A. H. Hoodi

July '73

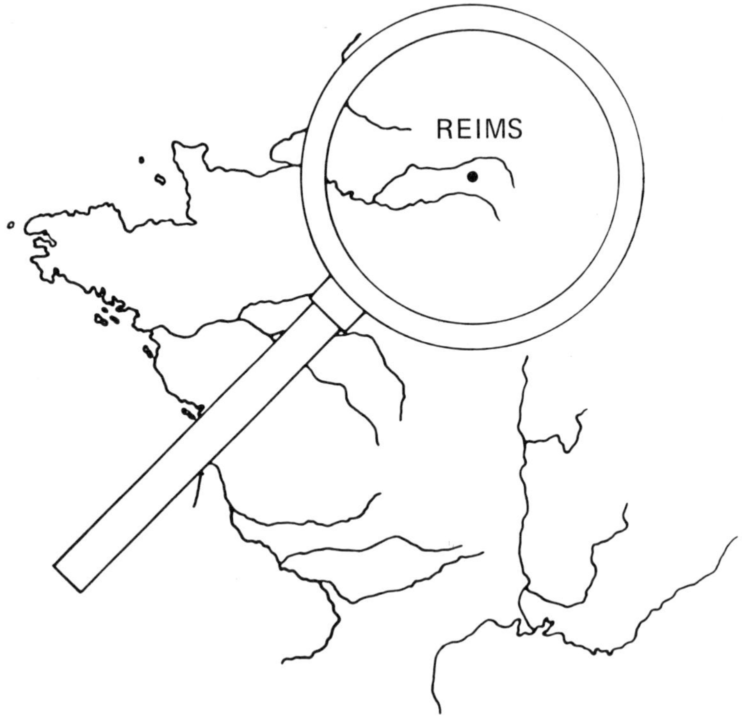

GAULE – CONQUÊTE RAMAN 1972

Advances in Raman Spectroscopy

Advances in Raman Spectroscopy

Volume 1

Proceedings of the Third International Conference on Raman Spectroscopy

University of Reims, France, September 1972

Editor: J. P. Mathieu
Laboratoire de Spectroscopie Cristalline et Moléculaire
Faculté des Sciences, Paris V, France

HEYDEN & SON LTD

London · New York · Rheine

Heyden & Son Ltd., Spectrum House, Alderton Crescent, London NW4 3XX
Heyden & Son Inc., 225 Park Avenue, New York, N.Y. 10017, U.S.A.
Heyden & Son GmbH, 4440 Rheine/Westf., Münsterstrasse 22, Germany

Library of Congress Catalog Card No. 73-76120

ISBN 0 85501 171 8

Printed in Great Britain by J. W. Arrowsmith Ltd., Bristol BS3 2NT

PREFACE

The third International Conference on Raman Spectroscopy was held at the University of Reims, France on September 8th–13th, 1972.

The Conference had the sponsorship of the Société Française de Physique and the Société de Chimie Physique.

Supporting grants have been received from:

Le Ministère de l'Education Nationale
Le Conseil Général de la Marne
La Ville de Reims
Le Conseil de l'Université de Reims
La Faculté des Sciences de Reims
La Société Coderg

The Organizing Commitee

M. Balkanski, Paris
L. Bernard, Reims
S. Bratos, Paris
M. Delhaye, Lille
R. Dupeyrat, Paris
R. Gaufrès, Montpellier
M. L. Josien, Paris
A. Kahane, Grenoble
R. Lafont, Montpellier
J. Lascombe, Bordeaux
J. P. Mathieu, Paris
A. Novak, Paris
B. Oksengorn, Paris
C. Troyanowsky, Paris

chose the following topics in the broad field of Raman Spectroscopy: Non-linear Phenomena, Flame Raman Effect, Resonance Raman Effect, Phase Transitions, Band Profiles in Gases and Liquids, New Techniques, Macromolecules and Biological Molecules, and Molecular Structures.

The proceedings of the first two International Conferences on Raman Spectroscopy (Ottawa 1969, chaired by Professor H. J. Bernstein; Oxford 1970, chaired by Professor D. A. Long) have not been published. We hope that this volume will help those who could not attend the Conference to survey the recent achievements of Raman Spectroscopy and will contribute to the development of this field of research.

Finally, we present the humorous introduction Professor Long gave to his plenary lecture.

'En tant que Président de la Second Conférence, tenue à Oxford en 1970, je suis un peu au courant du travail demandé par l'organisation de ces conférences. J'aimerais féliciter J. P. Mathieu, L. Bernard et tous les autres membres du comité: de leur choix de cette belle ville avec sa cathédrale magnifique et sa nouvelle université, pour notre réunion; du magnifique programme scientifique qu'ils nous ont proposé; et des délices gastronomiques qui nous attendent, à en croire le programme.

Le grand nombre de spécialistes en spectroscopie qui assistent à cette conférence témoigne également de l'excellence du programme. Cette véritable invasion de ce coin de la France par tant de spécialistes Raman rappelle une autre invasion, il y a bien des siècles, de ce qui fut alors la Gaule, et qui a été si délicieusement raconté dans les Chroniques de ce grand Gaulois, Astérix. Vous vous souviendrez sans doute de la carte qui se trouve en préface des Chroniques d'Astérix (par exemple "Les Lauriers de César", texte de Goscinny, dessins de Uderzo, Dargaud, Editeur, Neuilly-sur-Seine, 1972). En hommage à nos amis français, j'ai dressé une nouvelle carte dans le style d'Asterix pour marquer cette invasion moderne – cette conquête Raman.* J'ai un peu remanié les phrases célèbres qui ouvrent les Chroniques d'Astérix.

"Nous sommes en ~~50 avant~~ 1972 après Jésus-Christ.

Toute la ~~Gaule~~ ville de Reims est occupée par les ~~Romains~~ Ramans . . . Toute? NON! Un ~~village peuplé~~ comité d'organisation d'irréductibles Gaulois ~~résiste~~ aide encore et toujours ~~à~~ l'envahisseur!"

et je vous offre quelques "irréductibles Gaulois" de notre époque, que vous ne manquerez pas de reconnaître, j'en suis sûr.

"Mathieurix, le héros de ces avantures, a l'intelligence vive . . . Toutes les missions perilleuses lui sont confiées sans hesitation! Bernardix, le barde, toujours prêt à tout abandonner pour suivre Mathieurix dans une nouvelle aventure!" '

January 1973 J. P. Mathieu

* Voir le frontispice.

CONTENTS

SECTION I
NON–LINEAR PHENOMENA

1. SOME NON-LINEAR SPECTROSCOPIC PHENOMENA

D. A. Long
School of Chemistry, University of Bradford, Bradford, England

INTRODUCTION TO NON-LINEAR PHENOMENA

The most general relation between P, the induced dipole in a molecular system and E, the electric field producing it, is given by a power series

$$P = P_\alpha + P_\beta + P_\gamma + \ldots \tag{1}$$

$$= \alpha E + \tfrac{1}{2}\beta E^2 + \tfrac{1}{6}\gamma E^3 + \ldots$$

Here P and E are vectors and α, β and γ are tensors known as the polarizability, hyperpolarizability and second hyperpolarizability, respectively. These tensors measure the ease of distortion of the electron in a molecular system and their values depend on the nuclear and electron configurations. Typical equilibrium values are $\alpha = 10^{-24}$ cm^3, $\beta = 10^{-30}$ cm^3 esu^{-1} and $\gamma = 10^{-36}$ cm^7 esu^{-2}.

Since the contributions to the total induced moment P from the hyperpolarizability term P_β and the second hyperpolarizability term P_γ, depend on powers of the electric field strength higher than the first, these contributions are said to result from a non-linear interaction of the system with the electric field. Such interactions give rise to many novel effects now generally referred to as non-linear phenomena.

The production of significant non-linear interactions requires very high electric field strengths and in the optical frequency range these have only been available in recent years. For the contribution from P_β to reach 1% of P_α an electric field strength of 6×10^6 V cm^{-1} is required and the electric field strength must rise to 7.4×10^7 V cm^{-1} before the contribution from P_γ is 1% of P_α. However the electric field strengths associated with the conventional sources of radiation in the optical frequency range are many orders of magnitude less than those required for even these modest non-linear interactions. For example, the electric field strength associated with solar radiation at the earth's surface is only 0.14 V cm^{-1}, and the field strength associated with radiation from a typical Toronto mercury arc is only 27 V cm^{-1}.

The electric field strengths at optical frequencies necessary for significant non-linear interaction can only be produced in the focused beam of a giant pulse laser. As can be

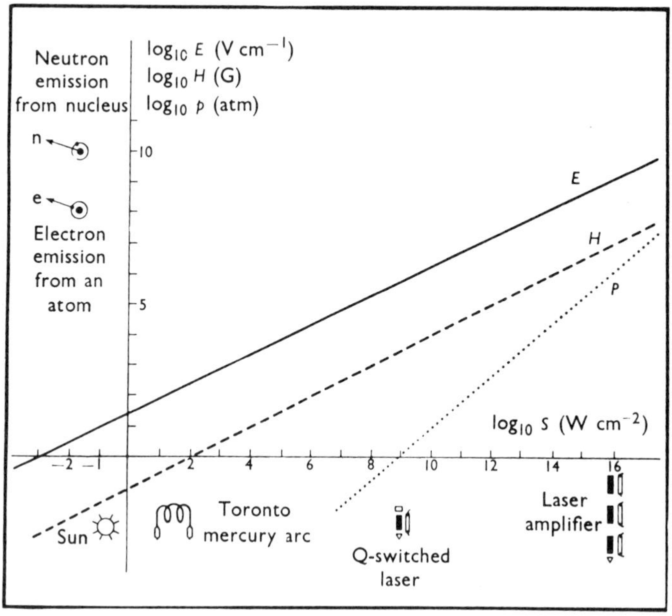

Fig. 1. Dependence of E, H and p on radiative flux density, S.

seen from Fig. 1, which shows in graphical form the linear relation between electric field strength in V cm^{-1} and power density in W cm^{-2}, an electric field strength of 6×10^6 V cm^{-1} requires a power density of 4×10^{10} W cm^{-2} and a field strength of 7.4×10^7 V cm^{-1} requires a power density of 7×10^{12} W cm^{-2}. A relatively modest giant pulse ruby laser with an output of 1 J in a pulse of 30×10^{-9} s duration and a focused beam diameter of 0.3 mm provides a power density of 4×10^{10} W cm^{-2}. Indeed, as can be seen subsequently, one problem in the investigation of non-linear optical phenomena is not so much to achieve an adequate power density, but to keep the power density below the level at which other phenomena, including some of a destructive nature, occur.

With such intense electric fields at optical frequencies there result new spectroscopic phenomena which are generically related to Rayleigh and Raman scattering. These novel phenomena[1] are capable of providing new information about molecules, information which has hitherto been regarded as spectroscopically inaccessible. Only some of the Rayleigh and Raman effects of non-linear origin are discussed in detail in this chapter. A reasonably thorough treatment of hyper-Rayleigh and Raman scattering and second hyper-Rayleigh and Raman scattering and a shorter account of the inverse Raman effect are given. The stimulated Raman effect is omitted. Although so far, very much more work has been reported in the literature relating to the stimulated Raman effect, it is the author's view that the *potential* value of hyper-Rayleigh and Raman scattering and the inverse Raman effect for the molecular spectroscopist is much greater. Perhaps this chapter will convince the reader of this and so justify this selection of specific topics within the broad region of non-linear spectroscopic phenomena.

Thus in subsequent sections, three treatments of non-linear light scattering of increasing completeness will first be presented: a qualitative pictorial treatment, a classical

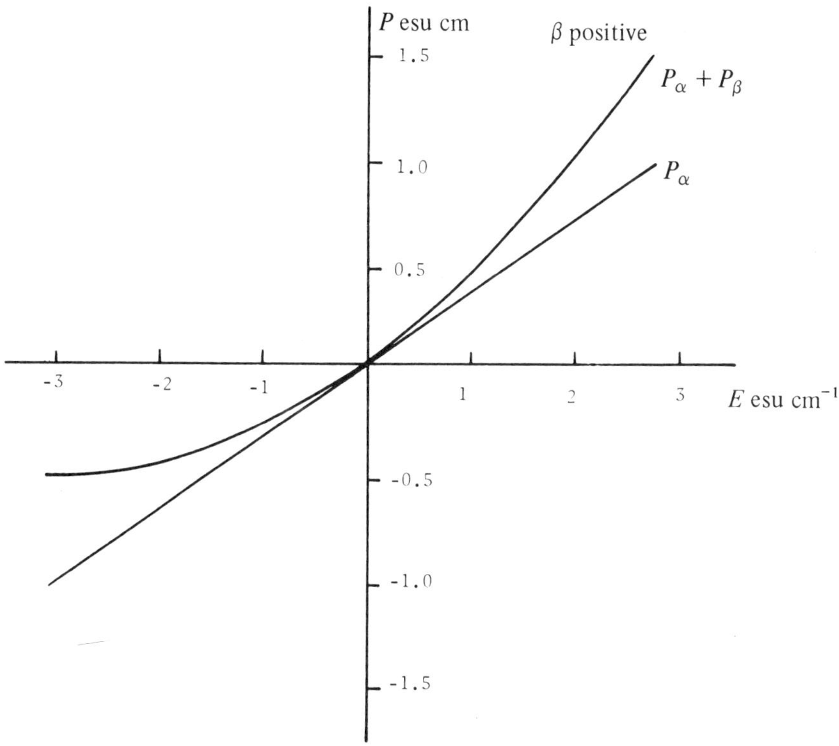

Fig. 2. Variation of the total induced dipole with electric field strength, for selected values of α, β (diagrammatic only).

treatment and a quantum mechanical treatment. Hyper-Rayleigh and hyper-Raman scattering will be treated in detail but for second hyper-Raleigh and Raman scattering just the essential results will be stated. In further sections on non-linear light scattering resonance phenomena, selection rules, experimental techniques and results will be discussed. Finally the inverse Raman effect will be briefly reviewed and experimental techniques and results discussed.

QUALITATIVE PICTORIAL TREATMENT OF NON-LINEAR LIGHT SCATTERING

Consider first a molecular system in which (a) the nuclear positions are fixed and so α and β are invariant, (b) the magnitude of the electric field is assumed to be sufficiently large for P_β to make a significant contribution to P_{total} and (c) β is positive. Figure 2 shows how, for such a system, P_α and $P_{\text{total}} = P_\alpha + P_\beta$ vary with the sign and magnitude of a static electric field. Note that for this case, where β is positive, for a given magnitude of the electric field, the magnitude of the induced dipole is greater when the electric field is positive than when it is negative. This asymmetry in the magnitude of the total induced dipole with change in sign of the field has important consequences when the electric field varies periodically with time.

Figure 3 shows the time dependence of an electric field E associated with radiation

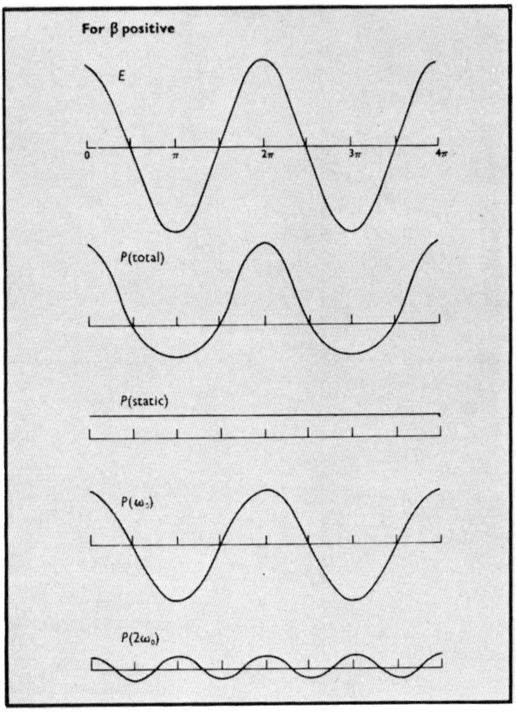

Fig. 3. Time dependence of electric field E, the total induced dipole P_{total} and its frequency components P_{static}, P_{ω_0} and $P_{2\omega_0}$ for β positive.

of frequency ω_0. Here ω_0 is a circular frequency defined by $\omega_0 = 2\pi c\nu_0$ where ν_0 is in cm^{-1}. If the interaction of the field with the system were only linear, the induced dipole would show the same time dependence as the field and Rayleigh scattering at ω_0 would result. However for the system we are considering, where a non-linear interaction through a positive hyperpolarizability β is involved, the time dependence of the total dipole P_{total} induced by the field of frequency ω_0 is as shown in Fig. 3. P_{total} now does not have the same time dependence as E. As Fig. 3 shows, P_{total} can now be resolved into three frequency components, P_{static}, P_{ω_0} and $P_{2\omega_0}$. Thus, whereas a linear interaction of radiation of frequency ω_0 with a system with fixed nuclei gives simply scattering at ω_0 (Rayleigh scattering) the non-linear interaction through β gives *in addition* a static field and scattering at $2\omega_0$ (hyper-Rayleigh scattering).

If instead of the hyperpolarizability β, the second hyperpolarizability γ is considered, then the total induced dipole $P_{\text{total}} = P_\alpha + P_\gamma$ can be resolved into two frequency components P_{ω_0} and $P_{3\omega_0}$. The P_{ω_0} dipole has, in this case, contributions to its amplitude from both E (through α) and E^3 (through γ). Thus for a system with fixed nuclei the non-linear interaction through γ results in a contribution to scattering at ω_0 (higher order Rayleigh scattering) and scattering at $3\omega_0$ (second hyper-Rayleigh scattering).

When nuclear motions are considered the polarizability α and the hyperpolarizabilities β and γ can vary with internuclear separation. This causes modulation of the Rayleigh, hyper-Rayleigh and second hyper-Rayleigh scattering. If the vibrational frequency is ω_M then we have

1. modulation of Rayleigh scattering at ω_0 through changes in the polarizability α to give Raman scattering at frequencies $\omega_0 \pm \omega_M$;
2. modulation of hyper-Rayleigh scattering at $2\omega_0$ through changes in the hyperpolarizability β to give hyper-Raman scattering at frequencies $2\omega_0 \pm \omega_M$; and
3. modulation of second hyper-Rayleigh scattering at $3\omega_0$ and the higher order normal Rayleigh scattering at ω_0 through changes in the second hyperpolarizability γ to give second hyper-Raman scattering at frequencies $3\omega_0 \pm \omega_M$ and a higher order contribution to normal Raman scattering at frequencies $\omega_0 \pm \omega_M$.

A CLASSICAL TREATMENT OF NON-LINEAR LIGHT SCATTERING

The ideas which have just been presented pictorially will now be expressed mathematically.[2] We shall consider for simplicity only the vibrational case.

The time dependence of the electric field E associated with the incident radiation can be written as

$$E = E_0 \cos \omega_0 t \tag{2}$$

where E_0 is the amplitude of the field, $\omega_0 = 2\pi c\nu_0$ and t is the time.

The polarizability α, and the hyperpolarizability β can be expressed as Taylor series in the normal coordinate Q associated with the vibrational frequency ν_M i.e.

$$\alpha = \alpha_0 + \left(\frac{\partial \alpha}{\partial Q}\right)_0 Q + \text{higher order terms} \tag{3}$$

$$\beta = \beta_0 + \left(\frac{\partial \beta}{\partial Q}\right)_0 Q + \text{higher order terms} \tag{4}$$

In the present treatment the higher order terms involving higher order derivatives of α and β and higher powers of Q will be neglected. These relate only to overtone and combination tones.

The time dependence of the normal coordinate Q can, in the harmonic oscillator approximation, be written as

$$Q = Q_0 \cos \omega_M t \tag{5}$$

where Q_0 is the amplitude of the normal coordinate and $\omega_M = 2\pi c\nu_M$.

If this relationship is introduced into the Taylor series expansions for α and β expressions are obtained for the time dependence of α and β for the molecular vibration frequency ω_M.

$$\alpha = \alpha_0 + \left(\frac{\partial \alpha}{\partial Q}\right)_0 Q_0 \cos \omega_M t \tag{6}$$

$$\beta = \beta_0 + \left(\frac{\partial \beta}{\partial Q}\right)_0 Q_0 \cos \omega_M t \tag{7}$$

If the expressions for the time dependence of α, β and E are introduced into the non-linear formula for the total induced dipole

$$P = \alpha E + \tfrac{1}{2}\beta E^2 \tag{8}$$

the following result is obtained

$$P = \left[\alpha_0 + \left(\frac{\partial\alpha}{\partial Q}\right)_0 Q_0 \cos\omega_M t\right] E_0 \cos\omega_0 t + \frac{1}{2}\left[\beta_0 + \left(\frac{\partial\beta}{\partial Q}\right)_0 Q_0 \cos\omega_M t\right] E_0^2 \cos^2\omega_0 t \tag{9}$$

On rearrangement and the use of well-known trigonometric relations it is found that the total induced dipole contains eight distinct frequency components. These are

1. $\omega = 0$: a d.c. field term

$$P_0 = \tfrac{1}{4}\beta_0 E_0^2 \tag{10}$$

2. $\omega = \omega_0$: Rayleigh scattering term

$$P_{\omega_0} = \alpha_0 E_0 \cos\omega_0 t \tag{11}$$

3 and 4. $\omega = \omega_0 \pm \omega_M$: Raman Stokes and anti-Stokes terms

$$P_{\omega_0 \pm \omega_M} = \tfrac{1}{2}E_0 Q_0 \left(\frac{\partial\alpha}{\partial Q}\right)_0 \cos(\omega_0 \pm \omega_M)t \tag{12}$$

5. $\omega = 2\omega_0$: hyper-Rayleigh scattering term

$$P_{2\omega_0} = \tfrac{1}{4}\beta_0 E_0^2 \cos 2\omega_0 t \tag{13}$$

6 and 7. $\omega = 2\omega_0 \pm \omega_M$: hyper-Raman Stokes and anti-Stokes terms

$$P_{2\omega_0 \pm \omega_M} = \tfrac{1}{8}E_0^2 Q_0 \left(\frac{\partial\beta}{\partial Q}\right)_0 \cos(2\omega_0 \pm \omega_M)t \tag{14}$$

8. $\omega = \omega_M$: molecular frequency term

$$P_{\omega_M} = \frac{1}{4}\left(\frac{\partial\beta}{\partial Q}\right)_0 Q_0 E_0^2 \cos\omega_M t \tag{15}$$

These follow the pattern expected from the pictorial treatment already presented. *Five* of the eight components arise from non-linear interaction with the electric field and are functions of E^2. *Two* of these non-linear terms involve the hyperpolarizability β and *three* the hyperpolarizability derivative $(\partial\beta/\partial Q)_0$.

This treatment may be readily extended to take into account terms in E^3 which involve the second hyperpolarizability γ. The frequency components of the induced dipole controlled by γ or its derivative are found to be

1. $\omega = 3\omega_0$: second hyper-Rayleigh term

$$P_{3\omega_0} = \tfrac{1}{24}\gamma_0 E_0^3 \cos 3\omega_0 t \tag{16}$$

2. $\omega = \omega_0$: higher order contribution to Rayleigh scattering

$$P_{\omega_0} = \tfrac{1}{8}\gamma_0 E_0^3 \cos \omega_0 t \tag{17}$$

3 and 4. $\omega = 3\omega_0 \pm \omega_M$: second hyper-Raman Stokes and anti-Stokes terms

$$P_{3\omega_0 \pm \omega_M} = \frac{1}{48}\left(\frac{\partial\gamma}{\partial Q}\right)_0 Q_0 E_0^3 \left\{\cos\,(3\omega_0 \pm \omega_M)t\right\} \tag{18}$$

5 and 6. $\omega = \omega_0 \pm \omega_M$: higher order contribution to normal Raman Stokes and anti-Stokes terms

$$P_{\omega_0 \pm \omega_M} = \frac{1}{16}\left(\frac{\partial\gamma}{\partial Q}\right)_0 Q_0 E_0^3 \left\{\cos\,(\omega_0 \pm \omega_M)t\right\} \tag{19}$$

A QUANTUM MECHANICAL TREATMENT OF LIGHT SCATTERING[3]

According to Placzek, P_{nk} the transition moment for a transition from the state k to the state n under the influence of a perturbing electric field E associated with electromagnetic radiation of frequency ω_0 is given by the matrix element

$$P_{nk} = \langle \Psi_n' | \mathrm{M} | \Psi_k' \rangle \tag{20}$$

where Ψ_n' and Ψ_k' are the time-dependent wave functions of the *perturbed* system in the states n and k respectively and M is the electric dipole moment operator.

The time dependence of the perturbing electric field associated with electromagnetic radiation of frequency ω_0 is now best expressed in the form

$$E = A \exp(-i\omega_0 t) + A^* \exp(i\omega_0 t) \tag{21}$$

For plane polarized radiation with, say $A_x \neq 0$ this corresponds to

$$E = 2A_x \cos \omega_0 t \tag{22}$$

Comparison with Eqn. (2) for E used in the classical treatment shows that $2A_x = E_0$.

Now Ψ_n', the time-dependent wave function of the perturbed system in the state n can be expressed in terms of the unperturbed time-dependent wave functions of the system, Ψ_n to a second-order by

$$\Psi'_n = \Psi_n + \Psi_n^{(1)} + \Psi_n^{(2)} \tag{23}$$

where
$$\Psi_n^{(1)} = \sum_r a_{rn}^{(1)} \Psi_r \tag{24}$$

and
$$\Psi_n^{(2)} = \sum_r a_{rn}^{(2)} \Psi_r \tag{25}$$

where the summation is over all states r of the system.

The first-order coefficients $a_{rn}^{(1)}$ are linear in A, the amplitude of the perturbing field, and the second-order coefficients $a_{rn}^{(2)}$ are quadratic in A. Substitution for Ψ'_n and Ψ'_k in P_{nk} gives to the second order of perturbation corrections

$$P_{nk} = \langle \Psi_n + \Psi_n^{(1)} + \Psi_n^{(2)} | \mathrm{M} | \Psi_k + \Psi_k^{(1)} + \Psi_k^{(2)} \rangle \tag{26}$$

Placzek considered only the terms in P_{nk} which involve the first power of A namely

$$P_{nk}^{(1)} = \langle \Psi_n^{(1)} | \mathrm{M} | \Psi_k \rangle + \langle \Psi_n | \mathrm{M} | \Psi_k^{(1)} \rangle \tag{27}$$

These terms control *inter alia* normal Raman scattering associated with a transition from a state k to a state n. Rayleigh scattering is included as a special case in which the initial and final states are the same ($n = k$).

The terms in P_{nk} involving A^2 control *inter alia* hyper-Raman scattering associated with a transition from a state k to a state n. These terms are

$$P_{nk}^{(2)} = \langle \Psi_n^{(1)} | \mathrm{M} | \Psi_k^{(1)} \rangle + \langle \Psi_n^{(2)} | \mathrm{M} | \Psi_k \rangle + \langle \Psi_n | \mathrm{M} | \Psi_k^{(2)} \rangle \tag{28}$$

Hyper-Rayleigh scattering is included as a special case in which the initial and final states are the same ($n = k$).

If the expressions for the perturbed wave functions had included third-order corrections which are cubic in A then P_{nk} would include terms in A^3 which would *inter alia* control second hyper-Rayleigh and Raman scattering.

Placzek proceeded to evaluate the first-order transition moment using standard perturbation theory. This procedure becomes increasingly cumbersome as one proceeds to higher order corrections. Stanton[5] has shown that the transition moment P_{nk} for a system subject to a perturbation can be expressed in an alternative form involving *only the unperturbed wave functions of the system* by making use of the density matrix. In its most general form the transition moment for a transition from state k to state n when the system is subject to a perturbation is given by

$$P_{nk} = \sum_{\mu,\nu} \langle \Psi_n | \rho^{(\mu)} \mathrm{M} \rho^{(\nu)} | \Psi_k \rangle \tag{29}$$

where μ and ν are orders of approximation of the density operator ρ.

The density matrix technique is not given in detail here. It is simply noted that the order of dependence of the transition moment on the electric field strength is given by $\mu + \nu$.

Thus normal Raman scattering is included in the first-order transition moment $P_{nk}^{(1)}$

which involves $\mu + \nu = 1$ (i.e. $\mu = 1, \nu = 0$ and $\mu = 0, \nu = 1$) namely

$$P_{nk}^{(1)} = \langle \Psi_n | \rho^{(1)} \mathrm{M} \rho^{(0)} | \Psi_k \rangle + \langle \Psi_n | \rho^{(0)} \mathrm{M} \rho^{(1)} | \Psi_k \rangle \tag{30}$$

Similarly hyper-Raman scattering is included in the second-order transition moment $P_{nk}^{(2)}$ which involves $\mu + \nu = 2$ (i.e. $\mu = 1, \nu = 1, \mu = 2, \nu = 0$ and $\mu = 0, \nu = 2$) namely

$$P_{nk}^{(2)} = \langle \Psi_n | \rho^{(1)} \mathrm{M} \rho^{(1)} | \Psi_k \rangle + \langle \Psi_n | \rho^{(2)} \mathrm{M} \rho^{(0)} | \Psi_k \rangle + \langle \Psi_n | \rho^{(0)} \mathrm{M} \rho^{(2)} | \Psi_k \rangle \tag{31}$$

Second hyper-Raman scattering would be included in the third-order transition moment $P_{nk}^{(3)}$ with $\mu + \nu = 3$.

The first-order transition moment contains terms with more than one frequency dependence. We shall restrict ourselves here and subsequently to the Raman-type terms. The term in $P_{nk}^{(1)}$ with the correct time-dependence for Raman scattering corresponds to the real dipole

$$C_{nk} \exp\left\{-i(\omega_{kn} + \omega_0)t\right\} + C_{nk}^* \exp\left\{i(\omega_{kn} + \omega_0)t\right\} \tag{32}$$

provided $\omega_{kn} + \omega_0 > 0$. Here $\omega_{kn} = \omega_k - \omega_n$, so that ω_{kn} is negative for Stokes Raman scattering (final state $\omega_n >$ initial state ω_k) and positive for anti-Stokes Raman scattering (final state $\omega_n <$ initial state ω_k). The condition that $\omega_{kn} + \omega_0 > 0$ is always satisfied for excitation frequencies in the visible region where $\nu_0 = \omega_0/2\pi c$ lies between 25,000 and 14,000 cm^{-1} and for rotation and vibration transitions where $|\nu_{kn}| = |\omega_{kn}/2\pi c|$ is less than 4,000 cm^{-1}.

The x component of the vector quantity C_{nk} has the form

$$[C_x]_{nk} = \sum_y [\alpha_{xy}]_{nk} A_y \tag{33}$$

where $[\alpha_{xy}]$ are components of a second order tensor for the transition $k \rightarrow n$.

That part of $P_{nk}^{(2)}$ with the correct time dependence for hyper-Raman scattering corresponds to the real dipole

$$B_{nk} \exp\left\{-i(\omega_{kn} + 2\omega_0)t\right\} + B_{nk}^* \exp\left\{i(\omega_{kn} + 2\omega_0)t\right\} \tag{34}$$

provided $\omega_{kn} + 2\omega_0 > 0$, a condition that again is always satisfied for excitation frequencies in the visible range and rotation and vibration transitions.

The x component of the vector quantity B_{nk} has the form

$$[B_x]_{nk} = \tfrac{1}{2} \sum_{x,y} [\beta_{xyz}]_{nk} A_y A_z \tag{35}$$

where $[\beta_{xyz}]$ are components of a third-order tensor.

By analogy there will emerge for the second-order hyper-Raman effect a real dipole of the form

$$D_{nk} \exp\left\{-i(\omega_{kn} + 3\omega_0)t\right\} + D_{nk}^* \exp\left\{i(\omega_{kn} + 3\omega_0 t)\right\} \tag{36}$$

The x component of the vector quantity D_{nk} will have the form

$$[D_x]_{nk} = \frac{1}{6} \sum_{yzw} [\gamma_{xyzw}]_{nk} A_y A_z A_w \tag{37}$$

where $[\gamma_{xyzw}]$ are components of a fourth-order tensor.

So far the quantum mechanical treatment has only given us explicitly the frequency dependence of the scattered radiation, information which was already available from the classical treatment. We gain a deeper insight into the factors controlling light scattering when we consider the nature of the tensor quantities α, β and γ.

We find that

$$[\alpha_{xy}]_{nk} = \frac{1}{\hbar} \sum_r \left[\frac{[\mathrm{M}_y]_{rk} [\mathrm{M}_x]_{nr}}{(\omega_{rk} - \omega_0 - i\Gamma_{rk})} + \frac{[\mathrm{M}_x]_{rk} [\mathrm{M}_y]_{nr}}{(\omega_{rn} + \omega_0 - i\Gamma_{rn})} \right] \tag{38}$$

where $[\mathrm{M}_y]_{nr}$ is the y component of the transition moment M_{nr} for a direct absorption or emission process involving the states n and r. That is

$$\mathrm{M}_{nr} = \langle \Psi_n | \mathrm{M} | \Psi_r \rangle \tag{39}$$

and so on. The terms $i\Gamma_{rk}$ relate to radiation damping and have been introduced phenomenonologically.

Similarly we find that

$$\begin{aligned}[\beta_{xyz}]_{nk} = \frac{2}{\hbar^2} \sum_{r,s} \Bigg[& \frac{[\mathrm{M}_y]_{nr} [\mathrm{M}_x]_{rs} [\mathrm{M}_z]_{sk}}{(\omega_{sk} - \omega_0 - i\Gamma_{sk})(\omega_{rn} + \omega_0 - i\Gamma_{rn})} \\ & + \frac{[\mathrm{M}_y]_{nr} [\mathrm{M}_z]_{rs} [\mathrm{M}_x]_{sk}}{(\omega_{sn} + 2\omega_0 - i\Gamma_{sn})(\omega_{rn} + \omega_0 - i\Gamma_{rn})} \\ & + \frac{[\mathrm{M}_x]_{ns} [\mathrm{M}_y]_{sr} [\mathrm{M}_z]_{rk}}{(\omega_{sk} - 2\omega_0 - i\Gamma_{sk})(\omega_{rk} - \omega_0 - i\Gamma_{rk})} \Bigg]\end{aligned} \tag{40}$$

The expression for $[\gamma_{xyzw}]_{nk}$ is still more complicated and we shall content ourselves with considering the general form of the terms in it.

The numerators will consist of products of four transition moments involving three intermediate states. That is they will be of the form

$$[\mathrm{M}]_{nr} [\mathrm{M}]_{rs} [\mathrm{M}]_{st} [\mathrm{M}]_{tk} \tag{41}$$

The denominators will involve products of three frequency terms which will include terms of the type $(\omega_{\mathbf{abs}} - \omega_0)$, $(\omega_{\mathbf{abs}} - 2\omega_0)$ and $(\omega_{\mathbf{abs}} - 3\omega_0)$ where $\omega_{\mathbf{abs}}$ refers to the frequencies associated with transitions of the type $n \longleftrightarrow r$, $r \longleftrightarrow s$, $s \longleftrightarrow t$, $t \longleftrightarrow k$.

We shall now examine the properties of these tensors in more detail.

The Raman tensor α

1. An 'intermediate' state r is involved. There must be non-zero transition moments in ordinary absorption (or emission) for the two pairs of states n and r and r and k, for *at least one intermediate state* r, if the tensor is not to be zero.
2. The frequency denominators can lead to 'resonance' if the excitation frequency ω_0 is close to an absorption frequency e.g. ω_{rn}.
3. The second order tensor is in general unsymmetric ($\alpha_{ij} \neq \alpha_{ji}$) with a maximum of nine distinct entries. However many systems satisfy conditions under which the tensor becomes symmetric ($\alpha_{ij} = \alpha_{ji}$) and there is then a maximum of only 6 distinct entries.
4. Many systems satisfy conditions under which the tensor can be regarded as a function of the nuclear coordinates. In such cases expansion in a Taylor series in the normal coordinate is justified.
5. The tensor transforms as $\langle \psi_n | xy | \psi_k \rangle$.

The hyper-Raman tensor β

1. Two 'intermediate' states r and s are involved. There must be non-zero transition moments for ordinary absorption (or emission) involving the three pairs of states n and r, r and s, and s and k for *at least one pair of intermediate states r and s* if the tensor is not to be zero.
2. Frequency denominators can lead to resonance if either the excitation frequency ω_0 or its first harmonic $2\omega_0$ is close to an absorption frequency e.g. ω_{sn}.
3. The tensor is in general unsymmetric with a maximum of 18 distinct components (not 27 since $A_xA_y = A_yA_x$). However many systems satisfy conditions under which the tensor becomes symmetric ($\beta_{ijk} = \beta_{jki}$) and there is then a maximum of only 10 distinct entries.
4. Many systems satisfy conditions under which the tensor can be regarded as a function of the nuclear coordinates. In such cases expansion in a Taylor series in the normal coordinate is justified.
5. The tensor transforms as $\langle \psi_n | xyz | \psi_k \rangle$.

The second hyper-Raman tensor γ

We shall note simply that:

1. *Three* intermediate states are involved.
2. Resonances with ω_0, $2\omega_0$ and $3\omega_0$ are possible.
3. The unsymmetric tensor would have 30 distinct entries and the symmetric tensor 15 distinct entries.

RESONANCE IN NON-LINEAR LIGHT SCATTERING

Attention has already been drawn to the fact that if one or more of the frequency denominators in a scattering tensor tends to zero the tensor will then become anomalously large and anomalously intense scattering will result.

Resonance phenomena are well recognized in the normal Raman effect and are being increasingly studied. Since higher order Raman effects are so much weaker than the normal Raman effect resonance could well be of great importance in facilitating their observation.[4] In the hyper-Raman effect, for example, resonance can occur in three ways; through an absorption band close to the incident frequency ω_0; through an absorption band close to $2\omega_0$; or through absorption bands close to both ω_0 and $2\omega_0$. We estimate that a 'single' resonance will lead to intensity enhancement by a factor of up to 10^6 and 'double' resonance to intensity enhancement by a factor of up to 10^{12}.

In the normal Raman effect some at least of the gain introduced by resonance at ω_0 is offset by absorption of both the incident radiation and the scattered radiation which lies close to ω_0. In the hyper-Raman effect, absorption of the scattered radiation which lies close to $2\omega_0$ can be avoided by using a single resonance in the region of ω_0.

In the second hyper-Raman effect resonances at ω_0, $2\omega_0$ and $3\omega_0$ are possible. This opens up a most interesting possibility. *The use of a single resonance at* $2\omega_0$ to enhance the intensity of the second hyper-Raman scattering around $3\omega_0$. The absorption at $2\omega_0$ will be well removed from both the exciting frequency ω_0 and the region close to $3\omega_0$ in which the second hyper-Raman frequencies are observed, so that no absorption losses will be present to offset the resonance enhancement. These interesting resonance experiments have yet to be performed but we hope to try some of them soon.

Table 1. Activity of fundamental vibrations in C_6H_6 (D_{6h}).

Symmetry species	μ	α	β	γ	Number of frequencies
A_{1g}		α		γ	2
A_{2g}		(unsymm.)		(unsymm.)	1
B_{1g}					0
B_{2g}				γ	2
E_{1g}		α		γ	1
E_{2g}		α		γ	4
A_{1u}					0
A_{2u}	μ		β		1
B_{1u}			β		2
B_{2u}			β		2
E_{1u}	μ		β		3
E_{2u}			β		2

SELECTION RULES

A detailed treatment of the selection rules for higher order Raman effects would be very extensive so only certain important general results are given and illustrated with particular examples. Detailed treatments of selection rules will be formed in the recent literature.[5]

The most important general result is that the selection rules for hyper- and second hyper-Raman spectroscopy are less restrictive than for normal Raman or infrared spectra. Thus many transitions previously regarded as spectroscopically inactive can be studied by higher order light scattering.

Consider first vibrational spectroscopy and let us take as a specific example the benzene molecule which belongs to the point group D_{6h}. The free molecule has 30 fundamental modes of vibration. There are 20 distinct vibration frequencies since 10 of the vibrations are doubly degenerate. Table 1 shows the symmetry classes to which the various components of α, β and γ belong. The assignment of tensor components to symmetry classes is based upon the transformation properties of the tensor. We see from Table 1 that there are 7 Raman active modes, $2A_{1g}$, $1E_{1g}$ and $4E_{2g}$, and 4 infrared active modes, $1A_{2u}$ and $3E_{1u}$. Nine modes are spectroscopically inactive using normal spectroscopic techniques. Six of these are active in the hyper-Raman effect, $2B_{1u}$, $2B_{2u}$ and $2E_{2u}$, and two more, $2B_{2g}$ are active in the second hyper-Raman effect. This leaves only the A_{2g} mode apparently inaccessible. However under conditions where the polarizability tensors are not symmetric, this mode is both normal Raman and second hyper-Raman active! It should also be noted that all the infrared active modes are also hyper-Raman active. Thus, by going to higher order light scattering effects it is possible to observe all the vibration frequencies of benzene, including those which are infrared active and inactive in the normal Raman effect.

It is useful to summarize certain general rules concerning selection rules in higher order vibrational Raman effects.

1. All infrared active modes are hyper-Raman active. Such modes are always polarized in the hyper-Raman effect. Hyper-Raman active modes which are infrared forbidden are always depolarized.
2. All Raman active modes are active in the second hyper-Raman effect.
3. In systems with a centre of symmetry all hyper-Raman active modes are ungerade and all second hyper-Raman active modes are gerade.

This discussion of selection rules will be concluded by commenting briefly on rotational hyper-Raman spectroscopy. It is well known that a spherically symmetric molecule like methane (point group T_d) shows no pure rotational infrared spectrum (no permanent dipole) and no pure rotational Raman spectrum (isotropic polarizability). However methane has an anisotropic hyperpolarizability so rotational transitions are hyper-Raman active. The selection rules are complicated and we shall simply note that in the most general case the permitted ΔJ values are $\Delta J = 0, \pm 1, \pm 2, \pm 3$ leading to *N, O, P, Q, R, S* and *T* branches. In specific cases there will of course be further restrictions, e.g. for a linear molecule $\Delta J = \pm 1, \pm 3, \Delta K = 0$ and $K = 0$.

THE EXPERIMENTAL OBSERVATION OF HYPER-RAYLEIGH AND HYPER-RAMAN SCATTERING

The detection of hyper-Raman scattering is unfortunately not easy. The problem can be gauged from the following. A giant pulse ruby laser operating at an optimum level for hyper-Raman scattering will deliver about 3×10^{17} photons of frequency ω_0 to the sample in a single pulse. The number of photons of frequency $2\omega_0 \pm \omega_M$ which the detector will receive in a typical experiment ranges from a few photons per laser pulse down to one photon every hundred laser pulses.

The first experimental observation of the hyper-Raman effect was reported in 1965 by Terhune, Maker and Savage of the Ford Research Laboratories, U.S.A. and was a considerable experimental triumph.[6] Since then they have continued to develop experimental techniques for hyper-Raman spectroscopy. Until recently all the reported observations of the hyper-Raman effect had been made in their laboratory.

There are at present two approaches to the detection of the hyper-Raman effect: the single channel and the multi-channel methods. Both these techniques for hyper-Raman spectroscopy were pioneered by Dr. Maker and his colleagues. In both methods the procedures for exciting the spectrum and dispersing the scattered radiation prior to detection are the same. Radiation from a giant pulse ruby laser is focused on to the sample and the

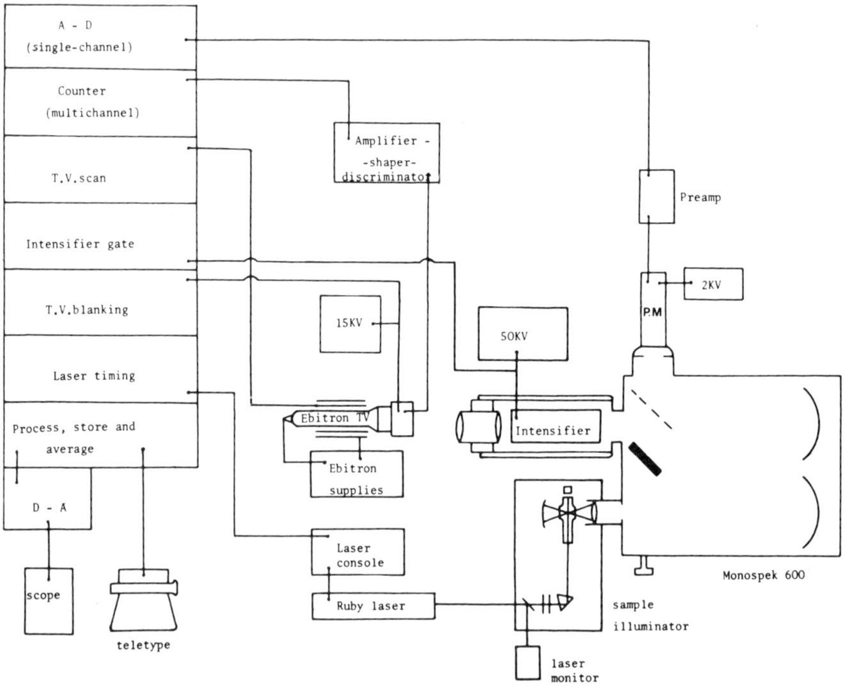

Fig. 4. Versatile single/multi-channel hyper-Raman instrument; block diagram.

radiation scattered at right angles to the laser beam is collected with a wide angle lens and fed to a grating system which disperses the radiation.

In single channel detection the grating dispersion system operates as a monochromator with an exit slit in front of the photomultiplier detector. With the monochromator set to pass radiation over a narrow frequency range the intensity of the scattered radiation in this range is found by averaging the number of photoelectrons produced over a large number of laser pulses – usually at least 50. This procedure is then repeated, spectral element by spectral element, over the whole frequency range of interest. This procedure is quite reminiscent of the point by point plotting of mirror galvanometer readings that was necessary for the measurement of infrared spectra fifty years ago. This method, while relatively simple is very time consuming and hard on the observer and the laser! Although the laser pulse has a duration of about 30×10^{-9} s the repetition rate is only one pulse per second, and so averaging over 50 laser pulses per spectral element and covering 1000 spectral elements would take 15 h and 50,000 laser shots! Nevertheless the Ford laboratories obtained many important results using this technique.

In the multi-channel method[7] the information from all the spectral elements over the desired spectral region is recorded for each laser shot. The grating dispersing system then operates as a spectrograph and the exit slit and photomultiplier are replaced by an image intensifier the photocathode of which is in the camera focal plane. For each laser shot there is produced on the output phosphor of the intensifier an intensified spectrum. This is scanned with a television camera which is optically coupled to the intensifier. In

Fig. 5. Photograph of versatile single/multi-channel hyper-Raman spectrometer at Bradford.

this way the optical information is converted into digital information and stored so that it can be combined with information from other laser shots and then averaged. Although each laser pulse lasts only 10^{-8} s there is signal storage of adequate duration at the intensifier–TV camera interface to permit the read off and processing of the information. This procedure is much more complicated but it achieves the same result as the single channel method in very much less time. Laser wear and tear is much less and fluctuations in laser output from shot to shot are unimportant. For example, with a laser pulse repetition rate of one per second and averaging over 50 laser pulses the whole spectral range covered by the intensifier cathode would be dealt with in 50 s.

Figure 4 shows a block diagram of a versatile hyper-Raman spectrometer we have constructed at Bradford.[8] Both single channel and multi-channel measurements can be made with this equipment. A specially adapted Digico Micro 16 P computer not only serves as a signal processer but also controls the whole sequence of operations during the

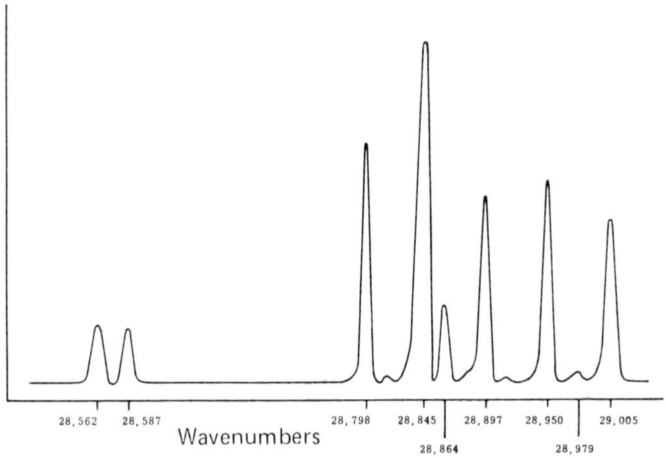

Fig. 6. Ultraviolet emission spectrum of neon bulb recorded with multi-channel spectrometer.

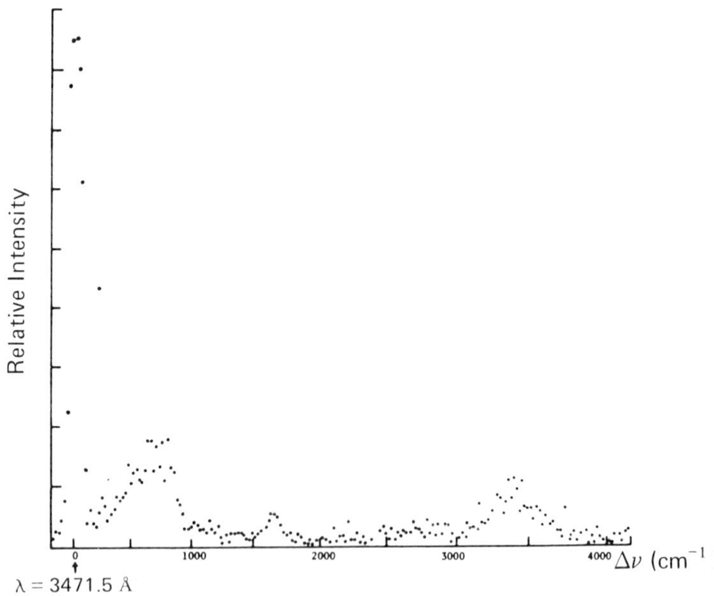

Fig. 7. Hyper-Rayleigh and Stokes hyper-Raman spectra of water.

course of recording a spectrum by either procedure. When operating in single channel mode the number of spectral elements and the number of laser shots per spectral element are inserted in the programme through the teletype. On initiation, the programme discharges the laser capacitor bank, and after a delay fixed by the programme, typically 10^{-3} s, opens the Q-switch device, a Pockels cell, so that a giant laser pulse is emitted. After further delay, typically 1.5×10^{-6} s also fixed by the programme, the photomultiplier load resistor voltage is interrogated by a sample-and-hold device, (sampling time typically

2×10^{-7} s), and the hold-voltage converted to digital form for storage and processing. This sequence of events is then repeated laser pulse by laser pulse and spectral element by spectral element.

In the multi-channel mode, the digital information produced by scanning the intensified spectrum on the output phosphor of the image intensifier with the television camera is accumulated in the computer. The information from each spectral element is stored in separate locations to await augmentation from subsequent shots. The total signal accumulation time per laser shot is about 25×10^{-3} s. The computer now controls not only the laser firing sequence, as for single channel operation, but also the timing of the television camera scans. In addition to avoid recording dark noise originating at times other than during the period of the giant laser pulse, the computer controls the gating of the intensifier stages.

At the end of an experiment the stored data can be presented in decimal integer form on the teletype, or displayed in analogue form on an oscilloscope or plotter. There is also provision to sample the computer store at any intermediate stages in the experiment and present the contents in analogue form on an oscilloscope. This 'quick-look-see' facility enables the efficacy of the averaging procedure in revealing weak features to be assessed as the experiment progresses.

Figure 5 shows a photograph of the complete hyper-Raman spectrometer. The image intensifier was custom built for us by Professor J. D. McGee at Imperial College, London, and the television camera is the Ebitron, a new camera recently developed by E.M.I. London. Figure 6 shows part of the emission spectrum of neon recorded with the multi-channel system at Bradford. This illustrates the present resolution of the multi-channel system.

EXAMPLES OF HYPER-RAYLEIGH AND HYPER-RAMAN SPECTRA

Figure 7 shows the hyper-Raman spectrum of water obtained at Bradford using the single channel technique. Figure 8 shows the hyper-Rayleigh and hyper-Raman spectrum of 1 : 2 dichlorethane also recorded at Bradford using the single channel technique. In the staggered configuration the free molecule has a centre of symmetry. It would therefore have a zero static hyperpolarizability β and would not give hyper-Rayleigh scattering. As can be seen from Fig. 8, hyper-Rayleigh scattering is observed from liquid 1 : 2 dichlorethane giving evidence of perturbations in the liquid state.

The remaining spectra presented here are taken from the work of Dr. Maker and his colleagues. Figure 9 shows the hyper-Raman spectrum of methanol obtained by the single channel technique. The infrared spectrum of methanol is also included in this figure and a careful comparison of the two spectra will show that all the infrared frequencies are present in the hyper-Raman spectrum although the intensities in the two spectra are different.[9]

Figure 10 shows the hyper-Raman spectrum of methane obtained by the multi-channel technique.[10] The low frequency band centred at about 100 cm^{-1} corresponds to the unresolved pure rotational hyper-Raman spectrum. The strong band at 3050 cm^{-1} corresponds to the $\nu_3(f_2)$ C–H stretching frequency which is observed in the infrared at 3020 cm^{-1}.

Figure 11 shows the hyper-Raman spectrum of ethane.[10] The hyper-Rayleigh line at $2\nu_0$ is absent since ethane has a centre of symmetry and so has a zero static hyperpolarizability. The band at 2950 cm^{-1} corresponds to the overlap of the C–H stretching frequencies $\nu_5(a_{2u})$ and $\nu_7(1u)$ which are observed in the infrared at 2954 cm^{-1} and 2994 cm^{-1}

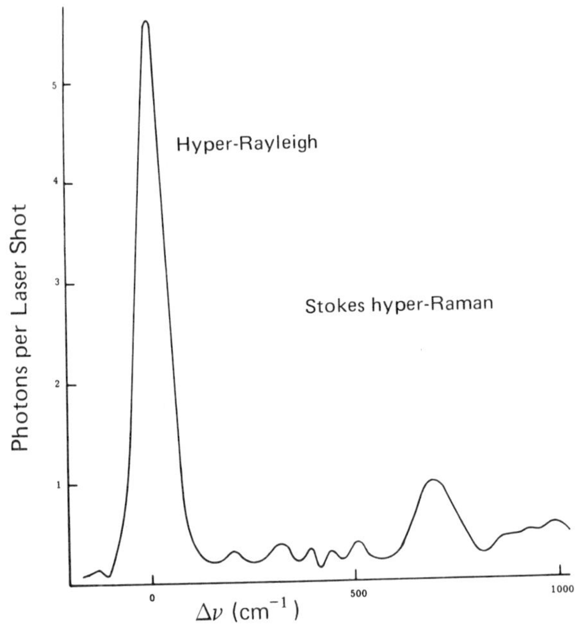

Fig. 8. Hyper-Rayleigh and Stokes hyper-Raman spectra of 1:2 dichloroethane.

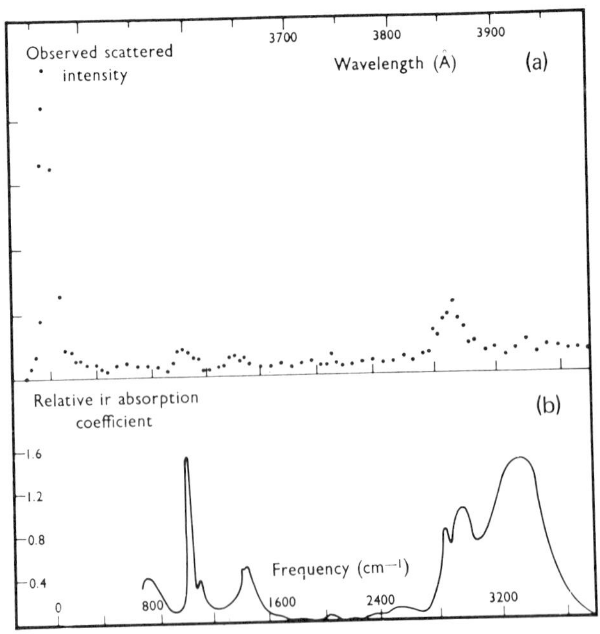

Fig. 9. (a) The hyper-Raman and (b) the infrared spectra of methanol, 0–3700 cm^{-1}.

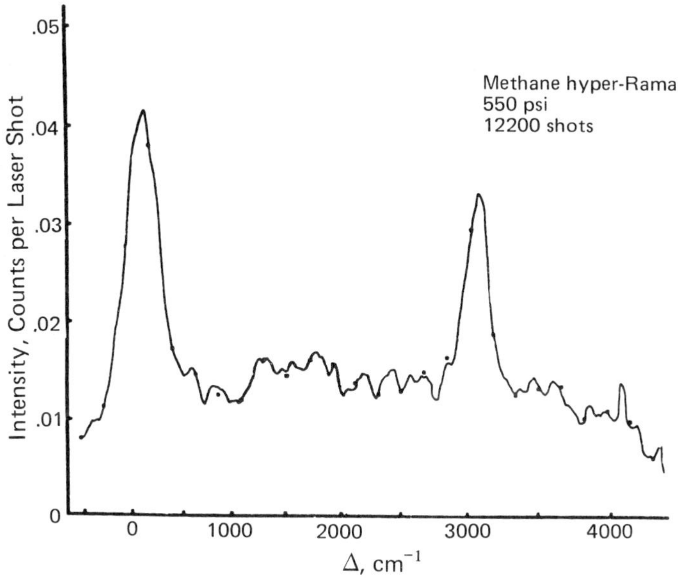

Fig. 10. The hyper-Raman spectrum of methane, under the conditions indicated. The horizontal scale is the hyper-Raman shift, $\Delta = 2\nu$ laser $- \nu$ scattered, and the vertical scale is the number of counts accumulated divided by the number of laser shots.

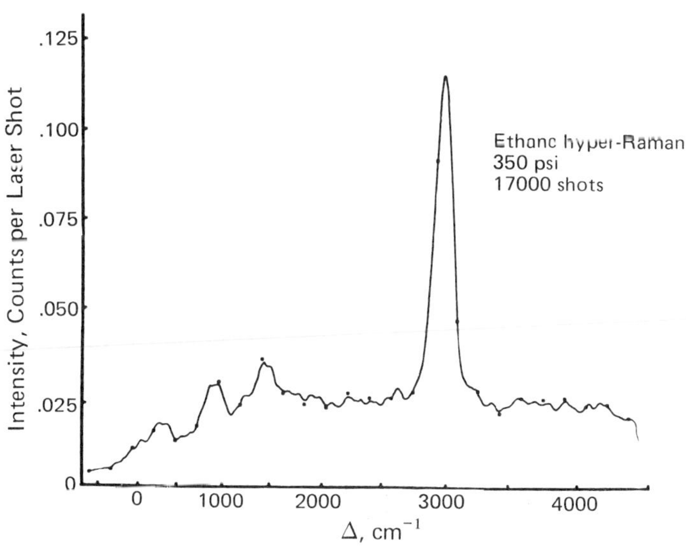

Fig. 11. The hyper-Raman spectrum of ethane (scales as in Fig. 10).

respectively. Other hyper-Raman bands with infrared counterparts appear at 1450 cm^{-1} and 900 cm^{-1}. Of special interest is the band at 300 cm^{-1} which can be assigned to the torsional vibration of A_{1u} symmetry which is infrared and normal Raman *inactive*.

Finally, Fig. 12 shows part of the hyper-Raman spectrum of a single crystal of

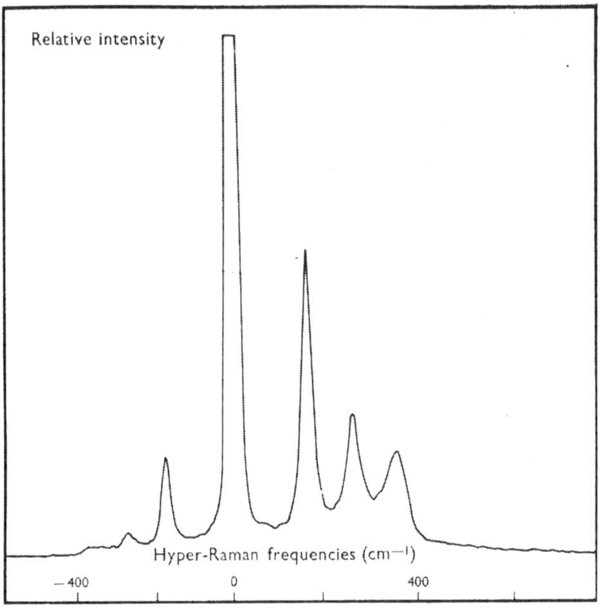

Fig. 12. Part of the hyper-Raman spectrum of a single crystal of ammonium chloride.

ammonium chloride obtained by the multi-channel technique. The band at 364 cm^{-1} on the Stokes side should be noted. It arises from a librational motion of the ammonium ion. It is not infrared or Raman active since the ammonium ion has no dipole moment and an isotropic polarizability.

THE INVERSE RAMAN EFFECT

The energy balance equation for the normal anti-Stokes Raman process is shown diagrammatically in Fig. 13. If this process is 'inverted' we have a formal description of what may be termed the *inverse* (anti-Stokes) Raman effect. The scattering molecule then absorbs radiation of frequency $\nu_0 + \nu_M$, which results in promotion of the molecule to a higher energy level, $h\nu_M$ energy units above its initial energy, and emission of radiation of frequency ν_0. From Fig. 14 we see that absorption of radiation of frequency $\nu_0 - \nu_M$ could result in decrease in the energy of the scattering molecule by $h\nu_M$ and emission of radiation at ν_0. This process is the inverse of the Stokes Raman effect.

The inverse Raman effect is another example of a non-linear Raman effect. It is found that absorptions at frequencies $\nu_0 \pm \nu_M$ only occur in the presence of radiation of very high power density at frequency ν_0. The necessary power density can only be achieved in the focused beam of a giant pulse laser. Thus the essential requirements for observation of the inverse Raman effect are simulataneous irradiation of the sample with intense radiation of frequency ν_0 *and* a strong continuum spanning the Stokes and anti-Stokes regions in which absorptions are expected. Ideally the continuum should span a region of about $\nu_0 + 3000\,cm^{-1}$ to $\nu_0 - 1000\,cm^{-1}$.

The theory of the inverse Raman effect is not treated here. However it should be

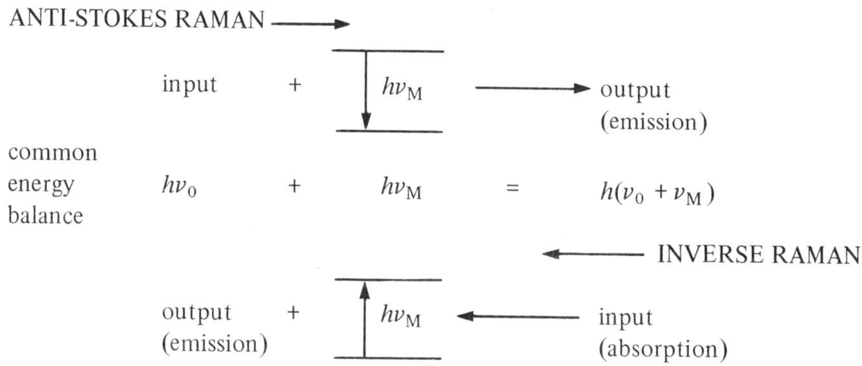

Fig. 13. Anti-Stokes, Raman and inverse Raman effects.

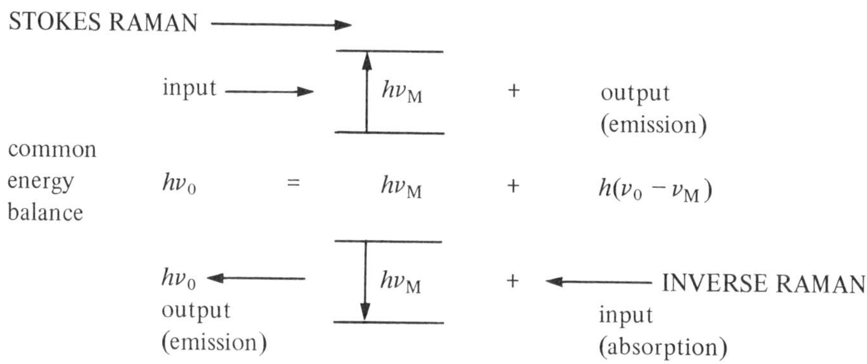

Fig. 14. Stokes Raman and inverse Raman effects.

noted that in principle the whole Raman spectrum is observable, in contrast to stimulated Raman emission, in which only one or two bands, namely those with the highest intensity per unit bandwidth, are observed. Furthermore, because of the high field strengths it is possible that the selection rules may be even less restrictive than for the normal Raman effect and bands active in the hyper- and second hyper-Raman effect may be observed. Since the necessary power density at frequency ν_0 can only be achieved with a giant pulse laser, the pulse duration of which ranges from nanoseconds to picoseconds, the inverse Raman effect offers the possibility of recording Raman spectra in extremely short times. The potential of this form of spectroscopy is thus very considerable. It could be particularly valuable for the investigation of short lived species and transient phenomena.

Unfortunately, although the first experimental observation of the effect was reported in 1964,[11] inverse Raman spectroscopy has not yet found any truly novel applications. The experimental effort has been devoted largely to devising improved methods of achieving a suitable continuum of adequate extent which is coincident with the intense laser pulse at ν_0 cm^{-1} as they pass through the sample. For photographic detection the continuum must also last no longer than the laser pulse. The remainder of this chapter is devoted to a brief review of progress in experimental techniques for inverse Raman spectroscopy.

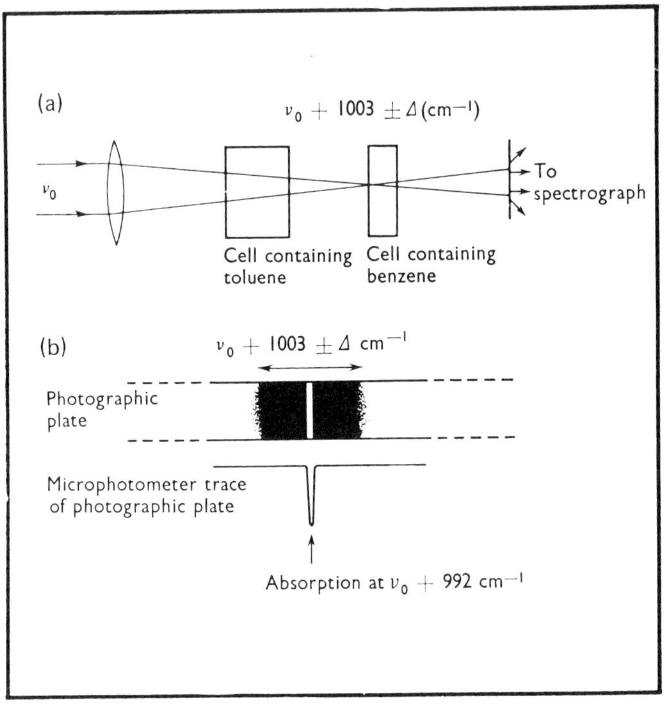

Fig. 15. (a) The initial experimental arrangement which Jones and Stoicheff[11] used to observe the inverse Raman effect and (b) a diagrammatic representation of part of the inverse Raman spectrum of benzene.

The inverse Raman effect was first reported by Jones and Stoicheff[11] in 1964. A typical example of their experimental arrangement is shown in Fig. 15. In general radiation from a giant pulse ruby laser is passed first into a cell containing a liquid with a molecular vibration frequency ν_M which is strong and sharp in the normal Raman effect. A broad, anti-Stokes stimulated Raman band at $\nu_0 + \nu_M \pm \Delta\,cm^{-1}$ is generated in this liquid and radiation in this frequency range passes together with the unchanged laser radiation at ν_0 into another liquid in a second cell. If the second liquid has a strong, Raman active molecular frequency within the range $\nu_M \pm \Delta\ cm^{-1}$ then absorption will be observed in the limited continuum. In Fig. 15 the liquid in the first cell is toluene which has a strong sharp Raman band at 1003 cm^{-1}. This generates a broad, anti-Stokes stimulated Raman band at $\nu_0 + 1003 \pm \Delta\ cm^{-1}$. Using this continuum of limited extent, Jones and Stoicheff observed absorption at $\nu_0 + 992.0\ cm^{-1}$ in benzene, (as shown diagrammatically in Fig. 15), $\nu_0 + 990.2\ cm^{-1}$ in pyridine and $\nu_0 + 917.8\ cm^{-1}$ in nitrobenzene. In the case of nitromethane since the absorption is about 90 cm^{-1} from the centre of the continuum ten laser shots were necessary. Using benzene in the first cell to generate a continuum centred at $\nu_0 + 992\ cm^{-1}$ the sharp benzene absorption at 992 cm^{-1} could be observed with benzene in the second cell. Using other appropriate continua of limited extent, Raman absorptions were also observed for bromoform, carbon disulphide, nitrobenzene, nitromethane, and carbon tetrachloride. However this intriguing procedure is clearly of limited applicability

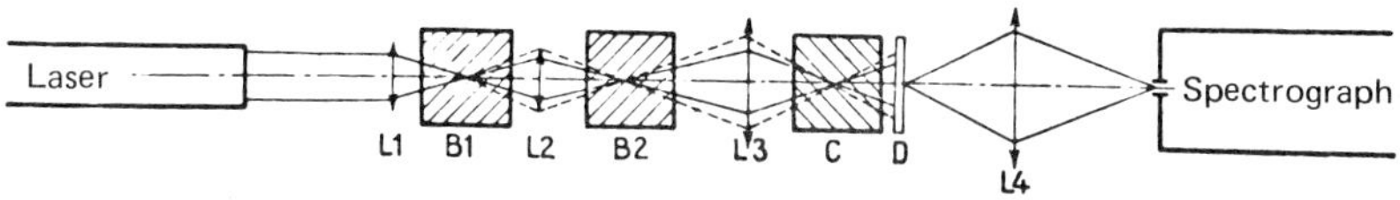

Fig. 16. Apparatus used by Dumartin, Oksengorn and Vodar.[12] (L = lens; B = high pressure cell; C = cell containing liquid to be studied).

since the continuum is so restricted in range.

In 1965, Dumartin, Oksengom and Vodar[12] described a technique for generating a continuum of greater extent. Their experimental arrangement is shown diagrammatically in Fig. 16. Radiation from a giant ruby laser was focused into a high pressure cell containing krypton. The emerging radiation was then focused into a second identical cell of krypton. The radiation from the second cell consisted of an intense continuum extending over both sides of the laser frequency together with ruby laser radiation of somewhat reduced power but still adequate for production of the inverse Raman effect. With this apparatus, inverse Raman bands on the anti-Stokes side at 668 cm^{-1}, 1216 cm^{-1} and 1389 cm^{-1} were observed in chloroform. These frequencies were assigned to the fundamentals ν_2 and ν_4 and the combination band $\nu_3 + \nu_5 + \nu_6$ respectively. In benzene an inverse Raman band was observed at 703 cm^{-1} and was interpreted as arising from the ν_7 fundamental, which is normally infrared and Raman inactive. Its appearance was attributed to relaxation of the selection rules in the presence of an intense electric field. In addition above a threshold in laser power these workers observed another type of Raman absorption appearing on the Stokes side, that is at frequencies less than ν_0. The special characteristics of this type of absorption led these workers to designate this absorption as a stimulated inverse Raman effect. They have discussed the underlying theory of this phenomenon but we are not treating this here.

In 1966, Duardo, Johnson and El-Sayed[13] described a variation of the original Jones and Stoicheff procedure for observing inverse Raman spectra. If a liquid happens to have two nearby strong Raman bands then, if the anti-Stokes stimulated Raman emission of one is sufficiently broad to overlap in frequency the position of the other band, the latter might be observed in absorption. These workers demonstrated this possibility using liquid acetonitrile which possesses two strong and sharp Raman bands, one at 2940 cm^{-1} (C–H stretch) and one at 2250 cm^{-1} (C≡N stretch). Using the second harmonic of a neodymium glass giant laser (output 5289 Å) they observed the CN band as an absorption in the broadened C–H band as shown in Fig. 17.

At the first International Raman Conference at Ottawa in 1969 Stoicheff and McLaren described[14] a new method for the production of the inverse Raman effect. Essentially the broad band fluorescence of a suitable dyestuff excited by the harmonic of the ruby laser frequency at 3470 Å is used as a source of the continuum. Rhodamine 6G (2.0×10^{-3} M in methanol) and rhodamine B (2.0×10^{-3} M in ethanol) absorb very strongly at 3470 Å and have intense fluorescence bands extending from 5600 Å–6800 Å. This is sufficient spectral breadth to cover Raman shifts from 300–3500 cm^{-1} on the high frequency side of the ruby frequency. The fluorescence lifetime is about 5 ns, sufficiently short to produce good coincidence with the laser pulse. The experimental arrangement is shown in Fig. 18. A ruby laser Q-switched by a rotating prism produced the stimulating radiation at ν_L in a single pulse with full width at half-maximum of 10 ns

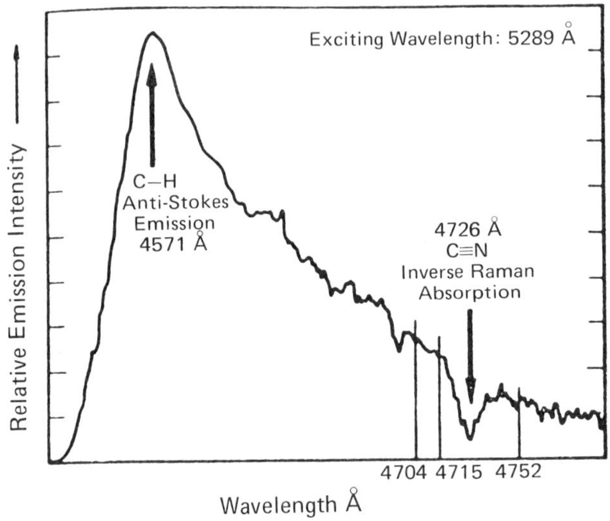

Fig. 17. Trace of the anti-Stokes side of a stimulated Raman emission spectrum of liquid acetonitrile excited by the second harmonic of the Nd^{3+} laser (at 5289 Å). The inverse Raman absorption of the C≡N stretching mode is observed on the tail of the diffuse C–H anti-Stokes emission line. (Duardo, Johnson and El-Sayed.[13])

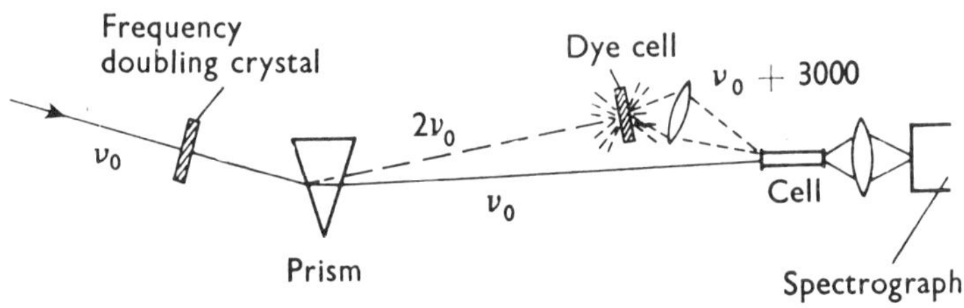

Fig. 18. The experimental arrangement with which McLaren and Stoicheff[14] observed the inverse Raman effect.

and with energy equal to 0.30 J. Approximately 3% of the laser emission was converted to $2\nu_L$ in a crystal of phase-matched KDP. Radiation at ν_L and $2\nu_L$ was then separated into two beams by a glass prism. The 3470 Å radiation was incident upon a 1 mm thickness of dye solution contained in a glass cell. The resulting normal fluorescence continuum along with the laser beam ν_L was then trapped by total internal reflection in a glass capillary cell containing the scattering liquid. This ensured spatial coincidence between the stimulating radiation and the continuum. The capillary cell had inner and outer diameters of 2.4 and 4.0 mm respectively, and was 15 cm long; the electric field inside the capillary was estimated to be 1.2×10^5 V cm^{-1}. The exit end of the cell was imaged on the slit of a grating spectrograph having a dispersion of 10 A mm^{-1}. Spectra were recorded on Kodak IF photographic plates and on Polaroid film. With a slit width of 100 μm, the intensity of the background continuum at larger frequency shifts was sufficient to record spectra in a single laser pulse; however, in the region of smaller frequency shifts (i.e.

Fig. 19. Apparatus for inverse Raman spectroscopy at Bradford.

longer wavelengths) several pulses were required to accumulate sufficient continuum intensity. Figure 19 shows a photograph of a similar apparatus constructed in our laboratories at Bradford University.

With this technique Stoicheff and McLaren[14] obtained inverse Raman spectra of diamond, liquid benzene and toluene. In benzene and toluene a number of lines were observed. In diamond a sharp band at 1332 cm^{-1} was observed from a cross sectional area of 5×10^{-2} cm^2 of crystal, 2.4 mm thick. Observations were also made of the effect of varying the state of polarization of the laser radiation and the continuum.

Stoicheff and McLaren concluded the published account of their work by suggesting that the inverse Raman effect could be extended to the ps range but using a mode locked laser for excitation. The necessary accompanying continuum of ps duration might be produced by a mode locked dye laser. Their prophecy was soon fulfilled. In 1971 Alfano and Shapiro of the General Telephone and Electronics Laboratories, Bayside, New York, described[15] successful experiments with ps inverse Raman spectroscopy. Alfano and Shapiro had shown in 1970 that, when a ps pulse of high intensity passes through a solid or liquid, self-focusing can occur and the pulse becomes self-phase modulated leading to spectral emission over the entire visible region. This continuum is suitable for inverse Raman spectroscopy. Its duration and the duration of the laser pulse is only 4 ps, that is 10^4 times shorter than the laser pulse and continuum used by Stoicheff and McLaren. Also the spectral region covered by the continuum is many thousands of wave numbers. This latest development in inverse Raman spectroscopy offers great potential for the study of very short lived species and the measurement of very short lifetimes.

The experimental procedure involved the use of a mode locked neodymium glass laser which emits ps pulses at 1.06 μm. These were converted to 5300 Å ps pulses by second harmonic generation in a KDP crystal. The 5300 Å pulses were then focused with a 20 cm focal length lens into a 5 cm sample of borosilicate BK–7 glass producing a self-phase modulated continuum. The beam measured 400 μm in diameter at the entrance and exit faces of the BK–7 glass and 350 μm in diameter at the centre. A fraction of the 5300 Å light was self-focused in the BK–7 glass into filaments, and the light emerging from the filaments was self-phase modulated and emitted in the region from 4000 Å–7000 Å. Since the dispersion is small in the BK–7 glass, the 5300 Å pulse, which still contained most of the energy, emerged from the glass coincident in time with the frequency swept part of the pulse. The swept pulse and the intense laser pulse passed into a sample of 5 cm length, and the emerging light was collected by a lens and imaged into either a 3/4 m Jarrell Ash f/6.5 or 1/2 m Jarrell Ash f/9 spectrograph. This experimental arrangement was chosen because the continuum generated in the glass is much more intense than any continuum generated in the samples themselves, so that the absorption takes place in the continuum generated in the BK–7 glass. Also the continuum generated by glass has a smooth intensity variation since there are no stimulated Raman lines emitted and no absorption in the visible region.

Each laser pulse had a power of 0.5 GW and 10^{-2} – 10^{-3} of the laser light energy emerged in the self-phase modulated continuum distributed over 10000 cm^{-1}. Most spectra were obtained with one laser shot. Appropriate filters were used to attenuate the direct laser radiation at 5300 Å. Spectra were recorded on Polaroid 3000 speed film. With this procedure, inverse Raman spectra on the anti-Stokes high frequency side of the laser frequency were obtained for carbon disulphide, benzene, liquid nitrogen, methanol and calcite.

An alternative method for the production of inverse Raman spectra involved allowing the ps pulse itself to generate the self-phase modulated continuum in the sample. The laser beam was focused directly into the sample and a ground glass screen placed after the sample. Radiation from the ground glass screen was imaged on to the grating of the spectrograph with a 5 cm focal length lens. Using this procedure Alfano and Shapiro observed inverse Raman spectra on the Stokes or low frequency side of the laser frequency in carbon disulphide, benzene and toluene. The results for benzene are particularly interesting. Stokes absorptions were observed at 507 cm^{-1}, 851 cm^{-1}, 1174 cm^{-1}, 1489 cm^{-1} and 1832 cm^{-1}. An emission at 992 cm^{-1} was also observed and this became very intense as the laser power was increased. The bands at 851, 1174 and 1489 cm^{-1} can be assigned to Raman active benzene fundamentals. The bands at 507 cm^{-1} and 1832 cm^{-1} could correspond to normally inactive benzene vibrations. There is also the possibility that some of these bands belong to a new species of very short life which can only be observed on a ps time scale.

CONCLUSION

Depuis toujours, l'homme a recherché la lumière et a trouvé en elle un joyeaux symbole de délivrance. Il y a plusieurs milliers d'années le prophète Isaie[16] *a parlé 'du peuple qui marchait dans les ténèbres (et qui) a vu une grande lumière'. Le spectroscopiste Raman a peut être marché longtemps dans l'ombre mais lui aussi a vu une grande lumière et j'espère vous avoir montré que ceci annonce sa delivrance!*

ACKNOWLEDGEMENTS

I would like to express my gratitude to Dr. L. Stanton, Dr. M. J. French (S.R.C.

research fellow) and Dr. A. G. Cram (S.R.C. research fellow), who have collaborated with me in recent years in work on non-linear phenomena. I would also like to thank Dr. P. D. Maker for keeping us informed of his work in this field and for the considerable help he gave me during the period he spent at Bradford University in 1971, as S.R.C. Senior Visiting Fellow. I am indebted to Professor J. D. McGee, F.R.S., of Imperial College, London, for advice and for supplying the image intensifier and to Mr. A. C. Dawe of E.M.I. London, for extensive help with the Ebitron TV camera.

For general technical assistance I am grateful to Mr. F. Bond, Chief Technician, and his workshop staff. I also thank Mr. A. Bolton, Mrs. J. K. Hillman and Mr. D. Farwell, who have prepared many of the drawings and photographs used in this chapter.

I also wish to acknowledge financial support received from the Science Research Council, the University of Bradford, and the Leverhulme Trust. Some of the work reported here was undertaken during my tenure of a Leverhulme Research Fellowship, 1970–1971.

Acknowledgements for figures

We acknowledge with thanks the permission of the authors and the journals concerned to reproduce the following figures: *Chem. Brit.* for Figs. 1, 3, 9, 12, 15 and 18; *Chem. Phys. Letters* and J. F. Verdieck, S. H. Peterson, C. M. Savage and P. D. Maker for Figs. 10 and 11; *Compt. Rend.* and S. Dumartin, B. Oksengorn and B. Vodar for Fig. 16; *Phys. Letters* and J. A. Duardo, F. M. Johnson and M. A. El-Sayed for Fig. 17.

The originals for Figs. 9 and 12 were supplied by P. D. Maker.

REFERENCES

1 D. A. Long, *Chem. Brit.* **7**, 108 (1971).

2 D. A. Long, *Essays in Structural Chemistry,* (Eds. D. A. Long, A. J. Downs and L. A. K. Staveley), Macmillan, London, 1971, Chapter 2.

3 D. A. Long and L. Stanton, *Proc. Roy. Soc.* **A318**, 441 (1970).

4 D. A. Long and L. Stanton, *Proc. Roy. Soc.* **A318**, 452 (1970).

5 S. J. Cyvin, J. E. Rauch and J. C. Decius, *J. Chem. Phys.* **43**, 4083 (1965), B. Fanconi, B. Tomlinson, L. A. Nafie, W. Small and W. L. Peticolas, *J. Chem. Phys.* **51**, 3993 (1969); B. Fanconi and W. L. Peticolas, *J. Chem. Phys.* **50**, 2244 (1969); J. H. Christie and D. J. Lockwood, *J. Chem. Phys.* **54**, 1141 (1971); L. Stanton, *J. Raman Spectry.* (1973) (in press).

6 R. W. Terhune, P. D. Maker and C. M. Savage, *Phys. Rev. Letters* **14**, 681 (1965).

7 C. M. Savage and P. D. Maker, *Appl. Opt.* **10**, 965 (1971).

8 D. A. Long and M. J. French, unpublished observation.

9 P. D. Maker, private communication.

10 J. F. Verdieck, S. H. Peterson, C. M. Savage and P. D. Maker, *Chem. Phys. Letters* **7**, 219 (1970).

11 W. J. Jones and B. P. Stoicheff, *Phys. Rev. Letters* **13**, 657 (1964).

12 S. Dumartin, B. Oksengorn and B. Voder, *C. R. Acad. Sci., Paris* **261**, 3767 (1965).

13 J. A. Duardo, F. M. Johnson and M. A. El-Sayed, *Phys. Letters* **21**, 168 (1966).

14 R. A. McLaren and B. P. Stoicheff, *Appl. Phys. Letters* **16**, 140 (1970).

15 R. R. Alfano and S. L. Shapiro, *Chem. Phys. Letters* **8**, 631 (1971).

16 *La Bible, Isaie,* ix, 2.

2. LES CHANGEMENTS DES SPECTRES RAMAN DANS UN FORT CHAMP LUMINEUX

M. A. Kovner, B. A. Medvedev, O. M. Parchkov
et S. K. Potapov
Université d'Etat de Saratov,
Saratov, USSR

ABSTRACT

The theory of several new effects arising in Raman spectra under the influence of a strong electric field of the light wave is proposed.

Stark effect. When the mean value of the Stark displacements in a disordered array of molecules is taken, one finds an asymmetric broadening of the Raman lines.

Fermi resonance. If the Raman scattering occurs on vibrational levels close enough, especially on the sub-levels of Fermi resonance, a new distribution of the intensities between the Fermi doublets must occur. The induced mechanical anharmonicity is also calculated and it is interesting to notice that it depends not only on the electric field of the light wave but also on the electrical anharmonicity of the molecule.

Modulation of the Raman line shape. If the field of the exciting radiation contains several modes, the Stark displacement can oscillate at the frequency of the beats between two neighbouring modes. Raman lines can thus appear with frequencies $\omega_S \pm n\pi$, with $n = 0, 1, 2 \ldots$ (ω_S = Stokes frequency, π = difference between the frequencies of two modes).

Enhancement of an infrared signal. When the stimulated Raman effect is excited in a non-stationary state, a population inversion between the levels of the normal vibration modes can be produced, even (in definite conditions) by an impulse of 10^{-9} s. This inversion can be used to enhance an infrared signal, e.g. the frequency difference $992 - 673 = 339$ cm^{-1} of benzene.

INTRODUCTION

Au fur et à mésure que l'intensité de la lumière excitante augmente, les paramètres de la substance éprouvent des changements considérables. C'est pourquoi on peut attendre une série entière des nouveaux effets pour la diffusion Raman dans un fort champ lumineux.

Dans ce travail nous étudions quatre effets de ce genre: le déplacement de Stark de la fréquence de la lumière diffusée, les changements des intensités des transitions entre les niveaux de la résonance de Fermi et en général entre les niveaux vibrationnels assez rapprochés, la structure discrète des lignes dans un cas d'un laser à plusieurs modes et l'émission infrarouge, dont la diffusion Raman est accompagnée. Tous les quatre effets sont étroitement liés entre eux et peuvent être étudiés à l'aide de la méthode de la matrice de densité dans ses diverses variantes en utilisant un hamiltonien effectif pour l'interaction entre le champ et la molécule

$$V = -\sum_{\mu\nu} \alpha_{\mu\nu} E_\mu E_\nu^* \tag{1}$$

où $\alpha_{\mu\nu}$ sont les composantes du tenseur de la polarisabilité dans un système d'axes fixes, et

$$\mathbf{E} = \mathbf{E}_L \exp(-i\omega t) + \mathbf{E}_S \exp(-i\omega_S t) \tag{2}$$

Avec cet hamiltonien, il est possible de formuler un système d'équations pour la matrice de densité S de sous-système des niveaux vibrationnels de l'état électronique fondamental

$$\dot{S}_{kl} + (\tau_{kl}^{-1} + i\omega_{kl}) S_{kl} = -\frac{i}{\hbar} \sum_j (V_{kj} S_{jl} - S_{kj} V_{jl}); k \neq l \tag{3}$$

$$\dot{S}_{ll} = -\frac{i}{\hbar} \sum_j (V_{lj} S_{jl} - S_{lj} V_{jl}) + \sum_j (T_{jl}^{-1} S_{jj} - T_{lj}^{-1} S_{ll}) \tag{4}$$

τ, T étant les temps transversal et longitudinal de la relaxation. En se basant sur la formule générale, qui détermine la liaison entre la polarisation et la matrice de densité, nous avons

$$P_\mu = \sum_{jkl} (\alpha_{kl}) \mu\nu S_{lk} (E_\nu + E_\nu^*) \tag{5}$$

En résolvant le systéme (3), (4), substituant les solutions dans (5) et en comparant le résultat avec le développement phénoménologique de la polarisation en série de Taylor par rapport au champ électrique, nous pourrons déterminer les susceptibilités Raman pour des cas concrets.

L'EFFET STARK

Soient deux niveaux vibrationnels 1 et 2 (avec les nombres quantiques 0 et 1) d'une vibration normale. Nous négligeons l'influence de la radiation diffusée sur cette vibration. Alors, dans les équations du type (3) on pourra conserver dans les coéffieients des élements non-diagonaux de la matrice S seulement le membre avec $|E_L|^2$ de (1). En outre, on se borne au cas d'une diffusion assez faible au point qu'elle ne change pas les populations des niveaux vibrationnels. Dans ce cas les élements diagonaux de S sont considérés comme fixés d'avance, et du système (3), (4) ne restera que l'équation unique

$$\begin{aligned}\dot{S}_{21} &= [\tau_{21}^{-1} + i\omega_{21} - \frac{i}{\hbar}(\alpha_{22} - \alpha_{11}) |E_L|^2 S_{21} \\ &= \frac{i}{\hbar} \alpha_{21} E_L E_S^* \exp(-i\omega t)(S_{11} - S_{22})\end{aligned} \tag{6}$$

La dérivation (6) faite nous avons pris pour simplifier que les champs électriques de la radiation laser et Stokes sont dirigés suivant l'axe z. Alors, la somme (1) ne consiste que d'un membre, l'indice z étant omis. Les cas plus généraux seront traités plus loin.

Cherchons la solution de l'équation (6) sous la forme

$$S_{21} = S_{21}^{\circ} \exp(-i\omega t), \omega = \omega_L - \omega_S \tag{7}$$

En substituant (7) dans (6), nous obtenons l'expression pour l'amplitude constante

$$S_{21}^{\circ} = \frac{\alpha_{21} E_L E_S^* (S_{11} - S_{22})}{\hbar [\omega_{21} - \omega - [(\alpha_{22} - \alpha_{11}) |E_L|^2]/\hbar - i\tau_{21}^{-1}]} \tag{8}$$

La substitution de (8), (7), (2) dans (5) nous permet de trouver la partie imaginaire de la susceptibilité

$$\psi''(\omega_L, -\omega_L, \omega_S) = -\frac{\alpha_{21}^2 \ \tau_{21}^{-1} (S_{11} - S_{22})}{\hbar[(\omega_{21} - \Delta\omega - \omega)^2 + \tau_{21}^{-2}]} \tag{9}$$

où le déplacement de Stark de la fréquence est

$$\Delta\omega = (\alpha_{22} - \alpha_{11}) \, |E_L|^2/\hbar \tag{10}$$

Pour l'évaluation numérique de $\Delta\omega$ nous développons α_{22} et α_{11} dans les séries par rapport à la coordonnée normale q. En utilisant les formules bien connues pour les élements matriciels de q et q^2, nous obtenons

$$\alpha_{22} - \alpha_{11} = \frac{\hbar}{2\omega_{21} M} \left(\frac{\partial^2 \alpha}{\partial q^2}\right)_0 \tag{11}$$

où M est la masse réduite de l'oscillateur.

Sidorov *et al.*[1] ont trouvé les deuxièmes dérivées de la polarisabilité par rapport aux coordonnées normales pour plusieurs substances, le calcul étant basé sur les valeurs experimentales des intensités des harmoniques. Ces valeurs laissent determiner la différence $\alpha_{22} - \alpha_{11}$ de l'ordre $10^{-25} - 10^{-26}$ cm^3. Alors, pour les champs de l'ordre $10^4 - 10^5$ CGSE nous avons calculé que $\Delta\omega$ est d'ordre de quelques cm^{-1}.

Dans les expressions (9) et (10) il faut exprimer les composantes des tenseurs de la polarisabilité par les cosinus directeurs entre les systèmes d'axes fixes et liés à la molécule. Faisant la moyenne pour les orientations chaotiques des molécules, nous pouvons présenter la susceptibilité sous la forme d'une intégrale par rapport aux angles d'Euler. Son calcul nous permet en conclure que le contour de ψ'' s'élargit et devient asymétrique. L'élargissement et l'asymétrie dépendent de E_L ainsi de l'anisotropie des tenseurs de la polarisabilité.

LA RESONANCE DE FERMI

Envisageons maintenant le système de trois niveaux vibrationnels 1,2 et 2′ avec les fréquences des transitions, satisfaisant à l'inégalité $\omega_{2'2} \ll \omega_{21}, \omega_{2'1}$. En admettant de nouveau que les champs de la lumière diffusée sont assez faibles devant E_L, nous pouvons poser nuls les élements diagonaux S_{22}, $S_{2'2'}$ et $S_{2'2}$. Dans le cas d'une lumière laser linéarement polarisée, on obtient deux equations du type (3)

$$\dot{S}_{21} + \left[\tau_{21}^{-1} + i\omega_{21} - i(\alpha_{22} - \alpha_{11}) \frac{|E_L|^2}{\hbar}\right] S_{21} - i\alpha_{2'2} \frac{|E_L|^2}{\hbar} S_{2'1}$$

$$= i\alpha_{2'1} E_L E_S^* \exp(-i\omega t) (S_{11} - S_{22}) \tag{12}$$

$$\dot{S}_{2'1} + [\tau_{2'1}^{-1} + i\omega_{2'1} - i(\alpha_{2'2'} - \alpha_{11}) \frac{|E_L|^2}{\hbar} S_{2'1} - i\alpha_{2'2} \frac{|E_L|^2}{\hbar} S_{21}$$

$$= i\alpha_{21} E_L E_S^* \exp(-i\omega t) (S_{11} - S_{22}) \tag{13}$$

Cherchons d'abord la solution du système d'équations homogènes à l'aide de la substitution suivante

$$S_{21} = a_{21} \exp(-i\omega t), S_{2'1} = a_{2'1} \exp(-i\omega t) \tag{14}$$

Celle-ci nous amène à deux équations homogènes pour les amplitudes

$$(\tilde{\omega}_{21} - \omega) a_{21} + \alpha_{2'2} \frac{|E_L|^2}{\hbar} a_{2'1} = 0$$

$$\alpha_{2'2} \frac{|E_L|^2}{\hbar} a_{21} + (\tilde{\omega}_{2'1} - \omega) a_{2'1} = 0 \tag{15}$$

où

$$\tilde{\omega}_{21} = \omega_{21} - (\alpha_{22} - \alpha_{11}) \frac{|E_L|^2}{\hbar} + i\tau_{21}^{-1} \tag{16}$$

La résolution de l'équation séculaire (ne tenant compte que des parties reélles des racines, parce que ce sont justement elles qui déterminent les fréquences déplacées nécessaires) nous donne deux racines

$$\omega^{(1,2)} = c^E \pm x^E \tag{17}$$

où

$$c^E = (\omega_{21} + \omega_{2'1})/2 - (\alpha_{22} + \alpha_{2'2'} - 2\alpha_{11}) \frac{|E_L|^2}{2\hbar} \tag{18}$$

$$x^E = (|p^E|^2 + |\Delta^E|^2)^{7/2} \tag{19}$$

$$p^E = \alpha_{2'2} \frac{|E_L|^2}{\hbar} \tag{20}$$

$$\Delta^E = (\omega_{2'1} - \omega_{21})/2 + (\alpha_{22} - \alpha_{2'2}) \frac{|E_L|^2}{2\hbar} \tag{21}$$

Les paramètres (19)–(21) sont analogues aux paramètres x, p, Δ, introduits par Placzek[2] pour la description de la résonance de Fermi ordinaire. Mais les paramètres nouveaux dèpendent du champ extérieur. La grandeur p^E répresente l'anharmonicité induite par le champ.

Pour calculer les élements matriciels de la polarisabilité, supposons tout concrètement que les niveaux 1,2,2′ sont les niveaux 00°0, 10°0 et 02°0 d'une molécule linéaire symétrique à trois atomes (CO_2, CS_2, N_2O), et en outre que les deux derniers niveaux sont les composantes de la diade de Fermi. Ensuite, en se servant des formules de Placzek pour les fonctions d'onde perturbées et développant la polarisabilité en série de Taylor par rapport aux coordonnées normales q_1, q_2 des vibrations symétriques de valence et de déformation, on obtient

$$\alpha_{11} = \alpha_0 + \left(\frac{\partial^2 \alpha}{\partial q_1^2}\right)_0 c_1^2/2 + \left(\frac{\partial^2 \alpha}{\partial q_2^2}\right)_0 c_2^2/2$$

$$\alpha_{22} = \frac{1}{2x} \left\{ (x + |\Delta|)\alpha_{10}^{10} + (x - |\Delta| \alpha_{02}^{02} \pm 2\sqrt{x^2 - \Delta^2}\, \alpha_{10}^{02} \right\}$$

$$\alpha_{2'2'} = \frac{1}{2x}\left\{(x-|\Delta|)\,\alpha_{10}^{10} + (x+|\Delta|)\,\alpha_{02}^{02} \mp 2\sqrt{x^2-\Delta^2}\,\alpha_{10}^{02}\right\}$$

$$\alpha_{2'2} = \frac{\sqrt{x^2-\Delta^2}}{2x}\left[\left(\frac{\partial^2\alpha}{\partial q_1^2}\right)_0 c_1^2 - 2\left(\frac{\partial^2\alpha}{\partial q_2^2}\right)_0 c_2^2\right] \tag{22}$$

Les élements matriciels de la polarisabilité non-perturbés par la résonance de Fermi ordinaire sont

$$\alpha_{10}^{10} = \alpha_0 + 3\left(\frac{\partial^2\alpha}{\partial q_1^2}\right)_0 c_1^2/2 + \left(\frac{\partial^2\alpha}{\partial q_2^2}\right)_0 c_2^2/2$$

$$\alpha_{02}^{02} = \alpha_0 + \left(\frac{\partial^2\alpha}{\partial q_1^2}\right)_0 c_1^2/2 + 5\left(\frac{\partial^2\alpha}{\partial q_2^2}\right)_0 c_2^2/2; \alpha_{10}^{02} = 0 \tag{23}$$

Les grandeurs b_i sont les amplitudes nulles et donc, selon Placzek, nous avons

$$\Delta = \nu_1 - 2\nu_2; x = \sqrt{\Delta^2 + 16p^2}, p = \frac{\beta_{12}}{4\pi}\sqrt{\frac{h}{(\nu_1 + 2\nu_2)^3}} \tag{24}$$

β_{12} est le coefficient du terme $q_1 q_2^2$ dans le développement de l'énergie potentielle par rapport aux coordonnées normales.

La substitution de (23) dans (22) et puis de (22) dans (17)–(21) donnera les formules finales pour les fréquences déplacées

$$\omega^{(1,2)} = (\omega_{21} + \omega_{2'1})/2 - \left[\left(\frac{\partial^2\alpha}{\partial q_1^2}\right)_0 c_1^2 + 2\left(\frac{\partial^2\alpha}{\partial q_2^2}\right)_0 c_2^2\right]\frac{|E_L|^2}{2\hbar}$$

$$\pm\left\{\frac{x^2-\Delta^2}{4x^2}\left[\left(\frac{\partial^2\alpha}{\partial q_1^2}\right)_0 c_1^2 - 2\left(\frac{\partial^2\alpha}{\partial q_2^2}\right)_0 c_2^2\right]^2 \frac{|E_L|^4}{\hbar^2}\right.$$

$$\left. + \left[(\omega_{2'1} - \omega_{21})/2 + \frac{\Delta}{x}\left[\left(\frac{\partial^2\alpha}{\partial q_1^2}\right)_0 c_1^2 - 2\left(\frac{\partial^2\alpha}{\partial q_2^2}\right)_0 c_2^2\right]\frac{|E_L|^2}{2\hbar}\right]^2\right\}^{7/2} \tag{25}$$

On voit de (25), que les fréquences, changées sous l'influence du champ, dépendent des paramètres de l'anharmonicité électrique de la molécule. Dans les cas les plus intéressants de CO_2 et CS_2 ces paramètres nous sont malheureusement inconnus. Cependant en utilisant les évaluations de Sidorov *et al.*[1] et en prenant pour le champ la valeur 10^4 CGSE nous obtenons pour les déplacements des fréquences les valeurs de l'ordre de quelques cm^{-1}.

A l'aide des équations (14)–(16) nous avons trouvé les solutions du système inhomogène (12), (13) pour les élements nondiagonaux de la matrice de dencité. Enfin en utilisant l'équation du type (5) nous avons obtenu pour les parties imaginaires des susceptibilitiés de Stokes les formules suivantes

$$\psi_{s1}'' = \left\{\sqrt{\frac{x^E + |\Delta^E|}{2x^E}}\sqrt{\psi_{s1}^{0''}} \pm \sqrt{\frac{x^E + |\Delta^E|}{2x^E}}\sqrt{\psi_{s2}^{0''}}\right\}^2 \tag{26}$$

$$\psi''_{s2} = \left\{ \mp \sqrt{\frac{x^E - |\Delta^E|}{2x^E}} \sqrt{\psi^{0''}_{s1}} + \sqrt{\frac{x^E + |\Delta^E|}{2x^E}} \sqrt{\psi^{0''}_{s2}} \right\}^2 \qquad (27)$$

où $\psi^{0''}_{s1}$, $\psi^{0''}_{s2}$ sont les parties imaginaires des susceptibilités pour le champ extérieur égal à zero. Il suit de (26), (27) que dans un cas général on peut s'attendre à des redistributions différentes des intensités parmi les lignes avec les fréquences $\omega^{(1)}$ et $\omega^{(2)}$. Ces intensités dépendent de differentes combinaisons de signes de Δ^E et x^E.

Il est à noter que dans la réf. 3 on a montré que la disposition des niveaux 100°0 et 02°0 dans la molécule de CO_2 est inverse de celle qui était acceptée jusqu'ici.

Passons maintenant au cas ou les niveaux 2,2′ ne présentent pas des composantes d'une diade de Fermi, mais sont des niveaux assez rapprochés. Ils peuvent appartenir à deux fréquences d'une même type de symétrie ou même de deux types de la symétrie différents. Comme on voit de (20), ces niveaux seront liés entre eux par le champ extérieur et entre ces niveaux surgit une résonance de Fermi induite. Il peut arriver, qu'en absence d'une telle interaction seulement la transition $2 \to 1$ est permise et se réalise. Mais, il suit de (26), (27) que la transition $2' \to 1$ peut acquérir une intensité notable sous l'influence du champ. C'est justement un cas pareil était observé dans la réf. 4 dont les auteurs ont constaté dans le spectre Raman de la paranitroaniline une redistribution des intensités entre les deux fréquences rapprochées pendant l'excitation de la diffusion Raman par la lumière d'un laser.

LA STRUCTURE DESCRETE DE LA RAIE RAMAN DANS UN CHAMP D'UN LASER A PLUSIEURS MODES

Si le champ d'un laser contient plusieurs modes, le déplacement de Stark de la fréquence des vibrations moléculaires peut osciller avec la fréquence des battements entre les modes voisins. Ces oscillations à leur tour modulent les vibrations moléculaires et dans le spectre Raman apparaissent des raies avec les fréquences $\omega_S + n\pi(-0, \pm 1, \pm 2$; π est la différence de deux modes voisins). Appliquons à une molécule à deux niveaux les équations (1) et (3). Dans le cas d'une diffusion faible l'équation (4) peut être négligée. Par contre à (2), présentons le champ électrique sous la forme

$$\mathbf{E} = \mathbf{E}_L \exp(-i\omega_L t) + \mathbf{E}'_L \exp(-i\omega'_L t) + \mathbf{E}_S \exp(-i\omega_S t) \qquad (28)$$

A l'aide du développement

$$\exp(ix \sin\theta) = \sum_{n=-\infty}^{\infty} J_n(x) \exp(in\theta) \qquad (29)$$

dans lequel J_n est la fonction de Bessel d'ordre n, nous trouvons la solution suivante de l'équation du type (3)

$$S_{21} = \frac{i}{\hbar} \alpha_{21} E^*_S \exp(-i\omega t)(S_{11} - S_{22})$$

$$\times \sum_{m,k=-\infty}^{\infty} \left\{ E_L \frac{J_m(x) J_k(x)}{\tau_{12}^{-1} - i(\omega - \varpi_{21} + k\pi)} \times \exp[-i(k-m)(\pi t + \rho)] \right.$$

$$+E_L' \frac{J_m(x)J_k(x)}{\tau_{12}^{-1} - i\,[\omega - \widetilde{\omega}_{21} + (k+1)\,\pi]} \exp\left\{- i\,[(k+1-m)\pi + (k-m)\rho]\right\}\Bigg\} \tag{30}$$

L'argument de la fonction de Bessel

$$x = 2(\alpha_{22} - \alpha_{11})\,|E_L E_L'|\,/\,\hbar\pi \tag{31}$$

ρ est la différence des phases de deux modes. Contrairement à (16), la fréquence déplacée sur le compte de l'effet de Stark

$$\widetilde{\omega}_{21} = \omega_{21} - (\alpha_{22} - \alpha_{11})\,(|E_L|^2 + |E_L'|^2)\,/\,\hbar \tag{32}$$

En prenant la moyenne des phases et admettant pour simplifier que $E_L \cong E_L'$, nous obtenons pour la polarisation de la fréquence de Stokes

$$\omega_s = \omega_L - \widetilde{\omega}_{21} + n\,\pi : P = -\frac{i}{\hbar}\alpha_{21}^2\,|E_L|^2 E_S \exp(-i\omega_S t)(S_{11} - S_{22})$$

$$\times \sum_{k=-\infty}^{\infty} \left\{ J_k^2(x) + J_{k-1}^2(x) \right\} \tau_{21} \tag{33}$$

La partie imaginaire de la susceptibilité de cette fréquence est

$$\psi'' = -\frac{i}{\hbar}\,\alpha_{21}^2\,(S_{11} - S_{22}) \sum_{k=-\infty}^{\infty} \left\{ J_k^2(x) + J_{k-1}^2(x) \right\} \tau_{21} \tag{34}$$

Il résulte de (34), que la modulation de Stark des vibrations moléculaires a pour conséquence l'apparition d'une structure discrète (quand $\pi > \tau_{21}^{-1}$) ou un élargissement flou (quand $\pi < \tau_{21}^{-1}$) de la ligne de la radiation diffusée.

Nous allons rechercher ensuite la déformation du contour de ψ'' sous l'influence du champ de laser. Admettant approximativement $\alpha_{22} - \alpha_{11} \cong 10^{-25} - 10^{-26}$ cm^3, $E_L, E_L' \cong 10^4 - 10^5$ CGSE, $\pi \cong 0,1$ cm^{-1}. Alors, $x \div 1 - 10$. Quand $\pi > \tau_{21}^{-1}$ ces chiffres correspondent à une situation possible dans les liquides N_2, O_2 et dans les gaz H_2, D_2, N_2, O_2, CH_4 ainsi que dans les liquides pour lesquelles la raie de la diffusion Raman stimulée est très étroite (par exemple, dans CS_2 la largeur $\sim 0,5$ cm^{-1}).

Le comportement de la susceptibilité est décrit par une fonction sans dimension

$$f(x) = \sum_{k=-\infty}^{\infty} [J_k^2(x) + J_{k-1}^2(x)]\,\pi\tau_{21} \tag{35}$$

et pour le cas $\tau_{21}^{-1} = 0.1\,\pi$ il est montré sur la Fig. 1. On y voit que l'augmentation de l'intensité de la radiation excitante et de paramètre x est accompagnée d'une égalisation des coéfficients de renforcement pour le nombre toujours croissant de composantes de la structure discrète de la radiation diffusée. Dans ces conditions le coefficient du renforcement maximal va s'abaïsser, ce que signifie la diminution de l'importance de la diffusion Raman stimulée. Une structure discrète de cette diffusion a été observée dans plusieurs travaux.[5] Sous l'excitation par champ d'un laser à deux modes ($\pi = 1,8$ cm^{-1})

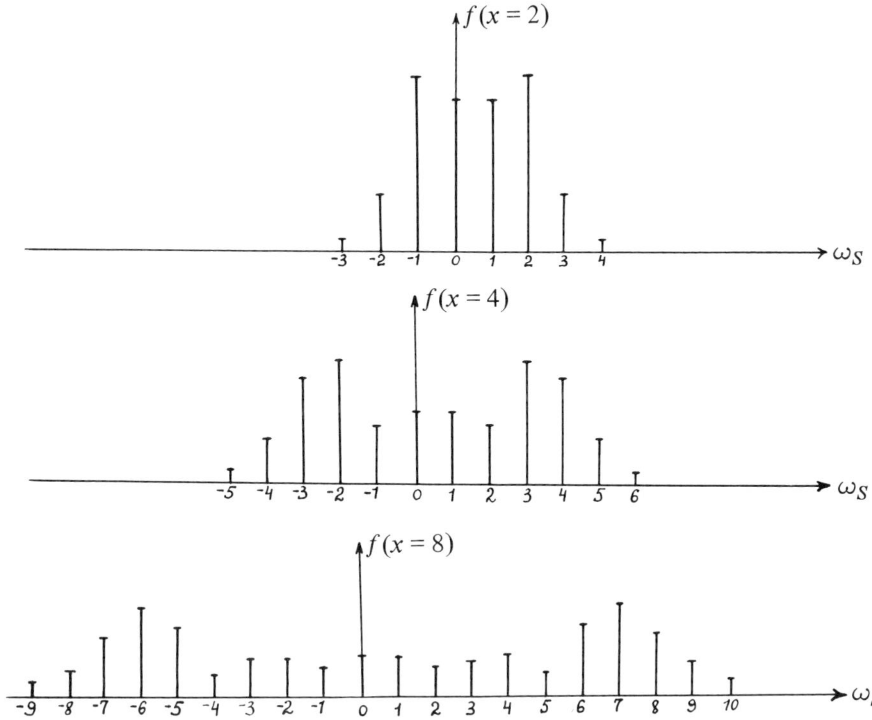

Fig. 1. La déformation du contour de la susceptibilité avec l'augmentation de l'intensité de la lumière excitante (l'échelle arbitraire). Le déplacement de la ligne centrale est déterminé par la composante constante du déplacement de Stark.

dans CS_2[6] on a observé une structure discrète avec une periode de 1,8 cm^{-1}. Elle peut être expliquée par le mécanisme que nous avons ici proposé et qui a pour origine la modulation de Stark des vibrations moléculaires.

LE RENFORCEMENT D'UN SIGNAL INFRAROUGE SUR LES NIVEAUX VIBRATIONNELS EXCITES DES MOLECULES PENDANT LA DIFFUSION RAMAN STIMULEE ET NONSTATIONNAIRE

La possibilité d'un renforcement d'un signal infrarouge dans les conditions d'une diffusion Raman stimulée stationnaire a été indiquée pour la première fois dans l'ouvrage.[7] Mais dans le cas d'une diffusion nonstationnaire, les processus de la relaxation sont éliminés; fait qui va créer des conditions plus favorables pour obtenir une inversion entre les niveaux excités. C'est pourquoi nous examinons le cas où la durée de la radiation excitatrice $\tau_L \ll T$. Sur la Fig. 2 est porté le schéma des niveaux actuels. Les niveaux 2 et 3 appartiennent à des vibrations normales différentes. La transition $1 \to 3$ est supposé être permise dans Raman et interdite dans l'infrarouge, tandis que la transition $2 \to 3$ possède des propriétés contraires. Prenons pour l'Hamiltonien de l'interaction entre le champ et la molécule la forme suivante

$$V = V - \hat{p}E_Q \exp(-i\,\omega_{23}\,t) \tag{36}$$

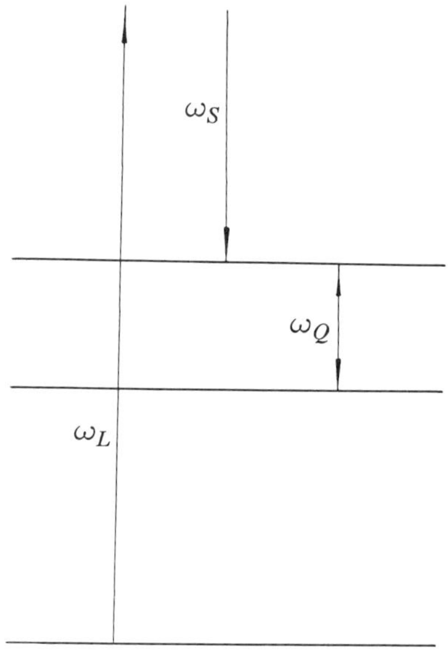

Fig. 2. Le système à trois niveaux. Les indices *L, S, Q* sont attribués à des champs de laser, Stokes et infrarouge.

où V est déterminé par (1), $\hat{p}$ est l'opérateur du moment dipolaire, E_Q est l'amplitude de champ de la radiation infrarouge.

Nous avons à résoudre les équations (3) et (4), en négligeant la deuxième somme dans (4) et en admettant que les champs E_L, E_S, E_Q dans le délais de l'action de l'impulse ne dependent pas du temps. La solution est recherchée par la méthode des approximations successives envers le champ E_Q. Dans l'approximation nulle ($E_Q = 0$) les solutions amènent aux formules suivantes pour les amplitudes R à des changements lents des élements matriciels de S

$$R_{12}^{(0)} = R_{23}^{(0)} = 0; \; R_{13}^{(0)} = A_{13} R_{11}^{e} \tau_{13} (1 - \exp(-t/\tau_{13}));$$

$$R_{22}^{(0)} = R_{22}^{e} \tag{37}$$

$$D^{(0)} = 2\,|A_{13}|^2 R_{11}^{e} \tau_{13} [t + \tau_{13} (\exp(-t/\tau_{13}) - 1)] + D^{e} \tag{38}$$

où $D = R_{33} - R_{22}$. La solution faite on a supposé que $R_{33} \ll R_{11} = R_{11}^{c}$, l'indice '$e$' indiquant la valeur d'équilibre. Les grandeurs A sont les suivantes

$$A_{13} = -\frac{i}{\hbar}\, \alpha_{13} E_L E_S^* = -A_{31}^*; \quad A_{23} = -\frac{i}{\hbar} P_{23} E_Q = -A_{32}^* \tag{39}$$

Supposons que[8] $\tau_{13}, \tau_{23} \ll \tau_I \ll T$. Alors, si $t = \tau_L$, il résulte de (38)

$$D^{(0)} = 2\,|A_{13}|^2 R_{11}^{e} \tau_{13} \tau_L + D^{e} \tag{40}$$

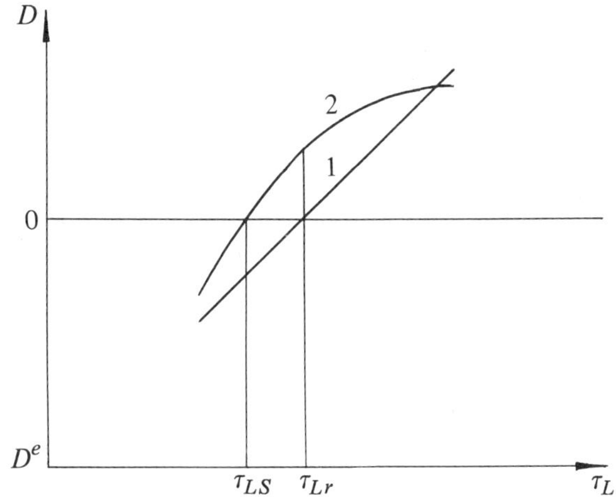

Fig. 3. La dépendance temporelle de la différence des populations dans la région $\tau_{13}, \tau_{23} \ll \tau_L \ll T$. Les courbes 1 et 2-les dépendances $D(t)$ en absence et en présence du champ infrarouge.

Nous déduisons une condition de seuil pour avoir une inversion entre les niveaux 3 et 2

$$|E_L\, E_S|_S \;=\; \hbar\alpha_{13}^{-1}\, \sqrt{|D^e|}\, (2\tau_L\, \tau_{13} R^e_{11})^{-1/2} \tag{41}$$

Si $\tau_L \cong 10^{-7}$s , $\tau_{13} \cong 10^{-12}$s , $\alpha_{13} \cong 10^{-25}\,\text{cm}^3, R^e_{11} \cong 1$, nous avons $|E_L E_S|_S \cong 10^6$ CGSE.

La première approximation donne le resultat $D^{(1)} = D^{(0)}$. Dans le deuxième ordre nous avons trouvé

$$D^{(2)} \;=\; D^{(0)} - 4|A_{23}|^2 R^e_{11}\, \tau_{13}\tau_{23}\tau_L^2 + 4|A_{23}|^2\, \tau_{23}\tau_L |D^e| \tag{42}$$

Le moment du temps, pour lequel les populations des niveaux 3 et 2 deviennent égales, est déterminé par l'expression

$$\tau_{L2} \;=\; \frac{|D^e|}{2|A_{13}|^2 R^e_{11}\,\tau_{13}} - \frac{|A_{23}|^2\; \tau_{23}|D^e|^2}{2|A^4_{13}|\; R^e_{11}\,\tau^2_{13}} \tag{43}$$

Le premier membre dans la partie droite de (43) caractérise le moment de la création d'une inversion si $E_Q = 0$.

Le renforcement du signal dans le canal 3–2 aura lieu, si la vitesse de pompage sur le niveau 3 dans las canaux 1–3 et 2–3 est moindre que la vitesse de pompage si $E_Q = 0$. Le moment du temps τ_{L2}, où commence le renforcement du signal infrarouge, peut être déduit de la condition $\dot{D}^{(0)} = \dot{D}^{(2)}$ et exprimé par la formule

$$\tau_{L2} \;=\; \frac{|D^e|}{2|A_{13}|^2 R^e_{11}\tau_{13}} \tag{44}$$

On peut voir de (43), (44) que le renforcement du signal dans le canal 2–3 est en retard relativement le moment de la création de l'inversion, le temps du retard étant donné par la différence

$$\tau_{Lr} - \tau_{LS} = |A_{23}|^2 \tau_{23} |D^e|^2 / 2|A_{13}|^4 R^e_{11} \tau^2_{13} \tag{45}$$

La cinétique du processus du mouvement des populations est qualitativement presentée sur la Fig. 3. Admettant $P_{23} \cong 10^{-19}$ CGSE, $|E_L E_S| \cong 10^6$ CGSE, $\tau_{23} \cong 10^{-10}$s., $|E_Q| \cong 10^{-1}$ CGSE, nous obtenons pour la différence $\tau_{Lr} - \tau_{LS}$ une valeur d'ordre 0,5.10^{-8} s, tandis que $\tau_{LS} \cong 10^{-7}$ s.

Le modèle mentionné peut être realisé dans une série de liquides, actives par rapport à la diffusion Raman stimulée. En particulier, pour le benzène les niveaux actifs peuvent être les niveaux, correspondant aux vibrations 992 et 673 cm^{-1}. Le renforcement aura lieu sur la fréquence de difference 319 cm^{-1}.

REFERENCES

1 N. K. Sidorov, L. S. Stalmachova et B. A. Eskin, *Opt. i Spektroskopiya* **25**, 194 (1968).
2 G. Placzek, *Rayleigh Streuung und Raman-Effekt,* Leipziger Vortrage, 1931.
3 G. Amat, *J. Pure Appl. Chem.* 383 (1969).
4 J. S. Bobovitch et A. V. Bortkevitch, *Opt. i Spektroskopiya* **25**, 763 (1968).
5 M. A. Bol'chov, G. V. Venkin, S. A. Jilkin et I. I. Nurminsky, *JETP* **58**, 3 (1970).
6 G. Gunter, K. H. Hellwege et G. Schaak, *Z. Physik.* **221**, 411 (1969).
7 V. S. Gorelik, V. A. Zoubov, M. M. Souschinsky et V. A. Tchirkov, *Pis'ma JETP* **4**, 52 (1966).
8 P. G. Krukov et V. S. Letochov, *Ouspechi Fisitch. Naouk* **99**, 169 (1969).

3. POSSIBILITIES AND LIMITS OF INVERSE RAMAN SCATTERING

A. Lau, W. Werncke, M. Pfeiffer, K. Lenz
and H.-J. Weigmann
Zentralinstitut für Optik und Spektroskopie
der Akademie der Wissenschaften der DDR, Berlin, DDR

In contrast to the usual Raman emission spectroscopy, inverse Raman scattering (IRS) is an absorption spectroscopy. The absorption lines appear within a continuous spectrum which is irradiated into the sample at the same time as the laser light.[1] The absorption lines lie on the anti-Stokes side of the laser frequency and come out from the conventional Raman transitions. The relationship of the IRS to the Raman emission

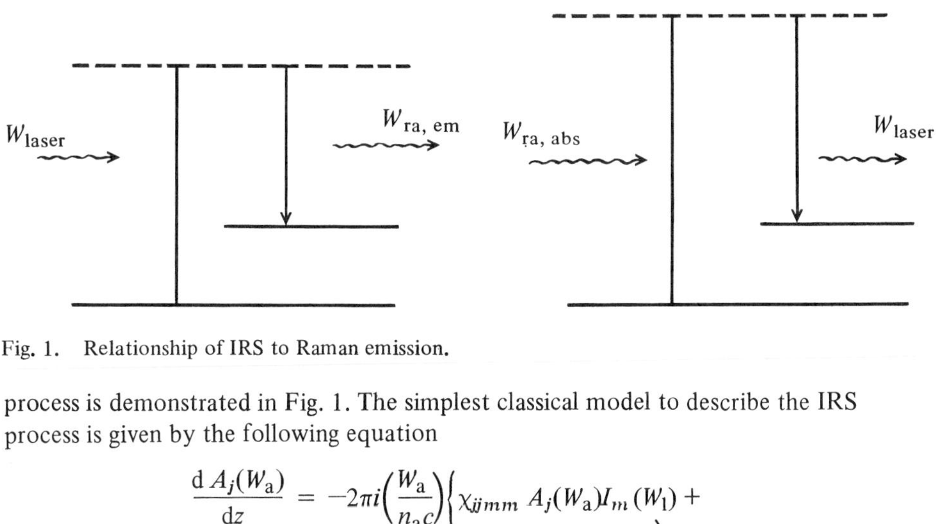

Fig. 1. Relationship of IRS to Raman emission.

process is demonstrated in Fig. 1. The simplest classical model to describe the IRS process is given by the following equation

$$\frac{\mathrm{d}\,A_j(W_a)}{\mathrm{d}z} = -2\pi i\left(\frac{W_a}{n_a c}\right)\left\{\chi_{ijmm}\,A_j(W_a)I_m(W_l) + \chi_{jjll}\,A_j(W_a)I_l(W_l)\right\} - a_j\,A_j(W_a) \qquad (1)$$

A = amplitude of the wave to be absorbed, W_a, W_l = anti-Stokes and laser frequencies respectively, $n_a c$ = diffraction coefficient and light velocity, χ = non-linear susceptibility in 3rd order and of rank 4 which is responsible for scattering efficiency and line width of normal Raman scattering, j, m, l = polarization components of waves and susceptibilities, a_j = absorption coefficient.

Discussion of the equation describing IRS shows the following facts:

1. if the laser radiation and the broad continuum pass the sample colinearly it is possible to get all normal Raman lines in absorption;
2. if the directions of the two radiations are non colinear during the interaction with the sample it is possible to get further information about vibrational frequencies which are forbidden in the case of the normal Raman scattering;[2]
3. using radiations which are plane-polarized it is useful to discuss the two cases (for isotropic media): (a) both radiations are polarized in parallel (both indices in each term of Eqn. (1) are equal), and (b) both radiations are polarized perpendicularly (both indices in Eqn. (1) are different).

The values for the absorption in the two cases are different for $\chi_{\parallel}, \chi_{\perp}$. With these different χ-values we get an expression

$$\rho = \frac{\chi_{\perp}}{\chi_{\parallel}}$$

which is equivalent to the degree of polarization of the normal Raman spectroscopy. From the connection between the χ-values and the mean value α and anisotropy δ of the polarizability ellipsoid, we get exactly the same expression for ρ as in normal Raman spectroscopy

$$\rho = \frac{3\delta^2}{45\alpha^2 + 4\delta^2}$$

Using pulsed laser radiation as the intense monochromatic source in connection with intense broad continua, the full information of the Raman spectrum will be obtained in a time as short as the duration of the laser pulse. For Q-switch lasers this means times in the range of 10^{-8} s. For mode-locked lasers it is possible to get this information in times of 10^{-11} to 10^{-12} s.

The limits for producing inverse Raman spectra are given by the qualities of the radiation fields and by the possibilities of the detection unit. For the laser intensity there exist two contradictory demands. Firstly, to make the absorption highly effective large laser intensities are necessary. In the range of a small absorption effect ($A = \Delta I_a / I_a < 1$) according to Eqn. (1) the relation

$$A = g N I_l l \tag{2}$$

holds. (N = mole %, I_l = laser intensity, l = length of interaction). High laser intensity is necessary because of the absorption coefficient g, which is essentially the imaginary part of χ is very small. A typical value is

$$g = 3.10^{-3} \left(\text{cm} \cdot \frac{\text{MW}}{\text{cm}^2}\ \text{mole}\right)^{-1}$$

Secondly, Eqn. (1) holds only for laser intensities below a value which gives generation of other non-linear effects, which destroy the visibility of the absorption.[3]

As to the necessary continuous spectrum, there appear to be two problems. First the continuum must have a minimum intensity for the absorption to surpass the spontaneous anti-Stokes emission at the same frequency, produced by the laser. Secondly it is rather difficult to produce a sufficiently intense continuum for such a short time interval which is given by the duration of the laser pulse. All together this results in

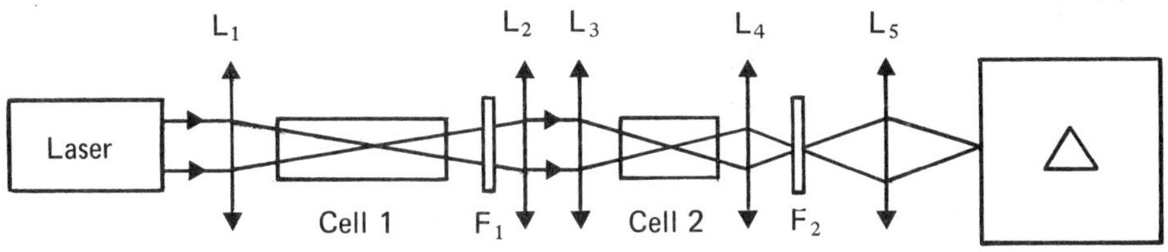

Fig. 2. Scheme of a 'two-cell' device. L_{1-5} are lenses, F a proper filter.

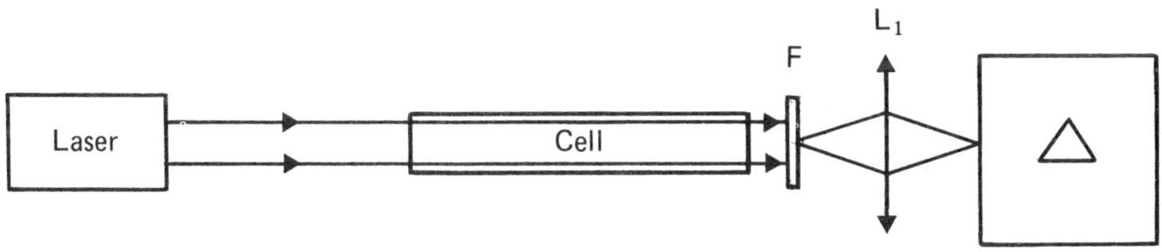

Fig. 3. Scheme of a 'one-cell' device.

Table 1. Detectability of different organic substances using the inverse Raman scattering in binary mixtures.

Substance	Solvent	Detectability (mole/l)
Pyridine	Toluene	0.005
Benzene	Toluene	0.01
Aniline	Toluene	1.0
Acetophenone	Toluene	1.0
Carbon disulphide	Chloroform	0.03

conditions for the two radiation fields which are difficult to perform.

There are two different experimental arrangements to get the inverse Raman scattering, the two-cell and the one-cell arrangement, which are shown in Figs. 2 and 3. In the 'two-cell' arrangement the first cell is taken to generate a proper continuum, e.g. laser plasma, fluorescence of dyes, broad anti-Stokes stimulated Raman scattering, phase modulation continua and so on. In the second cell the inverse Raman scattering is generated.[4]

In the 'one-cell' arrangement the sample itself generates a continuum suitable for the observation of the inverse Raman scattering by non-linear optical effects in the same cell.[4]

It has been shown experimentally, that all or nearly all Raman lines can be seen in the inverse Raman scattering. For the application of this effect to chemical systems the minimal amount of one component to be detectable in mixtures is very important. We have been able to get the results shown in Table 1.[4]

REFERENCES

1 W. J. Jones and B. P. Stoicheff, *Phys. Rev. Letters* **13**, 657 (1964).
2 L. D. Jevleva and T. J. Karagodova, *Opt. i Spektroskopiya* **23**, 991 (1967); S. Dumartin, B. Oksengorn and B. Vodar, *Compt. Rend.* **261**, 3767 (1965).
3 A. Lau, M. Pfeiffer *et al., Opt. Commun.* **4**, 228 (1971).
4 P. Gadow, A. Lau *et al., Opt. Commun.* **4**, 226 (1971).

4. ETAT DE POLARISATION D'UN FAISCEAU LUMINEAUX TRES INTENSE APRES TRAVERSEE D'UN MILEAU NON-LINEAIRE

X. Nguyen Phu, C. Desblancs et G. Rivoire

Institut Universitaire de Technologie,
Angers, France

ABSTRACT

Using Kielich's method of calculation for the study of the birefringence detected by a low intensity beam in a medium the molecules of which are oriented by an intense optical field, the following results have been obtained in the case when the field used for the measurement is the orientating field itself:

1. Calculation of the refractive index for a linearly polarized field and for a circularly polarized field.
2. Calculation of the variation of the orientation and of the ellipticity of an elliptically polarized field.
3. Calculation of the ratio of the thresholds of stimulated Raman effect obtained with linear and circular fields, assuming that the Raman threshold is identical with the self-focusing one.

INTRODUCTION

L'autofocalisation a été découverte expérimentalement en 1962 par Askaryan [1] et étudiée depuis lors par de nombreux auteurs. En particulier, Chiao, Garmire et Townes,[2] Talanov [3] et Kelley [4] ont montré que lors de la propagation d'un faisceau lumineux intense dans un milieu non linéaire dont l'indice de réfraction est de la forme $n = n_0 + B|E|^2$, (E champ électrique de l'onde lumineuse), le faisceau s'autofocalise dès que sa puissance est supérieure à une valeur P_a, appelée puissance de seuil d'autofocalisation avec

$$P_a = \frac{(1{,}22\,\lambda)^2 c^2}{64B}$$

Les causes possibles de variation de l'indice de réfraction avec le champ sont nombreuses, et dépendent du corps actif. Les principales causes étudieés jusqu'à présent sont l'orientation des molécules par le champ,[5] ou effet Kerr Optique, leur redistribution,[6,7] l'électrostriction,[8,9] l'effet électrocalorifique, et les divers phénomènes d'émission ou de diffusion, en particulier Raman [10] qui peuvent se produire dans le milieu actif. Pour les liquides, et dans les conditions d'excitation les plus fréquemment utilisées, en particulier avec des laser déclenchés, l'effet Kerr Optique joue souvent un rôle prédominant dans la variation d'indice.

Si le corps dans lequel se produit l'autofocalisation est également actif en Raman stimulé (c'est le cas de nombreux liquides comme le sulfure de carbone, le nitrobenzène, le benzène, etc.), les caractéristiques de l'effet Raman obtenu sont fortement perturbées par l'autofocalisation. En particulier l'existence d'un seuil d'effet Raman P_R dans un tel corps peut s'expliquer par l'existence d'un seuil d'autofocalisation P_a et l'on peut identifier P_R à P_a. L'état de polarisation du faisceau excitateur a une influence sur P_a; on peut calculer P_a, ce qui revient d'ailleurs à calculer B, pour divers états de polarisation, en particulier rectiligne et circulaire. Il est bien évident que le résultat du calcul dépendra des causes du phénomène d'autofocalisation. Nous nous sommes proposés de calculer le rapport $B_{\text{rect}}/B_{\text{circ}}$ dans l'hypothèse où les causes d'autofocalisation étaient conjoinement l'effet Kerr Optique, la déformation non linéaire des molécules et l'électrostriction. Nous avons pour cela pris la méthode utilisée par Kielich [11] pour le calcul de la variation d'indice d'une onde de mesure de faible intensité polarisée rectilignement, traversant un milieu soumis à un champ optique rectiligne intense. Cette méthode présente plusieurs avantages: d'une part elle permet d'obtenir B en fonction des caractéristiques moléculaires, d'autre part elle permet, par l'addition de divers termes dans le cheminement du calcul, de tenir compte ultérieurement d'autres effets comme la redistribution des molécules, les modifications des corrélations moléculaires. . .[12, 13]

Dans l'hypothèse $P_R = P_a$ nous avons

$$\frac{(P_R)_{\text{circulaire}}}{(P_R)_{\text{rectiligne}}} = \frac{B_{\text{rect}}}{B_{\text{circ}}}$$

Les valeurs mesurées pour le rapport des seuils Raman dans les 2 polarisations circulaire et rectiligne seront comparées aux valeurs calculées de $(B_{\text{rect}})/(B_{\text{circ}})$.

La méthode utilisée nous permettra également de calculer les modifications de polarisation subies par une vibration elliptique très intense au cours de la traversée d'un milieu non linéaire et de comparer les mesures aux calculs.

CALCUL DES VARIATIONS DE L'INDICE DE REFRACTION AVEC LE CHAMP D'UNE ONDE POLARISEE ELLIPTIQUEMENT

Théorie macroscopique

Considérons un milieu non absorbant macroscopiquement isotrope, de volume V contenant N molécules identiques, soumises à un champ optique intense $\mathbf{E}$ dont la fréquence est assez grande par rapport à toutes les fréquences naturelles des molécules. Le champ est supposé polarisé elliptiquement, et peut être décomposé en 2 vibrations rectilignes orthogonales

$$\mathbf{E} = \mathbf{E}^{(1)} + \mathbf{E}^{(2)} \tag{1a}$$

avec

$$\mathbf{E}^{(1)} = E_0^{(1)}\mathbf{e}^{(1)} \cos \omega t, \quad \mathbf{E}^{(2)} = E^{(2)}\mathbf{e}^{(2)} \sin \omega t \tag{1b}$$

$\mathbf{e}^{(1)}$ et $\mathbf{e}^{(2)}$ désignant deux vecteurs unitaires sur les axes de l'ellipse.

Le moment dipolaire total $\mathbf{M}$ sous l'action du champ local $\mathbf{E}^s$ est

$$M_\sigma = M_\sigma^0 + \left(\frac{\partial M_\sigma}{\partial E_\tau^s}\right)_0 E_\tau^s + \frac{1}{2}\left(\frac{\partial^2 M_\sigma}{\partial E_\tau^s\,\partial E_\nu^s}\right)_0 E_\tau^s E_\nu^s + \frac{1}{6}\left(\frac{\partial^3 M_\sigma}{\partial E_\tau^s\,\partial E_\nu^s\,\partial E_\rho^s}\right)_0 E_\tau^s E_\nu^s E_\rho^s + \ldots \tag{2a}$$

soit

$$M_\sigma = M_\sigma^0 + \left(\frac{\partial M_\sigma}{\partial E_\tau^s}\right)_0 [E_\tau^{(1)s} + E_\tau^{(2)s}] + \frac{1}{2}\left(\frac{\partial^2 M_\sigma}{\partial E_\tau^s \partial E_\nu^s}\right)_0 [E_\tau^{(1)s} + E_\tau^{(2)s}][E_\nu^{(1)s} + E_\nu^{(2)s}]$$

$$+ \frac{1}{6}\left(\frac{\partial^3 M_\sigma}{\partial E_\tau^s \partial E_\nu^s \partial E_\rho^s}\right)_0 [E_\tau^{(1)s} + E_\tau^{(2)s}][E_\nu^{(1)s} + E_\nu^{(2)s}][E_\rho^{(1)s} + E_\rho^{(2)s}] + \ldots \quad (2b)$$

avec

$$\mathbf{E}^s = \frac{n^2 + 2}{3}\mathbf{E} \quad (2c)$$

L'énergie interne U du système se développe en série de puissances de $\mathbf{E}^s$

$$U(\tau, \mathbf{E}) = U(\tau, 0) - M_\sigma^0 [E_\sigma^{(1)s} + E_\sigma^{(2)s}] - \frac{1}{2}\left(\frac{\partial M_\sigma}{\partial E_\tau}\right)_0 [E_\sigma^{(1)s} + E_\sigma^{(2)s}][E_\tau^{(1)s} + E_\tau^{(2)s}] - \ldots \quad (3)$$

En introduisant le tenseur de polarisabilité différentielle[11] $\Pi_{\sigma\tau}^{(1:2)}$ mesurant l'accroissement du moment dipolaire total (en présence de $\mathbf{E}^{(2)s}$) pour un accroissement unitaire de $\mathbf{E}^{(1)s}$ on obtient à partir de (2b)

$$\Pi_{\sigma\tau}^{(1:2)} = \left(\frac{\partial M_\sigma}{\partial E_\tau^s}\right)_0 + \left(\frac{\partial^2 M_\sigma}{\partial E_\tau^s \partial E_\nu^s}\right)_0 [E_\nu^{(1)s} + E_\nu^{(2)s}] + \frac{1}{2}\left(\frac{\partial^3 M_\sigma}{\partial E_\tau^s \partial E_\nu^s \partial E_\rho^s}\right)_0 [E_\nu^{(1)s} + E_\nu^{(2)s}][E_\rho^{(1)s} + E_\rho^{(2)s}] \quad (4)$$

$\Pi_{\sigma\tau}^{(1:2)}$ étant défini par

$$\delta M_\sigma = \Pi_{\sigma\tau}^{(1:2)} . \delta E_\tau^{(1)s}$$

Notons que

$$\Pi_{\sigma\tau}^{(1:2)} = \Pi_{\sigma\tau}^{(2:1)} = \Pi_{\sigma\tau}$$

L'indice de réfraction n_i du milieu pour la composante $\mathbf{E}^{(1)s}$ de l'onde en présence de l'onde totale $\mathbf{E}^s$ polarisée elliptiquement est donné par la relation de Lorentz–Lorenz

$$\frac{n_i^2 - 1}{n_i^2 + 2} = \frac{4\pi}{3}\left\langle \overline{\frac{1}{V}\Pi_{\sigma\tau} e_\sigma^{(i)} e_\tau^{(i)}}^{\,t} \right\rangle_{\mathbf{E}^s}, \quad i = 1, 2 \quad (5)$$

où le symbole $(\overline{\ldots}^{\,t})$ désigne la valeur moyenne sur plusieurs pédiodes d'oscillations de $\mathbf{E}$ et $\langle \Phi \rangle_{\mathbf{E}^{(s)}}$ désigne la moyenne statistique de la quantité Φ en présence du champ $\mathbf{E}^{(s)}$

$$\langle \Phi \rangle_{\mathbf{E}^{(s)}} = \frac{\int \Phi \exp\left[-\frac{\overline{U(\tau, \mathbf{E}^{(s)})}^{\,t}}{kT}\right] d\tau}{\int \exp\left[-\frac{\overline{U(\tau, \mathbf{E}^{(s)})}^{\,t}}{kT}\right] d\tau} \quad (6)$$

τ désignant les variables configurationnelles et les intégrales étant étendues à tout l'espace configurationnel.

En tenant compte de (1b), et en posant

$$V = V_0 + \frac{1}{2}\left(\frac{\partial^2 V}{\partial E^{2(s)}}\right) E^{2(s)}$$

ce qui revient à prendre en compte l'électrostriction – on trouve

$$\overline{U(\tau, \mathbf{E}^s)}^t = U(\tau, 0) - \frac{1}{4}\left(\frac{\partial M_\sigma}{\partial E_\tau}\right)_0 [E_0^{(1)^2 s} e_\sigma^{(1)} e_\tau^{(1)} + E_0^{(2)^2 s} e_\sigma^{(2)} e_\tau^{(2)}] - \dots \qquad (7)$$

$$\overline{\frac{1}{V}\Pi_{\sigma\tau} e_\sigma^{(1)} e_\tau^{(1)}}^t = \frac{1}{V_0}\left\{\left(\frac{\partial M_\sigma}{\partial E_\tau^s}\right)_0 e_\sigma^{(1)} e_\tau^{(1)} + \frac{1}{4}\left(\frac{\partial^3 M_\sigma}{\partial E_\tau^s \partial E_\nu^s \partial E_\rho^s}\right) [(E_0^{(1)s})^2 e_\sigma^{(1)} e_\tau^{(1)} e_\nu^{(1)} e_\rho^{(1)}\right.$$

$$\left. + (E_0^{(2)s})^2 e_\sigma^{(1)} e_\tau^{(1)} e_\nu^{(2)} e_\rho^{(2)}] - \frac{1}{4V_0}\left(\frac{\partial^2 V}{\partial E^2}\right)_0 \left(\frac{\partial M_\sigma}{\partial E_\tau}\right)_0 e_\sigma^{(1)} e_\tau^{(1)} [(E_0^{(1)s})^2 + (E_0^{(2)s})^2]\right\} \qquad (8)$$

Utilisons (7) et (8) dans la relation de Lorentz–Lorenz en tenant compte de (6) et (2c)

$$\frac{n_1^2 - 1}{n_1^2 + 2} - \frac{n_0^2 - 1}{n_0^2 + 2} = \frac{\pi}{3V_0}\left(\frac{n_0^2 + 2}{3}\right)^2 \left\{\left\langle\left(\frac{\partial^3 M_\sigma}{\partial E_\tau^s \partial E_\nu^s \partial E_\rho^s}\right)_0 e_\sigma^{(1)} e_\tau^{(1)} e_\nu^{(1)} e_\rho^{(1)}\right\rangle\right.$$

$$- \frac{1}{V_0}\left(\frac{\partial^2 V}{\partial E_s^2}\right)_0 \left\langle\left(\frac{\partial M_\sigma}{\partial E_\tau^s}\right)_0 e_\sigma^{(1)} e_\tau^{(1)}\right\rangle + \frac{1}{kT}\left[\left\langle\left(\frac{\partial M_\sigma}{\partial E_\tau^s}\right)_0 \left(\frac{\partial M_\nu}{\partial E_\rho^s}\right)_0 e_\sigma^{(1)} e_\tau^{(1)} e_\nu^{(1)} e_\rho^{(1)}\right\rangle\right.$$

$$\left.\left. - \left\langle\left(\frac{\partial M_\sigma}{\partial E_\sigma^s}\right)_0 e_\sigma^{(1)} e_\tau^{(1)}\right\rangle^2\right]\right\} (E_0^{(s)})^2 + \frac{\pi}{3V_0}\left(\frac{n_0^2 + 2}{3}\right)^2 \left\{\left\langle\left(\frac{\partial^3 M_\sigma}{\partial E_\tau^s \partial E_\nu^s \partial E_\rho^s}\right) e_\sigma^{(1)} e_\tau^{(1)} e_\nu^{(2)} e_\rho^{(2)}\right\rangle\right.$$

$$- \frac{1}{V_0}\left(\frac{\partial^2 V}{\partial E_s^2}\right)_0 \left\langle\left(\frac{\partial M_\sigma}{\partial E_\tau^s}\right)_0 e_\sigma^{(1)} e_\tau^{(1)}\right\rangle + \frac{1}{kT}\left[\left\langle\left(\frac{\partial M_\sigma}{\partial E_\tau^s}\right)_0 \left(\frac{\partial M_\nu}{\partial E_\rho^s}\right)_0 e_\sigma^{(1)} e_\tau^{(1)} e_\nu^{(2)} e_\rho^{(2)}\right\rangle\right.$$

$$\left.\left. - \left\langle\left(\frac{\partial M_\sigma}{\partial E_\tau^s}\right)_0 e_\sigma^{(1)} e_\tau^{(1)}\right\rangle \left\langle\left(\frac{\partial M_\nu}{\partial E_\rho^s}\right)_0 e_\nu^{(2)} e_\rho^{(2)}\right\rangle\right]\right\} (E_0^{(2)})^2 \qquad (9)$$

Dans l'expression (9) n_0 désigne l'indice de réfraction linéaire du milieu donné par

$$\frac{n_0^2 - 1}{n_0^2 + 2} = \frac{4\pi}{3V_0}\left\langle\left(\frac{\partial M_\sigma}{\partial E_\tau^s}\right)_0 e_\sigma^{(1)} e_\tau^{(1)}\right\rangle$$

et le symbole $\langle \ldots \rangle$ désigne la moyenne statistique (isotrope) en absence de champ excitateur

$$\langle \Phi \rangle = \frac{\int \Phi \exp\left[-\frac{U(\tau_1\, 0)}{kT}\right] d\tau}{\int \exp\left[-\frac{U(\tau_1\, 0)}{kT}\right] d\tau}$$

θr en absence de champ extérieur toutes les directions de $\mathbf{e}_1$ et $\mathbf{e}_2$ sont également probables de sorte que l'on peut écrire, en posant

$$n_i = n_0 + \Delta n_i, \quad i = 1, 2$$

$$\Delta n_1 = \frac{\pi}{2V_0\, n_0}\left(\frac{n_0^2+2}{3}\right)^4 \left\{\frac{1}{15}\left\langle\left(\frac{\partial^3 M_\sigma}{\partial E_\tau\, \partial E_\nu\, \partial E_\rho}\right)_0\right\rangle \Delta_{\sigma\tau\nu\rho} - \frac{1}{3V_0}\left(\frac{\partial^2 V}{\partial E^{2(s)}}\right)_0 \left\langle\left(\frac{\partial M_\sigma}{\partial E_\sigma^{(s)}}\right)_0\right\rangle\right.$$

$$\left. + \frac{1}{kT}\left[\frac{1}{15}\left\langle\left(\frac{\partial M_\sigma}{\partial E_\tau^{(s)}}\right)_0 \left(\frac{\partial M_\nu}{\partial E_\rho^{(s)}}\right)_0\right\rangle \Delta_{\sigma\tau\nu\rho} - \frac{1}{9}\left\langle\left(\frac{\partial M_\sigma}{\partial E_\sigma^{(s)}}\right)_0\right\rangle^2\right]\right\}(E_0^{(1)})^2 + \frac{\pi}{2V_0 n_0}\left(\frac{n_0^2+2}{3}\right)^4$$

$$\left\{\frac{1}{90}\left\langle\left(\frac{\partial^3 M_\sigma}{\partial E_\tau^s\, \partial E_\nu^s\, \partial E_\rho^s}\right)_0\right\rangle (10\delta_{\sigma\tau}\delta_{\nu\rho} - \chi_{\sigma\tau\nu\rho}) - \frac{1}{3V_0}\left(\frac{\partial^2 V}{\partial E^{2(s)}}\right)_0 \left\langle\left(\frac{\partial M_\sigma}{\partial E_0^{(s)}}\right)_0\right\rangle\right.$$

$$\left. + \frac{1}{kT}\left[\frac{1}{90}\left\langle\left(\frac{\partial M_\sigma}{\partial E_\tau^s}\right)_0 \left(\frac{\partial M_\nu}{\partial E_\rho^s}\right)_0\right\rangle (10\delta_{\sigma\tau}\delta_{\nu\rho} - \chi_{\sigma\tau\nu\rho}) - \frac{1}{9}\left\langle\frac{\partial M_\sigma}{\partial E_\sigma^s}\right\rangle^2\right]\right\} (E_0^{(2)})^2 \tag{10}$$

avec

$$\begin{aligned} \Delta_{\sigma\tau\nu\rho} &= \delta_{\sigma\tau}\delta_{\nu\rho} + \delta_{\sigma\nu}\delta_{\tau\rho} + \delta_{\sigma\rho}\delta_{\nu\tau} \\ \chi_{\sigma\tau\nu\rho} &= 3\delta_{\sigma\nu}\delta_{\tau\rho} + 3\delta_{\sigma\rho}\delta_{\tau\nu} - 2\delta_{\sigma\tau}\delta_{\nu\rho} \end{aligned} \tag{11}$$

Pour obtenir Δn_2 il suffit de permuter les indices 1 et 2.

On remarque, dans la formule (10), la présence de 3 types de contributions à la variation non linéaire de l'indice:

les termes en $\left(\frac{\partial^3 M_\sigma}{\partial E_\tau^s\, \partial E_\nu^s\, \partial E_\rho^s}\right)_0$ caractérisant l'hyperpolarisabilité

les termes en $\left(\frac{\partial^2 V}{\partial E_{(s)}^2}\right)_0$ caractérisant l'électrostriction

les termes en $\frac{1}{kT}$ caractérisant la réorientation

De la formule générale (10) on peut déduire les modifications subies par la vibration elliptique au cours de sa traversée dans le milieu non linéaire, ainsi que les indices de réfraction du milieu pour une onde rectiligne ou une onde circulaire. Pour obtenir l'indice dans ces deux derniers cas, il suffit de faire $\mathbf{E}^{(2)} = 0$ pour l'onde polarisée rectilignement et $E_0^{(1)} = E_0^{(2)}$ pour l'onde polarisée circulairement. En désignant I l'intensité de l'onde on a

$$(\Delta n)_{\text{rect}} = B_{\text{rect}} I = \frac{\pi}{V_0 n_0} \left(\frac{n_0^2 + 2}{3}\right)^4 [(B_{\text{déf}})_{\text{rect}} + B_{\text{electros}} + (B_{\text{réo}})_{\text{rect}}].I \tag{12}$$

$$(\Delta n)_{\text{circ}} = B_{\text{circ}} I = \frac{\pi}{2 V_0 n_0} \left(\frac{n_0^2 + 2}{3}\right)^4 [(B_{\text{déf}})_{\text{circ}} + 2B_{\text{électros}} + (B_{\text{réo}})_{\text{circ}}].I \tag{13}$$

où on a posé

$$\left.\begin{aligned}
B_{\text{électros}} &= -\frac{1}{8V_0} \left(\frac{\partial^2 V}{\partial E_{(s)}^2}\right)_0 \left\langle \left(\frac{\partial M_\sigma}{\partial E_\sigma^{(s)}}\right)_0 \right\rangle \\
(B_{\text{déf}})_{\text{rect}} &= \frac{1}{15} \Delta_{\sigma\tau\nu\rho} \left\langle \left(\frac{\partial^3 M_\sigma}{\partial E_\tau^{(s)} \partial E_\nu^{(s)} \partial E_\rho^{(s)}}\right)_0 \right\rangle \\
(B_{\text{réo}})_{\text{rect}} &= \frac{1}{45kT} \left[(\chi_{\sigma\tau\nu\rho} + 5\delta_{\sigma\tau}\delta_{\nu\rho}) \left\langle \left(\frac{\partial M_\sigma}{\partial E_\tau^{(s)}}\right)_0 \left(\frac{\partial M_\nu}{\partial E_\rho^{(s)}}\right)_0 \right\rangle - 5 \left\langle \left(\frac{\partial M_\sigma}{\partial E_\sigma^{(s)}}\right)_0 \right\rangle^2 \right] \\
(B_{\text{déf}})_{\text{circ}} &= \frac{1}{30} \left\langle \left(\frac{\partial^3 M_\sigma}{\partial E_\tau^{(s)} \partial E_\nu^{(s)} \partial E_\rho^{(s)}}\right)_0 \right\rangle (\Delta_{\sigma\tau\nu\rho} + 5\delta_{\sigma\tau}\delta_{\nu\rho}) \\
(B_{\text{réo}})_{\text{circ}} &= \frac{1}{90kT} \left[(\chi_{\sigma\tau\nu\rho} + 20\delta_{\sigma\tau}\delta_{\nu\rho}) \left\langle \left(\frac{\partial M_\sigma}{\partial E_\tau}\right)_0 \left(\frac{\partial M_\nu}{\partial E_\rho}\right)_0 \right\rangle - 20 \left\langle \left(\frac{\partial M_\sigma}{\partial E_\sigma^{(s)}}\right)_0 \right\rangle^2 \right]
\end{aligned}\right\} \tag{14}$$

La différence des indices de réfraction des 2 composantes $\mathbf{E}_1$ et $\mathbf{E}_2$ de la vibration elliptique est

$$n_1 - n_2 = \frac{\pi}{120 V_0 n_0} \left(\frac{n_0^2 + 2}{3}\right)^4 \left(\frac{1 - e^2}{1 + e^2}\right) \chi_{\sigma\tau\nu\rho}$$

$$\times \left[\left\langle \left(\frac{\partial^3 M_\sigma}{\partial E_\tau^{(s)} \partial E_\nu^{(s)} \partial E_\rho^{(s)}}\right)_0 \right\rangle + \frac{1}{kT} \left\langle \left(\frac{\partial M_\sigma}{\partial E_\tau^{(s)}}\right)_0 \left(\frac{\partial M_\nu}{\partial E_\rho^{(s)}}\right)_0 \right\rangle \right] . I_s \tag{15}$$

e désignant l'ellipticité de l'onde.

On remarque que dans la formule (15) interviennent seules la réorientation des molécules et l'hyperpolarisabilité.

A cette différence d'indices $n_1 - n_2$ correspondent les modifications suivantes de l'ellipse de vibration: une rotation des axes θ, et une modification de l'ellipticité qui passe de la valeur e à la valeur e' avec

$$\left.\begin{aligned} \tan\theta &= +\frac{2e}{e^2-1}\sin\phi; \quad e'^2 = \frac{e^2+1+\sqrt{(e^2-1)^2+4e^2\sin^2\phi}}{e^2+1-\sqrt{(e^2-1)^2+4e^2\sin^2\phi}} \\ \phi &= \frac{2\pi}{\lambda}\,l(n_1-n_2) \end{aligned}\right\} \qquad (16)$$

Théorie moléculaire

En désignant par $\mathbf{m}^{(p)}$ le moment dipolaire de la $p^{è}$ molécule, le moment dipolaire total du système contenant N molécules est égal à

$$\mathbf{M} = \sum_{p=1}^{N} \mathbf{m}^{(p)} \qquad (17)$$

avec

$$m_\sigma^{(p)} = \mu_{0\sigma}^{(p)} + \alpha_{\sigma\tau}^{(p)}(E_\tau + F_\tau) + \tfrac{1}{2}\beta_{\sigma\tau\nu}^{(p)}(E_\tau + F_\tau)(E_\nu + F_\nu) + \tfrac{1}{6}\gamma_{\sigma\tau\nu\rho}^{(p)}(E_\tau + F_\tau)(E_\nu + F_\nu)(E_\rho + F_\rho) + \ldots \qquad (18)$$

Négligeons en première approximation l'effet du champ moléculaire **F** (créé par $(N-1)$ autres molécules au centre de la $p^{è}$) sur les propriétés optiques des molécules. Les expressions (14) et la formule (15) deviennent

$$\left.\begin{aligned} B'_{\text{électros}} &= -\frac{1}{3V_0}\left(\frac{\partial^2 V}{\partial E^2}\right)_0 \left\langle \sum_{p=1}^{N} \alpha_{\sigma\sigma}^{(p)} \right\rangle = -\rho\alpha\left(\frac{\partial^2 V}{\partial E^2}\right)_0 \\ (B_{\text{déf}})_{\text{rect}} &= \frac{1}{15}\Delta_{\sigma\tau\nu\rho}\left\langle \sum_{p=1}^{N} \gamma_{\sigma\tau\nu\rho}^{(p)} \right\rangle \\ (B_{\text{réo}})_{\text{rect}} &= \frac{1}{45kT}\chi_{\sigma\tau\nu\rho}\left\langle \sum_{p,q=1}^{N} \alpha_{\sigma\tau}^{(p)}\alpha_{\nu\rho}^{(q)} \right\rangle \\ (B_{\text{déf}})_{\text{circ}} &= \frac{1}{30}(\Delta_{\sigma\tau\nu\rho} + 5\delta_{\sigma\tau}\delta_{\nu\rho})\left\langle \sum_{p=1}^{N} \gamma_{\sigma\tau\nu\rho}^{(p)} \right\rangle \\ (B_{\text{réo}})_{\text{circ}} &= \frac{1}{90kT}\chi_{\sigma\tau\nu\rho}\left\langle \sum_{p,q=1}^{N} \alpha_{\sigma\tau}^{(p)}\alpha_{\nu\rho}^{(q)} \right\rangle \end{aligned}\right\} \qquad (19)$$

et

$$n_1 - n_2 = \frac{\pi}{120 V_0 n_0}\left(\frac{n_0^2+2}{3}\right)^4\left(\frac{1-e^2}{1+e^2}\right)\chi_{\sigma\tau\nu\rho}\left[\left\langle\sum_{p=1}^{N}\gamma^{(p)}_{\sigma\tau\nu\rho}\right\rangle + \frac{1}{kT}\left\langle\sum_{p,q=1}^{N}\alpha^{(p)}_{\sigma\tau}\alpha^{(q)}_{\nu\rho}\right\rangle\right] I \tag{20}$$

Dans les expressions (19) nous avons fait une hypothèse supplémentaire en supposant que la trace des tenseurs $\alpha^{(p)}_{\sigma\tau}$, scalaire intrinsèque, est indépendante des variables configurationnelles

$$3\alpha = \alpha^{(p)}_{\sigma\sigma} = \alpha^{(q)}_{\sigma\sigma} = \ldots$$

alors

$$\left\langle\sum_{p=1}^{N}\sum_{q=1}^{N}\alpha^{(p)}_{\sigma\sigma}\alpha^{(q)}_{\tau\tau}\right\rangle = \left\langle\sum_{p=1}^{N}\alpha^{(p)}_{\sigma\sigma}\right\rangle^2$$

Restons dans la même hypothèse, ce qui revient à négliger l'effet de la redistribution, et transformons les expressions (19) et (20) en rapportant les tenseurs de polarisabilité et d'hyperpolarisabilité aux repères cartésiens orthonormés rigidement liés aux molécules. Soit $(\omega^{p}_{\sigma i})$ la matrice de passage du référentiel laboratoire au référentiel lié à la p^e molécule

$$\left.\begin{aligned}\alpha^{(p)}_{\sigma\tau} &= \omega^{(p)}_{\sigma i}\omega^{(p)}_{\tau j}\alpha^{(p)}_{ij}; \quad \gamma^{(p)}_{\sigma\tau\nu\rho} = \omega^{(p)}_{\sigma i}\omega^{(p)}_{\tau j}\omega^{(q)}_{\nu k}\omega^{(q)}_{\rho l}\gamma^{(p)}_{ijkl}\\ \alpha^{(p)}_{\sigma\tau}\alpha^{(q)}_{\nu\rho} &= \omega^{(p)}_{\sigma i}\omega^{(p)}_{\tau j}\omega^{(q)}_{\nu k}\omega^{(q)}_{\rho l}\alpha^{(p)}_{ij}\alpha^{(q)}_{kl}\end{aligned}\right\} \tag{21}$$

Du fait de l'orthonormalisation des bases nous pouvons poser

$$\omega^{(p)}_{\sigma i}\omega^{(q)}_{\sigma j} = \left|\begin{array}{ll}\delta_{ij} & \text{pour } p = q\\ \omega^{(pq)}_{ij} & \text{pour } p \neq q\end{array}\right. \tag{22}$$

où l'élément $\omega^{(pq)}_{ij}$ n'est autre que le cosinus de l'angle du i^{e} axe de la p^{e} molécule avec le j^{e} axe de la q^{e} molécule. Les formules (19) s'écrivent finalement, tenant compte du fait que

$$\alpha^{(p)}_{ij} = \alpha^{(q)}_{ij} = \alpha_{ij}$$

$$\left.\begin{aligned}B_{\text{électros}} &= -\rho\alpha\left(\frac{\partial^2 V}{\partial E^2}\right)_0 = \rho V_0\frac{\alpha}{4\pi}\left[\left(\frac{\partial n^2}{\partial p}\right)_T - (n_0^2-1)\beta_T\right]\\ (B_{\text{déf}})_{\text{rect}} &= \frac{1}{15}\rho V_0\gamma_{ijkl}\Delta_{ijkl} = \rho V_0\gamma\\ (B_{\text{réo}})_{\text{rect}} &= \frac{\rho V_0}{45kT}\left[6(\alpha_{ij}\alpha_{ij}-3\alpha^2) + \frac{\rho}{V_0}\alpha_{ij}\alpha_{kl}\iint(3\omega^{(pq)}_{ik}\omega^{(pq)}_{jl} + 3\omega^{(pq)}_{il}\omega^{(pq)}_{jk}\right.\\ &\qquad\left. - 2\delta_{ij}\delta_{kl})g^{(2)}(\tau_{pq})\,d\tau_{pq}\right]\end{aligned}\right\} \tag{23}$$

[Continuation of Eqn. (23)]

$$\left.\begin{aligned}(B_{\text{déf}})_{\text{circ}} &= \frac{\rho V_0}{6}(3\gamma + \gamma_{iijj}) \\ (B_{\text{réo}})_{\text{circ}} &= \frac{\rho V_0}{90kT}\left[6(\alpha_{ij}\alpha_{ij} - 3\alpha^2) + \frac{\rho}{V_0}\alpha_{ij}\alpha_{kl}\iint(\cdots)g^2(\tau_{pq})\mathrm{d}\tau_{pq}\right] = \tfrac{1}{2}(B_{\text{réo}})_{\text{rect}}\end{aligned}\right\}$$

Dans ces formules $\alpha = \alpha_{ii}/3$ et $\gamma = (\gamma_{iijj} + \gamma_{ijij} + \gamma_{ijji})/15$ désignent la polarisabilité moyenne et l'hyperpolarisabilité moyenne; β_T est le coefficient de compressibilité isotherme et $g^2(\tau_{pq})$, la fonction de corrélation des couples de molécules dont les variables configurationnelles sont notées par τ_{pq}. Notons que si la réorientation intervient seule dans l'autofocalisation le rapport des seuils $P_{\text{circ}}/P_{\text{rect}}$ sera égal à 4 (réf. 6) pour les molécules à symétrie non sphérique. La différence des indices de réfraction pour les deux composantes $\mathbf{E}_1$ et $\mathbf{E}_2$ d'une vibration elliptique s'écrit finalement

$$n_1 - n_2 = \frac{\rho\pi}{120n_0}\left(\frac{n_0^2+2}{3}\right)^4\left(\frac{1-e^2}{1+e^2}\right)(C^{(1)} + C^{(2)})I \tag{24}$$

avec

$$\left.\begin{aligned}C^{(1)} &= \chi_{ijkl}\left(\gamma_{ijkl} + \frac{1}{kT}\alpha_{ij}\alpha_{kl}\right) = 5(9\gamma - \gamma_{iijj}) + \frac{6}{kT}(\alpha_{ij}\alpha_{ij} - 3\alpha^2) \\ C^{(2)} &= \frac{1}{kT}\alpha_{ij}\alpha_{kl}\frac{\rho}{V}\iint(3\omega_{ik}^{(pq)}\omega_{jl}^{(pq)} + 3\omega_{il}^{(pq)}\omega_{jk}^{(pq)} - 2\delta_{ij}\delta_{kl})g^{(2)}(\tau_{pq})\,\mathrm{d}\tau_{pq}\end{aligned}\right\} \tag{25}$$

Les formules (23) et (24) permettant de calculer le rapport des seuils de puissances d'autofocalisation et la rotation des axes de l'ellipse de vibration sont très générales, valables pour tous les types de symétrie.

Cas de symétrie particuliers

Pour les cas de symétrie du type D_{4h}(e.g. C_4H_8), $C_{6h}(C_6Cl_6)$, $D_{6h}(C_6H_6)$ et $D_{8h}(CS_2)$ on a

$$\alpha_{ij} = \alpha_{11}\delta_{ij} + (\alpha_{33} - \alpha_{11})k_ik_j \tag{26}$$

où $\mathbf{k}$ désigne le vecteur unitaire de l'axe de symétrie. Remplaçons (26) dans les formules (23) en remarquant que

$$\alpha_{ij}\alpha_{kl}\chi_{ijkl} = 4(\alpha_{33} - \alpha_{11})^2 \tag{27}$$

$$\frac{\rho}{V_0}\alpha_{ij}\alpha_{kl}\iint(3\omega_{ik}^{(pq)}\omega_{jl}^{(pq)} + 3\omega_{il}^{(pq)}\omega_{jk}^{(pq)} - 2\delta_{ij}\delta_{kl})g^{(2)}(\tau_{pq})\mathrm{d}\tau_{pq} = 4(\alpha_{33} - \alpha_{11})^2J_A \tag{28}$$

avec

$$J_A = \frac{\rho}{2V}\iint (3\cos^2\theta_{pq} - 1)g^{(2)}(\tau_{pq})\,\mathrm{d}\tau_{pq} \tag{29}$$

où θ_{pq} est l'angle formé par les axes de symétrie de la $p^{\text{è}}$ molécule et de la $q^{\text{è}}$ molécule; nous obtenons finalement

$B_{\text{rect}} =$

$$\frac{\pi}{n_0}\left(\frac{n_0^2+2}{3}\right)^4\left\{\frac{\alpha}{4\pi}\left[2n_0\left(\frac{\partial n}{\partial\rho}\right)_T - (n_0^2-1)\beta_T\right] + \rho\left[\gamma + \frac{4}{45kT}(\alpha_{33}-\alpha_{11})^2(1+J_A)\right]\right\} \tag{30}$$

$B_{\text{circ}} =$

$$\frac{\pi}{n_0}\left(\frac{n_0^2+2}{3}\right)^4\left\{\frac{\alpha}{4\pi}\left[2n_0\left(\frac{\partial n}{\partial\rho}\right)_T - (n_0^2-1)\beta_T\right] + \frac{\rho}{4}\left[\gamma + \frac{\gamma_{iijj}}{3} + \frac{4}{45kT}(\alpha_{33}-\alpha_{11})^2(1+J_A)\right]\right\} \tag{31}$$

$$n_1 - n_2 = \frac{\rho\pi}{60n_0}\left(\frac{n_0^2+2}{3}\right)^4\left(\frac{1-e^2}{1+e^2}\right)\left[7\gamma\chi_\gamma + \frac{2}{kT}(\alpha_{33}-\alpha_{11})^2(1+J_A)\right] I \tag{32}$$

où $\chi_\gamma = (\gamma_{ijkl}\chi_{ijkl})/14\gamma$ désigne l'anisotropie de l'hyperpolarisabilité.

Pour le cas de symétrie sphérique on a

$$\alpha_{ij} = \alpha\delta_{ij}; \quad \gamma_{ijkl} = \tfrac{1}{3}\gamma\Delta_{ijkl} \tag{33}$$

Les expressions (30), (31) et (32) se simplifient et s'écrivent en notant que les effets de la réorientation sont nuls

$$(B_{\text{réo}})_{\text{rect}} = (B_{\text{réo}})_{\text{circ}} = 0$$

$$B_{\text{rect}} = \frac{\rho\pi}{n_0}\left(\frac{n_0^2+2}{3}\right)^4\left\{\gamma + \frac{\alpha}{4\pi}\left[2n_0\left(\frac{\partial n_0}{\partial\rho}\right)_T - (n_0^2-1)\beta_T\right]\right\} \tag{34}$$

$$B_{\text{circ}} = \frac{\rho\pi}{n_0}\left(\frac{n_0^2+2}{3}\right)^4\left\{\frac{2\gamma}{3} + \frac{\alpha}{4\pi}\left[2n_0\left(\frac{\partial n}{\partial\rho}\right)_T - (n_0^2-1)\beta_T\right]\right\} \tag{35}$$

$$n_1 - n_2 = \frac{\rho\gamma}{6n_0}\left(\frac{n_0^2+2}{3}\right)^4\left(\frac{1-e^2}{1+e^2}\right) I \tag{36}$$

Dans ce dernier cas si l'on suppose que l'effet de l'électrostriction est négligeable devant l'effet de déformation le rapport de seuil serait égal à

$$\frac{(P_a)_{\mathrm{circ}}}{(P_a)_{\mathrm{rect}}} = \frac{3}{2} \tag{37}$$

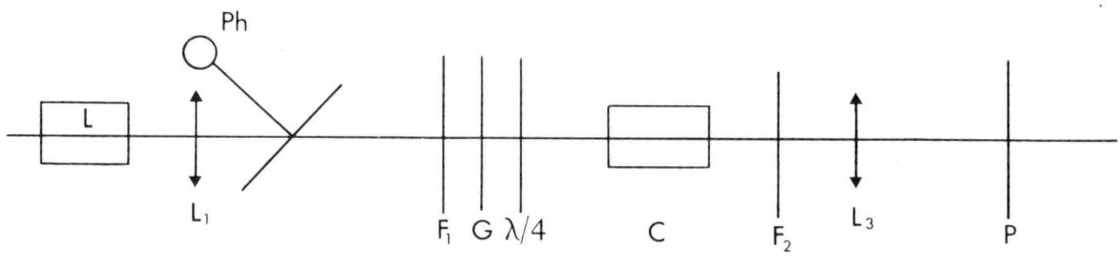

Fig. 1. L – laser déclenché par prisme tournant (20 MW, 20 ns). Ph – photodiode de contrôle de puissance du laser. L_1 – lentilles destinées à réduire la divergence du faisceau laser. F_1 – filtre neutre, F_2 – filtre sélectif. G – prisme de Glan. λ/4 – lame quart d'onde que l'on peut orienter par rapport à G de façon à obtenir une vibration rectiligne, circulaire ou elliptique. C – cuve contenant le liquide actif. P – plaque photographique ou photodiode.

EXPLOITATION DES RESULTATS ET COMPARAISON AUX MESURES

Description des expériences

Le montage utilisé pour la mesure du rapport des seuils Raman dans les polarisations rectiligne et circulaire est représenté sur Fig. 1.

Pour mesurer les modifications subies par l'ellipse de vibration après traversée de la cuve C, on rajoute après F_2 un prisme de Wollaston et on mesure les intensités W_1 et W_2 des faisceaux transmis.

On a travaillé en dessous du seuil Raman. Dans ces conditions, la rotation θ des axes de l'ellipse de vibration, et la variation $\Delta e = e' - e$ de l'ellipticité sont reliées à W_1 et W_2.

Les mesures de W_1 et W_2 nous permettent donc, en utilisant les relations (16) de déduire la différence $(n_1 - n_2)$ des indices avec lesquels se propagent les 2 axes de la vibration elliptique incidente, (les calculs sont faits à l'ordinateur, ce qui permet d'avoir en même temps les marges d'erreur).

Résultats concernant le rapport des seuils Raman en polarisation rectiligne et circulaire

Les résultats sont groupés dans le Tableau 1. Si l'on néglige les effets d'électrostriction et de déformation $(B_{\mathrm{rect}}/B_{\mathrm{circ}})_{\mathrm{calculé}} = 4$ pour les symétries axiales.

La présence du terme $\gamma_{iijj}/3$ au dénominateur de la formule (38) montre que la déformation des molécules conduit à $B_{\mathrm{rect}}/B_{\mathrm{circ}} < 4$. Pour CS_2 par exemple, la correction apportée par le terme $\gamma_{iijj}/3$ n'est que de 2/100, et ne peut donc expliquer l'écart entre le valeur mesurée 1,8 et la valeur calculée 4.

La présence des termes d'électrostriction conduit également à $(B_{\mathrm{rect}}/B_{\mathrm{circ}}) < 4$

$$\frac{B_{\mathrm{rect}}}{B_{\mathrm{circ}}} \simeq 4\left(1 - \frac{3B'_{\text{électros}}}{(B_{\mathrm{r\acute{e}o}})_{\mathrm{rect}}}\right)$$

Cependant pour les liquides, l'effet d'électrostriction n'a pas le temps de s'établir pendant

Tableau 1.

Corps actif	$\left(\dfrac{P_{r\,\mathrm{circulaire}}}{P_{r\,\mathrm{rectiligne}}}\right)_{\mathrm{mesuré}}$	$\left(\dfrac{B_{\mathrm{rectiligne}}}{B_{\mathrm{circulaire}}}\right)_{\mathrm{calculé}}$	
N_2 liquide	$1{,}2 \pm 0{,}2$	1,5 (on a admis $\alpha_{33} - \alpha_{11} = 0$)	(14)
CS_2	$1{,}8 \pm 0{,}2$		
$C_6H_5NO_2$	$2{,}5 \pm 0{,}1$	$\dfrac{\dfrac{\alpha}{4\pi}\left[2n_0\left(\dfrac{\partial n}{\partial \rho}\right)_T - (n_0^2 - 1)\beta_T\right] + \left[\gamma + \dfrac{4}{45kT}(\alpha_{33} - \alpha_{11})^2(1 + J_A)\right]}{\dfrac{\alpha}{4\pi}\left[2n_0\left(\dfrac{\partial n}{\partial \rho}\right)_T - (n_0^2 - 1)\beta_T\right] + \dfrac{1}{14}\left[\gamma + \dfrac{\gamma_{iijj}}{3} + \dfrac{4}{45kT}(\alpha_{33} - \alpha_{11})^2(1 + J_A)\right]}$	(38)
C_7H_8	$2{,}6 \pm 0{,}3$	$\simeq 4$ en négligeant la déformation et l'électrostriction	
C_6H_6	$2{,}4 \pm 0{,}1$		

Tableau 2.

Corps	Ellipticité e à l'entrée de la cuve	Rotation θ des axes de l'ellipse	Ellipticité e' à la sortie de la cuve	$\phi = \frac{2\pi l}{\lambda} \Delta n$ mesuré	Δn mesuré	Δn calculé
CS_2	3,2	1° (± 1°)	3,5 ± 0,3	3° (± 3°)	(3,5 ± 3,5) . 10^{-8}	3 . 10^{-8} on a admis $\alpha_{33} - \alpha_{11} = 50^{-23}$CGS et (17) $\gamma \ll \alpha_{33} - \alpha_{11}$
	11,5	1° (± 1°)	(1)			
$C_6H_5NO_2$	3,2	1° (± 1°)	3,4 ± 0,4	3° (± 3°)	(3,5 ± 3,5) . 10^{-8}	1,2 . 10^{-8} on a admis $\alpha_{33} - \alpha_{11} = 0{,}65 \times 10^{-23}$ et (17) $\gamma \ll \alpha_{33} - \alpha_{11}$
	11,5	1° (± 1°)	(1)			
N_2 liquide	11,5	0° (± 1°)	(1)	0° (± 3°)	0(± 3,5.10^{-8})	4 . $10^{25}\gamma$ (γ en CGS)

La fonction e' varie très rapidement au voisinage de $e = 11{,}5$, de sorte que les erreurs sur e' sont trop importantes pour que les mesures soient prises en considération.
Toutes les mesures ont été faites avec un champ excitateur $E = 2.10^6$ V/m.

la durée du pulse laser utilisé, de sorte que, même si $B_{\text{électros}}/B_{\text{réo}}$ est grand, l'électrostriction ne permet pas d'expliquer la différence entre la valeur théorique du rapport des seuils et la valeur mesurée.

Il faudra introduire dans les calculs l'effet de redistribution des molécules, l'effet non linéaire provenant de la variation du champ local ainsi que l'effet non linéaire dû au champ Raman. Ceci sera fait, pour la redistribution, au niveau des formules (14), dans lesquelles la polarisabilité sera fonction des variables configurationnelles par le jeu des interactions dipôle–dipôle . . , pour l'effet du champ local au niveau de la formule (18). La comparaison entre valeur théorique et mesurée du rapport des seuils doit permettre alors de donner l'importance de ces effets par rapport à l'effet Kerr Optique.

L'azote étant un liquide ne présentant pas de self focalisation aux puissances laser utilisées (nous l'avons constaté expérimentalement et le calcul le montre), on ne peut pas écrire

$$\frac{P_{R\,\mathrm{circ}}}{P_{R\,\mathrm{rect}}} = \frac{P_{a\,\mathrm{circ}}}{P_{r\,\mathrm{rect}}};$$

On ne peut donc comparer dans le Tableau 1 la valeur mesurée à la valeur calculée.

On remarque d'ailleurs qu'aux erreurs de mesure près, on peut considérer que le seuil Raman de l'azote n'est pas affecté par la polarisation du faisceau incident.

Résultats concernant les modifications d'ellipticité de l'onde laser au-dessous du seuil Raman

Les résultats sont groupés dans le Tableau 2. Pour les corps à symétrie axiale, Δn est dû essentiellement à l'effet Kerr optique. Avec les intensités utilisées, les rotations θ

mesurées pour ces corps sont faibles et conduisent à des valeurs de Δn de l'ordre de 10^{-8}, en bon accord avec les calculs.

Pour les corps à symétrie sphérique Δn est dû essentiellement à la déformation moléculaire, caractérisée par le paramètre γ; γ étant de l'ordre de 10^{-36}CGS, Δn est de l'ordre de 10^{-11}, donc non détectable dans nos expériences, ce qui est en accord avec nos observations. Pour des valeurs plus grandes du champ E, nous avons observé des rotations θ qui deviennent mesurables. En améliorant la précision sur les mesures angulaires, nous pourrons déduire de nos mesures, par un calcul très simple, les valeurs de l'hyperpolarisabilité pour les molécules à symétrie sphérique.

CONCLUSION

Les résultats obtenus montrent:

1. Les modifications d'indice de réfraction dans des corps du type CS_2 ou $C_6H_5NO_2$ sont essentiellement dues, pour des intensités inférieures aux seuil de self focalisation et d'effet Raman, à l'effet de réorientation des molécules.
2. Pour des intensités plus fortes, d'autres effets non linéaires entrent en jeu de façon importante (redistribution des molécules, effet du champ local, et création de champs, Raman par exemple).
3. L'on a obtenu une méthode de mesure de l'hyperpolarisabilité pour les molécules à symétrie sphériques, et une méthode de mesure de l'anisotropie pour les molécules à symétrie axiale.

REFERENCES

1 G. A. Askaryan, *Zh. Eksperim. i Teor. Fiz.* **42**, 1567 (1962); (traduction anglaise: *Sov. Phys. JETP* **15**, 1088 (1962).

2 R. Y. Chiao, E. Garmire et C. H. Townes, *Phys. Rev. Letters* **13**, 479 (1964).

3 V. I. Talanov, *Sov. Phys. JETP Letters* **2**, 138 (1965).

4 P. L. Kelley, *Phys. Rev. Letters* **15**, 1005 (1965).

5 P. D. Maker, R. W. Terhune et C. M. Savage, *Phys. Rev. Letters* **12**, 507 (1964).

6 F. J. McClung, Proceedings of the 1966 International Quantum Electronics Conference, Phoenix, Arizona, 1968; *IEGEJ Quant. Electron.* QE2, Nos 8, 9 (1966).

7 R. W. Hellwarth, *Phys. Rev.* **152**, 156 (1966).

8 Y. R. Shen, *Phys. Letters* **20**, 378 (1966).

9 B. Kasprowicz et S. Kielich, *Acta Phys. Polon.* **31**, 787 (1967).

10 V. S. Boutylkin, A. E. Kaplan et Y. G. Khronopulo, URSS Symposium of Non-Linear Optics, 1970.

11 S. Kielich, *Acta Phys. Polon.* **30**, 683 (1966).

12 S. Kielich, *Proc. Phys. Soc.* **90**, 847 (1967).

13 S. Kielich, *IEEEJ Quant. Electron.* **11**, 744 (1968).

14 Buckingham et J. A. Pople, *Proc. Phys. Soc.* **67a**, 905 (1955).

15 Y. R. Shen et Shaham, *Phys. Rev.* **163**, 224 (1967).

16 M. Pauthenier, *Ann. Phys.* **14**, 239 (1920); *C.R. Acad. Sci. Paris* 121 (1926).

17 I. L. Fabelinski, *Molecular Scattering of Light*, Plenum Press, New York, 1968.

5. EXCITATION OF THE HYPER-RAMAN EFFECT WITH A Nd-YAG LASER

J. A. Topp and H. W. Klöckner
Sektion Physik der Universität München,
München, Germany

INTRODUCTION

Usually the hyper-Raman effect is excited with a giant-pulse ruby laser and therefore detected near twice its frequency in the near ultraviolet region of the spectrum.[1–5] Some substances, however, show two-photon absorption and subsequent fluorescence in the ultraviolet which is stronger than the hyper-Raman effect and makes it unobservable. Therefore we decided to use a neodymium-YAG-laser for the excitation of hyper-Raman spectra in the hope of avoiding such effects in a greater number of substances.

APPARATUS

The laser is a Laser Associates Mod. 252 A emitting pulses at 1064·8 nm with an energy of 100 mJ, a pulse duration of 20 ns and a repetition frequency up to 20 Hz. In order to enhance the power of the exciting radiation (if necessary) a multiple reflection device was constructed (Fig. 1). The laser beam is reflected into a vertical direction by prism P1, focused into the sample by lens L1 (f = 30–50 mm) and made parallel again by an identical lens L2. The beam is then reflected by prism P2 and forms a second focus coincident with the first one. Because the beam is displaced from the optical axis of the lens system it can again be reflected by prism P3 to form two more foci. If the power enhancement by nearly a factor of four is not sufficient, prism P4 can be used to reflect the beam in itself and to obtain 8 traversals through the focus. One advantage of the system is that simple lenses can be used, because the beams can be adjusted to have equal distances from the optical axis so that spherical aberration is no problem.

In hyper-Raman excitation the power is often limited by the dielectric breakdown of the sample. In this case it is advantageous to use pulses of somewhat longer duration which can be produced in a longer laser cavity. In this arrangement the laser output mirror is removed and prism P2 serves as second laser reflector so that the sample is now inside the laser cavity. Obviously the beam must be adjusted to coincide with the optical axis of the lens system. The pulses now have a duration of 50 ns. The higher power level inside the laser cavity allows operation of the laser at lower input power to prolong the lifetime of the laser flashlamp. This arrangement is indicated in Fig. 2.

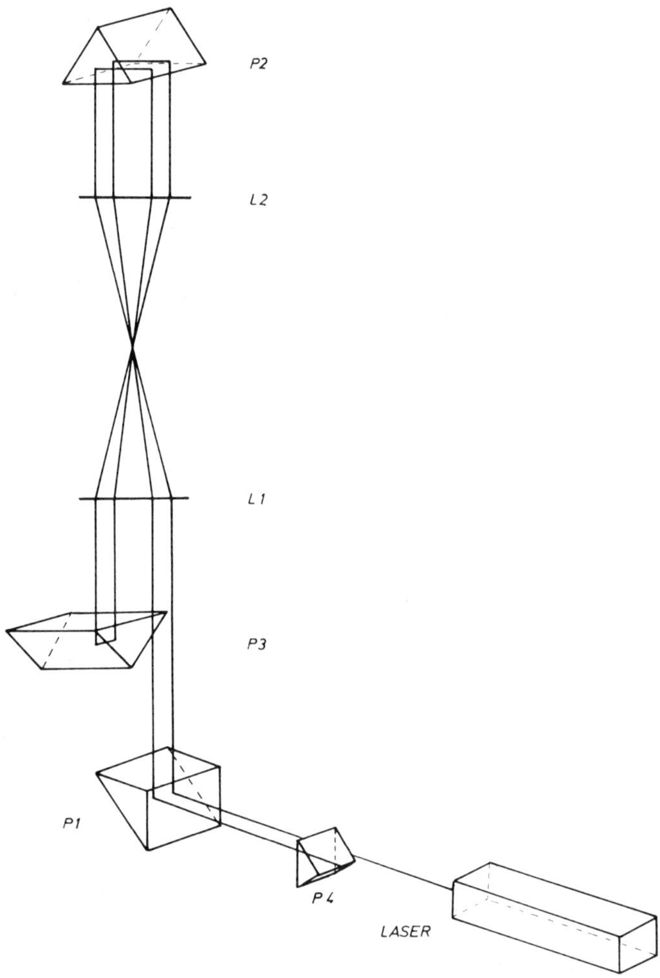

Fig. 1. Optical arrangement for multiple reflection of laser beam.

EXPERIMENTAL

The extremely low scattering cross section of the hyper-Raman effect requires that optimal use be made of the scattered light. After consideration of a rotational ellipsoid[6] we decided to use a rotational paraboloid as sample cell (Fig. 3) which allows the collection of light scattered from the focus F into a solid angle $> 2\pi$ and to transfer it to the spectrometer slit SP. This parabolic mirror has a focal length of 7·5 mm and a diameter of 43 mm.

A Jarrell-Ash 1 m Czerny-Turner monochromator equipped with a 600 $1\,mm^{-1}$ Bausch & Lomb grating is used for spectral dispersion. The photomultiplier RCA C 31000 E is insensitive to the infrared exciting radiation so that no stray light problems occur. The electronic detection system is schematically shown in Fig. 2. The dark current of the photomultiplier PM1 is reduced by cooling and by magnetic defocusing. The signal pulses from the photomultiplier are amplified 100-fold by amplifier V and are fed through

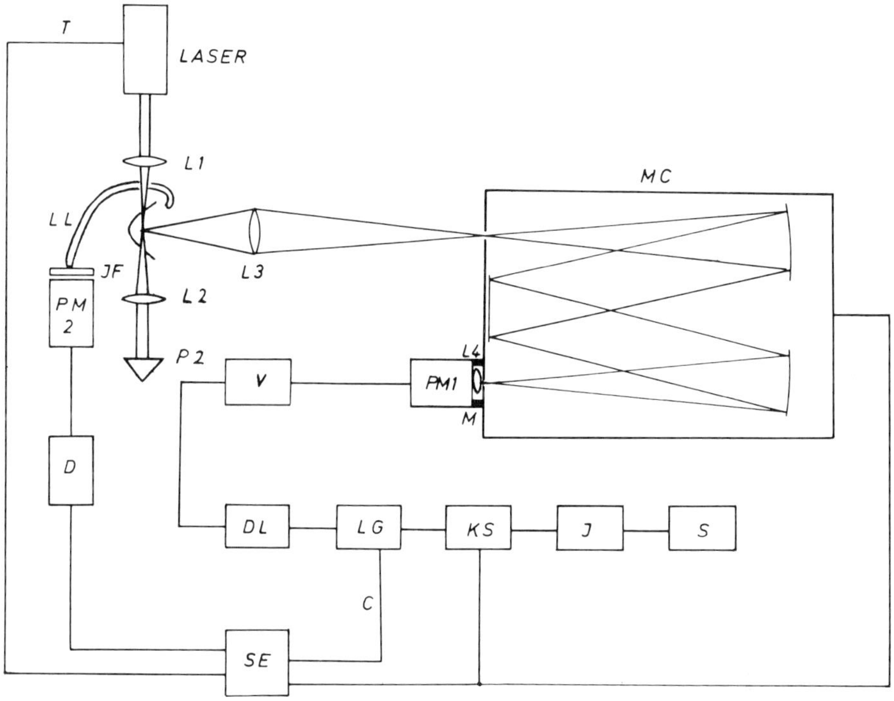

Fig. 2. Electronic detection system.

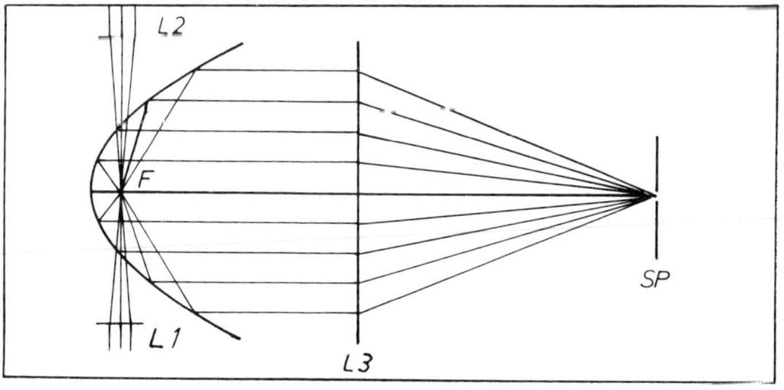

Fig. 3. Rotational paraboloid for stray light collection.

a delay line DL to the linear gate LG which is opened only at the time of laser emission by the control unit SE. The trigger pulse for the Pockels cell in the laser is used for triggering the control unit. The signal pulses transmitted by the linear gate are stored in the memory of a Geoscience Mod. 7000 Digital Processor KS.

When dielectric breakdown in the sample occurs for a particular laser pulse, the light flash is directed by a flexible light pipe LL to a second photomultiplier PM2 through an interference filter with a transmission maximum at a wavelength where no hyper-Raman line is expected. When the control unit SE receives a pulse from PM2 normalized by the discriminator D, the linear gate is not opened and the corresponding signal from PM1 is rejected. After storage of the signals from a number of laser shots which can be chosen between 1 and 4096 the control unit activates the stepping motor of the monochromator drive to the next spectral element and simultaneously switches the address of the storage system to the next channel. When a complete spectrum is recorded it can be read out directly by the recorder S after smoothing with a suitable integration time by the integrator I, or it can be transferred to a punched tape by a teletype for subsequent processing by a digital computer. With this apparatus the hyper-Raman spectrum of water could be reproduced.[1]

REFERENCES

1 R. W. Terhune, P. D. Maker and C. M. Savage, *Phys. Rev. Letters* **14**, 681 (1965).
2 P. D. Maker, *Physics of Quantum Electronics,* (Eds. P. L. Kelley, B. Lax and P. E. Tannenwald), McGraw-Hill, New York, 1966, p. 60.
3 J. F. Verdieck, S. H. Peterson, C. M. Savage and P. D. Maker, *Chem. Phys. Letters* **7**, 219 (1970).
4 C. M. Savage and P. D. Maker, *Appl. Opt.* **10**, 965 (1971).
5 D. A. Long, Plenary lecture at the Colloquium Spectroscopicum Internationale XVI, Heidelberg, October 1971.
6 A. Mooradian, *Science* **169**, 20 (1970).

6. QUANTUM THEORY OF THE RAMAN EFFECT AND NON-LINEAR OPTICAL PHENOMENA

D. F. Walls
University of Waikato,
Hamilton, New Zealand

INTRODUCTION

Non-linear optical effects arise as a result of the non-linear response of a medium to intense light fields. A successful method of analysis developed by Bloembergen[1,2] expands the polarization in powers of the electric field, the proportionality constant being a non-linear susceptibility tensor for the medium. Treating the electromagnetic field classically one obtains a set of coupled differential equations for the amplitudes of the interacting waves. The non-linear susceptibility of the medium is taken into account only phenomenologically as a coupling constant. A deficiency of the classical treatment of the electromagnetic field is the omission of the contribution of spontaneous emission to stimulated processes. The first quantum mechanical model for non-linear optical processes was proposed by Louisell *et al.*[3] They set up a phenomenological Hamiltonian for the interacting quantized field modes in a parametric amplifier. The non-linear susceptibility of the medium was included as a coupling constant. The pump field was treated as a constant classical variable in order to linearize the Heisenberg equations of motion for the mode operators. Numerical solutions to the non-linear operator problem have recently been presented.[4]

A microscopically correct Hamiltonian would contain the Bose operators of the light modes and the Fermi operators of the optically active electrons in the medium. It is the aim of this paper to present a nonperturbative method of solution of the Schrodinger equation for such a Hamiltonian.

HAMILTONIAN FORMULATION OF THE NON-LINEAR INTERACTION

The electric field operator for the free field may be expanded in terms of normal modes as in ref. 5.

$$E(\mathbf{r}, t) = i \sum_k (\tfrac{1}{2}\hbar\omega_k)^{1/2}(a_k(t)u_k(\mathbf{r}) - a_k^{\dagger}(t)u_k^*(\mathbf{r})) \tag{1}$$

where $a_k(t)$ and $a_k^{\dagger}(t)$ are the boson annihilation and creation operators for the kth mode. The effects of the linear polarizability in a non-dissipative isotropic medium may be accounted for by including a dielectric constant $\epsilon = 1 + 4\pi\chi$ where χ is the linear susceptibility of the medium. The mode functions are then taken to satisfy the wave equation

$$\left(\nabla^2 + \frac{\omega_k^2 \epsilon_k}{c^2}\right) u_k(\mathbf{r}) = 0 \tag{2}$$

with the orthonormality condition

$$\int (\epsilon_k \epsilon_l)^{1/2} u_k^*(\mathbf{r})\, u_l(\mathbf{r})\, \mathrm{d}^3 r = \delta_{kl} \tag{3}$$

The total Hamiltonian describing the interaction of the electromagnetic field with a non-linear medium may be expressed in the following form

$$H = H_0 + H_1 \tag{4}$$

where H_0 the Hamiltonian for the non-interacting system is

$$H_0 = \Sigma\, \hbar\omega_k a_k^\dagger a_k + H_{\text{medium}} \tag{5}$$

We consider the medium to consist of an ensemble of N f-level atoms with free Hamiltonian

$$H_{\text{medium}} = \hbar \sum_{i=1}^{N} \sum_{\alpha=1}^{f} \Omega_{\alpha i} c_{\alpha i}^\dagger c_{\alpha i} \tag{6}$$

where the $c_{\alpha i}$ are the fermion annihilation operators for the α level of the ith atom.

A general interaction Hamiltonian describing an $m + n$ photon process may be written as

$$H_1 = \hbar \sum_{i=1}^{N} \sum_{\alpha,\beta} \epsilon_{\alpha\beta}^{(m+n)} c_{\beta i}^\dagger c_{\alpha i} \prod_{j=1}^{m} E_j^{(-)}(\mathbf{r}_i) \prod_{k=1}^{n} E_k^{(+)}(\mathbf{r}_i) + \text{adjoint} \tag{7}$$

where $\epsilon_{\alpha\beta}^{(m+n)}$ is the matrix element for an $m + n$ photon transition, consisting of m emissions and n absorptions and an atomic transition from state α to state β. The term $E_k^{(+)}$ represents a single mode of the electromagnetic field.

$$E_k^{(+)}(\mathbf{r}_i) = [E_k^{(-)}(\mathbf{r}_i)]^\dagger = i(\tfrac{1}{2}\hbar\omega_k)^{1/2} u_k(\mathbf{r}_i)\, a_k \tag{8}$$

Substituting Eqn. (8) for $E_k^{(-)}$ and $E_k^{(+)}$ into Eqn. (7) and assuming an even distribution of atoms throughout the interaction volume V the interaction Hamiltonian becomes

$$H_1 \cong \sum_{\alpha,\beta} \mu_{\alpha\beta}^{(m+n)} \sum_{i=1}^{N} c_{\beta i}^\dagger c_{\alpha i} \prod_{j=1}^{m} a_j \prod_{k=1}^{n} a_k^\dagger + \text{adjoint} \tag{9}$$

where

$$\mu_{\alpha\beta}^{(m+n)} = \epsilon^{(m+n)} \prod_{j=1}^{m+n} i(\tfrac{1}{2}\hbar\omega_j)^{1/2} \int_V \mathrm{d}^3 r \prod_{j=1}^{m} u_j(\mathbf{r}) \prod_{k=1}^{n} u_k^*(\mathbf{r})$$

METHOD OF SOLUTION

We shall present here a method first used by Jaynes and Cummings[6] for the interaction of one electromagnetic field mode with one atom, but which proves to have a much greater applicability. If the interaction described by Eqn. (9) is perfectly energy conserving, the free Hamiltonian H_0^0 commutes with the interaction Hamiltonian H_1'. An energy mismatch in real transitions may be allowed for by choosing the interaction Hamiltonian as $H_1' = H_1 + H_0 - H_0^0$. It then follows that H_0^0 and H_1' commute. Thus a common set of eigenfunctions exists for H_0^0 and H_1'. This common set of eigenfunctions is the solution to the Schrodinger equation.

One proceeds by choosing as a basic set the eigenstates Ψ of the free Hamiltonian H_0^0. The Schrodinger equation for the modified interaction Hamiltonian assumes the form

$$H_1'\Psi = \hbar A\Psi \tag{10}$$

where A is the matrix representation of H_1' with dimension N the dimension of the Hilbert space of the systems studied.

The eigenstates and eigenvalues of the system are found by diagonalizing the matrix A. The probability amplitudes of the system being in a particular state at time t may be calculated exactly from the known eigenstates and eigenvalues.

APPLICATIONS IN NON-LINEAR OPTICS

The above method provides a powerful tool for studying problems involving the interaction of boson fields or boson and fermion fields in non-linear optics. It has its best applicability for systems where the dimension of the Hilbert space is small since an analytic solution is then feasible. Thus it is ideally suited for systems involving two or three fermion levels.

The technique has, however, been applied to studying coupled Stokes and anti Stokes radiation in Raman scattering from phonons.[7] The complexity of this problem is reduced by considering a one photon scattering process. By restricting the type of processes involved to those in which only one photon is present in an intermediate state at any time it is possible to obtain a solution in closed form to all orders of perturbation. The approximation involved in this restriction is analogous to the ladder approximation to the Bethe–Saltpeter equation.[8] The solutions predict a finite amount of Stokes and anti-Stokes emission for perfect energy matching in contrast to the results of classical calculations which omit spontaneous emission and laser depletion.[9]

A study of atomic coherence effects in Raman scattering reveals that a number of atoms acting in cooperation yield results that are markedly similar to those derived for scattering from phonons. This may be understood by observing that for large cooperation numbers the collective operators for the atoms obey boson commutation relations and hence the similarity with the phonon results.

For Raman scattering from atomic levels with definite parity one must consider a three level atomic system. A semi-classical analysis of Stokes generation from a three level system has been studied recently by Shimoda[10] who makes the assumption that the intensities of the laser and Stokes fields are constant over the interaction with a single molecule. Using the method outlined above a fully quantum mechanical treatment including the variation of the fields over the interaction with a single molecule is possible. We have shown that the semi-classical results of Shimoda are valid even in this fully quantum mechanical limit and extend his results to include the spontaneous Raman effect.[11]

Other applications in non-linear optics include two photon emission and absorption, parametric amplification, frequency up conversion and second harmonic generation.[12] Since the atomic operators are not removed by any linearization technique the effects of atomic cooperation are readily calculated. It is found, as well may be expected, that all the above non-linear optical processes are enhanced by cooperative atomic effects.

REFERENCES

1 N. Bloembergen, *Non-linear Optics*, Benjamin, New York, 1965.
2 J. A. Armstrong, N. Bloembergen, J. Ducuing and P. S. Pershan, *Phys. Rev.* **127**, 1918 (1962).
3 W. H. Louisell, A. Yariv and A. E. Siegman, *Phys. Rev.* **124**, 1646 (1961).
4 D. F. Walls and R. Barakat, *Phys. Rev.* **A1**, 446 (1970).
5 R. J. Glauber, *Phys. Rev.* **131**, 2766 (1963).
6 E. T. Jaynes and F. W. Cummings, *Proc. IEEE* **51**, 89 (1963).
7 D. F. Walls, *Z. Physik.* **237**, 224 (1970).
8 D. F. Walls, *Phys. Letters* **35A**, 345 (1971).
9 N. Bloembergen and Y. R. Shen, *Phys. Rev.* **137A**, 1787 (1965).
10 K. Shimoda, *Z. Physik.* **234**, 293 (1970).
11 D. F. Walls, *Z. Physik.* **244**, 117 (1971).
12 D. F. Walls, *J. Phys.* **A4**, 813 (1971).

SECTION II
NEW TECHNIQUES

7. NEW TRENDS IN RAPID LASER RAMAN SPECTROSCOPY

M. Bridoux, A. Chapput, M. Crunelle and M. Delhaye
Laboratoire Spectroscopie Raman, Université des Sciences et Techniques de Lille, Villeneuve d'Ascq, France

INTRODUCTION

The previous work of our laboratory has demonstrated the feasibility of recording in a very short time laser-excited Raman spectra of quickly evolving samples.[1,2] This is a fertile field of research for the study of photochemical reactions, of transient species and of excited states of molecules. This chapter reports on the recent improvements that have been found suitable to obtain well resolved Raman spectra and a good signal-to-noise ratio in the range of millisecond to nanosecond.

ELECTRON-OPTICAL ULTRAFAST SPECTROGRAPH

The experimental techniques that are based on the use of image intensifier photo tubes and high sensitivity electronic cameras have been developed mainly by Bridoux.[3] These techniques avoid the loss of information due to scanning by a monochromator. Through a careful study of the conditions of recording Raman spectra which are excited either by continuous lasers or by pulsed lasers, a very substantial gain of performance can be expected when using spectrographs fitted with such photoelectric tubes, instead of conventional scanning spectrometers.

The main recent improvements concerned the rejection of stray light by using two stages 'in tandem' grating spectrographs. The spectrographs employed in our earlier studies were constructed at the laboratory workshop. Our latest experiments were successfully made by using the new UF76 Coderg ultrafast spectrograph. The basic optical layout of this instrument is shown in Fig. 1. The radiation focused onto the entrance slit is dispersed in a Czerny-Turner configuration by a plane grating. In the focal plane of the second mirror is placed a wide adjustable slit which limits the band pass to the desired spectral range. This slit acts as an entrance slit for a second Czerny-Turner grating spectrograph. The dispersed spectra are focused onto the photocathode of an image intensifier phototube. The intensified spectral images which appear on the phosphor screeen of this tube are refocused by means of high aperture lenses onto the photocathode of a SEC television picture tube.

The photocathode transforms the dispersed spectra into an electronic image which is stored in memory in a low density KCl target. The information put in memory into

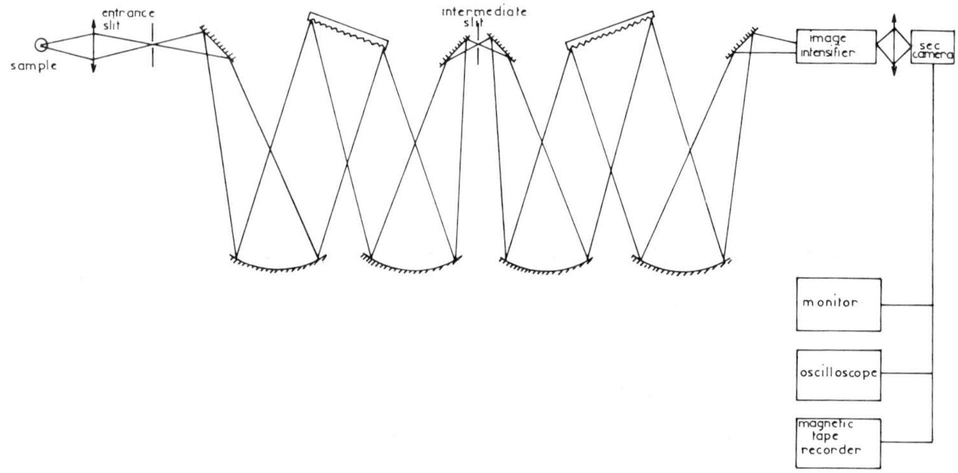

Fig. 1. Ultrafast spectrograph UF76 Coderg.

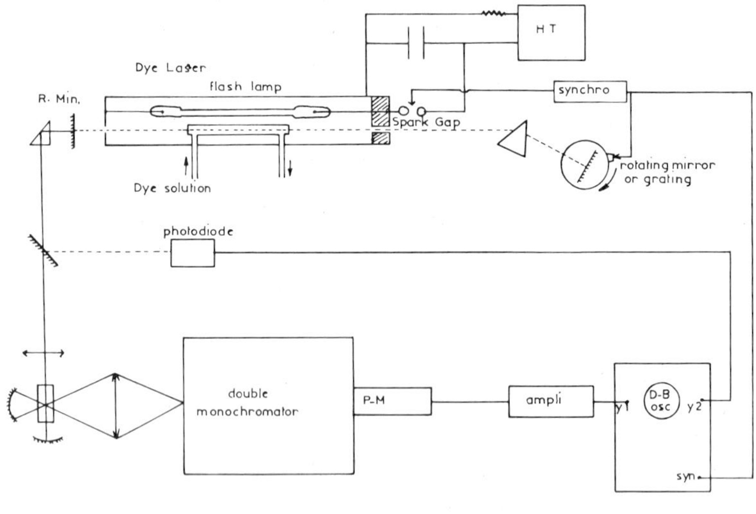

Fig. 2. 'Scanning Laser' used for Raman spectroscopy.

the target of the SEC camera is read by a beam of slow electrons as in classical television analyser tubes. The video signal picked up on the signal plate, then amplified, can be used to produce an image on a television monitor, to be analysed on an oscilloscope or transferred to a multichannel memory. The use of memory capability of the SEC camera allows the analysis of weak spectra, instantaneous recording of transient spectra, analysis of rapid evolution of a spectrum with a time resolution of the order of 10^{-6}s and simultaneous analysis of the different elements of a scattering sample optically focused along the spectrometer slit.

Four different gratings are mounted on turrets to provide quick interchange. Additive or subtractive combinations of gratings give a set of different dispersions.

TUNABLE DYE LASERS

Since the recent development of dye laser efforts have been directed to develop suitable techniques to obtain light sources which are tunable to a wide range of wavelength, without component change. Three different experimental arrangements were employed, using different spectrometers and detection systems.

Slow scanning

The intensity of the Raman scattering is measured at a fixed wavelength, whereas the laser is scanned. Either a continuously tunable dye laser which is pumped by an argon ion laser, or a high rate repetitively pulsed dye laser whose wavelength is varied between successive pulses may be used for such experiments.

The first report of such a technique was made by McNice[4] who described studies of gas scattering, by using a pulsed dye laser, tunable between 438 and 468 nm and narrow band interference filters. It is worth remarking that the performance of this arrangement may be seriously impaired by stray light, in the same manner as for conventional Raman spectrometers. That is why, in our experiments, the dye lasers were used in conjunction with a double monochromator. One feature of this technique, which we have not yet exploited, is the potential for observing the effect of resonance Raman by scanning the exciting radiation across an absorption band.

Rapid scanning

While the stability for a long time of the intensity of the radiation emitted by a tunable dye laser is a problem of considerable difficulty, the ability of a rapid scanning of wavelength during a single impulse suggested its use for observing transient phenomena by ultrafast spectroscopy.

An experimental arrangement for recording Raman spectra is shown in Fig. 2. The dye-cell is optically pumped by a flash lamp, which is synchronized with a wavelength selector placed inside the laser cavity. This wavelength selector may be either an interferometer, or a rotating grating or a variable polarization retardator.[5] The emitted wavelength is thus rapidly scanned during the pulse duration (typically from μs to ms). The laser beam is focused onto the sample and the scattered light is directed to a conventional Coderg PHO Raman spectrometer, whose scanning mechanism is stopped at a suitable wavelength. The signal output of the photomultiplier is processed by a wideband amplifier and presented to one channel of a dual channel oscilloscope. Into the other channel is fed the output of a photodiode which measures the intensity of the laser beam.

This display technique provides on one oscillogram both the variation of intensity of the exciting line during the laser pulse, and the variation of scattered light at a fixed wavelength as a function of time. The energy emitted in one pulse by dye lasers is typically of the order of 10–100 mJ and the peak power of the order of kilowatts. The substantial amount of power in the laser beam reduces the time constant that is needed to obtain a good signal-to-noise ratio to less than 1 μs. It is clear that the present arrangement makes more effective use of the pulsed laser as an exciting source than other scanning methods, which failed in recording the Raman spectra at high speed. This technique also avoids mechanical and optical problems which arise in the high speed motion of optical components in rapid scanning monochromators.

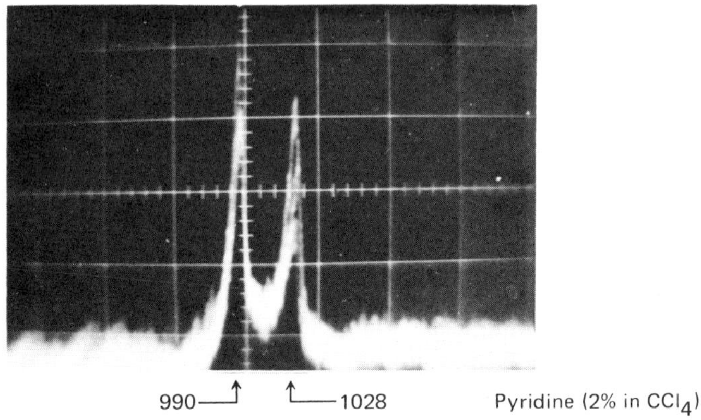

Fig. 3. Raman spectrum of dilute solution. Ultrafast spectrometer UF76; exciting line – 5471 Å; exciting energy – 7 mJ; scanning TV – 1 line.

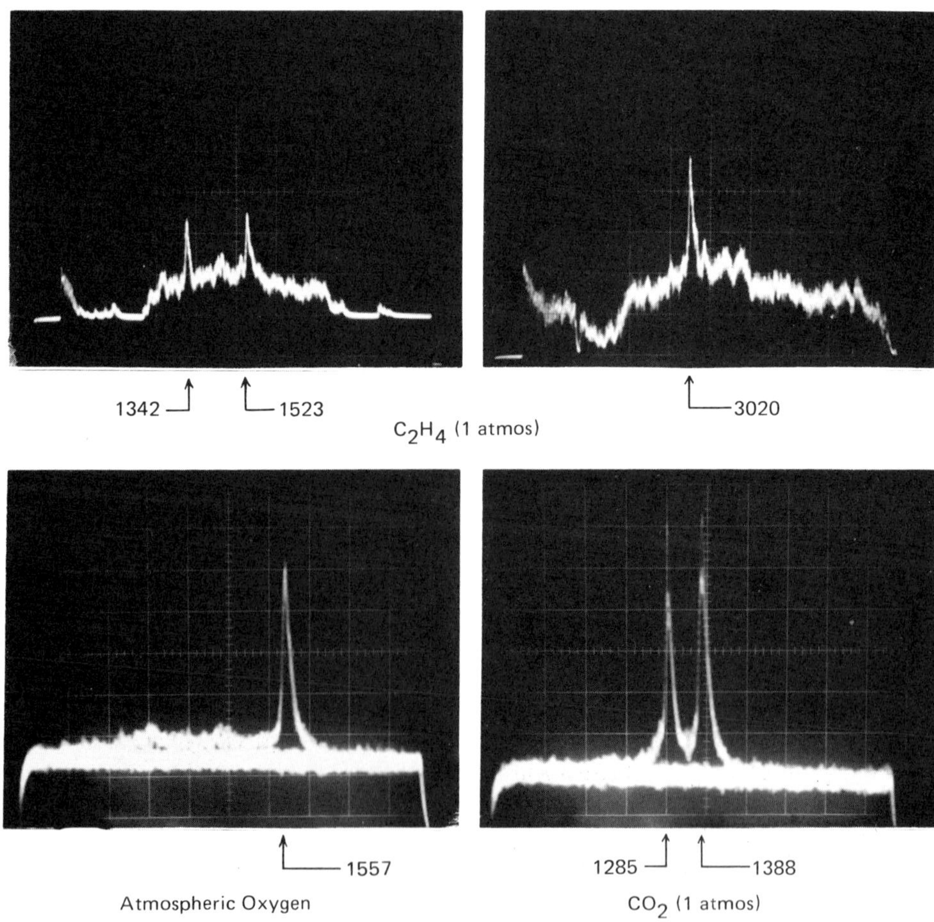

Fig. 4. Raman spectra of gases. Ultrafast spectrometer UF76; exciting line – 4880 Å; exciting energy – 8 mJ; scanning TV – 1 line.

Ultrafast scanning

A new solution to the problem of ultrafast Raman spectroscopy is offered by the conjunction of the excitation by means of a scanned pulsed dye laser with the recording by means of an electron-optical spectrograph. The laser source and the sample illuminator are similar to those described above. Instead of the monochromator, a double grating UF76 Coderg spectrograph analyses the Raman spectrum. In the focal plane of the spectrograph's camera is mounted an image intensifier phototube whose phosphor screen is optically coupled with a SEC vidicon. During the laser pulse the exciting wavelength varies as a function of time and simultaneously the spectral images are rapidly deflected in a direction perpendicular to the dispersion of the gratings. The deflection may be obtained either by an optical device placed between the sample and the entrance slit or by applying a magnetic field or an electric field to a suitable image intensifier tube.

A precise synchronization of laser scanning and deflection is necessary. If the illuminated part of the entrance slit is limited in height, time resolved spectra are displayed on the photocathode and are analysed by the SEC vidicon. The phototube is gated on only at the time the flash occurs. During a single laser pulse a considerable amount of information is stored on the target of the SEC phototube concerning the slopes of laser intensity, laser wavelength and of Raman spectra as functions of time.

CONCLUSION

The shortest intervals that have been handled by a device of this type is in the range of nanoseconds, which is far better than the limit due to the time constant needed in photomultiplier circuits in the case of a monochromator system.

Despite the early stages of this line of experiment there is, however, clear indication that it offers an attractive approach to the problem of recording Raman spectra in the sub-nanosecond range, where optical delay lines may be used as deflecting devices to obtain high precision time measurements. As an illustration of the operation of these systems Figs. 3 and 4 show several examples of ultrafast Raman spectra. These techniques and their use in providing new information on short lived molecules or radicals appear to be of value and wide applicability to a number of structural or kinetic problems.

ACKNOWLEDGEMENT

The authors wish to thank Dr. Lyndquist and M. Soep (University of Orsay) for helpful suggestions in the fabrication of the dye lasers.

REFERENCES

1 M. Delhaye, *Appl. Opt.* **7**, 2195 (1968).
2 M. Bridoux and M. Delhaye, *Nouv. Rev. d'Optique* **1**, 23 (1970).
3 M. Bridoux, *Rev. d'Optique* **8**, 389 (1967).
4 G. T. McNice, *Appl. Opt.* **11**, 699 (1972).
5 Lyndquist, Journées d'Etudes du GESEM sur les Sources et les Détecteurs en Spectroscopie Electronique Moléculaire, Orsay, October 1971.

8. RAMAN MATRIX ISOLATION SPECTROSCOPY

J. W. Nibler
Department of Chemistry, Oregan State University, Corvallis, Oregon, USA

INTRODUCTION

Since its introduction in the 1950's, infrared matrix isolation spectroscopy has proved an extremely valuable technique for the study of simple reactive molecules, radicals and high temperature species. Unfortunately, until quite recently, complementary and in some respects more informative Raman spectra of matrix isolated species were not accessible. In the last two years, however, considerable progress has been made in solving the formidable experimental problems of Raman matrix isolation spectroscopy, and the technique has evolved as a feasible and useful one. Among others, Shirk and Claassen[1] and particularly Ozin[2] and co-workers have pioneered in the development and exploitation of the technique.

We too, have had some success with the method, and the purpose of this chapter is to present some of our efforts at solving the experimental difficulties and to illustrate the value of the technique for a few of the chemical problems to which we have applied it.

EXPERIMENTAL

Figure 1 is a schematic of our cold cell and Raman illumination geometry. The rotatable 1 X 1 X 4 cm cold tip (aluminium, copper or platinum plated copper) is highly polished and is cooled to 12–14 K by an Air Products Displex refrigerator. The temperature is monitored by a hydrogen vapour thermometer and by a Au–Fe/chromel thermocouple imbedded just below the matrix surface. The cell has four optical ports, one being a CsI window to allow infrared scans using a homemade reflection assembly for a Beckman IR-7. For Raman scans, the cell is placed horizontally on the *x–y* translation stage of a Cary 82 spectrophotometer, and a 4880 or 5145 Å laser beam (50–750 mW–C.R.L. 52B argon ion laser) is focused on the sample tip. Sample positioning is quite important for maximum signal, and the tip angle and cell position must all be optimized. Use of N_2 as a matrix greatly facilitates this operation since the strong N_2 vibration can be used for alignment purposes.

The matrix gas is typically deposited at rates of 0.2–5.0 mM/h for periods varying from 1–30 h. The gas flow is monitored by a flowmeter and is introduced through a 3 mm O.D. copper tube terminating 3 cm from the cold tip. The sample (typically 0.2–1.0%)

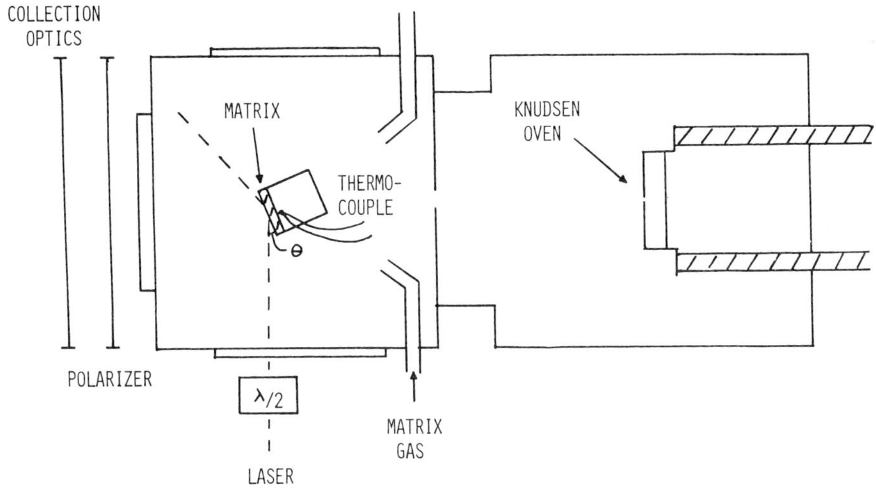

Fig. 1. Cold cell schematic and illumination geometry.

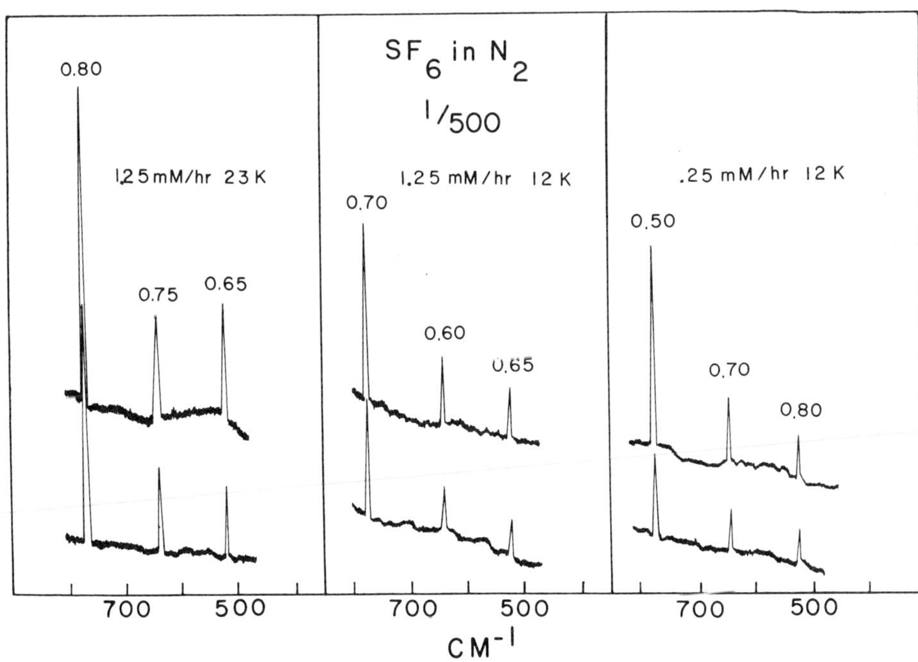

Fig. 2. Effect of deposition temperature and rate on the depolarization results for SF_6/N_2.

is either mixed initially with the matrix gas or else is introduced through a second jet. For studies of high temperature molecules, a separate molecular beam is generated by a resistively heated tantalum or graphite Knudsen oven situated 12 cm from the cold tip. The oven temperature is monitored by an optical pyrometer and is varied to give matrix ratios which minimize the formation of dimers. This operation is greatly facilitated by infrared examination of short duration depositions at four different temperatures, one for each face of the tip. Because of the greater sensitivity of the infrared measurement, the optimum isolation conditions can be quickly determined, and then extended deposits made for the Raman scan.

A variety of multiple reflection cold tips were tried in an attempt to enhance the Raman signal, but it was found that after a single pass through the matrix film the laser was randomly defocused and no gain was made on back reflection. From this and from other studies of the effect of the tip angle (θ) on the scattered intensity and depolarization ratios of Raman bands, it is clear that the sample can not be viewed as a 'thin film'. Indeed, for extended deposits sample thicknesses can exceed 1 mm. The ideal matrix seems to be a clear glassy film which is best obtained by long, slow deposits. The optimum tip angle seems to be about 20°.

One of the more pleasing experimental discoveries was our demonstration that depolarization measurements were feasible for glassy matrix films[3] and others have subsequently used this valuable tool to advantage in matrix studies.[4] We have investigated the conditions under which good depolarization results can be obtained and find, as indicated in Fig. 2, that low deposition temperatures and rates tend to give the most reliable results. We do not find a strong dependence on the tip angle θ and hence it appears legitimate to think of the process as 90° scattering from the bulk matrix.

EXAMPLES OF APPLICATIONS

We have examined a variety of systems using Raman matrix isolation spectroscopy and, due to space limitations, can only indicate here a few of the more interesting aspects of some of these.

Trans-formaldazine, $H_2C{=}N{-}N{=}CH_2$

This interesting molecule tends to polymerize quite readily so that Raman depolarization measurements on the liquid were not feasible, and hence its vibrational assignments were uncertain. For this reason, and our interest in its photolytic fragmentation products, infrared and Raman spectra were obtained for the matrix isolated molecule (Fig. 3). The absence of infrared-Raman coincidences is consistent with the *trans* structure, and the Raman spectra show that, as expected, most of the bands are polarized. An exception is the depolarized band at 851 cm^{-1}, a feature which had previously been assigned as the NN stretch. Since this mode should be strongly polarized, the polarized feature at 1205 cm^{-1} is a more logical choice and the 851 band is assigned as the B_g out of plane CH_2 twist.

It may be noted that careful examination of the matrix spectra revealed a number of features whose intensity varied depending upon deposition rate (e.g. Raman features at 3020, 2935, 1625, 1220, 1002 and 428 cm^{-1}). These are believed to be due not to polymers but to a room temperature rotamer of *trans*-formaldazine which is trapped during the rapid freezing process. Independent electron diffraction studies by Hedberg and coworkers at Oregon State University support this contention and indicate that the second rotamer has a *gauche*, not *cis* structure.

LaF_3 and PrF_3

A recent infrared matrix study[5] suggested that LaF_3 and most of the other lanthanide

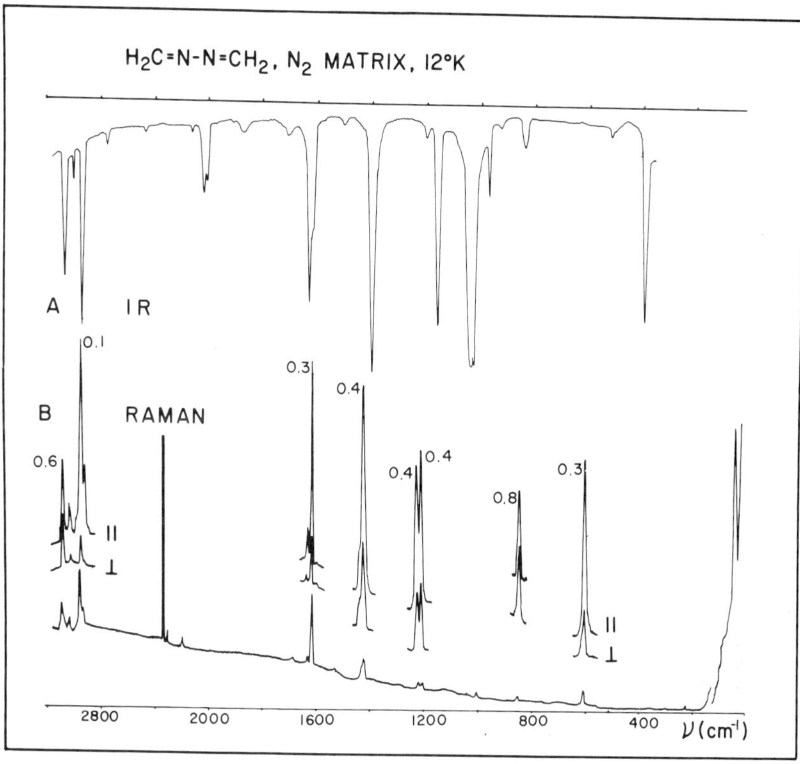

Fig. 3. Matrix spectra of *trans*-formaldazine.

trifluorides had planar D_{3h} structures. A notable exception was PrF_3 since the observation of *two* strong infrared stretches at 542 and 458 cm^{-1} seemed to require a pyramidal C_{3v} structure for this molecule. However, very recent electric deflection experiments by Falconer[6] give exactly the opposite structural conclusions (LaF_3 pyramidal, PrF_3 planar) and indicate that a number of the other trifluorides are pyramidal.

To help settle this question, we have obtained Raman matrix spectra for PrF_3 and LaF_3 as shown in Fig. 4. The LaF_3 results strongly support a pyramidal structure since the polarized Raman active symmetric stretch at 514 cm^{-1} is coincident with a weak infrared band. The Raman and infrared observation of both bends and their Raman depolarization values also require a C_{3v} structure. For PrF_3 however, the strong polarized Raman stretch at 526 cm^{-1} does not coincide with either of the infrared bands and only one Raman bending frequency is observed, results in complete accord with a planar D_{3h} structure. Thus the electric deflection results are confirmed, and it is seen that Raman matrix isolation spectroscopy can play a vital role in allowing positive structural conclusions for matrix isolated species.

C_2^-

Fluorescence is generally undesirable for Raman matrix isolation spectroscopy but it can sometimes be informative. A case in point is the fluorescence (not Raman) spectrum

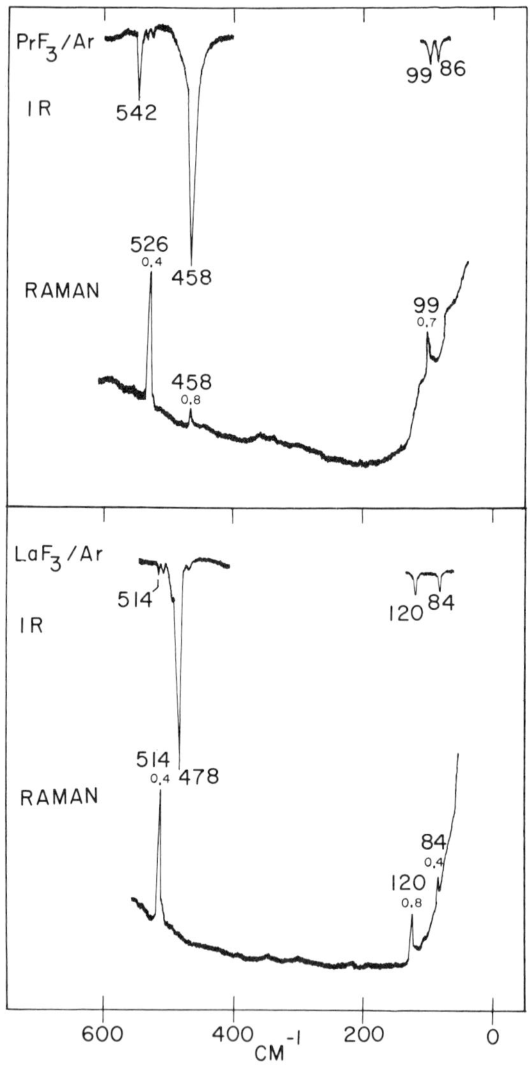

Fig. 4. Matrix spectra of PrF_3 and LaF_3.

shown in Fig. 5 for C_2^- formed by photolysis of C_2H_2. Superimposed on a broad fluorescence background are sharp features which are only slightly shifted from the Herzberg–Laqervist gas phase bands for C_2^-. Perhaps most interesting is the observation of a second progression series in argon, with a 1–0 transition on the anti-Stokes side of the laser line. Intensity studies clearly show that this series is due to secondary absorption of the laser radiation and hence one concludes that sizable concentrations of vibrationally excited C_2^- molecules must be present in the matrix. From steady state intensity considerations, a lower limit of 1 s can be deduced for the lifetime of the $\nu'' = 1$ vibrational state.[7]

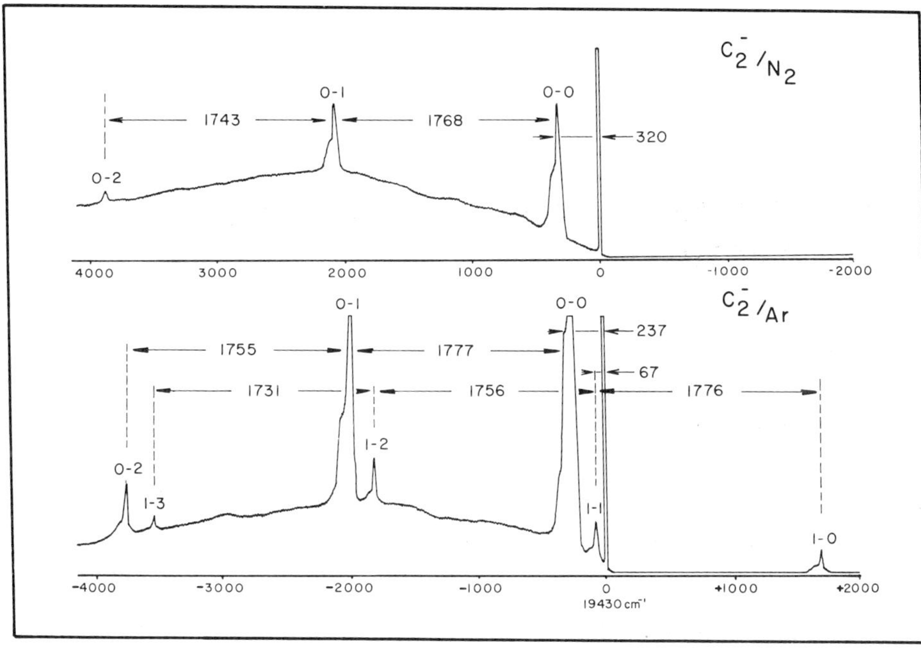

Fig. 5. Fluorescence spectra of C_2^-.

CONCLUSIONS

Raman matrix isolation spectroscopy has passed from the developmental stage and now must be viewed as a valuable new technique in the study of reactive molecules. The combined infrared-Raman matrix spectra and the Raman depolarization measurements will yield much more reliable conclusions regarding the structure and vibrational assignments of many species which cannot be easily studied by any other means.

REFERENCES

1 J. S. Shirk and H. H. Claassen, *J. Chem. Phys.* **54**, 3237 (1971).

2 D. Boal, G. Briggs, H. Huber, G. A. Ozin, E. A. Robinson and A. Vander Voet, *Nature (Phys. Sci.)* **231**, 174 (1971).

3 J. W. Nibler and D. Coe, *Raman Matrix Isolation Spectroscopy*, paper I 10, *26th Symposium on Molecular Structure and Spectroscopy*, Columbus, Ohio, June 1971; J. W. Nibler and D. Coe, *J. Chem. Phys.* **55**, 5134 (1971).

4 H. Huber, G. A. Ozin and A. Vander Voet, *Nature (Phys. Sci.)* **232**, 166 (1971).

5 R. D. Wesley and C. W. DeKock, *J. Chem. Phys.* **55**, 3866 (1971).

6 W. E. Falconer, Bell Laboratories, private communication.

7 V. E. Bondybey and J. W. Nibler, *J. Chem. Phys.* **56**, 4719 (1972).

9. SPECTROMETRE RAMAN MULTICANAL AUTOMATIQUE

M. Manfait, J. L. Beaudoin et L. Bernard
Laboratoire de Recherches Optiques,
Faculté des Sciences, Reims, France

ABSTRACT

Fully digital recording of Raman spectra is achieved with a Coderg PH0 tandem spectrometer used in synchronism with a multichannel analyser operating in the multiscaler mode. The associated logic device allows one to automatically obtain single or multi-scanned spectra (time-averaging method). A special device for compensating exciting source fluctuations is incorporated.

With the aid of a theoretical work by Ernst[1] we have found cases in which the method of accumulation of spectra improves the signal-to-noise ratio. Integrated spectra, difference spectra and rapid measurements of wavelengths are easily obtained thanks to the intrinsic possibilities of our device; a few representative examples are given.

INTRODUCTION

Lors de l'enregistrement d'un spectre Raman par comptage de photons, nous avons proposé[2] de substituer à l'ensemble chaîne de comptage-enregistreur analogique, un analyseur multicanal fonctionnant en mode multiéchelle. L'avantage de ce dispositif est d'éviter les distorsions, l'imprécision de lecture et la constante de temps inhérents à l'enregistrement analogique. En outre, l'analyseur multicanal apporte un certain nombre de possibilités nouvelles que nous allons préciser. Autour de cet organe, nous avons réalisé un ensemble de spectrométrie Raman laser permettant de réaliser automatiquement des spectres par accumulation de données, des spectres à énergie constante.

Le principe d'enregistrement est le suivant: le spectre à étudier est analysé en un nombre fini N de points et les N échantillons mis en mémoire. Le pas spectral p fonction de la largeur des fentes du spectromètre est déterminé à partir du théorème de Shannon. On a

$$p < \frac{\Delta\sigma}{2}$$

($\Delta\sigma$ largeur spectrale des fentes) ainsi, le spectre sera complètement déterminé en l'échantillonnant avec un pas spectral p inférieur ou égal à la demi-largeur des fentes.

DESCRIPTION DU SPECTROMETRE RAMAN AUTOMATIQUE

L'ensemble de spectrométrie se compose:

1. d'un spectromètre Coderg type PH0 muni d'un moteur pas à pas (un pas correspond

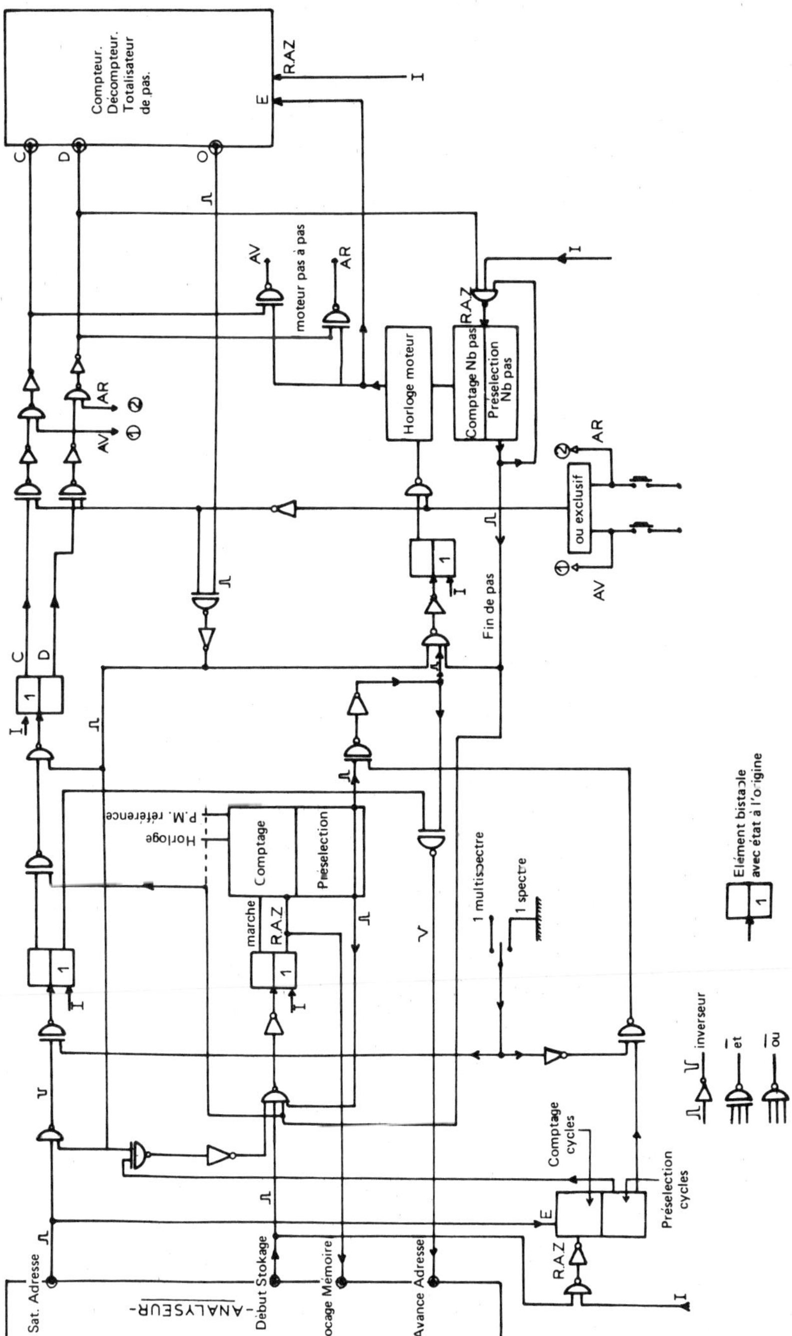

Fig. 1. Schéma de principe du spectromètre Raman multicanal automatique.

au balayage de O, 125 cm^{-1}).

2. d'un système de détection, d'acquisition et de traitement des données composé d'un photomultiplicateur EMI 9558QB fonctionnant en comptage de photons et d'un analyseur Intertechnique auquel sont associés une imprimante et un calculateur à bande magnétique.
3. d'un système électronique (Fig. 1) réalisé à partir de circuits intégrés qui assure d'une part un synchronisme rigoureux entre la rotation pas à pas du moteur et l'avance adresse dans l'analyseur (passage du canal n au canal $n + 1$), d'autre part l'enchaînement automatique des différents cycles lors de l'enregistrement d'un spectre, ces enregistrements pouvant être réalisés de différentes manières.

(a)Fonctionnement en mode normal. Le temps de comptage τ dans un canal de la mémoire est constant et prédéterminé. On a

$$\tau = T_0 - T_m$$

où T_0 est le pas temporel d'échantillonnage, et T_m le temps mort dû à différentes fonctions de servitude.

(b)Programmation et réalisation automatique des multipassages.

(c)Fonctionnement à énergie constante. Le temps de comptage τ dans un canal est variable; il dépend de l'amplitude des fluctuations d'intensité de source excitatrice.

POSSIBILITES INTRINSEQUES DU SYSTEME D'ACQUISITION ET DE TRAITEMENT DES DONNEES

Pour chaque mode de fonctionnement (normal, à énergie constante ou multipassage), les fonctions intrinsèques de l'analyseur et des organes périphériques permettent entre autre, de visualiser un même spectre sur plusieurs gammes de sensibilité, d'obtenir les résultats sous forme numérique décimale, d'enregistrer les spectres sur bande magnétique, d'effectuer des calculs sur le contenu de la mémoire.

Spectres intégrés

L'utilisation du calculateur intermédiaire RG23 permet le calcul instantané de la courbe intégrale d'un spectre ou d'une raie; il est alors facile d'effectuer des rapports d'intensités intégrées entre deux raies après normalisation de la valeur moyenne de leur ligne de base.

Pour tester cette méthode, nous avons enregistré les raies anti-Stokes et Stokes de vibration du Brome (319 cm^{-1}) et calculé le rapport des intensités intégrées (Fig. 2). Le résultat est identique à celui obtenu par des méthodes classiques.

Spectres par différence[3]

Cette technique utilisant conjointement l'analyseur et le calculateur permet par exemple, l'extraction du spectre d'un soluté par différence entre les spectres du solvant et de la solution. Dans des conditions expérimentales contrôlées, on peut procéder aux dosages rapides des constituants élémentaires d'une solution complexe à partir du spectre de cette solution et de ceux des constituants. On peut également mettre en évidence les effets d'interaction moléculaire (déplacement de raies, variation d'intensité) en comparant le spectre de la solution réelle à un spectre obtenu artificiellement par addition des spectres des constituants.

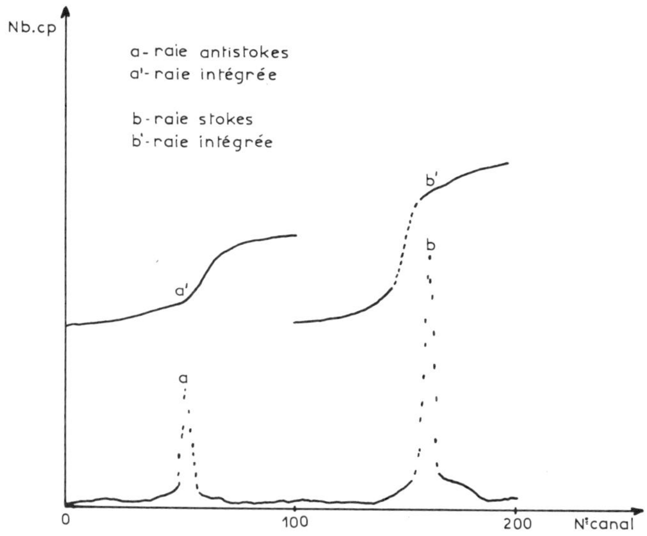

Fig. 2. Spectres intégrés. Calcul du rapport anti-Stokes/Stokes de la raie de vibration 319 cm^{-1} du brome.

Mesures des longueurs d'onde

L'enregistrement sans décalage en fréquence d'un spectre de référence (arc-au-fer, néon) par rapport au spectre expérimental, est permis grâce à un repositionnement correct du réseau au point d'analyse origine.[4]

La position d'une raie est repérée à l'aide du numéro de canal de la mémoire correspondant au maximum de cette raie. Le calcul des fréquences des raies Raman à partir des numéros de canaux de celles-ci et des numéros de canaux et longueurs d'onde correspondants aux raies de référence s'effectue sur ordinateur. L'utilisation d'un ordinateur Intertechnique Multi 8 en liaison avec l'analyseur permet d'effectuer certains calculs sur les spectres (fréquences et intensités, déconvolution, lissage) en temps réel.

Accumulation des données: influence sur le rapport signal sur bruit

Généralités. Le rapport S/B est défini comme étant égal à

$$\frac{S}{B} = \frac{\text{Valeur maximum du signal}}{\text{écart type du bruit}}$$

Proposons-nous d'additionner dans un même canal les échantillons de rang identique provenant d'enregistrements successifs du même spectre. En supposant la synchronisation parfaite, le signal s'additionne arithmétiquement à chaque analyse et le bruit tend vers une valeur moyenne. Supposons la fonction aléatoire $b_i(t)$ représentative du bruit stationnaire, ergodique et centrée. Après accumulation des données résultant de n balayages successifs, le signal et le bruit s'expriment respectivement par

$$S_{0_{\max}} = n\,S_{i_{\max}}$$

$$b_0(t) = \frac{1}{T_0} \sum_{x=0}^{n-1} \int_{t+xT}^{t+xT+T_0} b_i(\tau)\, d\tau$$

où T_0 est le pas temporel d'échantillonnage et T la période des balayages. Ce procédé de multipassage, couramment employé en R.M.N., en spectroscopie γ et Mossbauer, commence à être utilisé en spectroscopie optique. On considère souvent que le rapport S/B obtenu croît comme $\sqrt{n}$ où n est le nombre d'enregistrements successifs. Une étude mathématique due à Ernst[1] montre que la réalité est plus complexe. D'autre part, il ne faut pas oublier de comparer les résultats obtenus par accumulation des données à ceux obtenus en un seul enregistrement de même durée totale.

Etude de quelques cas particuliers: résultats.

(a) *Cas d'un bruit blanc.* L'amplitude d'un tel bruit est indépendante de la fréquence (bruits dans un photomultiplicateur). L'écart type après accumulation des données s'exprime par

$$\sigma = \sqrt{nk} \quad \text{avec} \quad k = A\left(\frac{1}{T_0} - \frac{1}{T}\right)$$

La loi en $\sqrt{n}$ est donc valable; cependant, un seul enregistrement à durée totale de mesure identique procure le même résultat.

(b) *Cas de bruits contenant des composantes périodiques.* Après accumulation des données, l'écart type est donné par

$$\sigma = \sigma_i\sqrt{n}$$

σ_i est l'écart type pour un balayage unique.

Si ce bruit contient des harmoniques de la fréquence de répétition des balayages $\omega_r = 2\pi/T$, ils ne seront pas atténués. Si ω_r ne correspond à aucune fréquence contenue dans le bruit, il y aura amélioration de S/B.

Considérons un bruit dont la bande de fréquence a pour demi-largeur $\Delta\omega$:

Si $\Delta\omega \gg 2\pi/T$, la loi en $\sqrt{n}$ reste valable.

Si $\Delta\omega \ll \dfrac{2\pi}{n_{max}T}$, S/B est indépendant du nombre de balayages.

(c) *Cas d'un bruit 'en* $1/f$*'.* Il s'agit d'un bruit dont l'amplitude augmente vers les basses fréquences; il apparaît par exemple dans les erreurs de positionnement et les oscillations du réseau au cours de sa rotation impulsionnelle ou bien dans les fluctuations des sources excitatrices.

Après accumulation des données, l'écart type est donné par

$$\sigma = \sigma_i\sqrt{n\mathrm{A} - \mathrm{B}}$$

La loi en $\sqrt{n}$ n'est pas rigoureusement vérifiée. Cependant, pour une même durée totale de mesure l'accumulation des données améliore S/B par rapport à un balayage unique. La Fig. 3 illustre l'amélioration de S/B lorsqu'on est en présence d'un bruit composite.

(d) *Domaine d'emploi de cette technique.* Nous avons essayé de déterminer expéri-

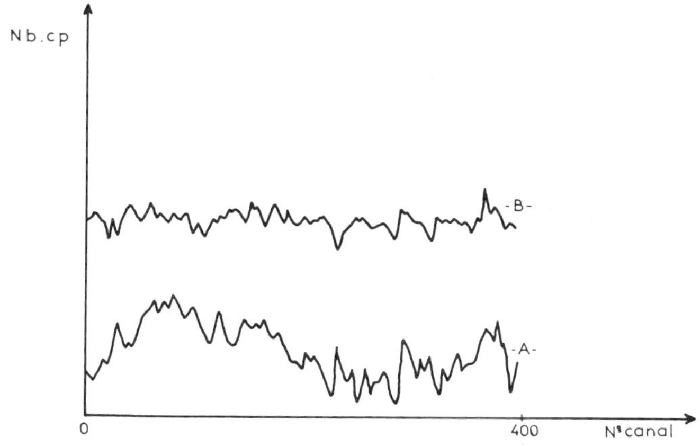

Fig. 3. Influence du multipassage dans le cas d'un bruit composite. A. 1 balayage $T_0 = 10$ s. B. 10 balayages $T_0 = 1$ s.

mentalement le domaine où l'emploi du multipassage apporte une amélioration par rapport à un enregistrement unique à durée de mesure identique.

Il en résulte de cette étude, que l'emploi du multipassage est recommandé d'une part, lorsqu'on est en présence de spectres de faible intensité nécessitant des durées d'analyses longues, d'autre part lorsque les sources excitatrices présentent des bruits de basses fréquences (sources conventionnelles et laser de grande puissance).

A titre d'exemple, nous présentons un multipassage sur la raie de vibration 2331 cm^{-1} de l'azote. Lors d'un balayage unique [Fig. 4(a)], l'amplitude du signal est pratiquement égale à celle du bruit. Un multipassage [Fig. 4(b)] améliore le rapport S/B de la détection. Si on compare à temps de comptage identique, on constate que la détection par multipassage [Fig. 4(b)] est supérieure à celle obtenue par monopassage [Fig. 4(c)].

Notons aussi que dans certaines conditions, cette technique permet de pallier l'effet d'une décomposition photochimique de l'échantillon.[3]

COMPENSATION DES VARIATIONS D'INTENSITE DE SOURCES ENSEMBLE A ENERGIE CONSTANTE

Si le flux de photons moyen diffusé par l'échantillon est proportionnel au flux de photons excitateurs, alors une mesure exacte de l'intensité Raman I impose que le nombre de photons excitateurs soit identique pour chacun des N points du spectre. Par suite, la durée de comptage τ dans le i ème canal doit satisfaire la relation

$$\int_{t_i}^{t_i+\tau_i} n(t)\,\mathrm{d}t = n_0 \forall_i \in (0, N) \tag{1}$$

où t_i est un instant quelconque, $n(t)$ le nombre de photons excitateurs par unité de temps à l'instant t, n_0 le nombre total prédéterminé de photons excitateurs par point d'analyse.

En fait, la grandeur mesurable est le nombre d'impulsions anodiques N issues du photomultiplicateur recevant n photons; dans certaines conditions on a

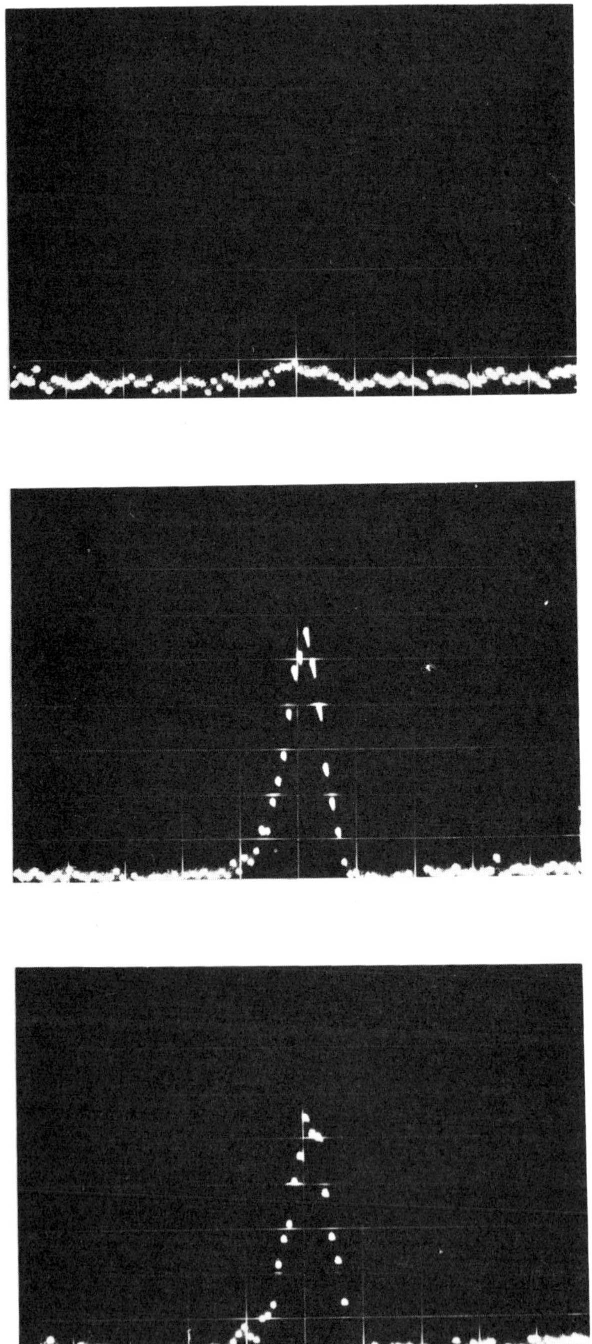

Fig. 4. Multipassage sur la raie de vibration 2331 cm^{-1} de l'azote. A. monopassage $T_0 = 0.1$ s, $\tau = 60$ ms. B. multipassage $n = 10$, $\tau = 600$ ms. C. monopassage $T_0 = 0.64$ s, $\tau = 600$ ms. Spectromètre Coderg $f = 4$ cm^{-1}, $p = 0.5$ cm^{-1}. T_0 période d'avance adresse. τ temps de comptage dans un canal.

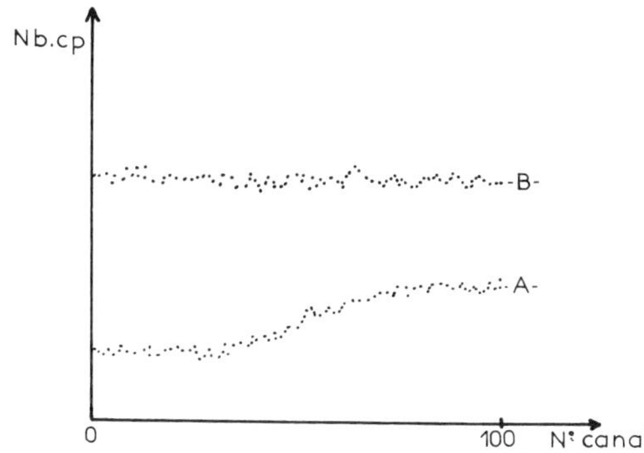

Fig. 5. Variation de l'intensité d'une raie Raman (760 cm^{-1} $CHCl_3$) en fonction de l'intensité de la raie excitatrice (4880 Å laser argon), 3 *h* de fonctionnement. A. sans contrôle et B. avec contrôle de l'intensité de l'excitatrice. Temps d'analyse 1000 s. $T_0 = 10$ s.

$$N(t) = K\,n(t) \tag{2}$$

en transposant dans la relation précédente, on obtient

$$\int_{t_i}^{t_i+\tau_i} N(t)\,\mathrm{d}t = N_0 \tag{3}$$

En pratique, il suffit donc de fixer N_0, c'est-à-dire le nombre d'impulsions que doit fournir le photomultiplicateur de référence, puis de mesurer et de stocker dans le canal *i* l'information Raman $I(t)$ pendant une durée τ_i satisfaisant la relation (3). Dans ces conditions, le résultat d'une mesure de l'intensité d'une raie Raman est indépendant des fluctuations de la source excitatrice. Notons que $N(t)$ doit être le plus grand possible, l'erreur relative sur la détermination de τ_i étant proportionnelle à $[N(t)]^{1/2}$

A titre d'exemple, nous présentons (Fig. 5) un enregistrement avec et sans contrôle de flux excitateur de l'intensité de crête d'une raie Raman (760 cm^{-1}) du chloroforme excité par la raie 4879,90 Å d'un laser argon ionisé: au bout de trois heures de fonctionnement, le laser a atteint sa stabilité maximum (± 1,5%), cependant, l'utilisation de notre montage compense ces fluctuations.

CONCLUSION

D'autres auteurs[5, 6] ont également proposé des dispositifs différents permettant l'application de l'accumulation des données à la spectroscopie optique. Nous avons mis en évidence les limites d'utilité de cette technique, de même que l'intérêt que présente l'emploi de l'analyseur multicanal et du calculateur associé pour le traitement des données.

REFERENCES

1 R. R. Ernst, *Rev. Sci. Instr.* **36**, 1689 (1965).
2 M. Manfait, *Thèse,* CNAM, Mai 1970.

3 M. Manfait, J. L. Beaudoin et L. Bernard, *Nouv. Rev. d'Optique Appliquée* **2**, 373 (1971).
4 L. Manfait, *Thèse 3° cycle*, Mars 1972.
5 E. D. Schmid, G. Berthold, H. Berthold et B. Brosa, *Ber. Bunsenges. Physik.* **75** (2), (1971).
6 C. Schwab, B. Meyer, A. Goltzene et S. Nikitine, *Rev. Phys. Appl.* **6**, N 4, (1971).

10. RAMAN INTENSITY STANDARDIZATION

S. G. Frankiss and W. Kynaston
Division of Chemical Standards,
National Physical Laboratory, Teddington, England

The substantial experimental advances in Raman spectroscopy during the last few years have extended the practical applications of the technique, which are expected to have increasing use in routine chemical analysis. The extent to which Raman spectroscopy contributes to molecular structure determination depends not only on suitable instrumentation, but also on the availability of documented spectra that are, as far as possible, independent of the measuring equipment. Wavenumber shifts and depolarization ratios of Raman bands can be conveniently measured using modern spectrophotometers, but in routine studies relative intensity data are usually not corrected for instrumental factors, and thus potentially useful information is lost. This problem has become more acute of late due to the multiplicity of sources, monochromators and detectors used to record Raman spectra.

Well-defined procedures for reducing observed intensities to a standard intensity scale have been described. These procedures require measurement of spectrophotometer sensitivity along with the density, refractive index and temperature of the sample, and the wavenumber shift and integrated intensity of the Raman band. In routine studies, however, it may not be generally practical to undertake all these measurements. Nevertheless, it can be readily shown that the relative integrated intensities of Raman bands in a given material, using excitation of a specified wavelength, are largely independent of the measuring equipment, provided that correction is made for the variation of spectrophotometer sensitivity with change of wavelength.

Spectrophotometer sensitivity is most reliably determined using radiation from a tungsten filament lamp of known spectral power distribution or colour temperature. Tungsten lamps, however, have short calibrated lifetimes, and should be used under carefully controlled experimental conditions. A more simple, though less accurate, method of measuring spectrophotometer sensitivity would appear to have some advantages in routine studies. We have therefore investigated the feasibility of using the measurement of integrated intensities of a few well-defined Raman bands in selected materials to determine spectrophotometer sensitivity.

Two requirements are necessary for spectrophotometer sensitivity to be determined from standard Raman bands. The spectrophotometer sensitivity must be a fairly smooth function of wavelength, and the other factors that may affect relative band intensities must be compensated for or eliminated.

Examination of gratings in several commercial Raman spectrophotometers has shown that the only discontinuities in the sensitivity curve arise from Wood's S-anomalies. The wavelength of these discontinuities are, however, readily calculated, and the magnitude of the discontinuities is easily observed. The Wood's S-anomalies for recently manufactured 1200 g/mm ruled gratings blazed at 5000 Å do not give unacceptable discontinuities in spectrophotometer sensitivity curve.

Several compounds have been examined for suitability as intensity standards, the most promising being CCl_4, CH_3CN, $C_6H_5CH_3$, and $C_6H_5COCH_3$.

Intensity data for both the pure compounds and binary mixtures of these compounds have been obtained. Spectrophotometer sensitivity may be conveniently measured using the ratios of the intensities in a given material. The calibration data have been determined for use over the temperature range 298 ± 10 K. A simple empirical procedure for correcting observed intensity ratios for possible errors due to optical divergence effects has been developed.

11. A RAMAN SPECTROMETER WITH SMALL ON-LINE DIGITAL COMPUTER FOR EXPERIMENTAL CONTROL AND DATA ACQUISITION

E. D. Schmid, G. Berthold, H. Berthold and B. Brosa*†

Institut für physikalische Chemie der Universität Freiburg, Freiburg, Germany

Our purpose is to measure the Raman intensities of weak bands to study intramolecular interactions. To do that in an efficient way we designed a system which consists essentially of a 6 W argon ion laser (Mod. 53 A from CRL), a sample illumination system, the Coderg Raman spectrometer PH0 with cooled multiplier and photon-counting, strip chart recorder, a decimal counter which serves as a monitor, and a computer system which has been interfaced to the spectrometer as shown in Fig. 1.

The computer system serves to operate the spectrometer, to acquire data from it on-line, to process the data, i.e. to obtain the area under a measured curve and to calculate the standard intensity S as defined by Bernstein and Allen.

The hardware of our LAB8 computer system includes a small PDP8L computer from Digital Equipment Corporation as the central processing unit. Core memory consists of 4K words of 12 bits. Some months ago we added another 8K of core memory and 2 magnetic tapes with a capacity of 184K of storage each. Results presented here do not yet make use of this expansion.

The interface consists of an AXO8-peripheral from DEC and a motor control unit which we have designed in TTL-logic. To the AXO8 we have added a binary counter which counts the signal pulses arriving from the photon counting. The AXO8 provides a facility for data storage of this digital input at the computer and performing digital-to-analog conversion on inputs from the computer for display on the oscilloscope. The timing for the real-time data acquisition is provided by a real-time crystal clock, operating with 10 kHz pulses. Furthermore it contains some set registers by which on a computer command logical powers of 3 V can be switched to create the motor step pulses and to change the stepping direction. Further input-output devices are the high-speed reader and punch, the $X-Y$ plotter and the teletype by which computer-operator communications are made.

On account of small core memory, the whole program has been divided in several sections. Each section must be read into core from paper tape before execution. These

---

* *Ber. Bunsenges. Physik. Chem.* **75**, 149 (1971).
† Presented by B. Brosa.

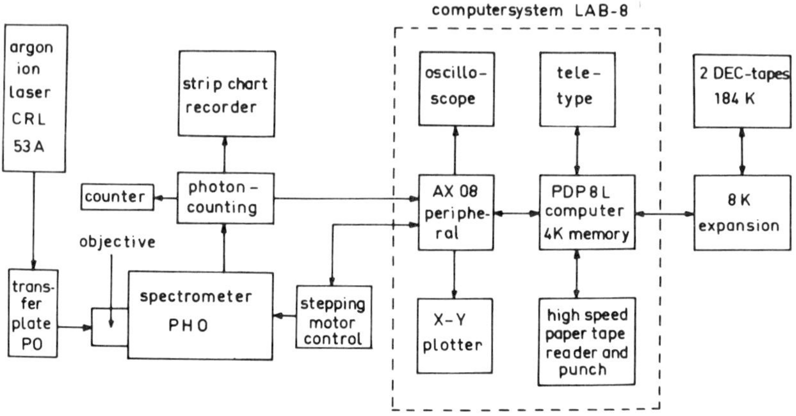

Fig. 1. Schematic diagram of the laser spectrometer computer system.

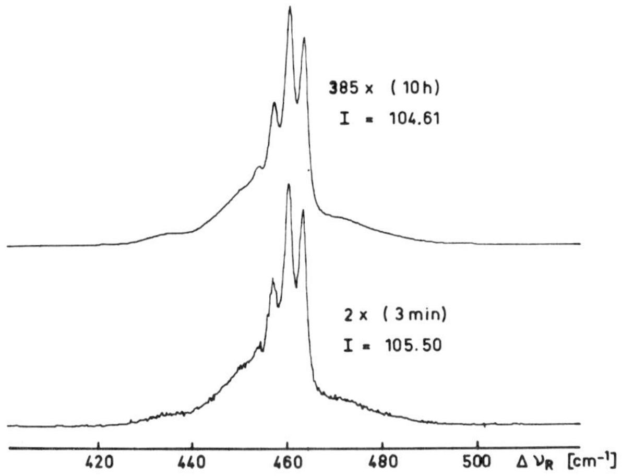

Fig. 2. The 459 cm^{-1} band of CCl_4 after 2 and 385 scans. Collection time 200 ms per point. Integral and resolution remain unchanged even after a great number of scans.

sections are

a) spectrometer control and data acquisition,
b) band intensity integration,
c) curve analysis of overlapping bands,
d) calculation of the standard intensity.

The program for the spectrometer control and data acquisition occupies half of the 4K memory and allows storage of up to 1024 data points, since they are stored in double precision form, giving a numerical range up to 2^{23}. To pick up a certain frequency range, the spectrometer must only be manually positioned at the starting wavenumber. Signal-to-noise enhancement can be obtained by multi-scan averaging. The following parameters

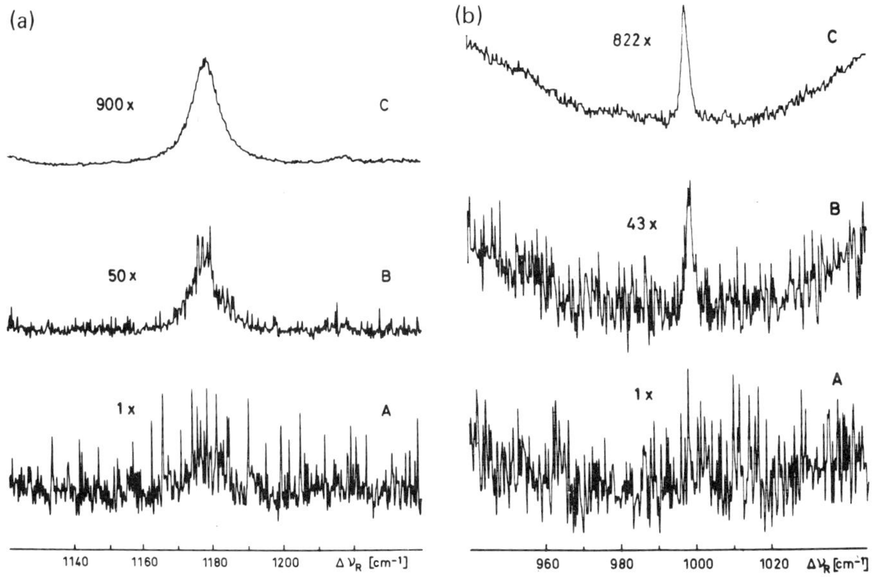

Fig. 3. Signal-to-noise enhancement by multi-scan averaging (a) 1178 cm^{-1} band of benzene in CCl_4 solution, concentration 3.5×10^{-6} mole/cm^3 (3×10^4 mole CCl_4/mole C_6H_6). (b) 992 cm^{-1} band of benzene in CCl_4 solution, concentration 2.3×10^{-7} mole/cm^3 (7.3×10^4 mole CCl_4/mole C_6H_6).

are entered on the teletypewriter:

1. The collection time ΔT (in ms) is the time during which for every data point signal pulses are counted. The range of ΔT is from 1–4096 ms, corresponding to a scanning speed from 625–1.5 cm^{-1}/min.
2. Number of data points to be taken (max. 1024).
3. Number of repetitive scans (1 . . . . max. 4096)
4. Number of steps between data points.

With 1024 data points we cover a spectral range of 128 cm^{-1}, if there is one step between two data points, since the stepping motor makes 8 steps per cm^{-1}. This spectral range can be expanded to 256, 384, 512 cm^{-1}, etc. by making 2, 3, 4, . . . or more steps between data points.

Data are only collected when the motor is stepping in a forward direction; in reverse the motor moves at the maximum speed of about 725 cm^{-1}/min. At any time a display request command can be given by setting a toggle switch on the operator console to one. This stops data acquisition at the end of the present sweep and displays the spectrum. Now 'CONTINUE' or 'END' may be chosen to continue or to finish the data acquisition. If 'END' has been chosen, the sum of the spectra is divided by the number of sweeps and then may be plotted on the X–Y plotter and/or punched on paper tape.

The main problem was to begin every scan exactly at the same wavenumber, to ensure that no resolution is lost on account of wavenumber displacements between the spectra of the various scans and that the integral remains unchanged during repeated scans (Fig. 2). This has been achieved by never removing motor and logical power during direc-

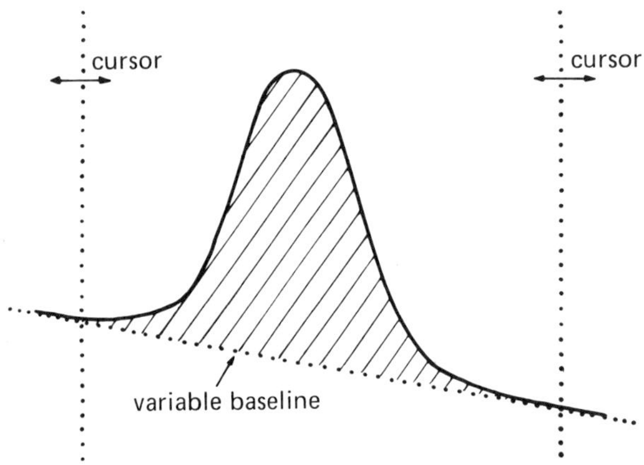

Fig. 4. Display of curve and integration limits on oscilloscope. Cursors and baseline can be moved by potentiometers.

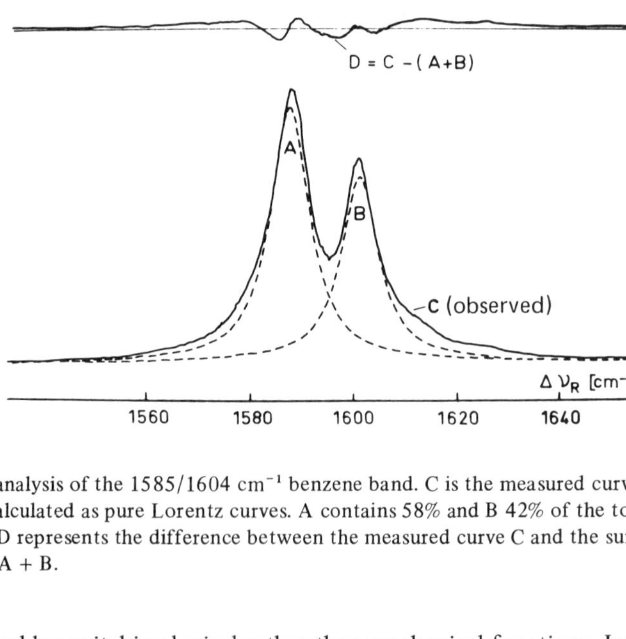

Fig. 5. Curve analysis of the 1585/1604 cm^{-1} benzene band. C is the measured curve, A and B have been calculated as pure Lorentz curves. A contains 58% and B 42% of the total area under the curve. D represents the difference between the measured curve C and the sum of the calculated curves A + B.

tion changes and by switching logical rather than mechanical functions. In Fig. 2 each curve consists of 512 data points. The spectral region covered is 128 cm^{-1}, since only 4 data points per cm^{-1} have been taken. The system has been working very reliably for 2 years even during long times of repetitive scanning, e.g. 10 h.

The advantage of multi-scan averaging is shown in Fig. 3(a) and (b). In Fig. 3(a) the 1178 cm^{-1} band of benzene has been measured in CCl_4 solution with a concentration of 3.5×10^{-6} mole/cm^3; in the same way Fig. 3(b) shows the 992 cm^{-1} band of benzene in CCl_4 solution with a concentration of 2.3×10^{-7} mole/cm^3. In each case 500 data points have been taken.

For band intensity integration we display the measured curve on the oscilloscope together with two vertical cursors, which determine the integration limits (frequency range) and which can be moved by potentiometers located on the AXO8.(Fig. 4). The surface to be integrated is determined by the curve, the two cursors and a baseline, the slope of which also is adjustable through two sense knobs but which is always a straight line. The integral is typed out as a decimal number on teletype.

In the case of overlapping bands one may sometimes be interested in the intensity of each line. We have tried to program a sort of visual curve analysis, which cannot replace a least square fit, but which already gives good results with only 4K core. We display the measured curve on the scope, then we calculate a linear combination of a Gaussian and Lorentzian curve, the peak of which is fixed by a moveable cross displayed on the scope. The position of the baseline can also be fixed on the scope. The standard deviation and the mixing factor are entered on the teletype. The calculated curve is then displayed to be compared with the measured one and can be subtracted from the measured curve if it fits well, otherwise a new Gauss/Lorentz curve may be calculated. This procedure may be continued with the difference curve, etc.; an example is shown in Fig. 5. Before calculating the standard intensities, spectral sensitivity corrections to relative band intensities are automatically achieved.

The expansion of core memory to 12K now makes it possible to store up to 8000 data points. Programs and curves will now be stored on magnetic tape which is faster and more convenient than storage on paper tape.

12. VIBRATIONAL DETERMINATION OF CRYSTAL STRUCTURES*

E. Silberman and J. R. Lawson
Fisk University, Nashville, Tennessee, USA
and
H. W. Morgan
*Oak Ridge National Laboratory,
Oak Ridge, Tennessee, USA*

INTRODUCTION

The basic approaches to a theory for the vibrational spectra of crystals were proposed by Bhagavantam and Venkatarayudu[1] in 1939, and by Halford[2] in 1946. These approaches were justified theoretically and further developed by Hornig[3] and Winston and Halford.[4] At that stage, if the structure of a crystal was known, it was possible to predict the number of its infrared and Raman active bands, as well as their fine structure and polarization. Experiments verified the early theory and promoted further developments.[5] We wish to discuss the use of this theory in the inverse problem, the determination of the structure of crystals through the measurement and interpretation of their vibrational spectra. For that purpose we will describe the theoretical limitations of the method and the experimental difficulties that must be overcome to make it useful. We will summarize the present status of the problem, including results of our recent experiments.

IDEAL METHOD

Using Halford's approach, with the experimentally verified assumption that the molecules or ions in a molecular crystal retain essentially the symmetry they have in the gas or liquid phases, the possible structures of a crystal are limited to those obtained by the following method:

i. use the character and correlation tables of the point groups[6] to determine all the possible site groups (subgroups of the molecular or ionic symmetry group);
ii. use the same tables to determine all the possible factor groups (supergroups of each site group).

In order to decide which is the correct factor group:

1. draw correlation diagrams[3] for each combination of the molecular, site and factor groups and determine for each case the splitting of the bands of the isolated molecule and the Raman or infrared activity and polarization of each component;
2. calculate the number, activity and polarization of the lattice vibrations;[7]
3. obtain the polarized near and far infrared spectra of an optically oriented single

* Research sponsored in part by U.S. Atomic Commission under contract with Union Carbide Corp.

crystal and the Raman spectra of the *same* oriented crystal for *all* the parallel and perpendicular scattering and polarization configurations. Assign the bands corresponding to the internal vibrations by the same methods used regularly for liquid or gas phase spectra. Although the experience is not as great, similar methods can be used to assign the lattice vibrations;[8]

4. compare the predicted spectra with the experimental spectrum; those which agree represent the only possible factor groups. It is our theory that their number will most often be just one. If it is not so, some non-spectroscopic information will be necessary to reduce the number to one. In any case, non-spectroscopic information will be necessary to determine the crystal group.

Practical limitations

The main obstacles to the application of the theoretically ideal method are:

a) the infrared and Raman instrumentation of the necessary range, resolution and sensitivity. The method may require measurements in the range 0–4000 cm^{-1} with 0.1 cm^{-1} resolution. In the infrared, the recently introduced Fourier Transform spectrometers may satisfy the requirement, but they are still too expensive to be widely used; there is no major sensitivity problem in the infrared. In Raman spectrometers, range and resolution are not difficult to satisfy, and the sensitivity necessary to measure very small samples is being rapidly reached with the availability of powerful laser sources. So, the instrumental problem may be solved with the present technology, but this became true so recently that is has not been yet fully utilized in any study of the kind we are discussing here.

b) the preparation of suitable crystals. Infrared spectrometers require samples of minimum dimensions 2 X 10 mm and thicknesses of a few microns. The area can be reduced by using beam condensers, but this increases the convergence of the beam and impairs the polarization measurements. The thickness cannot be increased, and that is the main difficulty when trying to cut a sample from a larger crystal. The alternative is to grow single crystal films, using techniques that we are going to mention shortly. In contrast, laser Raman spectrometers require more or less cubic samples with edges 1–2 mm long. It is now evident that the ideal condition of measuring a sample usable for both spectra is difficult to meet. This reduces the information contained in the data and it is another reason for which the method has not been fully exploited.

COMPLEMENTARY TECHNIQUES

The mentioned reduction of information can sometimes be compensated for by conclusions derived from other methods, such as X-ray diffraction, close-packing theory, crystallographic observations, optical dichroism, electrical or magnetic activity, etc. Vibrational spectra may provide crystallographic information impossible to determine with other methods, and in that direction the techniques involved become complementary; in fact, the collaboration leads frequently to the complete determination of the crystal group. A most typical example is contributed by the X-ray studies of crystals containing hydrogen atoms.

The close-packing theory is based on the assumption that the most stable configuration of the atoms in a crystal is the one which occupies the minimum volume. Although the theory cannot be strictly proved for the general case, it can be qualitatively justified.

Kitaigorodskii[9] has analysed the relations between close-packing and symmetry and deduced a table of the crystal groups which are compatible with the different kinds of packed structures (page 103 of Ref. 9). D'Alessio[10] has compiled a list of 540 molecular structures determined by X-rays and published by Wycoff,[11] and grouped them according to their crystal class and space group. Only three structures did not appear in Kitaigorodskii's table, and it turned out to be that they were early determinations made with only one crystallographic projection. Close-packing considerations should be expected to be less safe when oriented forces, such as dipole–dipole interactions between polar molecules, contribute noticeably to the crystal potential energy; but, even in some of these cases, experience[12] shows that it may help to choose the right structure.

Crystallographic measurements and optical dichroism can easily contribute to identify or eliminate cubic structures, and, in favourable cases, to narrow the choice or even identify the crystallographic class. In addition, they are essential in providing the directions of the optical axes in non-cubic crystals.

REVIEW OF PROGRESS

The question arises as to how much can be accomplished in view of the practical limitations of the method. Table 1 sums up a search of the literature on the subject, including the basic techniques used and the results obtained in each case. Vibrational analysis has been used to solve indeterminations left by the X-ray diffraction studies of HCl (II),[14, 25] HBr(II),[14] ethylene,[16, 38] methane(III),[18, 37] CS_2,[19] hydrazine,[26] thiophene,[27] cyclohexane(II),[30] $Cs_2Cr_2O_7$,[32] and cyclopropane;[33] these studies have not been listed. The following information and comments of Table 1 will help to understand the development of the method for the cases where X-ray results were not available.

The melting point or phase transition temperature is given in order to show that the method has mostly been useful for the study of substances which are solid only at low temperatures. Infrared spectra have been obtained mostly with polycrystalline films (P) made by condensing the vapour or freezing a liquid layer on a cooled infrared transmitting substrate. In some cases, however, controlled rate of deposit on alkali halide windows produced 'oriented' films (O) in which the crystallites have one or more axes parallel to preferred directions, usually perpendicular to the face of the substrate or coincident with its crystallographic axes; the effect seems to be quite common when samples are condensed and annealed on CsI windows.[10, 12, 40, 41] Swerdling and Halford[42] developed a technique for growing thin single crystals (S) which constitute almost ideal samples (except for uncontrollable orientation); the need for very accurate control of temperature gradients and flow-rates has limited the use of the method, but we think that the technique may become quite standardized with the availability of simple solid state controllers. The column of 'polarization measurements' shows that oriented films may permit the observation of preferential directions for the extinction of visible light between crossed polaroids and for the absorption of polarized infrared radiation; these observations may lead to the assignment of the polarization of the bands, as shown in the next section.

Raman spectra of large natural single crystals[43] and of powdered crystals[44] were reported very shortly after the discovery of the Raman effect[45] in 1928, but the technique remained, for more than 35 years, very inadequate for the measurement of solid samples in general. Only the recent introduction of the laser source has made it possible to obtain polarized Raman spectra of small crystals.[46]

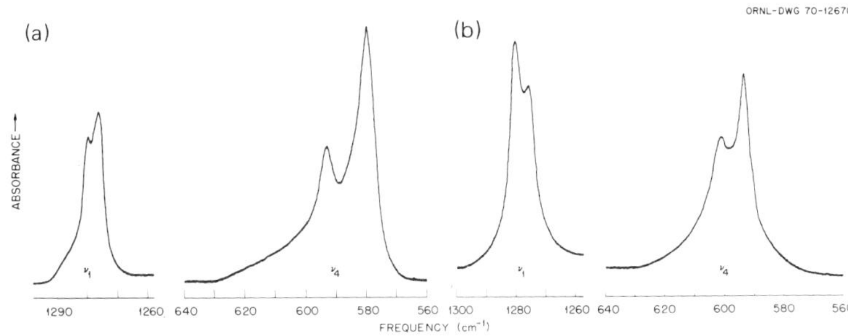

Fig. 1. Varying intensities of the components of ν_1 and ν_4 in the infrared spectra of solid $POCl_3$ at 80 K; (a) and (b) separate depositions.

Site and factor group symmetries listed in Table 1 are the ones compatible with the spectroscopic data. The crystal group has generally been obtained by close-packing considerations. In particular cases, reasonable hydrogen-bonding, observations of optical dichroism or approximate intensities calculated with the oriented gas model were used to select the most probable crystal structure.

It can be seen that in a majority of the cases the spectroscopic determination of the factor group has not been unambiguous. But, on the other side, the possibilities opened by modern instrumentation have not yet been exploited. In particular, no polarized Raman spectra have been obtained for the purpose of determining crystal structures and no advantage has been taken of the increased information that can be obtained from correlating the infrared and Raman spectra of the same sample.

The structure of $POCl_3$

As a test case of how much can be gained by using the above mentioned additional information, we have re-taken the study of $POCl_3$. Burgos, Huler and Silberman[40] (BHS) have shown that:

1. samples obtained by condensing the vapour on a CsI window may orient themselves in three different ways, in each of which one of the three components of the $\nu_4(E)$ is missing in the unpolarized infrared spectra. The two most frequent cases are reproduced in Fig. 1, where it can be noticed that the different orientations also give origin to changes in the intensities of the components of $\nu_1(A)$;
2. the spectroscopic data are compatible with three factor groups (C_{2h}, D_{2h}, D_{3d}) of which D_{2h} was chosen on the basis of close-packing considerations.

We tried to avoid the use of non-spectroscopic considerations by completing the spectroscopic data. For that purpose an infrared-Raman low temperature cell was constructed: it consisted of a conventional all-metal low temperature cell equipped with two CsI windows to which a third glass window was added at 90°; the fourth side was fitted with a nozzle through which the vapour was introduced at a rate controlled by a needle valve. While the samples were being deposited at about − 70°C, their thickness was estimated by following interferential colour changes under white light illumination. Films about 2 μm thick showed complete extinction of visible light between crossed polaroids for two perpendicular directions, which we shall call α and β. A Perkin-Elmer

Table 1. Vibrational studies on crystal structures.

Crystal (m.p. in °C)[a]	Infrared			Raman			Complementary information		Spectroscopic		Crystal group	Z	Ref. (year)
	Sample	Range	Polarization	Sample	Excitation	Polarization	Close-pack	Other	Site group	Factor group			
HCN (− 14)	P	N	x[b]					'reasonable'	C_{1h}	C_{4h}	$P4_2/m(C_{4h}^2)$	2	13(1950)
Cyclopropane (− 126.6)	S	N	x				x		C_s	C_{2v} D_{2h}	$Pmn2_1(C_{2v}^7)$	2	15(1961)
Allene (− 146)	S	N	x				x		C_i	monocl. orthor.	$P2_1/n(C_{2h}^5)$	4	17(1962)
Diacetylene (− 36.4)	S	N	x				x	H-bond	C_1 C_s	orthor.	$P22_12_1(D_2^3)$	4	20(1965)
α-Diborane (− 213)	S	N	x						C_1 C_i	monocl.			21(1965)
CH_2Cl_2 (− 96.7)	P	N					x		C_2	orthor.			23(1966)
CH_2Br_2 (− 52.8)	P	N					x		C_2	orthor.			23(1966)
CH_2I_2(I) (5.54)	P	N					x		C_{2v}	orthor. tetr. hex.	$Pmma(D_{2h}^5)$ $Pmmn(D_{2h}^{13})$	2 2	24(1966)
CH_2I_2(III) (6.01)	P	N					x		C_1		$P2_1/c(C_{2h}^5)$	4	24(1966)
CH_2Cl_2 (− 96.7)	P	N		P	H				C_2	D_{2h}		4	28(1967)
CH_2Br_2 (− 52.8)	P	N							C_2 C_s	C_{2h}		2	28(1967)

Cyclo-pentane (III) (− 150.7)	P	N						C_1	monocl.			29(1967)
CH_2Br_2 (− 52.8)	O	N	x	P	H	x		C_1	monocl. tricl.	$P2_1/c(C_{2h}^5)$ $P\bar{1}(C_i)$	4 4	31(1968)
Cyclo-pentane (III) (− 152)	O	N	x			x		C_1	C_{2h}	$P2_1/c(C_{2h}^5)$	4	34(1969)
Cis-1, 2-Di-chloroethyl-ene (− 80.5)	O	N		P	L	x		C_{2v}	D_{2h}	$Pmna(D_{2h}^5)$	2	10(1969)
Trans-1, 2-Dichloro-ethylene (− 50)	O	N		P	L	x		C_1	D_{2h} D_{3d} S_6	$P2_1/c(C_{2h}^5)$		10(1969)
Glyoxal	P	N		P	L	x		C_i	D_{2h}	$Pbca(D_{2h}^{15})$	4	35(1970)
(15)										$Pccn(D_{2h}^{10})$		
CH_2Cl_2 (96.7)	O	N	x	P	H	x	oriented gas	C_2	D_{2h}	$Pbcn(D_{2h}^{14})$	4	36(1970)
Glyoxal (15)	P	N F		P	L			C_i	C_{2h} D_{2h}			39(1971)
$POCl_3$ (2)	O	N	x[c]	P	L	x		C_1 C_s	C_{2h} D_{2h} D_{3d}	$Pnma(D_{2h}^{16})$		40(1971)
1, 1-Dichloro-ethylene (− 122.1)	O	N	x	P	L	x		C_1	D_2 C_{2v} D_{2h}	$Pca2_1(C_{2v}^5)$ $Pna2_1(C_{2v}^9)$	4 4	 41(1971)

Symbols: P = polycrystalline film (infrared), polycrystal or powder (Raman); S = single crystal;
O = oriented polycrystalline film; N = near infrared; F = far infrared; H = Mercury lamp;
L = laser.

[a] or higher transition temperature of corresponding phase.
[b] deduced from spectral changes with film thickness.
[c] deduced from band absences in some samples.

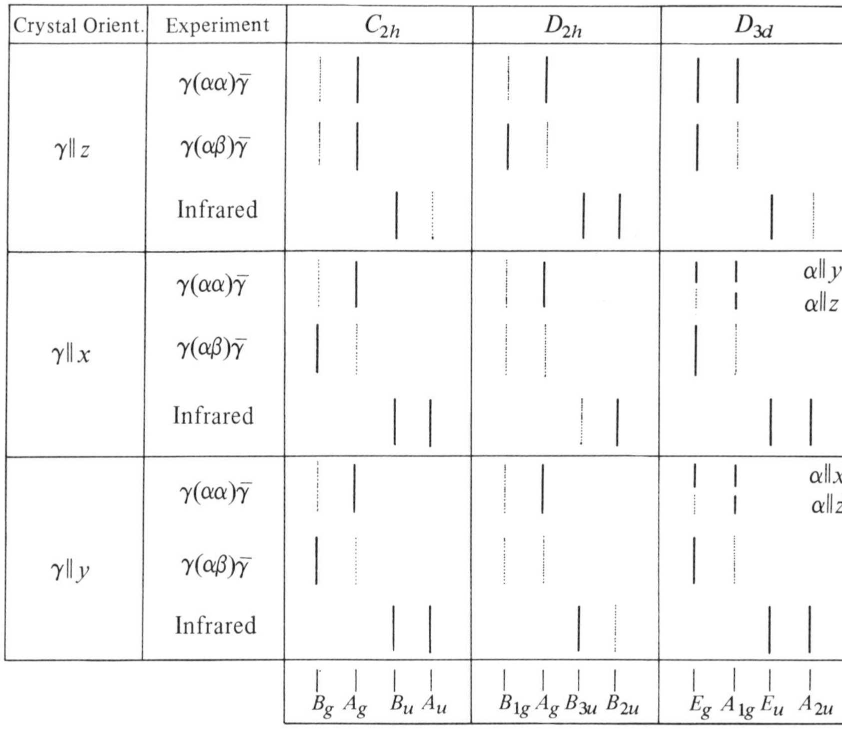

Fig. 2. Predicted activity of the components of the A_1 bands of a $POCl_3$ crystalline film for three possible factor group symmetries. Directions α and β in the film show extinction between crossed polaroids. γ is perpendicular to the film. Full lines: allowed. Dotted lines: forbidden.

521 spectrometer and a wire grid polarizer were used to measure $\nu_1(A)$ and $\nu_4(E)$ with infrared radiation polarized in the directions α and β.

The sample was then rotated until it became parallel to the glass window and cooled to liquid nitrogen temperatures. A Jarell-Ash 25-300 spectrometer was used to obtain Raman spectra; blue 5145 Å light from an Ar^+ laser was introduced through the glass window by way of a 1 mm^2 diagonal mirror and light scattered by the sample was collected at 180° with a 1 : 2, f = 100 mm lens system. With careful alignment, the laser light reflected by the entrance window and the sample was deflected back to the laser by the diagonal mirror, reducing noticeably the stray light. The cell was oriented so that the polarization of the laser light coincided with the direction α of the sample, and the scattered light was analysed with polaroids parallel or perpendicular to the polarization of the laser radiation, located in front of the entrance slit of the monochromator. The incident radiation intensity had to be limited to 500 mW to avoid melting the sample and good ν_1 bands were obtained with 2 cm^{-1} slits, which was more than sufficient resolution when the 22 cm^{-1} splitting of ν_1 is taken into account.

If we call γ the direction of the incident light (perpendicular to the surface of the film), our experiments can be described, with the notation of Damen, Porto and Tell,[46] as: 1) $\gamma(\alpha\beta)\bar{\gamma}$ when the polarization of the analyser was perpendicular to the laser polarization, and 2) $\gamma(\alpha\alpha)\bar{\gamma}$ when the analyser was parallel. Because of the behaviour of ν_4

in the infrared, we knew that the directions of α, β and γ coincided with optical axes of the crystal, although we could not yet identify their correspondence with the *x, y, z* conventional labelling.

The best results were obtained with a sample whose infrared spectrum was similar to the one shown in Fig. 1(b). Even if we did not obtain complete extinctions, the changes in intensity showed clearly the following polarizations.

1. Infrared ν_1: 1275 $cm^{-1} = \gamma$; 1280 $cm^{-1} = \alpha$.
   ν_4: 580 cm^{-1} = absent; 594 $cm^{-1} = \beta$; 603 $cm^{-1} = \alpha$.
2. Raman ν_1: $\gamma(\alpha\alpha)\bar{\gamma}$ only 1274 cm^{-1} active
   $\gamma(\alpha\beta)\bar{\gamma}$ both components inactive.

In order to interpret these results we calculated the theoretical structure of the *A* and *E* bands in the infrared and of the *A* bands in the Raman for the three factor groups compatible with BHS spectroscopic information. The predicted band structures are shown in Fig. 2 and they prove that just polarized Raman measurements of ν_1 can, at least in this case, distinguish among the three possible factor groups and give the orientation of the sample. Comparison with the results of our Raman measurement of ν_1 shows that $POCl_3$ has a D_{2h} factor group and that the sample was oriented with $\alpha = y$, $\beta = z$, $\gamma = x$ (or the orientation obtained interchanging *x* and *y*, which is spectroscopically equivalent). The infrared spectrum is compatible with this conclusion, and the bands can now be assigned as follows:

Infrared ν_1: 1275 (B_{1u}); 1280 (B_{2u})
Infrared ν_4: 580 (B_{3u}); 594 (B_{1u})
Raman ν_1: 1274 (A_g); 1296 (B_{1g})

Again, exchanging x–y will exchange B_{2u}–B_{3u} without major consequence.

CONCLUSIONS AND COMMENTS

The factor group of solid $POCl_3$ was deduced from the polarized infrared and Raman spectra of the same crystalline film, without resource to non-spectroscopic information. In fact, the measurement of the polarization of only one band in the Raman spectrum taken with a 180° scattering configuration was sufficient to resolve the indeterminations left by the unpolarized spectral information and to give the crystal orientation. Similar studies on the chloroethylenes and halogenated phosphoryls and thiophosphoryls are in progress and suggest that the method may be successful in most cases. We like to point out that far infrared measurements can add decisive information on the crystal structure, as it has been proved for hydrazine.[26]

Vibrational analysis, in addition to not being able to determine interatomic distances, cannot compete with X-ray diffraction as a general method for the determination of crystal structures, but it can:

1. be indispensable to complete inconclusive results of X-ray determinations, a very common case when the crystal contains hydrogen atoms;
2. give the instantaneous structure, as opposed to the time-averaged structure given by X-rays diffraction, when atoms can tunnel between equivalent positions in the crystal.[47] This information may be fundamental to the understanding of ferroelectric–paraelectric and other phase transitions.

REFERENCES

1 S. Bhagavantam and T. Venkatarayudu, *Proc. Indian Acad. Sci.* **9A**, 224 (1939).
2 R. S. Halford, *J. Chem. Phys.* **14**, 8 (1946).
3 D. F. Hornig, *J. Chem. Phys.* **16**, 1063 (1948).
4 H. Winston and R. S. Halford, *J. Chem. Phys.* **17**, 607 (1949).
5 T. H. Walnut, *J. Chem. Phys.* **20**, 58 (1952).
6 E. B. Wilson, J. C. Decius and P. C. Cross, *Molecular Vibrations*, McGraw-Hill, New York, 1955.
7 S. Bhagavantam and T. Venkatarayudu, *Proc. Indian Acad. Sci.* **37A**, 350 (1953).
8 For example, M. Ito, M. Suzuki and T. Yokoyama, in *Exiton, Magnons and Phonons in Molecular Crystals* (Ed. A. B. Zahlan), Cambridge Univ. Press, 1968, p. 1.
9 A. I. Kitaigorodskii, *Organic Chemical Crystallography*, Consultants Bureau, New York, 1961.
10 E. A. D'Alessio, Ph.D. Dissertation, Vanderbilt University, 1969.
11 R. W. G. Wyckoff, *Crystal Structures*, 2nd. ed., Interscience, New York, 1965.
12 E. Huler, A. Burgos and E. Silberman, *J. Mol. Structure* **12**, 212 (1971).
13 D. F. Hornig, *Discussions Faraday Soc.* 115 (1950).
14 D. F. Hornig and W. E. Osberg, *J. Chem. Phys.* **23**, 662 (1955).
15 Ch. Brecher, E. Krikorian, J. Blanc and R. S. Halford, *J. Chem. Phys.* **35**, 1097 (1961).
16 Ch. Brecher and R. S. Halford, *J. Chem. Phys.* **35**, 1109 (1961).
17 J. Blanc, Ch. Brecher and R. S. Halford, *J. Chem. Phys.* **36**, 2654 (1962).
18 G. B. Savitsky and D. F. Hornig, *J. Chem. Phys.* **36**, 2634 (1962).
19 H. Yamada and W. B. Person, *J. Chem. Phys.* **40**, 309 (1964).
20 I. Freund and R. S. Halford, *J. Chem. Phys.* **42**, 4131 (1965).
21 I. Freund and R. S. Halford, *J. Chem. Phys.* **43**, 3795 (1965).
22 A. Anderson and R. Savoie, *J. Chem. Phys.* **43**, 3468 (1965).
23 A. Le Roy, *C. R. Acad. Sci., Paris* **B 263**, 368 (1966).
24 M. P. Marzocchi, V. Schettino and S. Califano, *J. Chem. Phys.* **45**, 1400 (1966).
25 L. C. Brunel and M. Peyron, *C. R. Acad. Sci., Paris* **C 264**, 821 (1967).
26 F. G. Baglin, S. F. Bush and J. R. Durig, *J. Chem. Phys.* **47**, 2104 (1967).
27 J. Loisel and V. Lorenzelli, *Spectrochim. Acta* **23A**, 2903 (1967).
28 V. B. Kartha, *J. Mol. Spectry.* **24**, 368 (1967).
29 A. Le Roy, *C. R. Acad. Sci., Paris* **B 264**, 1087 (1967).
30 R. J. Obremski, Ch. W. Brown and E. R. Lippincott, *J. Chem. Phys.* **49**, 185 (1968).
31 M. P. Marzocchi, P. Manzelli, V. Schettino and S. Califano, *J. Chem. Phys.* **49**, 5438 (1968).
32 R. L. Carter and C. E. Bricker, *Spectry. Letters* **2**, 321 (1969).
33 J. B. Bates, D. E. Sands and W. H. Smith, *J. Chem. Phys.* **51**, 105 (1969).
34 V. Schettino, M. P. Marzocchi and S. Califano, *J. Chem. Phys.* **51**, 5264 (1969).
35 F. D. Verderame, E. Castelluci and S. Califano, *J. Chem. Phys.* **52**, 719 (1970).
36 M. P. Marzocchi and P. Manzelli, *J. Chem. Phys.* **52**, 2630 (1970).
37 R. Savoie and R. P. Fournier, *Chem. Phys. Letters* **7**, 1 (1970).
38 S. M. Blumenfeld, S. Paddi Redy and H. L. Welsh, *Can. J. Phys.* **48**, 513 (1970).
39 J. R. Durig and S. E. Hannum, *J. Cryst. Mol. Structure* **1**, 131 (1971).
40 A. Burgos, E. Huler and E. Silberman, *J. Mol. Structure* **9**, 283 (1971).

41 E. D'Alessio, E. Silberman and E. A. Jones, *J. Mol. Structure* **9**, 393 (1971).
42 S. Swedling and R. S. Halford, *J. Chem. Phys.* **23**, 2215 (1955).
43 G. Landsberg and L. Mandelstam, *Naturwissenschaften* **16**, 557 (1928).
44 A. C. Menzies, *Nature* **124**, 511 (1929).
45 C. V. Raman and K. S. Krishnan, *Nature* **121**, 501 (1928).
46 T. C. Damen, S. P. S. Porto and B. Tell, *Phys. Rev.* **142**, 570 (1966).
47 C. M. Hartwig, E. Wiener-Avnear and S. P. S. Porto, *Phys. Rev.* **B5**, 79 (1972).

13. A NEW METHOD FOR THE DETECTION OF RAMAN SCATTERING FROM ATMOSPHERIC POLLUTANTS*

W. Hayden Smith
Princeton University Observatory,
Princeton, New Jersey, USA

INTRODUCTION

The remote detection of atmospheric gases, particularly pollutants such as NO or SO_2, has been attempted in several ways in the last few years. Studies by Kobayasi and Inaba,[1] Inaba and Kobayasi,[2] Cooney,[3] and Leonard[4] using the vibrational Raman effect and pulsed lasers have provided data on the detection of the Raman shifted light with conventional techniques. The general conclusion of all these papers is that the method is marginal with the technology used by them. In no case has the ability to detect pollutant gases in the low p.p.m. range been confirmed for the type of sampling distance that will be required in field operations, say 300 m or more. The larger cross-sections of the rotational Raman transitions (usually two orders of magnitude larger than those for the vibrational-rotational Raman effect)[5, 6] would allow an immediate improvement in this situation. The difficulty of using rotational Raman scattering lies in the overlapping of the rotational transitions from the atmospheric gases in the region lying within about 100 cm^{-1} of the exciting line. For this reason only a very high resolution method could be used for an analysis of the rotational Raman effect. Ordinarily, this implies the use of a very large spectrometer which is not practical for applications outside the laboratory.

We show here how the interferometric method of Barrett and Myers[7] can overcome most of the above difficulties, and provide other operational improvements. Briefly, this consists of the use of a Fabry-Perot interferometer without narrow band blocking filters or monochromators, as is customary. Fabry-Perot interferometers typically have very high resolving power and transmittances of 70% or more in the visible and near ultraviolet. The details of the method with which the Fabry-Perot interferometer (FPI) can be used as a recurring transmission filter with windows at equal frequency intervals in wavenumber space (cm^{-1}) are given in the following discussion.

THEORY

The transmission function of an ideal Fabry-Perot interferometer $I(\Psi)$ can be described by an Airy function, or alternatively, the convolution of a non-periodic

* This research is supported by the Environmental Protection Agency under Grant R 800805.

Lorentzian line shape and Dirac comb.[8]

$$I(\Psi) = \frac{T^2}{1-2R\cos\Psi + R^2} = \frac{T^2}{1-R^2}\,\text{III}\,(\Psi/2)^*\left(\frac{2p}{p^2+\Psi^2}\right) \quad (1)$$

where $A = 1 - R - T$, and $\Psi = 4\pi\sigma d$, and A, R and T are the absorbance, reflectance and transmittance, $\sigma =$ the wavenumber frequency, $d =$ the plate spacing in cm and $p = -\ln R$.

Thus, we see that the transmission of a Fabry-Perot interferometer consists of a regular series of Lorentzian-shaped transmission windows, whose frequency spacing varies with the spacing of the etalon plates. This is illustrated schematically in Fig. 1(b).

The intensity transmitted by the FPI when illuminated by a spectrum of distribution $J(\sigma)$ is given by

$$F(d) = \int_{-\infty}^{\infty} J(\sigma)P(\sigma)\,\text{III}\,(\Psi/2)^*\left(\frac{2p}{p^2+\Psi^2}\right)\,d\sigma \quad (2)$$

where $P(\sigma) = (T^2/1 - R^2)$, d is the plate spacing and σ is the wavelength in cm^{-1}.

If we consider a specific example, a linear rigid rotor yielding a rotational Raman spectrum,[9] we have $J(\sigma)$ given by

$$J(\sigma) = \text{III}\,(\sigma/4B)^* S(\sigma) \quad (3)$$

where $S(\sigma)$ is given by the overlapped line profiles of the rotational lines at σ cm^{-1}, [Fig. 1(a)].

From the form of the transmission function of the FPI and the spectral distribution of the lines of the linear rigid rotor rotational Raman spectrum, we see that certain settings of the Fabry-Perot plates will pass all of the rotational lines simultaneously, namely, when the plate separations are $n/8\mu B$ cm, where B is the molecular rotational constant, n is an odd integer and μ is index of refraction of the medium between the plates. This has been demonstrated in the laboratory by Barrett and Myers[7] and is the essential property of the FPI that renders this scheme effective. Since for any given molecule, the rotational constants are unique quantities, it is clear that the identification of the emitters of the rotational Raman spectrum is made by the knowledge of the plate separation of the FPI at the point that transmits all the rotational Raman lines to produce this superposition, since this is a very accurate measure of the rotational constants. The plate separation can be measured precisely by using the laser lines available in an argon ion laser with very accurately known wavelengths, which give an interferometric accuracy.[7]

The ability of the FPI to separate the transmission maxima for various molecules depends primarily upon the resolving power of the FPI which is governed by the finesse $F = (\Delta\Psi/\delta\Psi)$, where $\delta\Psi$ is the half-width of the transmission profile and $\Delta\Psi$ is the fringe spacing in cm^{-1}, and by the interference order of the fringes that are responsible for the transmission $N \ldots N + J$.[8] For typical FPI's the finesse will be of the order 20–30, and the order of the fringes in excess of ten thousand. Resolving powers of the order of 500,000 are usually obtained. Rotational constants differing by more than a few parts in a thousand will then give separated transmission maxima.

The effect of including the centrifugal distortion is simply to broaden the maxima slightly and to make the position of the maximum dependent upon temperature, a fact

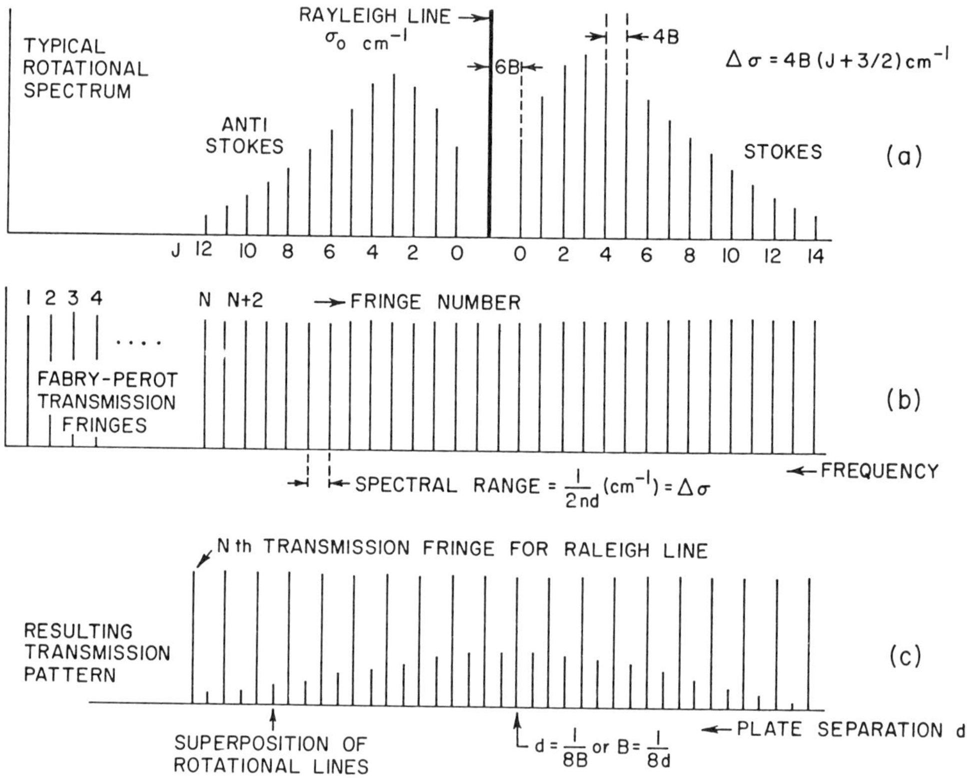

Fig. 1. (a) Schematic representation of a rotational Raman spectrum for a rigid linear molecule. The vertical lines represent the spectral lines and frequency is plotted along the horizontal scale. (b) Transmission profile (schematic) for a Fabry-Perot interferometer versus frequency. The frequency separation between adjacent transmission fringes may be continuously varied by changing the mirror separation of the FPI. (c) The resulting interferogram (schematic) which is obtained by scanning the FPI (changing d). The small lines, which alternate with the lines of constant amplitude (Rayleigh fringes), represent the superposition of all the rotational Raman lines.

that is of value as will be noted later. Symmetric rotors exhibit very similar transmission patterns to linear rotors[9] with an additional broadening due to the K effect on distortion. Asymmetric rotors differ sharply in that the levels can no longer be described by a simple term value formula, and thus do not transmit simultaneously. In the case of sulphur dioxide for example, where low or high J values exhibit line positions near one of the symmetric rotor limits, a unique and substantial transmission results. Spherical rotors of course, exhibit no pure rotational Raman effect to first order.

The throughput of the FPI is considerably greater than that for the spectrometer.[10] This is a purely geometrical advantage, which is characteristic of interferometers operating by beam division, and is independent of wavelength. The solid angle of the light beam that can be sent through the FPI is given by $\Omega = 2\pi/R$, where R is the resolving power. By contrast, a grating spectrometer has a solid acceptance angle for light of $\Omega = \beta/2R$, where β is the angular length of the slit. The gain in luminosity of the FPI over the spectrometer is then $G = 4\pi/\beta$. For the point sources considered here, β will not exceed 0.02 rad. The gain then is of the order of 10^2–10^3, for equal resolving power.

The laser radar range equation given in refs. 1–4 contains the following terms

$$N_p = N_i \sigma_r T_{1,2} \epsilon_{op} A_t R^{-2}$$

where $N_{p,i}$ are the number of Raman and laser photons received and emitted by the device, σ_r is the cross-section for the Raman process used, $T_{1,2}$ is the two-way atmospheric transmission, ϵ_{op} is the optical efficiency of the system, A_t is the optical receiver area, and R is the range at which the probed volume of the atmosphere lies. The conventional devices used earlier and the new technique discussed here are different only in the Raman cross-section and the efficiency of the optical system, in this case the gain in luminosity of the Fabry-Perot with respect to the conventional monochromator.

Although the difference in the Raman cross-sections has nothing intrinsically to do with the FPI, this gain is only practically realizable with the FPI, and so can be accounted in its favour. The other terms in the equation relating to laser output, range, gas density, and two-way transmittance to and from the observed region are the same in both cases. From the discussion in the theoretical section, we see that the luminosity of the FPI is of the order of magnitude of one hundred times that of the spectrometer, essentially due to the fact that there are no slits which are very wasteful of light. Transmittance of 70–75% are usual for FPI's employed in the visible. This is actually considerably superior to narrow band interference filters. The ratio of the total cross-section for the rotational Raman effect to that for the vibrational Raman effect is shown by Placzek[5] and Stansbury *et al.*[6] to be of the order of one hundred, usually somewhat more. The product of these two terms then represents the theoretical gain of the method over the use of the vibrational Raman effect and a conventional spectrometer. This amounts to a gain of $\sim 10^4$ in sensitivity.

The fact that the Fabry-Perot interferometer is the most luminous of spectroscopic instruments is well known[10] and requires no further verification here. In order to demonstrate the efficiency of the Fabry-Perot in superposing the individual rotational lines, we have computed the transmission patterns for NO under a range of experimental conditions. Two examples are shown in Fig. 2; these include the effect of rotational distortion. The patterns show that essentially all the intensity for the rotationally scattered Raman light is transmitted at the resonance positions indicated above. Similar patterns computed for NNO can be compared with the patterns found by Barrett and Myers[7] and are in complete agreement. The ratio of the Rayleigh to Raman scattered light for the NNO is very near to that expected for complete transmission of all the rotational Raman light, i.e. the ratio of the peak areas is about 3 or 4 to 1. This is very close to the expected ratio from the theory given by Manneback,[11] Placzek and Teller[12] and from the results of Stansbury *et al.*[6] Thus, we see that the total predicted gain is in fact realizable for linear rotors, and similarly for symmetric rotors. Asymmetric rotors require special consideration since the lifting of K degeneracy yields complex spectra, and will not be considered in detail here. It should be noted that this superposition gain is also fundamental to this scheme, in that the superposition process can give a signal gain for any type of spectrum that has regularly recurring sets of lines. It is applicable with reduced efficiency to vibration-rotation spectra as well.

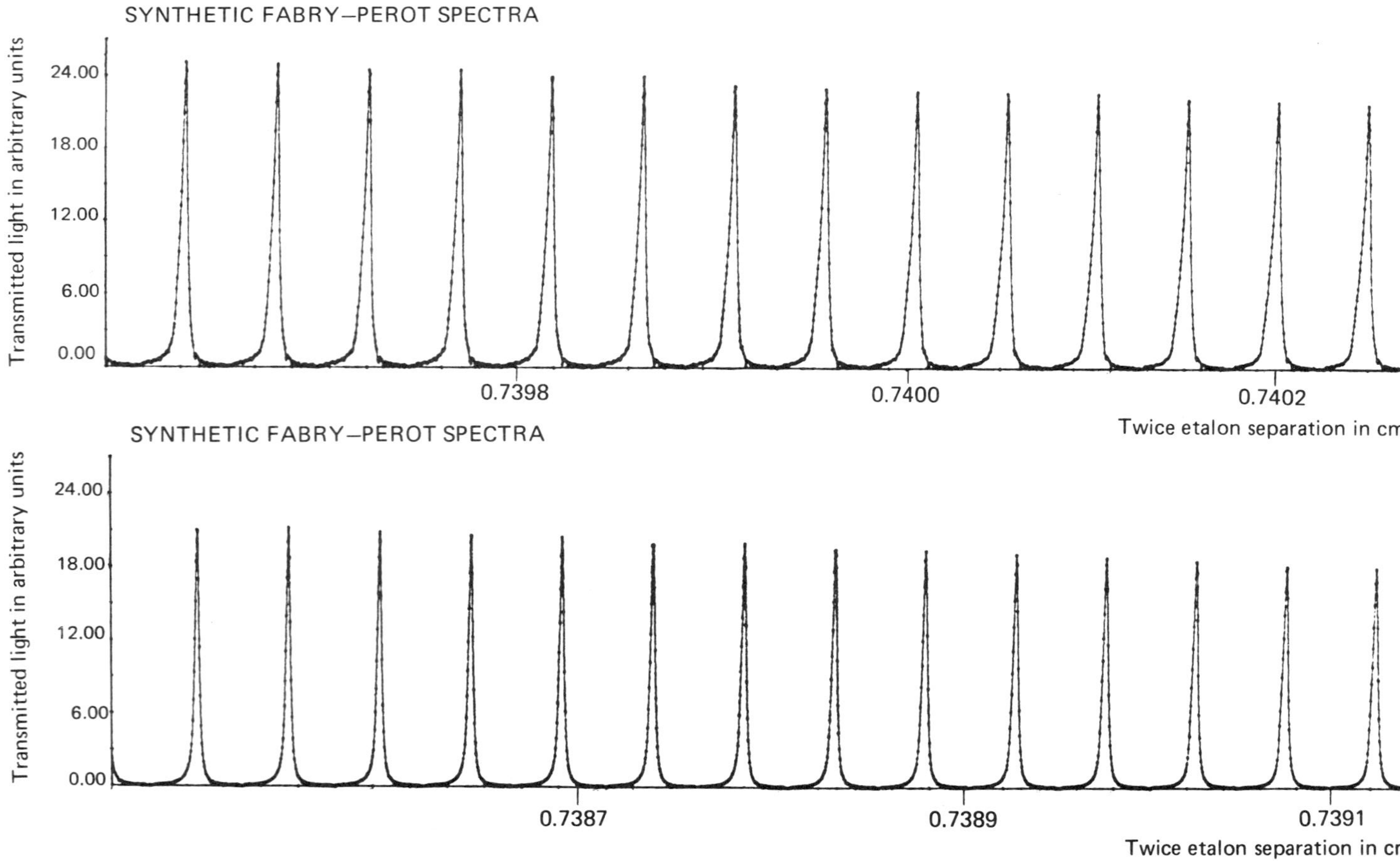

Fig. 2. Calcomp plots of the transmission fringe at a free spectral range of about $4B/3$ cm^{-1} for the molecule NO at 473 K and 273 K. The transmission maximum is located at the leftmost fringe. It will be noted from the plate spacings that the fringes have shifted about one maximum for each 20 K temperature change. The Rayleigh scattering is not introduced in these fringe patterns.

DISCUSSION

We have shown here that the use of a Fabry-Perot interferometer as a spectral filter having regularly recurring transmissions selected intervals in frequency should be of substantial value in improving the sensitivity for optical detection of atmospheric components. A gain of at least 10^4 can be expected from the foregoing remarks when these techniques are applied to linear or symmetric rotors, and somewhat less for asymmetric rotors. Thus by comparison with the experimental results of Leonard[4] we expect to be able to efficiently detect such molecules in concentrations of less than 100 p.p.m. at ranges of several hundred metres. This assumes that we limit ourselves to the other experimental considerations of Leonard. It is clear that the reduced imaging requirements of the Fabry-Perot compared with a conventional monochromator will allow the use of much cheaper and therefore larger light buckets for the collection of return radiation.

An important question for the applicability of the method is the ratio of rejection of unwanted light compared with the Raman light from the low concentration molecule being sought. Using a rather simple theory, it is possible to provide a good estimate of a minor component's detectability in the presence of larger amounts of the usual atmospheric gases. As is well known, the ratio of maximum to minimum transmission for a Fabry-Perot interferometer is given by[8]

$$T_{max}/T_{min} = 1 + (4F^2/\pi^2)$$

In our case, where we are adding the intensity of many rotational lines simultaneously, we effectively increase the transmitted intensity of minor constituent by a factor of more than fifty. Thus, for a finesse of thirty we have equal transmitted intensities for a random rotational Raman line from a major gas constituent occurring near a minimum of transmission for the FPI comb selected to be in resonance for the sought after minor constituent when the minor component is present in concentrations of about sixty p.p.m.

Naturally, every random line of a major component gas does not occur near the minimum of the transmission comb. In fact, far from its own resonance positions as is the case usually of importance for the measuring of minor constituents, essentially a constant background transmission occurs for each major component. For example, at the resonance for NO occurring at a free spectral range of $4B/3$ cm^{-1}, and for $F = 30$, rotational lines of N_2 and O_2 can only interfere seriously when their position lies within 0.075 cm^{-1} of an NO rotational line. At 300 K this occurs only once for O_2 ($J = 38$) and once for N_2 ($J = 20$) for appreciably populated levels. As shown by our calculated transmission patterns, change of the free spectral range by the half-width of a transmission brings other lines into the envelopes of a transmission maximum so that a background d.c. level results. This effect may be observed in the results of Barrett and Myers[7] where the Rayleigh maxima are very regular in intensity until a decrease is observed near a resonance since the rotational Raman light is being removed by the resonance effect from underneath the Rayleigh peak. Other abundant atmospheric gases such as H_2O and CO_2 produce a similar d.c. level away from their own resonances which is highly constant due to their greater density of lines.

The calculated transmission patterns given in Fig. 2 show that the temperature of the observed region may be measured due to the effect of the centrifugal distortion on the effective rotation constant for the molecule. In this case, we see that the maxima of transmission, located at the left side of each pattern shown, has shifted such that a change of position of one fringe occurs for each twenty degrees difference in temperature of the measured volume. Normally, one would probably just measure the temperature of the major constituents such as N_2, but minor constituents could be studied as well.

Some simple means of improving the detectivity of the method are readily arrived at. For example the reduction of the background scattered light at the unshifted frequency, and the limitation of the band of accepted frequencies to that region where the rotational lines occur will immediately enhance the sensitivity for a minor constituent. For our initial laboratory experiments, we have used the 5145 Å argon ion laser line. It is very convenient to use the coincidence of the I_2 absorption line at the same wavelength to substantially reduce the background light level (by about 1000). This, coupled with a broad band filter centred upon the 5145 Å line with a band pass of 100 Å, allows us to discriminate very well against much of the light that interferes with the detection of Raman rotational scattering. Such filtering is possible in part due to the large aperture of our instrument and its efficiency which wastes very little light.

The experimental realization of these theoretical considerations is underway at the present time in our laboratory, where we are investigating the various means of applying this promising technique to atmospheric studies.

ACKNOWLEDGEMENT

Mr. Harvey Liszt collaborated on the computer calculations discussed here briefly.

REFERENCES

1 T. Kobayasi and H. Inaba, *Proc. IEEE* **58**, 1568 (1970).

2 H. Inaba and T. Kobayasi, *Nature* **224**, 170 (1969).

3 J. A. Cooney, *Appl. Phys. Letters* **12**, 40 (1968).

4 D. A. Leonard, *Feasibility Study of Remote Monitoring of Gas Pollutant Emissions by Raman Spectroscopy*, AVCO Everett Research Laboratory, Final Report on Contract CPA 22–69–62 (November 1970).

5 G. Placzek, *Handbuch der Radiologie*, Vol. 6, part 2, Akademische Verlagsgesellschaft, Leipzig, 1934.

6 E. J. Stansbury, M. F. Crawford and H. L. Welsh, *Can. J. Phys.* **31**, 954 (1953).

7 J. J. Barrett and S. A. Myers, *J. Opt. Soc. Amer.* **61**, 1246 (1971).

8 W. H. Steel, *Interferometry*, Oxford University Press, Oxford, 1967.

9 G. Herzberg, *Infrared and Raman Spectra of Polyatomic Molecules*, Van Nostrand, Princeton, New Jersey, 1954.

10 P. Connes, *Ann. Rev. Astron. Astrophys.* 211 (1970).

11 C. Manneback, *Z. Physik* **62**, 224; **65**, 574 (1930).

12 G. Placzek and E. Teller, *Z. Physik* **81**, 209 (1933).

14. THE DEPOLARIZATION RATIO OF THE RAMAN BANDS OF THE VIBRATION OF DIATOMIC MOLECULES

W. Holzer and Y. Le Duff

Département des Recherches Physiques, Laboratoire associé au C.N.R.S., Université de Paris VI, France

INTRODUCTION

The depolarization ratio of the Raman band of the vibration of many diatomic molecules in the gas phase has already been studied with mercury excitation.[1] The precision of these results is, however, limited and the well defined geometry of laser excitation opens a new wide field of research.

Particular points of interest are as follows:

1. Extremely low depolarization ratios. The study of depolarizing effects due to various interactions of the molecule with the surrounding matter may be performed in these cases with very high precision. This has already been done for tetrahedral molecules.[2]
2. Changes in depolarization ratio when measured in gaseous or liquid phase.
3. Resonance effects on the depolarization ratio. These occur when excitation of the molecule is within the absorption band or close to it.[3,4]
4. Distinction between resonance fluorescence and resonance Raman effect by means of the depolarization ratio[4] and study of the intermediate transition region.[5,6]

In order to study the effects 1–3 precise gas phase results on the depolarization ratio of the Raman bands of a large number of molecules are necessary. The present investigation has been limited to diatomic molecules, in order to simplify a theoretical approach. The theoretical calculation of depolarization ratios is in general very difficult. Precise values have so far only been calculated for the molecules H_2^+ and H_2.[7] It is hoped that correlations found by systematic experimental studies may suggest suitable approximations and thus simplify the problem.

EXPERIMENTAL

The depolarization ratios have been measured by keeping the laser polarization fixed and analysing the scattered light (the only method to avoid large convergence errors).

By careful alignment of our conventional system (Spectra Physics 165 argon ion laser, Coderg PH1 spectrometer with photon counting, polarization scrambler, Polarex KSW68 foils), we were able to limit the leakage of the strong component into the weak one, to less than 3×10^{-4}. This value has been established by measuring the depolarization ratio of the ν_1 vibration of SF_6 (gas, 1 atm) to be $< 3 \times 10^{-4}$.

Table 1. Depolarization ratio of the rotational components of the *Q* branch of the Raman band of H_2 and D_2 (precision ± 0·001).

	$\rho_s(J)$				
	$J = 0$	$J = 1$	$J = 2$	$J = 3$	$J = 4$
H_2	<0·001	0·019	0·014	0·0135	
D_2	<0·002	0·020	0·0125	0·013	0·013

Table 2. Depolarization ratios of the vibrational Raman band of diatomic molecules (no resonance).

Gases	ρ_s	References
H_2	0·048 ± 0·005	this work
D_2	0·048 ± 0·005	this work
HF	0·12 ± 0·04	this work
N_2	0·075 ± 0·005	this work
NO	0·22 ± 0·04	this work
O_2	0·17 ± 0·01	this work
HCl	0·18	ref. 10
DCl	0·20	ref. 10
HBr	0·21	ref. 10
DBr	0·20	ref. 10
F_2	0·19	ref. 11
Cl_2	0·14	ref. 4
Cl_2	0·175 ± 0·01	this work
BrCl	0·22	ref. 4
Liquids	ρ_s	
N_2	0·19	ref. 12
O_2	0·09	ref. 12

RESULTS

Q branch of D_2 and H_2

We mention first the results on the depolarization ratio of the *Q* branch of the vibration of H_2 and D_2, because our results are principally different from former measurements.[8] The contributions from the rotational components $J = 0, 1, 2, 3 \ldots$ to the *Q* branch can be separated in these molecules and one should be able to observe the *J* dependence of the anisotropic part of the scattering intensity as calculated by Placzek and Teller.[9] The fact that the depolarization ratio of the $J = 0$ component is strictly zero is the most striking result. The results on the *Q* branch of H_2 and D_2 are given in Table 1. The data are in good agreement with theory; a detailed discussion will be given elsewhere. From ref. 9 and the data presented in Table 1 a value independent of $J(J = \infty)$ is easily derived to be

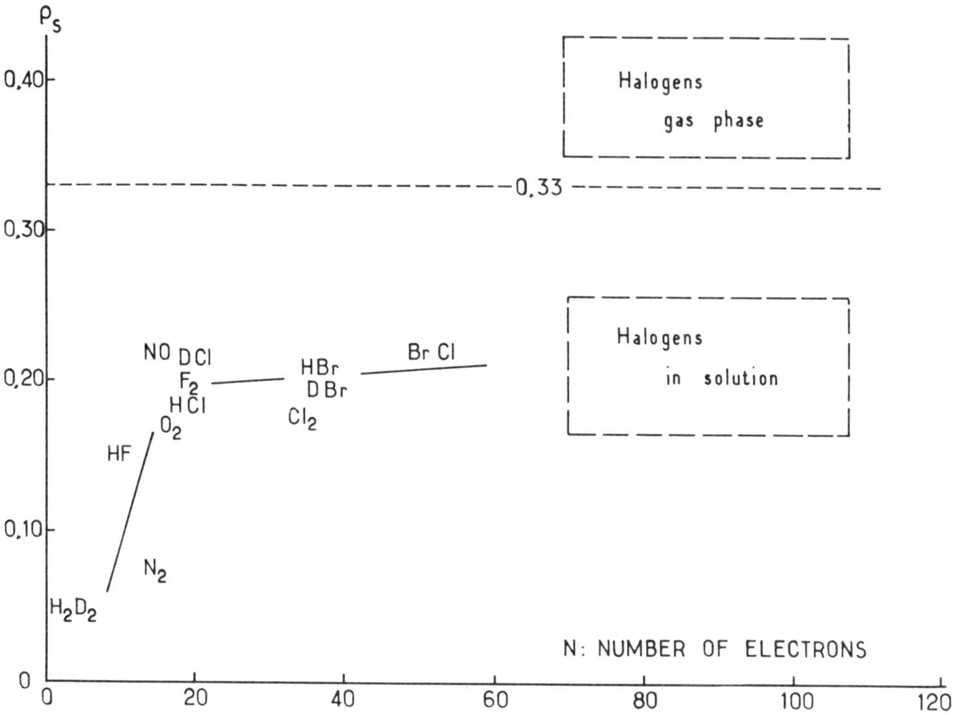

Fig. 1. Plot of the experimental depolarization ratio ρ_s of diatomic molecules as a function of N, the number of electrons of the molecule.

ρ_s(Q branch) $= 0{\cdot}0125 \pm 0{\cdot}001$ for H_2 and D_2. This gives $\rho_s = 0{\cdot}048$ for the total Raman band to be compared with the previous[8] value $\rho_s = 0.035$ ($\rho_n = 0.07$) which has been obtained indirectly from intensity measurements, neglecting the J dependence.

For molecules and temperatures, where the intensity contribution of levels with $J < 4$ to the total band is sufficiently small, the J dependence of the anisotropic part of the scattering intensity may be neglected.

Some results on diatomic molecules

The results of our measurements and of those of other authors are collected in Table 2. The values ρ_s are either measured directly or in cases where the Q branch is well isolated from the O, S branches, obtained from the depolarization ratio of the Q branch $\rho_{s,Q}$ by means of the relation:[9]

$$\rho_s = \frac{\rho_{s,Q}}{0{\cdot}25 + \rho_{s,Q}}$$

Results on ρ_s in liquid oxygen and nitrogen[12] are also included in Table 2.

Some results obtained under resonance excitation for halogen molecules

The discrete or continuous absorption bands of the heavier halogen molecules are

Table 3. Depolarization ratios of the vibrational resonance Raman band of the heavier halogen molecules.

Molecule	ρ_s	Wavelength of excitation (Å)	Reference
Br_2 gas	0·25	4880	4
	0·40	5017	13
	0·36	4965	13
	0·32	4880	13
	0·42	4765	13
	0·50	4579	13
	0·2	6943	14
Br_2 in $CHCl_3$	0·16 ± 0·01	6328	this work
I_2 gas	0·35	4880	4
	0·39	4965	13
	0·35	4880	13
	0·36	4765	13
	0·41	4579	
I_2 in $CHCl_3$	0·26	4765 to 5145	15
	0·23 ± 0·02	6328	this work
I_2 in CH_2Cl_2	0·26 ± 0·02	5145	this work
	0·26 ± 0·01	6328	this work
ICl gas	0·40	4880	4
ICl in CCl_4	0·23 ± 0·01	6328	this work
IBr gas	0·43	4880	4

extended over the complete visible region, thus either resonance fluorescence ($\rho_s = 0{\cdot}75$) or resonance Raman effect ($\rho_s \approx 0{\cdot}33$) is observed. We have tried to diminish resonance effects by working in solutions and with 6328 Å excitation. In this case the first overtone is usually too weak to be observed. The work in solution is difficult, since the concentration is limited by thermal lens effects and the solvent spectrum has to be subtracted very carefully. Our results are presented together with results in the literature in Table 3.

DISCUSSION

The trend that the depolarization ratio increases from light to heavy molecules (Table 2) is shown clearly in Fig. 1 where ρ_s has been plotted versus N, the total electron number of the molecules. The fact that the depolarization ratio is almost constant from $n = 15$ to $n = 50$ should be of high interest for a theoretical approach. For $n > 50$ resonance effects are taking over and non resonance data are no more available. The data for the heavier halogen molecules given in Table 3 are also indicated in Fig. 1. The gas data are slightly superior and the solution data slightly inferior to the theoretical value $\rho_s = 1/3$.

We will now discuss the results on the basis of the polarizability theory.[16] For diatomic molecules the trace and the anisotropy of the tensor of the polarizability

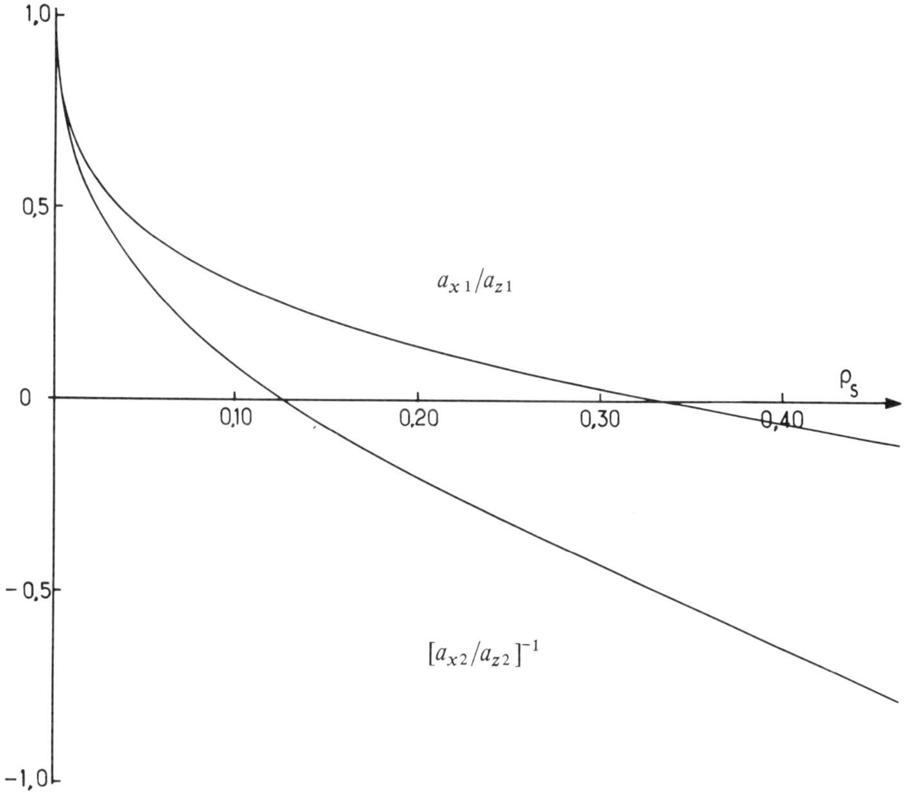

Fig. 2. The ratio of the principal polarizability derivatives as a function of the depolarization ratio ρ_s

derivative may be expressed in terms of the principal values of the tensor with respect to the three axes x, y and z: a_z and a_x equal a_y. Thus ρ_s may be calculated in terms of a_z and a_x.

We give here the ratio a_x/a_z as a function of ρ_s.

$$\frac{a_x}{a_z} = -\frac{2\rho_s + 1}{8\rho_s - 1} \pm \frac{1}{8\rho_s - 1} \cdot \sqrt{15\rho_s - 20\rho_s^2}$$

There exist two solutions and there is no simple physical argument to exclude one of them. The first solution a_{x1}/a_{z1} is plotted in Fig. 2 as a function of ρ_s. The second solution a_{x2}/a_{z2} is given as $(a_{x2}/a_{z2})^{-1}$ in order to facilitate the representation in the same plot.

We would like to point out the fact that the resonance values $\rho_s = 1/3$ (σ character) and $\rho_s = 1/8$ (π character) appear as smooth transition points when the ratio a_x/a_z passes from positive to negative values. Thus the polarizability theory which is in principle no more valid, gives qualitatively correct results. (Our representation in Fig. 2 is more convenient than the one given in ref. 16.)

The regions around $\rho_s = 0.33 \pm \epsilon$ and $\rho_s = 0{\cdot}125 \pm \epsilon$ are of special interest to us. It can be seen that the value $|a_{x1}/a_{z1}|$ is the same for $\rho_s = 0{\cdot}33 - \epsilon$ and $\rho_s = 0{\cdot}33 + \epsilon$, whereas only the sign of a_{x1}/a_{z1} changes. Similarly $|a_{x2}/a_{z2}|$ stays constant for $\rho_s =$

$0{\cdot}125 - \epsilon$ and $0{\cdot}125 + \epsilon$. Thus different values of ρ_s do not necessarily imply a change of the principal values $|a_z|$ and $|a_x|$. This may explain the obvious difference in ρ_s between the results in gas phase and in solution close to resonance $\rho_s \approx 0{\cdot}33$, as given in Table 2 and Fig. 2.

We discuss as an example ICl which has $\rho_s = 0{\cdot}40$ in the gas phase corresponding to a negative a_x/a_z, that is, the derivatives a_z and a_x are in antiphase. The lower value in solution $\rho_s = 0.23$ may then be rationalized by assuming the same $|a_z/a_x|$, but a_z and a_x to be in phase. The resonance may well determine the approximate region of the degree of depolarization ($\approx 0{\cdot}33$) while molecular structure and intermolecular interaction defines ϵ, both together result in the measured values $\rho_s = 0{\cdot}33 \pm \epsilon$.

The region $\rho_s = 0{\cdot}125$ should also be checked for similar effects. It is worthwhile to note, that this approach may be immediately extended to polyatomic molecules by the approximation of bond polarizability theory.

REFERENCES

1 W. F. Murphy, W. Holzer and H. J. Bernstein, *Appl. Spectry.* **23**, 211 (1969).
2 W. F. Murphy, W. Holzer and H. J. Bernstein, 23rd Symposium on Molecular Structure and Spectroscopy, Columbus, Ohio, 1968.
3 D. G. Rea, *J. Mol. Spectry.* **4**, 499 (1960); O. S. Mortensen, *Chem. Phys. Letters* **3**, 4 (1969).
4 W. Holzer, W. F. Murphy and H. J. Bernstein, *J. Chem. Phys.* **52**, 399 (1970).
5 I. R. Beattie, G. A. Ozin and R. O. Perry, *J. Chem. Soc.* **A**, 2071 (1970).
6 M. Berjot and L. Bernard, *Advances in Raman Spectroscopy,* Vol. 1 (Ed. J. P. Mathieu), Heyden & Son, London, 1973, p. 343.
7 J. Tang and A. C. Albrecht, *Raman Spectroscopy,* Vol. 2, (Ed. H. A. Szymanskii), Plenum Press, New York, 1970, p. 33.
8 T. Yoshino and H. J. Bernstein, *J. Mol. Spectry.* **2**, 213 (1958); D. M. Golden and B. Crawford, *J. Chem. Phys.* **36**, 1654 (1962).
9 G. Placzek and E. Teller, *Z. Physik* **81**, 209 (1933).
10 J. P. Perchard, W. F. Murphy and H. J. Bernstein, *Mol. Phys.* **23**, 535 (1972).
11 H. H. Claassen, H. Selig and J. Shamir, *Appl. Spectry.* **23**, 8 (1968).
12 M. Scotto, *J. Chem. Phys.* **49**, 5362 (1968).
13 M. Berjot, M. Jacon and L. Bernard, *C. R. Acad. Sci., Paris* **274B**, 404 (1972).
14 W. Kiefer and H. W. Schrötter, *J. Chem. Phys.* **53**, 1612 (1970).
15 O. S. Mortensen, *J. Mol. Spectry.* **39**, 48 (1971).
16 G. Placzek, *Handbuch der Radiologie,* Vol. 6, part 2, Akademische Verlagsgesellschaft, Leipzig, 1934.

SECTION III
PHASE TRANSITIONS

15. RAMAN STUDIES OF SOFT MODES AND STRUCTURAL PHASE TRANSITIONS

J. F. Scott

Bell Telephone Laboratories, Inc., Quantum Electronics Research Department, Holmdel, New Jersey, USA

and

*Department of Physics and Astrophysics,**
University of Colorado, Boulder, Colorado, USA

INTRODUCTION

The study of solid state phase transitions by light scattering began in 1939 with a study of the trigonal–hexagonal phase change in quartz.[1] Appropriately enough, this pioneering work was by C. V. Raman. Subsequently, some theoretical work followed (eventually culminating in Cochran's famous papers[2]), including that of Devonshire[3] in England, who described ferroelectric phase transitions in terms of phenomenological thermodynamic expressions. At about the same time Ginzburg[4] in the Soviet Union, applied Landau's earlier theory[5] of second-order phase transitions to Raman's experiments on quartz. Of special interest in Ginzburg's analysis was the prediction of 'central modes', centred at zero frequency shift,[6] which, like critical opalescence in fluid systems[7] would exhibit divergent cross sections at the phase transition temperature T_0. It was thought that these central modes had been directly observed in the Soviet Union,[8] but as subsequent work by Gammon,[9] Shapiro and Cummins[10] showed, the intense scattering in quartz at $T \approx T_0$ was due to micro-domain formation. While Shapiro and Cummins concluded that the domains were due to right- and left-handed Dauphine twins, it seems more likely that the effect is due to the coexistence of α- and β-quartz domains[11-13] over the ~ 3 K hysteresis region found to characterize this first-order phase transition.[10] Despite all the confusion in the literature about the possibility of 'central modes' or critical opalescence in solids, such phenomena have now been unambiguously observed by inelastic neutron scattering in both Nb_3Sn[14] and $SrTiO_3$,[15] and we shall discuss their origin in terms of Cowley's anharmonic theory,[16,17] as well as their detection via Raman spectroscopy.[18]

A second topic of great current interest is the possible deviation from classical or 'mean field' theories manifest in the occurrence[19] of exponents β differing from ½ in the order parameter and soft mode temperature dependences

$$\phi(T) = \phi_0(T_0 - T)^\beta \tag{1a}$$

$$\omega(T) = \omega_0(T_0 - T)^\beta \tag{1b}$$

*Permanent address. Work at Colorado supported by NSF grant GH–34681.

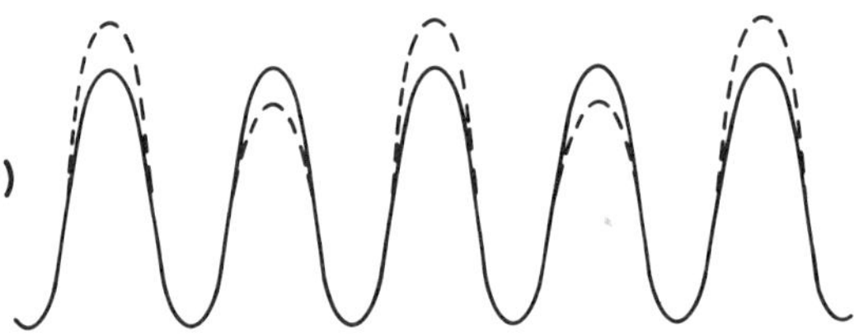

Fig. 1. Charge density $\rho(r)$ versus r for $T < T_0$ (dotted) and $T \geqslant T_0$ (solid).

We review in the theoretical section why $\beta = \frac{1}{2}$ theories are called 'mean field' or 'Molecular field' approximations and relate our soft mode studies to the older and slightly more respectable field of magnetic phase transitions. One point of sufficient importance to stress even in this introduction, however, is that despite the analogy with 'critical opalescence' the central modes we discuss have *no* necessary connection with 'critical exponents', such as $\beta = 1/3$ values determined in $SrTiO_3$. That is, central modes may be described within the confines of mean field theory.

We shall not attempt to provide a comprehensive review of soft mode spectroscopy in this chapter. Several good reviews have been published recently, and the reader is referred to the cited work by Cochran,[20] Worlock[21] and Fleury.[22] What does follow is a summary of work on a few selected prototypes of interest to me, and some comments to focus attention on problems of current interest, such as harmonic and anharmonic mode coupling,[23–31] in addition to the matters of critical exponents and central modes mentioned above.

Theory[32]

Landau's conditions for a second order displacive phase transition were defined in terms of symmetry requirements. In Fig. 1, we see a plot of charge density for a simple one-dimensional lattice. The solid line refers to $T > T_0$, and the dotted line to some $T < T_0$. It can be seen that below T_0 the periodicity of the lattice has changed, and that the unit cell has doubled from length A to 2A. If this kind of change is to occur continuously with temperature, Landau's first condition is that the charge density can be written as

$$\rho(r) = \rho_0(r) + \Delta\rho(r) \tag{2}$$

where $\rho_0(r)$ is a function invariant under the operations of the high-symmetry space-group, $\rho(r)$ is invariant under the operations of the low-symmetry space-group, *and* the symmetry group of $\rho(r)$ is a subgroup of $\rho_0(r)$. Physically, this implies that only infinitesimal displacements of ions are allowed.

Any function can be expanded in terms of the irreducible representations of the symmetry group characterizing the structure in question. Thus, we may write

$$\Delta\rho(r) = \sum_{i,j} c_{ij}\phi_{ij}(r) \tag{3}$$

where $\phi_{ij}(r)$ are the j-dimensional basis states for the ith irreducible representation. For example, in α-quartz there are three different irreducible representations (A_1, A_2, E), of which A_1 and A_2 are one-dimensional and E is two-dimensional. Thus the complicated double sum in Eqn. (3) contains only four terms

$$\sum_{i,j} c_{ij}\phi_{ij}(r) = C_1A_1(r) + C_2A_2(r) + C_3E_x(r) + C_4E_y(r) \tag{4}$$

although the functions $A_1(r)$ etc., may include many different combinations of normal modes of symmetry A_1, etc. Landau's second assumption is that changes corresponding to two different representations will set in at the same temperature only by accident. Thus, for usual systems

$$\sum_{i,j} c_{ij}\phi_{ij}(r) \longrightarrow \sum_j c_j\phi_j(r) \tag{5}$$

The difference between Landau's theory and Cochran's is that Cochran assumes not only that $\Delta\rho(r)$ can be expanded in terms of the irreducible representations ϕ_i (only one of which characterizes the structural change), but that in addition, the $\Delta\rho(r)$ can be expanded in terms of eigenvectors or normal coordinates of the lattice, and *only one of these* will characterize the symmetry change at T_0. The difference is that in general there will be more than one orthogonal normal mode of a given symmetry for a crystal. For example, in quartz there are four modes of Γ_1 symmetry. Landau's theory says only that the α–β transition, if second-order, cannot involve modes of two different symmetries; Cochran's theory says further that it cannot involve two modes of the same symmetry.

Cochran, Landau and Devonshire all assume that the free energy F near the transition temperature T_0, where the c_i in Eqn. (3) are small, can be expanded in a power series, e.g.

$$F = F_0 + \tfrac{1}{2}A\eta^2 + \tfrac{1}{3}B\eta^3 + \tfrac{1}{4}C\eta^4 + \ldots \tag{6}$$

where η is a normalized linear combination of the c_i.

The conditions for stability

$$\frac{\partial F}{\partial \eta} = 0 \quad \text{and} \quad \frac{\partial^2 F}{\partial \eta^2} > 0 \tag{7}$$

eliminate any linear term in η in Eqn. (6) and require that the coefficient A in Eqn. (6) be positive. The condition that F be a minimum at $\eta = 0$ eliminates the η^3 term (and all other odd powers) in Eqn. (6). If this latter condition is not satisfied, the phase transition cannot be second-order, for a finite η at $T = T_0$ implies finite c_i – which physically must be infinitesimal (Landau's third condition).

The above requirements yield (for all η)

$$0 = \frac{\partial F}{\partial \eta} = A\eta + C\eta^3$$

or

$$\eta^2 = \frac{-A}{C} \tag{8}$$

For $T > T_0$, $$0 < \left.\frac{\partial^2 F}{\partial \eta^2}\right|_{\eta=0} = A$$

Thus, A is positive. Usually it has been assumed that

$$A = A_0(T - T_0) \tag{9}$$

and that C is independent of temperature. This leads to the prediction that the order parameter η varies as

$$\eta = \eta_0(T_0 - T)^{1/2} \tag{10}$$

The assumption given in Eqn. (9) is not just heuristic. This particular temperature dependence follows from the assumption of infinite long-range forces. Because each particle is viewed under this assumption of infinite range forces, it can be treated as moving in an average potential or 'mean field' of all the other particles. This has been done classically for fluids in the Van der Waals theory, and for spins in magnetic systems as well.[33] In fluids and in magnetic systems, the mean field theory fails spectacularly, as shown beautifully by Heller and Benedek[34, 35] for ferromagnetic EuS and antiferromagnetic MnF_2. A theoretical derivation of the exponent $\beta = \frac{1}{2}$ in Eqn. (10) is given by Stanley.[33] Other critical exponents are derived from mean field theory in this reference as well. Because the interactions in fluids and magnetic solids are short-range, we expect the assumption leading to Eqn. (9) to fail. By comparison, the forces in ferroelectrics are largely Coulombic, and therefore of very long range. Thus we expect Eqn. (10) to hold for soft mode phase transitions in ferroelectrics. (Note that the α–β transition in quartz and the 106 K transition in $SrTiO_3$ are not of the ferroelectric type.) But in other systems we may find $\beta \approx 1/3$ typical of three-dimensional Ising and Heisenberg models.

The connection between the temperature dependence of the order parameter and a phonon frequency follows from the simple one-dimensional harmonic oscillator equation (η is a displacement; F, a potential) in linear[36] response

$$m\omega^2 = \frac{\partial^2 F}{\partial \eta^2} = A \tag{11}$$

[from Eqn. (8)] or

$$\omega = \omega_0 |T - T_0|^{1/2} \tag{12}$$

Thus we conclude that in ferroelectrics or other crystals in which the assumption of infinite range forces is reasonable, a mode having undamped frequency dependence given in Eqn. (12) may be expected.

The Curie–Weiss temperature dependence observed for the dielectric constant in many ferroelectrics follows immediately from Eqn. (12) if the harmonic oscillator model is assumed

$$\epsilon(\omega) = \epsilon_\infty + \frac{(\epsilon_0 - \epsilon_\infty)\,\omega_{TO}^2}{\omega_{TO}^2 - \omega^2} \tag{13}$$

The LO longitudinal mode frequency occurs where $\epsilon(\omega) = 0$ whereas $\epsilon(\omega) = \infty$ at ω_{TO}, the transverse mode frequency; this follows from Maxwell's equations. Hence

$$0 = \epsilon_\infty + \frac{(\epsilon_0 - \epsilon_\infty)\,\omega_{TO}^2}{\omega_{TO}^2 - \omega_{LO}^2} \tag{14}$$

or

$$\frac{\epsilon_0(T)}{\epsilon_\infty} = \left[\frac{\omega_{LO}}{\omega_{TO}(T)}\right]^2 \tag{15}$$

the famous Lyddane–Sachs–Teller equation.[37] It follows from Eqns. (12) and (15) that

$$\epsilon_0(T) = \frac{C}{|T_0 - T|} \tag{16}$$

the Curie–Weiss law exhibited by many ferroelectrics.

These macroscopic phenomenological descriptions of soft mode phase transitions have been related by Cowley[38] to microscopic anharmonic theory. The phase transitions are described as due to interactions between phonons. When phonons interact, the harmonic is broadened by a width Γ and shifted in frequency by an amount Δ, i.e., $\omega^2 = \omega_h^2 + \Delta$. In the lowest order theory, Δ will be given as proportional to the density of final states into which the phonon decays, e.g., $n_1 + n_2 + 1$ for a three-phonon interaction. At normal temperatures the Bose population factors n_1 and n_2 are larger than one and approximately $\frac{KT}{\hbar\omega_1} + \frac{KT}{\hbar\omega_2}$. Thus $\Delta = \mathrm{d}T$ where d is a constant. For a phonon which is unstable in the harmonic approximation, for example due to a double well of form $V(X) = -ax^2 + bx^4$, $\omega_h^2 = -a/m$ is negative. Let us call $-a/m = -\mathrm{d}T_0$. Then

$$\omega^2 = \omega_h^2 + \Delta = -\mathrm{d}T_0 + \mathrm{d}T = \mathrm{d}(T - T_0) \tag{17}$$

the equivalent of Eqn. (12) is derived. In this view, structures which are unstable in the harmonic approximation are stabilized by anharmonic decays, and a certain temperature T_0 must be reached before the decay is rapid enough to produce stabilization.

Why then do only transverse phonons go unstable in piezoelectric crystals and not LO's? We can see from Eqn. (15) that if an LO mode frequency went to zero at T_0 the low frequency dielectric constant would not diverge – i.e., the crystal would not become ferroelectric. But that does not explain why LO's do not soften. Perhaps the easiest answer is to consider that LO and TO phonon frequencies differ due to macroscopic Coulomb fields. We may write for a simple diatomic lattice[39]

$$\mu\omega_{TO}^2 = R_0 - \frac{4\pi z^2}{3V} \tag{18}$$

$$\mu\omega_{LO}^2 = R_0 + \frac{8\pi z^2}{3V} \tag{19}$$

where R_0 is the short-range potential common to both modes and the term in z^2 is the

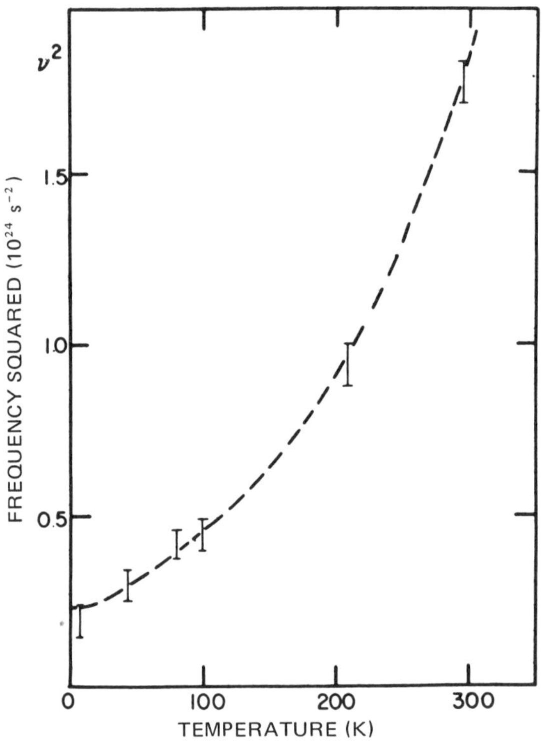

Fig. 2. SnTe soft mode ω versus T (refs. 40 and 41).

Lorentz field $\frac{4}{3}\pi P$ and the Lorentz-plus-macroscopic field $\frac{4}{3}\pi P - 4\pi P$. It is evident from Eqn. (18) above that the TO frequency can experience a sharp decrease at $T = T_0$ if the temperature variation of R_0 and z^2/V produce an accidental cancellation at some temperature T_0. However, since two positive quantities are added on the right side of Eqn. (19), the LO frequency can never go to zero. Thus, ferroelectricity requires no radical temperature dependence of bond strengths or short-range forces, only a subtle cancellation of long- and short-range effects.

EXPERIMENTAL

IV–VI Compounds

In Cochran's first paper on soft modes,[2] he discussed the unusual temperature dependence of the TO phonon in PbTe, which decreases in frequency with decreasing temperature. Subsequently, inelastic neutron scattering on SnTe at Chalk River[40] showed the presence of an extremely soft mode in that material. In both SnTe and PbTe, the TO phonon frequency remains positive even at 0 K; these materials are thus not ferroelectrics, but in SnTe the extrapolated Curie temperature is only a few degrees below absolute zero. This class of materials is of enormous theoretical interest because it has a simple diatomic rock-salt cubic structure in the paraelectric phase. They are thus more amenable to calculations than are most ferroelectrics. Gillis has treated this class of materials, and SnTe in particular,[41] by a self-consistent phonon field theory which has been quite

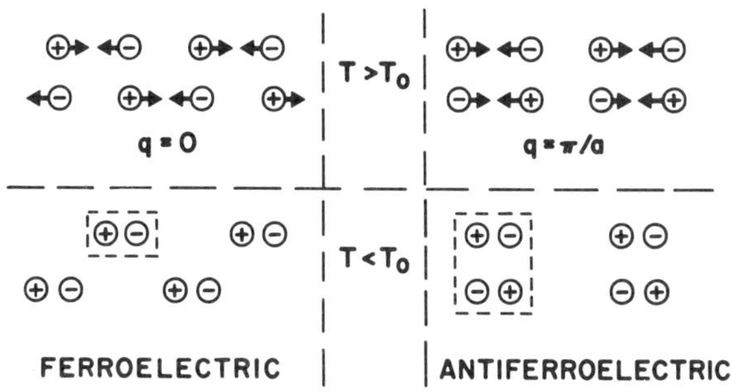

Fig. 3. GeTe soft mode diagram.

successful in describing the shape of the ω versus T soft-mode curve, shown in Fig. 2. Note that ω^2 is proportional to T over a reasonable range of temperature, but flattens out at very low temperatures. This behaviour may be predicted from Cowley's anharmonic theory, since the high temperature approximation

$$\Delta \propto (n_1 + n_2 + 1) \approx \frac{KT}{\hbar\omega_1} + \frac{KT}{\hbar\omega_2} \tag{20}$$

breaks down at low temperatures, yielding

$$n_1 + n_2 + 1 \approx 1 \tag{21}$$

and

$$\omega^2 = \omega_h^2 + \Delta \cong \text{const.} \tag{22}$$

rather than the Curie–Weiss form

$$\omega^2 = \omega_h^2 + \Delta = d(T - T_0). \tag{23}$$

Figure 3 shows a two-dimensional diagram of the soft modes in a diatomic lattice such as SnTe. Both soft ferroelectric modes and antiferroelectric modes are shown; although no real diatomic crystal has yet been found which undergoes an antiferroelectric phase transition, the distortions in SnS may approximate this case. More detailed two-dimensional modeling is given in an excellent review by Barker.[42]

Raman studies on this class of materials have been reported by Steigmeier and Harbeke,[43] who successfully analysed the soft modes in opaque GeTe, using 6328 Å He–Ne excitation. GeTe, unlike PbTe and SnTe, is ferroelectric at ambient temperatures. While its structure was previously known, no direct evidence of ferroelectricity had been reported, due to its high conductivity and the practical difficulty in making the usual capacitance measurements on such materials. In the paraelectric cubic form of GeTe, the TO photon is triply degenerate, of F_{1u} symmetry. This mode softens and becomes a static distortion at T_0, the transition temperature. Below T_0 the crystal is rhombohedral and the three-fold degeneracy is partly lifted, yielding a doubly degenerate Γ_3 (or E) symmetry mode and a nondegenerate Γ_1 (or A) symmetry mode. As the crystal is cooled much below T_0 the soft Γ_1 and Γ_3 components renormalize and increase in frequency.

Experimentally it is found that the phase transition is first order and that the Γ_1 frequency lies above the Γ_3 frequency. Thus, so far experiments on ferroelectric mixed $Sn_xGe_{1-x}Te$ crystals have not been successful.

The fact that the soft mode is Raman-active in the low-symmetry phase is a group theoretical requirement for all crystals with displacive, second-order phase transitions. Since the symmetry of the distortion in the lattice caused by the soft mode is the same as the symmetry of the soft mode (viewed in the high-symmetry phase), it follows that the soft mode symmetry in the low-temperature phase must be given by

$$\Gamma_j \times \Gamma_j = \Gamma_1 \tag{24}$$

If Γ_j is degenerate, as in GeTe, the product $\Gamma_j \times \Gamma_j$ will include more than one irreducible representation, but *always* includes Γ_1, and this totally symmetric representation transforms as a trace $(\alpha_{xx}, \alpha_{zz})$ Raman tensor for all crystal classes. Hence, any second-order displacive phase transition will exhibit a Raman-active soft mode in the *xx* and *zz* spectra. This fact was first explicitly discussed by Worlock.[44]

It is interesting to note, parenthetically, that experimentally *all* components of a degenerate soft mode have always been found to be Raman-active in the low-temperature, low-symmetry phase; however, no one has published a group theoretical proof that this should generally be so.

Perovskites

More effort, both theoretical and experimental, has been made on the study of perovskite phase transitions than on those of any other crystal class. We will discuss some of this work, not in chronological sequence, but in order of complexity.

$KTaO_3$. Potassium tantalate has the simple cubic 'perovskite' ABO_3 structure, with one formula group per primitive cell, and Ta ions at the centre of each cube, with K ions at the corner and oxygens at the centre of each cube face. As the temperature is lowered, the TO mode corresponding to Ta ion motion up-and-down relative to the K and O ions increases in amplitude and decreases in frequency. This mode is dipolar (infrared-active) and of triply-degenerate F_{1u} symmetry.[45] It becomes very soft at helium temperatures[46] but does not reach zero frequency, and, as with SnTe, no ferroelectric transition occurs. (By doping a few percent of Na into $KTaO_3$ or Ge into SnTe, the ferroelectric transitions can be produced at real temperatures.) As with SnTe, the Curie–Weiss temperature dependence for $\omega^2(T)$ fails[47] when

$$n_1 = \frac{1}{\exp(\hbar\omega_1/KT) - 1} \not\approx \frac{KT}{\hbar\omega_1} \tag{25}$$

Experimentally the same soft mode found in $KTaO_3$ is also observed in $SrTiO_3$,[48] $BaTiO_3$,[49] $PbTiO_3$[50] and various solid solutions, such as $K_xNa_{1-x}TaO_3$,[51] $Pb_{1-x}Ba_xTiO_3$[52] and $PbTi_{1-x}Zr_xO_3$.[53] The latter class of ferroelectric ceramics has important technological applications and has been studied most extensively by Burns at IBM.

These studies in $KTaO_3$ were facilitated by the extremely narrow line-width exhibited by the soft mode. Line-widths are an order of magnitude larger in $PbTiO_3$, $BaTiO_3$ and most of the mixed crystals, complicating analysis. Although the narrow line-widths

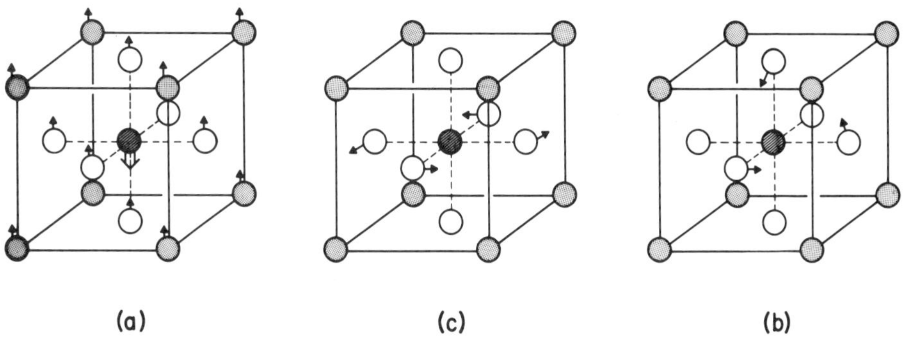

Fig. 4. ABO_3 mode diagrams. (a) ferroelectric mode ($KTaO_3$). (b) $LaAlO_3$ mode. (c) $SrTiO_3$ mode.

facilitated analysis in the cases of $KTaO_3$ and $SrTiO_3$, these experiments were difficult because the soft modes were not Raman-allowed ($KTaO_3$ and $SrTiO_3$ are paraelectric at all temperatures). To overcome this symmetry restriction, d.c. fields of a few kV/cm were pulsed onto the samples,[48] thus removing the inversion symmetry at each ion site and destroying the parity selection rule. Similar techniques could be applied to paraelectric SnTe, but the high conductivity of such semiconductors makes successful electric field-induced scattering unpromising.

$LaAlO_3$. The second kind of structural phase transition which has been found to characterize cubic ABO_3 perovskites is most simply illustrated by the case of $LaAlO_3$. The oxygen ions on the face centres of each cubic unit cell vibrate in the Γ_{25} or F_{2u} mode in a manner which approximates a rotation about a [111] axis of the cube. The sense of rotation is opposite for adjacent unit cells; thus, this mode lies at the Brillouin zone boundary in the cubic phase. At about 800 K this three-fold degenerate mode becomes unstable[54] and a second-order phase transition to a trigonal $D_{3d}^6(R\bar{3}c)$[55] structure occurs. This transition has been carefully studied by EPR,[56] inelastic neutron scattering,[54] X-ray[57] and Raman[58] techniques. Figure 4 shows the ionic displacements involved, and Fig. 5 plots soft mode frequencies versus T in the low symmetry phase. Because the unit cell doubles and the Brillouin zone halves at the phase transition, the soft modes which were described as 'zone boundary' in the cubic phase become zone-centre ($q = 0$) and Raman-allowed below $T_0 = 800$ K.

Soft modes associated with this cell-doubling phase transition have been observed by Raman spectroscopy in $PrAlO_3$, $NdAlO_3$,[58] and $SmAlO_3$,[59] and in $SrTiO_3$,[60, 61] where a related tetragonal distortion also occurs.

$SrTiO_3$. In addition to the soft ferroelectric mode discussed above in the section under $KTaO_3$, strontium titanate exhibits soft modes[62] associated with a unit cell-doubling phase transition at about 106 K. In fact, this crystal was the first in which a soft phonon at a Brillouin zone boundary was discovered. Because it is a more complicated situation than $LaAlO_3$, we discuss it out of chronological order.

The presence of a phase transition in $SrTiO_3$ near 106 K was made evident by Lytle's X-ray study in 1964 and by the studies of elastic coefficients by Bell and Rupprecht in 1963. Some earlier evidence for a phase transition was observed in dielectric studies by Grönicher. The space group of the crystal in the low-symmetry phase (I4/mcm) was not

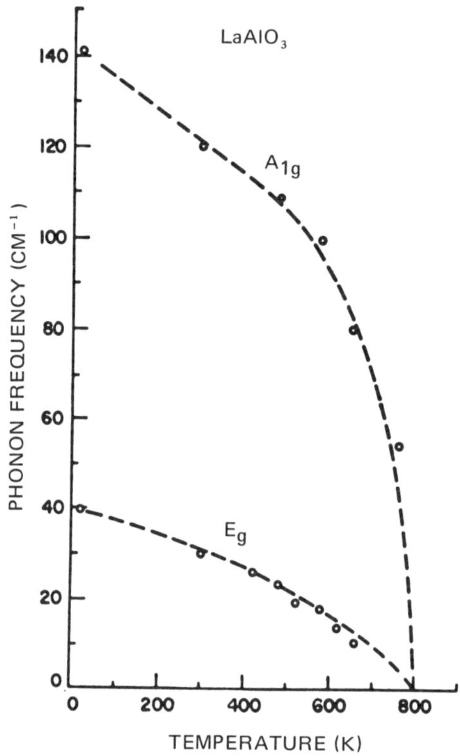

Fig. 5. ω versus T in $LaAlO_3$ (ref. 58).

correctly determined until Unoki and Sakudo's EPR study[62] in 1967. And in 1968, Fleury, Scott and Worlock[62] analysed soft modes and selection rules in the tetragonal phase, which verified Unoki and Sakudo's space group determination.

The ionic displacements in $SrTiO_3$ below 106 K correspond to the same soft $\Gamma_{25}(F_{1u})$ zone-boundary phonon as is responsible for the *trigonal* phase transition in $LaAlO_3$. However a different linear combination of the three components of this mode becomes unstable, because the quartic anharmonic terms are different from those in $LaAlO_3$. In $SrTiO_3$, it is the oxygen rotation about a [100] axis, and not a [111], which becomes unstable. This is shown in Fig. 4, and in Fig. 6 the soft mode frequency versus T is plotted.

Thomas and Müller[62] have provided a simple phenomenological theory which relates soft mode frequencies, anharmonic potentials, and the occurrence of trigonal and tetragonal transitions from the cubic perovskite structure.

They write an energy expression

$$E = E_0(T) + \tfrac{1}{2}a(T)(\Phi_x^2 + \Phi_y^2 + \Phi_z^2) + \tfrac{1}{4}b(T)(\Phi_x^4 + \Phi_y^4 + \Phi_z^4) + \tfrac{1}{2}\, c(T)(\Phi_x^2\Phi_y^2 + \Phi_y^2\Phi_z^2 + \Phi_z^2\Phi_x^2) \tag{26}$$

including terms up to quartic. (There are no cubic terms because the crystal retains inversion symmetry above and below the transition temperature.) Here Φ_x is the angle of

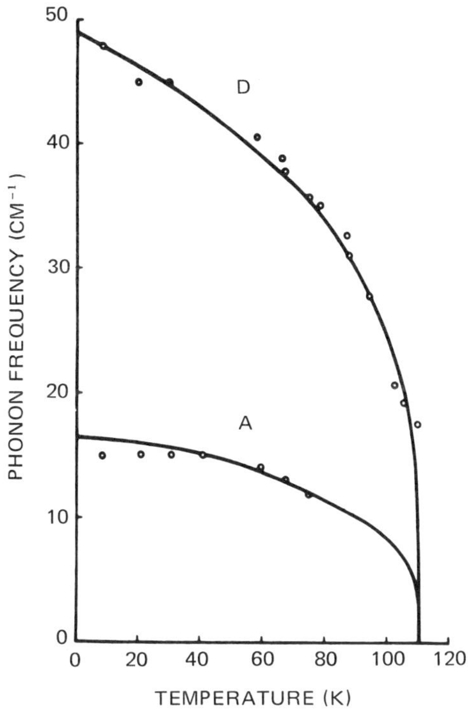

Fig. 6. ω versus T in $SrTiO_3$ (ref. 65).

displacement of the oxygen octahedra about the x axis [100] of the perovskite cube. a, b and c are potential coefficients assumed to vary slowly with T. If $0 < b < c$ the energy in Eqn. (26) has a minimum at $\Phi_x \neq 0$; $\Phi_y = \Phi_z = 0$; and the tetragonal distortion occurs. If $-2c < b < c$, the energy has a minimum at $\Phi_x = \Phi_y = \Phi_z \neq 0$, and the trigonal distortion occurs. The third phase transition possible group theoretically is $\Phi_x = \Phi_y \neq 0$, $\Phi_z = 0$, (which corresponds to an orthorhombic distortion in which the oxygen octahedra rotate about a [110] axis) and is found not to be stable thermodynamically.

The soft mode frequencies in both tetragonal and trigonal cases can be related to the quartic potential coefficients. For the tetragonal case

$$\frac{\omega_E^2(T)}{\omega_A^2(T)} = \frac{c-b}{2b} \tag{27}$$

and is approximately independent of T (as observed for $SrTiO_3$). In $SrTiO_3$, ω_E^2/ω_A^2 is very small (0.1) and shows that $SrTiO_3$ is very near the tetragonal–trigonal instability limit, i.e. $c = b$ in Eqn. (26). For this reason Burke and Pressley were able to drive $SrTiO_3$ trigonal by applying a small uniaxial stress along [111]. Similar stress experiments have been made by Müller and by Schawlow's group.[61] A complete phase diagram, shown in Fig. 7 is thereby obtained. A phenomenological theory including stress has been given in ref. 63, and in ref. 64. Pytte and Feder have given a microscopic theory which accounts for all details of the $SrTiO_3/LaAlO_3$ phase transitions *within the mean field approximation.*

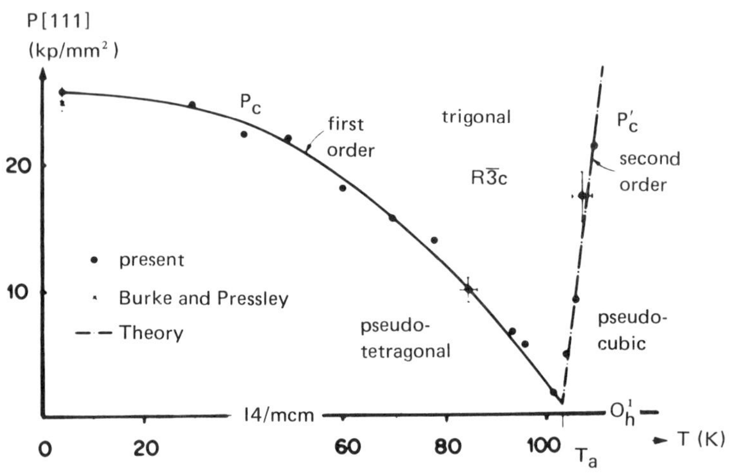

Fig. 7. Phase diagram for $SrTiO_3$ (ref. 61b).

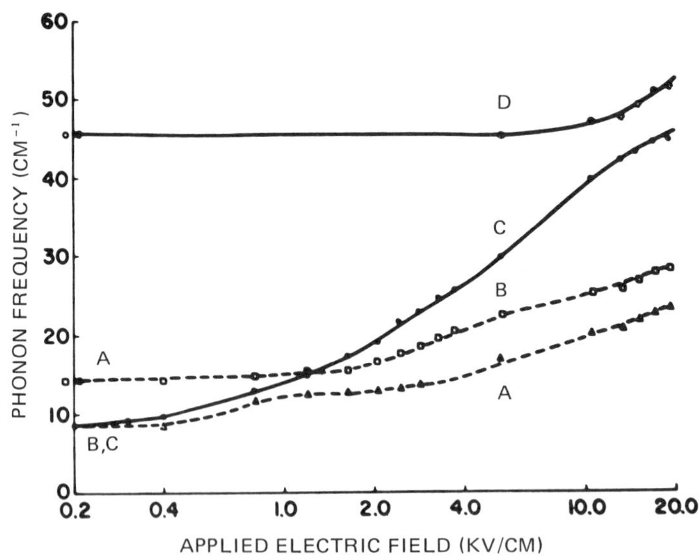

Fig. 8. Electric-field induced mode coupling in $SrTiO_3$ (ref. 65).

We should note that while $SrTiO_3$ has been driven trigonal by applying [111] stress, $LaAlO_3$ has *not* been driven tetragonal. The practical difficulty is that one would have to *pull* $LaAlO_3$, not squeeze it.

In the presence of an applied electric field the even and odd parity soft modes in $SrTiO_3$ couple. The simplest case is for $\hat{E}$ along the tetragonal axis. In this case A_{1g} and A_{2u} modes couple, and E_g and E_u couple, but no A/E coupling occurs. This is shown in Fig. 8 for $T \cong 10$ K. This is a very simple case of mode coupling in which no lineshape anomalies are manifest. The algebra of coupled harmonic oscillators is satisfactory, i.e.

$$\omega_\pm = \tfrac{1}{2}(\omega_1 + \omega_2) \pm \tfrac{1}{2}[(\omega_1 - \omega_2)^2 + 4W^2]^{1/2} \tag{28}$$

as discussed in ref. 65. More complex mode couplings are discussed in the KDP and $AlPO_4$ sections.

Critical phenomena in $SrTiO_3$ near T_0 = 106 K. Recently Müller and Berlinger[66] demonstrated unambiguously that over a temperature range of about 10 K (or 10% of T_0), the order-parameter in $SrTiO_3$, which is the angle of rotation of oxygen octahedra, varied not as

$$\phi = \phi_0(T_0 - T)^{1/2} \tag{29}$$

but as

$$\phi = \phi_0(T_0 - T)^{1/3} \tag{30}$$

In relating this observation to similar $\beta \approx 1/3$ values observed in magnetism and fluid phase transitions, the authors have placed solid state displacive phase transitions under the same 'universality' framework[67] which categorizes magnetic and liquid–gas phase changes. In this sense, Müller has made soft mode phase transitions respectable.

Raman experiments which reveal $\beta \approx 1/3$ exponents have been few. The accuracy of measurement over several decades of $(T_0 - T)$ is generally lacking. In $LaAlO_3$, extrapolation of $\omega = \omega_0(T_0 - T)^{1/2}$ curves to temperatures very near T_0 predicts[58] the same result as $\phi = \phi_0(T_0 - T)^{1/2}$ – namely, each yields[66] T_0 too high by about 20 K since the actual dependence of ϕ upon T is more rapid than predicted by $\beta = ½$. However, in $SrTiO_3$ careful measurement by Worlock and Olson[68] have shown that $\omega = \omega_0(T_0 - T)^{1/2}$ holds within 5 K of T_0. Thus it appears that ω is *not* generally proportional to the order parameter ϕ within the critical regime.

Two other measurements of β in soft mode expressions $\omega = \omega_0(T_0 - T)^\beta$ have been published. The first was by Scott,[69] who found $\beta = 0.31$ in quartz over a wide temperature range (~ 100 K). An independent measurement by Höchli and Scott[13] confirmed that the order parameter – an SiO_2 rotation angle – also varied as

$$\phi = \phi_0(T_0 - T)^{0.33} \tag{31}$$

over this range. And the value of the temperature range determined agrees well with Czaja's calculation[70] using the Ginzburg criterion[71]

$$\left(\frac{T_0 - T}{T_0}\right)_{\text{crit}} = \left(\frac{k}{\rho \Delta C_p}\right)^2 \frac{1}{l^6} \tag{32}$$

where ΔC_p is the specific heat jump at T_0 in the classical description, and l is the coherence length at absolute zero. The value of l obtained by Czaja is one unit cell length in quartz. The value of l obtained in $SrTiO_3$ from Eqn. (32) by Müller[72] is also one or two unit cell lengths. Thus we conclude physically that critical phenomena may be expected over a measurable temperature range (10–100 K) in systems with instabilities in short-range forces. It should be noted that neither the quartz transition nor the 106 K transition in $SrTiO_3$ involves ferroelectricity or long-range Coulomb forces. Consequently the report that ferroelectric SbSI exhibits $\beta = 1/3$ for its soft mode [73] must be viewed with surprise. Since this value has been obtained by only one of three independent groups [74, 75] working on SbSI, it must be viewed with some reservation yet.

In summary, values of $\beta = 1/3$ have been observed over wide temperature regions in solid state displacive phase transitions, but whether ω remains proportional to the order parameter in the critical temperature regime is an open question.

Central modes in $SrTiO_3$. The observation of a peak in the scattering spectra of $SrTiO_3$, centred at $\omega = 0$, near $T = T_0 = 106$ K was first made by Riste *et al.* via neutron scattering.[15] Raman studies of $SrTiO_3$ yield some indication of such a background[68] but nothing quantitative. A central mode has been reported by Steigmeier *et al.*[73] in SbSI but not yet confirmed by other independent publications on SbSI.[74, 75] While this phenomenon has a similarity to 'critical opalescence' in fluids, we wish to indicate that it follows from Cowley's anharmonic theory[16] and is not necessarily 'critical'.

Cowley has observed[16] that some three-phonon processes in which a soft optic mode emits and absorbs acoustic phonons cannot be treated by perturbation theory. (Recently, Silberglitt[76] has treated the four-phonon case with special application to $SrTiO_3$.) When these processes are included correctly, in the limit $\omega \to 0$, they yield a modification to the response function of the crystal. Instead of the usual expression

$$G(\omega) = [\omega_h^2 - \omega^2 + i\Gamma\omega]^{-1} \tag{33}$$

an expression

$$G(\omega) = \left[\omega_h^2 - \omega^2 + i\Gamma\omega + \frac{\alpha T}{1 + i\omega\tau}\right]^{-1} \tag{34}$$

is obtained. When the Raman spectrum is calculated from such a response function, it is found that an anomalous 'central mode' is present, whose intensity varies as $T \to T_0$. In addition, it follows that the soft mode frequency in such a system may obey a Curie–Weiss law

$$\omega^2 = \omega_0^2(T_0' - T) \tag{35}$$

but $T_0' \neq T_0$, the actual transition temperature. Further, the Lyddane–Sachs–Teller relation[37] will fail, since the dielectric constant $\epsilon(0)$ is measured at frequencies less than 10^{10} Hz, while the Raman response (or infrared) is measured at frequencies greater than 10^{11} Hz. The value 10^{11} Hz is a typical acoustic phonon lifetime τ and appears in the denominator in Eqn. (34). Thus Cowley's theory distinguishes crystal response on two time scales and is similar to the distinction in acoustic phonon theory between 'collisionless' and 'collision-dominated' regimes. All of the effects arising from Eqn. (34) above have been reported experimentally in ref. 17 for KDP structures.

Thus, the central mode is viewed as part of the response of the crystal near T_0, which, like the soft mode, arises from anharmonic effects.

$DyVO_4$: electronic soft modes. Dysprosium vanadate illustrates a class of Jahn–Teller systems in which the soft mode is electronic and not vibrational. $DyVO_4$, $DyAsO_4$ and $TbVO_4$ have all been studied by Harley, Hayes and Smith[77] at Oxford, and some of their data are reproduced in Fig. 9. A theory of electron–phonon interactions appropriate to

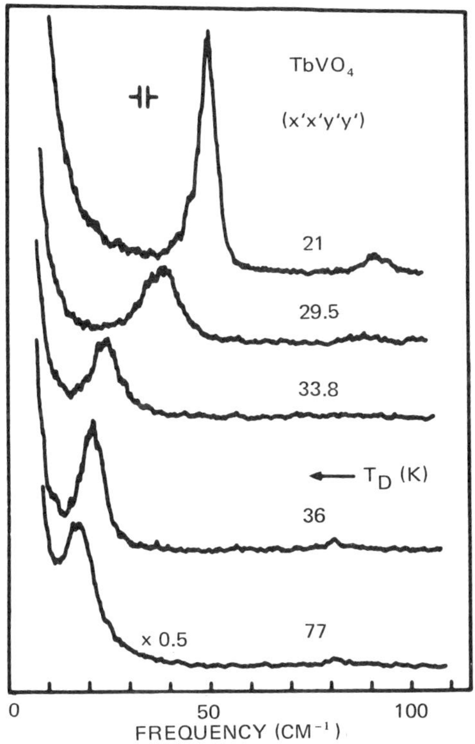

Fig. 9. Soft electronic mode (ref. 77).

$DyVO_4$ has been sketched by Elliott.[78] These studies will not be reviewed in detail here but serve to show that the order parameter or soft mode may in general be a phonon-like displacement, an electronic or even a magnetic parameter,[79] with the structural distortion in the latter cases arising from electron–phonon or magnon–phonon coupling.

$Gd_2(MoO_4)_3$ and $Tb_2(MoO_4)_3$. Gadolinium molybdate and its isomorph terbium molybdate are the first crystals to be studied in which a soft phonon at the zone boundary couples through internal strain to a zone centre acoustic phonon. The resulting 'antiferroelectrically-induced ferroelectricity' was explained theoretically by Pytte[80] shortly after weak ferroelectricity in $Gd_2(MoO_4)_3$ was reported by Cross *et al.*[81] Pytte's theory was confirmed by Axe *et al.* via inelastic neutron scattering[82] on $Tb_2(MoO_4)_3$. (Gd compounds cannot be used for neutron scattering, since Gd is an absorber.)

More recently, Cochran has provided[32] a simple model for a diatomic crystal which undergoes the same zone boundary optic phonon instability and induced ferroelectricity. It is important to note that Cochran's theory violates one of Landau's original requirements: that only a single order parameter be involved at T_0. It follows below that only one order parameter is explicitly temperature dependent in gadolinium molybdate-type systems, and this is the displacement which transforms like a zone boundary phonon; however, a second order parameter is coupled to the first via cross terms in the free energy. A temperature dependence is thereby induced in the second order parameter, which is the spontaneous polarization in the case of $Gd_2(MoO_4)_3$. Cochran assumes (to lowest order)

$$F = \tfrac{1}{2}A_1\eta_1^2 + \tfrac{1}{4}C_1\eta_1^4 + \tfrac{1}{2}A_2\eta_2^2 + H\eta_1^2\eta_2 \tag{36}$$

with $A_1 = a(T - T_0)$ as before and the other coefficients independent of T. Thus

$$0 = \frac{\partial F}{\partial \eta_2} = A_2\eta_2 + H\eta_1^2 = 0 \tag{37}$$

or

$$\eta_2 = -\frac{H}{A_2}\eta_1^2 \tag{38}$$

whence, substituting back

$$F = \tfrac{1}{2}A_1\eta_1^2 + \left(\tfrac{1}{4}C_1 - \tfrac{1}{2}\frac{H^2}{A_2}\right)\eta_1^4 \tag{39}$$

As we saw in the theoretical section, this free energy yields a phase transition with

$$\eta_1 = \mathrm{d}|T_0 - T|^{1/2} \tag{40}$$

(provided the coefficient of η_1^4 is positive). Thus, from Eqns. (38) and (40), we have

$$\eta_2 = \mathrm{d}'|T_0 - T| \tag{41}$$

i.e. a spontaneous polarization (η_2) is induced by a soft zone boundary optic mode displacement (η_1).

$NaNO_2$. Sodium nitrite is an excellent example of an order–disorder phase transition which has been interpreted by neutron scattering.[83] Above T_0, the NO_2 ions, which are V-shaped, flop slowly (very slowly!) back-and-forth between two possible orientations. Because this is a slow process involving a high potential barrier, it is a good case of quantum mechanical tunneling – better in fact, than the case of KH_2PO_4 which has so often been discussed with a tunneling formalism.

One of the most important things about $NaNO_2$ is that the very low frequency soft mode is 'extra', i.e. in addition to those predicted group theoretically (all of which have been studied and had their dispersion measured). This arises from the fact that group theory applies *only* to infinitesimally small displacements. Anharmonic oscillators can have two response peaks, one a narrow Lorentzian at frequency ω_0, and another broad response at very low frequencies, as long as large-amplitude motion is allowed.

Because the 'soft mode' or relaxation frequencies in $NaNO_2$ are so low (probably 10^8–10^{10} Hz or less) Raman spectroscopy is a waste of time and self-beating experiments should be attempted.

NH_4Cl and NH_4Br: order–disorder transitions. All of the crystals discussed so far except $NaNO_2$ are ordered in both high- and low-symmetry phases. What about true order–disorder transitions? Raman spectroscopy is a valuable tool for these systems as well,[84] *but* *there is no soft mode*, no optic phonon whose frequency tends toward zero as $T \rightarrow T_0$. In the ammonium halides the disorder involves rotational positions of the NH_4^+ ions. In the case of NH_4Cl the unit cell does not double in the ordered phase. Thus, it is directly analogous to $NaNO_2$ discussed above. Any 'soft mode' is 'extra', group theoretically, and lies at very low frequencies. Whether such 'modes' even exist is a semantic problem in NH_4Cl because that would imply a single relaxation time, which is not the case, experimentally.[85] Thus, Raman spectroscopy of NH_4Cl is not a good probe of the transition dynamics.

In NH_4Br, there are three ordered structures. One of the transitions involves a unit cell doubling in which the doubled cell has NH_4^+ ions randomly occupying two equivalent positions in the CsCl structure. In changing to an ordered tetragonal structure with *one* NH_4^+ ion per cell, the crystal halves its unit cell and doubles its Brillouin zone. No soft modes have been directly observed, but in analogue with $NaNO_2$, it should exhibit a very low frequency, temperature-dependent response – but at the zone boundary! Perhaps this could be studied by beating techniques in the disordered phase. It has not yet been analysed directly.

Raman spectroscopy does, however, yield direct information about the order parameter in NH_4Br. A measure of Raman *intensity* for modes described as zone boundary in the tetragonal phase (and nominally forbidden) is a measure of the degree of long-range disorder. This has been discussed in detail by Wang and Fleury.[84] It is shown that the 'forbidden' intensity correlates semiqualitatively with specific heat data, thus connecting microscopic and macroscopic parameters.

KH_2PO_4: Hydrogen-bonded ferroelectrics. The original discovery by Raman spectroscopy of a soft mode in KDP was of course in the well-known work by Kaminow and Damen.[86] They found an overdamped mode of B_2 symmetry in the paraelectric phase having a peak at $\omega = 0$, a width Γ of about 100 cm^{-1}, and a temperature-dependent harmonic frequency ω_h. Subsequently, many KDP isomorphs have been studied by many different groups and several reviews have been published.

A heavily damped soft mode now appears to be characteristic of most hydrogen bonded ferroelectrics. In fact, some of the most beautiful data I have seen are for *yy* scattering in thiourea in an unpublished 1970 thesis from Professor B. Schrader's group in Dortmund. Let me confine myself to what is new in KDP.

Firstly, the soft mode is rather complicated, since the K and P ions undergo an order–order transition at T_0, whereas the protons become disordered. (See Cochran[20] for a thorough review.) The new things in KDP structures are:

1. KDP also has an overdamped 'mode' of E-symmetry in the paraelectric phase.[87] This mode does not seem to be characterized by damped harmonic oscillator parameters.
2. This E-mode interacts strongly with the E (TO) phonon at 95 cm^{-1} yielding a line-shape asymmetry.[29]
3. ADP has both E- and B_2-symmetry[88] overdamped modes and these also interact with phonons.
4. When the phonon interactions are sorted out[28] it is found that Kaminow's temperature dependence for the soft mode in KDP is not correct. The uncoupled soft mode has a temperature dependence given by

$$\omega^2 = \omega_0^2 |T_0 - T| \quad \text{with} \quad \epsilon_0(T) = C|T_c - T|^{-1}$$

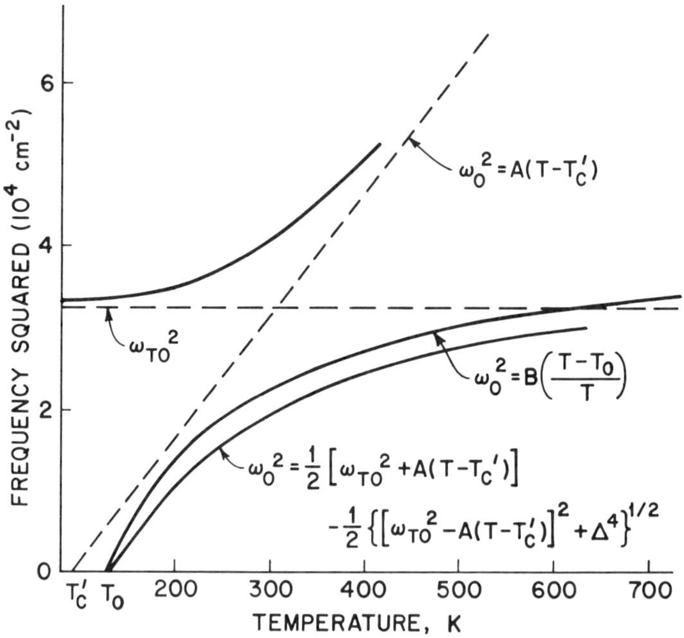

Fig. 10. ω_0 versus T in KDP (schematic) (ref. 29).

and

$$T_0 \neq T_c.$$

One possible explanation (Cowley's anharmonic mode-coupling theory[16]) for this $T_0 \neq T_c$ anomaly (which violates the Lyddane–Sachs–Teller relation) was mentioned in the beginning of this chapter. It is also possible that the linear, harmonic mode coupling model first used in KDP by Scott and Wilson[29] suffices to explain the $T_0 \neq T_c$ anomaly. This model was invoked to explain lineshapes. Following the theory of Maradudin and Fein[89] we wrote the Raman spectrum as

$$S(\omega) = [n+1]\ X''(\omega) \tag{42}$$

where n is the Bose population and $X(\omega)$ the susceptibility

$$X(\omega) = \sum_{ij} P_i P_j G_{ij}(\omega) \tag{43}$$

with $G_{ij}^{-1}(\omega)$ given as

$$G^{-1}(\omega) = \begin{bmatrix} \omega_1^2 - \omega^2 + i\Gamma_1\omega & \Delta^2 + i\Gamma_{12}\omega \\ \Delta^2 + i\Gamma_{12}\omega & \omega_2^2 - \omega^2 + i\Gamma_2\omega \end{bmatrix} \tag{44}$$

The interaction term Δ corresponds to a direct coupling. Γ_{12} corresponds to two modes coupling via the decay of each into a third state. By fitting data at several temperatures and requiring that all parameters have simple (or zero) temperature dependences, we were able to establish that in KDP structures the soft mode coupling with the lowest TO phonon has $\Delta \approx 100$ cm^{-1} and $\Gamma_{12} \approx 0$.

Figure 10 shows diagrammatically that this model predicts that the value T_0 extrapolated from the uncoupled soft mode frequency $\omega(T)$ will not equal T_c, which for a second-order transition, is just the transition temperature. Thus, both Cowley's nonlinear phonon coupling between a soft mode and two acoustic phonons, and our linear optic phonon model, predict $T_0 \neq T_c$. The high pressure studies of Peercy and Samara on KDP should clarify the situation.

Quartz and $AlPO_4$. SiO_2 was the first soft mode study.[1] SiO_2 and isomorphic $AlPO_4$ have beer analysed by the author in several papers.[13, 24, 26] The novel feature of these crystals is that the soft mode couples strongly with a two-acoustic-phonon state and produces a double-peaked spectrum. Ruvalds and Zawadowski have provided[25, 27] much of the theoretical analysis and have recently shown in unpublished work that the two-acoustic-phonon state must be a *bound* state, if the spectra are to be quantitatively interpreted. Such phonon bound states were first analysed by Cohen and Ruvalds.[90]

It is also of special interest that the uncoupled soft mode in quartz appears to exhibit the dependence

$$\omega = \omega_0 |T_0 - T|^{0.3}$$

CONCLUSION

We have shown that Raman spectroscopy and allied techniques have taken us from a macroscopic description of phase transitions, which in the case of ferroelectrics means specifying the macroscopic spontaneous polarization as the order parameter and discussing free energies by expansions about T_0, to a microscopic description in which the order parameter is a specific ionic displacement – often a rotation – and has soft mode frequency which is proportional to the magnitude of the displacement. Three new topics are not fully explained yet: central modes centred at $\omega = 0$ in the spectral repsonse at temperatures near T_0; the onset of critical phenomena, particularly $\beta = 1/3$ exponents in soft mode frequency expressions; and the coupled phase transitions involving phonon–phonon, phonon–electron, and phonon–magnon interactions in $Gd_2(MoO_4)_3$, $DyVO_4$, and UO_2.

ACKNOWLEDGEMENTS

The author wishes to thank the publishers of *The Physical Review, Physical Review Letters, Solid State Communications* and *Light Scattering Spectra of Solids*, as well as the authors of the cited references, for permission to reproduce some of the figures used in this review.

REFERENCES

1 C. V. Raman and T. M. K. Nedungadi, *Nature* **144**, 147 (1940).
2 W. Cochran, *Phys. Rev. Letters* **3**, 521 (1959); *Advan. Phys.* **9**, 387 (1960); *Advan. Phys.* **10**, 401 (1961).
3 A. F. Devonshire, *Phil. Mag.* **40**, 1040 (1949).

4 V. L. Ginzburg, *Usp. Fiz. Nauk* **38**, 490 (1949); *Zh. Eksperim. i. Teor. Fiz.* **19**, 36 (1949).

5 L. D. Landau, *Phys. Z. Sowjetunion* **11**, 26 (1937).

6 V. L. Ginzburg and A. P. Levanyuk, *J. Phys. Chem. Solids* **6**, 51 (1958); *Zh. Eksperim. i Teor. Fiz.* **39**, 192 (1960); *Usp. Fiz. Nauk.* **77**, 621 (1962).

7 T. Andrews, *Phil. Trans. Roy. Soc.* **159**, 575 (1969).

8 I. A. Iakovlev, T. S. Velichkina and L. F. Mikheeva, *Dokl. Akad. Nauk SSSR* **106**, 675 (1956); [translation, *Soviet Phys.-Dokl.* **1**, 91 (1956)]; *Kristallografiya* **1**, 91 (1956); I. A. Yakovlev and T. S. Velichkina, *Usp. Fiz Nauk.* **63**, 411 (1957), [translation, *Advan. Phys. Sci.* **63**, 552 (1957)].

9 R. W. Gammon, private communication cited in ref. 10.

10 S. M. Shapiro and H. Z. Cummins, *Phys. Rev. Letters* **21**, 1578 (1968).

11 U. T. Hochli, *Solid State Commun.* **8**, 1487 (1970).

12 V. K. Semenchenko and V. B. Baskakova, *Izv. Akad. Nauk SSSR* **5**, 766 (1969).

13 U. T. Hochli and J. F. Scott, *Phys. Rev. Letters* **26**, 1627 (1971).

14 G. Shirane and J. D. Axe, *Phys. Rev. Letters* **26**, 1803 (1971).

15 T. Riste, E. J. Samuelsen, K. Otnes and J. Feder, *Solid State Commun.* **9**, 1455 (1971).

16 R. A. Cowley, *J. Phys. Soc. Japan (Suppl).* **28**, S239 (1970).

17 R. A. Cowley, G. J. Coombs, R. S. Katiyar, J. F. Ryan and J. F. Scott, *J. Phys.* **C4**, L203 (1971).

18 E. F. Steigmeier, G. Harbeke and R. K. Wehner, *Light Scattering in Solids*, (Ed. M. Balkanski), Flammarion, Paris, 1971, p. 396.

19 K. A. Muller and W. Berlinger, *Phys. Rev. Letters* **26**, 13 (1971).

20 W. Cochran, *Advan. Phys.* **18**, 157 (1969).

21 J. M. Worlock, *Structural Phase Transitions and Soft Modes*, (Ed. E. J. Samuelsen *et al.*), Universitetsforlaget, Oslo, 1971, p. 329.

22 P. A. Fleury, *J. Acoust. Soc. Amer.* **49**, 1041 (1971).

23 A. S. Barker, Jr. and J. J. Hopfield, *Phys. Rev.* **135A**, 1732 (1964).

24 J. F. Scott, *Phys. Rev. Letters* **21**, 907 (1968).

25 J. Ruvalds and A. Zawadowski, *Phys. Rev.* **B2**, 1172 (1970).

26 J. F. Scott, *Phys. Rev. Letters* **24**, 1107 (1970).

27 A. Zawadowski and J. Ruvalds, *Phys. Rev. Letters* **24**, 1111 (1970).

28 R. S. Katiyar, J. F. Ryan and J. F. Scott, *Phys. Rev.* **B4**, 2635 (1971); *Light Scattering in Solids*, (Ed. M. Balkanski), Flammarion, Paris, 1971, p. 436.

29 J. F. Scott and C. M. Wilson, *Solid State Commun.* **10**, 597 (1972).

30 J. F. Scott, *Light Scattering in Solids*, (Ed. M. Balkanski), Flammarion, Paris, 1971. p. 387; J. F. Scott and R. S. Katiyar, *Ind. J. Pure Appl. Phys.* **9**, 950 (1971).

31 P. A. Fleury and P. D. Lazay, *Phys. Rev. Letters* **26**, 1331 (1971).

32 W. Cochran, *Structural Phase Transitions and Soft Modes*, (Ed. E. J. Samuelsen *et al.*), Universitetsforlaget, Oslo, 1971, p. 1.

33 H. E. Stanley, *Introduction to Phase Transitions and Critical Phenomena*, Oxford University Press, Oxford, 1971, especially Chapter 5.

34 P. Heller and G. B. Benedek, *Phys. Rev. Letters* **8**, 428 (1962).

35 P. Heller and G. B. Benedek, *Phys. Rev. Letters* **14**, 71 (1965).

36 For linear response theory applied to generalized soft modes, see H. Thomas, *Structural Phase Transitions and Soft Modes*, (Ed. E. J. Samuelsen *et al.*),

Universitetsforlaget, Oslo, 1971, p. 15.
37 R. H. Lyddane, R. G. Sachs and E. Teller, *Phys. Rev.* **59**, 673 (1941).
38 R. A. Cowley, *Advan. Phys.* **12**, 421 (1963).
39 W. Cochran, *Advan. Phys.* **9**, 387 (1960), Eqns. 1.12 and 1.13.
40 G. S. Pawley, W. Cochran, R. A. Cowley and G. Dolling, *Phys. Rev. Letters* **17**, 753 (1966).
41 N. S. Gillis, *Phys. Rev. Letters* **22**, 1251 (1969).
42 A. S. Barker, Jr., *Optical Properties of Solids* (Eds. S. Nudelman and S. S. Mitra), Plenum Press, New York, 1970.
43 E. F. Steigmeier and G. Harbeke, *Solid State Commun.* **8**, 1275 (1970).
44 J. M. Worlock, *Bull. Amer. Phys. Soc.* **14**, 368 (1969).
45 R. C. Miller and W. G. Spitzer, *Phys. Rev.* **129**, 94 (1963).
46 A. S. Barker, Jr., *Phys. Rev.* **145**, 391 (1966).
47 See H. H. Barrett, *Phys. Rev.* **178**, 743 (1969) and R. E. Nettleton, *Z. Physik* **220**, 401 (1969) for details of soft mode behaviour at low temperatures in perovskites.
48 P. A. Fleury and J. M. Worlock, *Phys. Rev. Letters* **18**, 665 (1967); *Phys. Rev. Letters* **19**, 1176 (1967); *Phys. Rev.* **174**, 613 (1968); J. F. Scott, P. A. Fleury and J. M. Worlock, *Phys. Rev.* **177**, 1288 (1969). This mode was first studied in detail in $SrTiO_3$ by R. A. Cowley, *Phys. Rev. Letters* **9**, 159 (1962); *Phys. Rev.* **134A**, 981 (1964).
49 G. Shirane, B. C. Frazer, V. J. Minkiewicz, J. A. Leake and A. Linz, *Phys. Rev. Letters* **19**, 234 (1967); M. DiDomenico, Jr., S. P. S. Porto, S. H. Wemple and R. P. Bauman, *Phys. Rev. Letters* **19**, 855 (1967); *Phys. Rev.* **174**, 522 (1968); P. D. Lazay and P. A. Fleury, *Light Scattering in Solids*, (Ed. M. Balkanski), Flammarion, Paris, 1971, p. 406.
50 G. Burns and B. A. Scott, *Phys. Rev. Letters* **25**, 167 (1970).
51 T. G. Davis, *J. Phys. Soc. Japan (Suppl.)* **28**, 245 (1970).
52 G. Burns and B. A. Scott, *Bull. Amer. Phys. Soc.* **16**, 415 (1971).
53 G. Burns and B. A. Scott, *Phys. Rev. Letters* **25**, 1191 (1970).
54 J. D. Axe, G. Shirane and K. A. Muller, *Phys. Rev.* **183**, 820 (1969).
55 B. Derighetti, J. E. Drumheller, F. Laves, K. A. Muller and F. Waldner, *Acta Cryst.* **18**, 557 (1965).
56 K. A. Muller, W. Berlinger and F. Waldner, *Phys. Rev. Letters* **21**, 814 (1968).
57 V. Plakhty and W. Cochran, *Phys. Stat. Sol.* **29**, K81 (1968).
58 J. F. Scott, *Phys. Rev.* **183**, 823 (1969).
59 J. F. Scott and J. P. Remeika, *Phys. Rev.* **B1**, 4182 (1970).
60 W. J. Burke and R. J. Pressley, *Solid State Commun.* **9**, 121 (1971).
61 (a) L. S. Wall, M. Rokni and A. L. Schawlow, *Solid State Commun.* **9**, 573 (1971). (b) K. A. Muller, W. Berlinger and J. C. Slonczewski, *Phys. Rev. Letters* **25**, 734 (1970).
62 H. Thomas and K. A. Muller, *Phys. Rev. Letters* **21**, 1256 (1968); H. Unoki and T. Sakudo, *J. Phys. Soc. Japan* **23**, 546 (1967); P. A. Fleury, J. F. Scott and J. M. Worlock, *Phys. Rev. Letters* **21**, 16 (1968).
63 J. C. Slonczewski and H. Thomas, *Phys. Rev.* **B1**, 3599 (1970); J. C. Slonczewski, *Phys. Rev.* **B2**, 4646 (1970).
64 E. Pytte and J. Feder, *Phys. Rev.* **187**, 1077 (1969); *Phys. Rev.* **B1**, 4803 (1970); see also J. Feder, *Structural Phase Transitions and Soft Modes* (Ed. E. J. Samuelsen

et al.) Universitetsforlaget, Oslo, 1971, p. 15.
65 J. M. Worlock, J. F. Scott and P. A. Fleury, *Light Scattering in Solids*, (Ed. M. Balkanski), Flammarion, Paris, 1971, p. 689.
66 K. A. Muller and W. Berlinger, *Phys. Rev. Letters* **26**, 13 (1971).
67 See for example, L. P. Kadanoff *et al.*, *Rev. Mod. Phys.* **39**, 395 (1967).
68 J. M. Worlock and D. H. Olson, *Light Scattering in Solids*, (Ed. M. Balkanski), Flammarion, Paris, 1971, p. 410.
69 J. F. Scott, *Phys. Rev. Letters* **16**, 907 (1968).
70 W. Czaja, *Phys. Kond. Mat.* **10**, 299 (1970).
71 V. L. Ginzburg, *Sov. Phys. Solid State* 2, 1824 (1960).
72 K. A. Muller, *Structural Phase Transitions and Soft Modes* (Ed. E. J. Samuelsen *et al.*), Universitetsforlaget, Oslo, 1971, p. 61.
73 E. F. Steigmeier, G. Harbeke and R. K. Wehner, *Structural Phase Transitions and Soft Modes*, (Ed. E. J. Samuelsen *et al.*), Universitetsforlaget, Oslo, 1971, p. 409; *Light Scattering in Solids*, (Ed. M. Balkanski), Flammarion, Paris, 1971, p. 396.
74 D. K. Agrawal and C. H. Perry, *Solid State Commun.* **8**, 225 (1970); *Light Scattering in Solids*, (Ed. M. Balkanski), Flammarion, Paris, 1971, p. 401.
75 M. Balkanski, M. K. Teng, M. Massot and S. M. Shapiro, *Light Scattering in Solids*, (Ed. M. Balkanski), Flammarion, Paris, 1971, p. 392.
76 R. Silberglitt, *Solid State Commun.* 1972, (to be published).
77 R. T. Harley, W. Hayes and S. R. P. Smith, *Solid State Commun.* **9**, 515 (1971).
78 R. J. Elliott, *Light Scattering in Solids*, (Ed. M. Balkanski), Flammarion, Paris, 1971, p. 354.
79 S. J. Allen, *Phys. Rev.* **166**, 530 (1968).
80 E. Pytte, *Solid State Commun.* **8**, 2101 (1970).
81 L. E. Cross, A. Fouskova and S. E. Cummins, *Phys. Rev. Letters* **21**, 812 (1968).
82 J. D. Axe, B. Dorner and G. Shirane, *Phys. Rev. Letters* **26**, 519 (1971).
83 J. Sakurai, R. A. Cowley and G. Dolling, *J. Phys. Soc. Japan* **28**, 1426 (1970), (extensive references are given).
84 C. H. Wang and P. A. Fleury, *Light Scattering in Solids*, (Ed. M. Balkanski), Flammarion, Paris, 1971, p. 651; C. H. Wang, *Phys. Rev. Letters* **26**, 1226 (1971).
85 J. H. Lunacek, *Ph. D. Thesis,* MIT, 1970.
86 I. P. Kaminow and T. C. Damen, *Phys. Rev. Letters* **20**, 1105 (1968).
87 B. Lavrencic, I. Levstek, B. Zeks, R. Blinc and D. Hadzi, *Chem. Phys. Letters* **5**, 441 (1970).
88 J. F. Ryan, R. S. Katiyar and W. Taylor, *J. Phys. (Paris), Suppl.* **C2**, 49 (1972).
89 A. A. Maradudin and A. E. Fein, *Phys. Rev.* **128**, 2589 (1962).
90 M. H. Cohen and J. Ruvalds, *Phys. Rev. Letters* **23**, 1378 (1969).

16. RAMAN SCATTERING OF POLARITONS IN UNIAXIAL AND BIAXIAL PIËZOELECTRIC CRYSTALS*

L. Graf, G. Schaack and B. Unger
Institut für technische Physik
der Technische Hochschule Darmstadt, Germany

In this contribution we report on some observations on the wavevector dependence of the frequencies and of the Raman scattering intensities of polar phonons (respectively polaritons). It is well known since the work of Poulet and Mathieu[1] that the anomalies observed in piëzoelectric crystals, where polar phonons are also Raman active, are the effect of an electrical field in the crystal due to the polar excitation of the lattice. The magnitude and the direction of this field change with the direction of the phonon wavevector k. The observed frequency shifts are a consequence of the interaction of this field with the polar phonons and the intensity anomalies are also caused by this interaction and, in addition, by an interaction between this field and the exciting light wave via a nonlinear susceptibility of the crystal.

We have studied the polar vibrations around 1100 cm^{-1} in orthorhombic $MgSO_4.7H_2O(D_2^4)$, which have their origin in the ν_3 states of the SO_4-tetrahedrons. With k along one of the orthorhombic axes we observe either $3B_1$-, $3B_2$-, or $3B_3$-states, which are split into two components with transverse or longitudinal polarization of the phonon. In total we observe six lines in one orientation. We have studied the variation of frequency and of the scattering cross section with k rotating in the three planes of symmetry at $T = 300$ K and at $T = 1.9$ K with the crystal immersed in superfluid helium.

As an example the angular dispersion in the yz plane at 1.9 K is shown in Fig. 1. The branches are symmetrical with respect to the binary axes. This implies that the dispersion branches have a horizontal slope on the axes, with the exception of accidental degeneracies between transversely and longitudinally polarized states of different symmetry where the point of inflection coincides with the axis. No crossings occur for a general k.[2,3] This symmetry does not hold for the scattering intensities as is roughly indicated by the shapes of the experimental points in the figure which represent different intensities. The continuous lines in the figure are results of a computer calculation of the angular dispersion according to existing theories (e.g. ref. 4).

The practical importance of these investigations lies in the fact that the good agreement between the calculations and the experimental points is a proof of the correct assignment of the symmetry and polarization of the transitions for k parallel to a symmetry

* Part of a project of the Sonderforschungsbereich 'Festkörperspektroskopie' Darmstadt-Frankfurt, financed by special funds of the Deutsche Forschungsgemeinschaft.

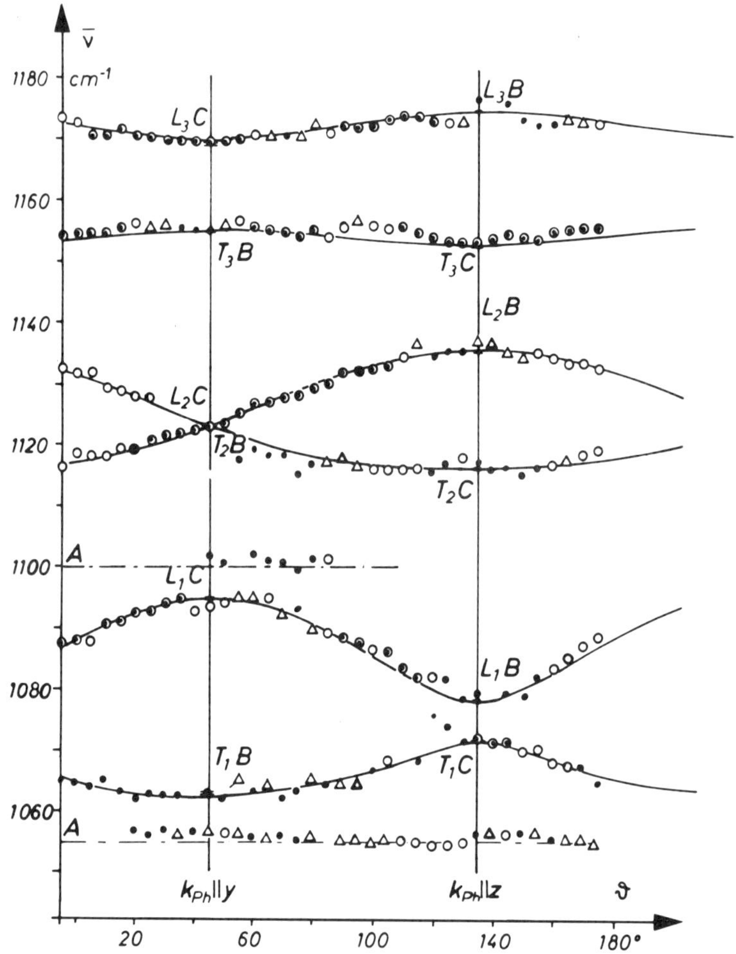

Fig. 1. Angular dispersion of polar phonons in the *yz* plane of $MgSO_4.7H_2O$ at $T = 1.9$ K. Full lines: calculated results; Intensities: double circles – very strong; open circles – strong; triangle – medium; full circles – weak.

axis. This check can be simplified by using two formulae, which we derived, for the curvature $d^2\omega/d\theta^2$ of a disperson branch on a symmetry axis (α) and the slope of two intersecting branches at an accidental degeneracy (β)

α) k in the $\alpha\beta$ plane, $k\|$ axis α:

$$\omega = \omega^T_{\beta,j}: \left(\frac{d^2\omega}{d\theta^2}\right)_{\omega^T_{\beta,j}} = \frac{4\pi}{\omega^T_{j,\beta}\cdot\epsilon^2_{\alpha,\infty}}\cdot(Z^T_{\beta,j})^2\cdot\epsilon^L_{\alpha}(\omega^T_{\beta,j}) \tag{1a}$$

$$\omega = \omega^L_{\alpha,i}: \left(\frac{d^2\omega}{d\theta^2}\right)_{\omega^L_{\alpha,i}} = \frac{-4\pi}{\omega^L_{\alpha,i}\cdot\epsilon^2_{\alpha,\infty}}\cdot(Z^L_{\alpha,i})^2\cdot\epsilon^T_{\beta}(\omega^L_{\alpha,i}) \tag{1b}$$

β)

$$\left(\frac{d\omega}{d\theta}\right)_{\omega^T_{\beta,j}=\omega^L_{\alpha,i}} = \pm\frac{2\pi}{\epsilon_{\alpha,\infty}\cdot\omega^L_{\alpha,i}}\cdot(Z^L_{\alpha,i})\cdot(Z^T_{\beta,j}). \tag{2}$$

Table 1.

	Calculated	Experimental	Dimension
$\left[\frac{d^2\omega}{d\theta^2}\right]_{\omega^L = 1095\,cm^{-1}}$	$-8.19\cdot10^{-3}$	$(-9.5 \pm 2)\cdot10^{-3}$	$cm^{-1}\ deg^{-2}$
$\left[\frac{d\omega}{d\theta}\right]_{\omega = 1123\,cm^{-1}}$	$\pm\ 0.188$	$\pm\ (0.15 \pm 0.03)$	$cm^{-1}\ deg^{-1}$

The following abbreviations have been used for the effective charges Z of branches i, j at frequencies $\omega^T_{\beta,j}$ or $\omega^L_{\alpha,i}$ (transverse or longitudinal, dipole moment parallel α or β, total number of polar transitions:N_α, N_β)

$$(Z^T_{\beta,j})^2 = \frac{\epsilon_{\beta,\infty}}{4\pi} \cdot \frac{\prod_k^{N_\beta} (\omega^{2L}_{\beta,k} - \omega^{2T}_{\beta,j})}{\prod_{k \neq j}^{N_\beta} (\omega^{2T}_{\beta,k} - \omega^{2T}_{\beta,j})}$$

$$(Z^L_{\alpha,i})^2 = \frac{\epsilon_{\alpha,\infty}}{4\pi} \cdot \frac{\prod_k^{N_\alpha} (\omega^{2L}_{\alpha,i} - \omega^{2T}_{\alpha,k})}{\prod_{k \neq i}^{N_\alpha} (\omega^{2L}_{\alpha,i} - \omega^{2L}_{\alpha,k})}$$

with $\epsilon_{\alpha,\infty}$ the high frequency d.c. in α-direction. The dielectric functions $\epsilon(\omega)$ are

$$\frac{\epsilon^T_\beta(\omega)}{\epsilon_{\beta,\infty}} = \prod_k^{N_\beta} \frac{(\omega^2 - \omega^{2L}_{\beta,k})}{(\omega^2 - \omega^{2T}_{\beta,k})} = \frac{\epsilon_{\beta,\infty}}{\epsilon^L_\beta(\omega)}$$

Thus the knowledge of the phonon frequencies in the region of interest with k in a symmetry direction is sufficient for a first view on the dispersion of phonons with an oblique wavevector in uniaxial and in a plane of symmetry of orthorhombic crystals. In Table 1 some values for the curvatures and the slopes of dispersion branches taken from Fig. 1 are given and compared with calculated values from Eqns. (1) and (2).

In Fig. 2 some Raman spectra of this crystal in the ν_3-region are shown taken at $T = 300$ K and at $T = 1.9$ K. We see the lowering of the line widths and the corresponding gain in resolution at low temperatures and the rapid change of scattered intensity with the direction of k.

Figure 3 gives the experimental results of the intensity variation of some dispersion branches in the yz plane together with the theoretical interpretation (full lines). This interpretation starts with the scattering intensity of a pure transverse phonon Q_j. This intensity is solely determined by $(\partial\alpha/\partial Q_j)$, the derivative of the polarizability α with respect to the normal coordinate Q of vibration j. This contribution is due to the deformation of the lattice. For an oblique phonon, two effects happen:

1. For a general k vector, the eigenstate of the phonon has changed from its form for the pure transverse state; in general this is a mixture of all transverse polar phonons. This

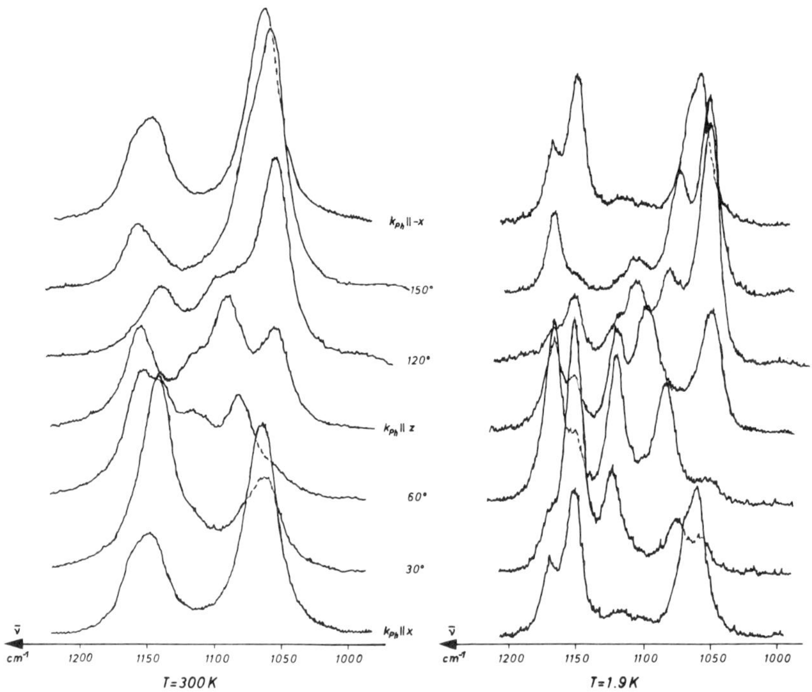

Fig. 2. Raman spectra of $MgSO_4 7H_2O$ in the x,z plane at two temperatures and different angles of the phonon wavevector with respect to the crystalline axes.

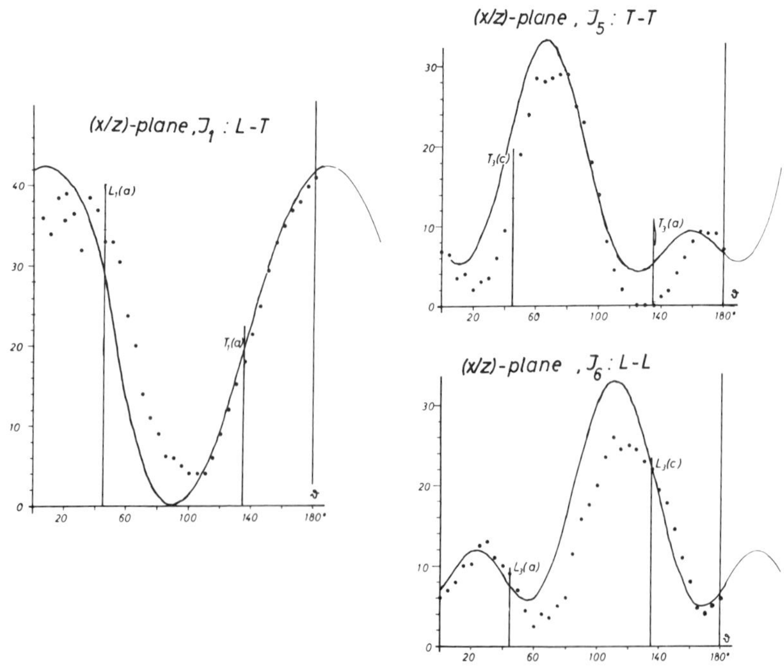

Fig. 3. Intensity variation of some dispersion branches in $MgSO_4.7H_2O$ in the x,z plane. Full lines: theoretical results.

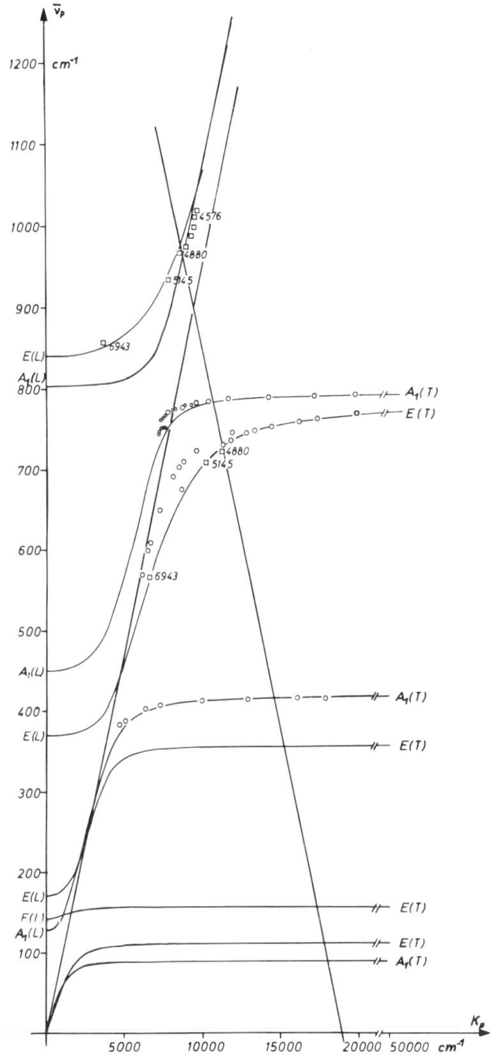

Fig. 4. Polariton dispersion ($k_p \approx \omega_p/c$) and angular dispersion ($k_p \gg \omega_p/c$) in $KBrO_3$ at $T = 300$ K. Full lines: calculated results.

mixing can be calculated with the theories on angular dispersion.[4] Consequently, the scattered intensity is a superposition of the deformation contributions ($\partial\alpha/\partial Q_j$) of all transverse branches.

2. For phonons with at least a partial longitudinal polarization the macroscopic electric field, which vanishes for transverse polarization, gives rise to another source of intensity due to the nonlinear susceptibility. The macroscopic electric field and the laser light mix and generate the Stokes radiation with the difference frequency of both waves. We have determined in relative units the size of the derivatives $\partial\alpha/\partial Q_j$ and the contribution of the macroscopic field via the nonlinear susceptibility d from the scattered intensity of each transverse and longitudinal phonon and have interpolated

the intensities of the oblique phonons, using the information provided by the theory of angular dispersion. The rather irregular behaviour of the intensity dependence is the result of a complicated interference between the different intensity contributions.

In Fig. 4 the angular dispersion of all polar phonons of $KBrO_3$ is shown. $KBrO_3$ is a uniaxial crystal of C_{3v}-symmetry, with one molecular unit per elementary cell, which has very interesting polar and nonlinear properties. The region below 200 cm^{-1} covers the lattice modes, above 350 cm^{-1} we find the internal modes of the bromate ion. The branches at 150 cm^{-1}, 350 cm^{-1}, 450 cm^{-1} connect the transverse with the corresponding longitudinal polarization of the same type of symmetry with only a small admixture of other modes. This mixing of states is large for branches which connect different types of symmetry with the same polarization (near 120 cm^{-1} and 800 cm^{-1}). Otherwise the branches would show a crosspoint, which is forbidden by symmetry and avoided by repelling of the two interfering branches. The branches near 800 cm^{-1} demonstrate this effect very clearly.

On the left of Fig. 4 the calculated and the observed frequencies in the polariton region are shown, where k has such a size that the electric field connected with the polar phonon can propagate as an electromagnetic wave. For k parallel to the optic axis we have two systems of polariton branches which arise from phonons of A_1- and E-symmetry and which may cross each other because the symmetry is preserved.

In the low frequency region there are no experimental points because of imperfect crystals. The two straight lines shown in the figure mark the smallest polariton k vectors, which can be generated for a given polariton frequency and laser wavelength due to the optical properties of the crystal. The line with the positive slope in the origin refers to the case where both the incident and scattered radiation belong to the ordinary or extraordinary wave; the line with the negative slope at $\omega = 0$ to the case where one beam is extraordinarily polarized; the other is ordinarily polarized. As the experimental points show, the proper choice of the wavelength of the exciting laser and of the polarization of the exciting and scattered light beams permit the studies of these polariton branches over a wide range of wave vectors and wavelengths.

Two features deserve further consideration: at 760 cm^{-1} we observe a splitting of the A_1-polariton branch. This splitting must be caused by a polar two-phonon state, which is coupled with the photon component of the polariton. This two-phonon excitation gains intensity by its coupling with the nearby one-phonon A_1-state and near the cross over of the polariton with the two-phonon-state we observe a characteristic exchange of intensity between the two branches and the two–phonon state. As a consequence we observe here a complicated excitation of the crystal, comprising a hybrid (one-phonon)–(two-phonon)-component and a photon component.

On the uppermost branch the phonon-like component of the polariton is decreasing with respect to the photon-like component with growing distance on the wavenumber axis from the E-phonon with highest energy. Following this branch with increasing k and ω, we observe a continuous transition from the classical Raman process to parametric fluorescence. We change from the field of lattice dynamics into the field of nonlinear optics. This change is also manifested by the fact that the intensity of this scattering phenomenon is depending more and more on the size of the nonlinear susceptibility d with increasing separation from the phonon branches. At present we are preparing a quantitative study of the polariton cross section and line shape in this frequency area.

REFERENCES

1 H. Poulet, *Ann. Phys. (Paris)* **10**, 908 (1955).

2 C. H. Hartwig, E. Wiener-Avnear, J. Smit and S.P.S. Porto, *Phys. Rev.* **B3**, 2078 (1971).

3 E. Anda, *Solid State Commun.* **9**, 1545 (1971).

4 B. Unger and G. Schaack, *Phys. Status Solidi* **48**, 285 (1971).

17. THE RAMAN SPECTRUM OF AMMONIUM BISULPHATE

J. W. Arthur, D. J. Lockwood and W. Taylor
Department of Physics, University of Edinburgh, Edinburgh, Scotland

INTRODUCTION

Ammonium bisulphate (NH_4HSO_4 or AHS) has three crystalline phases. At room temperature the space group is $P2_1/c$, but at 270 K there is a second-order phase transition to P_c and below this at 154 K there is a first-order transition to P_1. At the upper transition there is a sharp dielectric anomaly, below which the crystal becomes ferroelectric.[1] This anomaly, however, is much smaller than in the classic ferroelectrics KDP and $BaTiO_3$ where strong low frequency Raman scattering in the form of temperature dependent 'Rayleigh wings' has been observed[2] and interpreted variously in terms of an overdamped phonon, which in KDP is coupled with a hydrogen bond order-disorder mechanism.

The aim of the present work was to determine if scattering from a similar mechanism might be observed in AHS. It would appear that the NH_4^+ ions are not involved in the upper transition, since a similar transition exists in RHS ($RbHSO_4$) at 258 K; but there is evidence that they have a high degree of rotational freedom and take part in the lower transition.[3] Recent structural work on AHS by Nelmes[4] indicates large thermal parameters for one of the symmetrically inequivalent types of sulphate ion in the room temperature structure. Nelmes has fitted this to a model in which these ions are disordered between two orientations.[5] In passing to the ferroelectric phase the centre of inversion is lost, and the disordered ions become ordered.[6] It is possible that this ordering mechanism is associated with a ferroelectric mode as postulated by Cochran in his lattice dynamical theory of ferroelectrics.[7] It is difficult, however, to regard the sulphate order-disorder motion as a phonon, and it may be more appropriate to interpret the spectra in terms of molecular dipole reorientation.

The positions of the H^+ ions with respect to the SO_4^{2-} ions have not been determined in the ferroelectric phase, but it is clear that they are already ordered at room temperature.[4,5] In addition to this, deuteration does not appreciably affect the transition temperatures.[8] Although this seems to indicate that the hydrogens play no direct part in the transition, there may be some involvement, as suggested by infrared[9] and NMR[3] via H bonding between sulphate groups along the *b* axis.

Earlier Raman work on AHS has been carried out by Bazhulin *et al.*[10] using mercury lamp excitation, but no low frequency or extensive polarization data were reported.

GROUP THEORY

AHS has eight formula units in the primitive cell giving 264 normal modes at $q = 0$. Considering the internal molecular vibrations of NH_4^+ and HSO_4^- separately, there are 93 external lattice modes, excluding acoustic modes. In the room temperature phase the modes are classified according to the irreducible representations of the point group $C_{2h}(2/m)$

$$\begin{array}{ll} \text{Translatory modes:} & 12A_g + 12B_g + 11A_u + 10B_u \\ \text{Rotatory modes:} & 12A_g + 12B_g + 12A_u + 12B_u \end{array}$$

The remaining internal modes transform as:

$$42A_g + 42B_g + 42A_u + 42B_u$$

The 'g' modes are Raman active, whereas the 'u' modes are infrared active. If the Y axis is oriented along the crystallographic b axis, the Raman tensor has the form:

$$\begin{pmatrix} a & & d \\ & b & \\ d & & c \end{pmatrix} \sim A_g \qquad \begin{pmatrix} & e & \\ e & & f \\ & f & \end{pmatrix} \sim B_g$$

For infrared polarizations, $P_y \sim A_u$, and $P_x, P_z \sim B_u$.

In the ferroelectric phase, the modes transform under the irreducible representations of C_s, and the compatibility relations are

$$\begin{array}{l} C_{2h}(A_g, B_u) \longrightarrow C_s(A') \\ C_{2h}(A_u, B_g) \longrightarrow C_s(A'') \end{array}$$

The centre of inversion is no longer present and all modes are simultaneously Raman and infrared active.

The 9 internal modes of the NH_4^+ ion and the 12 internal modes of the HSO_4^- ion give rise to 72 and 96 modes respectively in the crystal. In the room temperature phase these modes transform as

$$\begin{array}{ll} NH_4^+: & 18A_g + 18B_g + 18A_u + 18B_u \\ HSO_4^-: & 24A_g + 24B_g + 24A_u + 24B_g \end{array}$$

The compatibility relations can be applied to give the results for the ferroelectric phase.

A rotatory motion of the SO_4^{2-} ions, as associated with the upper phase transition, transforms as B_u above, and A' below the transition. It is therefore to be expected that a related soft ferroelectric mode, if it exists, would appear in the Raman spectrum only below the transition in the *xx, yy, zz* or *xz* polarizations, unless there is a relaxing of selection rules above the transition caused by the disorder situation in the lattice.

RESULTS AND DISCUSSION

The Raman spectra of AHS were recorded using conventional techniques. The

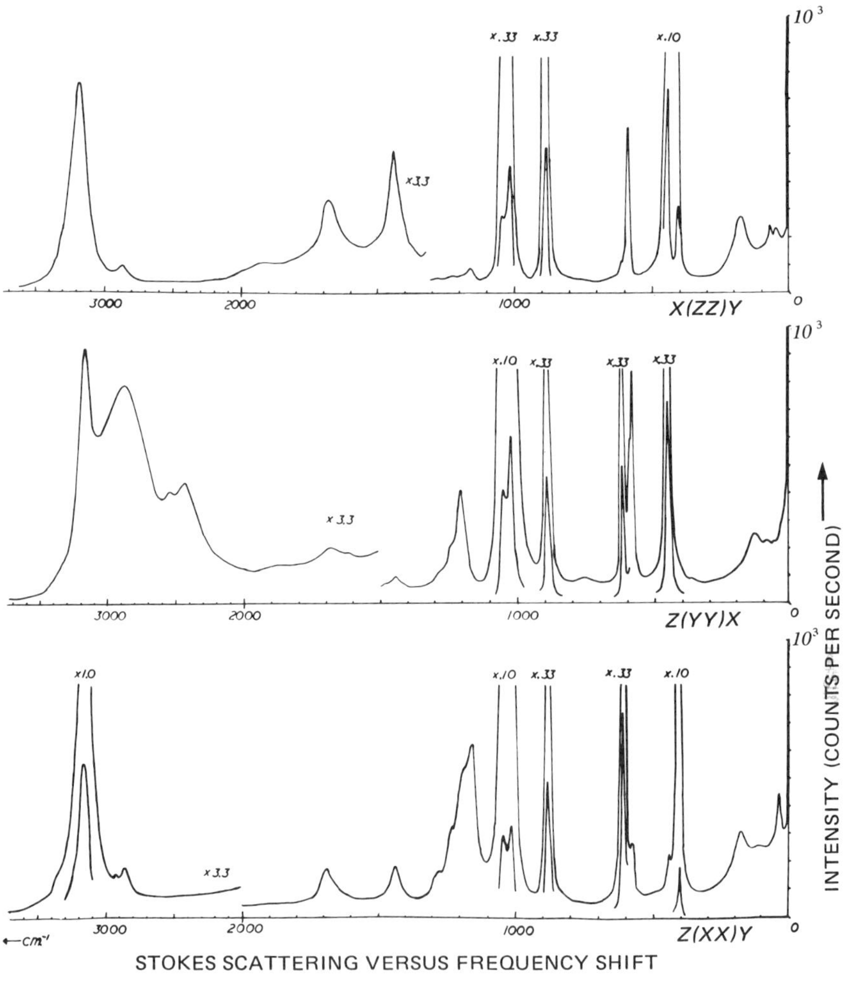

Fig. 1. Stokes scattering at room temperature.

excitation used was the 488 nm line of the argon-ion laser, and the spectrometer slit width was about 3 cm^{-1}. The sample was a single crystal of approximate dimensions 4 mm × 5 mm × 6 mm, with *X, Y* and *Z* directed along the *a, b* and *c* axes of the pseudo-orthorhombic cell. The spectra are displayed in Figs. 1–3, and the frequencies and assignments are presented in Tables 1 and 2. All the internal frequencies of the SO_4^{2-} ion reported by Bazhulin *et al.*[10] were observed, but the frequencies obtained here are higher by as much as 9 cm^{-1}. The assignments were based on free-ion frequencies and the results of other works.[9,11]

Although group theory predicts a doubling in the number of Raman active lattice modes, no significant change takes place on passing through the 270 K transition. It is possible that, as in $BaTiO_3$, the disordered nature of the lattice breaks the expected symmetry requirements.[12] On the other hand, comparison of the Raman spectra with the infrared spectra of Schutte and van Rensburg[9] shows that the frequencies of the internal vibrations are apparently exclusive, still consistent with a centre of inversion. In

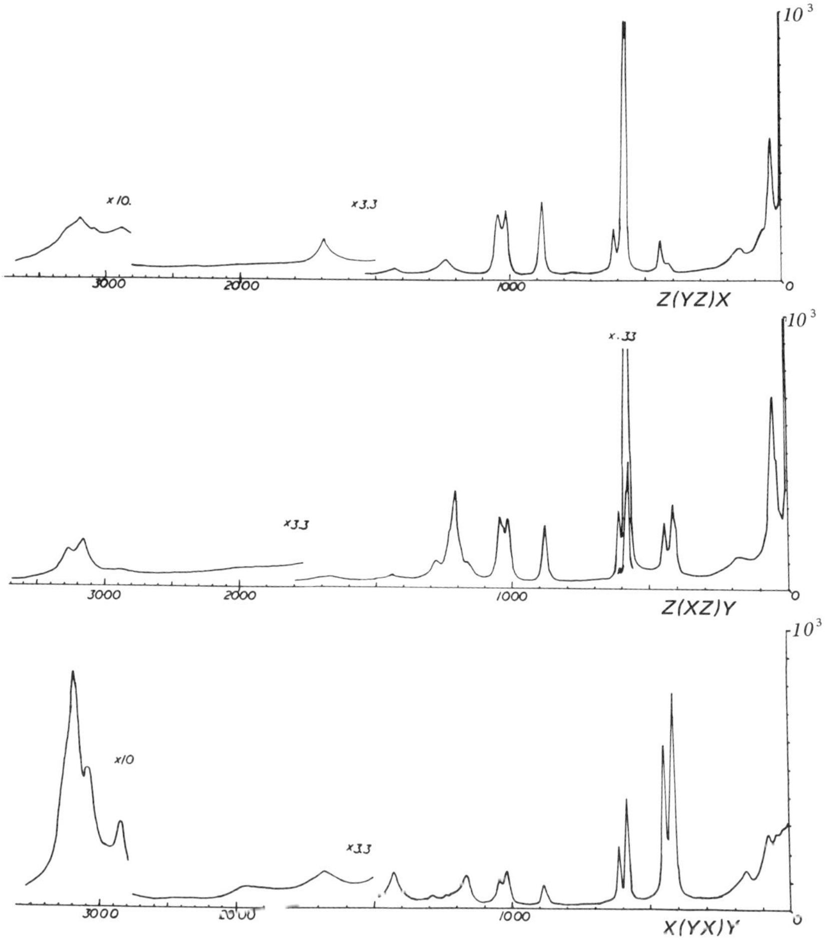

Fig. 2. Stokes scattering at room temperature.

the ferroelectric phase, the deviation of the structure from one having a centre of inversion is quite small, suggesting that most of the modes which are infrared active at room temperature become only weakly Raman active in the ferroelectric phase and are not easily seen.

In the low frequency region, principal features of interest are the prominent 'Rayleigh wings', especially obvious in the *yy* and *xy* spectra. These Rayleigh wings exhibit some temperature dependence; between 290 K and 170 K the *xy* wing loses intensity in an approximately linear fashion, but, perhaps more interestingly, the *yy* wing appears to change its profile, losing intensity and simultaneously revealing a small temperature dependent peak around 40 cm^{-1}. The *yy* spectrum is compatible with modes of A' symmetry, this being the expected symmetry of a ferroelectric mode involving SO_4^{2-} rotatory motions as mentioned above. It is possible, however, that the Rayleigh wings result from nearly-free NH_4^+ rotations, but some distortion of the ion from T_d symmetry would then be necessary to allow Raman activity. It is not clear that these wings can be accounted for by a soft mode at the upper transition, particularly as they change little

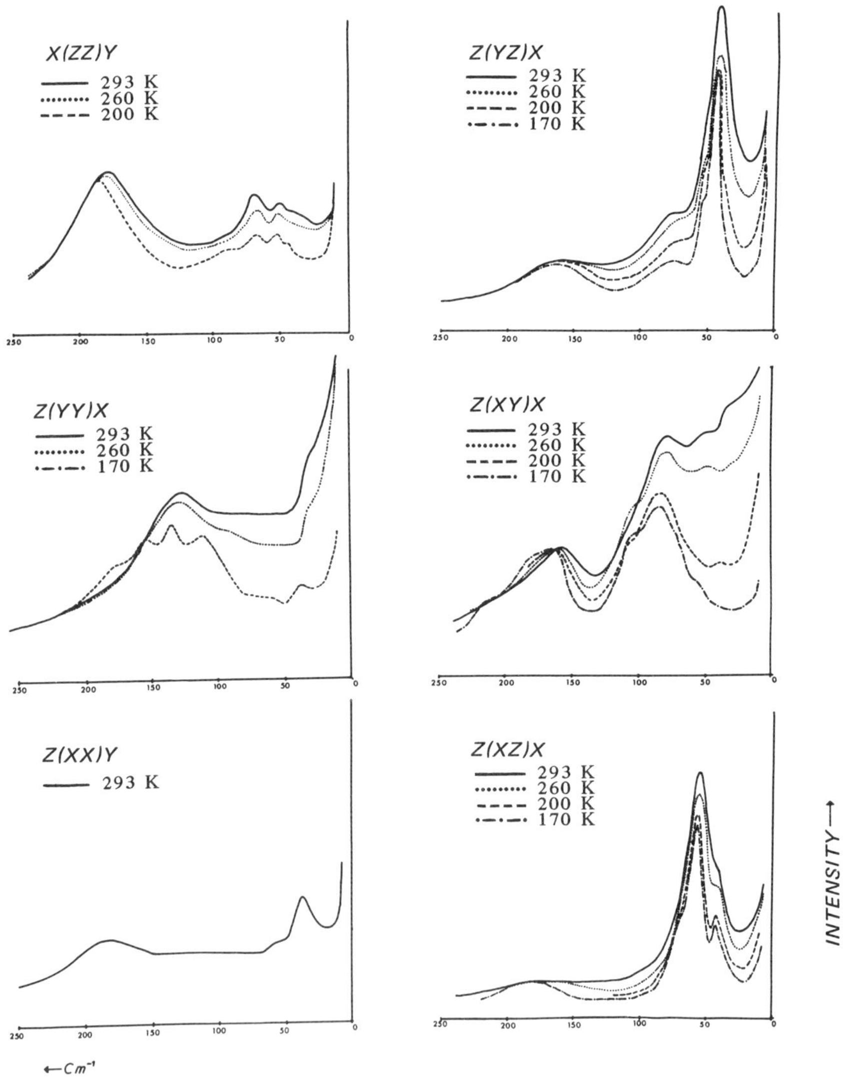

Fig. 3. Low frequency scattering.

at this point. Some additional temperature dependence is found in the frequency of some strong, low-frequency peaks. In the *yz* spectrum, for example, a peak shifts out from 40 cm^{-1} at 290 K to 45 cm^{-1} at 170 K.

In the high frequency region, a feature of interest is the very broad 2870 cm^{-1} peak in the *yy* spectrum ($\Delta\nu \approx 600$ cm^{-1}). This is presumably the O–H stretch band. Some broadening might be expected in this band because of the role played by the hydrogens in the O–H – – – O bonding which couples the SO_4^{2-} ions along the *b* axis. Unfortunately, the behaviour of this 2870 cm^{-1} peak could not be clearly observed in the ferroelectric phase because it is obscured by intensity from a strong background which appeared in the region 300–3500 cm^{-1}

Table 1. Internal molecular frequencies at room temperature, cm^{-1}.

XX ν	*XX* $\Delta\nu$	*YY* ν	*YY* $\Delta\nu$	*ZZ* ν	*ZZ* $\Delta\nu$	*XZ* ν	*XZ* $\Delta\nu$	*XY* ν	*XY* $\Delta\nu$	*YZ* ν	*YZ* $\Delta\nu$	
404 s	13			404 s	16	410 sh						
						419		419	18	418		ν_2 SO_4^{2-}
445		446 s		443 vs	15	448		449	14	446	16	
578		575				576 s				575 s		
		582		586	16	583 sh		585		583 s		ν_4 SO_4^{2-}
609 s	10	611 s	11	612 wsh		611		615		615	19	
		750 vw										S-O-H torsion
881 s	18	887 s	18	883 s		880	17	883 w		882	19	ν_1 SO_4^{2-}
1014 vs		1017 vs		1014 s		1017		1017		1018		ν_3 SO_4^{2-}
1043 vs		1046 vs		1043 s		1044		1047		1046		
1155				1160 w		1160 w		1168				ν_4 SO_4^{2-}
1188 sh		1188				1213						
1230 sh		1235 sh		1220 sh		1230 sh		1230 wsh		1235 w		S-O-H bend
1275 sh		1280 sh		1275 wsh		1281 w		1280 w				
1438		1445 w		1438	95	1435 w		1435		1430 w		ν_4 NH_4^+
1630 w		1630 vw		1630 vw						1635 vw		
1685		1680 w		1680		1685 w		1685 w		1680 w		ν_2 NH_4^+
1900 vw,b		1860 vw,b		1900 vw,b				1900 vw,b				$\nu_1 + \nu_3 SO_4^{2-}$
		2423 w						1420 vw				
		2540 w										
2856 w		2870	~ 600	2860 vw		1870 vw		2855 w	80	2890 vw		O-H str. and
2920 vw								3095 wsh				2 ν_4 NH_4^+
3160	~ 95	3160	~ 100	3175	~ 140	3180 w		3210 w		3190 vw		ν_1 and ν_3 NH_4^+
						3280 w				3260 w		
				3360 wsh								2 ν_2 NH_4^+

s: strong. w: weak. vw: very weak. b: broad. sh: shoulder.

Table 2. Lattice vibrational frequencies (cm^{-1}).

	XX	*YY*			*ZZ*			*XZ*				*XY*				*YZ*			
Temperature	293	293	260	170	293	260	200	293	260	200	170	293	260	200	170	293	260	200	170
			20 sh	21 sh															
		31 sh	32 sh	38 w	33 w														
	37											36 w	36 w	30 w	40 w				
							42 w	40 sh	40 sh	41	42					40 ns	41 ns	44 ns	45 ns
					48 w	48 w	51 w					51 w	50 w	57 w	57 w		54 sh	54 sh	55 sh
								54 ns	55 ns	56 ns	58 ns								
	59 w			63 w															
					68	68	67			70 sh	75 sh								
																75	75	75	74
							90 w					80	78	84	87				
			112	109									110 sh	110 sh	106 sh				
		130 b	130	133															
			150	155															
												158 b	163 b	163 b	163 b	163 b	163 b	165 b	166 b
	182 b			180 sh	179 bs	180 bs	184 bs	175 b	175 b	*	185 b				178 sh				

w: weak. s: strong. n: narrow. b: broad. sh: shoulder.
* not measured

When the temperature is reduced from 290 K to 160 K, the profile of the ν_3 SO_4^{2-} band changes distinctly as the high frequency component shifts out in frequency and becomes clearly resolved into an unusually sharp intense peak. This feature was also reported by Bazhulin *et al.*,[10] and as they suggest, it is likely evidence of increasing distortion in the sulphate ions. No similar feature is apparent in other bands.

Although much of interest has been seen in the present work, more detailed temperature studies and work on isomorphous materials are required before a precise understanding of the relation between the ferroelectric properties of AHS and its Raman spectra can be reached.

ACKNOWLEDGEMENTS

The authors wish to acknowledge helpful discussions with Dr. Richard Nelmes. The work was supported by the Science Research Council.

REFERENCES

1 R. Pepinsky, K. Vedam, S. Hoshino and Y. Okaya, *Phys. Rev.* **111**, 1508 (1958).
2 I. P. Kaminow and T. C. Damen, *Phys. Rev. Letters* **20**, 1105 (1968). A. Pinczuk, W. Taylor, E. Burstein and I. Lefkowitz, *Solid State Commun.* **5**, 429 (1967). M. Didomenico, S. Wemple, S. P. S. Porto and R. P. Bauman, *Phys. Rev.* **174**, 522 (1968).
3 S. R. Miller, R. Blinc, M. Brenman and J. S. Waugh, *Phys. Rev.* **126**, 528 (1962).
4 R. J. Nelmes, *Acta Cryst.* **B27**, 272 (1971).
5 R. J. Nelmes, *Acta Cryst.* (to be published).
6 R. J. Nelmes, *Ferroelectrics* (to be published).
7 W. Cochran, *Advan. Phys.* **9**, 387 (1960); **10**, 401 (1961).
8 M. Kasahara and I. Tatzusaki, *J. Phys. Soc. Japan* **29**, 1392 (1970).
9 C. J. H. Schutte and D. J. J. Van Rensburg, *J. Mol. Structure* **9**, 77 (1971).
10 P. A. Bazhulin, T. P. Myasnikova and A. V. Rakov, *Sov. Phys.–Solid State* **5**, 1299 (1964).
11 B. H. Torrie, C. C. Lin, O. S. Binbrek and A. Anderson, *J. Phys. Chem. Solids* **33**, 697 (1972).
12 M. P. Fontana and M. Lambert, *Solid State Commun.* **10**, 1 (1972).

18. RAMAN SCATTERING INVESTIGATION OF THE COOPERATIVE ORDER–DISORDER PHASE TRANSITION IN AMMONIUM CHLORIDE AND AMMONIUM BROMIDE

Y. Ebisuzaki

Department of Chemistry, University of California, Los Angeles, California, USA

INTRODUCTION

Extensive studies on NH_4Cl show a strong volume dependence of the temperature for the cooperative order-disorder transition.[1] This suggests a relationship between volume and the ordering process of the NH_4^+ ion. Thus a knowledge of the volume and temperature dependence of the vibrational spectra would be most useful for a better understanding of this transition.

This chapter reports the most recent results on the volume and temperature dependence of the libration and translational motion of the NH_4^+ ion and changes of Raman activity due to the ordering process. The existence of the direct cooperative phase transition in NH_4Br also is reported.

EXPERIMENTAL

At low temperatures, both NH_4Cl and NH_4Br have ordered cubic phases in which the NH_4^+ ions have parallel orientations. The fundamental modes and their representations are

$$\nu_1(A_1) + \nu_2(E) + \nu_3(F_2) + \nu_4(F_2)$$

[internal modes of the NH_4^+ ion] and

$$\nu_5(F_2) + \nu_6(F_1)$$

[lattice modes]. Only the lattice modes are of immediate interest. These are the translational motion of the NH_4^+ against the Cl^- sublattice, ν_5, and the librational motion about the four-fold axes of the cube, ν_6. In the ordered phase, the librational mode is the only Raman-inactive fundamental, and it must be observed as a two-phonon process. However, in the disordered phase, NH_4Cl shows a complex spectrum with all fundamentals being Raman active.

Figure 1 shows the polarization spectra of the lattice modes for several scattering geometries of the disordered phase of NH_4Cl (296 K, 1 atm). The internal modes also are

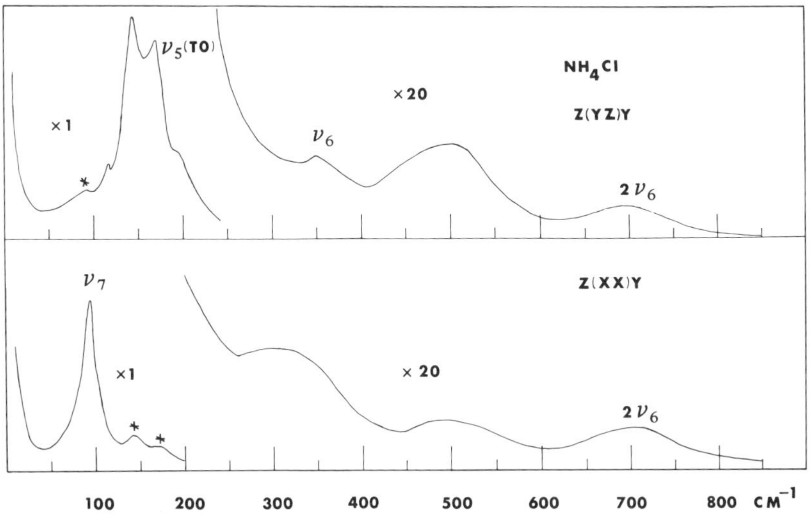

Fig. 1. Right-angle Raman scattering for the disordered phase (1 atm, 296 K) of NH_4Cl in two scattering geometry (see ref. 5 for notation). The asterisks indicate the spill-over of scattering from peaks active in other scattering geometry.

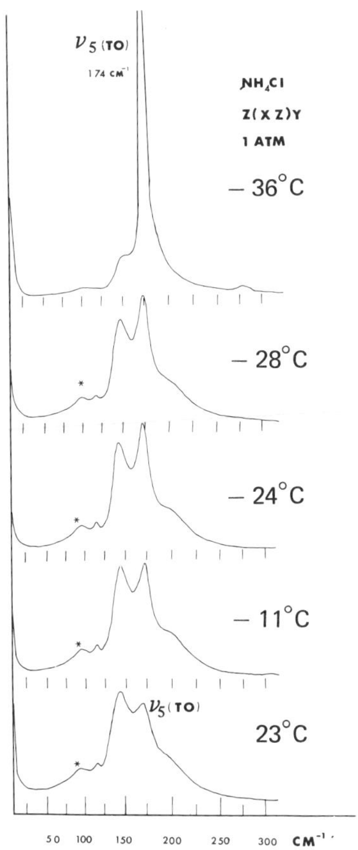

Fig. 2. Raman scattering intensities for modes centred around ν_5 (TO). The asterisks indicate peaks attributed to spill-over of scattering from other phenomena.

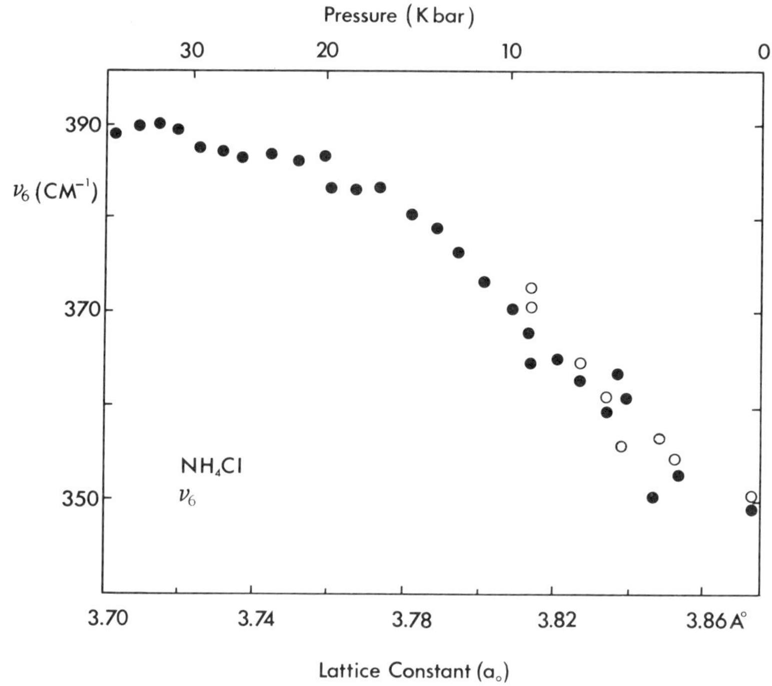

Fig. 3. The dependence of the librational frequency on pressure and lattice constant (a_0). Frequencies obtained from zone centre and as a two phonon process are indicated by open and solid circles, respectively.

characterized by well-defined polarizations; however, some modes such as the librational fundamental are active in all scattering geometries. The spill over from modes inactive in the particular scattering geometry (designated by asterisks) is quite small for both internal and lattice modes.

Figure 2 shows Raman peaks in the ν_5(TO) region with off-diagonal components in the polarization tensor. Of special interest is the observation that the total integrated intensity between 50 and 250 cm^{-1} remains constant while the intensity of ν_5(TO) itself depends strongly upon temperature and pressure. As the transition to the ordered phase[2] (−30.6°C, at 1 atm) is approached isobarically from higher temperatures, ν_5(TO) increases in intensity relative to others. The frequencies of ν_5(TO) and other modes shown in Fig. 2 are very volume sensitive in the disordered phase with large Grüneisen constants. The Grüneisen constant for ν_5(TO) is 3.1 and 2.5 for the disordered and ordered phases respectively.

Observation of ν_6 at the centre of the Brillouin zone (350 ± 15 cm^{-1} at 1 atm and 296 K) and as a two phonon process, $2\nu_6$, indicates that even in the disordered phase, the librational branch is flat within the precision of the experimental results. In the disordered phase, the frequency of ν_6 is very sensitive to change in the lattice constant (a_0); this is shown in Fig. 3. No differences can be distinguished between the frequency shift for the librational mode obtained as a fundamental and as a two phonon process.

The volume dependence or mode Grüneisen constants (γ_i) for the librational mode

Table 1. Modes which become Raman inactive upon 'cooperative disorder-order' transition in NH_4Cl and NH_4Br.

NH_4Cl (cm^{-1})[a]	NH_4Br (cm^{-1})[a]
96 (ν_7) [$Z(XX)Y$]	56 (ν_7) [$Z(XX)Y$]
119[b] [$Z(YZ)Y$, $Z(XZ)Y$]	80 [$Z(YZ)Y$, $Z(XZ)Y$]
144[b] [$Z(YZ)Y$, $Z(XZ)Y$]	141[b] [$Z(YZ)Y$, $Z(XZ)Y$] (low energy shoulder appearing on this peak in phase V)
198[b] [$Z(YZ)Y$, $Z(XZ)Y$]	180[b] [$Z(YZ)Y$, $Z(XZ)Y$]
310[c] [$Z(XX)Y$]	(found at low T in phase V weak, broad, unresolved)
350 (ν_6) [$Z(XX)Y$, $Z(XZ)Y$, $Z(XZ)Y$]	

[a] Frequencies are given at 1 atm and 296 K. [ ] Indicate scattering geometry in which the mode is Raman active.
[b] Modes fitted to an uncoupled damped harmonic oscillator.
[c] Frequency has a large uncertainty.

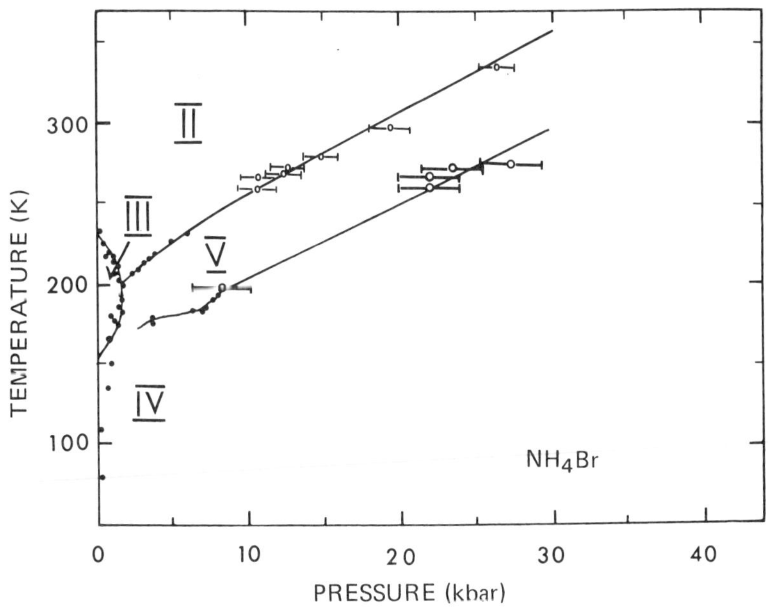

Fig. 4. Phase diagram for NH_4Br. Open circles with error bars represent data obtained in the present study; solid circles represent the literature values (ref. 4).

are 1.30 and 0.29 for the disordered and the ordered phase respectively. The magnitude of γ_i for the disordered phase indicates that there is a fair amount of anharmonicity associated with the librational mode. This anharmonicity probably arises from repulsive interactions between hydrogens on adjacent NH_4^+ ions. The large frequency increase in ν_6 as the transition temperature is approached isobarically from room temperature cannot be accounted for with the room temperature Grüneisen constant. That is the self-energy or explicitly temperature dependent shift is larger than the volume shift.

The phase diagram[4] for NH_4Br is shown in Fig. 4. Both II–V and V–IV transitions are disorder-order transitions with slopes of 5.0 ± 0.8 and 5 ± 1 K/kbar respectively. Experimental evidence for identifying V–IV transition as the cooperative disorder-order phase transition consists in the fact that at this transition, the Raman spectrum shows the same characteristic changes as at the disorder-order transition in NH_4Cl. This is true in the change of the number of Raman active peaks (Table 1), intensity increase of ν_5 (TO) and decrease of the Grüneisen constants upon ordering. The II–V transition is accompanied by a decrease of the Grüneisen constants, with typical changes being: 3.2 to 0.75 (ν_7); 0.72 to 0.46 ($2\nu_6$); and 0.036 to 0.003 ($\nu_2(E)$). These decreases of the γ's suggest that there must be some type of ordering which leads to a decrease in anharmonicity. This is especially true for the librational overtone ($2\nu_6$) where the orientation-dependent part of the force constant between the NH_4^+ ions would be expected to decrease upon ordering.

The interesting implication of the studies on NH_4Br is that both NH_4Br V and NH_4Cl II are ordered. Also, the possible existence of another high temperature phase in NH_4Cl is suggested; and this aspect is now being explored.

The present Raman data give no indication of a disorder-order transition mechanism involving a soft or other low-energy mode. However, the experimental data indicate the presence of volume- and temperature-dependent anharmonicity associated with the librational mode in the disordered state. There are also modes coupled to the translation motion of the NH_4^+ ion in the disordered phase, but their role in the phase transition is not known.

ACKNOWLEDGEMENT

The author wishes to express her appreciation to Dr. M. F. Nicol for encouragement and for the use of the equipment. She also wishes to acknowledge helpful discussions with Dr. R. S. Katiyai. This work is supported in part by the U. S. Atomic Energy Commission.

REFERENCES

1 P. W. Bridgman, *Phys. Rev.* **38**, 182 (1931); P. Nissilä and I. Poyhönen, *Phys. Letters* **33A**, 345 (1970); C. W. Garland and B. B. Weiner, *Phys. Rev.* **13**, 1634 (1971); C. W. Garland and R. Renard, *J. Chem. Phys.* **44**, 1130 (1966).

2 V. Hovi, *Proc. Int. Conf. Sci. Technol. of Non-Metallic Crystals*, New Delhi, India, 1969, p.67.

3 H. G. Smith, J. G. Traylor and W. Richard, *Phys. Rev.* **181**, 1218 (1969).

4 Yukiko Ebisuzaki, (to be published); C. W. Garland and R. A. Young, *J. Chem. Phys.* **49**, 5282 (1968); P. W. Bridgman, *Phys. Rev.* **38**, 182 (1931); C. C. Stephenson and H. E. Adams, *J. Chem. Phys.* **20**, 1658 (1952); R. Stevenson, *J. Chem. Phys.* **34**, 1757 (1961).

5 T. C. Damen, S. P. S. Porto and B. Tell, *Phys. Rev.* **142**, 570 (1966).

19. SOFT MODES IN THE FERROELECTRIC AND ANTIFERROELECTRIC CRYSTALS OF THE KDP-TYPE

B. Lavrenčič, I. Levstek, B. Žekš and R. Blinc
"J. Stefan" Institute, University of Ljubljana, Ljubljana, Yugoslavia

INTRODUCTION

Since the classical paper of Kaminow and Damen [1] the groups at Baltimore, Edinburgh and Ljubljana have tried to understand better the phase transition which occurs in the KDP-type series of compounds by means of the light scattering. It has been found:

1. The main driving mechanism for the ferroelectric transition in KH_2PO_4 is due to a highly damped, T-dependent coupled proton-lattice TO, $q = 0$ mode which pushes an acoustic mode to zero slope before it itself becomes unstable.[2]
2. The Raman spectrum shows remarkable changes on deuteration as expected from the four proton cluster picture not only in the B_2-type spectrum but also in the E-type, the width of the 'ferroelectric wing' decreasing by a factor of ten.[3]
3. A deviation from the Lyddane-Sachs-Teller (LST) relation is observed in KDP,[4] KDA [5,6] and CsDA.[6] This deviation was interpreted as the coupling of the soft mode with a central mode, i.e. either a low lying mode [7] or a continuum peaked at zero frequency.[6]
4. A soft mode was observed in the ferroelectric phase whose frequency is proportional to the spontaneous polarization;[8] the B_2 spectrum below T_c has been recently remeasured[4,9,10] and the new feature at 153 cm^{-1} has been found around 84 K.

In this chapter we wish to report some new experimental as well as theoretical results, hopefully to throw more light on the nature of the phase transition in the KDP-type crystals. In particular we present some evidence that the soft mode becomes in the undeuterated compound underdamped below T_c.

THEORY

It is well established by now that the following Hamiltonian describes reasonably well the static and the dynamic properties of KH_2PO_4 [11]

$$\mathcal{H} = -2\Gamma \sum_i S_x^i - \tfrac{1}{2} \sum_{i,j} J_{ij} S_z^i S_z^j \tag{1}$$

where Γ is the tunnelling matrix and J_{ij} is the coupling between protons and i is the index of bond. From this Hamiltonian one can derive in the molecular field approximation

(MFA) the following expression for the inverse susceptibility

$$\chi_q^{-1} = \frac{4\Gamma}{\tanh \beta\Gamma} - J_q \tag{2}$$

where $\beta = 1/k_B T$ and $J_q = \sum_j J_{ij} \exp [iq(R_j - R_i)]$. When we lower the temperature the Fourier component with wave vector q_0 for which J_{q_0} is maximum goes soft first. Thus the transition temperature is given by

$$4\Gamma - J_{q_0} \tanh \beta_c\Gamma = 0, \;\; \beta_c = 1/k_B T_c \tag{3}$$

From the symmetry properties of the Brillouin zone we know that the maximum value of J_{q_0} occurs at the special point, most likely at $q_0 = 0$ or $q_0 = \pi/a$, i.e. at the zone centre or zone edge. The former case we call the ferroelectric one and the latter the antiferroelectric one. Only transitions with $q_0 = 0$ or π/a can be, or nearly be of the second order.[12]

From the Hamiltonian, Eqn. (1), one can derive the eigenfrequency of the motion.[11] One must solve the dynamic equation

$$\frac{d\langle S^i\rangle}{dt} = -i\langle[S^i, \mathcal{H}]\rangle_t \tag{4}$$

in the random phase approximation (RPA). The result for the transverse excitation which couples to the light wave is[11] for $T > T_c$

$$\omega^2(q) = \Gamma(4\Gamma - J_q \tanh \beta\Gamma) = A(T - T_c) \tag{5}$$

From Eqns. (3) and (5) one sees immediately that the temperature where $\omega(q = 0) \to 0$, is when $T \to T_c$.

For $T < T_c$ we distinguish two cases:

1. Ferroelectrics, where

$$\omega^2(q) = (J_0 \langle S_z\rangle)^2 + (2\Gamma)^2 \left(1 - \frac{J_q}{J_0}\right) \tag{6a}$$

This mode is critical with $q = 0$.

2. Antiferroelectrics. Here we have to take into account that the transition involves the doubling of the unit cell. Thus, the low temperature phase contains two oppositely oriented spins per unit cell. Therefore, for each wave vector there are two quasi transverse mode frequencies

$$\omega^2(q) = (J_q \langle S_z\rangle)^2 + (2\Gamma)^2 (1 - J_0/J_q) \tag{6b}$$

$$\omega^2(q) = (J_q \langle S_z\rangle)^2$$

The first of the two frequencies is finite at all temperatures. The second displays the critical behaviour. The two frequencies are equal for $q = 0$. Here the $\langle S_z\rangle$ means the sublattice polarization.

Raman measurements are one of the few methods available to determine the temperature variation of the sublattice polarization of an antiferroelectric. Also it is worth noting that in the case of antiferroelectrics the soft zone edge mode of the paraelectric phase always correlates with the soft totally symmetric zone centre mode of the low temperature phase. This is the extension of the Worlock's theorem for ferroelectrics.[13]

Let us consider now the coupling of the quasi-spin system with a phonon field. The Hamiltonian now becomes

$$\mathcal{H} = -2\Gamma \sum_i S_x^i - \tfrac{1}{2} \sum_{i,j} J_{ij} S_z^i S_z^j + \tfrac{1}{2} \sum_q (P_q P_{-q} + \Omega_q^2 Q_q Q_{-q}) - \sum_{i,q} F_q S_z^i Q_q \exp(iq \cdot R_i) - \sum_q \rho_q Q_q E_{-q} \exp(i\omega t) \tag{7}$$

where Q_q, P_q are respectively the phonon coordinates and momenta, Ω_q is the phonon frequency, F_q is the proton–lattice coupling constant and ρ_q and E_q are the Fourier components of the density and electric field respectively. This Hamiltonian was first solved by Kobayashi[14] and recently extended to the ferroelectric phase by us.[15,16] We have found the following expression for $\langle Q_q \rangle$

$$\langle Q_q \rangle = \frac{NF_q}{\Omega_q^2} \langle S_z \rangle \delta_{q,0} \tag{8}$$

Here is seen the relation between $\langle S_z \rangle$ which describes the ordering of the proton system with the $\langle Q_0 \rangle$ which describes the displacement and thus the spontaneous polarization. The Hamiltonian in Eqn. (7) can be transformed to that of Eqn. (1) by replacing J_q with

$$\tilde{J}_q = J_q + N|F_q|^2/\Omega_q^2$$

The effect of the proton–lattice coupling is merely to change the proton–proton coupling constant J_q. However, the $\tilde{J}_q$ might have the maxima at different values of q_0 from J_q thus drastically changing the star of the soft mode. All this could explain why KH_2PO_4 is ferroelectric and $NH_4H_2PO_4$ (with almost identical proton network) is antiferroelectric.

The above RPA treatment neglects damping which is known to be large in KDP-type ferroelectrics. When the fluctuations around the molecular field are small as compared to the molecular field itself, Bloch type damping terms can be introduced into the equations of motion.[15] This seems to be a good approximation below T_c. It predicts that the soft mode will become underdamped in the ferroelectric phase. In the ferroelectric phase, however, the fluctuations are large as compared to the molecular field and the Bloch approach becomes inadequate.

Until now we have considered the coupling of the proton with the well defined phonon mode. The influence of all low frequency modes on the proton system can be described[16] by a randomly fluctuating internal field $H(t)$. For such a system the Hamiltonian reduces to

$$\mathcal{H} = -2\Gamma \sum_i S_x^i - \tfrac{1}{2} \sum_{i,j} J_{ij} S_z^i S_z^j - H(t) \sum_i S_z^i \tag{9}$$

where $H(t)$ is a randomly fluctuating internal field, describing the influences of the surround-

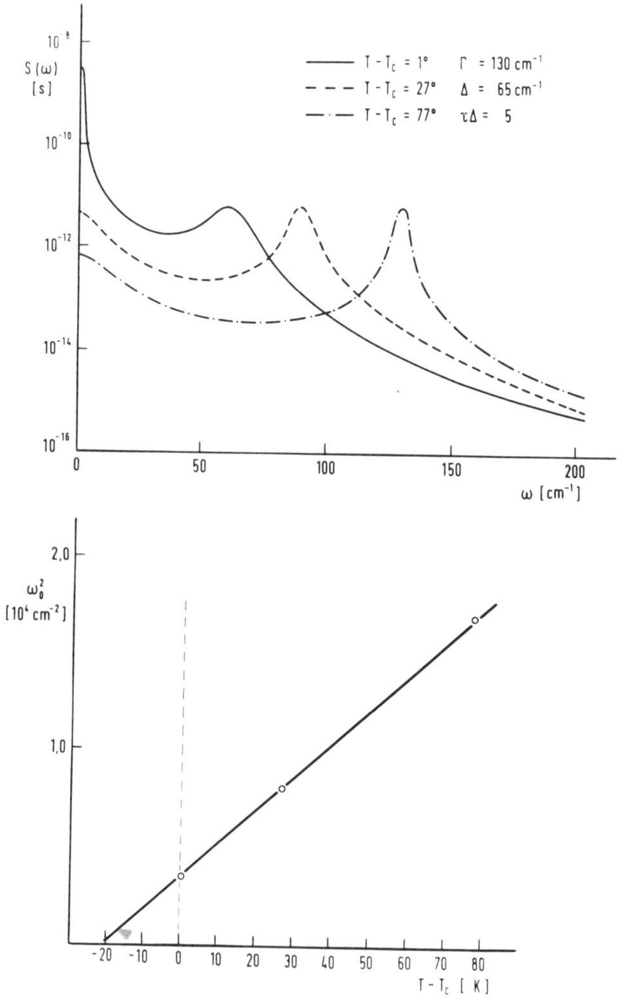

Fig. 1. The Raman scattering instensity $S(\omega)$ as calculated from the Hamiltonian, Eqn. (9). The parameters Γ, Δ and τ were estimated (upper figure). The peak frequency squared versus temperature shows the deviation from the LST relation.

ing on the protons. The equations of motion become stochastic. For $H(t)$ we conjecture Markoffian and Gaussian process

$$\langle H(0)H(t)\rangle = \Delta^2 \exp(-t/\tau) \tag{10}$$

The details of the calculation [16] will be omitted here and will be presented elsewhere.[17] The calculated Raman intensities are shown in Fig. 1. The wing, i.e. the central mode behaves critically as T_c is approached, the soft mode peaks do not. The peaks are shown to exhibit a deviation from the LST relation.

RESULTS AND DISCUSSION

We have measured the Raman spectra of KH_2AsO_4 (KDA) and RbH_2PO_4 (RbDP)

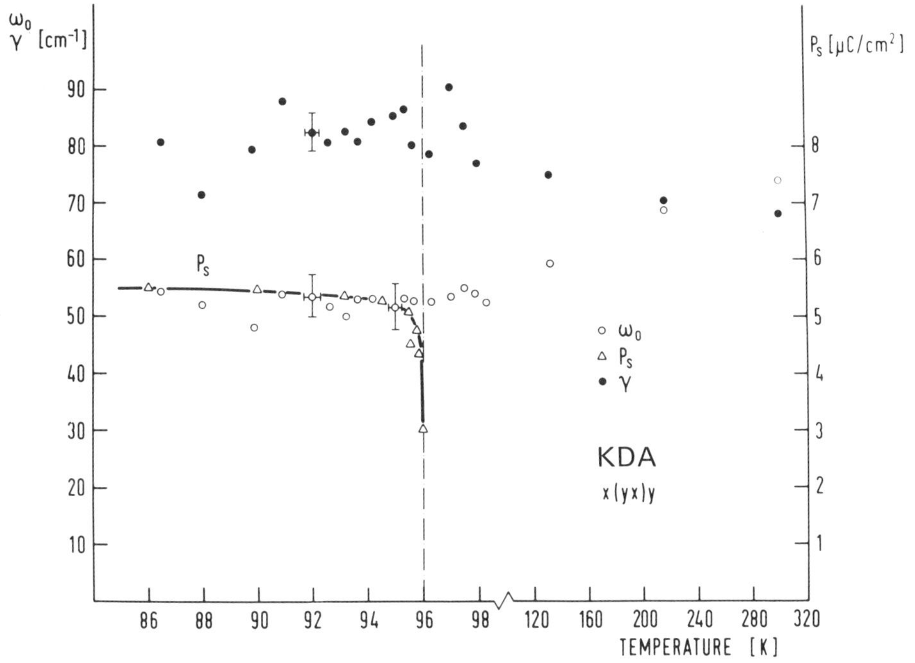

Fig. 2. Temperature dependence of the parameters ω_0 and γ as a function of temperature is given for KH_2AsO_4 (KDA) in both phases and ω_0 is compared with the spontaneous polarization.

above and below T_c (96 K and 147 K, respectively). In KDA a broad wing was observed in the vicinity of the excitation line. If the 'ferroelectric' wing is fitted to the usual damped harmonic oscillator formula

$$\chi(\omega) = \frac{\chi^*(0)\,\omega_0^2}{\omega_0^2 - \omega^2 + 2i\gamma\omega} \tag{11}$$

it yields three parameters $\chi^*(0)$, ω_0 and γ as the function of temperature, shown in Figs. 2 and 3. ω_0 is identical with $\omega(q = 0)$ in Eqn. (5) and is indeed proportional to the spontaneous polarization.

However, if we plot T/τ ($\tau = 2\gamma/\omega_0^2$) versus temperature, the mode does not approach zero according to Eqn. (5) or in other words, ω_0 is not proportional to the static reciprocal susceptibility, i.e. to the LST relation. Also, $\chi^*(0)$ does not follow the LST relation as expected. The same phenomenon was observed in CsDA,[7] RbDP and RbDA.[17] If we plot the deviation from the LST relation, $T_c - T_0$, as a function of the cation atomic mass R in the RH_2AsO_4 isomorphs, Fig. 4, we get a straight line through the origin. We feel that this linear dependence is an important clue in explaining the LST deviation.

Let us consider some antiferroelectrics as they are shown in Raman or infrared spectroscopy. Only a few investigations have been done so far.[18,19] We have measured the Raman spectra of $NH_4H_2AsO_4$ in the $x(yx)y$ geometry and the results have been compared with Perry and Agrawal's[18] results. The frequencies follow the Curie–Weiss law with T_0 far below T_c. The correlation with the dielectric data appears to be rather poor,[19] except

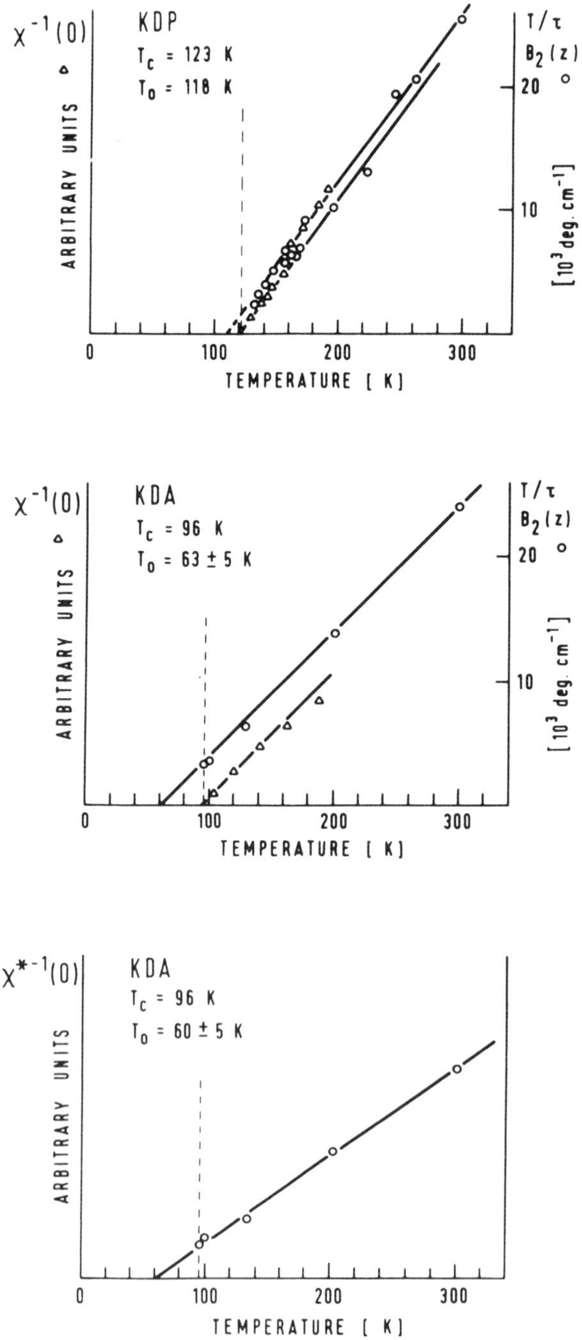

Fig. 3. The T/τ is shown as a function of temperature for KH_2PO_4 (KDP) and KH_2AsO_4 (KDA) and is compared with the inverse static susceptibility $\chi(0)^{-1}$. For the KDA we also plot the inverse static susceptibility $\chi^*(0)^{-1}$ as given by light scattering, see Eqn. (11).

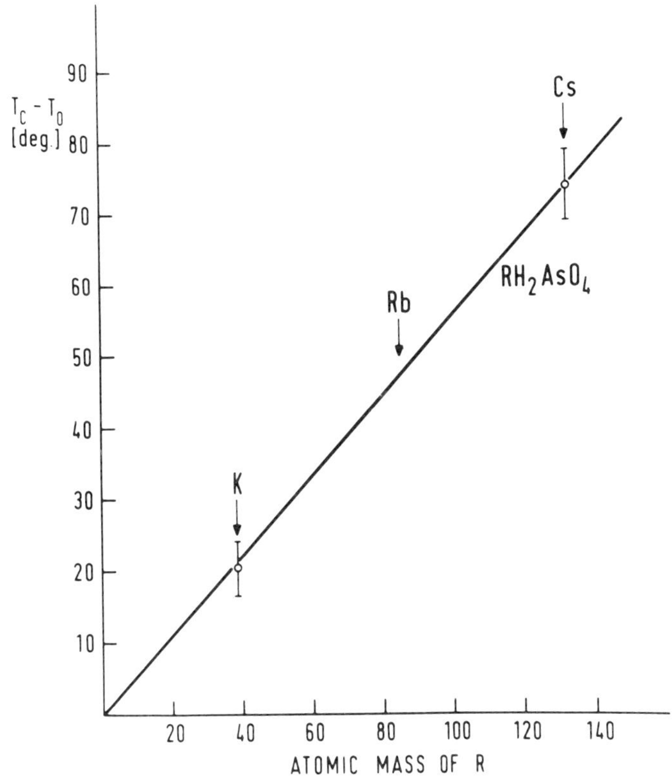

Fig. 4. Dependence of $T_C - T_0$ [5,6] on the atomic mass of R. The calculation of the rubidium point is in progress.

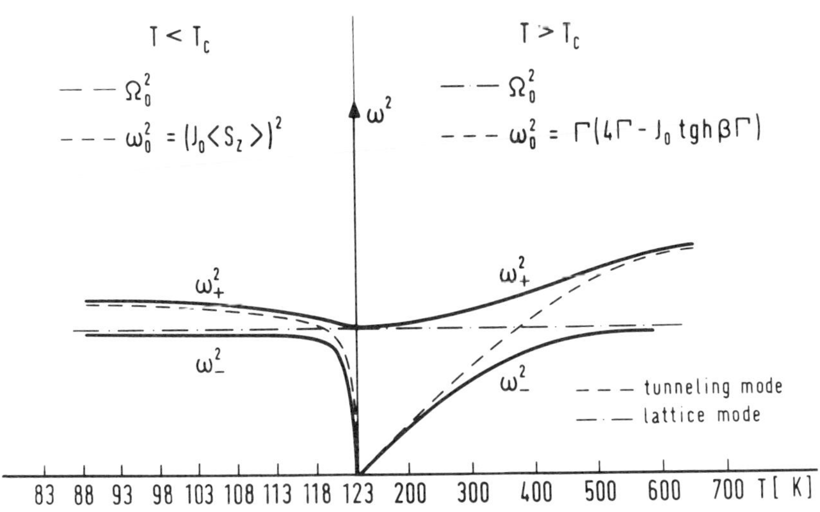

Fig. 5. Proton–phonon interaction in KDP (schematically). The temperature scales above and below 123 K are different. The ω_- and ω_+ are proton-like and phonon-like coupled modes respectively.

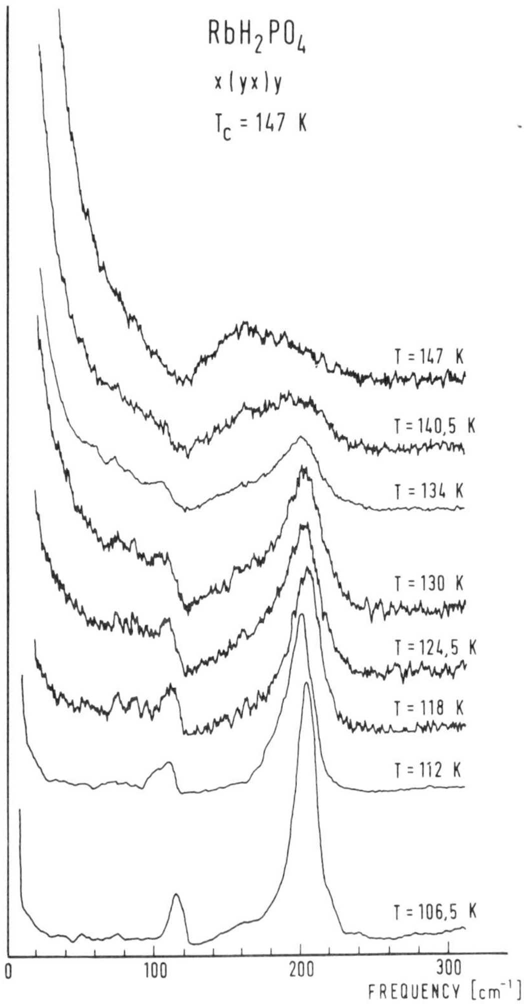

Fig. 6. *x*(*yx*)*y* Raman spectra of RbH_2PO_4 (RbDP) in the ordered phase. Notice the similarity with the equivalent KDP data (reference 4, Fig. 3).

for the *E* symmetry modes where the frequency as well as the inverse susceptibility stay constant until T_c, and then they drop. Rather interestingly, the same effect was seen in ferroelectric KH_2PO_4.[4] It now remains for a skilled experimenter to find the way of bringing the ADP or ADA single crystal undamaged through the transition into the ordered phase where the nature of the soft mode can be elucidated by Raman scattering.

Let us now consider the problem of the proton coupling with the lattice. Scott and Wilson[10] have recently considered the proton–phonon interaction in KDP. They interpreted the paraelectric results as a coupling of the bare protonic system described by $\omega(q)$ with the lattice as described by Ω_q [see Eqns. (5) and (7)]. The ordered phase spectra (where a new feature appears at 153 cm^{-1} around 84 K) were interpreted as an interaction of the 'ferroelectric' wing with a lattice mode of (paraelectric) A_1 symmetry. We feel that the latter interpretation is basically incorrect. We have measured the Raman spectra of the deuterated KDP 50 K below T_c. There is no mode seen at 153 cm^{-1} and there is no reason why that mode should not be there if it were of 'nonprotonic' origin as Scott and Wilson suggested. Our belief is that what we see at 84 K is an underdamped soft mode $\omega(0)$ coming out of the wing. We propose a different picture of the proton–phonon interaction in KDP (Fig. 5). Additional arguments for this model may be derived from the observations of the temperature dependence of the lattice mode which behaves as indicated in Fig. 5 and from the total Raman cross section which follows Curie–Weiss law.

Until now we were unable to calculate the parameters Γ and J_0 from the Raman measurements alone. This is possible now. If we assume that the value of the ferroelectric mode frequency at 84 K is 153 cm^{-1} and that this result is virtually the same at $T = 0$ (it is at 20 K[20]), then by combining Eqns. (3) and (6), using the result from MFA, namely

$$J\langle S_z\rangle_{T=0} = (\tfrac{1}{4}J_0^2 - 4\Gamma^2)^{1/2}$$

we arrive at the estimates, $2\Gamma = 140$ cm^{-1} and $J_0 = 415$ cm^{-1} which are reasonable. The intensity of this mode disappears on deuteration as its eigenvectors have no component in the direction of spontaneous polarization.[17]

We have recently measured the Raman spectra of RbH_2PO_4 below T_c, (Fig. 6). They also show the same general behaviour as KDP in spite of the deviation from the LST relation. In particular the soft mode again becomes underdamped in the ordered phase. Here we found $2\Gamma = 400$ cm^{-1} and $J_0 = 830$ cm^{-1}.

REFERENCES

1 I. P. Kaminow and T. C. Damen, *Phys. Rev. Letters* **20**, 1105 (1968).
2 E. M. Brody and H. Z. Cummins, *Phys. Rev. Letters* **21**, 1263 (1968).
3 B. Lavrenčič, I. Levstek, B. Žekš and D. Hadži, *Chem. Phys. Letters* **5**, 441 (1970).
4 C. M. Wilson and H. Z. Cummins, *Light Scattering in Solids*, (Ed. M. Balkanski), Flammarion, Paris, 1971, p. 420.
5 V. Smolej, *Thesis*, University of Ljubljana, 1971 (unpublished).
6 R. A. Cowley, G. J. Coombs, R. S. Katiyar and J. F. Scott, *J. Phys. Chem.* **4**, 1203 (1971); R. S. Katiyar, J. F. Ryan, and J. F. Scott, *Phys. Rev.* **B4**, 2635 (1971); R. S. Katiyar, J. F. Ryan and J. F. Scott, *Light Scattering in Solids*, (Ed. M. Balkanski), Flammarion, Paris, 1971, p. 436.
7 R. Blinc, B. Lavrenčič, I. Levstek, R. Pirc and B. Žekš, *Proton–Phonon Coupling and the Anomalous Temperature Dependence of the Ferroelectric Mode in KH_2AsO_4*, (to be published).
8 I. Levstek and B. Lavrenčič, *Second Int. Conference on Raman Spectroscopy*, Oxford, Sept. 1970 (unpublished).
9 T. Shigenari and Y. Takagi, *J. Phys. Soc. Japan* **31**, 312 (1971).
10 J. F. Scott and C. M. Wilson, *Solid State Commun.* **10**, 597 (1972).
11 R. Braut, K. A. Müller and H. Thomas, *Solid State Commun.* **4**, 507 (1966).
12 C. Haas, *Phys. Rev.* **140A**, 863 (1965).
13 B. Lavrenčič and R. Blinc, *Physica Status Solidi* **49**, 119 (1972).
14 K. Kobayashi, *J. Phys. Soc. Japan* **24**, 497 (1968).
15 B. D. Silverman, *Phys. Rev. Letters* **25**, 107 (1970); B. Lavrenčič, I. Levstek, R. Blinc and B. Žekš, *Light Scattering in Solids*, (Ed. M. Balkanski), Flammarion, Paris, 1971, p. 424.

16 B. Žekš, *Ph.D. Dissertation*, University of Ljubljana, 1972 (unpublished).
17 R. Blinc, B. Lavrenčič, I. Levstek and B. Žekš,(to be published).
18 C. M. Perry and D. K. Agrawal, *Phonons*, (Ed. M. A. Nusimovici), Flammarion, Paris, 1971, p. 342.
19 J. F. Ryan, private communication.
20 D. K. Agrawal and C. H. Perry, *Light Scattering in Solids*, (Ed. M. Balkanski), Flammarion, Paris, 1971, p. 429.

20. RAMAN SPECTRA OF CRYSTALS OF THE KDP-TYPE IN THEIR PARAELECTRIC AND FERROELECTRIC PHASES

H. Hammer

Sektion Physik der Universität München,
München, Germany

INTRODUCTION

Several experimental groups have been studying intensively the ferroelectric phase transitions of KDP-type crystals with inelastic light scattering techniques for some years (see, for example, ref. 1). Though theoretical efforts have been made to explain some peculiarities of the Raman spectra of these crystals above and below the transition points, the spectra are not at all well understood up to now. Most of these experimenters were interested in the anomalous line shapes and temperature dependences of the low frequency regions of the spectra of $B_2(z)$- and $E(x)$ or $E(y)$-modes in the paraelectric phases and just below the transition temperatures.

We want to point out additional peculiarities observed in the Raman spectra above and below the ferroelectric transition points. These effects have not been taken into account up to date and cannot be explained in a simple way. We have found the following effects by studying the crystals KDP, KD*P (90%, 80%, and 50% deuteration), KDA and RDP in the temperature range from 300 K to about 100 K or 10 K, respectively.

Our experimental arrangement was a Spectra Physics argon ion laser model 165 with an output of about 1W for the 5145 Å line, a Jarrell Ash 1 m double monochromator and a photon counting system. For the experiments at low temperatures we used a cryostat from Coderg for liquid nitrogen and a bath cryostat from Oxford Instruments for liquid helium. The size of the oriented single crystals was about 5 × 7 × 5 mm. All the spectra were taken with the spectral slit width of 1 cm^{-1} (35 μm for 5145 Å) to get high resolution.

EXPERIMENTAL

Most of our experimental data are taken from RDP spectra, because this is a less well studied crystal of the KDP family. One of the remarkable features of the Raman spectra of KDP-type crystals are the asymmetric line shapes of some bands, i.e. the B_2 phonon line at 393 cm^{-1} for RDP for $T > T_c$ or the A_1 phonon lines at 349 cm^{-1} and 918 cm^{-1} for $T < T_c$. The feet of these Raman lines remain asymmetric even at about 10 K.

Another interesting observation is the temperature dependence of the intensities of some Raman lines with respect to the neighbouring bands. In the $x(yy)z$ spectrum a new

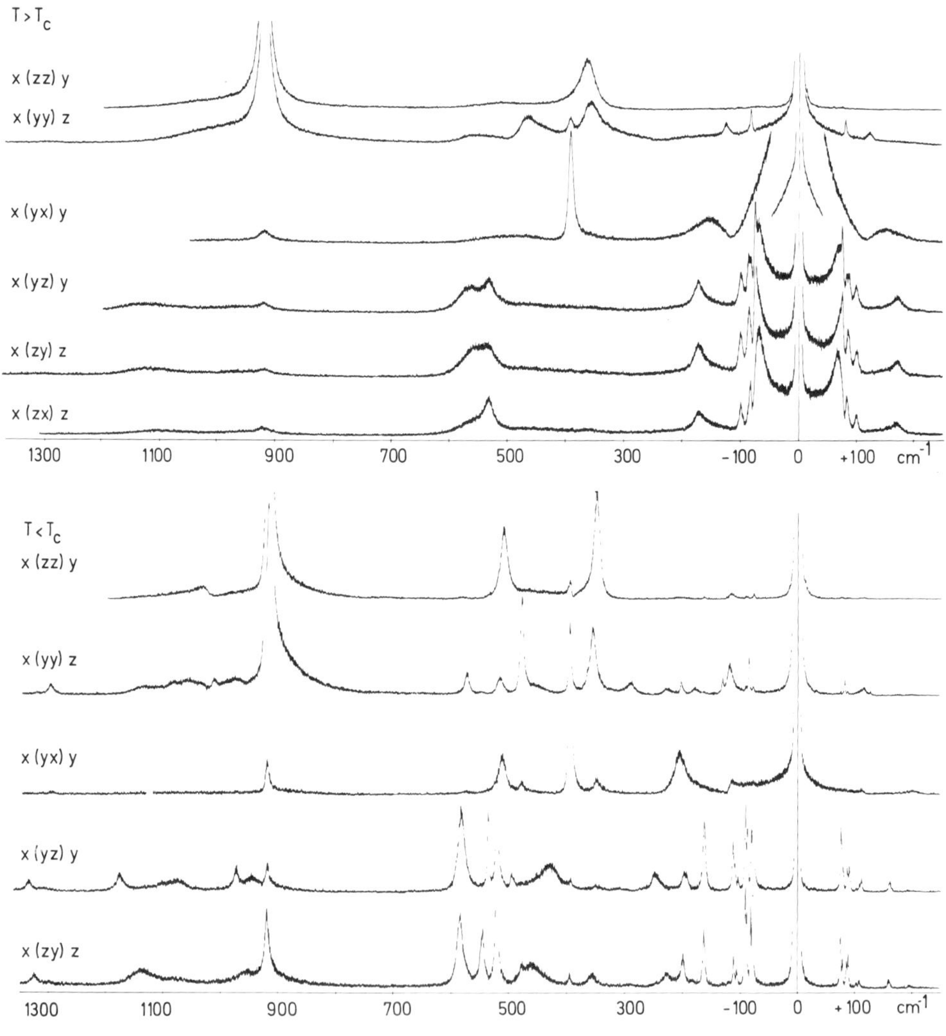

Fig. 1. Raman spectra of RbH_2PO_4 above and below T_C; exciting line 5145 Å, 1 W, slit width 1 cm^{-1}, scanning speed $\nu = 20$ cm^{-1}/min, time constant 1 s.

line appears at 130 cm^{-1} below T_c which increases in intensity with decreasing temperature compared with the line at 140 cm^{-1}. While the intensity of this line increases continuously down to helium temperatures, in the B_1 or B_2 spectra, for instance, a distinct temperature dependence of the intensity appears only near the transition point. We are concerned with the low frequency region and the group at 500 cm^{-1}.

We have observed not only temperature dependent changes of the intensity or frequencies, but a strong frequency dependence from the observation or polarization direction, too, by changing the y axis and the z axis.

In the $x(yy)z$ spectrum above T_c not only are the additional B_1 phonon lines compared with $x(zz)y$, but also the A_1 mode at 355 cm^{-1} in $x(zz)y$ is shifted to 363 cm^{-1} in $x(yy)z$. A corresponding effect has been registered for $T < T_c$, where frequency shifts are

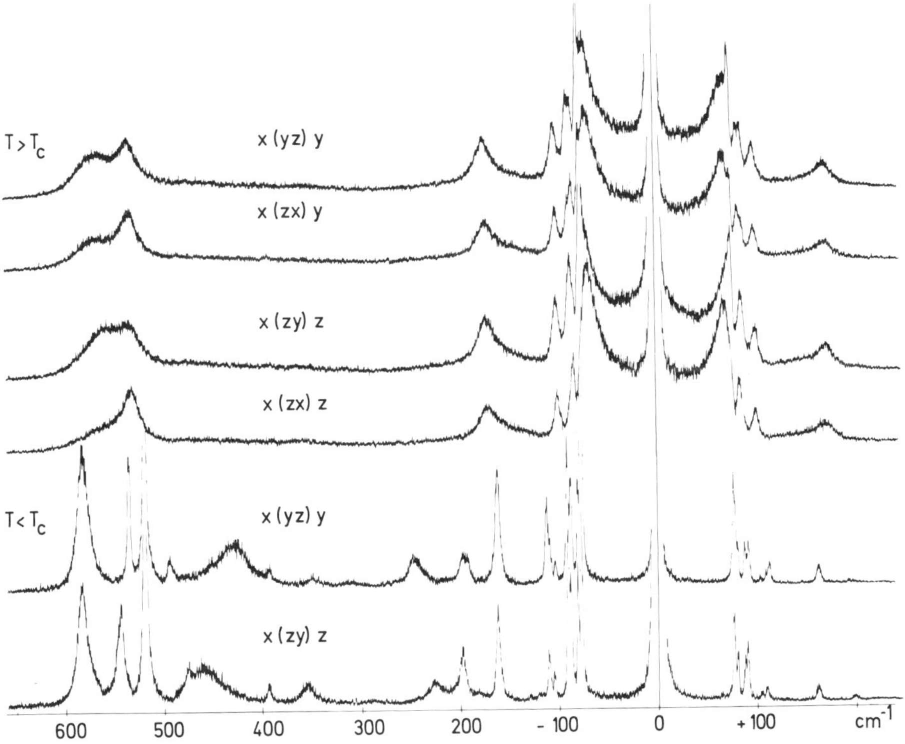

Fig. 2. *E* and *B* modes of RbH_2PO_4 in different directions of polarization and observation for *T* above and below T_c.

observed from 349 cm^{-1} for $x(zz)y$ to 354 cm^{-1} for $x(yy)z$, and from 510 cm^{-1} to 514 cm^{-1}, respectively (Fig. 1). The $E(x)$ and $E(y)$ spectra for $T > T_c$ are different for the four possible arrangements of observation, $x(yz)y$, $x(zx)y$, $x(zy)z$ and $x(zx)y$, and the B_1 or B_2 spectra for the two possible arrangements, $x(yz)y$ and $x(zy)z$ for $T < T_c$.

Figure 2 shows the different spectra above and below T_c. There are differences at the bands at 70 cm^{-1}, at 85 cm^{-1} and in the 500 cm^{-1} region. The broad band at 70 cm^{-1}, corresponding to the well known 'continuum' of KDP[1,2] has different intensities and line shapes in the four different spectra. Next, the line at 85 cm^{-1} is split into two lines for the observation in the *y* direction. In the 500 cm^{-1} region the very broad bands show rather different line shapes in the four spectra. Below T_c we have observed distinct frequency shifts: the line at 535 cm^{-1} in the $x(yz)y$ spectrum is shifted to 545 cm^{-1} in the $x(zy)z$ spectrum; the one at 427 cm^{-1} to 458 cm^{-1} and the line at 248 cm^{-1} to 226 cm^{-1}, respectively. Furthermore, the intensity of the line of lowest frequency at 78 cm^{-1} for $x(zy)z$ shows about twice the intensity of the line for $x(yz)y$.

But these frequency shifts are not only caused by the different direction of observation. We have done a nearly forward scattering experiment with a KDP crystal and have found comparable effects only by changing $x(zy)x$ and $x(yz)x$, shown for $T < T_c$ in Fig. 3, where the scattering angle decreases from the bottom to the top of the figure (from 5° to 2°).

The line shape of the band at 70 cm^{-1} with the foot reaching to the exciting line

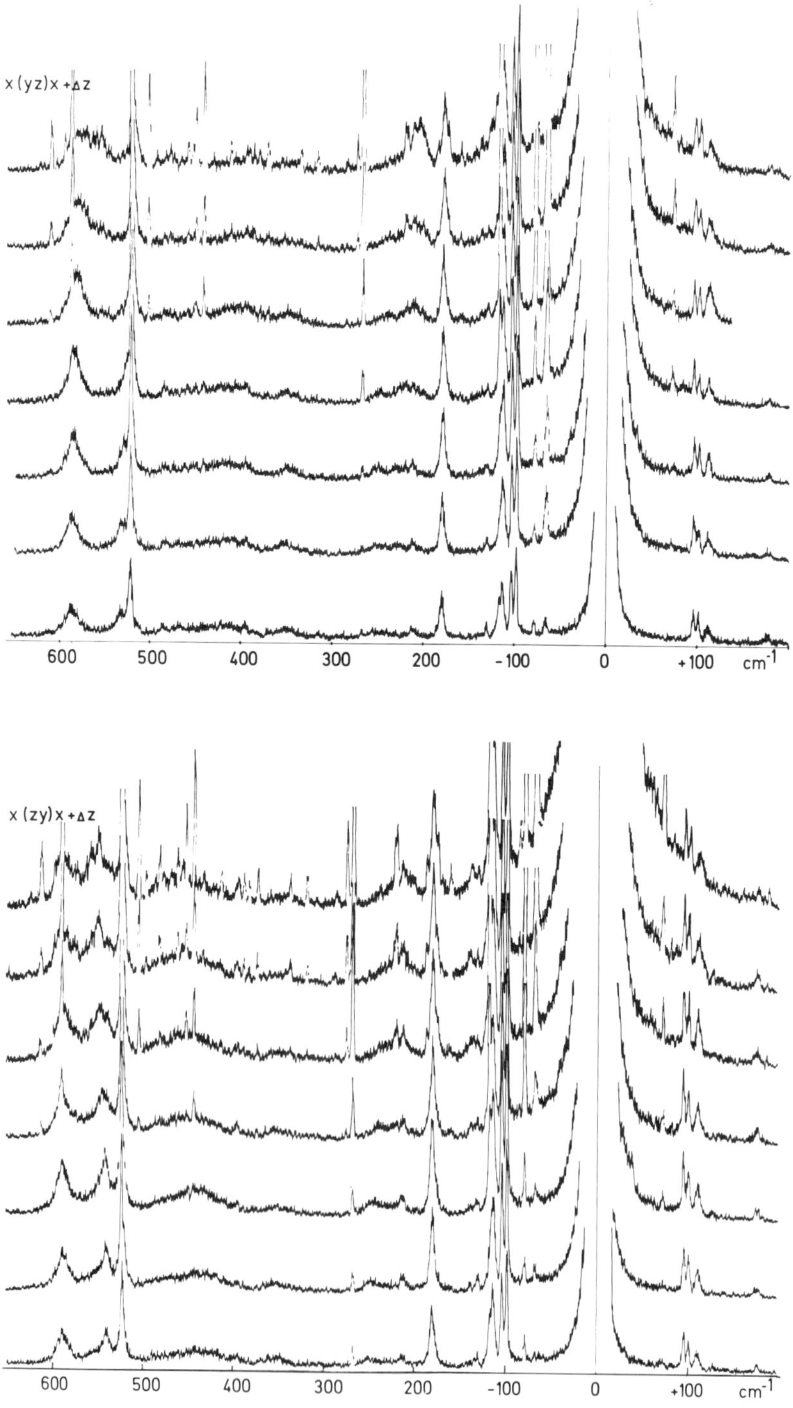

Fig. 3. Nearly forward scattering experiments in KH_2PO_4 in the low temperature phase. Parameter is the scattering angle changing from about 5° to about 2° (spectral slit width 2 cm^{-1}).

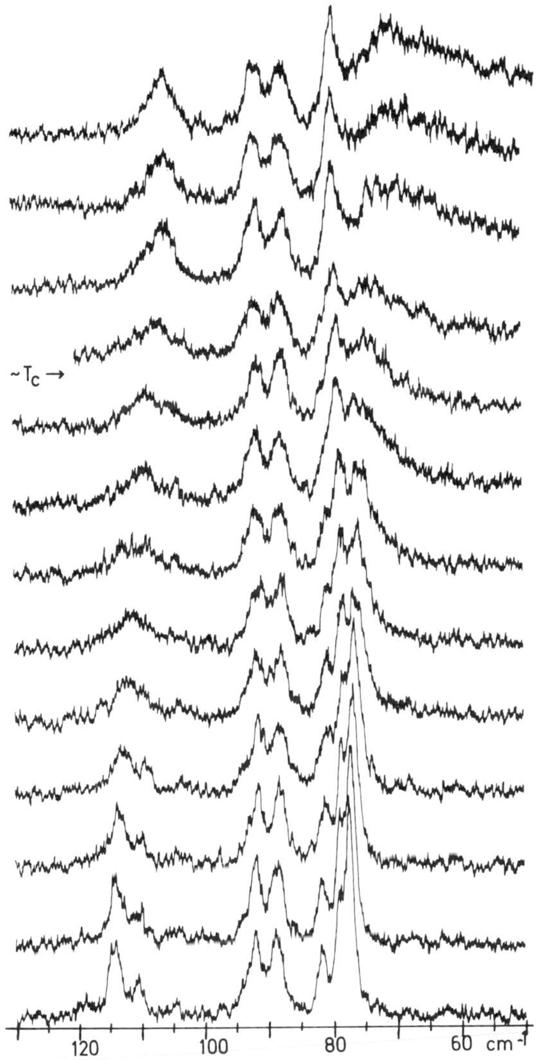

Fig. 4. $x(yz)y$ spectra of RbH_2PO_4 in the low frequency region above and below T_c.

has been studied close to the transition temperature. Figure 4 shows the change of the low frequency $x(yz)y$ spectrum in the temperature range from 150 K to 70 K. Near the transition point the temperature was changed in steps of about 1°. One can see the vanishing of the broad foot and the sharpening of the line continuously. At the transition temperature the line at 78 cm^{-1} splits into two indicated by the not yet resolved shoulder. With decreasing temperature the line at 74 cm^{-1} sharpens more and more, increases in intensity and is shifted by about 3 cm^{-1} to higher frequencies. In the spectrum at the bottom of Fig. 4 one can see only the new line at 81 cm^{-1} caused by the lifted degeneracy and a shoulder of the new line now at 78 cm^{-1}.

Finally we want to pay attention to the frequency range from 1500 cm^{-1} to 3500 cm^{-1}

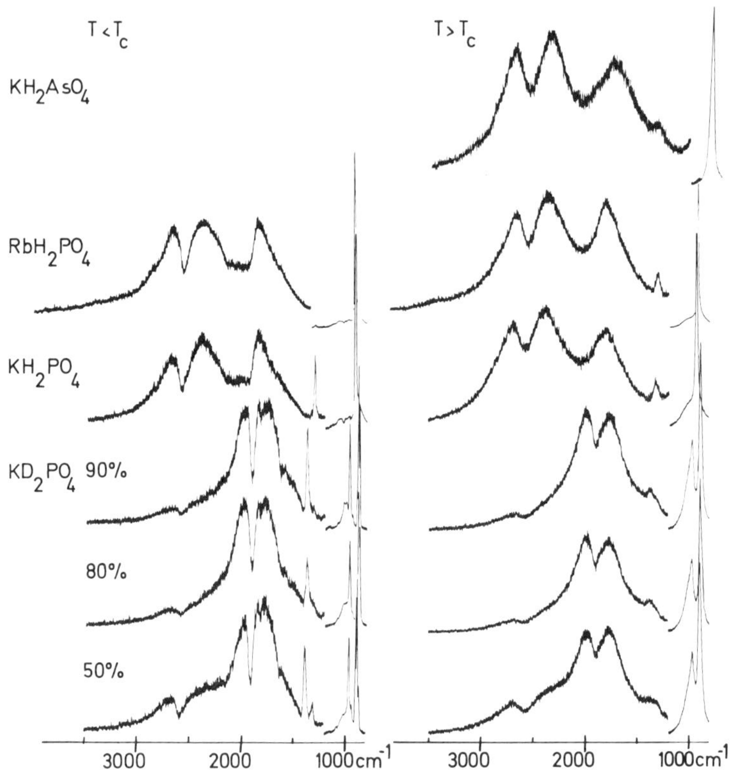

Fig. 5. High frequency region of the phonon spectra of RbH_2PO_4, KH_2PO_4, KD_2PO_4 (90%, 80%, 50% deuteration) above and below T_C; slit width 3 cm^{-1}, scanning speed 200 cm^{-1}/min, time constant 0.25 s.

These bands have been assigned to O–H vibrations[3] for RDP and RDA crystals. We have studied these bands for KDP, RDP and KD*P above and below T_c and found, contrary to ref. 3 that these vibrations have B_1 symmetry above T_c. Furthermore we found these bands to be strongly influenced by deuteration, mainly below T_c, where rather peculiar line shapes are observed (Fig. 5).

CONCLUSIONS

We have listed some peculiarities in the Raman spectra of crystals of the KDP-family in their paraelectric and ferroelectric phases. These effects are not yet taken into account in theoretical treatments for explaining the phonon spectra of these crystals. It might be possible that the coupling of proton motions to the phonons plays an important role on these effects.

ACKNOWLEDGEMENTS

I wish to thank Prof. Haussühl, Cologne, for the oriented single crystals and the Deutsche Forschungsgemeinschaft for the apparatus and the Bundesministerium für Bildung und Wissenschaft for the cryostat from Oxford Instruments.

REFERENCES

1 *Light Scattering in Solids* (Ed. M. Balkanski), Flammarion, Paris, 1971, p. 415.

2 J. F. Scott and C. M. Wilson, *Solid State Commun.* **10**, 597 (1972).

3 E. V. Chisler, J. T. Savatinova and V. Yu. Davydov, *Sov. Phys. Solid State,* **13**, 1339 (1971); *Sov. Phys. Solid State* **13**, 1635 (1972).

21. TEMPERATURE DEPENDENT RAMAN SCATTERING IN SODIUM NITRATE*

A. D. Prasad Rao, R. S. Katiyar and S. P. S. Porto
Departments of Physics and Electrical Engineering, University of Southern California, Los Angeles, California, USA

INTRODUCTION

Raman scattering in $NaNO_3$ had been studied as early as 1938.[1,2] Since then a number of workers have tried to explain various anomalies in this crystal and a fair account of this work has been reported by Rousseau *et al.*[3] By cooling the crystal Rousseau *et al.* could give a definite assignment of low frequency vibrations belonging to E_g symmetry. These E_g vibrations also appear in the A_{1g} orientation which according to Porto[4] and co-workers is due to the higher birefringence in these crystals. Chisler[5] has studied the temperature dependence of spectrum of $NaNO_3$ and tried to correlate the structure of the 1385 cm^{-1} mode of E_g symmetry to the phase transition which occurs at $T_\lambda = 548$ K. We have studied the Raman spectrum of single crystals of $NaNO_3$ from room temperature to a few degrees above the transition temperature. Our studies confirm the earlier observations of Chisler.[5] We observed, however, at least two new modes having A_{1g} symmetry. The temperature dependence of these modes show that they are intimately connected with the structural phase transition in this crystal.

Sodium nitrate belongs to the space group $D_{3d}^6(R\bar{3}c)$ with two formula units per primitive cell at room temperature. Group theoretical calculations show that the 27 long wavelength optical modes of vibrations are distributed among various symmetry species as follows

$$1A_{1g} + 2A_{1u} + 3A_{2g} + 3A_{2u} + 4E_g + 5E_u$$

The normal modes corresponding to species A_{1g} and E_g are active in Raman effect and those of A_{2u} and E_u are infrared active. The polarizability tensors for A_{1g} and E_g are given below

$$A_{1g}: \begin{pmatrix} a & 0 & 0 \\ 0 & a & 0 \\ 0 & 0 & b \end{pmatrix} \quad E_g^{(1)}: \begin{pmatrix} c & 0 & 0 \\ 0 & -c & d \\ 0 & d & 0 \end{pmatrix} \quad E_g^{(1)}: \begin{pmatrix} 0 & -c & -d \\ -c & 0 & 0 \\ -d & 0 & 0 \end{pmatrix}$$

* Supported by National Science Foundation.

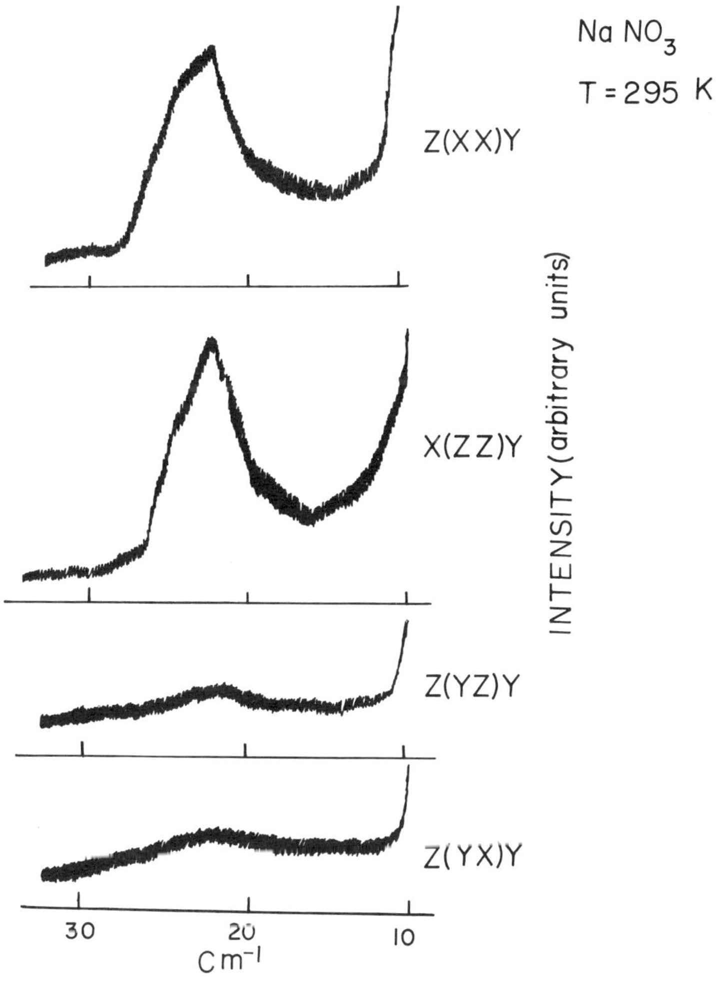

Fig. 1. Polarization measurements of the 22 cm^{-1} mode in $NaNO_3$ at room temperature. The leakage in the *yz* and *yx* polarizations could be due to the relatively high birefringence of the crystal.

If we assume tight binding among the atoms of a NO_3 ion then the spectrum of $NaNO_3$ can be divided into an external and an internal vibration region. In the external vibration region NO_3 ions move as a whole against the Na ions and the translatory type of lattice vibrations of the following symmetry species are possible

$$A_{2g}(NO_3, z) + A_{1u}(Na, z) + 2A_{2u}(Na, NO_3, z) + E_g(NO_3, x, y) + 3E_u(Na, NO_3, x, y)$$

where the notation in the parentheses indicates the type of ions involved in the vibration and the direction of their motion. In addition to this translatory type of vibrations, there would be a rotatory type of vibrations due to the NO_3 ion in the external frequency

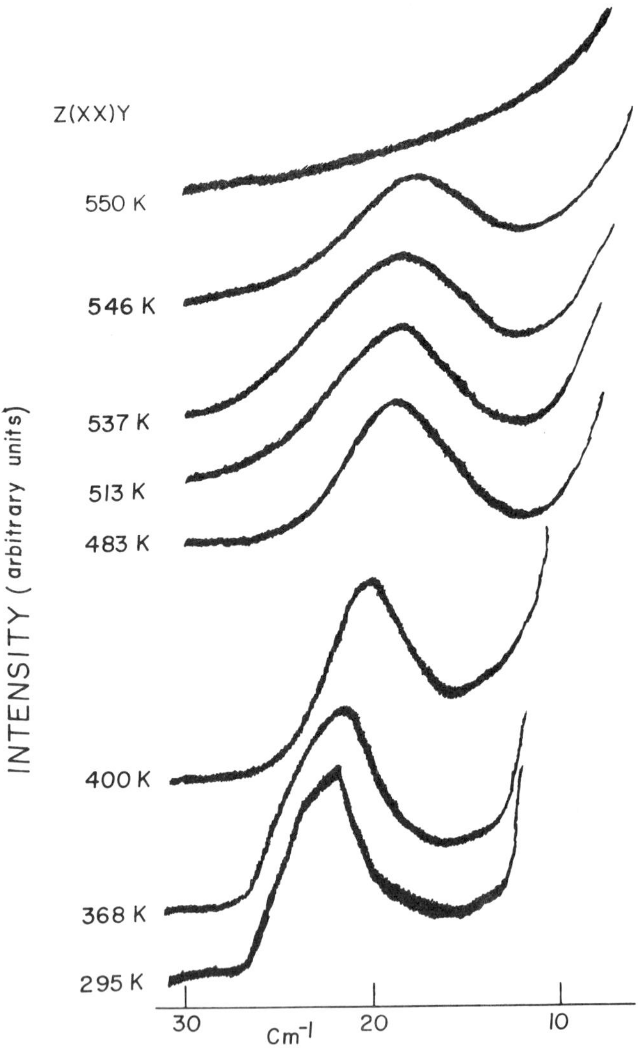

Fig. 2. Temperature measurements of the 22 cm^{-1} mode. The shift in frequency is about 5 cm^{-1} and the half width increases by 4 cm^{-1}. The transition point is 548 K.

region; these are distributed as follows

$$(A_{2g} + A_{2u})\,(R_z) + (E_g + E_u)\,(R_x, R_y)$$

where the notation in the parentheses indicates the direction about which the NO_3 ion librates.

In the high temperature phase the crystal has $D_{3d}^5(R\bar{3}m)$ symmetry with one formula unit per primitive cell. This would involve the reorientation of the NO_3 ion in this phase. Two theories exist. According to one NO_3 groups are rotating freely around the trigonal axis.[7] The other theory considers a two-position disorder model[8] in which NO_3 groups

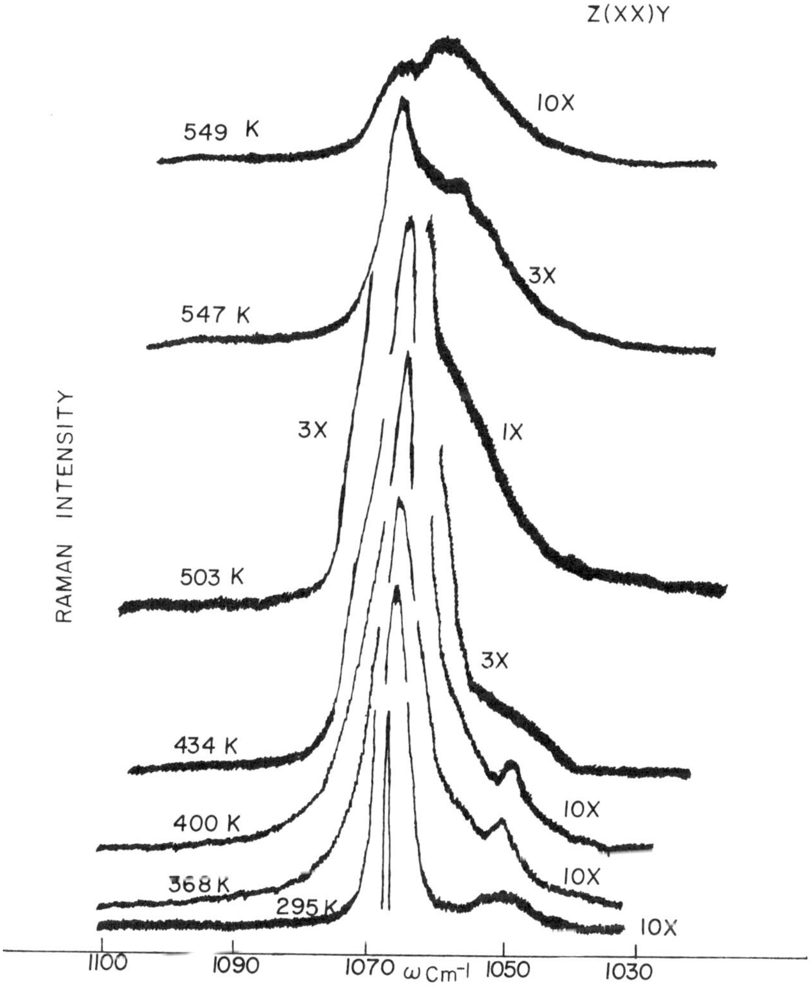

Fig. 3. The internal modes 1068 and 1053 cm^{-1} belonging to the A_{1g} representation with temperature. Fitted to an uncoupled damped harmonic oscillator model. Intensity scale changes are marked by 1 X, 3 X, etc. The drawing is not to scale exactly.

undergo hindered rotation.

EXPERIMENTAL

Large single crystals of $NaNO_3$ were grown from the solution by slow evaporation technique and were oriented with x and z axes as the crystallographic axes of a trigonal system, and the y axis was chosen perpendicular to these axes. The crystal was placed in a furnace equipped with a temperature controller with an accuracy of ±0.5°.

Raman experiments were performed using a double monochromator using the lines 5145 and 4880 Å of the argon ion laser. The spectra were checked against the presence of any ghosts or fluorescence. Some of the spectra thus recorded are shown in Figs. 1, 2 and 3. The various polarization studies reported in Fig. 1 insure that the mode at 22 cm^{-1} corre-

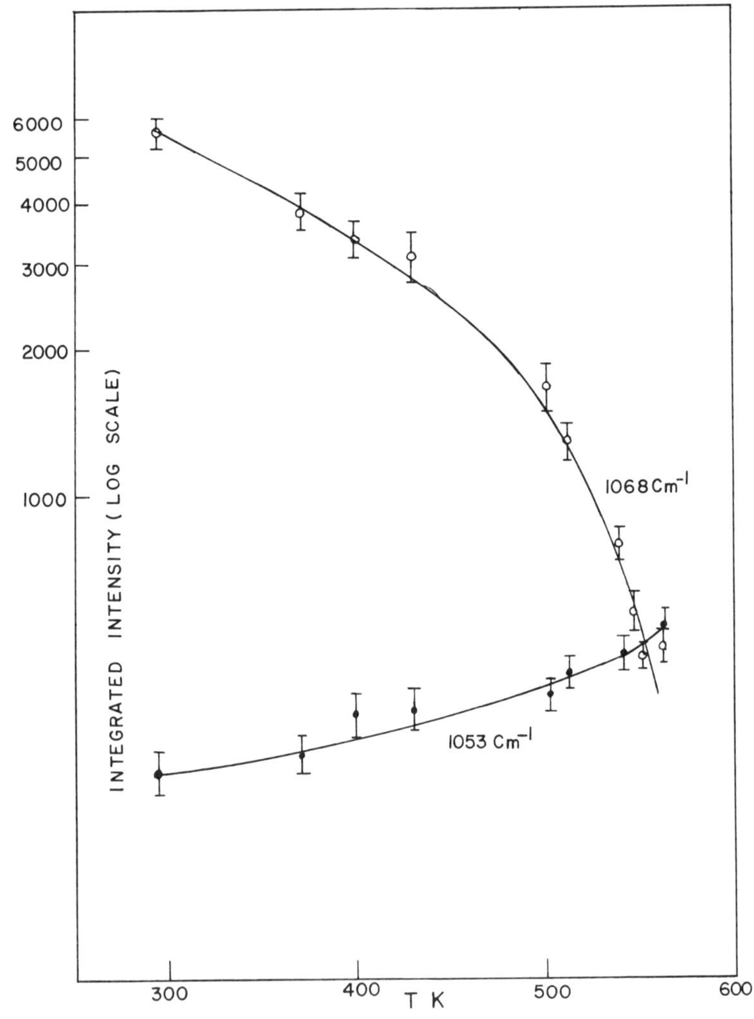

Fig. 4. Integrated intensity with temperature of the 1053 and 1068 cm^{-1} modes. Logarithmic plot.

sponds to A_{1g} symmetry. Group theory however does not permit the presence of such a mode in the low frequency region. In Fig. 2 we have followed the temperature dependence of this mode and it is interesting to know that this mode disappears at the transition temperature. In the internal mode region we observed two phonons belonging to the A_{1g} symmetry. The intensity ratio of these two modes at room temperature is 1 : 30. However, as one approaches the transition temperature the intensity of the weaker phonon increases and that of the stronger phonon decreases, and at the transition temperature both phonons have almost the same intensity (Fig. 4). The spectra in the E_g orientation are the same as reported by earlier workers.[3,5] In Fig. 5 we report the temperature dependence of all the modes.

In order to compare the Raman spectrum of $NaNO_3$ with that of calcite (which has identical structure at room temperature), we have carried out similar investigations on

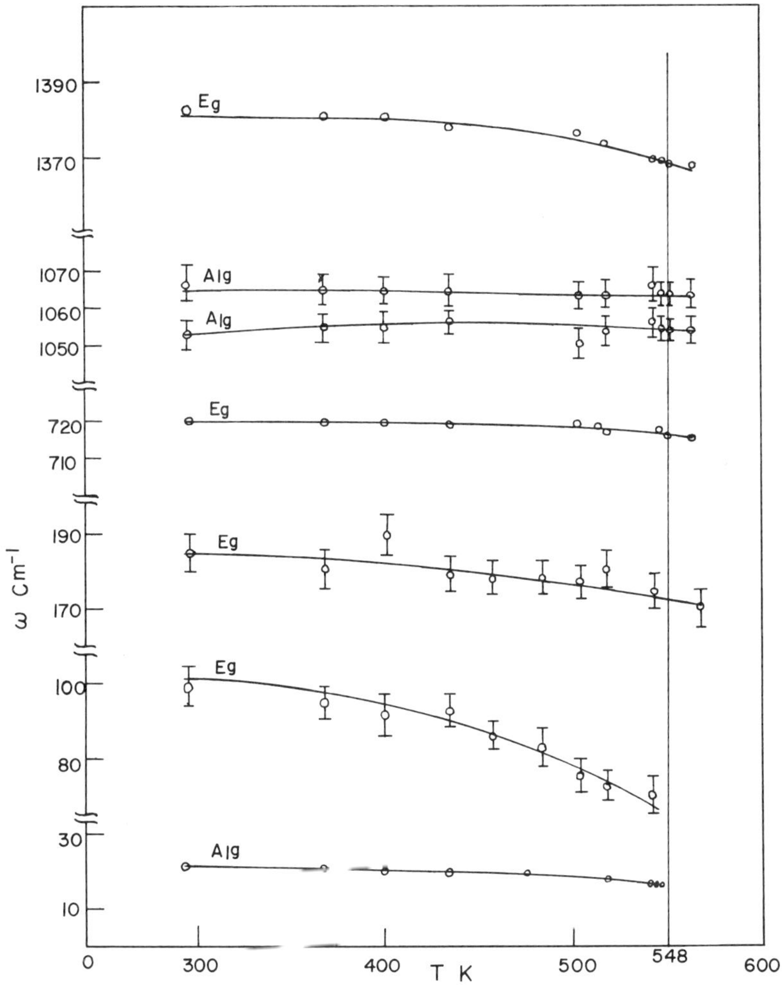

Fig. 5. Raman frequencies with temperature in the external and internal mode regions. The modes at 98, 185 cm^{-1} are fitted to a coupled damped harmonic oscillator model. The modes at 1053 and 1068 cm^{-1} are fitted to an uncoupled damped harmonic oscillator model.

calcite. The frequencies corresponding to A_{1g} symmetry in calcite are 26 cm^{-1}, 1086 cm^{-1} and 1071 cm^{-1}, which compare well with those observed in $NaNO_3$. Later two modes were also seen by Krishnamurti.[9]

DISCUSSION

The presence of extra phonons in these crystals indicates that there is some kind of distortion present at the site of the NO_3 ion. Since only the nitrate ions which are active in Raman effect vibrate in the symmetry species, only the distortion of the nitrate ion could possibly activate the modes which are otherwise forbidden in Raman effect. The phonons which are inactive in both Raman and infrared spectra belong to A_{2g} and A_{2u} symmetry.

This totally different picture of the infrared and Raman spectra indicates that the crystal possesses the centre of symmetry and therefore the A_{1u} modes could not be Raman active. Modes belonging to A_{2g} symmetry would become Raman active if the two-fold axes at the site of the nitrogen atom were not present. This is only possible if the nitrogen atoms are pushed out of the NO_3 plane along the z axis in such a way that they still preserve the centre of symmetry at the sodium ion. This indicates some kind of order-disorder in which the nitrogen atom moves up and down the NO_3 plane. This kind of distortion would reduce the instantaneous symmetry of the crystal to C_{3i} with the symmetry C_3 at the site of the nitrogen atom. If this happens, then all the modes belonging to A_{2g} symmetry become Raman active and those belonging to A_{1u} symmetry would be infrared active. Thus the weakly observed modes at 1053 cm^{-1} (Fig. 3) and 22 cm^{-1} (Fig. 1) belong to A_{2g} symmetry. The mode at 22 cm^{-1} decreases in frequency but only slightly as the transition temperature is approached, where it disappears abruptly (Fig. 2). This indicates that this is a librational mode in which the nitrate ion librates about the z axis. The amplitude of libration increases as the temperature is raised where it crosses the potential barrier to become a free rotor. The change in ratios of the intensities of the modes at 1053 cm^{-1} and 1068 cm^{-1} with temperature indicates that the state of disorder increases with temperature. The spectra in general indicate that the transition is gradual and it is complete at 548 K. This may also explain the low frequency mode in the region of 15–30 cm^{-1} observed by Schroeder *et al.*[10] in their infrared spectrum as belonging to A_{1u} symmetry. This mode however involves the motion of the sodium ions against the nitrate ions along the z axis and is not a librational mode.

CONCLUSION

Our Raman measurements on calcite and sodium nitrate reveal that there are at least two extra modes belonging to A_{1g} symmetry which could not be accounted for by group theory assuming D_{3d}^6 structure of the crystal. We believe that this structure is true only on the average and the instantaneous structure of the crystal is C_{3i}. Such a disorder in their structure is produced due to the displacement of the nitrogen atom out of the NO_3 plane along the z axis. In this situation the modes belonging to A_{2g} symmetry become Raman active and those belonging to A_{1u} symmetry become infrared active. The disappearance of the 22 cm^{-1} mode at the transition temperature indicates free rotation of the NO_3 ion about the z axis in the high temperature phase of $NaNO_3$. The possibility of hindered rotation in this phase is however not ruled out,[9] and for this one has to look for an overdamped mode in the low frequency region.

REFERENCES

1 P. Pattabhiramayya, *Proc. Ind. Acad. Sci.* **A7**, 229 (1938).
2 T. M. K. Nedungadi, *Proc. Ind. Acad. Sci.* **A8**, 397 (1938).
3 D. L. Rousseau, R. E. Miller and G. E. Leroi, *J. Chem. Phys.* **48**, 3409 (1968).
4 S. P. S. Porto, J. A. Giordmaine and T. C. Damen, *Phys. Rev.* **147**, 608 (1966).
5 E. V. Chisler, *Sov. Phys.–Solid State* **11**, 1032 (1969); *Fizika Tverdoga Tela* **11**, 1272 (1969).
6 F. C. Kracek, E. Posnjack and S. B. Hendricks, *J. Amer. Chem. Soc.* **53**, 3339 (1931).
7 Y. Sato, K. Gesi and Y. Takagi, *J. Phys. Soc. Japan* **19**, 449 (1964).
8 Y. Shinnaka, *J. Phys. Soc. Japan* **19**, 1281 (1964).
9 D. Krishnamurti, *Proc. Ind. Acad. Sci.* **46A**, 183 (1957).
10 R. A. Schroeder, C. E. Weir and E. R. Lippincott, *J. Res. Natl. Bur. Std.* **A66**, 407 (1962).

22. THE RAMAN SPECTRUM OF SINGLE CRYSTAL $PbTiO_3$

R. A. Frey*
Physics Department, Fisk University,
Nashville, Tennessee, USA

INTRODUCTION

$PbTiO_3$ belongs to the perovskite family comprising $BaTiO_3$, $KNbO_3$ and $KTaO_3$ as some of the outstanding members. All these materials are ferroelectrics, having one or more phase transitions. However, $PbTiO_3$ is exceptional in several aspects: its tetragonal distortion is the largest of all, as is the spontaneous polarization in the tetragonal phase as well as the second harmonic generation coefficients. Likewise, the thermodynamic quantities are a maximum for $PbTiO_3$ at the transition point. It is obvious that these outstanding properties are intimately related to the large tetragonal distortion $c/a = 1.063$.[1]

The transition at decreasing temperature from the cubic (paraelectric) phase to the tetragonal (ferroelectric) phase near 495°C was first reported in 1950.[2] The tetragonal phase is stable through a wide temperature range. Two further phase changes in the powder near −100 and −150°C have been detected in X-ray and dielectricity measurements[3] though their effect is minute. However, reliable data on single crystals have not been available until recently, as good crystals are difficult to grow. Since Cochran's paper on lattice dynamics and crystal instability,[4] extensive Raman studies have been carried out on $BaTiO_3$ in the search for soft modes accompanying the phase transition. The results are difficult to interpret as the spectra are dominated by second order Raman scattering obscuring the first order features. In $PbTiO_3$, due to the large value of c/a, it is expected that the Raman spectrum at room temperature is predominantly of first order, making the observation of soft phonons promising. Existing Raman data on single crystal $PbTiO_3$ [5] are not complete and are partially contradictory to infrared reflectance measurements,[6] especially as to the dominating role of either short or long range forces in the crystal.

Crystal samples were obtained from J. P. Remeika, and the method of growth has been described by the same author.[7] The multi-domain crystals were made single-domain by combined heat treatment and axial pressure or applied d.c. field. The crystals were then cut and polished giving crystals having typical sizes of $1 \times 1 \times 0.6\ mm^3$.

---

* Present address: Department of Physical Chemistry, Swiss Federal Institute of Technology, Zürich, Switzerland.

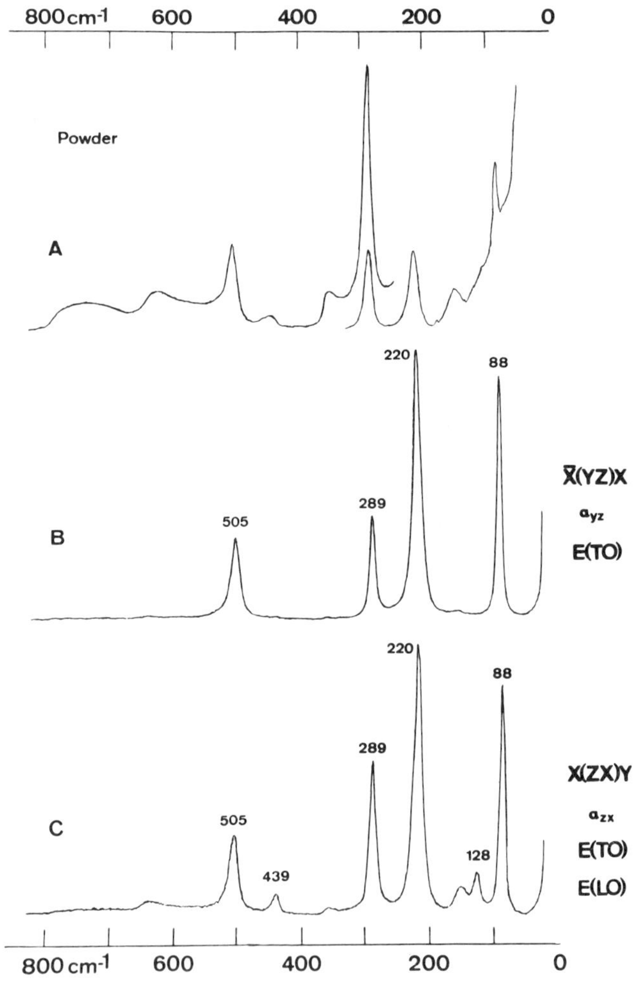

Fig. 1. Raman spectra of $PbTiO_3$. (a) Powder spectrum. (b) and (c) E-type phonons in the single crystal.

RAMAN SPECTRA

As there are five atoms per unit cell in the tetragonal phase, $PbTiO_3$ has fifteen vibrational degrees of freedom in the $k = 0$ limit, delineated as $3F_{1u} + F_{2u}$ in the cubic phase (the acoustic F_{1u} phonon not being included). In the tetragonal phase, still keeping the $k = 0$ limit, the branches decompose according to $3A_1 + 3E + B_1 + E$. All phonons are Raman active and all but B_1 are infrared active. This fact of simultaneous Raman and infrared activity has important consequences: some of the phonons are accompanied by long range electric fields which shift their frequencies away from the $k = 0$ values and give rise to some frequency variation with the direction of phonon propagation.[8] This makes the classification of the phonons in the tetragonal class C_{4v} irrelevant in a rigorous sense. However, experimental evidence suggests, as will be discussed later, that phonons propagating along the a and c axes are completely degenerate. It is for this reason that the phonon classification according to C_{4v} is retained. Since the phonons derived from F_{1u} are both Raman and infrared active, the electric field accompanying the longitudinal phonon will raise its frequency above that of the transverse phonon as expressed by the Lyddane–Sachs–Teller relation[9]

$$\frac{\epsilon^0}{\epsilon} = \left(\frac{\omega_l}{\omega_t}\right)^2 \tag{1}$$

Consequently, the Raman spectrum may show phonons according to

$$3A_1(\mathrm{TO}) + 3A_1(\mathrm{LO}) + 3E(\mathrm{TO}) + 3E(\mathrm{LO}) + B_1 + E(\mathrm{TO}) + E(\mathrm{LO})$$

Furthermore, for E(TO) phonons with $k \neq 0$ the energy may be different for propagation along the a and the c axis. All splittings and degeneracy liftings are related to the action of long range electrostatic forces and to short range anisotropic forces,[8] the former causing the TO–LO splitting, the remaining splittings being due to the latter forces. Thus from the observed splittings some conclusions may be drawn as to the role of these forces in the crystal.

Raman spectra were recorded on a Cary 81 spectrometer using a He–Ne laser for excitation. Spectra were obtained at right angle scattering as well as with forward and backward scattering. The notation of Damen *et al.*[10] is adapted to describe the geometrical arrangement.

Figure 1(a) shows the Raman spectrum of $PbTiO_3$ powder. The spectrum does not show any polarization effect and offers little information towards the assignment of the phonons. Single crystal spectra are given in Figs. 1(b) and 1(c). The backscattering spectrum $\bar{x}(yz)x$ is due to the polarizability tensor component α_{yz}. In this particular geometry, only E(TO) phonons travelling along the $\bar{x}$ axis should be observed. These are designated in the spectrum (b) by their respective Raman shifts. The 90° scattering spectrum $z(yz)x$, showing E(TO) phonons propagating in the xz plane at an angle of 45° to the z axis, is identical to the (b) spectrum. The conclusion drawn from this is that the frequency of a particular E(TO) phonon does not depend on the direction of propagation. The spectrum $x(zx)y$ as displayed in trace (c) is due to E(TO) and E(LO) phonons propagating in the xy plane. The E(TO) phonons are present at their proper locations. Additional lines, assigned to E(LO) phonons are at 439 and 128 cm^{-1} and, by intensity comparison with spectrum (b), also at 289 cm^{-1}. The additional weak features cannot be assigned to any E(LO) phonon; the fourth E(LO) phonon thus could not be observed. The longitudinal and transverse E phonons at 289 cm^{-1} derived from the F_{2u} branch are degenerate; the same situation was observed in $BaTiO_3$.[11]

Phonons of symmetry A_1 are shown in Fig. 2. Spectra (a) and (b) are due to the polarizability tensor components α_{zz} and α_{xx} respectively. Three A_1(TO) phonons are expected in these spectra. They are denoted in spectrum (a) by their frequencies. As the lines are comparatively weak, some extra bands are present, partly due to leakage of improperly polarized light and partly due to deviation from ideal alignment of the crystal. Furthermore, the crystals are far from being perfect owing to the presence of lattice defects. The treatment necessary to get single-domain crystals may cause considerable internal stress, particularly if 90° domains are involved, and may lead to various lattice defects which in turn may cause depolarization of the light. The strong line at 289 cm^{-1}

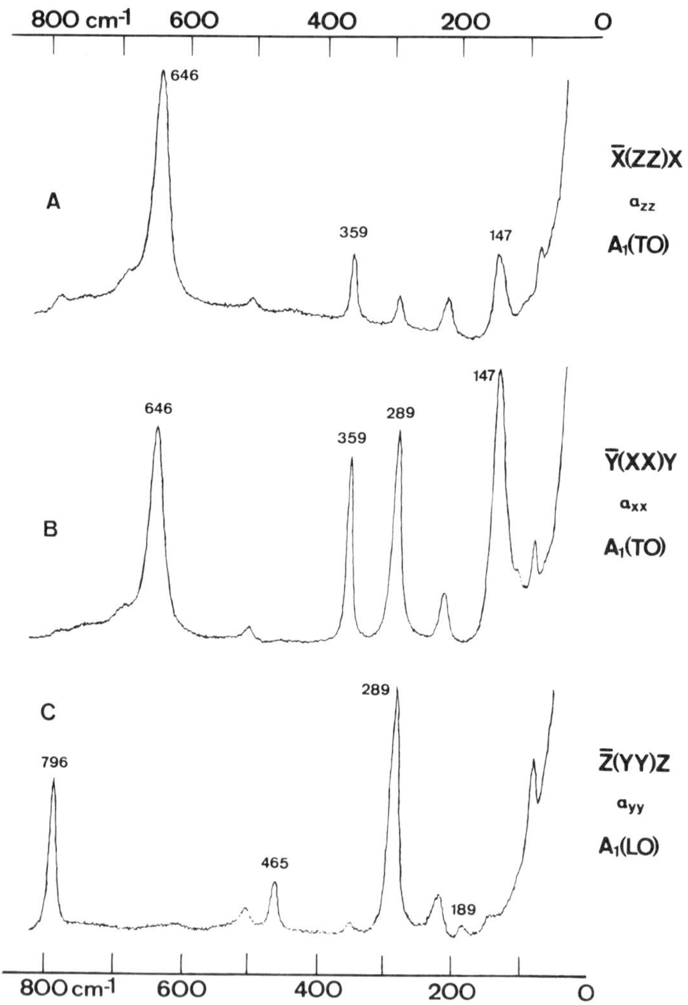

Fig. 2. Raman spectra of $PbTiO_3$. A_1-type phonons in the single crystal.

Table 1. Lattice modes in $PbTiO_3$.

		Ref. 6	Ref. 5	Quasi modes			Ref. 6	Ref. 5
E(TO1)	88	88	89	109	A_1(TO1)	147	150	126[a]
E(LO1)	128	130	128	161	A_1(LO1)	189	170	185[a]
E(TO2)	220	220	220	310	A_1(TO2)	359	360	359
E(LO2)	439	440		451	A_1(LO2)	465	415	465
E(TO3 + LO3) + B_1	289	290	290					
E(TO4)	505	505	506	559	A_1(TO4)	646	640	645
E(LO4)		680	749[a]	773	A_1(LO4)	796	790	796

[a] Calculated from quasi mode spectra.

originates from the B_1 phonon, active in the (b) spectrum. The A_1(LO) phonons are presented in trace (c). The three expected phonons are at 796, 465 and 189 cm^{-1}. Again the B_1 phonon manifests itself by a prominent line at 289 cm^{-1}.

The observed Raman lines are compiled in Table 1 together with previously reported Raman and infrared data. The general agreement is good; the present results complete the Raman data and suggest a new assignment for the A_1(TO1) phonon, in accordance with infrared reflectance results. The line at 147 cm^{-1} ascribed to a second order Raman line in ref. 5 does not exhibit the strong temperature dependence of the intensity as commonly observed for second order lines. Table 1 also gives the frequencies of quasi modes observed in $z(yy)x$ and $x(zx)z$ spectra. In these spectra the phonons travel at an angle of 45° to the z axis with the polarization vector lying in the xz plane. Consequently they are not phonons of pure symmetry or pure TO or LO phonons, but are related to them by a formula due to Loudon[8]

$$\frac{\omega_{A_1}^2(\mathrm{TO})\,\epsilon_{\parallel}^0 - \omega_q^2\,\epsilon_{\parallel}}{\omega_{A_1}^2(\mathrm{TO}) - \omega_q^2}\cos^2\theta + \frac{\omega_E^2(\mathrm{TO})\,\epsilon_{\perp}^0 - \omega_q^2\,\epsilon_{\perp}}{\omega_E^2(\mathrm{TO}) - \omega_q^2}\sin^2\theta = 0 \qquad (2)$$

where θ is the angle between the z axis and the direction of propagation of the phonon. In the limit of dominating long range forces, the phonons are essentially TO or LO phonons and Eqn. (2) reduces to

$$\begin{aligned} \omega_q^2 &= \omega_{A_1}^2(\mathrm{TO})\sin^2\theta + \omega_E^2(\mathrm{TO})\cos^2\theta \\ \omega_q^2 &= \omega_{A_1}^2(\mathrm{LO})\cos^2\theta + \omega_E^2(\mathrm{LO})\sin^2\theta \end{aligned} \qquad (3)$$

For short range forces dominating over the long range forces, the phonons are predominantly A_1 or E phonons and Eqn. (2) gives

$$\begin{aligned} \omega_q^2 &= \omega_{A_1}^2(\mathrm{TO})\sin^2\theta + \omega_{A_1}^2(\mathrm{LO})\cos^2\theta \\ \omega_q^2 &= \omega_E^2(\mathrm{TO})\cos^2\theta + \omega_E^2(\mathrm{LO})\sin^2\theta \end{aligned} \qquad (4)$$

These formulae are derived for crystals having only one infrared active mode[8] and do not necessarily hold for $PbTiO_3$. Furthermore, Eqn. (1) has to be used in a modified form, given by Cochran *et al.*[12]

$$\prod_i \frac{\omega_{A_1}(\mathrm{LO}_i)}{\omega_{A_1}(\mathrm{TO}_i)} = \left(\frac{\epsilon_{\parallel}^0}{\epsilon_{\parallel}}\right)^{1/2} ; \prod_i \frac{\omega_E(\mathrm{LO}_i)}{\omega_E(\mathrm{TO}_i)} = \left(\frac{\epsilon_{\perp}^0}{c_{\perp}}\right)^{1/2} \qquad (5)$$

In order to elucidate the role of short range and long range forces, the observed A_1–E and TO–LO splittings are collected in Table 2. Looking at the magnitude of the splittings it is obvious that the influence of both short range and long range forces are comparable. Tentatively, Eqns. (3) and (4) are applied to the observed quasi modes; the results are given in Table 3. These results suggest that the quasi mode spectra can be better understood if the long range forces are assumed to dominate. Moreover, going back to spectra (b) and (c) in Fig. 1, one may see that the E(TO) modes appear at exactly the same positions in both spectra although the phonons travel in different directions. In other words the phonons are either pure TO or LO phonons whatever their direction of

Table 2. Mode splitting in $PbTiO_3$.

Short range force splitting $E–A_1$		Long range force splitting TO–LO	
TO1	59	$E(1)$	40
LO1	61	$A_1(1)$	42
TO2	139	$E(2)$	219
LO2	26	$A_1(2)$	106
TO4	141	$E(4)$	
LO4		$A_1(4)$	150

Table 3. Quasi modes in $PbTiO_3$.

Calculated Eqn. (3)	Calculated Eqn. (4)	Observed
121	110	109
161	169	161
298	348	310
452	416	451
579		559
	725	773

propagation may be. This again is characteristic for the action of long range electrostatic forces.

We thus arrive at a fairly surprising conclusion that short range forces have comparable effects to the long range forces as far as the $A_1–E$ splitting is concerned. They are, however, insensitive with respect to the direction of phonon propagation. One is therefore tempted to use Eqn. (3) to compute the unobserved E(LO4) mode with the help of the quasi mode at 773 cm^{-1}. The result is 748 cm^{-1} for E(LO4). Using Eqn. (5) and the values of $\epsilon_{\parallel} = 7.081$ and $\epsilon_{\perp} = 7.134$ from a recent publication by Singh *et al.*,[13] one obtains for the static clamped dielectricity constants $\epsilon_{\parallel}^{0} = 29.8$ and $\epsilon_{\perp}^{0} = 185$.

PHASE TRANSITIONS

The Raman spectrum of $PbTiO_3$ was observed as the temperature was increased towards the transition point at 495°C. All Raman lines were more or less lowered to some extent except for the E(TO1) mode, which showed a drastic reduction in the Raman shift. At 485°C its position was near 58 cm^{-1}. At higher temperatures it escaped detection due to low peak height and large band width, but there is strong evidence that this mode represents a soft phonon, accompanying the phase transition.

Particular attention was paid to the low temperature region. However, with cooling rates not exceeding 0.3°/min as suggested in ref. 3, no phase change was indicated in the Raman spectrum near −100°C, even at a spectral slit width of 0.8 cm^{-1}. This is not necessarily contradictory to the observation reported in ref. 3, as these measurements were made on powder samples. It is conceivable that the volume/surface ratio may have a critical influence on crystal stability, especially for small particles as those present in the powder. Moreover, the domain structure in powder particles and its influence on phase stabilization are completely unknown. We thus may say that low temperature phase changes in $PbTiO_3$ powder are possible but the single crystal has a stable tetragonal phase down to 77 K.

REFERENCES

1 H. D. Megaw, *Proc. Phys. Soc. (London)* **58**, 10 (1946).
2 G. Shirane, S. Hoshino and K. Suzuki, *Phys. Rev.* **80**, 1105 (1950).
3 J. Kobayashi *et al., Phys. Rev. Letters* **103**, 830 (1956).
4 W. Cochran, *Advan. Phys.* **9**, 387 (1960).
5 G. Burns and W. A. Scott, *Phys. Rev. Letters* **25**, 167 (1970).
6 N. E. Tornberg and C. H. Perry, *J. Chem. Phys.* **53**, 2946 (1970).
7 J. P. Remeika and A. M. Glass, *Mater. Res. Bull.* **5**, 37 (1970).
8 R. Loudon, *Advan. Phys.* **13**, 423 (1964).
9 R. H. Lyddane, R. G. Sachs and E. Teller, *Phys. Rev.* **59**, 673 (1941).
10 T. C. Damen, S. P. S. Porto and B. Tell, *Phys. Rev.* **142**, 570 (1966).
11 M. DiDomenico *et al., Phys. Rev.* **174**, 522 (1968).
12 W. Cochran and R. A. Cowley, *J. Phys. Chem. Solids* **23**, 447 (1962).
13 S. Singh, J. P. Remeika and J. R. Potopowicz, *Appl. Phys. Letters* **20**, 135 (1972).

23. RAMAN ACTIVE EXTERNAL PHONONS AND THE PHASE TRANSITIONS IN SODIUM AND THALLOUS AZIDES

Z. Iqbal
Explosives Division, Feltman Research Laboratory, Picatinny Arsenal, Dover, New Jersey, USA

INTRODUCTION

The metal azides form a group of halide-like crystals with a linear anion, which exhibit a variety of second order structural transformations.[1] A study of the Raman active external modes (where external phonons involve the centre of mass oscillations of the ions), in particular the librational oscillation of the azide ion in such crystals, can provide useful information regarding the microscopic nature of these transitions. With this aim in view we have collected together data on the Raman active librational mode frequencies in a whole series of metal azides, in an attempt to correlate this frequency with the ionic character and the thermodynamic stability of the crystals at atmospheric pressure. Following this, temperature dependent Raman data on two crystals, sodium azide (NaN_3) and thallous azide (TlN_3), which undergo a low temperature phase transition, are discussed and a possible mechanism of the transition process is indicated.

EXPERIMENTAL

The Raman spectra were excited with the 6471 Å (200 mW) line of an argon-krypton mixed gas laser. The light scattered at 90° was analysed by a Perkin-Elmer LR 3 double monochromator coupled to a photoelectric detection system. Single crystals used had dimensions of approximately $3 \times 3 \times 3$ mm^3, were grown by slow cooling and/or evaporation from aqueous solutions and were oriented by X-ray diffraction. In the case of lithium, strontium and caesium azides polycrystalline samples were used for recording the Raman spectra. The variable temperature measurements were made with a conventional cold finger type dewar in which the temperature was varied by means of a nichrome wire heating element. The temperature was measured by a pre-calibrated chromel-alumel thermocouple which was pressed against the side of the sample. The accuracy of the temperature measurements was estimated to be ± 1 K.

RESULTS AND DISCUSSION

Librational mode and ionicity in different metal azides

The metal azides can be divided into the largely ionic alkali and alkaline earth salts, and the semi-ionic heavy metal salts. The degree of ionicity in the lattice can be considered

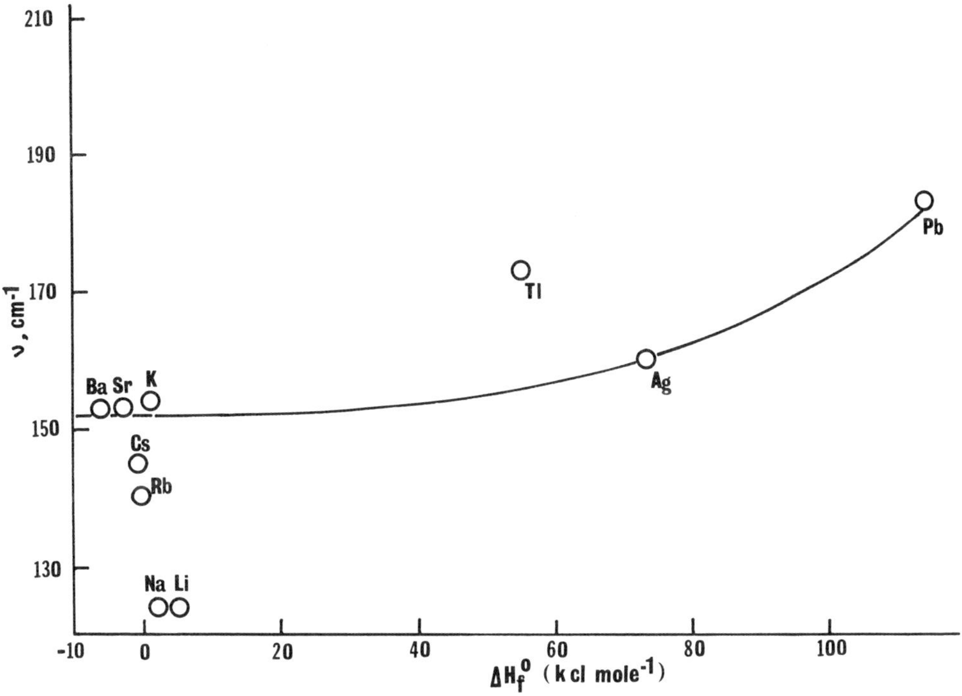

Fig. 1. Plot of the heat of formation (ΔH_f°) against the azide ion librational frequency, for a number of metal azides. The metal ions are indicated in the figure.

to be proportional to the heat of formation (ΔH_f°) of the salt. This follows from the Born-Haber cycle, according to which ΔH_f° of an $A^+B_3^-$ lattice is equal to $[(I+L)-(U+2RT)] + [\Delta H_f^\circ(\text{radical}) - E]$, where I and L respectively are the ionization potential and the sublimation heat of A, U is the lattice energy, and ΔH_f° (radical) and E are respectively the heat of formation and the electron affinity of the B_3 radical. The experimental values obtained for the metal azides[2] show the expected trend as one goes from the ionic to the semi-ionic azides.

The low frequency Raman spectra of a number of metal azides have been recorded while the results for the K, Rb, Ba and Pb salts have been taken from previously published work.[3-5] The peak which can be assigned to a largely librational motion in each of the azide lattices, was then selected. One criterion used in selecting these peaks is the relatively large Raman intensity of the librational mode compared with the translational modes lying in the same region of the spectrum. In the case of KN_3 structures, the B_{1g} component was chosen since the E_g component undergoes anharmonic coupling with the E_g translational mode, to a fairly large degree.[1] The selected values of the librational frequencies of a number of metal azides are plotted against the respective ΔH_f° values, as shown in Fig. 1. It can be seen that the selected librational mode increases monotonically with increasing ΔH_f°. When compared with the recent observation[5] that one component of the bending internal vibration of the azides decreases in frequency with increasing ΔH_f°, it may be suggested that the heavy metal azide lattices are more 'molecular' as a result of the decreased separation between the internal and external modes. It can also be seen from

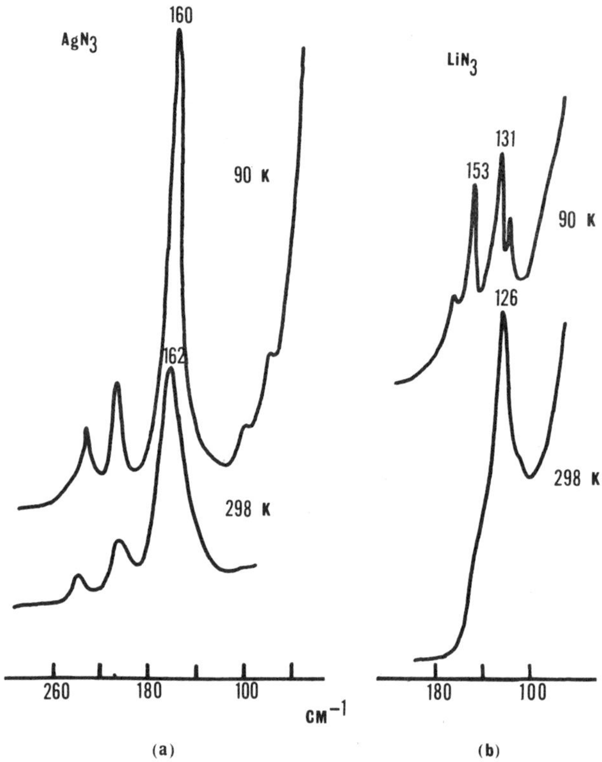

Fig. 2. Low frequency Raman spectra of (a) AgN_3 (b) LiN_3 at 90 and 298 K.

Fig. 1 that the librational mode frequencies of the Li, Na, Rb, Cs and Tl salts do not fall on the curve shown in the figure. This is an interesting result because all these azides except LiN_3 are known to undergo second order phase transitions at low temperatures. The deviation of the librational mode frequencies of some of the azides from the curve shown in Fig. 1 probably results from increased anharmonic contributions to the vibrational potential energy of the crystal concerned. This apparently also results in the decreased thermodynamic stability of the particular azide lattice at atmospheric pressure. The idea of increased anharmonic coupling finds support in the increased width of the B_{1g} librational mode as one goes from the K to the Cs salt in KN_3 type structures.[3] Also if the librational mode spectra of LiN_3 and AgN_3 at 298 and 90 K, shown in Fig. 2, are compared it can be seen that a librational mode component in LiN_3, whose librational mode component at 298 K does not fall on the curve shown in Fig. 1, undergoes an unusually large temperature shift. A relatively large temperature dependence is a reflection of appreciable anharmonic interaction coupled to the thermal expansion of the lattice in question.

Phase transition in NaN_3

At 298 K, NaN_3 has a rhombohedral structure (phase I), which can be described as a distortion of the sodium chloride lattice. The space group is D^5_{3d} and there is one

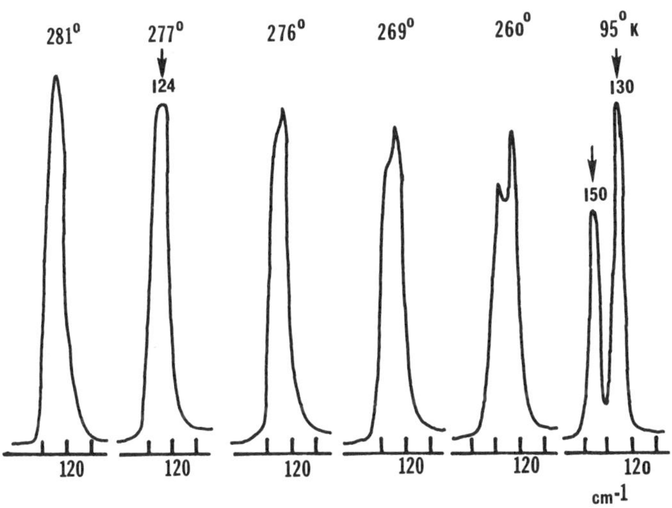

Fig. 3. Low frequency Raman spectrum of single crystal NaN_3 as a function of temperature.

formula unit in the primitive unit cell. The linear and symmetric azide ion is aligned along the hexagonal *c* axis or the (111) direction of the rhombohedron. A phase transition at a transition temperature (T_c) between 286 and 295 K[6–8] has been observed, in which the rhombohedral unit cell distorts to a monoclinic structure (phase II), which incidently is isomorphous with the stable phase of LiN_3. Phase II belongs to the space group C^3_{2h} with one formula unit in the primitive cell. Diffraction results[7] indicate that phase II involves a shear of the Na^+ and N_3^- ion layers, which is coupled with a tilting of the N_3^- ion from the hexagonal *c* axis by an angle of 12.3° at 163 K.

In phase I there is one doubly degenerate E_g Raman active external vibration, which is of the librational type. Factor group considerations indicate that the mode degeneracy breaks down in phase II, giving rise to two Raman active modes of A_g and B_g symmetry. This does indeed occur as can be seen from the Raman spectra of NaN_3 at different low temperatures shown in Fig. 3. At 103 K, two peaks at 130 and 150 cm^{-1} are observed in the Raman spectrum. The separation between the peaks progressively decreases as T_c is approached. Near T_c, which the present results indicate to be between 277 and 281 K, the separation between the two peaks almost disappears. The value of T_c obtained here is somewhat lower than those reported previously. It was observed that there is some sluggishness in the transition process when the temperature is lowered rather slowly in the vicinity of T_c, which may account for differences in the value of T_c obtained by various workers.

Phase transition in TlN_3

TlN_3 has a body centred tetragonal structure at 298 K (phase I), which is isomorphous with the structure of KN_3 and can be described as a distortion of the CsCl lattice. The space group is D^{18}_{4h} and there are two formula units in the primitive unit cell. At a temperature of 233 K, determined by differential thermal analysis,[9] a phase transition has been observed, which low temperature powder diffraction results[10] indicate to be orthorhombic (phase II). However, the space group of the structure has not yet been

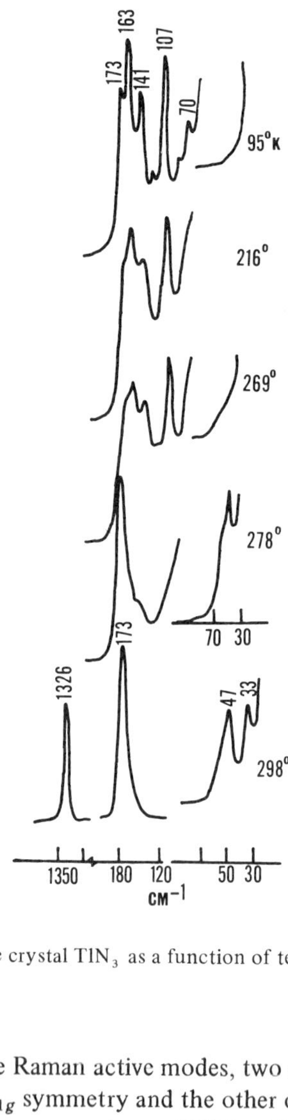

Fig. 4. Raman spectrum of single crystal TlN_3 as a function of temperature.

established.

In phase I there are three Raman active modes, two of E_g and one of B_{1g} symmetry. Two of these modes, one of B_{1g} symmetry and the other of E_g symmetry, are librational in character. The Raman spectrum of TlN_3 at 298 K is shown in Fig. 4. Polarized data could not be obtained due to appreciable birefringence and the fact that the crystals grew only on complex (112) type planes. The assignment of the peaks at 33 and 47 cm^{-1} to E_g symmetry motions and the peak at 173 cm^{-1} to the B_{1g} mode was therefore made by analogy with KN_3.[11] The E_g mode at 47 cm^{-1} cannot be correctly described as a librational mode since it undergoes strong anharmonic interaction with the translational E_g mode.[1]

At low temperatures, interesting changes in the Raman spectrum occur, which clearly indicate that a phase transition has taken place. New features of note are the appearance of strong peaks at 107 and 141 cm^{-1}, the shift of the strong librational peak from 173 cm^{-1} to 163 cm^{-1} and the disappearance of the peaks at 33 and 47 cm^{-1} (*cf.* Fig. 4). The downward frequency shift of the librational peak is in line with the obser-

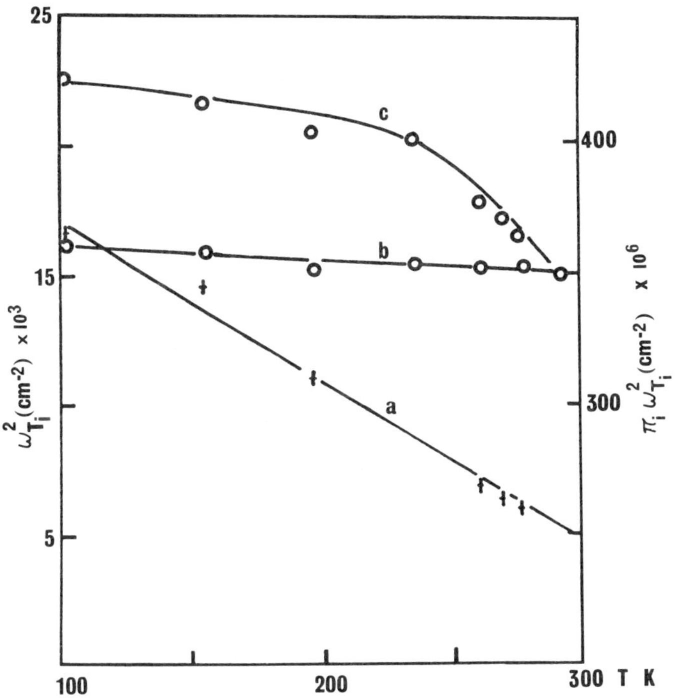

Fig. 5. Plot of the squares of the librational frequencies and the product of the squares of the two frequencies against temperature, for NaN_3.

vation (*cf.* Fig. 1) that the librational mode frequency in TlN_3 is anomalously high. Another interesting feature is the correspondence of the low temperature Raman peaks at 70, 107, 141 and 163 cm^{-1} with the infrared peaks of TlN_3 at 90 K.[11] This observation when coupled with the fact that the infrared inactive internal symmetric stretching mode of phase I becomes weakly active in phase II,[11] suggests a possible breakdown of the rule of mutual exclusion in phase II. Hence it is likely that phase II has a non-centrosymmetric space group of C_{2v} symmetry. A non-centrosymmetric unit cell may result either from a distortion of the Tl^+ ion from the body centred position in the cell or from a distortion of the symmetric azide ion to $C_{\infty v}$ symmetry followed by a parallel arrangement of the ions. Large distortions of the azide ion can occur in the lattices of the heavy metal azides as shown by diffraction results (*cf.* ref. 1). It is therefore likely that phase II consists of a parallel arrangement of distorted azide ions.

In slow warming up experiments it was found that T_c is 278 K, a value which is appreciably higher than that obtained by the thermal method. However, when the crystal is warmed up rapidly, a value of 225 K is obtained, which is close to that previously observed in the thermal measurement. This suggests that some sluggishness in the phase transition analogous to that in NaN_3, may be involved.

Temperature dependence of Raman active phonons and the nature of the phase transitions

The temperature dependence of the Raman active modes can determine whether

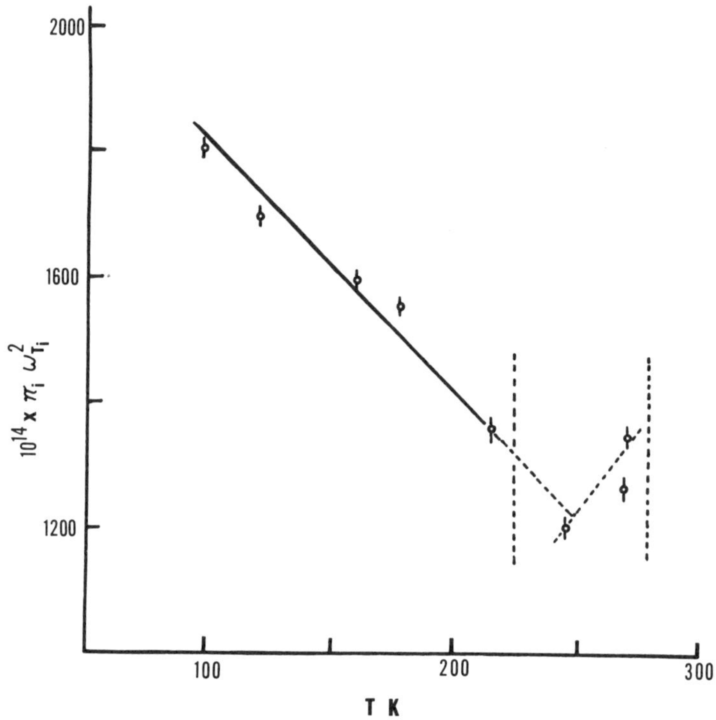

Fig. 6. Plot of the product of the squares of the frequencies at 107, 141, 163 and 173 cm^{-1} (95 K values against temperature, for TlN_3). Dotted lines indicate region where sluggishness occurs in the transition process.

the transition is displacive (single ion potential well) in character or of the order-disorder (double or multiple ion potential well) type. It has been shown[12] that an external transverse vibrational mode can become unstable in the vicinity of T_c due to the near cancellation of the short range forces by the long range field of the lattice oscillation. In the limit of long wavelengths the square of the frequency of such a mode would therefore vary linearly with temperature, with the mode frequency tending to zero near T_c. Such a relationship has been found to apply to the displacive transition in barium titanate.[13] However in the case of the order-disorder phase transition in ferroelectrics like sodium chlorate and sodium nitrite[14] and 'quasi' antiferroelectrics like sodium nitrate[14] and potassium thiocyanate,[15] more than one 'soft' mode is probably involved and it appears that a general relationship in which the product of the frequencies squared varies linearly with temperature, is followed. One might therefore speculate that a combined 'soft' mode approach is applicable to ionic crystals like sodium chlorate in which the anion has strong intra-ion covalent bonding, while a single 'soft' mode approach applies to perovskite crystals in which the anion contains a transition metal atom and only partly covalent character. The former type of crystals undergoes phase transitions by a largely order-disorder mechanism while the latter class depends on a displacive process of the type found in barium titanate.

The frequency squared against temperature plots for the two Raman active modes of NaN_3 (phase II) are shown in Fig. 5. It can be seen that the higher frequency mode shows a greater temperature dependence but does not follow a linear relationship. However, if the product of the squares of the two Raman active modes is plotted against temperature as is shown in the same figure, a sensibly linear plot is obtained. One might therefore suggest, in view of the above discussion, that microscopically the phase transition proceeds through a disordering process with increasing temperature. The order parameter involved is likely to be associated with the position of the azide ion along the hexagonal *c* axis of the unit cell.

In the case of TlN_3, the Raman active modes of phase II show a fairly small but similar temperature dependence. If the product of the squares of the four Raman active mode frequencies at 107, 141, 163 and 173 cm^{-1} (95 K values) are plotted against temperature as shown in Fig. 6, a sensibly linear relationship is obtained up to a value close to T_c. Above this temperature, the graph appears to reverse its slope. The temperature dependence indicates that an order-disorder type mechanism is also applicable in the case of TlN_3. This would suggest that the transition is likely to involve a slight distortion of the azide ion rather than a displacement of the Tl^+ ion, followed by the increased ordering of the anion dipoles with decreasing temperature.

REFERENCES

1 Z. Iqbal in *Structure & Bonding,* Vol. 10, (Ed. P. Hemmerich *et al.*), Springer, Berlin, 1972, p. 25.
2 P. Gray and T. C. Waddington, *Proc. Roy. Soc.* **235A**, 481 (1956).
3 Z. Iqbal, *J. Chem. Phys.* **53**, 3763 (1970).
4 Z. Iqbal and M. L. Malhotra, *J. Chem. Phys.* **55**, 528 (1971).
5 Z. Iqbal, W. Garrett, C. W. Brown and S. S. Mitra, *J. Chem. Phys.* **55**, 4528 (1971).
6 G. E. Pringle and D. E. Noakes, *Acta Cryst.* **16**, A192 (1963).
7 G. E. Pringle and D. E. Noakes, *Acta Cryst.* **B24**, 262 (1968).
8 R. B. Parsons and A. D. Yoffe, *Acta Cryst.* **20**, 36 (1966)
9 A. D. Yoffe in *Developments in Inorganic Nitrogen Chemistry,* Vol. 1, (Ed. C. B. Colburn), Elsevier, New York, 1966, p. 72.
10 O. Haase, private communication (1971).
11 Z. Iqbal and M. L. Malhotra, (to be published).
12 W. Cochran, *Advan. Phys.* **9**, 387 (1960).
13 M. DiDomenico, S. H. Wemple, S. P. S. Porto and R. Bauman, *Phys. Rev.* **174**, 522 (1968).
14 A. D. P. Rao, R. S. Katiyar and S. P. S. Porto, *Phys. Rev. Letters,* **28**, 665 (1972).
15 Z. Iqbal, *J. Chem. Phys.* (submitted).

24. THE LASER RAMAN SPECTRA OF SEVERAL PEROVSKITE ZIRCONATES

A. E. Pasto and R. A. Condrate, Sr.
Division of Ceramic Engineering and Science, S.U.N.Y. College of Ceramics, Alfred University, Alfred, New York, USA

INTRODUCTION

The close relationship between the ferroelectric effect and lattice dynamics for the perovskite-structured materials has led to extensive investigation of their infrared and Raman spectra. Vibrational spectra have been obtained and interpreted for single crystals and solid solutions of perovskite titanates. Further studies have also been conducted with the tantalates and niobates. However, very little work has been done with the perovskite zirconates. A few papers have been published on the infrared adsorption and reflection spectra of some powdered zirconates,[1–3] but no Raman data are currently available in the literature. This lack of data is unusual since Raman spectra can give useful information concerning the structures of these materials, which may possess ferroelectric or antiferroelectric properties. We have attempted to fill this gap by obtaining and interpreting the Raman spectra of available powders or single crystals of several perovskite zirconates ($BaZrO_3$, $SrZrO_3$, $CaZrO_3$ and $PbZrO_3$). Also, perturbations on the Raman spectra of zirconate powders due to impurities are discussed. Finally, the results of a temperature-dependence study on the Raman spectra of a single crystal of $PbZrO_3$ are presented and discussed.

EXPERIMENTAL PROCEDURES

Preparation of materials

Zirconate powders were prepared by methods described in the literature.[4,5] Powdered samples were also available commercially from the Titanium Alloy Division of National Lead Industries, Inc. and Cornell-Dubilier Electronics, Inc. Small single crystals of $PbZrO_3$ were grown from a $PbCl_2$ flux.[6] The solid reaction mixture was heated at 900°C for 12½ h in a tightly covered platinum crucible. Crystals were picked from the resultant product. Larger single crystals were obtained from Dr. W. R. Cook, Jr. of Gould Laboratories and Dr. B. A. Scott of IBM Thomas J. Watson Research Center. Samples were characterized by wet chemical, X-ray diffraction and spectrographic methods.

Raman spectra

The Raman spectra were obtained using a Spex Model 1401 double monochromator with a Coherent Radiation Laboratory Model 54 argon ion laser. Light was detected by a thermoelectrically cooled ITT FW-130 photomultiplier and amplified by photon counting electronics. All spectra illustrated in the figures were obtained in the Stokes region of the blue laser line (4880 Å). Similar Raman spectra were also observed in the anti-Stokes region of the blue line, and in the Stokes region of the green laser line (5145 Å). The instrument was calibrated using benzene and CCl_4 samples. Precision was ± 1 cm^{-1}, and wavenumber was ± 3 cm^{-1}.

Raman spectra of the powders could be investigated by preparing pellets of the pure materials in a sample press. However, the densely packed surface of small crystallites cause excessive scattering close to the exciting line, obscuring any Raman peaks in that region (less than 100 cm^{-1}). This problem was eliminated by using a different sample preparation technique. A drop of Duco Cement was smoothed across the surface of a glass-slide and some of the powder was sprinkled until the glue was completely covered. Another slide was laid on top of the powder and pulled lengthwise across it, flattening the surface. The slide was suspended with the flat face of the powder at 45° to the beam. The Raman spectra of these samples could be observed to within 25 cm^{-1} of the exciting line.

Large multi-domain single crystals of $PbZrO_3$ were mounted on goniometers. The oriented crystals were placed on a mounting platform in the Raman spectrophotometer for spectral investigation. Raman spectra were also obtained for these crystals at high temperatures, using a specially designed alumina tube furnace. Low temperature Raman spectra were observed using a cell described by Bates.[7]

RESULTS AND DISCUSSION

The relation of the crystal structures of perovskite zirconates to their vibrational spectra

The ideal perovskite structure is cubic and is illustrated in Fig. 1. Divalent *A*-cations are placed in cube corners with *B*-cations such as Zr at body centres, and *O*-anions at face centres. $BaZrO_3$ possesses such cubic symmetry. Its space group is given by O_h^1-*Pm*3*m* with one molecular unit per cell. Both $SrZrO_3$ and $CaZrO_3$ are distorted from ideal perovskite symmetry to an orthorhombic symmetry. In these cases, the space group is D_{2h}^{16}-*Pnma* with four molecular units per cell. $PbZrO_3$ possesses lower orthorhombic symmetry with a space group of C_{2v}^8-*Pba*2 and eight molecular units per cell.

The number of allowed Raman fundamental modes can be predicted using the factor group method described by Fateley.[8] The results are collected in Table 1 for the perovskite structures mentioned above. Factor group analysis predicts that no fundamental modes are expected in the Raman spectra of $BaZrO_3$. A similar analysis for $SrZrO_3$ and $CaZrO_3$ predicts 24 possible Raman-active fundamental modes. Because of the centre of symmetry in their crystal structures, none of these bands are also infrared active. $PbZrO_3$ possesses an orthorhombic structure with no centre of symmetry. Factor group analysis predicts the possibility of 117 Raman active modes. 89 of these modes are also infrared active due to a lack of the centre of symmetry. The high temperature phase of $PbZrO_3$ possesses the ideal perovskite structure. Therefore, no first order Raman bands should be observed for this phase.

The interpretation of the Raman spectra of $BaZrO_3$

The Raman spectrum of a $BaZrO_3$ powder is illustrated in Fig. 2. This spectrum exhibits only one very weak band at ~ 123 cm^{-1}. However, the factor group selection

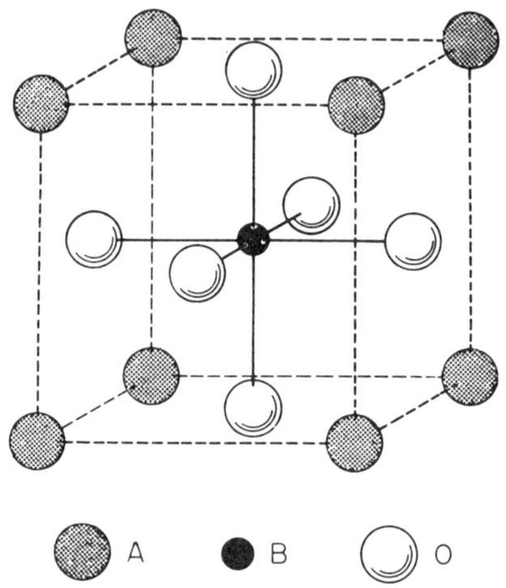

Fig. 1. The ideal perovskite structure.

Table 1. Spectral predictions for various perovskite structures.

Factor group	Symmetry species of fundamentals	Number of Raman fund.	Number of IR fund.	Number of coincidences
O_h	$3F_{1u} + F_{2u}$	0	3	0
D_{2h}	$7A_g + 5B_{1g} + 7B_{2g} + 5B_{3g} + 8A_u + 9B_{1u} + 7B_{2u} + 9B_{3u}$	24	25	0
C_{2v}	$27A_1 + 28A_2 + 31B_1 + 31B_2$	117	89	89

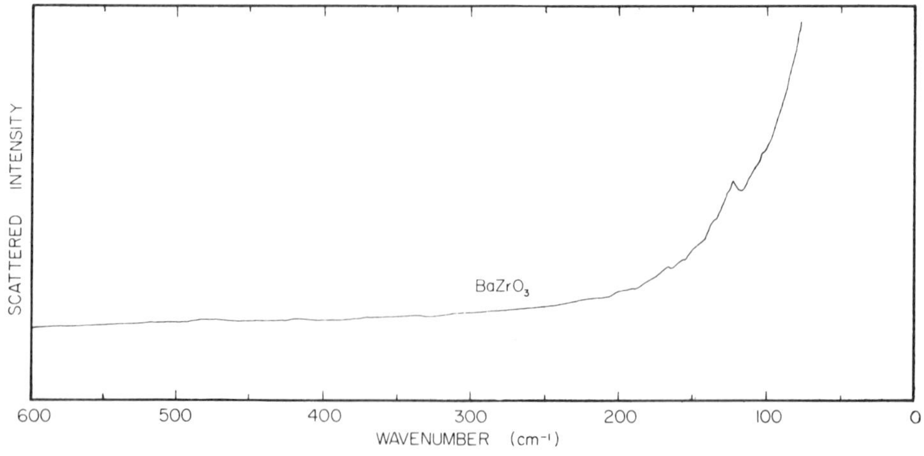

Fig. 2. The Raman spectrum of a $BaZrO_3$ powder.

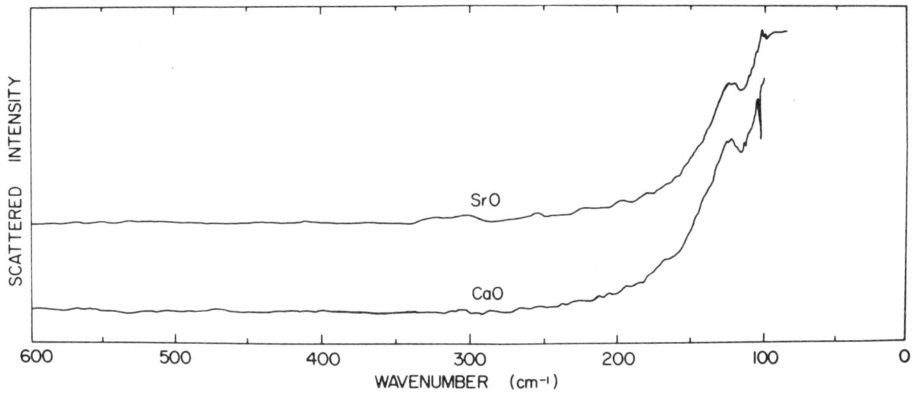

Fig. 3. The Raman spectra of SrO and CaO powder.

rules do not allow any first order Raman scattering. Since this band is observed at the same shifted wavenumber position in the Stokes and anti-Stokes regions of both the 4880 and 5145 Å exciting lines, it must be due to Raman scattering rather than a fluorescence effect. Therefore, this band must be a second order band. The broadness of the band suggests that it is second order. A band appears in the same location in the Raman spectra of CaO and SrO (see Fig. 3), neither of which should possess first order Raman spectra under their factor group O_h. This further strengthens the assignment of the band for $BaZrO_3$ to a second order mode. Also, $CaTiO_3$ exhibits a Raman spectrum with a second order band near 123 cm^{-1}. Therefore, we can conclude that $BaZrO_3$ does not possess any first order spectra as is predicted by factor group analysis.

The interpretation of the Raman spectra of $SrZrO_3$

The Raman spectra of $SrZrO_3$ powders containing increasing amounts of impurity are illustrated in Fig. 4. Factor group analysis predicted that the fundamental modes are Raman active. The observed Raman bands are very sensitive to the concentrations of the impurities present in the materials. The orthorhombic structure of $SrZrO_3$ is very close to a cubic structure. Therefore, very slight changes in the concentrations of impurities may drastically alter the structure of these materials, along with their spectra. It appears that increasing the concentration of Al and Si in the material distorts the structure towards cubic symmetry. Tentative band assignments for Sample I containing the least impurity are given in Table 2.

Interpretation of the Raman spectra of $CaZrO_3$

The Raman spectra for various $CaZrO_3$ powder samples are illustrated in Fig. 5. All spectra for $CaZrO_3$ are more intense than any spectra for $SrZrO_3$. This observation is expected because of the larger orthorhombic distortion in the $CaZrO_3$ lattice. No Raman bands are observed in these spectra from the starting materials, monoclinic ZrO_2 and $CaCO_3$. The degree of nonstoichiometry, concentrations of the impurities present and method of heat-treating the sample influences the observed spectra. The MY IV $CaZrO_3$ sample possesses nearly ideal stoichiometry. This powder which was prepared by firing at 1300°C for 10 h yields 13 Raman bands. The sample was refired for 10 h at 1300°C. The

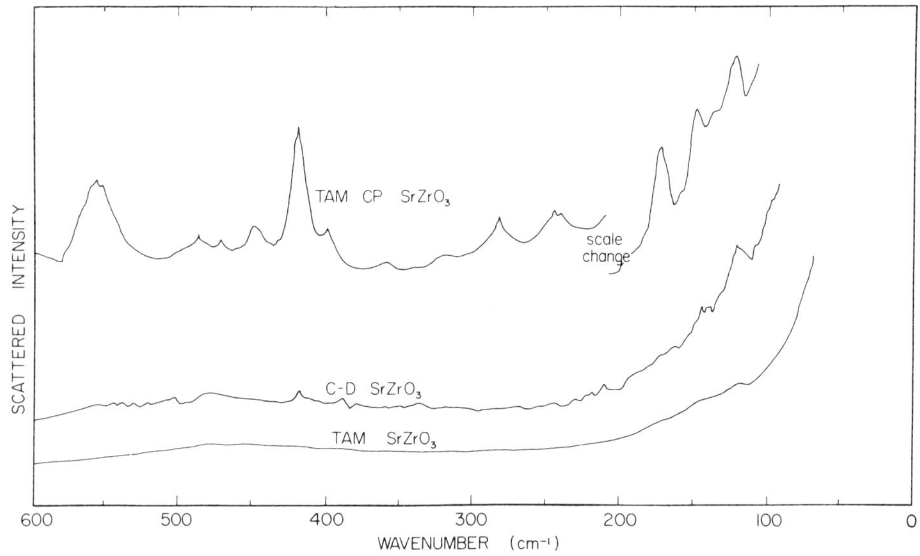

Fig. 4. The Raman spectra of several $SrZrO_3$ powders.

Table 2. Tentative band assignments for $SrZrO_3$, $CaZrO_3$ and $PbZrO_3$ (cm^{-1}).

$SrZrO_3$	$CaZrO_3$	$PbZrO_3$	Band assignments
556	551	560	Zr–O stretching
483	473	535	Zr–O stretching
470		508	Zr–O stretching
448	444	350	Torsional mode
418	430	336	Torsional mode
405	362	332	Torsional mode
358	315	302	Torsional mode
320			Torsional mode
		288	Pb–O group mode
283	293	238	Zr–O bending
245	268	212	Zr–O bending
	230		Zr–O bending
	218		Zr–O bending
173	186		Lattice soft mode
150	150		Lattice soft mode
		140	Pb–O group mode
123	124		Second order mode
		75	External mode
		60	External mode
		55	External mode
		49	External mode
		39	External mode

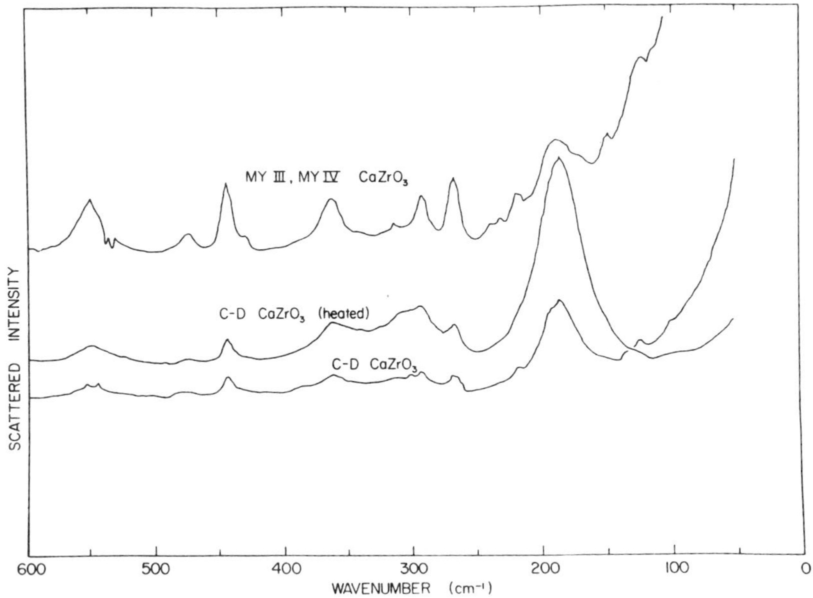

Fig. 5. The Raman spectra of several $CaZrO_3$ powders.

resultant powder displays a Raman spectra that is identical to that of the original material, indicating that the material is at structural equilibrium. Tentative band assignments for $CaZrO_3$ are made in Table 2 using this sample.

The CD $CaZrO_3$ sample is highly non-stoichiometric with the Ca/Zr ratio deviating by approximately 10% from the ideal ratio. The spectrum of this sample significantly differs from that of the previous sample. Furthermore, refiring this sample at 1300°C for 10 h will cause significant changes in the spectra (see Fig. 5). It appears that the initial CD sample was not a pure single phase powder, but it composed of a stoichiometric $CaZrO_3$ phase along with another zirconium-containing phase. Heat treatment of the sample changed the phases present. Raman spectra are sensitive to these changes.

Interpretation of the Raman spectra of $PbZrO_3$

The Raman spectra of a multi-domain single crystal of $PbZrO_3$ with nearly ideal stoichiometry are illustrated in Fig. 6. The Raman spectrum of the powder of this crystal is similar to it. The tentative band assignments for this crystal are given in Table 2. The bands at 288, 140 and 75 cm^{-1} are selectively assigned to PbO group modes. These bands appear at similar locations as those observed for yellow PbO (see Fig. 7). The bands at 60, 55, 49 and 39 cm^{-1} are assigned to external modes. The Raman spectra of $PbZrO_3$ are also sensitive to the degree of nonstoichiometry, the amount of impurities present and the preparation process. Also, focusing the laser at full power on the crystal destroys the sample. A white cloud of PbO is generated about the crystal during exposure to the laser beam. The Raman spectrum observed for the solid after laser damage is that of monoclinic zirconia.

A temperature dependence study of the Raman spectra of $PbZrO_3$ was conducted to obtain structural information and to attempt to identify the so-called antiferroelectric

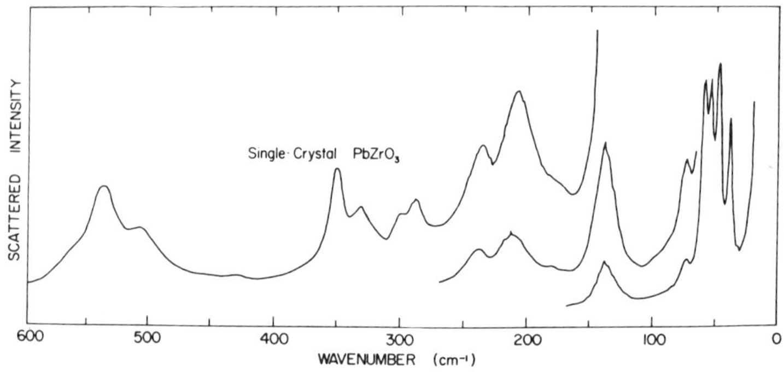

Fig. 6. The Raman spectra of a $PbZO_3$ crystal.

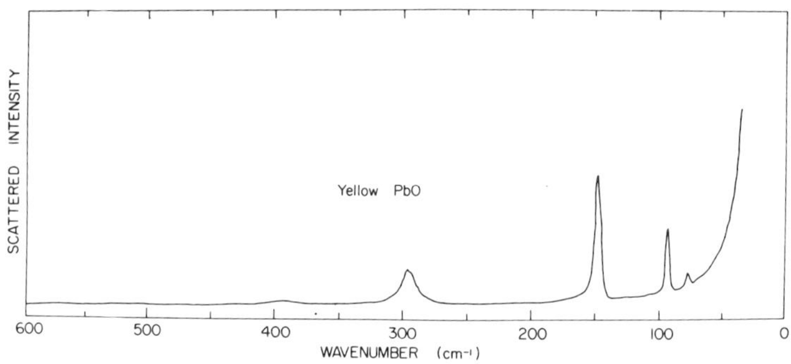

Fig. 7. The Raman spectra of yellow PbO.

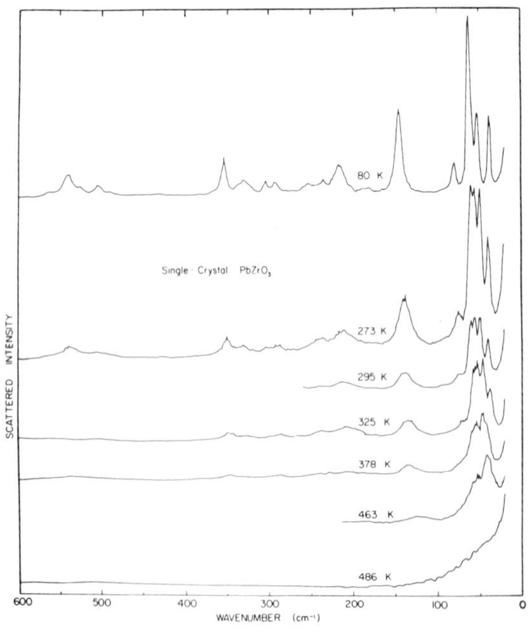

Fig. 8. The Raman spectra of a $PbZrO_3$ single crystal at several temperatures.

or ferroelectric modes (Fig. 8). The intensities of all Raman bands for $PbZrO_3$ decreased continuously as the temperature is increased. The band intensities approach zero as the Curie temperature (505 K) is approached. This behaviour is expected for first order Raman modes of a material undergoing a transition to the cubic phase. This study does not disclose any large wavenumber shift for any band. Therefore, we conclude that either the ferroelectric and antiferroelectric modes leave no discernible traces in the orthorhombic phase of $PbZrO_3$ or they occur at lower wavenumbers than can be readily observed with our Raman spectrophotometer (below 20 cm^{-1}). The latter result might be expected because the lead atom is heavy and the force constants involved are small. The soft modes for $SrTiO_3$ that relate to this transition occur below 50 cm^{-1}. Further investigation is necessary.

REFERENCES

1 A. F. Iatsenko, *Bull. Acad. Sci. USSR (Phys. Series)* **22**, 1450 (1958).

2 C. H. Perry, D. J. McCarthy and G. Rupprecht, *Phys. Rev.* **138**, 1537 (1965).

3 T. Dupuis and M. V. Lorenzelli, *C. R. Acad. Sci., Paris* **264B**, 1019 (1967).

4 R. S. Roth, *J. Res. Nat. Bur. Std.* **58**, 75 (1957).

5 B. A. Scott, IBM Thomas J. Watson Research Center, Yorktown Heights, New York (private communications).

6 M. L. Sholokhovish, O. P. Kramarov and V. I. Varicheva, Segnetoelektriki, Rostovsk. -na-Donu Gos. Univ., Nauchn, -Issled. Fiz. -Mat. Inst., Sb. Statei, 31 (1961).

7 J. B. Bates, *Appl. Spectry.* **25**, 380 (1971).

8 W. G. Fateley, N. T. McDevitt and F. F. Bentley, *Appl. Spectry.* **25**, 155 (1971).

25. ETUDE STRUCTURALE ET DYNAMIQUE DES HEXAFLUORURES DE SOUFRE, SELENIUM ET TELLURE

M. Gilbert et M. Drifford

Service de Chimie-Physique, Centre d'Etudes Nucleaires de Saclay, Gif-sur-Yvette, France

ABSTRACT

SF_6, SeF_6 and TeF_6 have been studied by laser Raman spectroscopy between +25°C and −190°C. All of these compounds show a body centred cubic plastic phase. The study of the shape and width of Raman lines as a function of temperature, in both the liquid and plastic phases allows one to follow the evolution of molecular rotational motion. The evaluation of vibrational relaxation is discussed. Finally the results obtained by vibrational spectroscopy are compared with NMR measurements of ^{19}F relaxation.

The solid–solid transition (body centred cubic, orthorhombic) is studied and Raman spectra of the orthorhombic phase are analysed and discussed.

INTRODUCTION

Nous présentons l'étude structurale et dynamique par diffusion Raman laser des hexafluorures de métalloides SF_6, SeF_6, TeF_6 et quelques résultats obtenus sur MoF_6. Ces hexafluorures présentent une phase «plastique» caractérisée par des mouvements de réorientation moléculaire. L'étendue du domaine d'existence de la phase plastique dépend de façon remarquablement simple de la distance élément–fluor (Fig. 1).[1]

Nous avons précisé:

1. la structure des deux phases solides;
2. les mouvements de réorientation moléculaire dans les phases liquides et plastiques.

Cette étude vibrationnelle fait suite aux mesures de RMN faites par Rigny et Virlet[1,2] dans les différentes phases liquides et solides.

ETUDE STRUCTURALE DES HEXAFLUORURES DE SOUFRE, SELENIUM, TELLURE ET MOLYBDENE

Méthode expérimentale

Les hexafluorures SF_6, SeF_6 et TeF_6 sont liquides à la température ambiante. Le groupe de symétrie ponctuelle de ces molécules est O_h (autour de l'atome métalloidique placé au centre, les fluors occupent les six sommets d'un octaèdre régulier).[3] Ces composés ont une tension de vapeur importante à la température ambiante, ils sont très réactifs et extrêment toxiques. SeF_6 et TeF_6 en provenance de l'Allied Chemical Company sont purifiés par sublimation fractionnée. Des capillaires de quartz de diamètre égal à

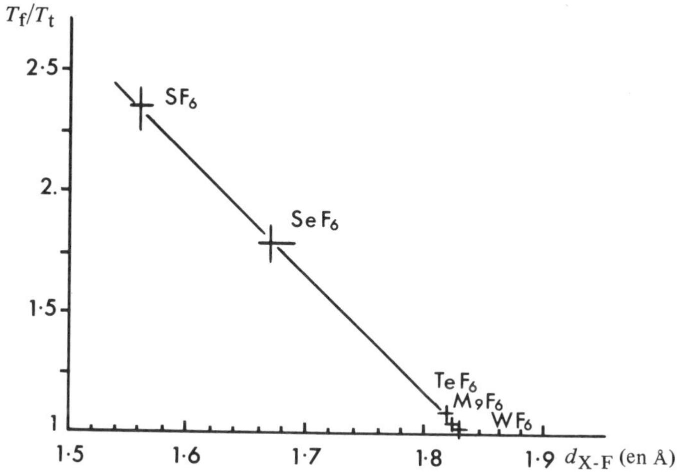

Fig. 1. Le rapport $T_f/T_t = T(\text{fusion})/T(\text{transition solide–solide})$ est fonction linéaire de la distance élément fluor d_{X-F} dans la molécule XF_6.

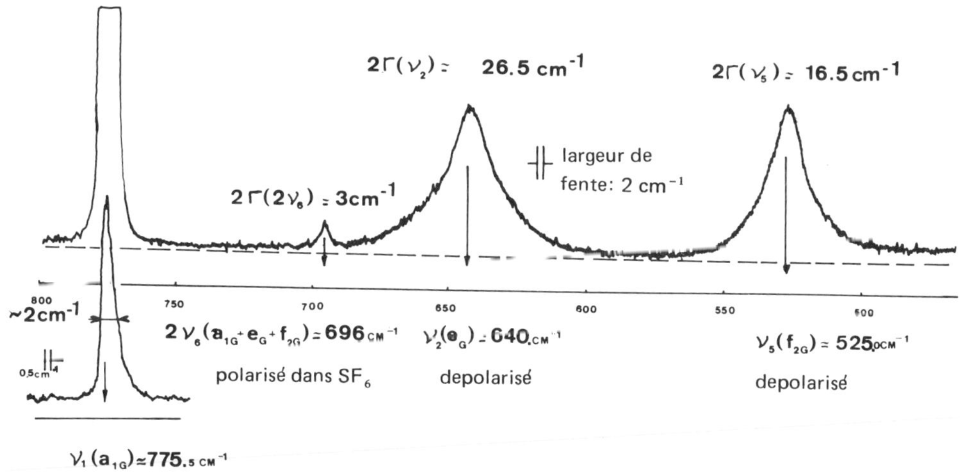

Fig. 2. Spectre Raman de SF_6 liquide à la température ambiante ($T \sim 23°C$).

1,4 mm sont remplis par piégeage du gaz à la température de l'azote liquide, puis scellés et conservés à 77 K.

L'étude de la diffusion Raman est réalisée sur un spectromètre Coderg avec un laser Spectra-Physics 125 Å He–Ne de puissance égale à 60 mW, ou un laser argon ionisé 165 de puissance 1 W. Un cryostat adaptable sur la platine porte-échantillons permet l'étude des composés entre 297 K et 80 K. Le capillaire est placé dans un porte-échantillon en cuivre réglable sous vide et à basse température par rapport au faisceau laser. Le refroidissement du porte-échantillons est obtenu par circulation de gaz froid issus de l'azote

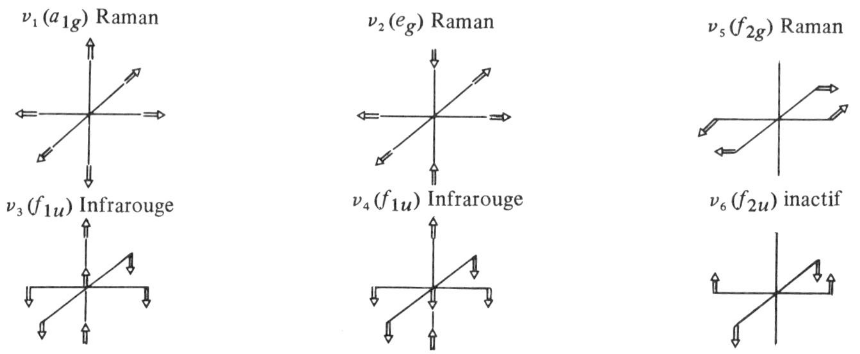

Fig. 3. Modes normaux de vibration de XF_6, symétrie, activité.

liquide placé dans un dewar sous le spectromètre. L'une des extrémités de la canalisation du cryostat plonge dans le gaz liquéfié, tandis que l'autre est relié à une pompe à circulation. La régulation de la température est obtenue par un thermocouple qui agit sur le débit de la pompe.

La précision des mesures est de l'ordre de 1 cm^{-1} pour les fréquences Raman, et 0,2 cm^{-1} pour les décompositions des raies. La différence de température entre celle de l'échantillon et celle du porte-échantillons directement lue est d'environ 5°C.

Résultats et interprétation

Phase liquide. Les spectres présentent à température ambiante les trois raies actives en Raman d'une molécule MF_6 (M = S, Se, Te, Mo)de symétrie ponctuelle O_h (Fig. 2).[4]

Singulet – Totalement polarisé ($\rho = 0$)

Doublet, Triplet – Dégénérés, dépolarisés ($\rho = 3/4$)

Les modes normaux de vibration sont représentés Fig. 3.

Tableau 1.

Frequences (cm^{-1})	$\nu_1 (a_{1g})$	$\nu_2 (e_g)$	$\nu_5 (f_{2g})$
SF_6	774	640	525
SeF_6	707,5	657,5	405
TeF_6	696	668,4	314,5
MoF_6	743,5	647	326

Le Tableau 1 donne les fréquences des trois raies Raman des hexafluorures en phase liquide.

Phases solides. *A la solidification* des hexafluorures le spectre Raman n'est pas modifié (Fig. 4). Les fréquences des trois raies actives en Raman (ν_1, ν_2, ν_5) ne sont pratiquement pas modifiées, seules leurs largeurs subissent une discontinuité à la transition.

Des études anterieures en diffraction X[6] ont montré que ces composés cristallisent

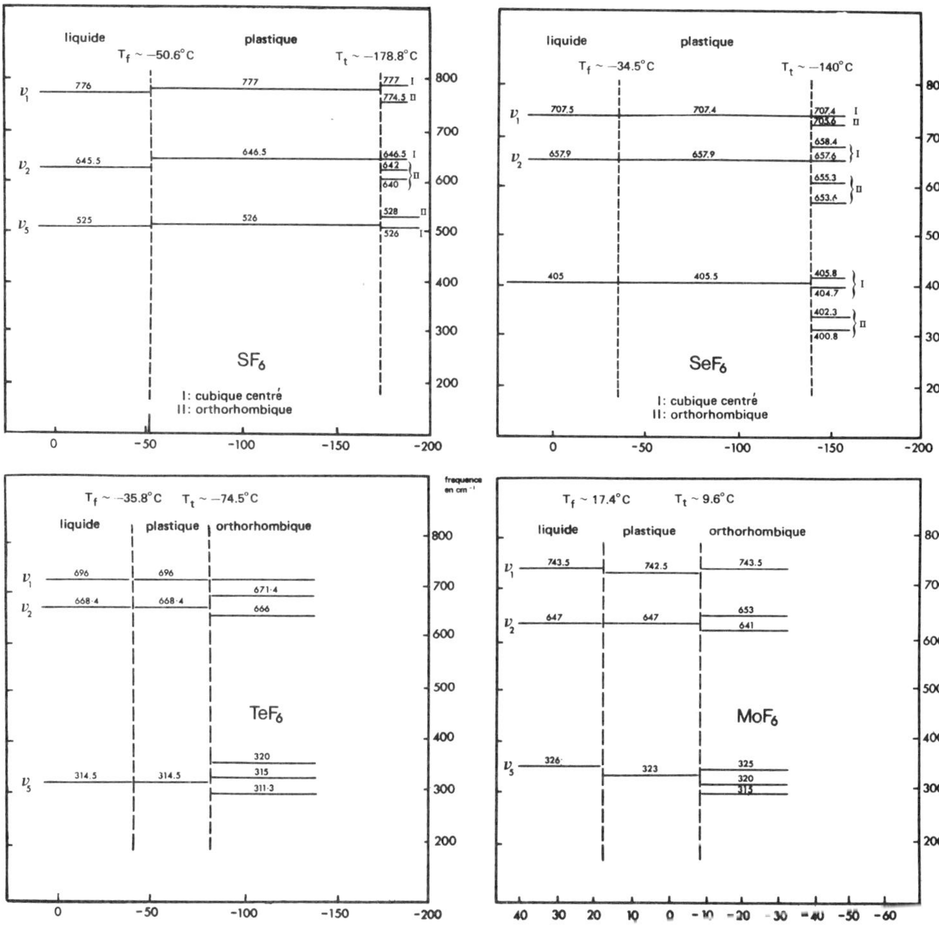

Fig. 4. Fréquences des raies Raman dans les trois phases condensées des hexafluorures SF_6, SeF_6, TeF_6 et MoF_6.

dans une structure cubique centrée. Les atomes de S, Se et de Te occupent les noeuds et le centre de la maille et les atomes de fluor sont sur les arêtes de la maille. Cette maille élémentaire contient deux molécules et les paramètres cristallins sont respectivement de 5,79 Å pour SF_6, 5,99 Å pour SeF_6 et 6,32 Å pour TeF_6. Le motif élémentaire d'un tel réseau de Bravais peut être décrit avec un rhomboèdre dont les axes principaux correspondent aux trois directions (111) du cube.

La symétrie ponctuelle de la molécule MF_6 est toujours O_h et le motif élémentaire contient une seule molécule. Le spectre Raman est donc identique en nombre de raies à celui du liquide. De plus les fréquences Raman ν_1, ν_2, ν_5 n'évoluent pas en fonction de la température (Fig. 4). Généralement, il existe des modifications de fréquences des modes normaux des molécules entre l'état liquide et l'état solide. Dans le cas particulier des hexafluorures de sélénium et de tellure, des considérations thermodynamiques[3] confirmées par des études de RMN[2] ont montré que ces phases cubiques peuvent être considérés comme des cristaux «plastiques». En effet, les mouvements de rotation des molécules y sont très rapides et sont également accompagnés d'une lente diffusion. Dans ces conditions, il est logique que les fréquences Raman de ce solide «plastique» soient pratiquement égales à celles du liquide (les interactions intermoléculaires étant faibles).

Au-dessous du point de transition solide–solide (Fig. 4) les spectres Raman sont modifiés et semblent compatibles avec les résultats expérimentaux de la diffraction des rayons X; cependent leur interprétation est différente.[6]

Dans TeF_6 et MoF_6, toutes les dégénérescences des raies sont levées à la transition: la symétrie ponctuelle des molécules peut être D_{2h} en accord avec la cristallographie et correspondre à une structure orthorhombique (Fig. 4).

Dans SeF_6 et SF_6 (Fig. 4) les spectres Raman des phases basse température sont plus complexes. Il semble qu'il y ait coexistence de deux phases – l'une cubique centré correspondant à la phase plastique et l'autre de plus basse symétrie, par exemple, orthorhombique.

La cinétique de transformation semble extrèmement lente et seules des études à 4 K permettront d'élucider ce problème.

ETUDE DYNAMIQUE DES HEXAFLUORURES DE SOUFRE, SELENIUM ET TELLURE EN PHASE LIQUIDE ET PLASTIQUE

A la transition liquide–plastique, les largeurs des raies de diffusion Raman présentent une discontinuité importante caractéristique d'une modification de la dynamique réorientationnelle. Nous allons donc exploiter les propriétés des spectres Raman dans l'étude des mouvements moléculaires de ces composés.

Effet de la relaxation de vibration et de la réorientation sur les raies Raman – généralités[7, 8]

Si la lumière incidente est polarisée selon la direction i, le spectre de la lumière diffusée inélastiquement avec la polarisation j est

$$I_{ij}(\omega) \propto \int_{-\infty}^{+\infty} \exp(-i\omega t) \langle \alpha_{ij}(0)\, \alpha_{ij}(t) \rangle \, dt$$

où α_{ij} est un élément du tenseur de polarisabilité, et $\langle \alpha_{ij}(0)\, \alpha_{ij}(t) \rangle$ sa fonction d'autocorrélation.

On sépare la polarisabilité moléculaire en ses parties: isotrope $(\alpha\delta_{ij})$ et anisotrope $(\beta_{ij} = \alpha_{ij} - \alpha\delta_{ij})$ où $\alpha = \frac{1}{3}(\alpha_{11} + \alpha_{22} + \alpha_{33})$ – l'une invariante dans une rotation moléculaire, l'autre étant un tenseur irréductible de rang 2.

Introduisons les notions de spectre isotrope

$$I_{\text{iso}}(\omega) = \int_{-\infty}^{+\infty} \langle \alpha(0)\, \alpha(t) \rangle \exp(-i\omega t) \, dt$$

et anisotrope

$$I_{\text{anis}}(\omega) = \int_{-\infty}^{+\infty} \langle \beta_{ij}(0)\, \beta_{ij}(t) \rangle \exp(-i\omega t) \, dt \qquad \text{avec } i \neq j$$

selon la direction de polarisation d'observation on obtient deux types de spectres

$$\begin{cases} i = j & I_{ij} \equiv I_{\parallel}(\omega) = I_{\text{iso}}(\omega) + \frac{4}{3} I_{\text{anis}}(\omega) \\ i \neq j & I_{ij} \equiv I_{\perp}(\omega) = I_{\text{anis}}(\omega) \end{cases}$$

I_{iso} *ne reflète que la relaxation de vibration* comme

$$G_{\text{iso}} = \langle\alpha(0)\,\alpha(t)\rangle = \left|\frac{\partial\alpha}{\partial q}\right|^2 \langle q(0)\,q(t)\rangle = \left|\frac{\partial\alpha}{\partial q}\right|^2 G^q_{\text{vib}}$$

(q: coordonnée normale de vibration) puisque α est invariant dans une rotation moléculaire.

I_{anis} *est sensible à la relaxation de vibration et aux mouvements de réorientation*; la fonction d'autocorrélation correspondante s'écrit

$$G_{\text{anis}} = \langle\beta_{ij}(0)\,\beta_{ij}(t)\rangle = \left\langle \frac{\partial\beta_{ij}(0)}{\partial q}\frac{\partial\beta_{ij}(t)}{\partial q}\,q(0)q(t)\right\rangle$$

$$= \left\langle \frac{\partial\beta_{ij}(0)}{\partial q}\frac{\partial\beta_{ij}(t)}{\partial q}\right\rangle\langle q(0)\,q(t)\rangle = \left|\frac{\partial\beta}{\partial q}\right|^2 G^{(2)}_{\text{rot}}G_{\text{vib}} \qquad (1)$$

où $\partial\beta_{ij}/\partial q$ est un élément d'un tenseur irréductible de rang 2 (dépendant donc des rotations) dont la fonction d'autocorrélation normalisée à 1 $G^{(2)}_{\text{rot}}$ se reduit à

$$G^{(2)}_{\text{rot}} = \frac{\langle Y^m_{(2)}(0)\,Y^{m*}_{(2)}(t)\rangle}{\langle|Y^m_{(2)}|^2\rangle}$$

($Y^m_{(2)}$ = harmonique sphéricale; m = quelconque), pour une réorientation isotrope des molécules. La troisième égalité (Eqn. (1)) suppose que les variables d'orientation et les coordonnées normales de vibration ne sont pas corrélées. Avec cette hypothèse, I_{anis} est le produit de convolution de deux spectres

$$I_{\text{anis}} = I_{\text{vib}} \times I^{(2)}_{\text{rot}}$$

$$I_{\text{iso}} = I_{\text{vib}}$$

La largeur à mi-hauteur de I_{anis} est dans l'approximation Lorentzienne

$$2\Gamma_{\text{tot}} \sim 2\Gamma_{\text{vib}} + 2\Gamma_{\text{rotation}}$$

Pour un mode polarisé ($I_{\text{iso}} \neq 0$ et $I_{\text{anis}} \neq 0$) on connait la largeur de vibration de I_{anis}, qui est celle de I_{iso}. On sépare ainsi facilement les effets de vibration et de rotation sur la largeur de I_{anis}.

Application au spectre des hexafluorures

Le mode «de respiration» totalement polarisé $\nu_1(a_{1g})$ constitue la partie isotrope du spectre (ayant la symétrie de $\alpha_{11} + \alpha_{22} + \alpha_{33}$ il contribue uniquement à la modulation de la partie isotrope de la polarisabilité $\bar{\bar{\alpha}}$).
Les modes dépolarisés ($\rho = 3/4$)

$$\nu_2(e_g) \longrightarrow \alpha_{xx} + \alpha_{yy} - 2\alpha_{zz}, \alpha_{zz} - \alpha_{yy}$$

$$\nu_5(f_{2g}) \longrightarrow \alpha_{xy}, \alpha_{yz}, \alpha_{zx}$$

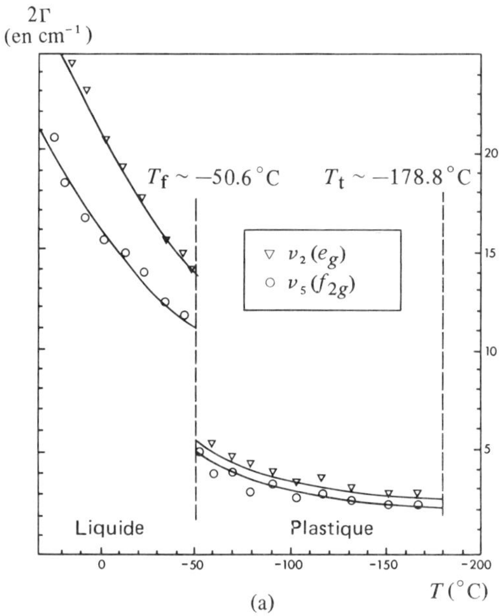

(a)

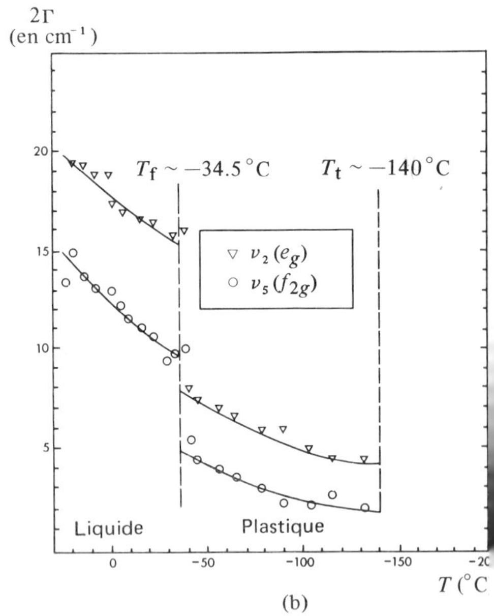

(b)

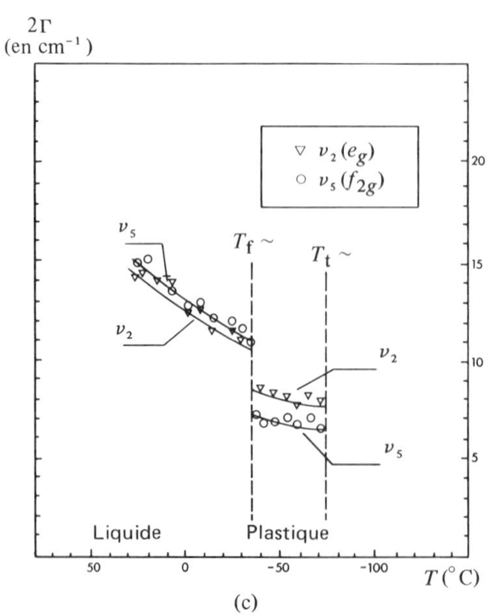

(c)

Fig. 5. Evolution des largeurs des modes depolarisés ν_2 et ν_5 avec la température.

participent à la modulation de la partie anisotrope de $\bar{\bar{\alpha}}$ et les raies ν_2 et ν_5 sont donc la partie I_{anis} du spectre. Comme ν_2 et ν_5 sont complètement dépolarisés ($I_{iso} = 0$), on ne peut estimer simplement:

- leurs largeurs de vibration respectives
- leur largeur de rotation:

$$2\Gamma = \frac{2}{2\pi\tau^{(2)}c} \text{ (en cm}^{-1}\text{)}$$

avec:

$$\tau^{(2)} = \int_0^\infty \mathrm{Re} \frac{\langle Y_2^m(0)\, Y_2^{m*}(t)\rangle}{\langle |Y_2^m|^2\rangle} \mathrm{d}t$$

en comparant I_{iso} et I_{anis} comme on peut le faire sur les modes polarisés.

Evolution du mode isotrope ν_1 avec la température. La largeur du mode isotrope est très faible; de l'ordre de quelques cm^{-1}, (largeur à mi-hauteur $2\Gamma \simeq 2$ cm^{-1}). Cette raie semble fortement dissymétrique à la température ambiante (Fig. 1). On constate une légère décroissance de la largeur avec la température et la dissymétrie disparaît (bande chaude). La largeur est donc généralement plus faible en phase plastique, mais le pouvoir de résolution du spectromètre ne permet pas de suivre l'évolution complète. Pour SF_6 $2\Gamma_{(\text{liquide})} = 2$ cm^{-1}, $2\Gamma_{(\text{point de fusion})} = 1{,}7$ et $2\Gamma_{(\text{transition plastique orthorhombique})} = 1{,}58$ cm^{-1}. Pour TeF_6 $2\Gamma_{(\text{liquide})} = 2{,}75$ cm^{-1} et $2\Gamma_{(\text{point de fusion})} = 2{,}05$ cm^{-1}.

Evolution des largeurs des modes dépolarisés avec la température. L'évolution des largeurs des deux modes dépolarisés avec la température est indiquée sur la Fig. 5.

On note:

1. Une discontinuité très nette des largeurs des modes à la transition liquide–plastique, cette discontinuité étant variable avec les hexafluorures.
2. Une différence également variable suivant les composés entre les modes ν_2 et ν_5 – à la fois dans la phase liquide et plastique. En admettant une réorientation isotrope, il semble que ces différences soient dues à la fois à une relaxation vibrationnelle différente et à l'influence du couplage de Coriolis tout au moins pour le mode ν_5. Une estimation grossière semble cependant indiquer une valeur de $2\Gamma_{\text{Coriolis}} \sim 0{,}25$ cm^{-1} pour TeF_6 liquide, donc une contribution négligeable.
3. Des variations thermiques généralement plus importantes en phase liquide, qu'en phase plastique et variables avec les éléments.

Pour séparer les contributions de relaxation vibrationnelle et orientationnelle dans ces modes dépolarisés, nous avons suivi la méthode de Rakov[9] qui peut être décrite par une formule empirique de la forme

$$2\Gamma = 2\Gamma_0 + 2\Gamma_1 \exp(-T_0/T)$$

$2\Gamma_0$ représente la largeur «résiduelle» ou plus simplement la contribution vibrationnelle.

Les résultats pour la phase plastique sont indiqués sur le Tableau 2.

Pour SF_6 et SeF_6 en phase plastique, la relaxation vibrationnelle déduite de la loi de Rakov est faible, et de l'ordre de grandeur de celle déduite à partir du mode isotrope.[10]

L'étude des harmoniques en phase liquide peut éventuellement permettre une séparation des contributions vibrationnelle et orientationnelle.

L'harmonique $2\nu_6$ est le plus intense dans les spectres Raman des phases liquides. $2\nu_6$ $(a_{1g} + e_g + f_{2g})$ est polarisé et le degré de polarisation dépend des intensités relatives des transitions a_{1g} et $e_g + f_{2g}$. Dans ce cas a_{1g} contribue à I_{iso}, tandis que e_g et f_{2g} contribue à I_{anis}.

On observe que pour SF_6, $2\nu_6$ est polarisé: $2\Gamma \sim 2{,}5$ cm^{-1} alors que $2\nu_6$ est depolarisé pour TeF_6, $2\Gamma(2\nu_6) \cong 16 \pm 3$ cm^{-1}. Dans TeF_6 on constate que $2\Gamma(2\nu_6)$ est identique à $2\Gamma(\nu_2)$ et à $2\Gamma(\nu_5)$ de sorte que les largeurs intrinsèques de ces modes sont du même ordre de grandeur et sont également faibles par rapport aux largeurs dues aux réorientations.

L'harmonique $2\nu_5$ a été observé dans TeF_6, il est polarisé et sa largeur est de $2\Gamma(2\nu_5) \simeq 7{,}5 \pm 1$ cm^{-1}. En admettant que l'harmonique $2\nu_5$ a une largeur intrinsèque double de celle de ν_5 on obtient $2\Gamma(\nu_5) \cong 3{,}75 \pm 1$ cm^{-1}.

Le résultat semble compatible avec l'extrapolation de Rakov et de l'ordre de grandeur de la largeur du mode isotrope.

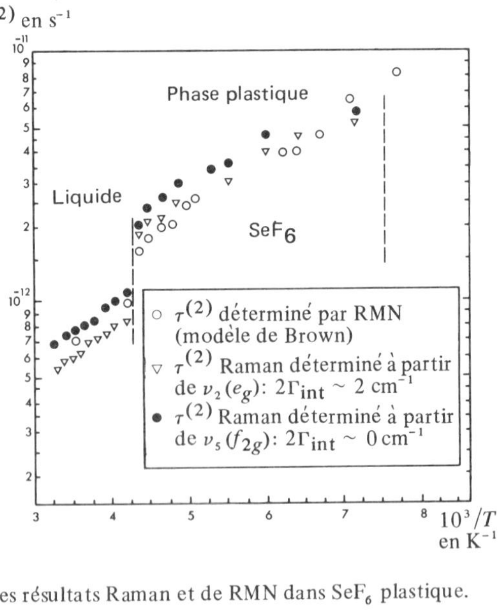

Fig. 6. Comparaison des résultats Raman et de RMN dans SeF_6 plastique.

Modèle de réorientation moléculaire

L'étude des formes de raies et des fonctions de corrélation peut préciser le type de modèle.[8] La comparaison des profils des bandes infrarouge et Raman qui sont respectivement les transformées de Fourier des fonctions de corrélation des harmoniques sphériques $Y_1^m(\theta, \phi)$ et $Y_2^m(\theta, \phi)$ peut également caractériser le modèle de réorientation.[10]

Enfin, les études de relaxation nucléaire sur ^{19}F permettent de mesurer T_1 (temps de relaxation longitudinal). Dans un modèle de réorientation par saut de grands angles avec bloquage du moment angulaire entre les sauts (modèle de Brown) il est possible de déduire un intervalle moyen entre sauts T_R. De même à partir des résultats de diffusion

Tableau 2.

		SF_6	SeF_6	TeF_6
$2\Gamma_0$ (cm^{-1})	ν_2 (e_g)	2,7	3,6	7,4
	ν_5 (f_{2g})	2,2	1,4	6

Raman, on déduit le temps $\tau^{(2)}$ de corrélation des harmoniques sphériques $Y_2^m(\theta, \phi)$. La comparaison des valeurs de $\tau^{(2)}$ obtenues par la RMN et la diffusion Raman (Fig. 6) permet de déduire un angle de saut de l'ordre de 50°.[11]

Indépendamment de ce modèle, on peut constater que les largeurs de rotation de ν_2 et ν_5 ont une valeur minimum qui est de 0,5 cm^{-1} pour SF_6, 0,65 cm^{-1} pour SeF_6 et 0,45 cm^{-1} pour TeF_6 à la transition solide–solide.

CONCLUSION

Cette étude préliminaire en diffusion Raman de la dynamique des hexafluorures de métalloîdes a montré:

1. Une discontinuité des temps de réorientation à la transition liquide–plastique.
2. Une largeur de vibration des modes dépolarisés faible par rapport aux largeurs de rotation.

La dispersion des fréquences dues à des environnements moléculaires différents est donc faible, ce qui est confirmé par un déplacement négligeable des fréquences des raies entre gaz, liquide et phase plastique $\Delta\nu \sim 2$ cm^{-1}.

L'anharmonicité du potentiel intramoléculaire, la présence de bandes «chaudes», l'éventuelle différence de largeurs intrinsèques entre les phases liquide et plastique ne permettent de caractériser complètement ce type de relaxation. Des études de potentiel intermoléculaire sont en cours pour essayer d'expliquer les différentes valeurs de largeurs de vibration des raies Raman.

On obtient des temps de réorientation de $0,5 \times 10^{-12}$ s à température ambiante pour SF_6 et SeF_6 liquide. En phase plastique, près de la transition solide–solide, les temps sont de l'ordre de 8×10^{-12} s.

Un accord convenable entre les mesures de RMN et de Raman pour un modèle de sauts avec bloquage du moment angulaire entre les sauts. Ce modèle semble valable pour décrire les phases plastiques et sans doute plus approprié que le modèle d'Hubbard.

Des études en infrarouge et diffraction de neutrons sont en cours pour essayer de préciser plus en détail les mouvements moléculaires dans ces cristaux plastiques.

REFERENCES

1 J. Virlet, Thèse de Doctorat des Sciences, Paris, 1972.

2 J. Virlet et P. Rigny, *Chem. Phys. Letters* **6**, 4377 (1970).

3 B. Weinstock et G. L. Goodman, in *Advances in Chemical Physics*, Vol. IX, Wiley, 1965.

4 H. H. Claassen *et al.*, *J. Chem. Phys.* **53**, 341 (1970)

5 J. Michel, M. Drifford et P. Rigny, *J. Chim. Phys.* **31**, 671 (1970).

6 M. Drifford et M. T. Boucraut, *Appareillage et caractérisation des composés minéraux*, Vol. 4, Masson, 1970, p. 115.

7 R. G. Gordon, *J. Chem. Phys.* **43**, 1307 (1965).
8 S. Bratos et E. Maréchal, *Phys. Rev.* **A4**, 1078 (1971).
9 A. V. Rakov, *Proc. Lebedev Phys. Inst.* **27A**, 111 (1965).
10 P. Rigny, M. Drifford et J. Virlet, *J. Chim. Phys.* **32**, 65 (1971).
11 D. A. Pinnow *et al.*, *J. Chem. Phys.* **49**, 1 (1968).
12 J. Virlet et M. Drifford, *Chem. Phys. Letters* 8, 347 (1972).

26. RAMAN DEPOLARIZATION RATIO IN THE SMECTIC AND NEMATIC LIQUID CRYSTALS

B. Lavrenčič and S. Lugomer
"J. Stefan" Institute, University of Ljubljana, Ljubljana, Yugoslavia

Liquid crystals are substances with properties between solids and (isotropic) liquids, i.e. exhibiting some properties of both.[1] When a (thermotropic) liquid crystal melts, it forms optically an anisotropic mesophase which after further heating changes into an ordinary liquid. The liquid crystalline mesophases are designated with names such as nematic, smectic A, B, C, . . . , cholesteric. Some substances acquire the mesophase upon mixing with water. The latter are called lyotropic and are of biological interest. For example, a solution of the tobacco mosaic virus (TMV) in water is known to form a mesophase.

From a microscopic point of view, the mesophase indicates the onset of a long range order parameter in one or two dimensions. Structurally, the liquid crystals are made of long, rod-like molecules which orient themselves along an axis. We shall consider a series of liquid crystals, 4,4′-bis-(alkoxy) azoxybenzenes with a chemical formula:

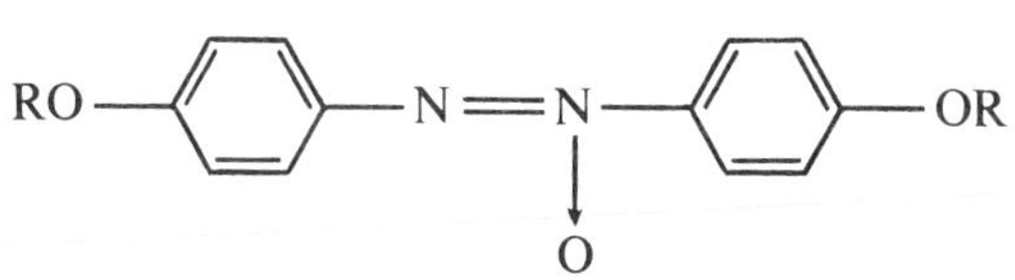

(1)

$$R = C_7H_{13}$$

The phase boundaries for the series are shown in Fig. 1.

So far, only a few Raman spectra have been taken from any liquid crystal.[2] Amer and Shen have thoroughly studied the series of 4,4′-bis-(alkoxy) azoxybenzenes up to the sixth atom in the alkyl chain.[3] Since the seventh member of this series, heptyloxyazoxybenzene (HOAB) is the first in the series to exhibit, beside a nematic, also a smectic C mesophase, it appeared interesting to study Raman spectra. In order to deduce more of the structural information we have oriented the sample in a magnetic field and have measured the polarized and depolarized components. This work appears to be the first of its kind. The scattering geometry is seen in the insert of Fig. 2. The laser beam enters the sample polarized parallel to the magnetic field (~ 1000 G). The scattered light is

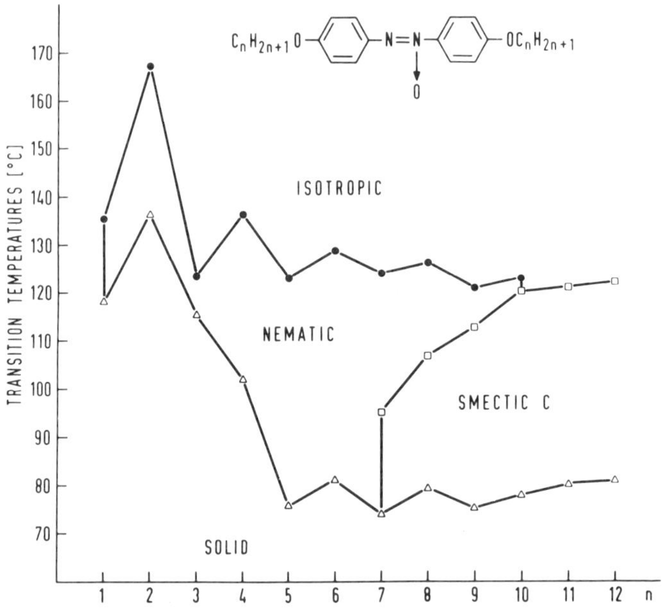

Fig. 1. The phase transitions in the series 4,4′-bis-(alkoxy) azoxybenzene as a function of alkyl length. HOAB belongs to $n = 7$.[1]

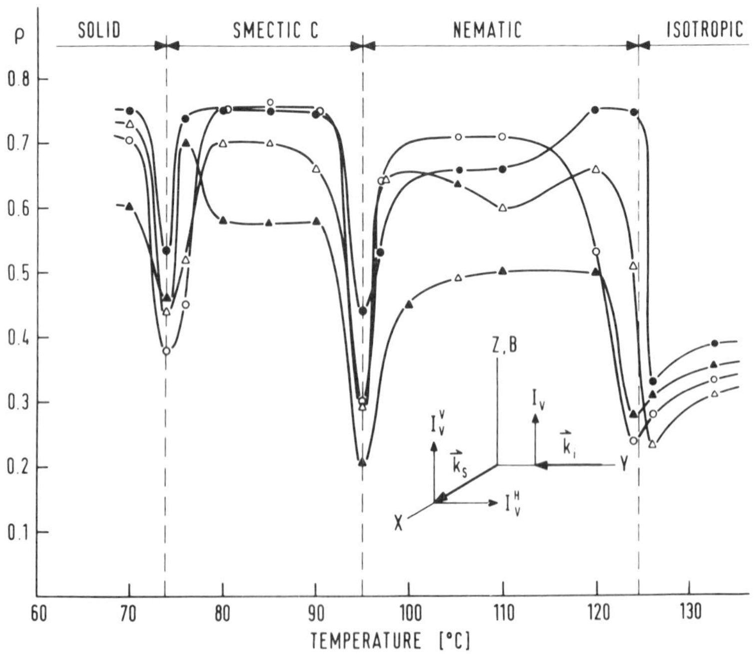

Fig. 2. The depolarization ratio, $\rho = I_V^H / I_V^V$, is shown as a function of temperature in all four phases of HOAB for the lines 1267 cm^{-1} (△), 1459 cm^{-1} (●), 1489 cm^{-1} (○) and 2925 cm^{-1} (▲). The insert shows the scattering geometry, z axis is parallel with the magnetic field B.

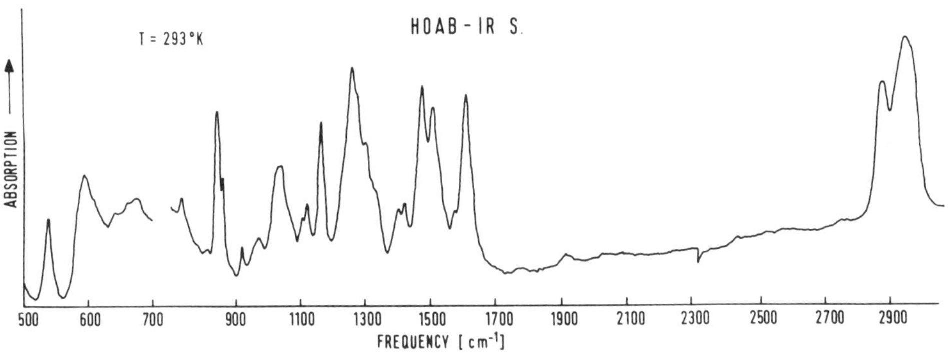

Fig. 3. Room temperature infrared spectrum of HOAB.

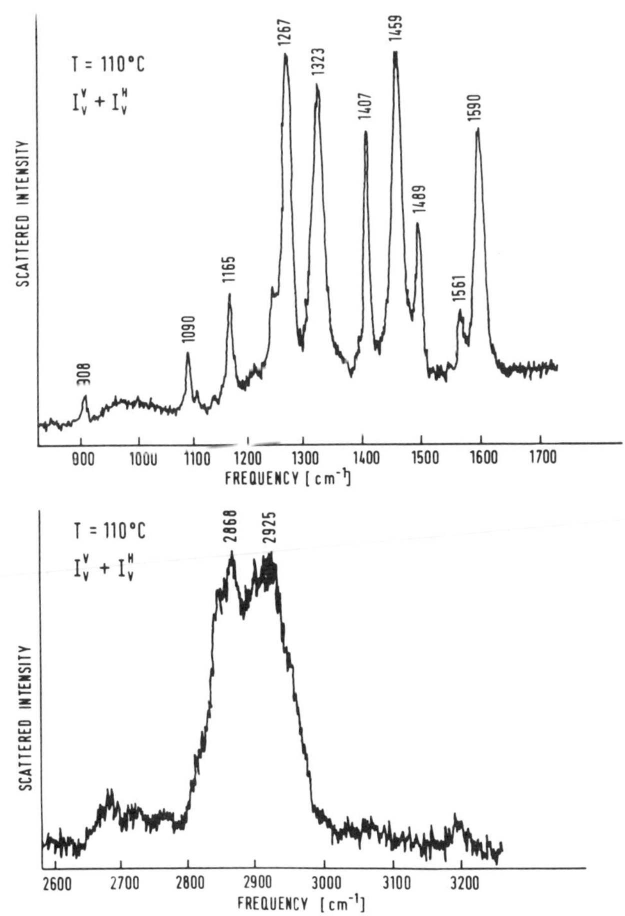

Fig. 4. The VV + VH spectra at 110°C (nematic phase) are shown for two regions of interest.

analysed either parallel to the scattering plane (I_V^H) or perpendicular to it (I_V^V).

We have measured the infrared and Raman spectra of HOAB in the solid, smectic, nematic and isotropic phases. The infrared spectra (Fig. 3) and Raman spectra (Fig. 4) show the same peaks as were found by Amer and Shen in the shorter alkyl chains. For several peaks the depolarization ratio $\rho = I_V^H / I_V^V$ has been plotted as the function of temperature (Fig. 2). Rather large drops in ρ are observed around all phase transitions (respectively at 74°C, 95°C and 124°C). The corresponding infrared lines do not change. Such a change can be expected from the fact that the molecules in the mesophase are not allowed to rotate freely.

The case of isotropic averaging of the molecular polarizability was treated by Nafie *et al.*[4] In the liquid crystal their steps can be followed by replacing the $\overline{\cos^2 \theta}$ or $\overline{\cos^4 \theta}$ where θ is the angle between the molecular axis and the preferred direction, with the ensemble averages $\langle \cos^2 \theta \rangle$ and $\langle \cos^4 \theta \rangle$ and to average all other angles over a sphere. The latter quantities are evaluated by using Maier and Saupe's mean field theories,[5] using the distribution function $f(\cos \theta)$. Thus, the depolarization ratio is a function of the second and fourth powers of $\cos \theta$ averaged over the ensemble and therefore of temperature and the order parameter.[6] The details of this calculation will be presented elsewhere.

The new feature of our results is that as far as the depolarization ratio is concerned, the immediate neighbourhood of the solid–smectic as well as smectic–nematic point behaves as if the system becomes isotropic. It may well be that the fluctuations and rearrangements near the above transitions force the system into a quasi liquid state. The same phenomenon has been found by Blinc and Pirš of this laboratory, in the proton NMR spectra of HOAB which show a sharp quasi liquid line within three degrees of the smectic–nematic transition.

REFERENCES

1 G. H. Brown, J. W. Doane and V. D. Neff, *CRC Review of the Structure and Physical Properties of Liquid Crystals,* Butterworths, London, 1971.

2 R. Freymann and R. Servant, *Ann. Phys. (Paris)* **20**, 131 (1945); A. S. Zhdanova, L. F. Morozova, G. V. Peregudov and M. M. Sushinskii, *Opt. i Spektroskopiya* **26**, 206 (1969); B. J. Bulkin and F. T. Prochaska, *J. Chem. Phys.* **54**, 635 (1971); N. M. Amer, Y. R. Shen and H. Rosen, *Phys. Rev. Letters* **24**, 718 (1970); W. J. Borer, S. S. Mitra and C. W. Brown, *Phys. Rev. Letters* **27**, 379 (1971).

3 N. M. Amer and Y. R. Shen, *J. Chem. Phys.* **56**, 2654 (1972).

4 L. Nafie, P. Stein, B. Fanconi and W. L. Peticolas, *J. Chem. Phys.* **52**, 1584 (1970).

5 W. L. McMillan, *Phys. Rev.* **A4**, 1238 (1971).

6 P. G. de Gennes, *Phys. Letters* **30A**, 454 (1969).

27. RAMAN STUDIES OF LIQUID CRYSTALS

G. Vergoten

Laboratoire de Spectroscopie Raman, Université des Sciences et Techniques de Lille, Villeneuve d'Ascq, France

and

Laboratoire de Physique, UER de Pharmacie, Université du Droit et de la Santé, Lille, France

INTRODUCTION

The classification of matter into solids, liquids and gases is not absolute. There are many substances, having properties intermediate between those of an isotropic liquid and a solid, which can exist in well-defined phases.

Unlike most organic solids which have sharp melting points at which the crystal undergoes a change to an isotropic liquid however, certain compounds pass through an intermediate stage upon heating. They melt to a liquid, often cloudy, which exhibits many properties usually associated with the crystalline state. At a higher temperature, the cloudy liquid suddenly changes to a clear, isotropic one. This class of compounds is known as 'liquid crystals'. Depending on the precise nature of these phases, they are termed 'nematic', 'smectic', or 'cholesteric'. Of these, the nematic phase has been found to be of greatest interest. The nematic phase is a true liquid, which takes the shape of its container. It differs from a normal liquid in that the molecules are oriented to a certain degree. The weak intermolecular attraction resulting in the observed orientation is favoured by a rod shaped molecular geometry.

PRESENTATION OF NEMATIC MBBA

N.(*p*.methoxy-benzylidene)-*p*-n-butylaniline (MBBA) (Fig. 1) which has been recently synthesized,[1] is a rare example of nematic liquid crystal at room temperature and has widespread applications to display devices.

A large number of investigations of its physical properties have been undertaken. But one of the major problems associated with the use of substituted benzylidene-aniline type liquid crystals is their relative instability to cleavage or reaction of the anil linkage. MBBA melts at 20°C to a mobile, turbid, pale yellowish liquid with all the properties of a nematic phase; its clarification point is approximately 44°C, but many workers have noted the presence of ubiquitous impurities in this material which influence the reproducibility of phase transition temperatures and their electrooptic properties.

EXPERIMENTAL

Raman spectra were obtained using a Coderg PH1 double monochromator equipped

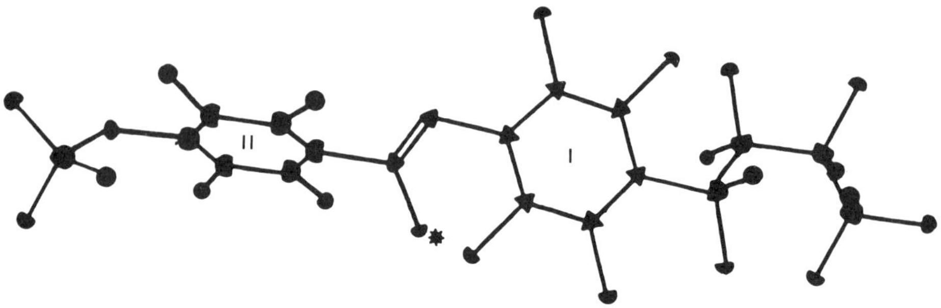

Fig. 1. Molecular model of MBBA.

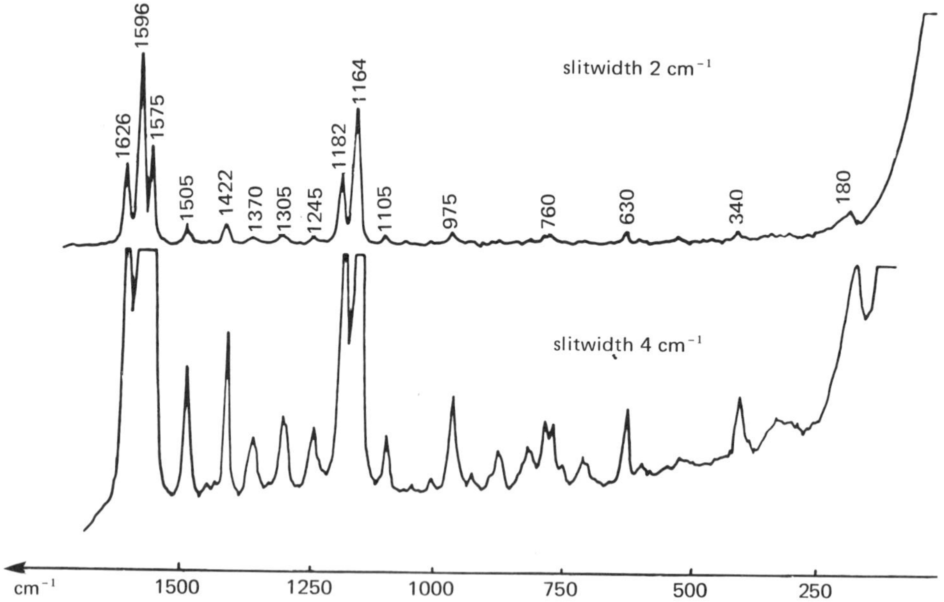

Fig. 2. Raman spectrum of MBBA in the isotropic liquid phase.

with 1800 lines/mm gratings, used in the first order. A coherent radiation laboratories Model 52 ionized krypton laser (647·1 nm) was used with transfer optics which yield 60 mW power at the sample.

The spectral slitwidths were 1, 2 and 4 cm^{-1}. Spectra were scanned at 50 or 100 cm^{-1}/min. A spectral range of 1700 cm^{-1} from the exciting frequency was investigated in the solid, nematic and isotropic liquid states.

RAMAN SPECTRA OF MBBA IN THE SOLID, NEMATIC AND ISOTROPIC STATES

The Raman spectrum of MBBA in the isotropic phase (Fig. 2) shows five intense sharp lines. In addition, there are a number of broad and weaker bands. In this spectrum

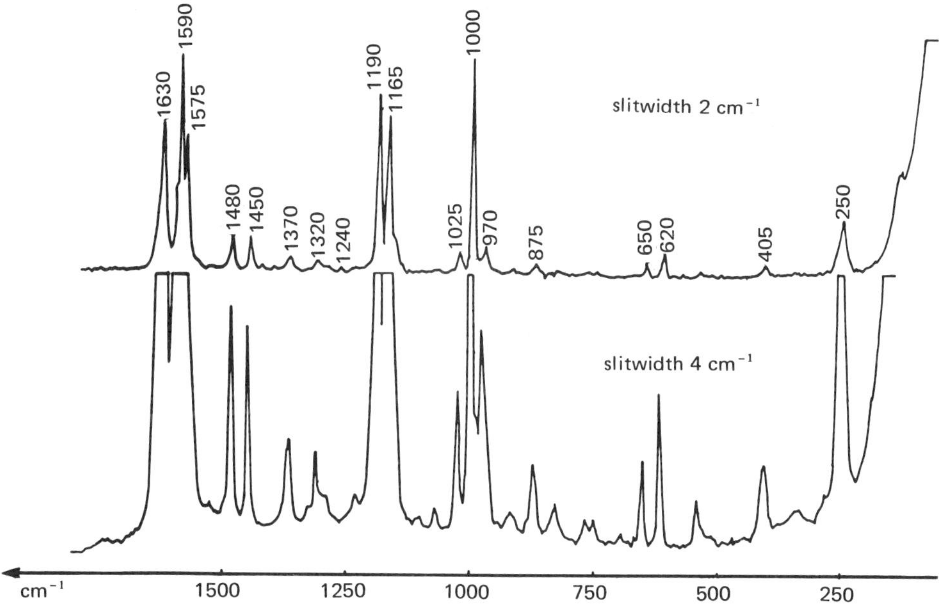

Fig. 3. Raman spectrum of liquid BA.

we find several lines belonging to the benzylidene-aniline (BA) the infrared spectrum of which is known.[2,3] We have also recorded the Raman spectrum of BA (Fig. 3) which is not mesomorphic and have undertaken a normal coordinate treatment for this compound. We report the observed frequencies, relative intensities and assignments of MBBA and the observed and calculated frequencies of BA (Table 1)

The spectral range can be divided into two regions of interest: 0–150 cm^{-1} and 150–1700 cm^{-1}: this division is based on the response of these regions to the phase transitions. We first consider the spectral region 150–1700 cm^{-1}. As the compound changes from the solid to nematic phase, all lines decrease quasidiscontinuously at the transition temperature and become broader. No lines disappear. These features indicate that the solid-to-nematic transition is a true first order one. There is no difference between the spectra in the nematic and the isotropic liquid states. This phase transition is still of first order, but weak. The intermolecular interaction does not change considerably when going from the nematic state to the liquid one.

Let us now consider the spectral region 0–150 cm^{-1}. Low frequency Raman scattering has been used here to probe phase transitions and short range ordering. The low frequency Raman spectrum of MBBA is reported in the solid state (Fig. 4). The assignments, which we present (Table 2), are tentative and are in need of verification when single crystal, crystallographic data and far infrared spectra become available.

In the solid phase of a liquid crystal, the molecules are bound by orientational forces so strong that the crystal melts before the orientations become disordered. Thus the lattice periodicity breaks down before orientational correlations between neighbouring molecules are lost. In undergoing the transition from solid to nematic, most of the low frequency Raman lines disappear, with the exception of the presence of two broad lines in the nematic phase of MBBA (Fig. 5). We have used for the recording of the low frequency

Table 1. Observed frequencies, relative intensities and assignments of MBBA. Observed and calculated frequencies of BA.

MBBA			BA
Frequency (cm^{-1})	Relative intensity	Assignments (Wilson's notation)	Observed frequency
180	6		
315	1		250
340	2		
410	4	16 a	405
			540
630	5	6 b	620
			650
720	2		
760	2		750
775	4		770
790	4		
825	3		825
886	3	δ (C–H)*	875
934	1	17 b	920
975	7	δ 5	970
			1000
1014	1	18 a (I and II)	1025
1105	4		
1164	70	9 a (I and II)	1165
1182	37	ν (ϕ–N)	1190
1245	5	ν (ϕ–C)	1240
1305	5		1320
1370	4	δ (C–H)*	1370
1422	12	19 b (I and II)	1450
1503	9	19 a (I and II)	1480
1575	53	8 a (II)	1575
1596	100	8 a (I)	1590
1626	41	ν (C=N)	1630

Raman spectrum in the nematic state, a spectrometer equipped with a premonochromator as described by Wallart.[4]

We did not observe the presence of a non-propagating (relaxational) soft mode shifting toward zero frequency in approaching the solid-nematic transition, as described by Bulkin and Prochaska,[5] in the spectrum of *para*azoxyanisol, and as theoretically suggested by Kobayashi.[6]

Concerning the two broad lines observed in the spectrum of MBBA in his nematic phase, there must exist relationships between these low frequencies and the Frank elastic coefficients k_{11}, k_{22}, k_{33} associated with the nematic phase and which correspond to, respectively, the splay, torsion, and flexion modes of distortion. What is the significance of

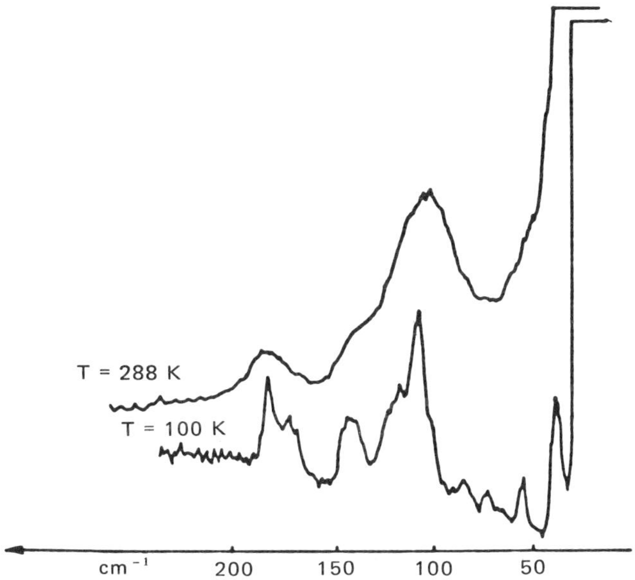

Fig. 4. Low frequency Raman spectrum of MBBA in the solid state.

Table 2. Frequencies and assignments of the lattice Raman lines for solid MBBA.

	$\bar{\nu}$ (cm^{-1})			Assignment
100 K	194 K	250 K	288 K	
38	37·5			$2R + 2T$
55	54·5	53		
71·5				
85·4	84·5			
107	104·5	102	97	R_z (axis of elongation)
119	113·5	113	105	
142	140	137	134	T
179	179			Internal modes
		180	180	
181	181			

these bands in the low frequency Raman spectra of the nematic phase? Let us consider the nematic-isotropic liquid phase transition from a group theoretical point of view. A phase transition can be defined as follows.

A continuous group of symmetry in the disordered phase is broken in the ordered one. When the symmetry of the continuous group is broken, there will appear a set of collective modes in the condensed phase. When the isotropic liquid undergoes nematic transition, the rotational symmetry O_3 is broken into $\bar{O}$ which is the unidimensional continuous group of rotational symmetry. The collectives modes which will appear in the

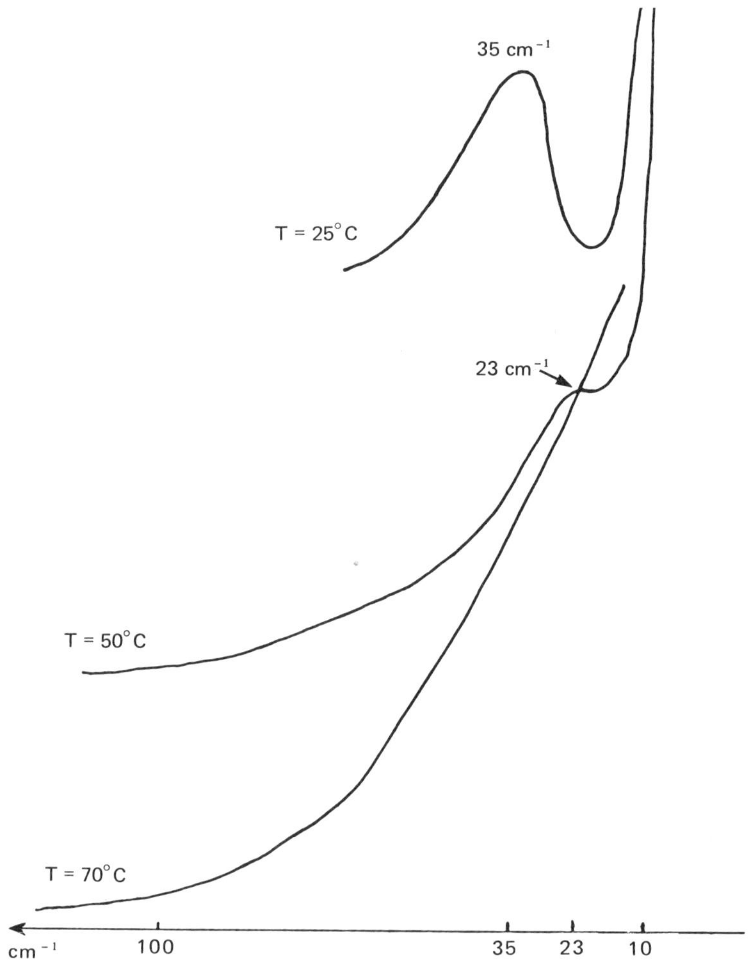

Fig. 5. Low frequency Raman spectra of MBBA in the nematic and isotropic phases.

nematic phase can be described as classical orientational waves. The broad lines disappear in the isotropic phase but we have also observed a low frequency Raman line just above the clarification point which disappears at higher temperature.

Obviously the isotropic nematic phase transition displays typical features of a second order phase transition. When approaching the transition temperature from the high temperature side, there is a pronounced pretransitional ordering phenomenon.

CONCLUSION

We are continuing our Raman studies of liquid crystalline materials in two directions: it will be possible to define an orientational order parameter from polarization measurements on homogeneously oriented samples of MBBA. We also use the nematic materials as solvents to give a solute molecule an anisotropic environment.

ACKNOWLEDGMENT

The normal coordinate treatment of benzylidene aniline was performed with the help of Pr. Fleury.

REFERENCES

1 H. Kelker and B. Scheurle, *Angew. Chemie (Int. Ed.)* 8, 884 (1969).
2 H. Shindo and B. Umezawa, *Chem. Pharm. Bull.* **10**, 492 (1962).
3 B. Ostrowska and A. Tramer, *Acta Phys. Polom.* **33**, fasc. 1 (7), 14 (1967).
4 F. Wallart, *Thesis*, 1970.
5 B. J. Bulkin and F. T. Prochaska, *J. Chem. Phys.* **54**, 635 (1971).
6 K. K. Kobayashi, *Mol. Cryst. (Liq. Cryst.)* **13**, 137 (1971).

28. MISE EN EVIDENCE PAR SPECTROSCOPIE RAMAN D'UN CHANGEMENT DE STRUCTURE MOLECULAIRE DE L'ETHYLENEDIAMINE A L'ETAT SOLIDE

H. Brusset, Nguyen Quy Dao et M. Jouan

Institut de Chimie, Ecole Centrale des Arts et Manufactures, Chatenay-Malabry, France

ABSTRACT

The determination of the crystalline structure of solid ethylenediamine at −60°C previously performed in our laboratory, has shown that this molecule appears in the *trans* form. The Raman spectra of ethylenediamine and of N,N,N',N' tetradeuteroethylenediamine or ethylenediamine-d_4 have been recorded at temperatures varying between room temperature and that of liquid nitrogen. We have seen that the two amines undergo a change of phase. The transition temperature has been evaluated at about −80°C for ethylenediamine by DTA. The first form, 'low temperature', which corresponds to single crystals studied by X-ray diffraction presents a spectrum which conforms to the *trans* configuration, with regard to selection rules. The second form, 'very low temperature', is very difficult to obtain, especially with en-d_4. The presence of water in particular is sufficient to prevent transformation. The change of phase is, moreover, often incomplete, and in fact a mixture of the two forms is generally obtained. The 'very low temperature' spectra can be distinguished from the 'low temperature' spectra by visible disappearances of peaks, appearances of new bands and splitting of peaks, particularly for the skeleton bands.

The skeleton bands are assigned as follows:

	'Low temperature' form		'Very low temperature' form	
	en	en-d_4	en	en-d_4
δ (C–N)	1109	1036	1121, 1128	1938, 1053
δ (C–C)	1054	1051	1055	1054
δ (N–C–C)	469	430	471, 479	431, 445
τ (N–C–C–N)	–	–	186	178

INTRODUCTION

Notre travail se situe dans le cadre des études menées au laboratoire sur les composés de solvatation formés par divers sels avec des solvants organiques et est orienté plus particulièrement vers l'étude des liaisons entre sel et solvant, à l'état solide, par spectroscopies infrarouge et Raman. Ainsi, nous nous intéressons aux vibrations dans les composés formés par l'éthylènediamine avec les halogénures de lithium, à la suite des études structurales effectuées au laboratoire sur les composés cristallins de formule LiCl-2 en, LiBr-2 en, LiI-3 en.[1,2]

Pour l'étude spectroscopique de ces composés de solvatation, il nous a semblé souhaitable de nous appuyer sur une connaissance aussi complète que possible du solvant pur. C'est pourquoi nous avons voulu connaître les spectres de vibration de l'éthylènediamine à l'état pur.

La principale étude spectroscopique que nous ayons trouvée est celle de Sabatini et Califano[3] qui ont enregistré les spectres infrarouge de l'éthylènediamine et de l'éthylènediamine-d_4(ND_2-CH_2-CH_2-ND_2) à l'état vapeur, liquide et solide avec, dans ce dernier cas, étude de la polarisation des bandes. Ces auteurs, en utilisant des données Raman sur l'éthylènediamine liquide provenant d'autres sources, ont proposé une attribution des bandes basée sur la configuration *cis* pour la molécule.

Leurs conclusions sont donc en contradiction avec les résultats obtenus récemment au laboratoire par Jamet-Delcroix[2,4] sur la structure cristalline de l'éthylènediamine solide à $-60°$C.

D'autres auteurs ont également publié des spectres infrarouge[5] et Raman[6] de l'éthylènediamine liquide et de sa solution aqueuse. Cependant, à notre connaissance, aucune étude n'a été faite par spectroscopie Raman à l'état solide, ni sur l'éthylènediamine-d_4.

L'absence de ces données Raman, ainsi que de données infrarouge en dessous de 400 cm^{-1} constitue à notre avis un réel handicap pour une étude poussée des vibrations dans la molécule. C'est la raison pour laquelle nous nous sommes proposé d'étudier le spectre Raman de l'éthylènediamine solide, à basse température.

Nos résultats, obtenus en refroidissant progressivement de l'éthylènediamine (T_f = 11°C) et de l'éthylènediamine-d_4 depuis la température ordinaire jusqu'à la température de l'azote liquide nous ont montré que ces deux isotopomères présentent une allotropie à très basse température. Pour exposer ce phénomène, il nous paraît nécessaire de nous appuyer sur des résultats précis à 'basse température'.

Nous exposerons donc dans ce qui suit tout d'abord l'attribution des bandes que nous proposons pour le spectre Raman de l'éthylènediamine et de l'éthylènediamine-d_4 à 'basse température', sous la configuration *trans*, puis la mise en évidence d'une nouvelle phase 'très basse température' pour laquelle nous donnerons une hypothèse de structure.

Dans la suite de l'exposé, nous désignerons par 'BT' la phase 'basse température' et 'TBT' la phase 'très basse température' d'autre part, sauf indication contraire, les valeurs valables pour l'éthylènediamine-d_4 (ou en-d_4) seront indiquées entre parenthèses derrière les valeurs correspondantes pour l'éthylènediamine ordinaire (ou 'en').

CONDITIONS EXPERIMENTALES

Produits

Nous avons utilisé de l'éthylènediamine 'Merck' distillée deux fois sur sodium, indifféremment sous azote ou sous vide, et de l'éthylènediamine-d_4 obtenue suivant le procédé décrit par Gray et Lord:[4,7] la deutération est effectuée sur le chlorhydrate et l'éthylènediamine-d_4 est ensuite régénérée par chauffage vers 200–250°C sous vide, en présence de chaux désséchée à 1000°C pendant 48 heures. Contrairement aux auteurs précédents, nous avons ainsi recueilli une amine contenant une quantité d'eau lourde non négligeable, pouvant atteindre une mole d'eau par mole d'amine [ainsi que nous le verrons à propos de *l'Influence de l'hydratation* (p. 235), les spectres de l'éthylènediamine (en-d_4) et du monohydrate en, H_2O (en-d_4, D_2O) présentent, en particulier à l'état solide, des différences très nettes; on peut d'ailleurs remarquer que le monohydrate solide est effectivement un composé défini. Il est donc possible d'évaluer le degré d'hydratation quand il est inférieur à 1 mole d'eau par mole d'amine]; nous l'avons donc déshydraté par chauffage à reflux sur sodium suivi d'une distillation. Cette opération est répétée une ou deux fois jusqu'à ce que le spectre ne nous montre plus de variation par distillation.

Enregistrement des spectres

Les spectres ont été enregistrés sur un spectromètre Raman Laser Coderg PH1, en utilisant la raie à 488,0 nm d'un laser à argon Spectra-Physics modèle 141. Pour les mesures à basse température, nous avons utilisé une cellule à température variable Cryocirc (Coderg) refroidie avec de l'azote liquide. La température est mesurée à l'aide d'un thermocouple dont la soudure froide est plongée dans l'azote liquide et la soudure chaude se trouve dans un bloc métallique sur lequel est fixé un tube capillaire de 1 mm de diamètre contenant le liquide. A cause des phénomènes de gradient de température et d'inertie thermique, il n'est pas possible de mesurer de manière précise, à l'aide de ce dispositif, la température au point de focalisation du faisceau laser; il faudrait, en fait, introduire dans le liquide, à l'intérieur du capillaire, un thermocouple qui risque alors d'être attaqué par l'éthylènediamine et de perturber le phénomène. Nous nous sommes donc contentés de relever l'indication du thermocouple, ce qui donne des valeurs relativement reproductibles; pour des mesures de température de changement de phase, nous avons préféré effectuer, en collaboration avec Jamet-Delcroix, des mesures par analyse thermique différentielle.

SPECTRE RAMAN DE L'ETHYLENEDIAMINE A 'BASSE TEMPERATURE', SOUS LA CONFIGURATION *TRANS* (EN 'BT')

Rappel de la structure de l'en 'BT' à 60° C[2,4]

L'en 'BT' cristallise dans le système monoclinique de groupe spatial $P2_1/c(C^5_{2h})$. Les caractéristiques de maille sont:

$$a = 5{,}65 \pm 0{,}05 \text{ Å}$$
$$b = 7{,}14 \pm 0{,}02 \text{ Å}$$
$$c = 5{,}50 \pm 0{,}05 \text{ Å}$$
$$\beta = 125^\circ\ 35' \pm 40'$$
$$d_{\text{th}} = 1{,}11$$

Il y a deux molécules NH_2-CH_2-CH_2-NH_2 dans la maille. L'en 'BT' se présente sous la configuration *trans* de symétrie C_{2h}. Le centre de gravité de la molécule se trouve à un centre d'inversion du groupe spatial. La Figure 1 représente la structure projetée parallèlement à l'axe a. Il est à remarquer que la distance entre les atomes d'azote N_6–N_8 reliés par pont hydrogène est de 3,22 ± 0,10 Å. Ceci correspond à une faible liaison hydrogène N–H.....N.

Dénombrement des vibrations dans la molécule

Les divers types de vibrations dans la molécule sous la configuration *trans* de symétrie C_{2h}, à l'état isolé, peuvent être classés suivant leur type de symétrie, en utilisant les résultats de la théorie des groupes (Tableau 1). Ainsi en spectroscopie Raman, nous devons en principe observer 15 raies, soit 9 de symétrie A_g et 6 de symétrie B_g. Les 9 vibrations de type A_g comportent trois vibrations du squelette et trois vibrations corres-

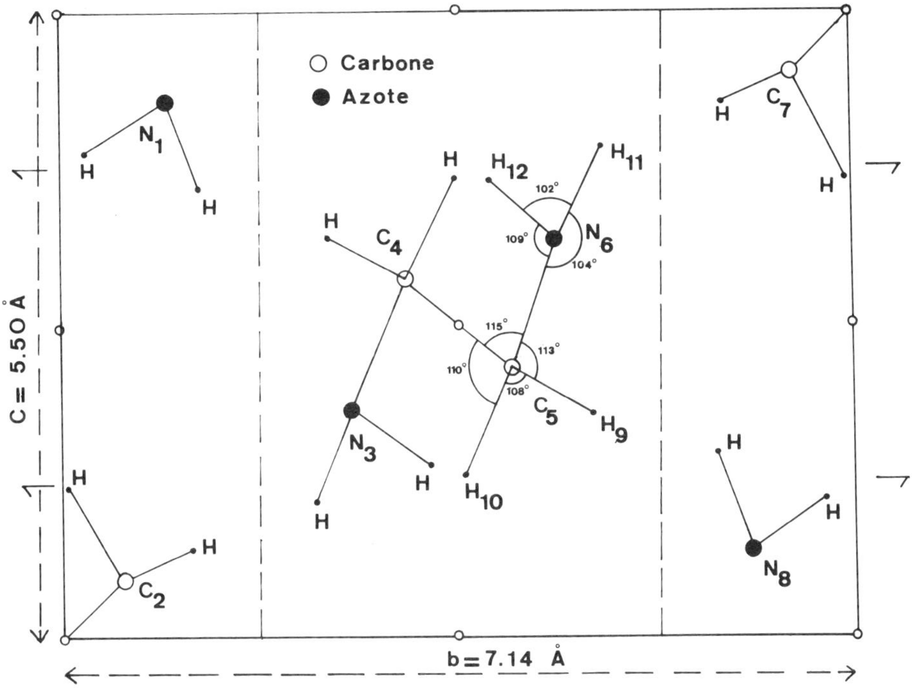

Fig. 1. Structure cristalline de l'éthylènediamine à l'état solide à $-60°C$, projetée parallèlement à l'axe a.

Tableau 1. Types de symétrie des vibrations de la molécule d'éthylènediamine sous sa conformation *trans* à l'état isolé.

C_{2h}	Squelette	CH_2 et NH_2	Activité
$9A_g$	ν(C–C) ν(C–N) δ(N–C–C)	ν_s δ w	Raman
$6B_g$		ν_a ρ t	Raman
$7A_u$	τ(N–C–C–N)	ν_s ρ t	Infrarouge
$8B_u$	ν(C–N) δ(N–C–C)	ν_a δ w	Infrarouge

pondant à chacun des deux groupes CH_2 et NH_2.

Les 6 vibrations de type B_g, sont dues aux groupes CH_2 et NH_2: trois vibrations chacun. Dans le cristal, le groupe facteur est C_{2h} et la molécule occupe le groupe de site C_i. Comme il y a deux molécules dans la maille, les effets dynamiques peuvent se manifester par un dédoublement des fréquences internes.[8] Le Tableau 2 indique la table de corrélation pour le cristal d'éthylènediamine.

Nous avons basé notre proposition d'attribution des fréquences Raman de l'éthylènediamine sur ce dénombrement et sur l'effet isotopique.

Tableau 2. Table de corrélation pour le cristal d'éthylènediamine

Symétrie de la molécule C_{2h}	Groupe de site C_i	Groupe facteur C_{2h}
A_g (9 vib. + 1R) → A_g	A_g	→ A_g (15 vib. + 3 lib.)
B_g (6 vib. + 2R) → A_g	A_g	→ B_g (15 vib. + 3 lib.)
A_u (7 vib. + 1T) → A_u	A_u	→ A_u (15 vib. + 2T' + T)
B_u (8 vib. + 2T) → A_u	A_u	→ B_u (15 vib. + T' + 2T)

Attribution des bandes

Les Figures 2(b) et 3(b) nous montrent les spectres Raman de l'en 'BT' et de l'en-d_4 'BT'. Les échantillons étudiés, obtenus en refroidissant très lentement l'éthylènediamine contenu dans le capillaire, se présentent sous la forme d'un solide transparent et dont le spectre présentait de très nets effets de polarisation, ce qui indiquait la présence d'un monocristal. Malheureusement celui-ci est orienté de manière quelconque dans le capillaire,[2] ce qui empêche d'étudier systématiquement la polarisation des bandes et nous prive donc d'un moyen précieux pour l'attribution des bandes du spectre à 'basse température'.

A titre indicatif, nous avons montré sur Figs. 2(a) et 3(a) les spectres Raman à l'état liquide. Nous pouvons remarquer que les spectres correspondant à l'état solide présentent des pics à la fois beaucoup plus nets et différents de ceux que l'on obtient à l'état liquide.

Fréquences de vibrations correspondant au groupe NH_2. Elles ont été attribuées essentiellement sur la base des déplacements isotopiques.

Dans la région des vibrations de valence du groupe $NH_2(ND_2)$, cinq bandes peuvent être observées. Comme il ne doit y en avoir théoriquement que quatre, la cinquième est donc une bande de combinaison.

Les bandes à 3250 et 3336 cm^{-1} (2470 et 2494 cm^{-1}) correspondent aux vibrations de valence antisymétrique. D'après les spectres partiellement polarisés, nous pouvons considérer que la bande à 3250 (2470 cm^{-1}) appartient au type B_g.

Les bandes à 3170 et 3198 cm^{-1} peuvent être attribuées aux vibrations de valence symétrique $\nu_s(NH_2)$; les bandes correspondantes $\nu_s(ND_2)$ se trouvent à 2345 et 2400 cm^{-1}

Il reste donc un épaulement à 3230 (2425 cm^{-1}) dont nous pensons qu'il est dû à la combinaison: $2\delta(NH_2)$ soit $2 \times 1623 = 3246$ cm^{-1} ou $2\delta(ND_2)$ soit $2 \times 1216 = 2432$ cm^{-1} pour l'isotopomère deutéré.

La fréquence de déformation $\delta(NH_2)$ ($\delta(ND_2)$) se trouve dans la région habituelle à 1623 (1216 cm^{-1}). Le rapport des fréquences (1, 33) correspond assez bien à la valeur prévue. Il reste 4 petites bandes qui, par substitution isotopique se déplacent vers les basses fréquences.

$$970\ \text{cm}^{-1}\ (799\ \text{cm}^{-1})\ \text{rapport}\ \frac{\nu(\text{NH}_2)}{\nu(\text{ND}_2)} = 1{,}25$$

$$892\ \text{cm}^{-1}\ (742\ \text{cm}^{-1})\ \text{rapport}\ \frac{\nu(\text{NH}_2)}{\nu(\text{ND}_2)} = 1{,}20$$

$$545\ \text{cm}^{-1}\ (441\ \text{cm}^{-1})\ \text{rapport}\ \frac{\nu(\text{NH}_2)}{\nu(\text{ND}_2)} = 1{,}23$$

$$427\ \text{cm}^{-1}\ (314\ \text{cm}^{-1})\ \text{rapport}\ \frac{\nu(\text{NH}_2)}{\nu(\text{ND}_2)} = 1{,}37$$

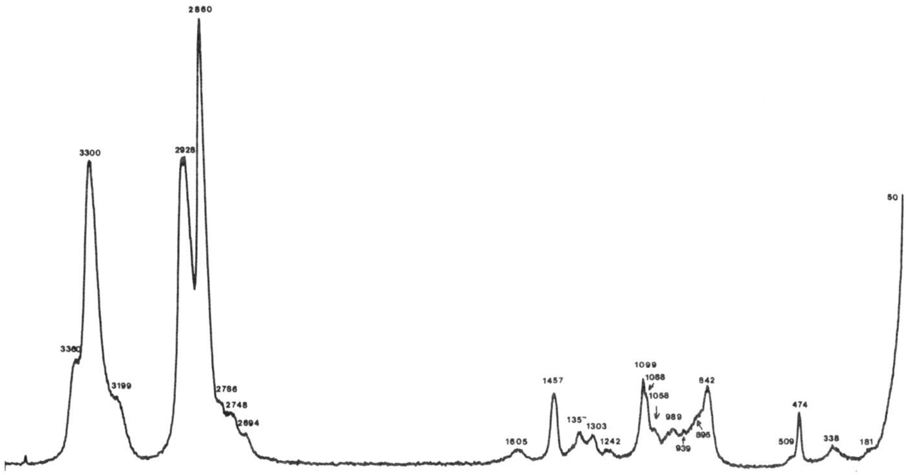

Fig. 2(a). Spectres Raman de l'éthylènediamine à l'état liquide.

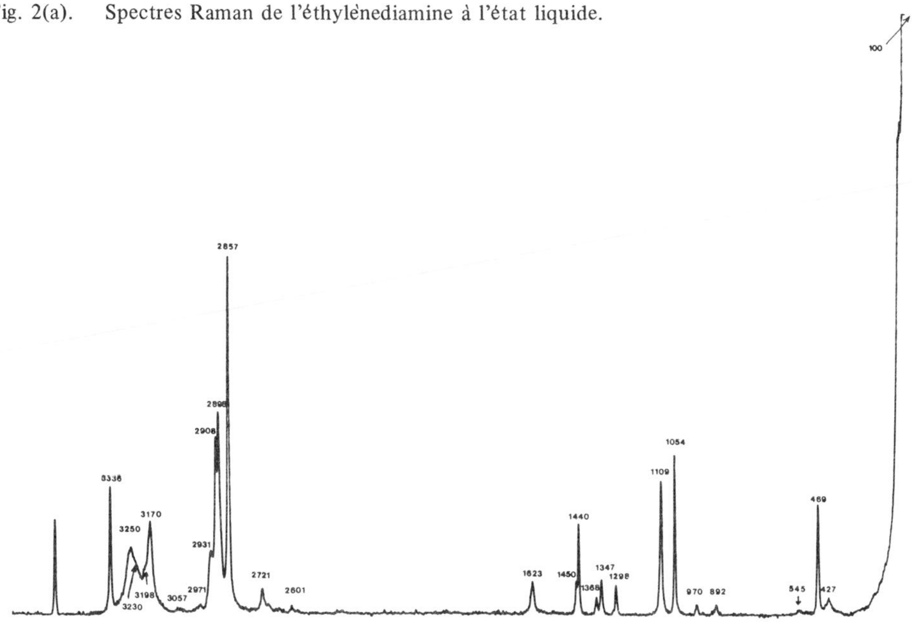

Fig. 2(b). Spectres Raman de l'éthylènediamine sous la conformation *trans* (en-BT).

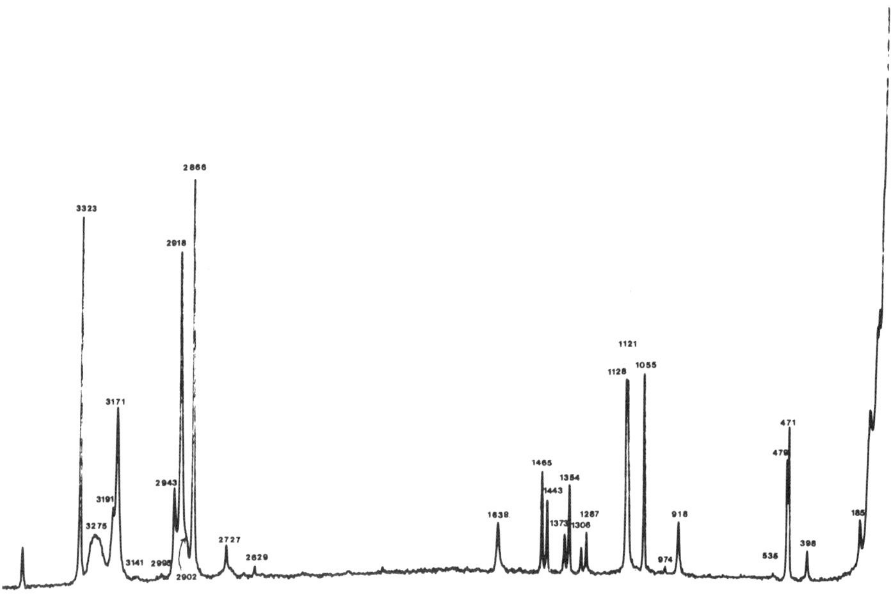

Fig. 2(c). Spectres Raman de l'éthylènediamine sous la forme en 'TBT'.

La bande à 545 cm^{-1} n'apparaît plus dans l'éthylènediamine hydraté. Nous pensons qu'il s'agit d'une bande de combinaison. Les trois autres bandes peuvent être attribuées respectivement aux vibrations ρ_w (wagging) à 970 cm^{-1}; ρ_t (twisting) à 892 cm^{-1} et ρ_r (rocking) à 427 cm^{-1}.

Fréquences de vibration correspondant au groupe CH_2. Les bandes à 2931 (2934 cm^{-1}) et 2908 (2909 cm^{-1}) peuvent être considérées comme dues aux vibrations de valence antisymétrique $\nu_{as}(CH_2)$; les deux autres à 2857 (2862 cm^{-1}) et 2898 (2897 cm^{-1}) correspondant aux vibrations de valence symétrique.

La fréquence de déformation $\delta(CH_2)$ à 1440 (1438 cm^{-1}) présente un épaulement à 1450 (1447 cm^{-1}) qui disparaît dans le monohydrate et peut donc être dû soit à l'effet cristal conformément à ce qui est prévu par la table de corrélation, soit à une bande de combinaison. Nous avons attribué la bande à 1298 (1316 cm^{-1}) à la torsion $\rho_t(CH_2)$.

Pour la fréquence ρ_w, on peut observer deux fréquences à 1368 et 1347 cm^{-1} correspondant à 1 seule bande à 1376 cm^{-1} dans l'en-d_4. Nous pensons que la présence de 2 bandes dans l'en serait due à une résonance de Fermi avec une bande de combinaison.

Enfin la vibration $\rho_r(CH_2)$ qui apparaît vers 720 cm^{-1} en infrarouge n'a pas été observée en spectroscopie Raman.[10]

Fréquences de vibration correspondant au squelette N–C–C–N. Il nous reste à attribuer 3 bandes du squelette de la molécule. La déformation symétrique se trouve à 469 (429 cm^{-1}). La bande à 1054 (1051 cm^{-1}) ne se déplace pratiquement pas par substitution isotopique. Nous l'avons donc attribué à l'élongation C–C.

La dernière bande à 1109 (1036 cm^{-1}) enfin, correspond à la vibration de valence C–N. Le déplacement isotopique est en bon accord avec le modèle approché de l'oscillateur diatomique: $\nu(C{-}NH_2)/\nu(C{-}ND_2) = 1{,}07$.

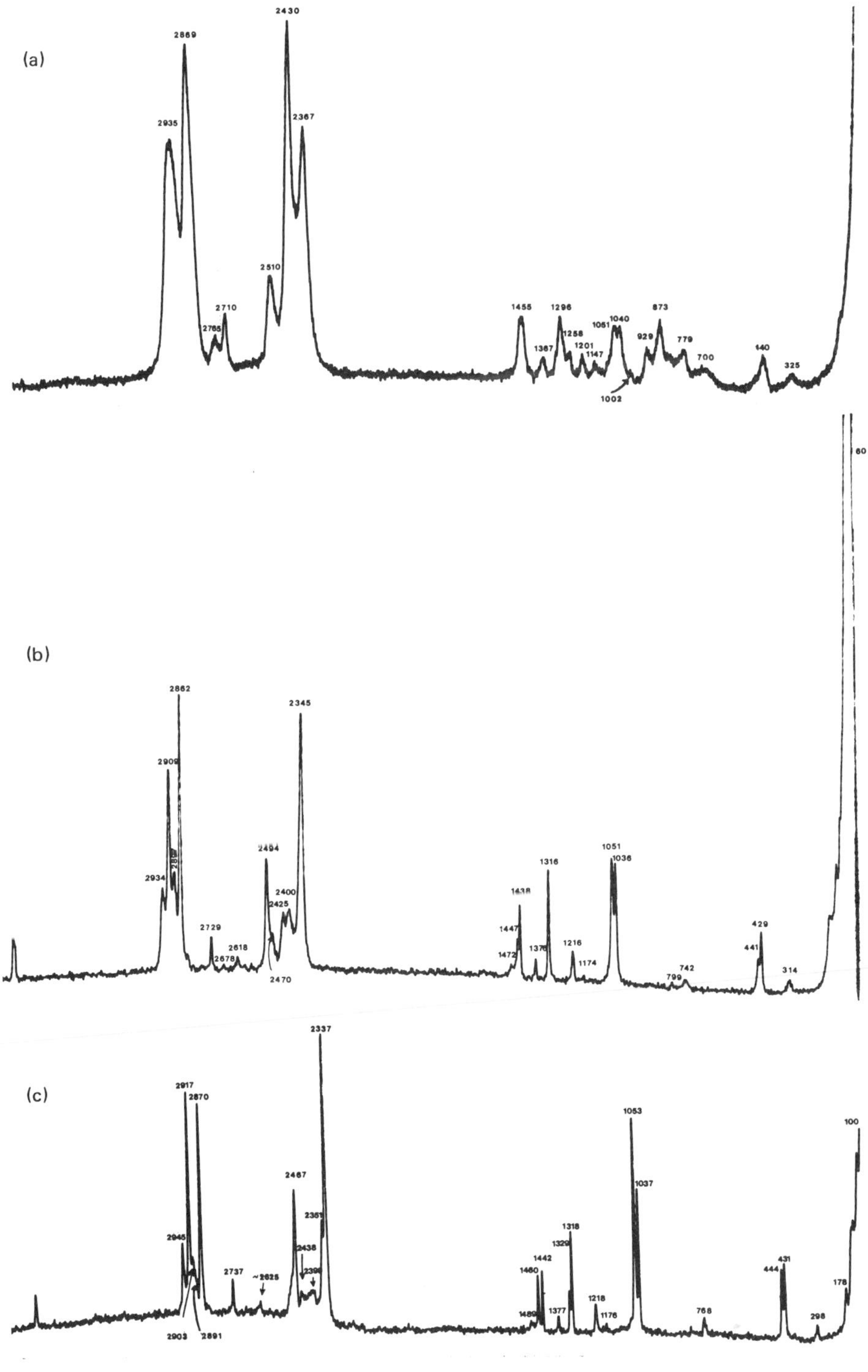

Fig. 3. Spectres Raman de l'éthylènediamine-d_4 sous ses diverses formes. (a) A l'état liquide. (b) Sous la forme en 'BT'. (c) Sous la forme en 'TBT'.

Tableau 3. Attribution des bandes à basse température ('BT') de l'en et de l'en-d_4

Fréquences en cm^{-1}		Attribution	
en	en-d_4	en	en-d_4
3336		$\nu_{as}(NH_2)$	
3250		$\nu_{as}(NH_2)$	
3230		$2\delta(NH_2)$	
3198		$\nu_s(NH_2)$	
3170		$\nu_s(NH_2)$	
3057		bande de combinaison	
2971		bande de combinaison	
2931	2934	$\nu_{as}(CH_2)$	$\nu_{as}(CH_2)$
2908	2909	$\nu_{as}(CH_2)$	$\nu_{as}(CH_2)$
2898	2897	$\nu_s(CH_2)$	
2857	2862	$\nu_s(CH_2)$	
2721	2729	bande de combinaison	bande de combinaison
	2678		bande de combinaison
	2618		bande de combinaison
2601		bande de combinaison	
	2494		$\nu_{as}(ND_2)$
	2470		$\nu_{as}(ND_2)$
	2425		$\nu = 2\delta(ND_2)$
	2400		$\nu_s(ND_2)$
	2345		$\nu_s(ND_2)$
1623		$\delta(NH_2)$	
	1472		
1450	1447		
1440	1438	$\delta(CH_2)$	$\delta(CH_2)$
1368			
1347	1376	$\rho_w(CH_2)$	$\rho_w(CH_2)$
1298	1316	$\rho_t(CH_2)$	$\rho_t(CH_2)$
	1216		$\delta(ND_2)$
	1174		
1109		élongation C–N	
1054	1051	élongation C–C	
	1036		élongation C–N
970	799	$\rho_w(NH_2)$	$\rho_w(ND_2)$
892	742	$\rho_t(NH_2)$	$\rho_t(ND_2)$
545	441		
469	429	déformation symétrique du squelette	
427	314	$\rho_r(NH_2)$	$\rho_r(ND_2)$

Il reste enfin quelques bandes vers 2721 (2729 cm^{-1}) qui ne changent pratiquement pas par la substitution isotopique; nous pensons que ce sont des bandes de combinaison dues au groupement CH_2.

Le Tableau 3 indique en résumé les fréquences observées et leur attribution pour les deux variétés isotopiques à 'basse température'. Cette attribution que nous considérons comme provisoire devra être complétée par une comparaison avec les spectres infrarouge de l'éthylènediamine 'BT'. Elle sera confirmée, enfin, par une analyse en coordonnées normales portant sur la molécule complète.

Nous pouvons remarquer, dans le cas présent, que les effets dynamiques du cristal n'affectent que les fréquences faisant intervenir les groupements CH_2 et NH_2. D'autre part, l'interprétation que Sabatini et Califano ont donnée de leurs spectres infrarouge est à revoir puisque ces auteurs se sont basés sur une conformation *cis* pour la molécule. Il y a en particulier incompatibilité avec nos résultats à propos de l'attribution des bandes C–C et C–N du squelette qu'ils situent respectivement à 980 (973 cm^{-1}) et 1104 et 1065 (1003 et 1067 cm^{-1}).

CHANGEMENT DE STRUCTURE DE L'ETHYLENEDIAMINE A TRES BASSE TEMPERATURE ('TBT')

Mise en évidence d'une nouvelle phase

C'est au cours de l'enregistrement de spectres de l'éthylènediamine à très basse température que nous avons obtenu le changement de phase. En effet, par cristallisation lente du liquide dans le capillaire, nous avons assez souvent réussi à obtenir un solide translucide présentant les caractéristiques de polarisation d'un monocristal. En continuant à refroidir lentement, nous avons pu alors observer un changement de phase visible à l'oeil nu, car il provoque la rupture du cristal initial en une masse polycristalline.

Cette nouvelle phase 'TBT' présente un spectre assez différent de celui de l'éthylènediamine 'BT', avec dédoublement de pics, déplacement de pics et apparition de nouveaux pics [Fig. 2(c)]; il y a de plus disparition des effets de polarisation ce qui correspond bien à un échantillon polycristallin. En réchauffant lentement, on retrouve le spectre 'basse température' en passant par une superposition des deux spectres (Fig. 4).

Le changement de phase, par refroidissement, est subit et total quand le solide 'BT' était monocristallin, ainsi que nous avons pu le constater en particulier au cours d'un enregistrement; par contre, il peut être progressif et incomplet si le solide 'BT' est polycristallin.

Nous avons également mis ce phénomène en évidence avec l'en-d_4 (Fig. 3(c)). Le diagramme de diffraction de rayons X de cette nouvelle phase à l'état de poudre a pu être enregistré par Jamet-Delcroix.[11] Elle a pu ainsi montrer que les clichés de Debye–Scherrer obtenus à $-190°C$ et à $-20°C$ présentent de nombreuses raies dont la répartition en intensité et en position diffèrent très nettement surtout au centre du diagramme; ceci prouve encore que nous sommes en présence de deux phase distinctes. La température de changement de phase a été évaluée par analyse thermique différentielle: elle se situe vers $-80°C$.[2]

Influence de l'hydratation

Nous avons refroidi également des mélanges en, xH_2O avec $x = 1/2, 1, 3/2$. Le spectre obtenu pour $x = 1$ est différent de celui pour l'éthylènediamine pure. Le spectre correspondant à $x = 1/2$ est la superposition des spectres pour $x = 0$ et $x = 1$. Par contre

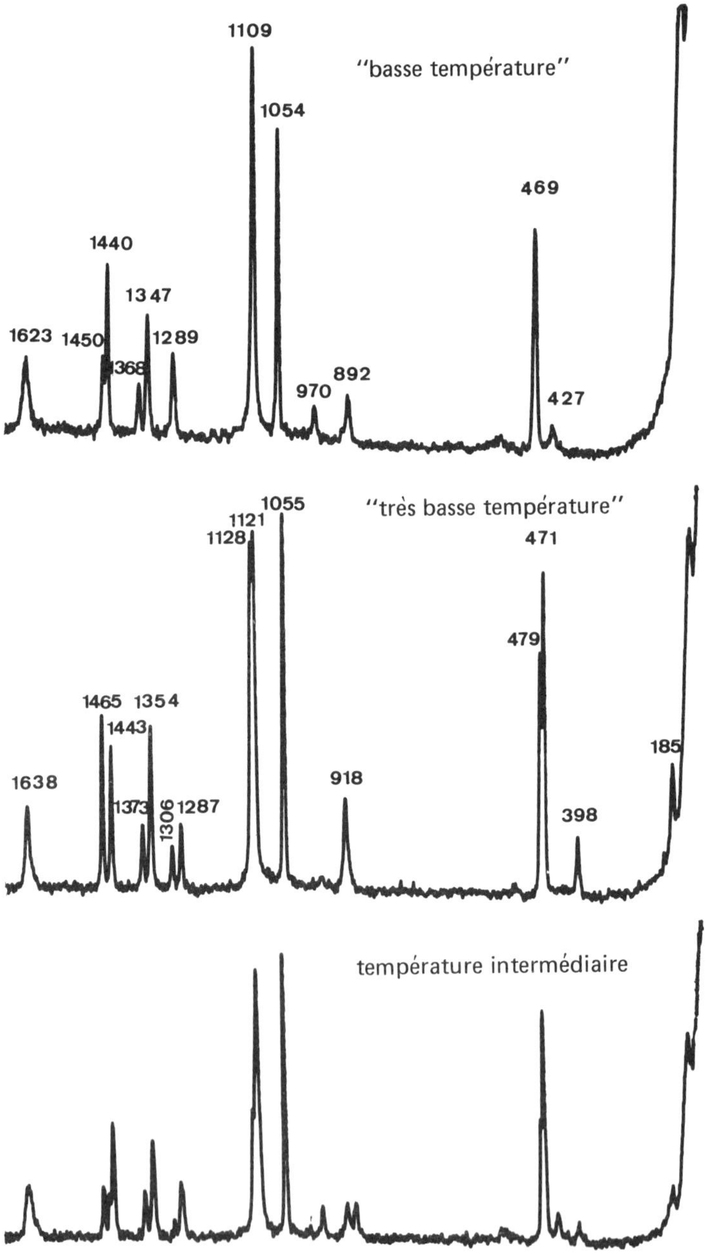

Fig. 4. Comparaison des spectres Raman de l'éthylènediamine aux diverses températures.

le spectre pour $x = 3/2$ ne diffère de celui pour $x = 1$ que par l'augmentation d'intensité de certaines bandes, en particulier C–N et le déplacement d'une bande large vers 3000 cm^{-1}. Ce phénomène que nous avons pu mettre également en évidence pour l'en-d_4 permet donc un dosage d'eau dans l'en et l'en-d_4; nous nous en sommes servis en particulier pour évaluer la pureté de l'en-d_4 avant et après distillation sur sodium.

Nous avons observé d'autre part qu'une présence d'eau empêche la formation de la phase 'TBT' de l'en (en-d_4). Il semble donc que la présence d'impuretés et en particulier d'eau gêne le changement de phase à très basse température.

Analyse du spectre Raman de l'en 'TBT'

Pour l'éthylènediamine, l'effet du changement de phase sur les bandes est assez net: les deux bandes correspondant aux élongations symétriques C–N, et à la déformation symétrique N–C–C–N laissent place chacune à deux bandes, tandis que celle qui correspond à l'élongation C–C reste inchangée. De plus, une bande apparaît à 185 (178) cm^{-1}; on l'attribue en général à la torsion N–C–C–N. On la retrouve d'ailleurs aisément dans l'éthylènediamine liquide alors qu'elle n'apparaît pratiquement pas dans l'isotopomère deutéré. Il semble donc qu'il y ait apparition des bandes du squelette qui étaient interdites en Raman dans la symétrie C_{2h}.

Dans le cas de l'éthylènediamine-d_4, le phénomène est moins clair; nous pensons cependant que l'augmentation de l'intensité des bandes à 1051 et 441 cm^{-1} correspond à l'apparition de bandes de type B correspondant aux bandes de type A situées à 1036 et 429. En nous limitant à l'examen des seules vibrations du squelette nous pouvons déjà en déduire que l'en (l'en-d_4) change de structure à très basse température.

Ce changement de structure cristalline entraîne une rupture des monocristaux en masse polycristalline ainsi qu'un changement du diagramme de Debye–Scherrer; il est lié un changement de structure moléculaire visible en particulier par spectroscopie Raman. Actuellement, il ne nous est par contre pas possible par spectroscopie Raman de trancher entre une conformation *gauche* ou *trans.* Un tel renseignement doit pouvoir être obtenu d'une part par analyse du spectre infrarouge de la forme 'TBT' à condition que les effets dynamiques du cristal ne lèvent pas l'interdiction prévue par les règles de sélection, d'autre part par un calcul en coordonnées normales portant sur la molécule complète.

Il est peu probable qu'une étude de diffraction par des monocristaux puisse être effectuée car ces derniers ne peuvent pas être obtenus par refroidissement du liquide.

CONCLUSION

Nous avons donc étudié le spectre Raman de l'éthylènediamine et de l'éthylènediamine-d_4 à basse température. Par refroidissement du solide stable à basse température nous avons pu mettre en évidence un changement de phase. Nous avons proposé une attribution des bandes pour la forme 'basse température' en nous basant sur la structure, sur la variation des spectres par deutération et en utilisant la variation des spectres lors ce changement de phase.

Actuellement, parallèlement aux études spectroscopiques des vibrations dans les composés de solvatation, nous essayons de confirmer l'attribution de bande de l'en sous sa confirmation *trans* par une analyse en coordonnées normales. D'autre part, afin d'étendre les résultats obtenus avec l'en à très basse température nous avons entrepris l'étude des éthylènediamines substituées.

REFERENCES

1 S. Jamet-Delcroix, *J. Chim. Phys.* **64**, 601 (1967).
2 S. Jamet-Delcroix, *Thèse*, à soutenir à Paris en 1972.
3 A. Sabatini et S. Califano, *Spectrochim. Acta* **16**, 677 (1960).

4 S. Jamet-Delcroix et H. Gillier-Pandraud, *Compt. Rend.* **274C**, 771 (1972).
5 L. Segal et F. V. Eggerton, *Appl. Spectry.* **15**, 116 (1961).
6 S. A. S. Ghazanfar, J. J. Edsall et D. V. Myers, *J. Amer. Chem. Soc.* **86**, 559 (1964).
7 A. P. Gray et R. C. Lord, *J. Chem. Phys.* **26**, 690 (1957).
8 H. Poulet et J. P. Mathieu, *Spectres de Vibration et Symétrie des Cristaux*, Gordon & Breach, Paris, 1970, p. 314.
9 G. Herzberg, *Molecular Spectra and Molecular Structure*, Van Nostrand, Princeton, 1968, p. 229.
10 M. Tobin, *Laser Raman Spectroscopy*, Wiley-Interscience, New York, 1971, p.98.
11 S. Jamet-Delcroix, communication personnelle.

29. RAMAN SPECTROSCOPY OF ORIENTATIONAL DISORDER

T. Geisel, W. Dultz and W. Gebhardt
Fachbereich Physik, Universität Regensburg, Regensburg, Germany

INTRODUCTION

A large number of crystals containing molecules show orientational phase transitions. In some of these systems (e.g. ammonium halides) the molecules are randomly distributed between two equivalent orientations when the crystal is in the disordered phase. In the ordered phase the molecules become parallel or antiparallel ordered. The disorder above the transition temperature T_λ destroys the translational symmetry of the lattice and induces first order Raman scattering which includes phonons with wavevector $\mathbf{k} \neq 0$. At lower temperatures increasing order changes the spectrum drastically.

In this chapter we discuss NH_4Br as an example of an orientational phase transition. Raman measurements on NH_4Br have already been done by other authors.[1-4] We found a quite interesting temperature dependence of the intensity of one band, that has not yet been reported. Furthermore we succeeded in giving a complete assignment of the observed bands, that allows us to interpret the temperature behaviour of the bands with a detailed knowledge of the modes involved. In addition we give a theoretical treatment for the temperature dependence of the scattered intensity, which can also be applied to other systems undergoing orientational phase transitions. It is valid for modes that are only weakly perturbed by the disorder.

STRUCTURES AND λ-TRANSITION IN NH_4Br

We have studied the λ-transition in NH_4Br from phase II to phase III. Phase II has a disordered CsCl-structure (space group O_h^1). The disorder consists in the orientation of the NH_4^+-tetrahedra (Fig. 1). The two possible orientations above $T_\lambda = 235$ K may be described by a pseudo-spin variable σ_l that takes on two possible values corresponding to an Ising model. Below T_λ the NH_4^+-tetrahedra become ordered (phase III). In the x and y direction the ordered arrangement is *abab* . . . , in the z direction the arrangement is *aaa* . . . or *bbb* . . . This configuration causes a tetragonal distortion in the z direction. Two possible Wigner-Seitz cells of this structure are shown in Fig. 2.

The unit cell of the ordered phase contains two NH_4Br molecules and the space group is D_{4h}^7. The Wigner-Seitz cell as shown in Fig. 3 is doubled in volume and rotated by 45° about the z axis. The halving and rotation of the corresponding Brillouin zone is

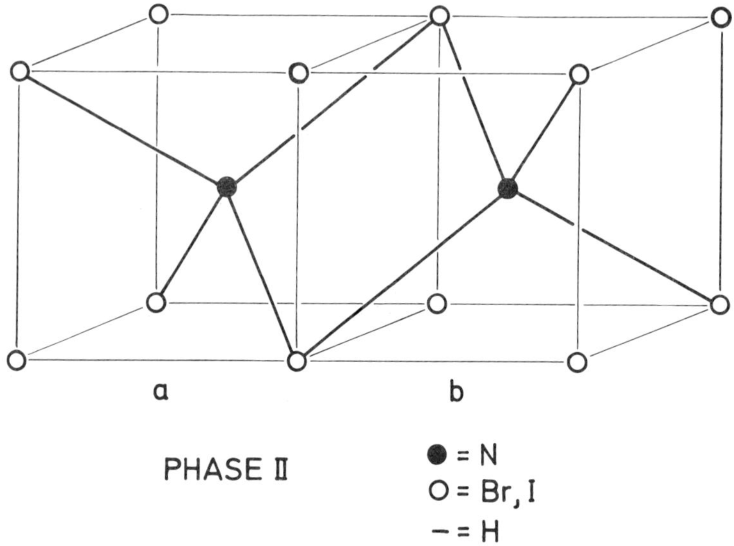

Fig. 1. The two possible orientations of the NH_4^+ ions in phase II.

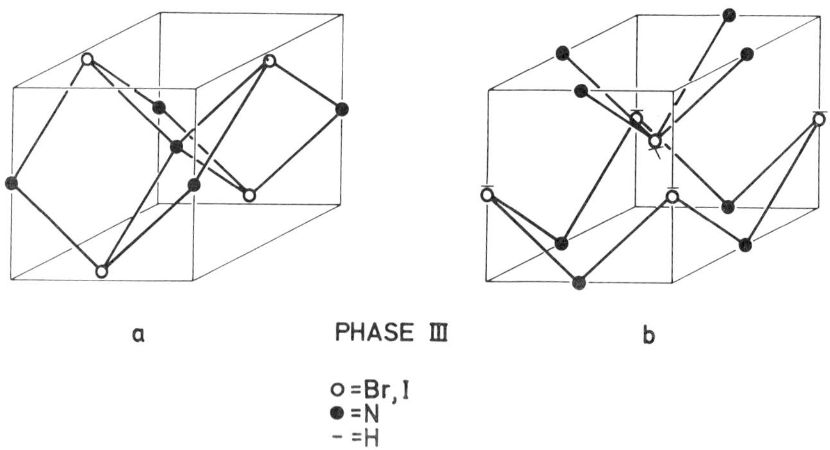

Fig. 2. Two Wigner-Seitz cells for phase III.

shown in Fig. 4. It is evident that M-point phonons in phase II appear as Γ-point phonons in phase III, whereas Γ-point phonons of phase II remain at the Γ-point in phase III. In Table 1 we give the irreducible representations of the relevant modes in the two phases and the correspondence to phonon branches.

The representations of the lattice vibrations in phase II were taken from Warren,[5] those for phase III have been calculated by the factor group method. Perry and Lowndes[2] obtained a B_{2g} mode in phase III instead of the B_{1g} mode. They used a coordinate system corresponding to the crystal axes of phase II. Our result refers to the crystal axes of phase III.

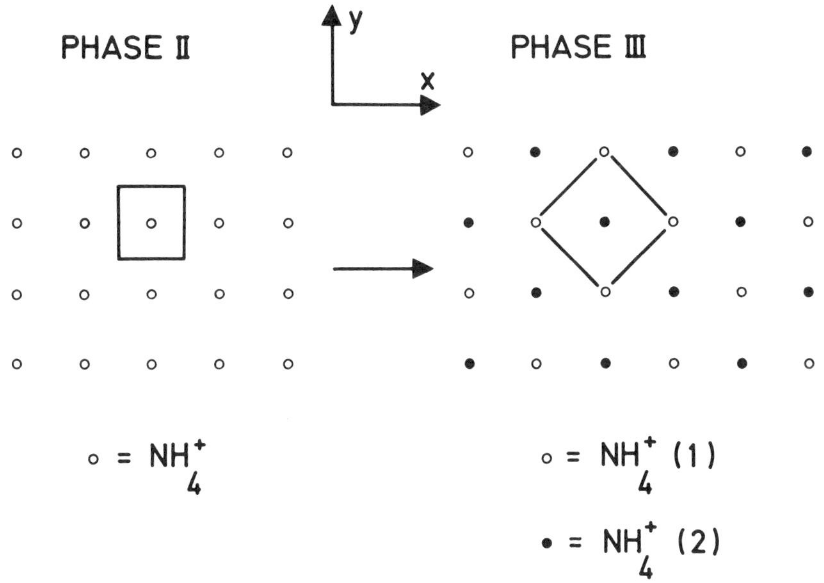

Fig. 3. The changing of the Wigner-Seitz cell in the λ-transition.

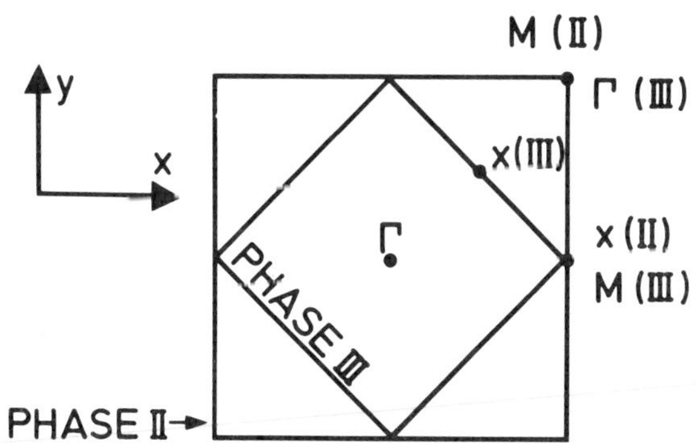

Fig. 4. The Brillouin zones of phase II and phase III.

The correspondence between phase II and phase III phonons shown in Table 1 can be easily understood by the following arguments: in phase II the Γ-point F_{1u} optical phonons are split by long range electrostatic forces in a LO phonon and doubly degenerate TO phonons. It follows from the compatibility relations that in phase III this LO phonon is of type A_{2u} and the TO phonons are of type E_u. In both phases two doubly degenerate and two non degenerate modes remain to be considered. Since the degeneracy of these modes is not changed by the phase transition, the E_u modes of phase II must correspond to the E_g modes of phase III. The assignment of the remaining 1-dimensional representa-

Table 1. Symmetry properties of the phonons.

	Phase II		Phase III	
Γ-point (O_h)	opt.	$\Gamma_{15} = F_{1u}$	$\nearrow A_{2u}$, $\searrow E_u$	opt. modes Γ-point (D_{4h})
M-point (D_{4h})	LO	$M_5' = E_u$	$\longrightarrow E_g$	
	TO	$M_2' = B_{1u}$	$\longrightarrow B_{1g}$	
	LA	$M_5' = E_u$	$\longrightarrow E_g$	
	TA	$M_4' = A_{2u}$	$\longrightarrow A_{1g}$	

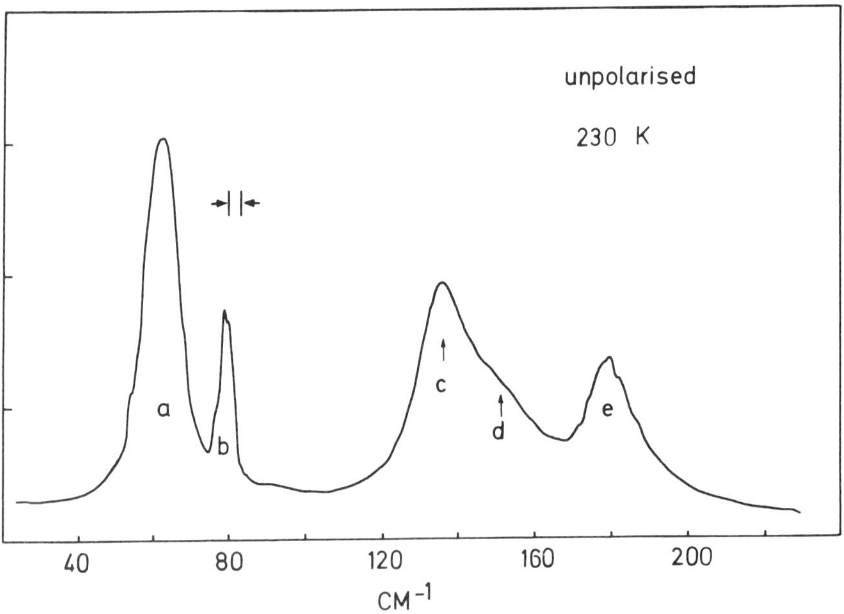

Fig. 5. The external modes of NH_4Br at 230 K.

tions can only be proved by a symmetry consideration of the particle displacements. The assignment of the phonon branches of phase II (Table 1, column 1) has been given by Warren.[5]

EXPERIMENTAL RESULTS AND BAND ASSIGNMENT

The measurements have been taken with a conventional Raman setup with photon counting technique. We used an evaporation cryostat of high temperature stability ($\Delta T \leqslant 0.05$ K). The temperature was measured by a thermocouple.

The low frequency Raman spectra of NH_4Br at various temperatures are shown in Figs. 5, 6 and 7. We assigned the peaks of the Raman spectra using the results of polarization measurements and taking into account the selection rules in accordance with

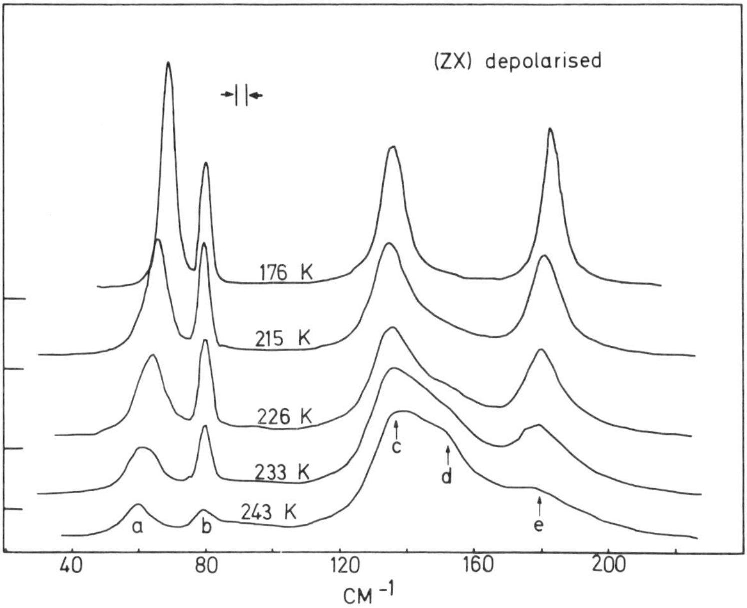

Fig. 6. The external modes of NH_4Br above and below T_λ.

Table 1. Furthermore we compared the energetic position of the measured peaks with the relative position of the phonon branches.

We will first consider phase III. The four observed peaks *a, b, c, e* (Fig. 6) correspond to four modes which are *M*-point modes in phase II. The only polarized band in phase III (band *a* in Fig. 7) belongs to the A_{1g} mode which arises from the TA (110) phonon of phase II. Thus band *b* belongs to the E_g mode corresponding to the LA (110) phonon of phase II. If we consider the relative energetic position of the remaining bands we come to the conclusion, that the *c*-band and the *e* band coincide with the B_{1g} mode arising from TO (110) and the E_g mode from LO (110) respectively. For the B_{1g} mode the polarizability tensor has only diagonal components. Its appearance in the depolarized spectrum can be explained by the domain structure.

We are now able to give the assignment of all low frequency Raman peaks in phase II. The strongest contributions are obtained from the same modes as in phase III, since short range order is found in a broad temperature range above T_λ. These vibrations are *M*-point modes in phase II. Therefore they are only Raman active in the disordered phase because the translational symmetry is broken. We obtain the following assignment by comparison of phase II with phase III spectra: TA for band *a*, LA for band *b*, TO for band *c* and LO for band *e*. Table 2 gives their frequencies at 256 K. It is reasonable to assume for the reststrahl frequency ν_r : $\nu_{TO}\ (110) \leqslant \nu_r \leqslant \nu_{LO}(110)$. Perry and Lowndes[2] found $\nu_r = 154\ cm^{-1}$ in good agreement with the frequency of our *d* band.

We shall now discuss the bands *a* and *b* in more detail. The temperature dependences of the integrated intensities of the two bands are shown in Figs. 8 and 9. Measurements on band *a* have already been reported by Wang.[4] In the case of band *b* the signal-to-noise

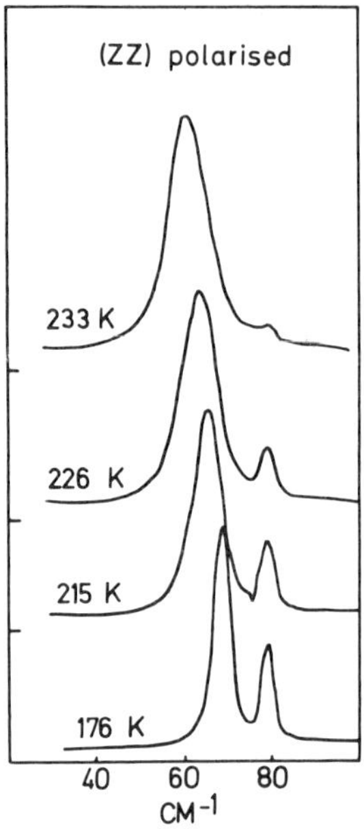

Fig. 7. Bands *a* and *b* in the polarized spectrum.

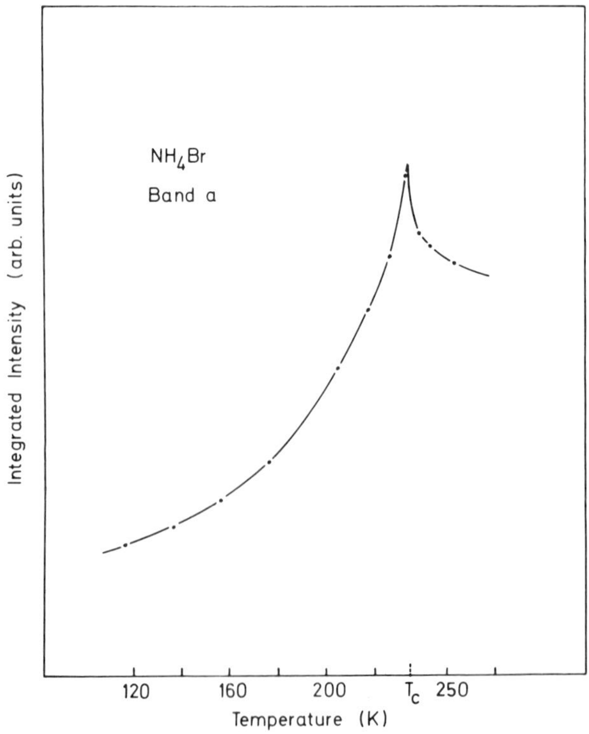

Fig. 8. Temperature dependence of the integrated intensity of band *a* in the polarized spectrum.

Table 2. Low frequency Raman bands at 256 K.

Band	*a*	*b*	*c*	*d*	*e*
Main contributions	TA (110)	LA (110)	TO (110)	TO	LO(110)
Frequency (cm^{-1})	58	80	134	154	177

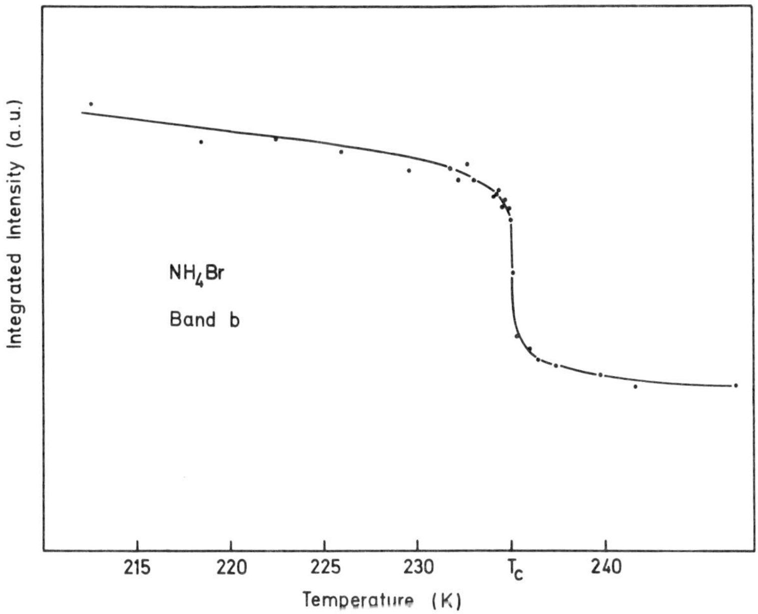

Fig. 9. Temperature dependence of the integrated intensity of band *b* in the depolarized spectrum.

ratio was small above T_λ. We therefore subtracted completely the background intensity. The most surprising result is the striking difference in the temperature dependence of band *a* and band *b*. This phenomenon will be further discussed in the next section.

THEORY

In the following discussion it will be shown that the Raman intensity is related to the correlation function of the orientation of nearest neighbours. We will first deal with the case of phonons which are weakly perturbed by the disorder. This means that phonon states and pseudo spin states may be separated.

We will describe the orientation of the νth molecule in the lth unit cell by variables $\sigma_{l,\nu}$. In general this may be a set of Euler angles. In our special case the $\sigma_{l,\nu}$ correspond to Ising-spins, since only two orientations are possible.

The scattered Raman intensity is given by[6,7]

$$I(\omega) = \frac{\omega_i^4}{2\pi c^3} \sum_{\alpha\beta\gamma\lambda} n_\alpha n_\beta i_{\alpha\gamma,\beta\lambda}(\omega) E_\gamma^- E_\lambda^+ \qquad (1)$$

where ω_i is the frequency of the incident light, ω is the frequency shift, **n** is a unit polarization vector of the scattered light and $\mathbf{E}^+$ and $\mathbf{E}^-$ are the amplitudes of the positive and negative frequency components of the electric field of the incident radiation. $i_{\alpha\gamma,\beta\lambda}(\omega)$ is given by

$$i_{\alpha\gamma,\beta\lambda}(\omega) = \sum_{\{\sigma\},\{\sigma'\}}\sum_{v,v'} \frac{\exp[-\beta(E_v + E\{\sigma\})]}{Z_v \cdot Z\{\sigma\}} \times < \{\sigma\}, \nu | P_{\beta\lambda} | \nu', \{\sigma'\} > < \sigma', \nu' | P^+_{\alpha\gamma} | \nu, \{\sigma\} > \times \delta\left[\omega - \frac{1}{\hbar}(E_v - E_{v'})\right] \tag{2}$$

The $P_{\alpha\beta}$ are components of the electronic polarizability tensor operator,[6] $|\nu>$ and $|\nu'>$ are phonon eigenstates and E_v the corresponding energy levels. Z_v is the vibrational partition function. $\sum_{v'} \delta(\omega - 1/\hbar (E_v - E_{v'}))$ is introduced to account for more than one possible transition with frequency ω. In addition to the conventional formula we averaged in Eqn. (2) over the orientational configurations $\{\sigma\}$. $P_{\alpha\beta}$ may be expanded in powers of the components of the nuclear displacements $u_\mu(l\kappa)$ (lth cell, κth molecule)

$$P_{\alpha\beta} = P^{(0)}_{\alpha\beta} + \sum_{l\kappa\mu} P_{\alpha\beta,\mu}(l\kappa,\{\sigma\}) u_\mu(l\kappa) + \ldots \tag{3}$$

The first term in Eqn. (3) gives rise to Rayleigh scattering and will not be considered here. It follows from Eqns. (2) and (3)

$$i_{\alpha\gamma,\beta\lambda}(\omega) = \sum_{\substack{l\kappa\mu \\ l'\kappa'\mu'}} < P_{\beta\lambda,\mu}(l\kappa\{\sigma\}) P_{\alpha\gamma,\mu'}(l'\kappa'\{\sigma\}) >_{\{\sigma\}} \times I_{\mu\mu'}(l\kappa, l'\kappa', \omega) \tag{4}$$

with

$$I_{\mu\mu'}(l\kappa, l'\kappa', \omega) = \frac{1}{2\pi} \int_{-\infty}^{\infty} dt \exp(-i\omega t) < u_\mu(l\kappa, t)\, u_{\mu'}(l'\kappa', 0) >_v \tag{5}$$

We write the ionic displacements as

$$u_\mu(l, \kappa) = \sum_{s\mathbf{k}} \left(\frac{\hbar}{2Nm_\kappa \omega_{s\mathbf{k}}}\right)^{1/2} \exp(i\mathbf{k}\mathbf{R}_l)\, \epsilon_{\mu s\mathbf{k}\kappa} (b^\dagger_{s-\mathbf{k}} + b_{s\mathbf{k}}) \tag{6}$$

Here s is the phonon branch, **k** the phonon wave vector and $\omega_{s\mathbf{k}}$ the corresponding frequency. The $\epsilon_{s\mathbf{k}\kappa}$ are the phonon's polarization vectors. Using Eqn. (6) we find for the Stokes component

$$I_{\mu\mu'}(l\kappa, l'\kappa', \omega) = \frac{[< n(\omega) >_v + 1]\hbar}{2N\omega\sqrt{m_\kappa m_{\kappa'}}} \sum_{s\mathbf{k}} \epsilon_{\mu s\mathbf{k}\kappa} \epsilon_{\mu' s\mathbf{k}\kappa'} \times \delta(\omega + \omega_{s\mathbf{k}}) \exp[i\mathbf{k}(\mathbf{R}_l - \mathbf{R}_{l'})] \tag{7}$$

So far the theory is quite general. In order to get a unique description for both phases of NH_4Br, we shall use the lattice of phase II (CsCl structure). The pseudo spins on different sublattices are defined with opposite signs. This convention leaves the ferromagnetic Ising formalism unchanged. Thus at sufficiently low temperatures the spins on both sublattices take on the value + 1.

LA (110) mode

The above equations simplify drastically, if special cases are considered. Since for band b only the LA (110) mode is important, $\sum_{\mathbf{k}} \delta(\omega - \omega_{s\mathbf{k}})$ in Eqn. (7) yields a constant factor $\rho(\omega_{LA(110)})$, the density of states. In this mode the NH_4^+ ions remain fixed ($\epsilon_{\mu LA(110)NH_4} = 0$). The Br^- ions are displaced in the (110) or $(1\bar{1}0)$ direction. The phase factor is $\exp(ik\mathbf{R}_l) = (-1)^{l_1 + l_2}$, where l_1 and l_2 are the x and y components of l. We also use the fact that band b has been observed in (zx) polarization and obtain for the scattered intensity

$$I(\omega) = \frac{\omega_i^4 |E_z|^2 \hbar[<n(\omega)>_v + 1]}{2\pi N c^3 \omega m_{Br}} \sum_{ll'} (-1)^{l_1 + l_2 - (l'_1 + l'_2)} \times <P_{zx,x}(l\kappa,\{\sigma\}) P_{zx,x}(l'\kappa,\{\sigma\})>_{\{\sigma\}} \rho(\omega_{LA(110)}) \quad (8)$$

Since the spin independent contributions of $P_{zx,x}(l\kappa, \sigma)$ yield no first-order Raman scattering, we need only consider the spin dependent contributions which we may write in a first approximation as

$$P_{zx,x}(l\kappa,\{\sigma\}) = \eta\sigma_l$$

η is a proportionality constant. It follows

$$I(\omega) \propto \sum_{ll'} (-1)^{l_1 + l_2 - (l'_1 + l'_2)} <\sigma_l \sigma_{l'}> \quad (10)$$

Now for $|l - l'| \gg 1$ and $T \neq T_c$

$$<\sigma_{l_1 l_2 l_3} \sigma_{l'_1,l'_2,l'_3}> \approx <\sigma_{l_1,l_2,l_3} \sigma_{l'_1+1,l'_2,l'_3}>$$

These two correlations are weighted with opposite signs in Eqn. (10), so that we can restrict the summations in Eqn. (10) to nearest neighbour correlations. The result is then

$$I(\omega) = \frac{\omega_i^4 |E_z|^2 \hbar[<n(\omega)>_v + 1]\eta^2}{2\pi c^3 \omega m_{Br}} \sum_{\delta} <\sigma_l \sigma_{l+\delta}> \rho(\omega_{LA(110)}) \quad (11)$$

The integrated intensity of band b shown in Fig. 9 thus gives a direct measurement of nearest neighbour correlations at the λ-transition in NH_4Br. Our measurements on this band may be used for comparison with model calculations. The theory outlined will yield similar results in all cases where special modes become Raman active due to an orientational phase transition. Its application however is limited to those modes which are only weakly perturbed by the disorder.

TA (110) mode

The above assumption cannot be made for the TA (110) mode. Its displacement configuration anticipates the lattice distortion of phase III. In phase II the positions of the Br^- ions coincide with the cube edges. In phase III the Br^- ions are shifted in $+z$ and $-z$ direction (Fig. 2). For a qualitative discussion we introduce a short-range order parameter $<S>$ describing the anion displacement in z direction. The TA (110) phonon displaces the Br^- ions in $\pm z$ direction and leaves the NH_4^+ ions fixed. The particle displacement by the TA (110) phonon can therefore be described as

$$u_z(l) \propto f(S_l - \langle S_l \rangle) = f(\Delta S_l) \tag{12}$$

and the correlations $\langle u_z(l, t)\, u_z(l, 0)\rangle$ are functions of $\langle \Delta S_l(t)\, \Delta S_{l'}(0)\rangle$. The averaging is over the S-states. The scattered intensity depends on the Fourier transform of these correlation functions

$$I_{zz}(l, l', \omega) = \frac{1}{2\pi} \int_{-\infty}^{\infty} dt \exp(-i\omega t) \langle \Delta S_l(t)\, \Delta S_{l'}(0) \rangle \tag{13}$$

From this qualitative discussion we may draw the following important results: The a band Raman scattering is obviously critical fluctuation scattering. The TA (110) mode is strongly excited near T_λ (Fig. 8). Its configuration corresponds to the lattice distortion in phase III. Therefore it plays an active role in the mechanism of the λ-transition. However, it must be noted that it is not a soft mode as Wang[4] has already observed. The reason for this may be the fact that the local potential is a function of the molecular orientation.

REFERENCES

1 C. H. Wang and P. A. Fleury, *Light Scattering Spectra of Solids,* Springer Verlag, New York, 1969, p. 651.

2 C. H. Perry and R. P. Lowndes, *J. Chem. Phys.* **51**, 3648 (1969).

3 J. R. Durig and D. J. Antion, *J. Chem. Phys.* **51**, 3639 (1969).

4 C. H. Wang, *Phys. Rev. Letters* **26**, 1226 (1971).

5 J. L. Warren, *Rev. Mod. Phys.* **40**, 38 (1968).

6 M. Born and K. Huang, *Dynamical Theory of Crystal Lattices,* Clarendon Press, Oxford, 1954.

7 A. A. Maradudin, *Solid State Physics*, Vol. 19, (Eds. F. Seitz and D. Turnbull), Academic Press, New York, 1966, p. 2.

SECTION IV
FLAME RAMAN EFFECT

30. LASER RAMAN AND RESONANCE FLUORESCENCE SPECTRA OF FLAMES

P. J. Hendra and C. J. Vear
Department of Chemistry, Southampton University, England
and
J. J. Macfarlane,
N.G.T.E., Pyestock, Hampshire, England

INTRODUCTION

The study of the components within a flame, the mechanisms of reaction and the thermodynamics of the various processes involved is relatively difficult. The conventional methods of following the dynamics of reacting systems are frequently inapplicable, but with modification, analyses of these essentially flowing systems have been achieved. The normal methods are the extraction of small quantities of gas from the system with a probe and the analysis for components by vapour-phase chromatography or mass spectrometry[1] and the use of spectroscopic techniques.[2] Although invaluable as a means of monitoring materials within the combustion zone the use of probing techniques is questionable as the introduction of a probe must inevitably disturb the gas flow whilst its temperature and chemical composition will affect the reactions close to it. The use of cooled probes of sufficiently small diameter will minimize these effects but some flame quenching may well occur, while fractionation of the quenched, cooled products can occur within the probe, so producing a distorted picture of materials present at the selected point inside the flame. Emission or absorption spectroscopic techniques are both in principle applicable and do not suffer from these defects, but are in turn limited in their spatial resolution properties. Emission or absorption can occur at any point in the flame but the process assayed by the spectrometric system will be the sum total of all that lies within the optical cone of acceptance of the optical system. Methods have been devised for improving the effective spatial resolving properties of these techniques but usually involve the use of atypical combustion conditions (e.g. the use of low pressure environments to increase the thickness of the reaction zone relative to the assayed volume or by resorting to specially designed burners producing flat flames). In this chapter we wish to show how laser excited spectra, (both Raman and fluorescent) can yield information on flame processes. Although the technique will never rival mass spectrometry in sensitivity of detection, it offers an optical probing method with excellent spatial resolution properties.

THEORY

Before describing Raman experiments on flames it is relevant to briefly discuss one or two properties of flames and the chemistry involved. Simple flames may be divided

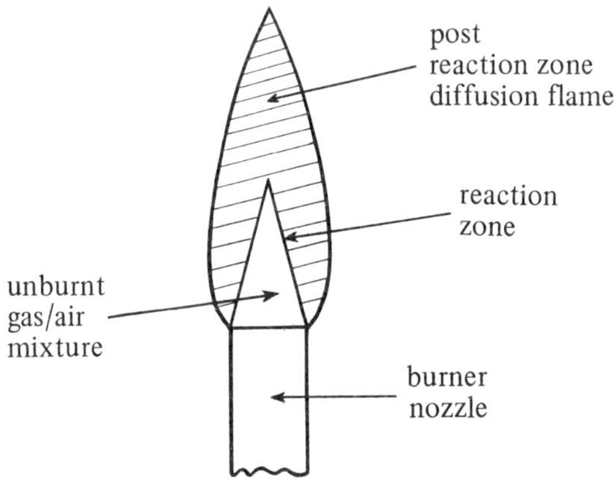

Fig. 1. Typical pre-mixed flame geometry.

into two main classes – pre-mixed (of which the fully aerated bunsen flame is a familiar example) and diffusion flames (e.g. a 'luminous' bunsen flame). In the former case the reacting gases are crudely mixed in a system devised to produce a mixture of controlled composition and are then allowed to mix fairly intimately before burning. Clearly the design of the system is important here but for laboratory work it is normal to construct a burner so that complete mixing is guaranteed before combustion. The gases are then flowed through an orifice and burn in a familiar form – the principal reactions occurring within a thin conical section normally apparent in an air–gas flame as a transparent incandescent film between the unreacted species and the more diffuse products (Fig. 1). In the diffusion flame the reactants are allowed to flow into each other, the flame then occurring at the mixing interface, its size and form being governed by diffusion processes, there being then a less well-defined reaction zone. It is known that in the common hydrocarbon/air (or oxygen) case that the hydrocarbon is pre-heated immediately prior to entering the reaction zone and that in a fuel-rich mixture a considerable number of C_2 and other radical fragments are generated within the reaction zone and are subsequently rapidly oxidized. The C_2 radicals give rise to electronic emission spectra in the visible part of the spectrum (the Swan bands) whilst in rich flames incomplete oxidation results in the formation of soot particles which are hot enough to act as intense black-body radiators throughout the visible and infrared. In many technologically important systems – gas-turbines, coal and oil fired furnaces, internal combustion reciprocating engines etc. the flame is still more complex because of evaporation of liquid fuels, turbulence and in some cases deliberate stratification of the flow patterns inside the combustion system. Thus, it is clear that a complete study of the reactions involved in these systems ideally requires a method which does not perturb the flame equilibrium and has excellent spatial resolution. Laser Raman spectroscopy is a technique having these properties, but is difficult to apply to flame systems due to the intense background emissions present.

EXPERIMENTAL

Early experiments attempted at Southampton involving excitation of laser Raman spectra within flames, used a commercial laser Raman spectrometer (Spex 1401, although some preliminary studies were undertaken using a single monochromator instrument constructed at Southampton) with conventional 90° viewing optics, and a Spectra Physics Model 165 argon ion laser (producing 1.5 W at 488.0 nm and 514.5 nm). The laser was interrupted by a mechanical chopper at 700 Hz, the output from the photomultiplier (cooled ITT FW 130) being fed to a A.I.M. 'Minilock' (MLS 249A) phase sensitive detector. The latter enabled us to discriminate between the flame emissions and the laser excited signal. Photon counting electronics were, however, used in estimates of total flame emission. The flame system was of the pre-mixed burner type and used a variety of hydrocarbon fuels plus air or oxygen.

Although early experiments did not seem to yield any meaningful results, refinement of the viewing optics/spectrometric system gave good spectra from stoichiometric or slightly fuel-rich oxygen/ethylene flames at one atmosphere. From a study of these spectra we surmized that they must be due to laser excited resonance fluorescence of the species C_2 in the flame. Our findings, broadly outlined were that the signal

1. was proportional in intensity to laser output
2. was mixture dependent and
3. varied in strength throughout the flame.

This is as expected for Raman spectra, *but* the signal

4. was relatively strong compared to the anticipated Raman band intensity
5. when measurements were made using the spatial resolution property of the method it became apparent that the signal emanated *only* from the primary reaction zone
6. varied in intensity and characteristic form using different laser lines as sources
7. was of a complex, band-like structure.

In Fig. 2 a series of spectra demonstrating points (6) and (7) are given. A closer high resolution study of these bands [Fig. 2(c)] yielded data consistent with our original diagnosis that the species was C_2 formed in the primary reaction zone, i.e. in the volume where its concentration is highest under the stated conditions.[2, 3] A detailed analysis of these spectra is not relevant to this chapter, but one point of interest may be mentioned here. The emission spectrum of C_2 in the region of interest is that of the well known Swan bands.[2, 3] Comparison of this pure emission spectrum with the laser induced transitions yields a much simpler result in the latter case. In our opinion the resonant fluorescence bands excited by the laser are due to transitions from a very low lying electronic state of C_2 i.e. $^3\pi_u \leftrightarrow {^3\pi_g}$. Thus, a series of bands appears shifted to the red of the exciting wavelength (514.5 nm) due to the $\nu' = 0, \nu'' = 1$ [designated (0, 1)] etc. transitions while a further series appears shifted to the blue due to the (1, 0,), (2, 0) .... transitions. The latter series is quite strong due to the relatively high concentrations of vibrationally excited molecules in the reaction zone. Comparison of the intensities of relevant bands makes it possible to determine the temperature at the point of focus of the laser. Corrections must be applied for instrument response and the particular transitions observed, but a simple uncorrected temperature from a consideration of the (0, 1) and (1, 0) band intensities has yielded a figure of 2600 K, a figure not inconsistent with that widely accepted[4] for this type of flame. Whether or not there is equipartition of vibrational energy in this region is dubious, so that a false 'vibrational' temperature may ensue from this type of study. The importance of this distortion may be checked, however, by a detailed analysis of the rotational lines within the fluorescence bands to give a 'rotational temperature'. This will be a better figure as translational $\leftrightarrow$ rotational energy transfer at atmospheric pressure will almost certainly lead to a Boltzman-type rotation energy distribution. This

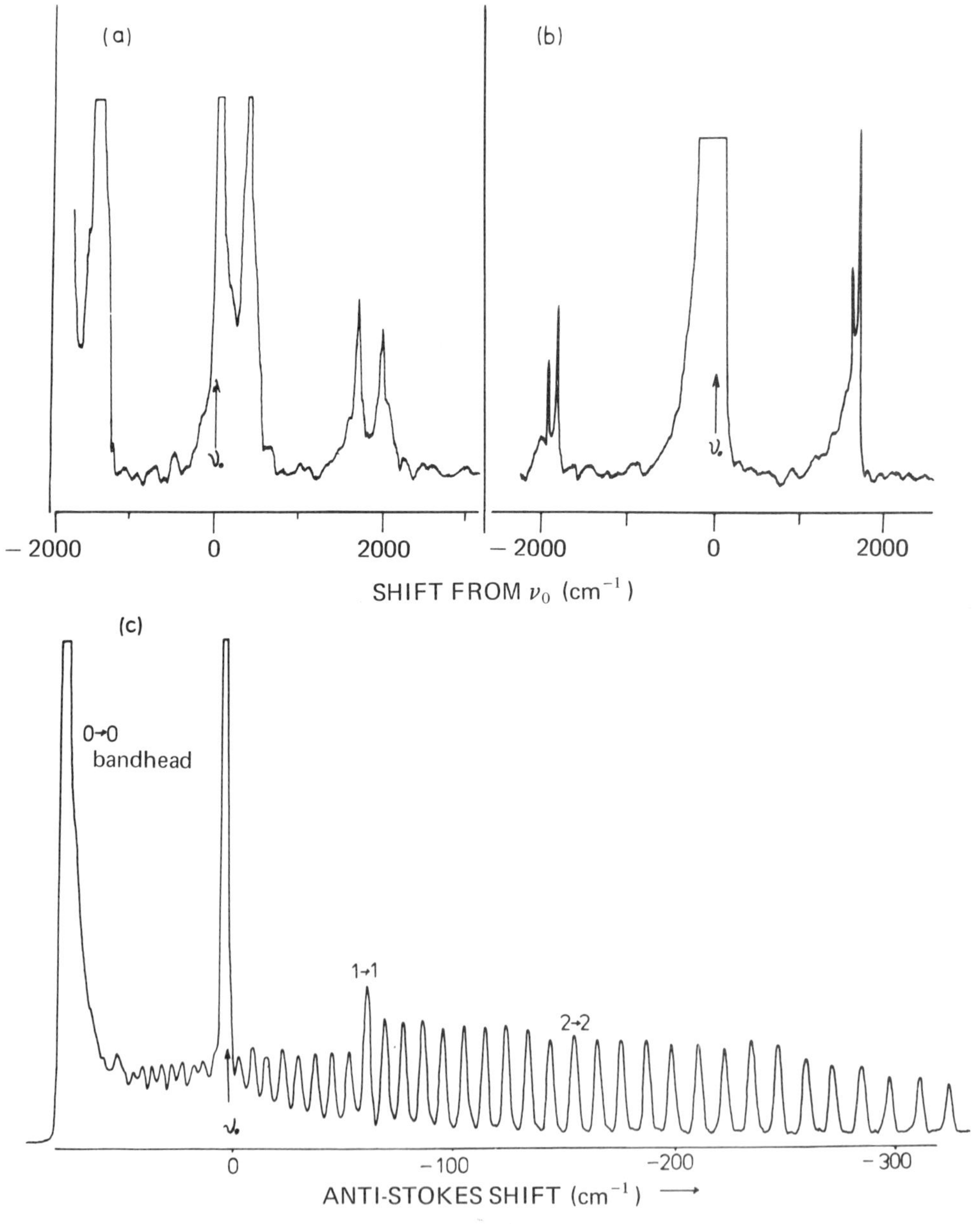

Fig. 2. (a) Low resolution resonance fluorescence spectrum of C_2 excited by $\lambda_0 = 501.7$ nm radiation. (b) Low resolution resonance fluorescence spectrum of C_2 excited by $\lambda_0 = 514.5$ nm radiation. (c) High resolution study of the origin (0, 0) band of C_2 excited by $\lambda_0 = 514.5$ nm.

theme is expanded in greater detail elsewhere, where more consideration is given to the energy exchange problem.

Following these experiments more investigations were undertaken to obtain laser

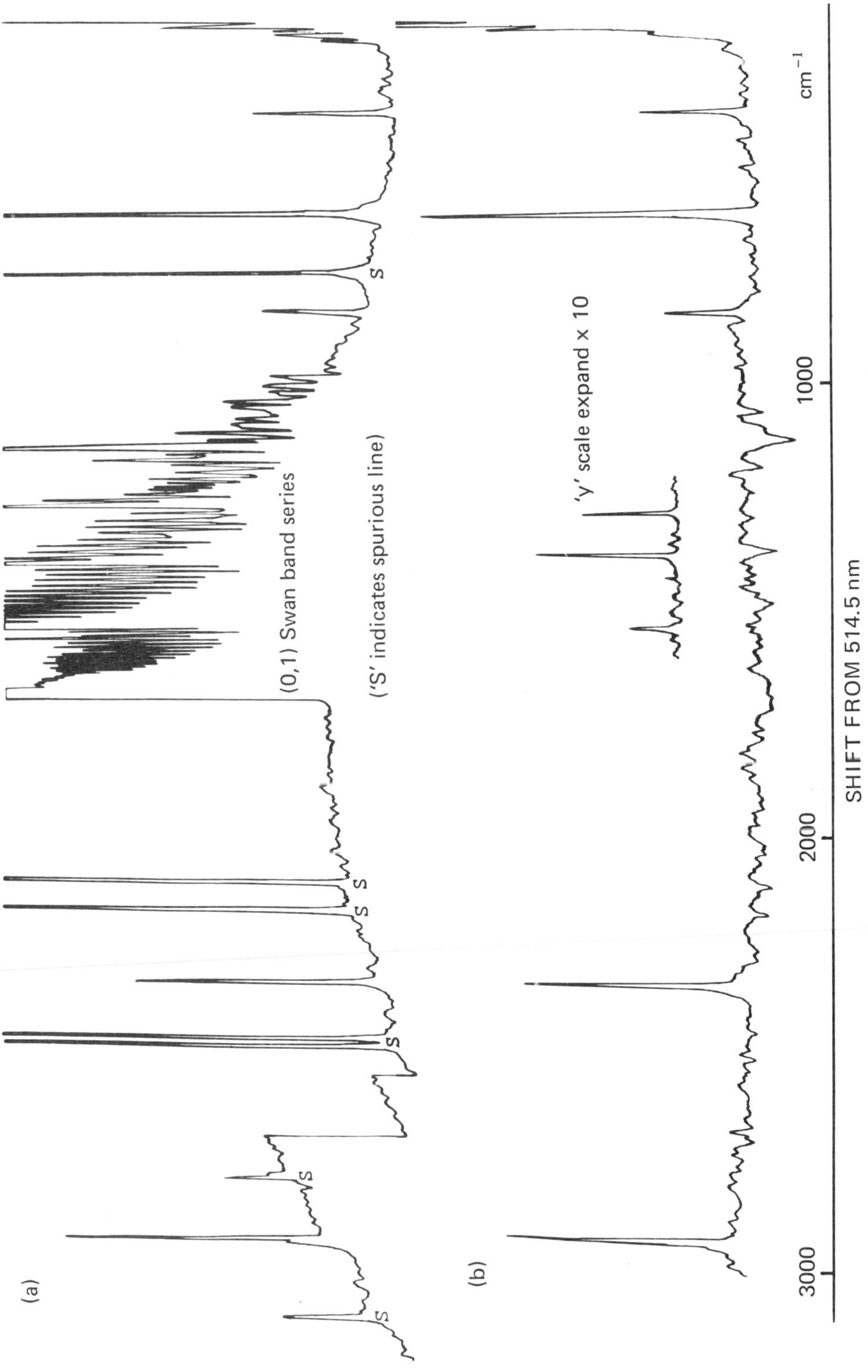

Fig. 3. (a) Total emissions from domestic gas/air diffusion flame. (b) As in (a) using phase sensitive detection.

Raman spectra from less emissive systems. A study of domestic gas*/air pre-mixed and diffusion flames was found to yield vibrational Raman spectra characteristic of CO_2, N_2, O_2, CH_4 and rotational lines of H_2. Relative intensities of the bands were shifted from the unlit gas/air mixture and varied with mixture strength and position within the flame envelope. With suitable burner geometry and optical design it is possible to avoid problems arising from laser focus shifts due to refractive index gradients in the gas flow.

The spectra (Fig. 3) appear to be those due to a mixture of unburnt gas and products. The presence of nitrogen in air/fuel flames can be used in principle as an internal intensity standard, enabling assays of O_2/N_2 and CO_2/N_2 to be made, giving an estimate of air/fuel ratio. Our experiments suggest that measurements of this type are feasible, but burner design must be modified in pre-mixed flame systems to reduce air entrainment in the post-reaction zone gases.

Temperature measurements are also possible using Raman spectroscopy, the temperature being determined at the well defined point of observation without disturbing gas flow. Three methods are possible in principle, i.e. from

1. The observation of maxima in pure rotational band envelopes. This is often not possible due to the large number of components in the gas mixture. If the system contains hydrogen, the very widely spaced H_2 rotational lines can be used to give a crude temperature estimate.
2. The comparison of Stokes and anti-Stokes vibrational band intensities.
3. The observation of vibrational 'hot' bands.

Of these options (2) is most attractive as one may monitor the relative populations of the ground and first vibrationally excited states exclusively, which is of importance if molecules are formed in vibrationally excited states in the reaction process. The technique would, however, be difficult as for most common systems the anti-Stokes band will be weak, presenting difficulties in accurate determinations of intensity ratios. Method (3) is the technique most commonly used to date, being applied to temperature estimates in N_2 in an electric discharge[7] and to measurements in O_2/H_2 and air/H_2 flames.[8] Such flames are easy to investigate using Raman spectroscopy as there is no visible emission from the reaction zone and one effectively looks at a hot stream of reactants and products. Observations of hot bands is suspect as their relative intensities can be shifted by formation of vibrationally excited molecules. In all of these temperature measurement methods allowance has to be made for corrections in bandwidths due to temperature/pressure effects and for instrument response. Our experiment is continuing at present with a range of other hydrocarbon/air flame systems.

CONCLUSION

From the preceding statements we suggest that both laser excited fluorescent and Raman spectra are potentially powerful tools for the investigation of burning systems under a variety of conditions. Since the light observed has its source in a volume of approximately cylindrical form about 20 μm diameter $\times$ 100 μm long at the laser focus,[6] this optical probe is capable of great spatial resolving power. The resonance fluorescence phenomenon, limited at present to the C_2 species is, furthermore, highly sensitive, the spectra already recorded being due to an estimated concentration of 10^{16} molecules/ml or less, and from the quality

* The domestic gas used had the following quoted analysis – H_2 54.3%; CO 1.5%; O_2 0.0%; C_2H_6 0.6%; N_2 1.0%; CO_2 12.4%; CH_4 30.1%; C_3H_8 0.1%; by volume.

of the spectra a reduction of about three orders of magnitude on this signal would still give meaningful results. Raman spectra in this context are less sensitive but vibrational spectra can be recorded at partial pressures of 1/100 atm within the flame for CO_2 and N_2 for example, while attempts to improve sensitivity are at present under active investigation.

All early experiments on flames were conducted using a chopped laser of high power and phase-sensitive detection to select the a.c. component of the total signal for the photomultiplier. This method is simple but is not the most satisfactory for this type of work. A new system based on a pulsed laser, a specially designed system comprising a double monochromator, blue-green sensitive photomultiplier and a box-car detector is under development. The laser, a Laser Associates instrument is a Q-spoiled Nd^{3+} in YAG device operating with an in-cavity frequency doubler to yield microsecond pulses of radiation at 3,200 p.p.s. with a wavelength of 5300 Å. Each pulse has a height of $>$ 20 W. The detector system is based on a Brookdeal Model 415/425A instrument and is so designed that a timed gate opens for a controlled short interval at the instant of laser pulsing. Hopefully, the high-energy pulsed output of the laser will improve the sensitivity of the method and since the electronic gate is closed for most of the time the detected noise should be minimized.

Our conclusion is that these experiments show that laser excited spectra are capable of giving composition and/or temperature profiles throughout the burning gas flow without any perturbation of the system.

ACKNOWLEDGEMENTS

The authors would like to thank Professor I. R. Beattie for access to the spectrometer and laser, and the National Gas Turbine Establishment, Pyestock (Ministry of Defence) for advice and financial support.

REFERENCES

1 G. J. Minkoff and C. F. H. Tipper, *Chemistry of Combustion Reactions*, Butterworths, London, 1962, see for example p. 271 *et seq.*
2 A. G. Gaydon, *The Spectroscopy of Flames*, Chapman & Hall, London, 1957.
3 J. D. Shea, *Phys. Rev.* **30**, 825 (1927).
4 A. G. Gaydon and H. G. Wolfhard, *Flames*, Chapman & Hall, London, 1970, p. 296.
5 A. G. Gaydon and H. G. Wolfhard, *Flames*, Chapman & Hall, London, 1970, see for example p. 196.
6 J. J. Barrett and N. I. Adams, *J. Opt. Soc. Amer.* **58**, 311 (1968).
7 L. Y. Nelson, A. W. Saunders, A. B. Harvey and G. O. Neely, *J. Chem. Phys.* **55**, 5127 (1971).
8 M. Lapp, L. M. Goldman and C. M. Penney, *Science* **175**, 1112 (1972).

31. RAMAN BAND SHAPES FOR FLAME GASES

M. Lapp
General Electric Company, Corporate Research and Development Schenectady, New York, USA

INTRODUCTION

In earlier work,[1] some of the general temperature-sensitive features of Raman scattering from flames were explored. Stokes *Q*-branch fundamental vibrational band profiles for nitrogen and oxygen were theoretically fit as a function of temperature, with a view toward diagnostic measurements. This series of bands consists of the ground state band (scattering from the ground vibrational level) and the upper state or 'hot' bands (scattering from excited vibrational levels).

The promising results obtained have led to further detailed study of the shapes of Raman bands under flame conditions. We present here new preliminary experimental data and analyses of nitrogen and hydrogen band profiles obtained for flames whose temperatures were also determined by thermocouple measurements, thereby 'calibrating' the system. In addition, theoretically calculated spectra are given from 300 K to 3500 K. These profiles indicate the spectral spread of scattered intensity as the temperature is increased.

The fact that the Raman technique under discussion measures both vibrational and rotational excitation temperatures (and does so with good spatial resolution) indicates that a substantial use for the method should develop.

Additional fundamental interest exists in Raman scattering band shape experiments in flames, since they offer a way of obtaining fundamental molecular data (such as anharmonic molecular constants[2]) not easily available at lower temperatures. Thus, species not present in significant concentrations at lower temperatures may be studied, as well as excited vibrational and rotational states of molecules which are not significantly populated at lower temperatures.

EXPERIMENTAL TECHNIQUE

The basic experimental apparatus has been described previously.[1] New additions to the equipment include the capability for data logging by paper tape, the use of accurate flow metering techniques, and utilization of fine wire thermocouples for flame temperature measurements. A continuous wave argon ion laser source of about 1 W at the flame position at 4880 Å was used for all the results reported here.

Horizontal hydrogen-air flames were produced on a 2.5 cm diameter water-cooled

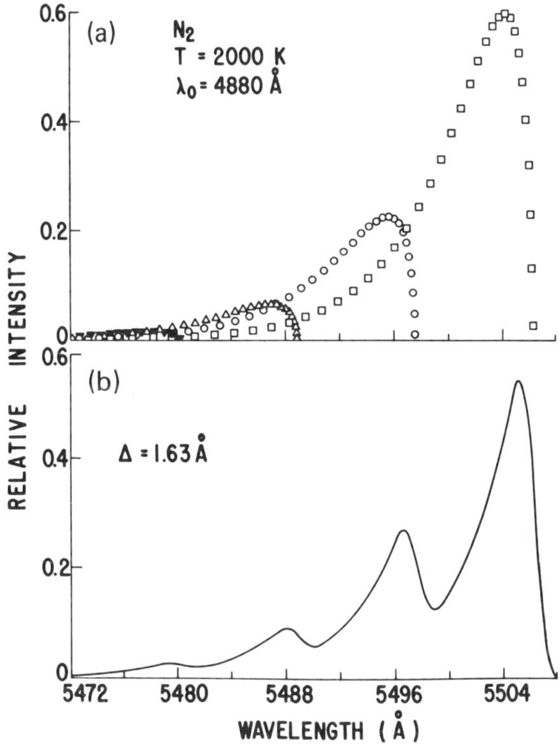

Fig. 1. Calculated Stokes Q-branch fundamental intensity at 2000 K for nitrogen. (a) Alternate 'strong' line intensities. Four bands are shown. The square data points correspond to the ground state band; the circular points to the first upper state band; etc. (b) Triangular slit function – convoluted profile, where Δ is the spectral slit width (FWHM).

porous plug burner[3] which was placed 2.0 cm away from another water-cooled porous plug connected to a vacuum line. The gas flows were monitored and made steady for these experiments through use of critical flow orifices and regulators. The temperatures of the flames were measured as a function of position by means of fine thermocouples made in our laboratory. These thermocouples[3] were made of 0.0005 in diameter wires of Pt–Pt 10% Rh, coated with quartz to prevent catalytic heating. The thermocouples were moved throughout the flame with an accurate vernier manipulator using, as a reference position locator, a finely-machined metal cone which could be placed reproducibly on the burner head. When the burner assembly was then placed in the test position in front of the spectrometer, the burner could be accurately located in this same reference position by placing it so that the laser beam just touched the cone tip. (By observing slight attenuation of the laser beam with a power meter at the position of the laser dump, this positioning could be accomplished with high sensitivity.)

In principle, the thermocouple measurement position and laser Raman scattering position could be made coincident by imaging the laser beam on the thermocouple junction. However, it is experimentally difficult to accomplish this. In addition, the thermocouples are relatively fragile, and it was therefore found to be advisable to calibrate the flame before embarking on the scattering measurements.

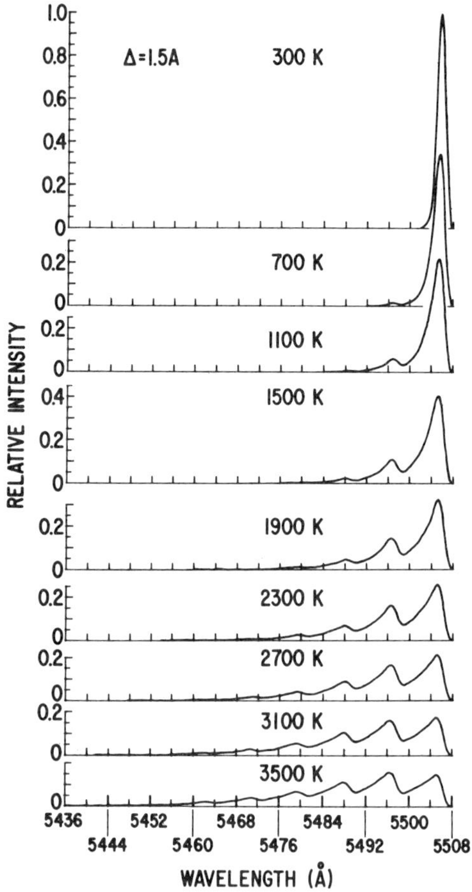

Fig. 2. Calculated Stokes Q-branch fundamental intensities from 300 K to 3500 K for nitrogen.

Two different flames were studied. First, for nitrogen data, a steady stoichiometric hydrogen-air flame was utilized (37.5 ml/s hydrogen and 88.8 ml/s air) for which 65% of the product gases was nitrogen. The flame was found to be about 1575 K at its centre point (i.e. half way between the burner head and the vacuum plug), including a rough 50 K correction for thermocouple radiative losses. Since the image of the spectrometer entrance slits at the flame scattering position was about 5 mm high, an estimate was made of the temperature variation along this zone. This was found to be about 16 K. The reproducibility of the thermocouple data was about ± ½ K.

Secondly, for hydrogen data, a less steady fuel-rich (four times stoichiometric) hydrogen-air flame was used (79.3 ml/s hydrogen and 47.0 ml/s air) for which about 51% of the product gases was hydrogen. This flame was found to be about 1390 K at its centre point, including a rough 40 K correction for radiative losses. Here, the variation of temperature with position along the slit image was much more severe, being roughly 110 K over a 5 mm vertical zone. Furthermore, the reproducibility was significantly poorer, being roughly ± 3 K. This flame, coloured red from the emission of water vapour vibration–rotation bands,[4] was affected more severely by diffusion from the ambient atmosphere than the preceding flame, which undoubtedly contributed to its less reproducible characteristics. It had, however, the virtue of a high hydrogen content.

THEORETICAL PREDICTIONS

For the diatomic molecules considered in these experiments, Eqn. (3) of ref. 1 can be used for the calculation of the Stokes Q-branch fundamental series ($\nu + 1 \leftarrow \nu$) profiles. This intensity relation neglects the small depolarized contribution for our cases. The profiles calculated in ref. 1 in this fashion were used to fit experimental profiles in order to determine the scattering-gas temperature. Therefore, all calculated profiles were normalized to the experimental curve peaks. Such normalized profiles are also presented here for purposes of temperature determination (see Figs. 5 and 7), but for other purposes, it is desired to calculate profiles for a given molecule at various temperatures while maintaining the spectral intensity differences at these temperatures (i.e. not normalizing each profile at the peak intensity). This is the case, for example, if it is desired to calculate the relative intensities at different temperatures obtained through use of a monochromator or filter designed to isolate a spectral portion of the Q-branch. We point out here that Eqn. (3) of ref. 1 can be used for these calculations if the vibrational partition function Q_{vib} given in ref. 1 following Eqn. (3)

$$Q_{\text{vib}} \approx [1 - \exp(-hc\omega_e/kT)]^{-1} \tag{1}$$

is multiplied by the factor $\exp[-(hc/kT)G(0,0)]$, where $G(0,0)$ is the zero-point energy. Here

$$G(0,0) = (1/2)\omega_e - (1/4)\omega_e x_e + (1/8)\omega_e y_e$$

where ω_e, $\omega_e x_e$ and $\omega_e y_e$ are vibrational constants defined by Eqn. (1) of ref. 1. The exponential factor compensates for the fact that the term value $G(\nu,J)$ defined by Eqn. (1) of ref. 1 relates to the energy above zero rather than above the zero-point energy.

Furthermore, substitution of ω_0 for ω_e in Eqn. (1) for Q_{vib}, where

$$\omega_0 \approx \omega_e - \omega_e x_e + (3/4)\omega_e y_e$$

results in a slightly more accurate calculation of Q_{vib}. Here, ω_0 is the coefficient of the linear ν term in the term value expression

$$G_0(\nu) = \omega_0 \nu - \text{const} \times \nu^2 + \ldots\ldots$$

Finally, if comparisons of Q-branch intensities are to be made between different molecules, account must be taken of the absolute value of the nuclear spin statistical weight g_I. This may be accomplished by substitution of g_I for the relative factor η in Eqn. (3) of ref. 1, and, for Σ states of homonuclear molecules, multiplication of the value of $Q_{\text{rot}} \approx kT/2hcB_e$ given in ref. 1 following Eqn. (3), by the factor $(2I + I)^2$. Here, I is the nuclear spin quantum number.

In analogy with the previous comment on Q_{vib}, substitution of B_0 for B_e in the

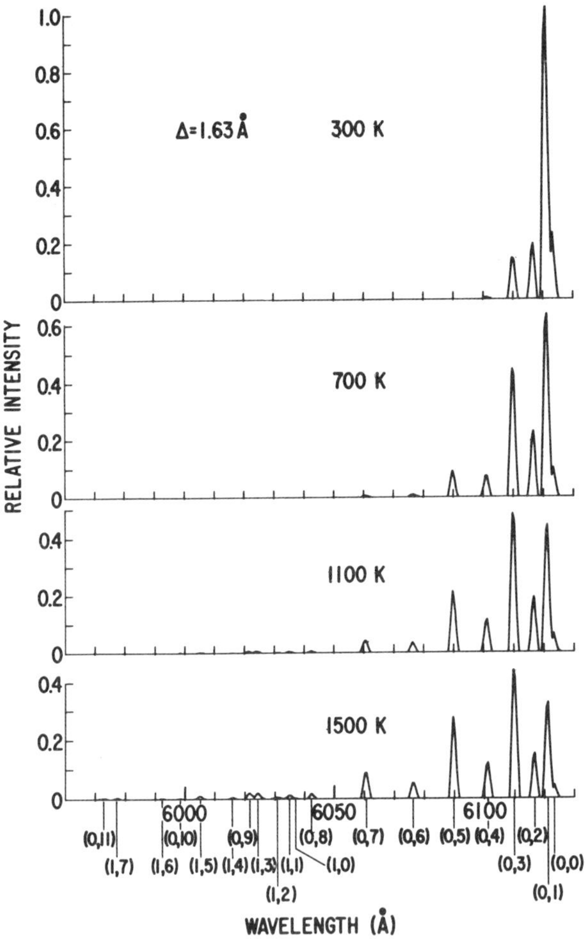

Fig. 3. Calculated Stokes Q-branch fundamental intensities from 300 K to 1500 K for hydrogen. The vertical lines along the wavelength axis indicate the positions of the various (v, J) vibration–rotation lines, where the parenthetic notation corresponds to the lower level quantum numbers.

relation for Q_{rot} results in a slightly more accurate calculation of Q_{rot}. Here

$$B_0 \approx B_e - \alpha_e/2$$

where α_e is the coefficient of the vibration–rotation interaction term in the term value $G(v,J)$ given in Eqn. (1) of ref. 1.

In Fig. 1(a) are shown the calculated nitrogen Q-branch vibration–rotation lines at 2000 K, different symbols denoting the various fundamental bands. The nitrogen spectrum has alternately 'strong' and 'weak' lines because of nuclear spin degeneracy, but only the 'strong' lines are shown here for clarity. In Fig. 1(b) is shown the spectrometer slit-convoluted intensity appropriate for the instrument used in our work, *viz.* a triangular slit function of full width at half maximum (FWHM) $\Delta = 1.63$ Å corresponding to 300 μm

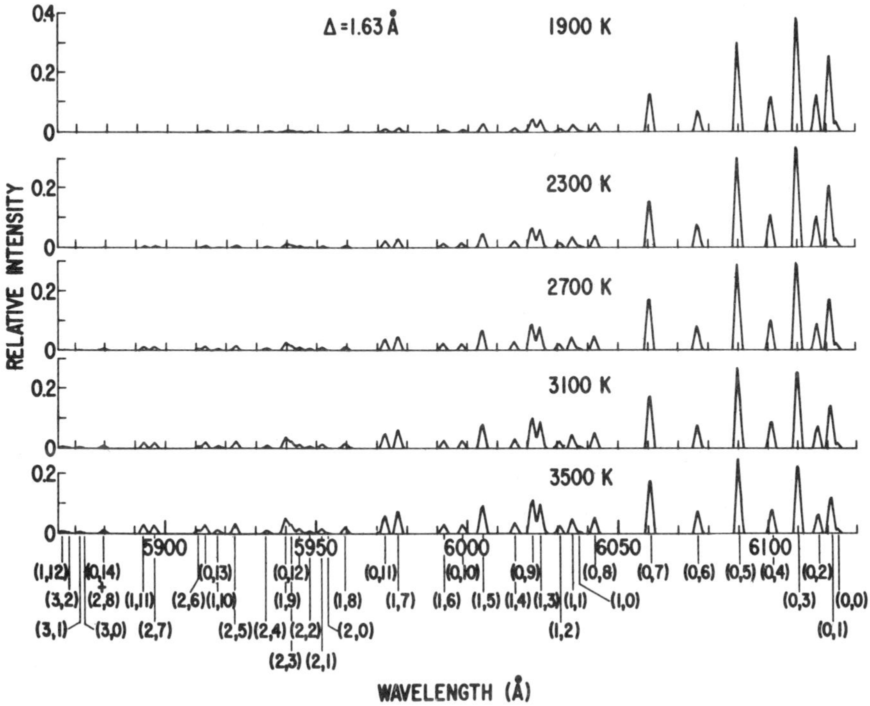

Fig. 4. Calculated Stokes Q-branch fundamental intensities from 1900 K to 3500 K for hydrogen. The vertical lines along the wavelength axis indicate the positions of the various (v, J) vibration–rotation lines, where the parenthetic notation corresponds to the lower level quantum numbers

entrance and exit slits on a Spex 1400-II double spectrometer.

Use of the type of data shown in Fig. 1(a) permits estimation of appropriate bandwidths for measuring the various rotational and vibrational excitation temperatures possible. In general, vibrational temperatures are indicated by the integral of intensity for particular bands (i.e. the ground state band or any specific upper state band), while rotational temperatures are indicated by the profile on the short wavelength side of each band via the influence of the vibration–rotation interaction. Thus, in principle, it is possible to determine vibrational excitation temperatures for any pair of vibrational levels and rotational excitation temperatures associated with any vibrational level.

For general comparative purposes, the nitrogen profile has been calculated from 300 K to 3500 K in Fig. 2 for $\Delta = 1.5$ Å. Here, the relatively broad spectral width at elevated temperatures becomes quite apparent.

The profiles for hydrogen are very different from those for nitrogen, since the individual vibration–rotation lines of the Q-branch for light molecules are spread far apart because of their very large vibration–rotation interaction constant, α_e. (For hydrogen, α_e is over 10^2 times larger than the value for nitrogen.) In Fig. 3 are shown calculated shapes for hydrogen from 300 K to 1500 K, while in Fig. 4 are shown the profiles for 1900 K to 3500 K. We note that the first vibration–rotation line of the first upper state band in hydrogen does not appear for a *longer* spectral interval [starting

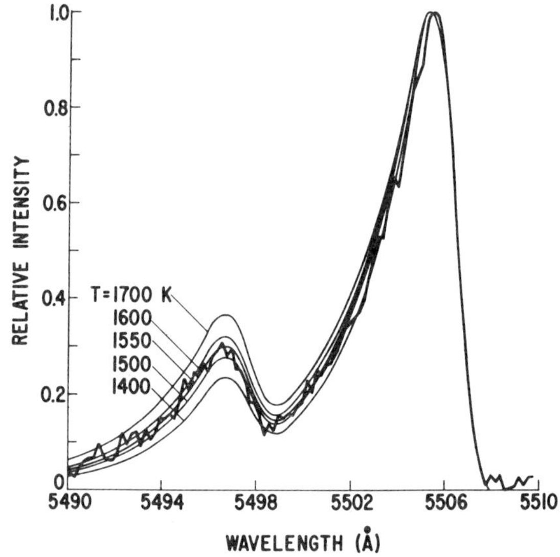

Fig. 5. Experimental Stokes Q-branch fundamental profile (heavy curve) for nitrogen for stoichiometric hydrogen-air flame ($2H_2 + O_2 + 3.73N_2$). The lighter curves are theoretically calculated and normalized to the peak value of the experimental ground state band. The thermocouple-measured temperature was approximately 1575 K, and the spectral slit width $\Delta = 1.63$ Å.

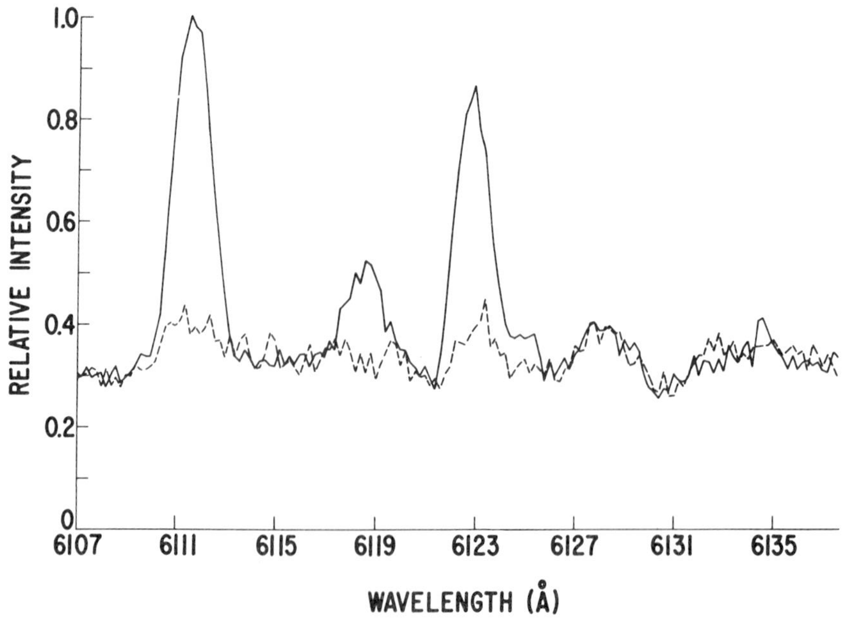

Fig. 6. Experimental Stokes Q-branch fundamental profile (solid curve) for hydrogen for four times stoichiometric hydrogen-air flame ($8H_2 + O_2 + 3.73N_2$). The dashed curve is a subsequently run emission spectrum for this slightly luminous flame. The thermocouple-measured temperature was approximately 1390 K, and the spectral slit width $\Delta = 1.62$ Å.

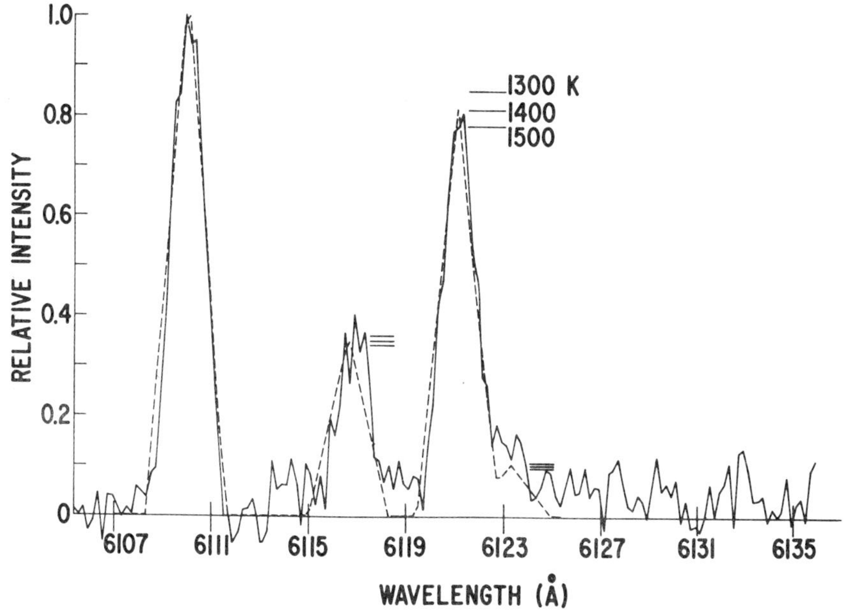

Fig. 7. Experimental Stokes Q-branch fundamental profile (solid curve) for hydrogen for the conditions of Fig. 6. Here, the emission spectrum (dashed curve of Fig. 6) has been subtracted from the scattering-plus-emission spectrum (solid curve of Fig. 6). A theoretically-calculated profile for hydrogen at 1400 K (dashed curve) has been added, and has been normalized to the experimental peak intensity for the $J = 3$ line. Keeping the peak intensities normalized for the $J = 3$ line, profiles were calculated at 1300 K and 1500 K. The corresponding peak intensities of the $J = 0$, 1 and 2 lines are indicated here by horizontal bars.

from the (0, 0) position, where the parenthetic notation corresponds to the lower level quantum numbers (v, J)] than that corresponding to the *entire* wavelength scale of all the nitrogen data plotted in Fig. 2.

EXPERIMENTAL RESULTS

The profile of nitrogen observed from the stoichiometric flame at a thermocouple-measured temperature of about 1575 K is shown in Fig. 5. Superimposed on the data are several theoretically-calculated profiles, normalized at the peak of the ground state band. The experimental profile can be fitted to an equilibrium temperature of about 1550 K, to within an uncertainty of about 50 K. This profile was obtained by data logging techniques, recording the number of counts every 10 s (i.e. every 0.2 Å for the spectrometer scan speed used).

If the combustion gases were not in equilibrium, a curve based upon a single temperature would then not fit all the data. For example, Nelson *et al.*[5] have published a nitrogen Q-branch profile obtained from 200 torr of nitrogen in a flowing electric discharge. This gas is vibrationally hot and rotationally cold, and the results offer an interesting comparison to our equilibrium work.

The profile of hydrogen obtained from the four times stoichiometric hydrogen-air flame at a thermocouple-measured temperature of about 1390 K is shown in Fig. 6, where the data-logged trace shown as the solid curve contains the Raman scattering data, and the dashed curve is a subsequently run emission spectrum of this luminous flame. The emission spectrum has been subtracted from the emission-plus-scattering spectrum in Fig. 7, and a theoretically-calculated (dashed) curve added for hydrogen at 1400 K. Keeping the $J = 3$ calculated peak intensity normalized to the experimental peak intensity, profiles were also calculated for other temperatures. The peak values for $J = 0$, 1 and 2 at 1300 K, 1400 K and 1500 K are indicated by the appropriate horizontal lines in this figure. The accuracy of temperature measurements for the hydrogen data shown here is not good for two main reasons: (1) the flame is somewhat unsteady and non-isothermal, and (2) ratios of the vibration–rotation line intensities shown here are not particularly sensitive to the temperature over this temperature range. Other ratios utilizing higher rotational lines are more sensitive for this range. However, the profile presented here is indicative of the type of data and the required treatment for temperature estimates utilizing light molecules. The relatively wide spectral intervals between vibration–rotation lines for these molecules suggests that, with proper choice of bandpass, interference filters could eventually be used for temperature determinations with greater ease than would be the case for heavier molecules.

CONCLUSIONS

Calculated Raman scattering Q-branch fundamental profiles for nitrogen and hydrogen have been given for 300 K to 3500 K. Experimental nitrogen and hydrogen profiles for hydrogen-air flames have also been shown. The nitrogen data have been sufficiently well fitted theoretically as a function of temperature that a new diagnostic tool for temperature measurement is suggested. This method is capable of determining vibrational and rotational excitation temperatures as well as equilibrium values.

ACKNOWLEDGEMENTS

The author is grateful to Dr. John Moore and Mr. Harry Horton for their contributions to the thermocouple temperature measurements, and to Dr. C. M. Penney for his contributions to the experimental program. This work has been performed in connection with a program supported in part through Project SQUID of the Office of Naval Research.

REFERENCES

1 M. Lapp, L. M. Goldman and C. M. Penney, *Science* **175**, 1112 (1972).

2 R. Gaufrès and S. Sportouch, *Compt. Rend.* **272**, 995 (1971).

3 W. E. Kaskan, *Sixth Symposium (International) on Combustion,* Reinhold, New York, 1957, p. 134.

4 A. G. Gaydon, *Spectroscopy of Flames,* Chapman and Hall, London, 1957, pp. 79, 90, 24

5 L. Y. Nelson, A. W. Saunders, Jr., A. B. Harvey and J. O. Neely, *J. Chem. Phys.* **55**, 5127 (1971).

SECTION V
MACROMOLECULES & BIOLOGICAL MOLECULES

32. RAMAN SPECTRA OF CHAIN CONFORMATIONS

J. A. Koenig
Case Western Reserve University,
Cleveland, Ohio, USA

VIBRATIONAL SPECTROSCOPY OF THE α-HELICAL CONFORMATION

It has been established that many polypeptides and proteins exist in the α-helical conformation. In the α-helix, there are 18 peptide residues in five turns and intrachain hydrogen bonds are found between each pair of every third neighbours. The factor group for the helix is a $\bar{C}_{18/5}$ group. The character table for the $\bar{C}_{18/5}$ factor group is shown in Table 1. The infrared active molecular vibrations are found in the A and E_1 representations corresponding to phase differences of 0 and $2\pi/3.6$, respectively. The A and E_1 vibrations are parallel and perpendicularly polarized with respect to the helix axes respectively, in the infrared.

The Raman scattering tensors for the α-helix have been derived.[1] The Raman active modes are found in the A_1, E_1 and E_2 representations. The A modes are polarized in the Raman while the E_1 and E_2 modes are depolarized. For an oriented helix and polarized radiation, it is possible to isolate each of the A_1, E_1 and E_2 contributions.

Raman polarization measurements are designated by the following notation $X(zz)Y$. The first capital letter before the parenthesis (in this case X) designates the direction of the light propagation. The last capital letter designates the direction of viewing of the scattered light. The first letter inside the parentheses which is written in lower case (in this case z) denotes the polarization direction of the incident laser light while the second lower case letter in the parentheses denotes the polarization direction of the scattered light. With the $X(zz)Y$ spectral measurement, the A modes are isolated while the $X(zx)Y$ spectral measurement isolates the E_1 modes and the $Z(yx)Y$ spectral measurement gives rise to the E_2 modes. The $Z(xx)Y$ spectral measurement gives rise to both A and E_2 vibrations. These measurements have been made for the Raman spectrum of poly-L-alanine (PLA).[1]

Poly-L-alanine is the simplest polypeptide having the α-helical conformation since polyglycine assumes other helical conformations. It is appropriate to examine the vibrational properties of this α-helical polypeptide in some detail in order to understand the spectroscopic implications. The vibrational properties of α-poly-L-alanine have been investigated by infrared spectroscopy,[2,3] Raman spectroscopy,[1,4,5] and normal coordinate analysis.[5-7]

The optical frequencies represent frequencies for particular values for θ on the

Table 1. Group character table for $C_{18/5}$ factor group.

$C_{18/5}$	E	${C_{18/5}}^1$	${C_{18/5}}^2$	${C_{18/5}}^3$	…	${C_{18/5}}^{16}$	${C_{18/5}}^{17}$	Infrared	Raman
A	1	1	1	1		1	1	z	$z^2; x^2+y^2-i(xy-yx),$ $x^2+y^2+i(xy-yx)$
ϵ_1	1	$e^{i5\pi/9}$	$e^{i10\pi/9}$	$e^{i15\pi/9}$	…	$e^{i80\pi/9}$	$e^{i85\pi/9}$	$x+iy$	$(xz+iyz, zx+izy)$
ϵ_2	1	$e^{i10\pi/9}$	$e^{i20\pi/9}$	$e^{i30\pi/9}$	…	$e^{i160\pi/9}$	$e^{i170\pi/9}$		$[x^2-y^2+i(xy+yx)]$
ϵ_3	1	$e^{i15\pi/9}$	$e^{i30\pi/9}$	$e^{i45\pi/9}$	…	$e^{i240\pi/9}$	$e^{i255\pi/9}$		
⋮	⋮	⋮	⋮	⋮		⋮	⋮		
$B=\epsilon_9$	1	$e^{i45\pi/9}$	$e^{i90\pi/9}$	$e^{i135\pi/9}$		…			
⋮	⋮	⋮	⋮	⋮					
$\epsilon_3^*=\epsilon_{15}$	1	$e^{i75\pi/9}$	$e^{i150\pi/9}$	$e^{i225\pi/9}$		…			
$\epsilon_2^*=\epsilon_{16}$	1	$e^{i80\pi/9}$	$e^{i160\pi/9}$	$e^{i240\pi/9}$		…			$[x^2-y^2-i(xy+yx)]$
$\epsilon_1^*=\epsilon_{17}$	1	$e^{i85\pi/9}$	$e^{i170\pi/9}$	$e^{i255\pi/9}$		…		$x-iy$	$(xz-iyz, zx-izy)$
A	1	1	1	1				z	$z^2; (x^2+y^2, xy-yx)$
E_1	2	$2\cos(5\pi/9)$	$2\cos(10\pi/9)$	$2\cos(15\pi/9)$		…		x, y	$(xz, xz; yz, zy)$
E_2	2	$2\cos(10\pi/9)$	$2\cos(20\pi/9)$	…					$(x^2-y^2; xy+yx)$
E_3	2	$2\cos(15\pi/9)$	$2\cos(30\pi/9)$	…					
⋮	⋮	⋮	⋮						
E_4	2	$2\cos(40\pi/9)$	$2\cos(80\pi/9)$	…					

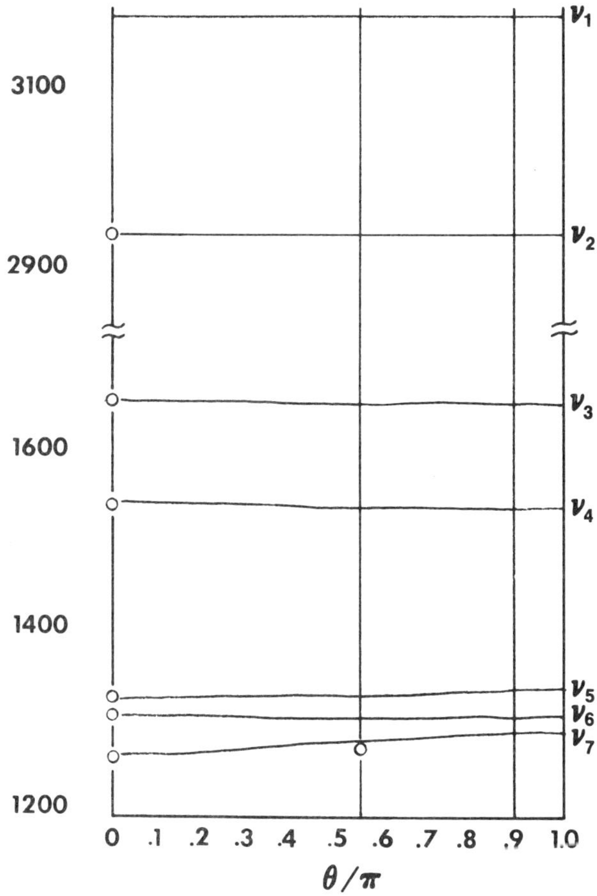

Fig. 1. Dispersion curves for poly-L-alanine (high frequency range).

phonon dispersion curve where θ is the phase difference between vibrations of equivalent atoms in adjacent repeat units. The frequencies of the active modes for the α-helix occur at $\theta = 0°$ (A species); $\theta = 100°$ (E_1 species) and $\theta = 160°$ (E_2 species). The E_2 frequencies are at $\theta = 160°$ because of periodicity of the Brillouin zone. The phonon dispersion curves for α-PLA are shown in Figs. 1 and 2.[5] The intercepts of the vertical lines at the values of $\theta = 0$, 100 and 160° give the theoretical frequencies which should correspond to the infrared and Raman data.

The dispersiveness of the curves is an indication of the coupling of the particular normal mode within adjacent repeat units. When the frequency curves are flat as a function of θ, the vibrations are localized within the chemical repeat unit indicating that the motions of the adjacent unit have no effect on the frequency. Usually, the internal motions correspond to bond stretching or angle deformations of substituents on the backbone of the polymer chain. These modes correspond to the higher frequency portions of the spectrum. The external modes, that is, those showing change in frequency with phase angle, usually involve skeletal motions and occur at lower frequencies.

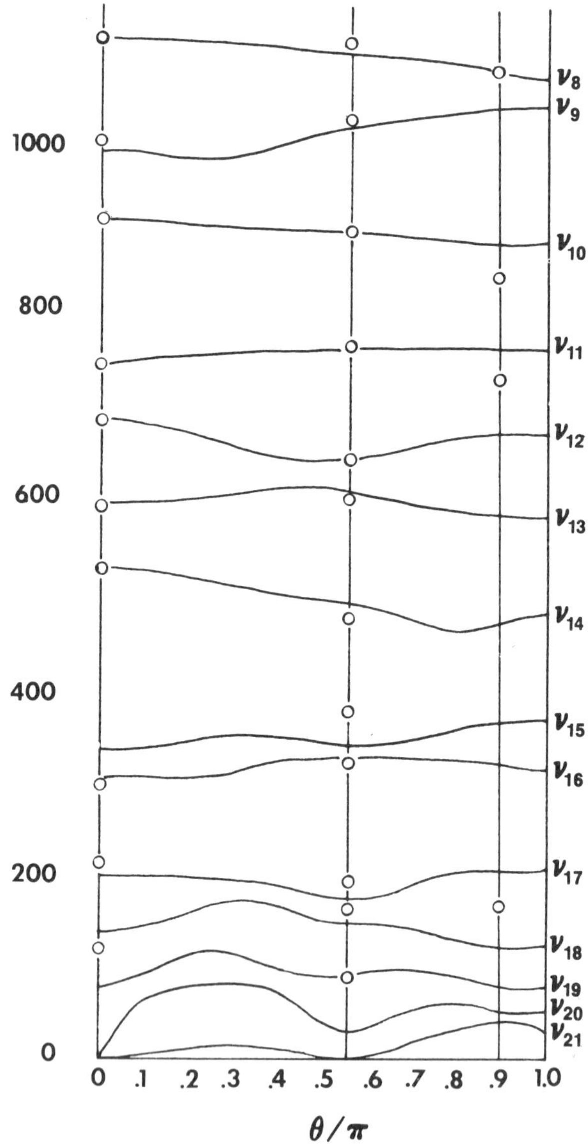

Fig. 2. Dispersion curves for poly-L-alanine (low frequency range).

It has been well established that the *trans*-CONH group undergoes highly localized vibrations and exhibits characteristic bands at: 3280 cm^{-1} (amide A), 3090 cm^{-1} (amide B), *ca*. 1650 cm^{-1}(amide I), 1550 cm^{-1} (amide II) and 1300–1250 cm^{-1} (amide III), 627 cm^{-1} (amide IV), 725 cm^{-1} (amide V), 600 cm^{-1} (amide VI) and 206 cm^{-1} (amide VII). The amide A and B frequencies are believed to arise from Fermi resonance between the first overtone of the amide II vibration and the N–H stretching vibration. In the Raman spectra, the amide A and B lines are usually marked by water lines and aromatic C–H stretching lines. Lord and Yu[8] suggest that the amide A line can be observed for aqueous solutions of proteins if the spectrum is obtained immediately after dissolution in deuterium oxide since the exchange rate of the peptide hydrogens is slow. The amide I vibration is primarily C–O stretching (80%), with some contributions due to the C–N stretching and the N–H in-plane bending. The amide I in the α-helix shows very little splitting of the A and E_1 components and the experimental resolution of the two components has not been made in the Raman so only a single line is observed at 1651 cm^{-1}. The polarized infrared of poly-γ-benzyl L-glutamate shows a parallel band at 1650 cm^{-1} and a perpendicular band at 1652 cm^{-1}.

The amide II mode is primarily the N–H in-plane deformation (65%) with a C–N stretching contribution. This mode in the α-helix gives an intense perpendicular infrared band (1545 cm^{-1}) but only a weak broad Raman line at 1549 cm^{-1}. The calculated frequencies of the amide II modes show very little splitting but it has been suggested that the parallel infrared mode at 1515 cm^{-1} in poly-benzyl-L-glutamate is the A component of the amide II mode.

The amide III mode is calculated at 1305 cm^{-1} for N-methyl acetamide and at 1328 (A), 1333 (E_1) and 1340 (E_2) mode for α-PLA. Fanconi *et al.*[5] assign the line at 1339 cm^{-1} in the Raman for poly-L-alanine to this mode while Koenig and Sutton[4] suggest the 1310 cm^{-1} line. These latter two lines decrease to 1329 cm^{-1} and 1293 cm^{-1} respectively upon N-deuteration. However, Raman lines at 1275 cm^{-1} and 1262 cm^{-1} disappear entirely upon N-deuteration and the α-helix amide III mode for α-PLA could be in this region of the spectrum.

The amide IV line was assigned to O=C–N in-plane bending, amide V to N–H out-of-plane bending, amide VI to C=O out-of-plane bending and amide VII to torsion about the C–N band. All of these amide modes are expected to be weak in the Raman effect.

The remainder of the vibrational modes are not easily classified because of the large interaction with several different internal coordinates. The Raman lines at 909 and 889 cm^{-1} may be the C-methyl stretching frequency which should be fairly localized. There is usually a strong Raman line in this frequency range for most peptides.

The low frequency normal modes (below 500 cm^{-1}) are mainly skeletal motions and are highly phase-dependent. These frequencies are expected to be highly sensitive to polymer conformation but they will also depend on the residue, so clear cut conformational assignments seem unlikely. Fanconi *et al.*[5] assign the lines at 371 and 378 cm^{-1} of α-PLA as characteristic of the α-helix, but no Raman lines are found in this region for poly-L-leucine. This difference between alanine and leucine illustrates the sensitivity of these modes to their environment.

Raman spectra of α-helical polypeptides (amide modes)

The Raman spectra of four polypeptides in the α-helical form have been reported: poly-L-alanine,[4] (α-PLA), poly-γ-benzyl-L-glutamate,[9] (α-PBLG), poly-L-lysine,[10] (α-PLL·HCl), poly-L-leucine,[9] (α-PLL). A strong Raman line at 1652 cm^{-1} for α-PLL, α-PBLG, α-PLL·HCl and α-PLA is the amide I mode. Since this mode is not experimentally resolvable in the Raman spectra, the A, E_1 and E_2 modes are apparently common at 1652 cm^{-1}.

In amide II region, a weak, broad Raman line is found at 1547 cm^{-1} for α-PLL, 1549 cm^{-1} for α-PLA and only broad diffuse scattering resulting in this region in α-PBLG. This weak line probably corresponds to the A mode since a strong perpendicular infrared band is found near 1540 cm^{-1} for these polymers. One would be hard pressed to deter-

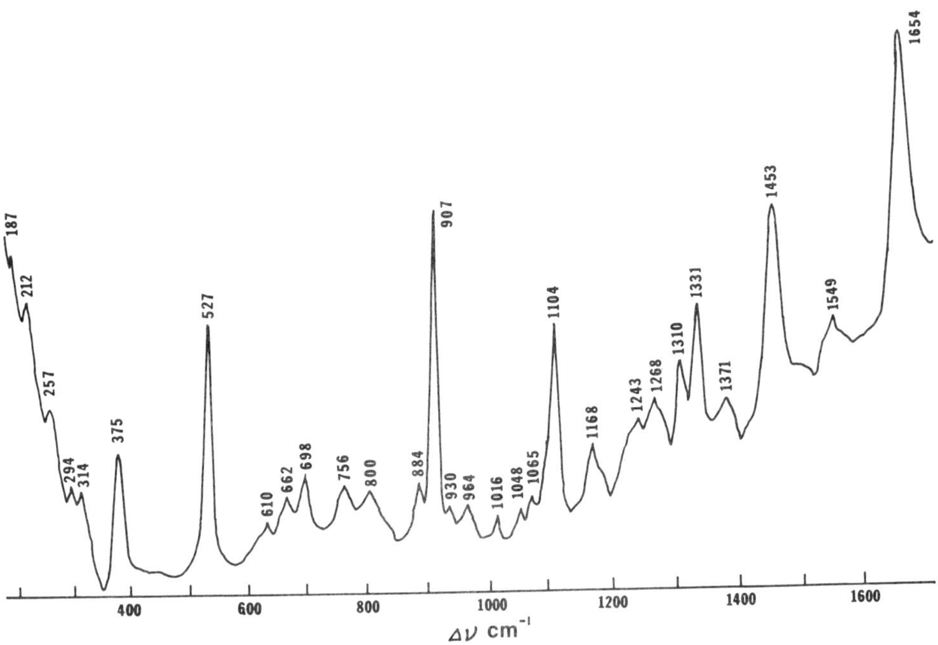

Fig. 3. Raman spectrum of poly-L-alanine.

mine that the infrared and Raman frequencies are experimentally different in this region.

In the amide III region, a Raman line is observed at 1320 cm^{-1} for α-PLL and α-PLL·HCl, at 1336 cm^{-1} for α-PBLG and two lines at 1339 and 1310 cm^{-1} for α-PLA. In this frequency range of 1200–1300 cm^{-1} region, the C_α–H deformation is also expected.

The remaining lower frequency amide modes (IV, V, VI, VII) are difficult to assign from the Raman effect due to their inherent weaknesses and the presence of strong C–C motions from the skeleton of the α-helix.

The Raman spectrum of poly-L-alinine is shown in Fig. 3. Assignments are given in Table 2.[5] The methyl modes are easily identifiable and in several cases coincide with the observations in the infrared and Raman spectra of poly-L-valine. The expected doubling of the methyl modes due to the A and E species is rarely observed indicating that the intrachain interactions for the methyl residue are small.

VIBRATIONAL SPECTROSCOPY OF THE β-CONFORMATION

Another common chain conformation of polypeptides is the so-called β-conformation in which the polypeptide chain is very nearly fully extended. An antiparallel pleated sheet accommodates the hydrogen bonding.

Let us consider polyglycine I as an example of the β-conformation. According to X-ray investigations the molecule of polyglycine I belongs to the factor group C_{2v}. Therefore, the 26 normal modes of the chain are grouped into nine A_1 (polarized Raman and infrared); four A_2 (Raman) and nine B_1 (Raman and infrared) and four B_2 (Raman and infrared) modes. The crystal of polyglycine I has symmetry D_2 and due to the interchain hydrogen bonding in the crystal, each of the internal modes of the polyglycine I mole-

Table 2. Observed and calculated frequencies of α-poly-L-alanine.

	Frequencies (cm^{-1})			Assignments[a]
Phonon Curve No.	Raman	Infrared	Calculated	
1	3308		3184 (A, E_1, E_2)	$r_{NH}(96)$
	2998	2988		Methyl asymmetric stretch
	2950	2938		Methyl symmetric stretch
2	2892	2883	2939 (A, E_1, E_2)	$r_{C'H}(100)$
3			1657 (A)	$r_{CO}(71), r_{CN}(17)$
	1659	1657($\parallel$)		
3			1655 (E_1, E_2)	$r_{CO}(72), r_{CN}(18)$
	1587			
	1563			
4			1545 (A)	$\phi_{NH}(65), r_{CN}(38)$
4	1543	1545	1542 (E_1)	$\phi_{NH}(69), r_{CN}(32)$
4			1541 (E_2)	$\phi_{NH}(68), r_{CN}(33)$
	1528			
	1462	1457($\parallel$)		Methyl asymmetric deformation
	1405			
	1390	1384($\perp$)		Methyl symmetric deformation
5			1340 (E_2)	$r_{CC'}(34), r_{CN}(25), \phi_{NH}(20)$
5	1339	1330($\parallel$)	1333 (E_1)	$r_{CC'}(21), r_{CN}(17), \phi_{HC'C}(16), \phi_{NH}(17)$
5			1328 (A)	$\phi_{HC'C}(37), \phi_{HC'N}(16), r_{CO}(14), \phi_{NH}(10)$
6	1309	1308($\parallel$)	1309 (A, E_1, E_2)	$\phi_{HC'M}(57), \phi_{HC'N}(41)$
7			1290 (E_2)	$\phi_{HC'C}(57), \phi_{HC'M}(35)$
7	1275	1274($\perp$)	1282 (E_1)	$\phi_{HC'C}(46), \phi_{HC'M}(20), r_{CC'}(15)$
7	1262	1262($\parallel$)	1266 (A)	$r_{CC'}(32), \phi_{HC'C}(28), r_{CN}(20)$
	1167	1167($\parallel$)		Methyl rocking
8			1112 (A)	$r_{C'N}(61), r_{C'C}(20)$
8	1106	1107($\perp$)	1093 (E_1)	$r_{C'N}(48), r_{C'C}(12)$
8	1072		1076 (E_2)	$r_{C'N}(45), r_{CN}(10)$
	1058	1050($\perp$)		Methyl mode
9			1034 (E_2)	$r_{CN}(17), r_{C'C}(15), r_{CO}(10)$
9	1017		1012 (E_1)	$r_{CC'}(18), r_{C'N}(16)$
9	994		984 (A)	$\phi_{CO}(19), r_{CC'}(13), \phi_{C'NC}(12), r_{CN}(10)$
10	909	906($\parallel$)	911 (A)	$r_{C'M}(64)$
10	889	893($\perp$)	896 (E_1)	$r_{C'M}(50), \pi_{CO}(14), \phi_{MC'C}(10)$
10	842		883 (E_2)	$r_{C'M}(57), \phi_{MC'C}(15)$
11		772	771 (E_1)	$\pi_{NH}(28), r_{C'M}(24), \pi_{CO}(20)$
11	757		752 (A)	$\pi_{NH}(44), \pi_{CO}(18)$
11	728		769 (E_2)	$\pi_{NH}(34), \pi_{CO}(30), r_{C'M}(10)$
12	695	686	691 (A)	$\pi_{CO}(29), \phi_{CC'N}(17), \phi_{NCC'}(13)$
12			673 (E_2)	$\phi_{CO}(38), \pi_{NH}(11)$
12	662	653	652 (E_1)	$\phi_{CO}(28), \pi_{NH}(12)$

(cont.)

Table 2 *(cont.)*

Phonon Curve No.	Raman	Infrared	Calculated	Assignments[a]
	Frequencies (cm^{-1})			
13	610	610	611 (E_1)	$\pi_{NH}(30), \pi_{CO}(29)$
13		595	599 (A)	$\pi_{NH}(31), \pi_{CO}(15), \tau_{CN}(13)$
13			583 (E_2)	$\pi_{CO}(26), \pi_{NH}(24), \tau_{CN}(11)$
14	531	526	530 (A)	$\phi_{CO}(36), r_{C'M}(15), \phi_{NC'M}(10)$
14			493 (E_1)	$\phi_{CO}(21), \phi_{NC'M}(15), \tau_{CN}(10)$
14	469		472 (E_2)	$\pi_{CO}(25), \pi_{NH}(20)$
15			363 (E_2)	$\phi_{NC'M}(62)$
15			339 (E_1)	$\phi_{CO}(19), \phi_{MC'C}(16), \phi_{C'NC}(12)$
	378	371		
15			334 (A)	$\phi_{MC'C}(35), \phi_{C'NC}(12), \pi_{CO}(12)$
16		325	324 (E_1)	$\phi_{NC'M}(45), \phi_{MC'C}(16)$
	314			
16			313 (E_2)	$\phi_{C'NC}(20), \phi_{CO}(21), \phi_{CC'N}(17)$
16	294	287	304 (A)	$\phi_{NC'M}(35), \phi_{MC'C}(22), \phi_{CC'N}(11)$
17			201 (E_2)	$\phi_{MC'}(41), \phi_{NCC'}(16)$
	214			
17			200 (A)	$\phi_{CC'N}(28), \tau_{CN}(10), \tau_{CC'}(15)$
17	194	190	173 (E_1)	$\phi_{MC'C}(20), \phi_{NCC'}(13)$
18	165	167	147 (E_1)	$\phi_{C'NC}(19), \tau_{CC'}(15), \phi_{NCC'}(12)$
18	159		125 (E_2)	$\tau_{C'N}(34), \tau_{CN}(16), \pi_{NH}(13)$
18		120	137 (A)	$\phi_{NCC'}(20), \phi_{C'NC}(19), \phi_{NC'M}(15)$
19			90 (E_1)	$\tau_{CC'}(26), \tau_{C'N}(23), \phi_{CC'N}(14)$
19			80 (A)	$\tau_{CC'}(31), \tau_{C'N}(23), \phi_{CC'N}(14)$
19			79 (E_2)	$\tau_{CC'}(19), \phi_{CC'N}(12)$
20			53 (E_2)	$\tau_{CC'}(47), r_{N..O}(10)$
21			40 (E_2)	$r_{N..O}(34), \tau_{CN}(21), \tau_{CC'}$
20			26 (E_1)	$\tau_{CC'}(33), \tau_{CN}(18), \tau_{C'N}(14)$

[a] Potential energy distribution among diagonal elements of the force matrix: r = bond stretch; ϕ = angle bend; π = out-of-peptide-plane bend; τ = torsion.

cule is expected to split into two vibrations. The results are summarized in Table 3.

With the antiparallel chain pleated sheet, a unit cell contains four peptide groups; two groups from one chain and two from an adjacent chain. In this case, adjacent chains may move in-phase or out-of-phase ($\delta' = 0, \pi$) as well as the intramolecular phasing ($\delta = 0, \pi$).

Since the infrared and Raman modes of the crystal arise from in- or out-of-phase motions of corresponding groups in various cells, it is more convenient and easier to understand the splitting of the frequencies in terms of intrachain and interchain interactions. Miyazawa[7] has derived the appropriate equations

Table 3. Analysis of the β-conformation of polyglycine.

Molecule	Crystal	Activity
C_{2v}	$(\delta, \delta')D_2$	
A_1	$A_1(0, 0)$	Raman
	$B_1(0, \pi)$	Raman, Infrared
B_1	$B_2(\pi, 0)$	Raman, Infrared
	$B_3(\pi, \pi)$	Raman, Infrared
A_2	$A_1(0, 0)$	Raman
	$B_1(0, \pi)$	Raman, Infrared
B_2	$B_2(\pi, 0)$	Raman, Infrared
	$B_3(\pi, \pi)$	Raman, Infrared

$$\nu(0, 0) = \nu_0 + D_1 + D_1'$$
$$\nu(0, \pi) = \nu_0 + D_1 - D_1'$$
$$\nu(\pi, 0) = \nu_0 - D_1 + D_1'$$
$$\nu(\pi, \pi) = \nu_0 - D_1 - D_1'$$

where D_1 and D_1' are the intra- and interchain interaction terms respectively. These equations are useful in making vibrational assignments for the β-form.

Although all four frequencies are expected in the Raman, only one Raman line at 1665 cm^{-1} is found in the oligomers of polyglycine in the amide I region. A single line at 1674 cm^{-1} is reported for polyglycine I but this frequency appears to lie on the upper edge of the experimental limit when compared with other measurements.[11] Two of the three infrared lines are observed at 1685 and 1636 cm^{-1}. Using these frequencies, the three parameters, ν_0, D_1 and D_1' can be calculated for polyglycine I.[12]

The amide II mode is weak but two lines are observed at 1564 and 1515 cm^{-1} in the Raman spectra of polyglycine I and a single strong line at 1517 cm^{-1} in the infrared. In the Raman spectra of polyglycine I in the amide III region, lines are observed at 1252, 1234, and 1219 cm^{-1} and weak infrared lines at 1243, 1230 and 1214 cm^{-1}. The lines at 1234 and 1230 and at 1219 and 1214 cm^{-1} are the same within experimental error. Therefore all four modes are observed. The 1252 cm^{-1} mode is assigned to the infrared inactive $A_1\nu(0, 0)$ mode as it is the only mode of the four not observed in the infrared spectrum. This amide III mode will probably have different frequencies for different β-structures since the interaction terms can have a large range of values for this mode.

The Raman spectra of the β-form is considerably different from the α-form and very little difficulty is expected in differentiating between these two conformations.

The Raman spectra of several polypeptides in the β-form have been obtained including poly-L-valine (β-PLV), poly-L-serine (β-PLS), polyglycine (PGI) and oligomers of poly-L-alanine (β-PLA). In the amide I region, one observes a strong Raman line at

1666 cm^{-1} for β-PLV, at 1665 cm^{-1} for (PGI), at 1663 cm^{-1} for β-PLA and at 1668 cm^{-1} for β-PLS. These modes have the same frequency within the limits of experimental error. It is most likely that this unique Raman line is the $A_1\nu(0, 0)$ mode, The infrared bands at 1685 and 1625 cm^{-1} have been established as characteristic of the β-structure and the Raman line at 1666 cm^{-1} falls into the same category.

The amide II region is rather weak and diffuse in the Raman effect for the β-conformation, and peak positions are difficult to determine. The characteristic infrared frequencies in the 1530 cm^{-1} and 1555 cm^{-1} are apparent in these polymers. It would appear that the amide II mode is of little value in the Raman effect for the β-structure because of its broad weak character.

The amide III region of the Raman spectra is rich with lines in the β-form. For β-PLV, three lines at 1291, 1276, and 1231 cm^{-1} are found. For the oligomers of β-PLA, lines are observed at 1268, 1250, and 1231 cm^{-1}. For β-pentaglycine, lines are observed at 1296, 1252, and 1234 and for PGI at 1295, 1255, 1234 and 1220 cm^{-1}. For PGI Small, Fanconi and Peticolas assigned the 1255 cm^{-1} mode to the CH_2 twisting mode and the 1234 and 1220 cm^{-1} modes to the amide III. A single line is observed at 1235 cm^{-1} in β-PLS. It is apparent that this region is rich in lines and is highly diagnostic of the β-conformation. The frequencies are slightly variable because of the extensive coupling but this frequency range should be inspected when β-conformation is suspected and the lines at 1230 to 1235 cm^{-1} seems most diagnostic.

VIBRATIONAL SPECTROSCOPY OF OTHER HELICAL FORMS OF POLYPEPTIDES

In addition to the α-helical conformation, polypeptides can assume other helical conformations. Polyglycine can exist as a 3_1 helix (PGII) as well as the β-form. The molecule in this conformation has factor group symmetry isomorphous with the $C(2\pi/3)$ point group, and the chain vibrations can be classified as symmetric modes ($A: \delta = 0$) and degenerate asymmetric modes (E: $\delta = 2\pi/3$). Normal coordinate analysis and the phonon dispersion for PGII has been reported.[11] The A modes are polarized in the Raman and have parallel infrared dichroism while the E modes are depolarized and have perpendicular infrared dichroism. The primary differences between the infrared and Raman spectra of the polypeptide in the 3_1 helix conformation are in the intensities of the modes; that is, some modes which are weak or absent in the Raman spectra may appear strong in the infrared spectra and vice versa. The symmetric A modes are expected to be the strongest in the Raman spectra with the E modes strongest in the infrared spectra.

VIBRATIONAL SPECTROSCOPY OF THE RANDOM COIL CONFIGURATION

It is possible for polypeptides to exist in the random coil conformation as well as the extended β and folded helical conformations. Poly-L-glutamate and poly-L-lysine have been prepared in the random coil conformation. In the random coil conformation, the frequency shifts due to the interchain interactions as well as the intrachain interaction will average zero. This will result in broad Raman lines indicative of liquid scattering. In addition, the frequency shifts arising from coupling of modes from units having identical structure and energies will be small. This latter effect may result in little change in frequency but will cause a substantial increase in the extinction and scattering coefficients in the infrared and Raman, respectively. Infrared studies indicate that the amide I band is at

1655 cm^{-1} for the random coil conformation whose peptide groups are hydrogen bonded with others. The amide II appears at 1535 cm^{-1} in the infrared for the random coil. Unfortunately, these infrared frequencies are similar to the α-helix in the amide I region (1656 cm^{-1} for random and 1650 cm^{-1} for α-helix) and the amide II band is weak for the random coil and is often interfered with by the band due to the ionized carboxyl mode.

The random coil conformation is apparently as elusive for detection by Raman spectroscopy as with infrared. The major differences involve a small lower frequency shift in the amide I region and higher frequency shift and a greatly enhanced intensity of the amide III mode for the coil compared to the α-helix.

RAMAN SPECTRA OF AQUEOUS SOLUTIONS OF POLYPEPTIDES

Although considerable information exists about the structure of the various polypeptides in the solid state, much less information is available about the confirmation of the polypeptides in aqueous solutions. The reasons for this lack of information are primarily due to the lack of proper instrumental techniques for the determination of the conformation on a molecular level. The overall size and shape of the molecule can be measured by light and X-ray scattering techniques, but more detailed information is not available. As a result, many questions remain unanswered. Does a structural change occur when a polypeptide is dissolved in water? What structural changes are induced by environmental differences such as pH, ionic strength, temperature and mixed solvents?

Raman studies can attempt to answer many questions of this type due to the sensitivity of the Raman effect to conformational differences and the minimal interference of the Raman scattering of water. Some difficulties arise since the levels of concentration required for quality Raman spectra are generally higher than encountered in previous work and the natural system, so questions of the concentration dependence of the results can justifiably be asked. However, with the continued increase in laser power and spectrophotometric detection and amplification, appropriately low concentrations of polypeptides can be studied and the concentration dependence of the structural changes can be studied. However, at this time, Raman spectroscopy represents the technique offering one of the fastest, easiest and most detailed structural information for the study of biopolymers in aqueous solutions.

The scattering due to water interferes with the measurement of the amide I mode, but studies of deuterium oxide solutions are not encumbered with this difficulty. So when it is desirable to study the amide I region in the Raman effect, D_2O should be used. The other Raman lines due to water at 800, 450 and 175 cm^{-1} are less of a problem because of their low intensities.

The signal-to-noise ratios of the Raman spectra obtained from aqueous solutions are not as high as for solids or liquids. The basis of these differences are probably the optical inhomogeneities of the aqueous solutions. These inhomogeneities act as Rayleigh scattering centres and the exciting line is broader in frequency than for liquids and the higher level of scattered light in the monochromator results in a lower signal-to-noise ratio. In addition, the desire to operate at the lowest possible concentration consistent with good Raman spectra is also reflected in the lower signal from the solution so a higher amplification factor is generally used.

The first Raman spectrum of an aqueous solution of a polypeptide, poly-L-lysine, was reported by Edsall and coworkers.[1] The spectrum was obtained photographically and was extremely weak. For poly-L-lysine in the random coil, a conformational change is reflected by the shift of the amide III mode from 1216 cm^{-1} to 1243 cm^{-1} in going from the solid state to aqueous solution.

At room temperature and pH 10, it is believed that poly-L-glutamic acid assumes a random coil form with its carboxyl side groups ionized and the polypeptide backbone hydrated. The amide I mode appears at 1665 cm^{-1} and is quite broad.[13] In the amide III region, Lord and Yu[8] report only one broad line at 1250 cm^{-1}.

The Raman spectra of PGA in aqueous solutions of different pH show spectral shifts indicative of conformational transitions. The amide I modes in D_2O at a pD of 4.5 and 10.5 occur at 1656 cm^{-1}. The methylene wagging and twisting modes and the amide III occur at 1353, 1313 and 1249 cm^{-1} at pH 11.4 but shift to 1338, 1305, and 1238 cm^{-1} at lower pH. The C–C stretch of the C–C–N unit shifts from 949 cm^{-1} to 931 cm^{-1} upon lowering the pH.

CONCLUSION

The Raman spectra of synthetic polypeptides appears to be useful in the determination of the conformation of the chain and the nature of the interaction between chains. This background information is essential to the interpretation of the spectra of protein.

REFERENCES

1 B. Fanconi, B. Tomlinson, L. A. Nafie, W. Small and W. L. Peticolas, *J. Chem. Phys.* **51**, 3993 (1969).

2 C. H. Bamford, A. Elliott and W. E. Hamby, *Synthetic Polypeptides*, Academic Press, New York, 1956, Chapter 5.

3 K. Itoh, T. Nakahara and T. Shimanouchi, *Biopolymers* **6**, 1759 (1968).

4 J. L. Koenig and P. L. Sutton, *Biopolymers* 8, 167 (1969).

5 B. Fanconi, E. Small and W. L. Peticolas, *Biopolymers* **10**, 1277, (1971).

6 K. Itoh and T. Shimanouchi, *Biopolymers* **9**, 383 (1970).

7 T. Miyazawa, K. Fukushima, S. Sugano and Y. Masuda, *Proteins*, Vol. II, Academic Press, New York, 1967, p. 557.

8 R. C. Lord and Nai-Teng Yu, *J. Mol. Biol.* **50**, 509 (1970).

9 J. L. Koenig and P. Sutton, *Biopolymers* **10**, 89 (1971).

10 J. L. Koenig and P. Sutton, *Biopolymers* **9**, 1229 (1970).

11 E. W. Small, B. Fanconi and W. L. Peticolas, *J. Chem. Phys.* **52**, 4369 (1970).

12 M. Smith, A. G. Walton and J. L. Koenig, *Biopolymers* **8**, 29 (1969).

13 J. L. Koenig and B. Frushour (unpublished).

33. SPECTRES RAMAN A LASER DU POLY-(TEREPHTALATE D'ETHYLENE) ET QUELQUES AUTRES POLYMERES

J. Derouault et P. J. Hendra
Department of Chemistry, Southampton University, England
et
M. E. A. Cudby, G. Fraser, J. Walker et H. A. W. Willis
I.C.I. (Plastics) Ltd., Welwyn Garden City, Herts., England

ABSTRACT

The laser Raman spectra of a number of polymers were investigated. The use of high pressure techniques is of great value for correlation splitting (Davydov effect) analysis. The results show that some typical features of 'pseudo-crystalline structure' appear on the Raman spectra of some higher polyolefines and copolymers. The anisotropic properties of Raman scattering from polymers are used for determination of chain orientation in bulk specimens.

INTRODUCTION

Le spectre de vibration d'un polymère contient[1] des informations sur:

1. Les caractéristiques vibrationnelles des cristallites du polymère et, en particulier, l'état de la molécule polymérique dans son environnement cristallin. Par conséquent le spectre sera caractéristique de la chaîne influencée par sa symétrie de site, c'est-à dire par le nombre de chaînes et leur orientation relative à l'intérieur de la cellule unitaire.
2. Les vibrations des espèces polymériques se trouvant dans les régions non cristallines et dans les séquences correspondant aux plis des chaînes.
3. 'Impuretés' du système, telles que groupes terminaux, branches latérales et additifs.
4. Les propriétés directionnelles des chaines par rapport à l'orientation du faisceau laser incident et à la direction d'observation.

Dans cette communication, nous décrivons une série d'expériences réalisées à Southampton sur des polymères cristallins sous pression, sur des homopolymères amorphes, sur des copolymères et sur quelques systèmes orientés.

ECLATEMENT PAR EFFET DE CORRELATION (DAVYDOV)

Si la cellule unitaire d'un matériau cristallin contient au moins deux espèces identiques, le nombre de bandes du spectre de vibration augmente à cause des interactions entre les composants de la cellule unitaire. Considérons par exemple le polyéthylène; dans ce cas, la cellule unitaire contient deux chaînes et sa symétrie est isomorphe du groupe $D_{2h} = V_h$. Par suite, le mouvement de valence symétrique (Fig. 1) de la chaîne isolée prend les formes indiquées dans la Fig. 2 dans les domaines cristallins. Donc, si les interactions moléculaires sont définies, on peut attendre un doublet au lieu d'une seule bande Raman

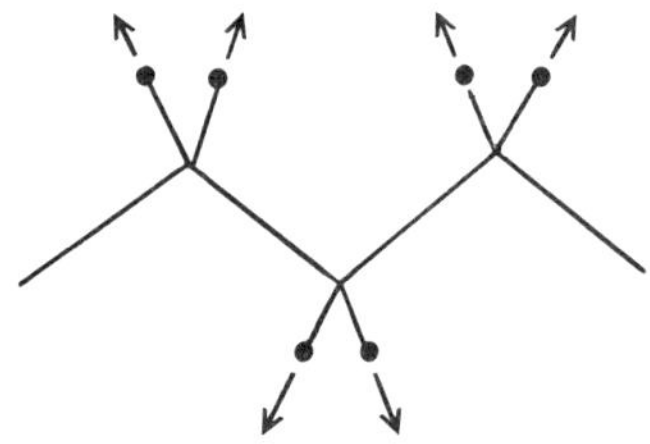

Fig. 1. Valence $\nu(CH_2)$ symétrique dans la chaine isolée de polyéthylène.

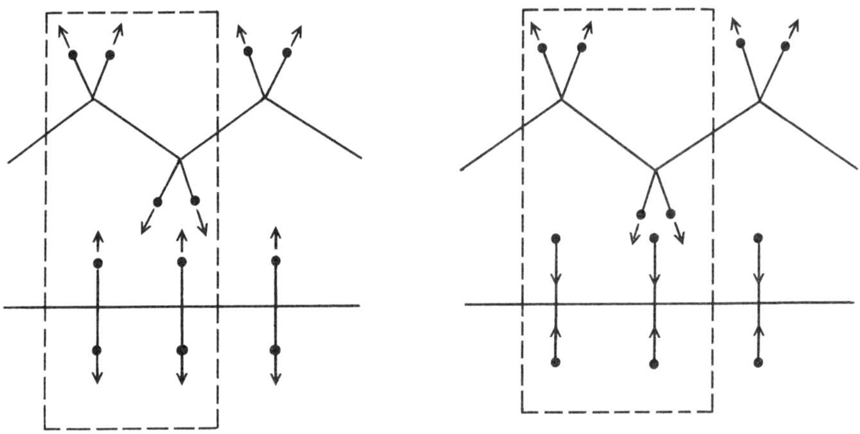

Fig. 2. Combinaison en phase et en opposition de phase de la valence $\nu(CH_2)$ symétrique dans la maille du polyéthylène.

correspondant à ce mouvement. Dans son travail maintenant classique sur le polyéthylène, Koenig a étudié des composés de haute cristallinité refroidis à 73 K et a montré que certaines bandes se dédoublent comme prévu. De plus, le système complexe de bandes situé vers à $\Delta\nu = 1440$ et $1415\ cm^{-1}$ a été expliqué par un effet de corrélation important.[2]

Le refroidissement d'un échantillon entraîne une légère réduction de la largeur de bande à mi-hauteur (déjà très faible en effet Raman – souvent 3 ou 4 cm^{-1} environ) ainsi qu'une diminution de l'intensité des bandes chaudes dans la région des basses fréquences. Ce dernier effet, bien qu'utile expérimentalement, peut prêter à confusion. De plus des changements de structure peuvent être induits par l'abaissement de la température.

A Southampton, nous avons appliqué la technique des hautes pressions à ce problème. La pression ne change ni la largeur des bandes ni l'intensité des bandes chaudes mais simplement maintient les molécules rapprochées, et donc augmente l'effet de corrélation. De plus, cet effet peut être suivi sur une gamme de pressions étendue. Nous trouvons que certaines bandes du polyéthylène se dédoublent sous l'influence de la pression, l'éclatement étant d'environ 1 cm^{-1}/Kbar. Des valeurs de cet ordre de grandeur ont été trouvées par ailleurs pour des déplacements de fréquence des modes de réseau de sels d'oxyanions minéraux.[3]

Le polypropylène dans sa forme isotactique α a quatre chaînes par cellule unitaire et son spectre Raman devrait également présenter des éclatements par effet de corrélation. La symétrie de la cellule unitaire est telle que chacun des modes du groupe linéaire (mode identifié par sa coordonnée dans une chaîne considérée comme isolée) doit donner lieu à deux bandes Raman pour l'échantillon cristallin. Il est probable que l'éclatement est très faible et ne sera donc pas résolu mais que l'anisotropie de la lumière diffusée et en particulier les rapports de dépolarisation seront la résultante des différents comportements anisotropiques des composantes du multiplet de corrélation. Ainsi, pour de tels polymères, les mesures anisotropiques effectuées avec des résolutions normales peuvent être inutilisables, ou du moins prêter à confusion.

Nous avons enregistré les spectres Raman du polypropylène de haute cristallinité α à la température ambiante, à 90 K et à 4 K. Les deux derniers spectres sont essentiellement identiques. La comparaison avec les résultats infrarouges (*y* compris le dichroïsme – qui n'est pas affecté par l'effet de corrélation de la même façon que le spectre Raman) et avec l'analyse en coordonnées normales de Tadokoro *et al.*[4] nous a permis de proposer une explication détaillée du spectre et de montrer que les bandes Raman à 1255/1220 ($A + E$) 1170 (A), 1154 (E), 1103 (E), 1000 (A), 974 (A), 940 (E), 841 (A), 809 (E), 528 (E), 400 (A), 318 (E) et 250 (A) sont dues à des modes singulets du groupe de symétrie linéaire. La plupart des autres bandes, si ce n'est toutes, sont des multiplets. Les singulets présentent soit un éclatement, soit une modification de la forme de bande, soit un changement de fréquence (ou tout à la fois) quand la température s'abaisse, par exemple la bande de classe E située à 528 qui a une largeur à mi-hauteur de 10 cm^{-1} à 25C donne à 90K ou à 4K deux bandes séparées par 4,5 cm^{-1}.

Une chaîne de polypropylène isolée appartient au groupe de symétrie linéaire C_3, qui n'a aucune proprieté en commun avec le groupe d'espace C_{2h}^6. Dans ces conditions, la dégénérescence des vibrations de classe '*e*' du groupe linéaire est levée. En fait, les paires '*e*' donnent lieu à deux vibrations fondamentales dans le groupe d'espace, elles-mêmes composées de quatre fréquences influencées par l'effet de corrélation. Deux sont actives en Raman et une des deux autres en infrarouge (B_u). Les vibrations de classe '*a*' éclatent en un doublet Raman et un singulet infrarouge dans le cristal. En résumé nous avons:

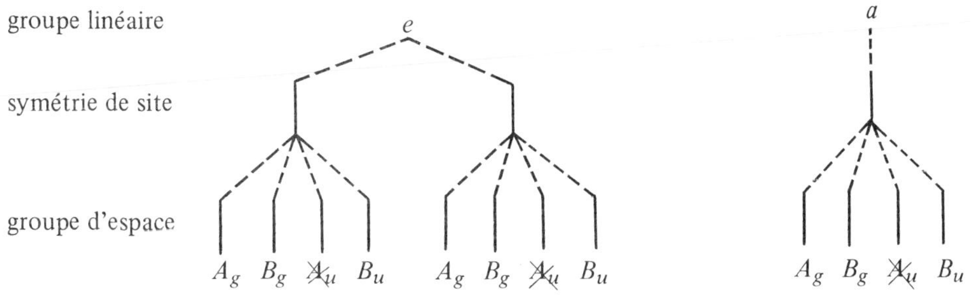

Nous avons obtenu l'éclatement de quelques vibrations de classe '*e*', actives en Raman ce qui montre que la symétrie est effectivement modifiée en passant du groupe linéaire au groupe d'espace.

Nous espérons mettre en évidence des éclatements par effet de corrélation en utilisant la méthode des hautes pressions pour le polypropylène α-isotactique.

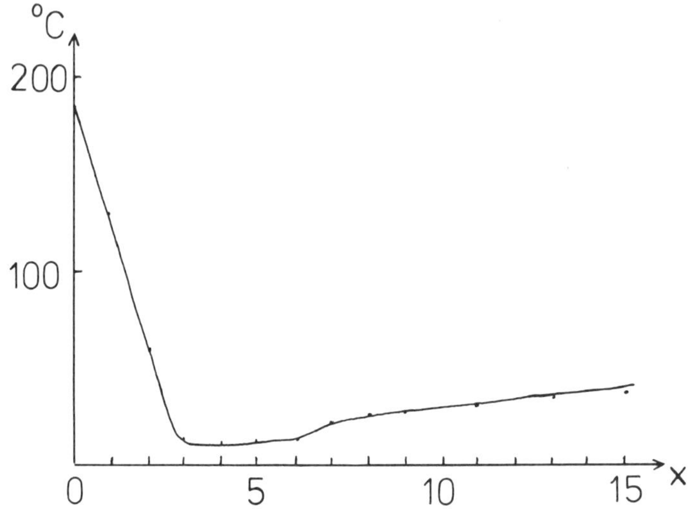

Fig. 3. Point de fusion des paraffines normales $\left[-CH_2-\underset{\underset{CH_3}{|}}{\underset{(CH_2)_x}{\underset{|}{CH}}}-\right]_n$ en fonction de *x*.

LE SPECTRE RAMAN DES POLYOLEFINES D'ORDRE PLUS ELEVE

Nous avons récemment éxaminé les spectres infrarouges et Raman des polyoléfines linéaires isotactiques dérivés du polypropylène. Les points de fusion surprenants caractéristiques de ces matériaux ont reçu seulement un commencement d'explication; ainsi les résultats de rayons X par Turner-Jones *et al.*[5] suggèrent que le squelette de ces polymères est hélicoïdal et que la chaîne latérale est plane dans quelques cas: l'intéraction chaîne latérale–chaîne latérale accroît alors le degré de cristallinité, ce qui explique la variation du point de fusion (Fig. 3). Il semble cependant difficile d'expliquer pourquoi le poly [3-méthylbutène-1] fond à 310°C et le poly [4-méthylpentène-1] à 250°C alors que les espèces non méthylées correspondantes ont des points de fusion très bas. Nous avons enregistré les spectres Raman sur de larges intervalles de température et avons observé une amélioration nette de la qualité spectrale pour les espèces fondant le plus bas quand on les refroidit à des températures cryogéniques. Cela signifie qu'un type d'interaction chaîne–chaîne régulière (sinon une véritable cristallinité) s'établit à froid. Cette propriété a également été observée pour le polypropylène isotactique; lorsqu'on chauffe l'échantillon, le spectre commence à se détériorer *au-dessous* du point de fusion mais conserve quelques traits caractéristiques de la structure cristalline légèrement *au-dessus* du point de fusion et s'évanouit dans un spectre diffus qui n'est pas sans ressembler à celui du polypropylène atactique à haute température. La comparaison des spectres des polyoléfines d'ordre élevé avec celui du polyéthylène nous a convaincus que certaines chaînes latérales doivent avoir la structure zig-zag plane et que ces chaînes s'associent entre elles en s'orientant d'une façon comparable à celle du polyéthylène. Le spectre du poly(octadécène-1) par exemple présente des bandes Raman avec des fréquences convenables et un profil caractéristique d'un polyéthylène dans le même état d'association que la cellule unitaire du polyéthylène cristallin.* Par suite on peut attendre que le point de fusion s'élève, ce qui est en fait le cas.

Pour le polyhexène, nous pensons que l'arrangement zig-zag plan ne peut pas s'effectuer et que l'association entre chaînes latérales n'est certainement pas du même type que dans le polyéthylène, à cause de la petite longueur de la chaîne et de l'effet stérique des groupes méthyles. Dans les polyoléfines inférieures, la structure hélicoïdale du squelette et l'interaction entre chaînes et groupes latéraux sont la cause du maintien de la cristallinité et des points de fusion élevés.

SPECTRES RAMAN DES COPOLYMERES

L'examen des spectres Raman d'une série étendue de polyéthylènes, polyoléfines supérieures, et aussi quelques copolymères séquencés suggère que l'apparence du spectre est très sensible, en général et en détail, au type d'interactions chaîne–chaîne mis en jeu. Si l'interaction est faible, ou désordonnée, le profil de bande dans la région 1450 cm^{-1} se modifie de façon significative, à cause probablement de la rupture des interactions entre les multiplets de corrélation et de modifications des résonances de Fermi.

La présente étude concerne les copolymères d'éthylène et de propylène pour une large gamme de concentrations en polyéthylène (3 – 80%). Elle conduit à ces conclusions:

1. Dans les polyalkènes, le propylène se polymérise en des fragments de conformation isotactique; et même pour 15% seulement d'éthylène, dans un système dont on sait qu'il est séquencé, les fragments éthyléniques sont de longueurs suffisante pour pouvoir former des zones cristallines analogues à celle du polyéthylène lui-même. Cette caractéristique a été démontrée par Snyder[6] qui a comparé les spectres Raman de paraffines comportant une ou deux chaînes par cellule unitaire.
2. Dans les copolymères désordonnés (éthylène, propylène) nous trouvons que les bandes associées aux fragments polyméthyléniques ne sont que très peu caractéristiques d'associations interchaînes. Les élastomères propylène-éthylène, (c'est-à-dire des échantillons dans lesquels la tacticité du polypropylène est incontrôlée) ont été étudiés par Schreier et Peitcher;[7] ces auteurs trouvent également une corrélation phénoménologique entre la forme des spectres Raman et la cristallinité estimée par rayons X.

INFLUENCE DE L'ORIENTATION SUR LES SPECTRES RAMAN DES POLYMERES

La diffusion Raman est normalement anisotropique et l'anisotropie peut être mesurée quantitativement par l'utilisation de sources polarisées, d'analyseurs entre l'échantillon et le spectromètre, ainsi que le contrôle approprié de l'orientation.[8] L'anisotropie dépend de l'arrangement géométrique utilisé et de la symétrie de la coordonnée normale donnant lieu à la vibration.

Théoriquement, l'analyse du comportement anisotropique permet donc de définir de façon sûre la symétrie du mode correspondant à une bande donnée. Il existe de nombreux exemples de cette approche appliquée à des cristaux mais dans le domaine des polymères, des problèmes expérimentaux importants se posent et seulement quelques applications ont été réalisées à ce jour (Polyéthylène: Carter,[9] Hendra;[10] Poly(téréphtalate d'éthylène): Derouault et Hendra[11]). Les complications principales sont:

---

* Des structures de ce type ont été etablies pour le poly(décène-1) et les dérivés de poids moléculaires plus élevés analogues.[5]

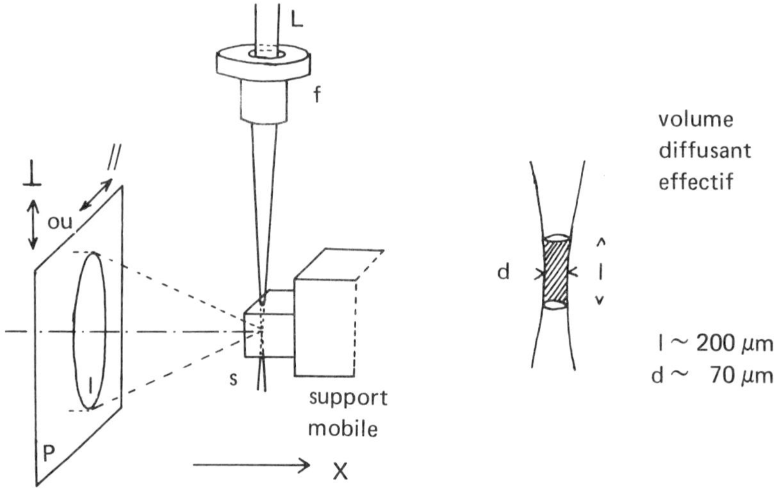

Fig. 4. Arrangement expérimental. L = faisceau laser; f = objectif de focalisation; S = échantillon; P = analyseur; l = objectif de transfert.

1. Perte de polarisation de la lumière due à des discontinuités de l'index de réfraction à l'intérieur de l'échantillon.
2. Le manque de résolution des multiplets de corrélation entraîne souvent un comportement hybride des bandes.
3. La distorsion du vecteur électrique d'une radiation à l'intérieur d'un échantillon par effet dichroïque.

Au cours de l'analyse du comportement anisotropique de specimens orientés, le vecteur électrique de la source est toujours en coïncidence avec l'un des axes optiques de l'échantillon; de cette façon, l'effet dichroïque (3) ne perturbe pas les mesures. Un échantillon très convenable est constitué par un film épais (350 μm) de téréphtalate d'éthylène étiré uni-axialement (rapport 4,5:1) car le degré de cristallinité est très élevé et la cellule unitaire ne comporte qu'une seule chaîne, ce qui exclut la présence de multiplets de corrélation (2). Malgré l'imperfection de l'orientation des chaînes et la perte (1) de polarisation due aux inhomogénéités, notre analyse montre que:

1. Les tenseurs de polarisabilité dérivée se classent en deux types
   (a) les éléments diagonaux sont les plus élevés: mode A_g
   (b) les éléments rectangles sont les plus élevés: mode B_g

et ainsi, quelques points de l'attribution antérièure sont confirmés ou complétés.
2. L'élément diagonal du tenseur des bandes A_g correspondant à la variation de polarisabilité le long de l'axe de la chaîne est toujours le plus élevé (l'élément correspondant du spectre de fluorescence également).

Cependant on observe encore un effet anisotropique sur le spectre d'un échantillon dont l'axe optique forme un angle quelconque avec le vecteur électrique du faisceau laser. En principe, si les propriétés anisotropiques du matériau étudié sont connues, de tels spectres doivent contenir une information sur l'orientation de la molécule examinée par rapport aux axes de références expérimentaux.

Nous avons par suite enregistré le spectre de quelques échantillons commerciaux d'orientation inconnue pour le poly(téréphtalate d'éthylène) et poly(4-méthylpentène-1). L'utilisation de l'arrangement géométrique (Fig. 4) met en évidence des modifications dans l'orientation des molécules à l'interieur du matériau avec une résolution spatiale de 25–30 μm et avec une reproductibilité de la position quelque peu meilleure que cette valeur. Nous avons montré de plus que, lorsqu'on coupe un échantillon de poly(téréphtalate d'éthylène) avec une lame fine, l'orientation des chaînes est modifiée à cause de la déformation mécanique et que cette réorientation s'étend jusqu'à environ 150 μm de l'arête d'un film de 350 μm d'épaisseur. Ces mesures sont réalisables sur des polymères transparents d'un volume assez important, et donc permettent de mettre en évidence des modifications à l'intérieur de fibres, de films et de specimens moulés.[12]

Une première tentative d'analyse détaillée de la diffusion Raman produite par un échantillon dont les axes moléculaires forment des angles quelconques avec le vecteur électrique du faisceau laser a été proposée par Bower et Ward dans une communication à un congrès récent.[13] Leur analyse suppose cependant que l'échantillon est homogène et parfaitement orienté.

Deux problèmes se presentent pour interpréter nos résultats. Il est nécessaire de tenter de prévoir l'influence de l'effet dichroïque sur le faisceau laser incident et sur la lumière Raman émergente, et il faut analyser les données en termes d'orientation par rapport aux axes de référence expérimentaux. Le premier de ces problèmes semble impossible à résoudre car, par définition, les échantillons ne sont pas homogènes et l'influence du dichroïsme est donc impossible à prévoir théoriquement. L'analyse du spectre à différentes pénétrations dans l'échantillon doit cependant permettre de résoudre le problème expérimentalement de proche en proche.

L'analyse des données de diffusion pour déduire une orientation nécessite des renseignements sur au moins deux bandes Raman dont on sait par avance qu'elles sont dues à des vibrations de symétrie différentes.

D'éventuels progrès dans ce domaine seront rapportés verbalement au cours de la réunion.

REFERENCES

1 M. J. Gall et P. J. Hendra, *Spex Speaker* **16**, 1 (1971).
2 F. J. Boerio et J. L. Koenig, *J. Chem. Phys.* **52**, 3425 (1970).
3 M. Nicol et J. F. Durana, *J. Chem. Phys.* **54**, 1436 (1971).
4 H. Tadokoro *et al., J. Chem. Phys.* **42**, 1432 (1965); W. L. Peticolas *et al., Appl. Phys. Letters* **18**, 87 (1971); M. Tasumi et T. Shimanouchi, *J. Chem. Phys.* **43**, 1245 (1965).
5 A. Turner-Jones, *Makromol. Chem.* **71**, 1 (1964).
6 R. Snyder, *J. Mol. Spectry.* **31**, 464 (1969).
7 G. Schreier et G. Peitcher, *Z. Anal. Chem.* **258**, 199 (1972).
8 T. R. Gilson et P. J. Hendra, *Laser Raman Spectroscopy*, Wiley, London, 1970.
9 V. Carter, *J. Mol. Spectry.* **34**, 356 (1970).
10 M. J. Gall, P. J. Hendra *et al., Spectrochim. Acta* **28A**, (1972).
11 P. J. Hendra, J. Derouault, V. Ellis, H. A. Willis et M. E. A. Cudby, *Industria i Chimica,* (sous presse); *European Polymer J.* (à paraître).

12 J. Derouault, P. J. Hendra, H. A. Willis et M. E. A. Cudby, *Chem. Commun.* (sous presse).
13 I. M. Ward et D. I. Bower, Communication présenteé à la réunion *Radiation Scattering by bulk polymers*, Societé Européanne de Physique, Strasbourg, May 1972.

34. THE RAMAN EFFECT FROM NUCLEIC ACIDS

W. L. Peticolas
Department of Chemistry, University of Oregon, Eugene, Oregon, USA

INTRODUCTION

The nucleic acids are the molecules which carry and transmit genetic information used in the synthesis of the cellular proteins. A great deal has been written about these materials and we will review here only those features necessary for an interpretation of the Raman spectra. Nucleic acids may be divided into two distinct classes depending upon the structure of their constituent sugars. Ribonucleic acid (RNA) contains D-ribose while deoxyribonucleic acid (DNA) contains 2′-deoxy-D-ribose. The general primary bonding of the nucleic acid chain consists of alternating sugar and phosphate groups bonded by means of ester linkages at the 3′ and the 5′ hydroxy positions of the sugar. The backbone of the nucleic acid chain is thus a copolymer of a ribose sugar and phosphoric acid. Attached to each of the sugar groups in the one position is a nitrogenous base. Normally this base will consist of one of the following: adenine (A), guanine (G), cytosine (C), uracil (U) or 5-methyluracil called thymidine (T). It is a general rule that U is found only in RNA and T is found only in DNA. Sometimes instead of guanine, it is more convenient to study hypoxanthine which is identical to guanine except that it lacks the pendant $-NH_2$ group. Polymers containing hypoxanthine are often much easier to work with than those containing guanine because the guanine base interacts very strongly both with itself and with other bases.

In living cells, the nucleic acids contain all four types of bases arranged in an informational array in order to direct the biological processes of the cells. However, model nucleic acids can be made in which only one type of base is attached to each sugar in the chain. These materials are called polynucleotides. For a given base, the polynucleotide has the following names: adenine, polyadenylic acid or polyA; uracil, polyuridylic acid or polyU; thymidine, poly-d-thymidylic acid or polyT; guanine, polyguanylic acid or polyG; cytosine, polycytidylic acid or polyC; hypoxanthine, polyinosinic acid or polyI. The small 'd' in poly-d-thymidylic acid reminds us that with thymidine we normally have only the deoxyribose sugar.

When polynucleotides are placed in an aqueous environment they are capable of undergoing a great number of mutual interactions which give rise to a number of special conformations and ordered structures. Indeed one of the major applications of Raman

Table 1. Structure of helical polynucleotides from X-ray diffraction measurements.

Polymer	Form	810–815 cm^{-1} Raman band	Pitch Å	Residues per turn	Translations per residue	Rotation per residue	Tilt angle
Na DNA fibre humidity = 75%	A	yes	28.2	11	2.55	32.7°	70°
Na DNA fibre humidity = 92%	B	no	34.6	10	3.46	36°	90°
Li DNA fibre humidity = 66%	B	no	33.7	10	3.37	36°	88°
Li DNA fibre humidity = 30%	C	no	31.0	9.3	3.32	39°	84°
poly d(A–T)	B	no	33.4	10	3.34	36°	90°
Na RNA poly rI · poly rC poly rA · poly rU	A	yes	30.9	11	2.73	32.7°	70°

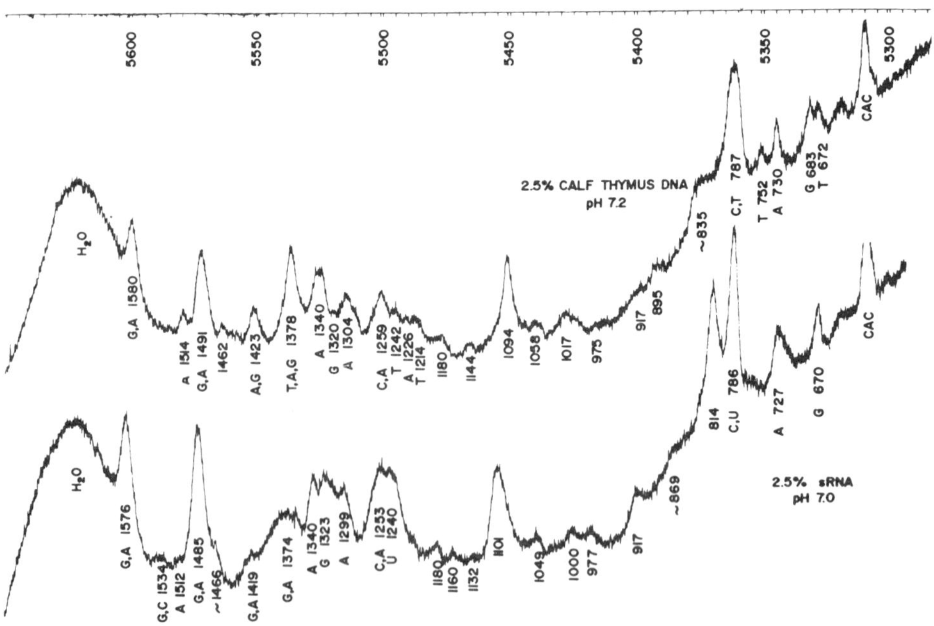

Fig. 1. Raman spectra of calf thymus DNA in 2.5% aqueous solution at pH 7.2 (upper) and yeast transfer RNA in a 2.5% aqueous solution at pH 7 (lower).

spectroscopy to the study of nucleic acids is the illustration of these structures and the forces which stabilize them. Several limiting structures can be distinguished:

1. completely disordered or random coil structures,
2. single stranded, self-stacked and partially helical (but non-hydrogen-bonded) structures, and

3. multistranded hydrogen-bonded helical stuctures.

Among the latter, the double stranded (duplex) helices are the most prevalent and the most important.

The actual geometry of the double helices has been worked out in detail from X-ray analysis of the fibres[1, 2] of nucleic acids. The conformational geometries of the double stranded helices can be classified into three general classes which have been designated as the A-form, the B-form and the C-form. Table 1 gives the geometry of these helical structures as determined by X-ray diffraction. The presence or absence of a strong Raman band at 810–815 cm^{-1} is also noted. This band will be discussed in the next section.

RAMAN SPECTRA OF DNA AND RNA – IDENTIFICATION OF THE A- AND B-FORMS BY LASER RAMAN SCATTERING

We will begin our discussion of the Raman spectra of nucleic acids with the comparison of the Raman spectra of RNA and DNA in dilute solution with the spectra of DNA fibres where the geometry of the helical conformation is known. In this section we will only be interested in the most fully formed double helical structures of DNA and RNA. Although RNA almost always exists in living cells as a single stranded linear polymer, this single chain folds back upon itself to form double stranded duplex helical regions. Such double helical structures form in single stranded RNA molecules spontaneously in aqueous solution at neutral pH and room temperature. Similarly, DNA exists in aqueous solution as a double stranded duplex helix, under ordinary conditions.

Figure 1 shows a comparison of the Raman spectrum of aqueous solutions of a purified yeast t-RNA mixture and calf thymus DNA.[3, 4] These spectra are typical of these polymers in their helical forms in aqueous solution. The region to which we wish to draw attention in these spectra is the region from 750 cm^{-1} to 850 cm^{-1}. In particular it should be noticed that in RNA there is a sharp band at 814 cm^{-1} and in DNA a band at 835 cm^{-1}. The 814 cm^{-1} band was first discovered to be present in all ribonucleic acid structures[3,5–7] including homoribopolymers when these structures are in an ordered helical form. When raising these solutions of ribonucleic acid materials above their melting point it was found that this sharp, distinct band at 814 cm^{-1} moved to a broader weaker diffuse band at 797 cm^{-1}.[3] Since this band is present in all ordered ribonucleic acid structures and since it is highly polarized it has been assigned to the phosphate–sugar backbone and particularly to the symmetric phosphate diester stretch[3, 6] although this assignment is in disagreement with the calculations and Raman spectra of simple phosphate diesters.[8] This band is never found in aqueous solutions of DNA or deoxy-homopolymers even when these polymers are known to be in a helical form. This fact is illustrated by the Raman spectrum of helical calf thymus DNA shown in Fig. 1, where the band at 810–815 cm^{-1} is obviously absent.

Recently it has been found that the reason for the presence of this band at 810–815 cm^{-1} is due to the fact that this vibration arises from the phosphate–sugar backbone only when the backbone conformations of the nucleic acid is in the A-form.[4] This conclusion can easily be established by an examination of Fig. 2. This figure shows the spectra of a low salt fibre sodium DNA at 98% and 75% relative humidity. At the former humidity, the helix is known to be of the B-form while in the latter it is known to be of the A-form. The resemblance between the solution spectra of Fig. 1 and the fibre spectra of Fig. 2 is striking and shows quite conclusively that in aqueous solutions RNA is always in the A-form and DNA is always in the B-form. It also shows that the band at 810–815 cm^{-1}

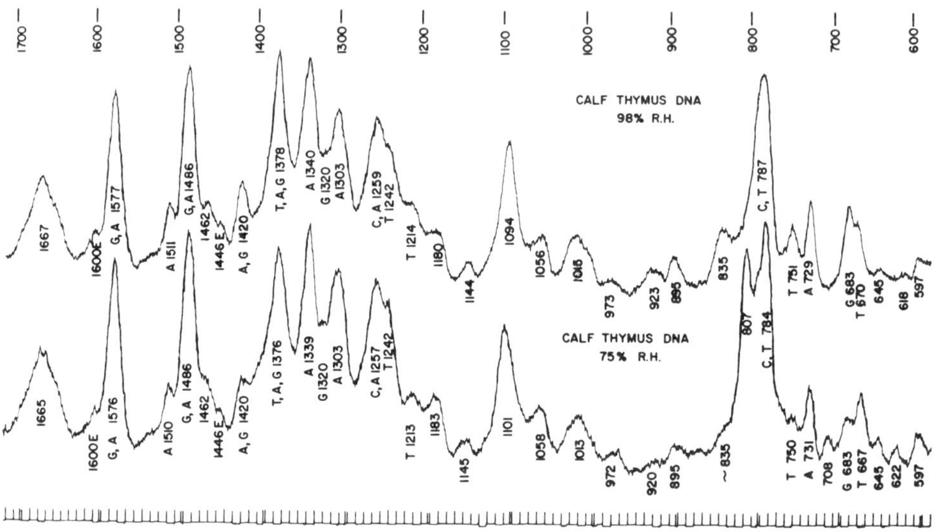

Fig. 2. Raman spectra of a fibre of the sodium salt of calf thymus DNA at 98% (upper) and 75% (lower) relative humidity. These spectra are completely reversible with changing humidity.

Table 2. The ratio of the peak height of that Raman band at 810–815 cm^{-1} to the Raman band at 1100 cm^{-1} as a measure of the fraction of the backbone bonds in the A-type conformation in nucleic acids.

Molecule	T (°C)	pH	$\frac{I\,(810–815\ \text{cm}^{-1})}{I\,(1098\ \text{cm}^{-1})}$	% A-type conformation[d]
PolyA–PolyU[a]	15	7.0	1.7	100
PolyA–PolyU[a]	65	7.0	0	0
PolyG–PolyC	95	7.0	1.4	82
PolyI–PolyC	25	7.0	1.5	88
PolyC	18	6.4	1.7	100
PolyC	22	7.0	1.2	70
PolyC	75	7.0	0	0
PolyG	19	7.0	0.6	35
PolyI	9	7.0	1.2	70
PolyA[b]	23	7.0	1.4	82
PolyA[b]	80	7.0	0	0
PolyU[b] (Mg^{++})	0	7.0	1.4	82
PolyU[a] (Mg^{++})	20	7.0	0	0
t-RNA[a]		7.0	1.6	94
DNA[c,e]	—	—	1.64	96

[a] Ref. 3.
[b] Ref. 5.
[c] Ref. 9.
[d] A-form as described in refs. 1 and 2.
[e] Fibre at low humidity of 75%.

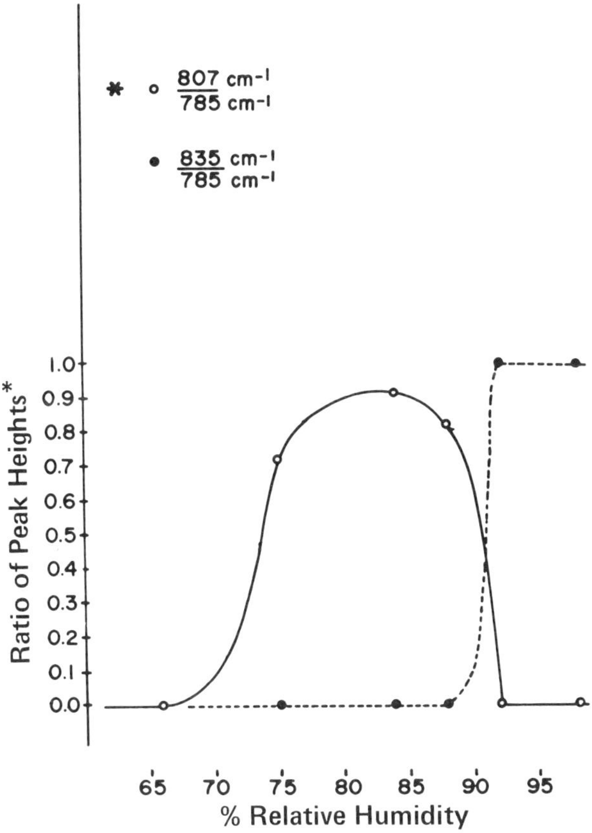

Fig. 3. Intensity ratio of the Raman bands of DNA fibres as a function of the relative humidity of the air surrounding the fibre. The 807 cm^{-1}/785 cm^{-1} ratio gives approximately the amount of the A-form while the 835 cm^{-1}/785 cm^{-1} ratio gives an estimate of the amount of the B-form in the fibre.

may reasonably be used as a measure of the amount of A-type structure a nucleic acid or polynucleotide can possess.

Table 2 gives an estimate of the fraction of the backbone structure which is of the A-form for a large number of single and double chain polynucleotides and nucleic acids.[9] This estimate is obtained from the ratio of the 810–815 cm^{-1} band to the conformationally independent phosphate band at 1100 cm^{-1}.

Figure 3 shows the ratio of the 807 cm^{-1} band in the DNA fibre to the 785 cm^{-1} band as a function of the relative humidity of the fibre sample. This ratio may be taken as an estimate of the relative amount of the helix in the A-form. We see that as the relative humidity is raised past 75% the polymer is entirely in the A-form but that this form decreases above 90% relative humidity. Similarly the ratio of the 855 cm^{-1} band to the 785 cm^{-1} band may be taken as a measure of the B-form and we see that the B-form begins to occur as the humidity increases above 90%.

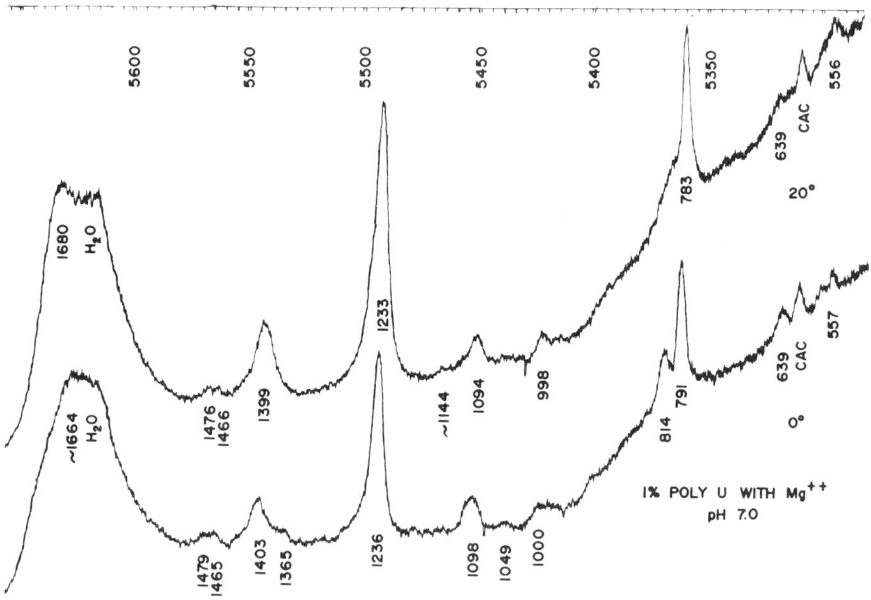

Fig. 4. Raman spectra of a 2% aqueous solution of polyU in the presence of Mg^{++} ions at 0° and 20°C. The decrease in the 1236 cm^{-1} band is due to base stacking interactions while the increase in the 814 cm^{-1} band is due to ordering of the single chain helix into the A conformation.

CONFORMATION DEPENDENCE OF THE RAMAN SCATTERING INTENSITIES FROM POLYNUCLEOTIDES AND NUCLEIC ACIDS

In the previous section we have seen how the frequencies of the backbone vibrations change with the type of helix. In this section, we wish to show changes which occur upon disruption of all or part of the order in a helical structure. There are three specific types of changes in order which can each be seen separately by Raman spectroscopy. These are

1. changes in the backbone disorder,
2. changes in the base stacking interactions, and
3. changes in the number of Watson–Crick hydrogen bonded base pairs.

Due to the interactions between the electronic states of the aromatic, highly polarizable nitrogenous bases, it is to be expected that there will be changes in the intensities of the Raman bands.[3, 5] If the Raman bands are more strongly coupled to the low lying electronic levels of the base molecules, then one would expect a decrease in the Raman intensity upon increasing order of the helix leading to greater base–base interactions. This phenomenon has been called Raman hypochromism.[3, 5] A striking example is the self stacking of the single chain polynucleotide, polyU which is shown in Fig. 4. Here the 1236 cm^{-1} band is considerably decreased in intensity when the polymer is cooled to 0°C at pH 7 in the presence of Mg^{++} ions.[3] Note also the presence of the 814 cm^{-1} band at low temperature. This shows that the base stacking interactions which lead to changes in the electronic and Raman intensities also lead to single chain helix formation of the A-form. It has been found that at higher temperatures and in the absence of Mg^{++} ions the Raman spectrum is remarkably invarient to temperature change showing that the observed inten-

Table 3. Raman active vibrations of nucleic acid bases which show substantial changes in Raman intensity upon increasing order of the nucleic acid structure at neutral pH.

Base	Frequency wavenumbers	Intensity change with increasing order
Adenine	725	decreases
	1304	decreases
	1380	decreases
	1520	decreases
Uracil	790	decreases
	1240	decreases
Guanine	670	increases
	1485	decreases
	1580	decreases
Cytosine	790	decreases due to frequency shift
	1256	decreases at pH < 6.8
	1547	decreases at pH < 6.8

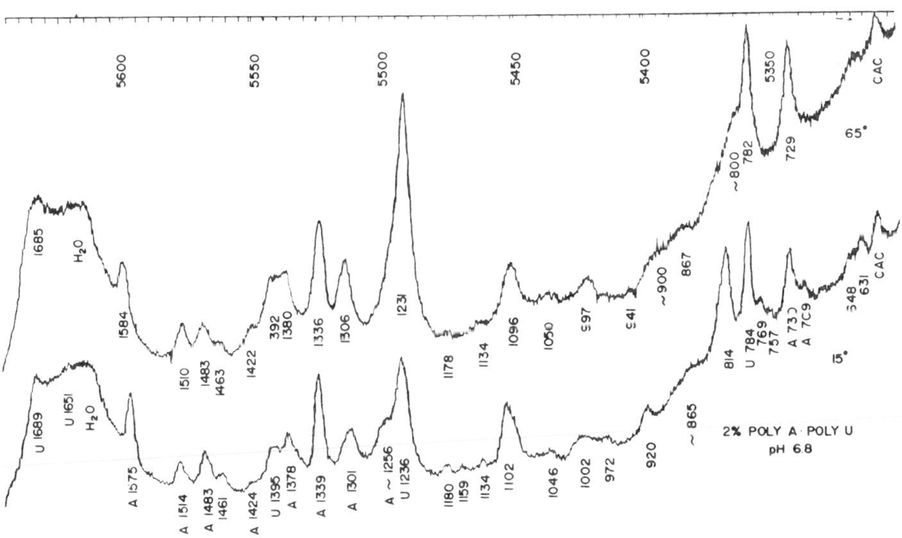

Fig. 5. Raman spectra of 2% solution of polyA–polyU in H_2O at pH 6.8, below and above the melting point of the double helical complex.

sity changes in the Raman spectrum are due primarily to the change in polymer conformation, since polyU is known not to form ordered structures under these conditions.[3, 5]

Many of the Raman bands from the bases are sensitive to stacking interactions and Table 3 gives a list of the prominent Raman bands which change intensity because of base stacking.

Figures 5–8 show the Raman spectra of various double helical complexes, polyA–polyU, polyC–polyG and polyC–polyI. The polyC–polyG double helix is so completely

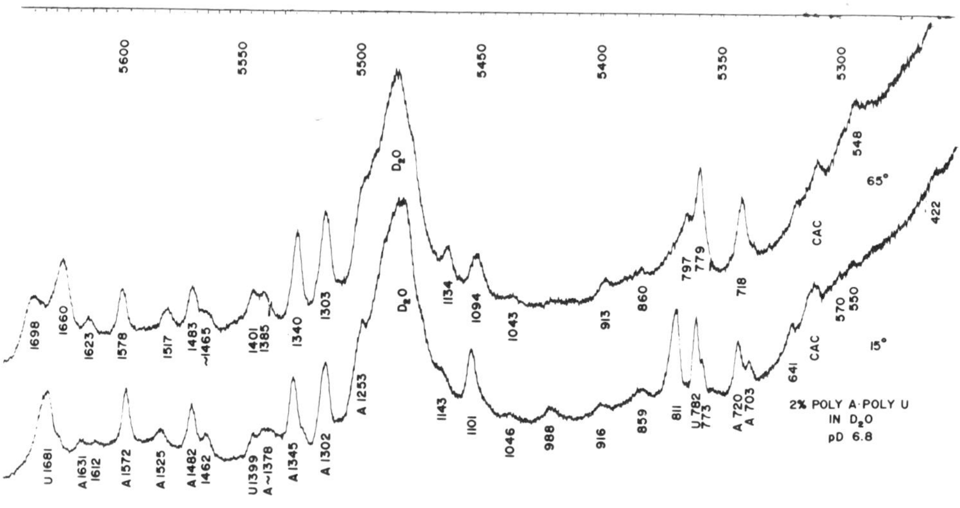

Fig. 6. Raman spectra of 2% solution of the polyA–polyU in D_2O at pD 6.8, below and above the melting point of the double helical complex.

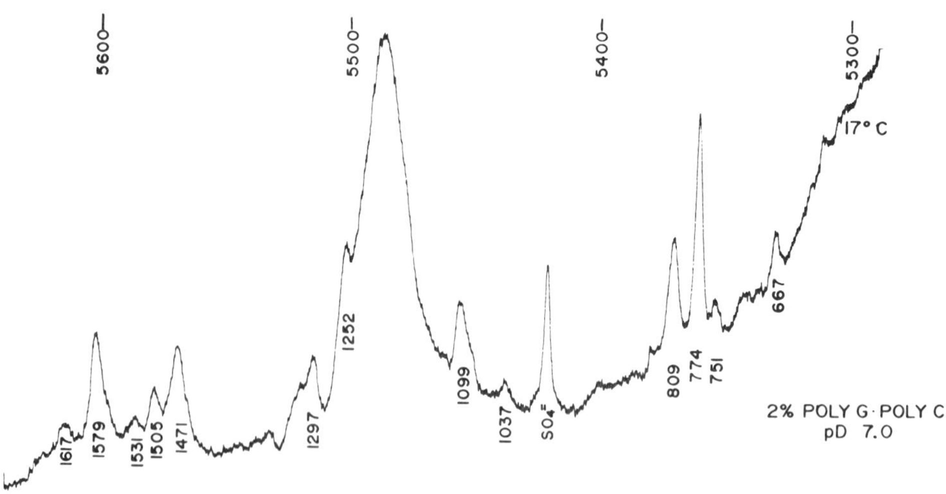

Fig. 7. Raman spectrum of a 2% solution of the polyG–polyC helical complex at pD 7.0. This double helix is so stable that it does not change between 0° and 90°C.

stable that its spectrum does not change in the slightest between 0° and 90°C. However the other two duplex helices melt at an easily accessible temperature and for these helices both the helical and dissociated forms are given. In each figure, the intensity changes listed in Table 3 can be observed.

Finally, it should be noted that there are two cases where the amount of Crick–Watson base pairing can be quantitatively obtained from Raman spectra. Figure 6 shows

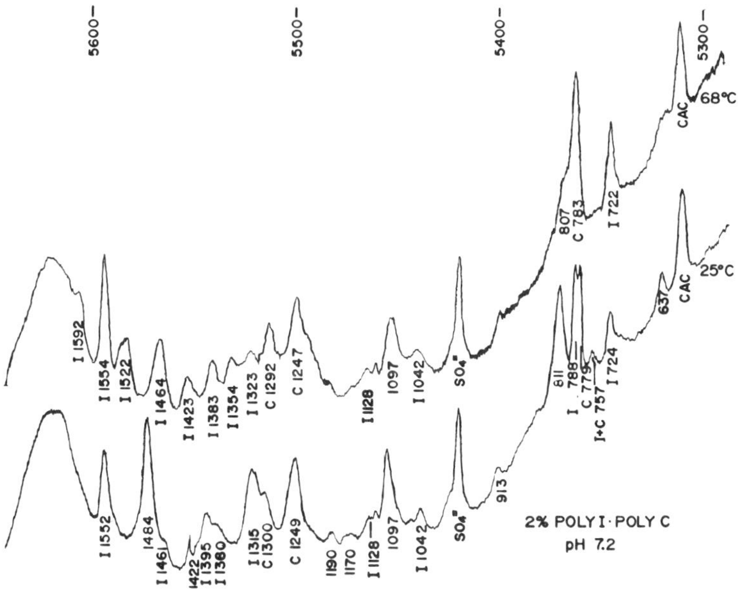

Fig. 8. Raman spectrum of a 2% solution of the polyI–polyC below and above the melting point of the double helical complex, in H_2O pH 7.2.

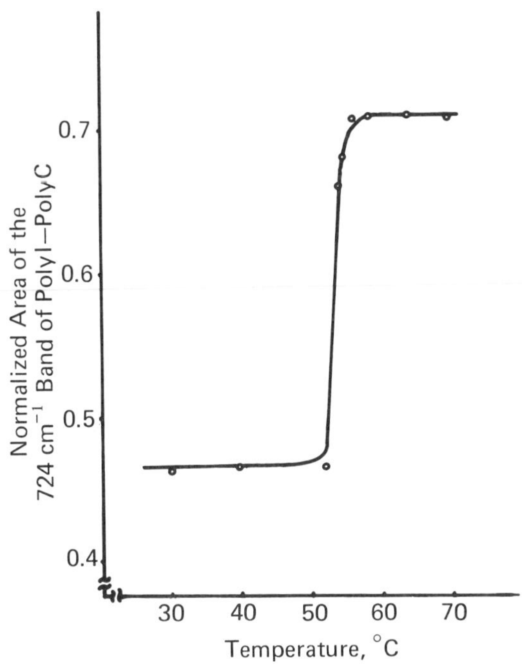

Fig. 9. The intensity of the 724 cm^{-1} vibration of the hypoxanthine base of polyI in the polyI–polyC helical complex. The sharp melting of this double helix at 55°C is evident from the increase in the Raman intensity from this ring vibration.

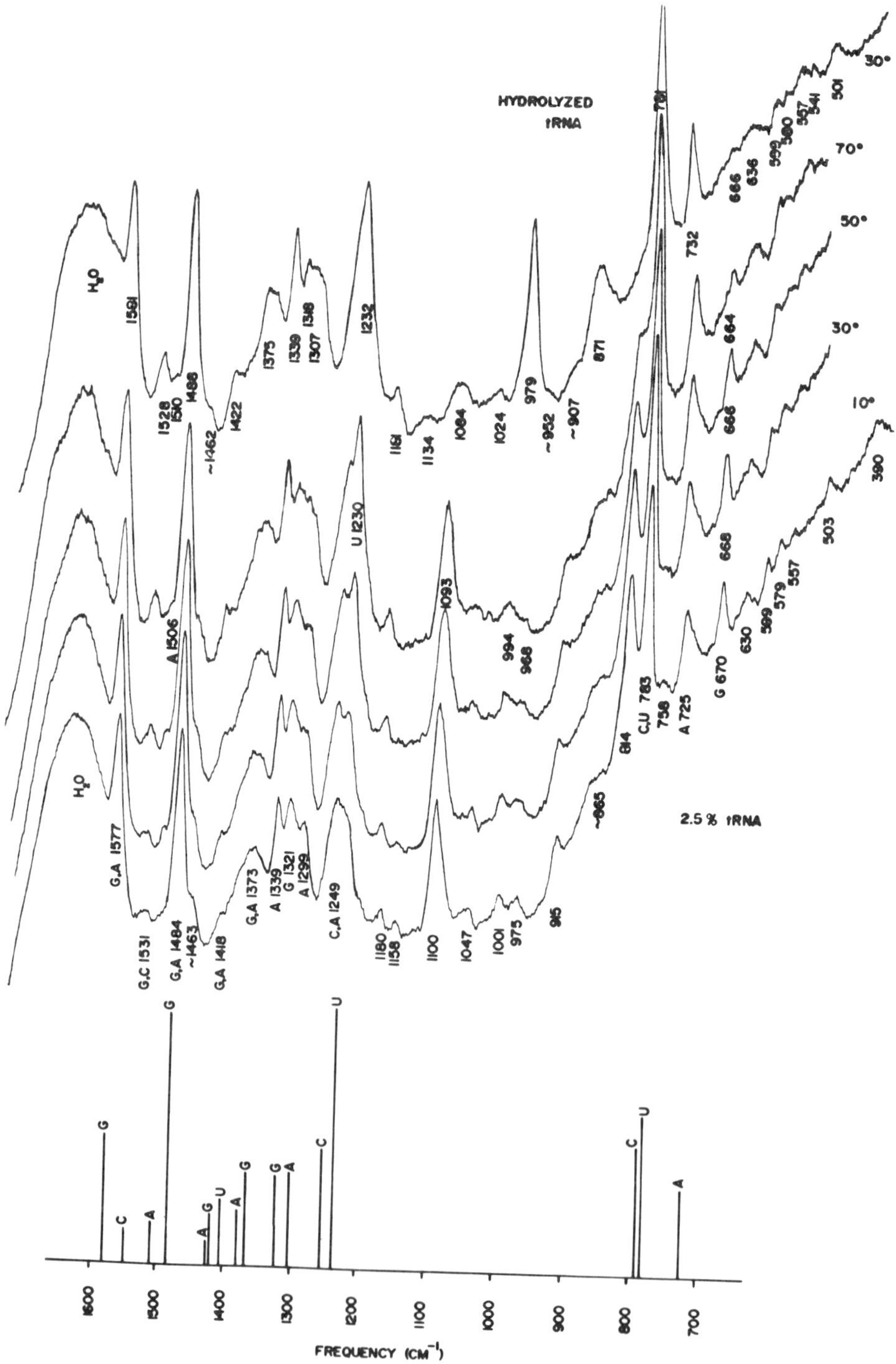

Fig. 10. The spectra of a purified mixture of transfer RNA's from yeast as a function of temperature and alkaline hydrolysis.

the spectrum of polyA–polyU in D_2O. In this solvent the frequencies of the two carbonyl groups of uracil are split into 1698 and 1660 cm^{-1} when the uracil is unhydrogen-bonded. This splitting of the carbonyl frequencies is not observed in H_2O and this may be seen by computer subtraction of the water spectrum of polyA–polyU.[10] Apparently the NH vibration is coupled to the two C=O vibrations in H_2O but the ND vibration which is at a much lower frequency is not. However, as may be seen in Fig. 6 the decoupled carbonyl vibrations in uracil in D_2O become recoupled when the bases are hydrogen-bonded, so that only a single frequency at 1681 cm^{-1} is observed. By taking a linear sum of these two spectra one can fit partially double helical structures and determine the number of Watson–Crick base pairs of the type U=A.

In the polyI–polyC helix, the 1454 cm^{-1} band of the hypoxanthine base shifts to 1484 cm^{-1} upon base pair formation.[9] This Raman band has been assigned to the poly-amide II vibration of the *cis*-amide group of the hypoxanthine base. Again using this pair of Raman bands one can determine the number of H=C base pairs formed. Unfortunately there appear to be no strong Raman bands from which one can determine the number of C=G base pairs. This is particularly unfortunate because C=G is the most common base pair in nucleic acids and the hypoxanthine base appears extremely rarely if at all in living cells.

When a double helical structure melts, it is possible to follow these Raman intensity changes as a function of the solution temperature throughout the melting region and to obtain a melting curve for each type of change which we have discussed. Melting curves for the unstacking of A, the unstacking of U, the disordering of the backbone chain and the dissociation of the hydrogen-bonded base pairs have all been recorded and published elsewhere.[3] Figure 9 shows the melting of the polyI–polyC helix as measured by the Raman hypochromism of the 724 cm^{-1} band of polyI. The sharp melting behaviour of this helix at about 55°C is easily seen.

Finally we wish to show a detailed melting of transfer RNA in Fig. 10. As an example of the completely disordered structure, the spectrum of hydrolysed t-RNA is also shown. Here, as the temperature is raised and the double helical sections are melted out, all of the types of changes discussed can be observed. However since there is little change in the guanine bands it appears that this base restacks upon melting. A detailed analysis of this set of spectra will be published elsewhere.[11]

REFERENCES

1 R. R. Davis, *Ann. Rev. Biochem.* **36**, 321 (1967).
2 S. Arnott, *Progr. Biophys. Mol. Biol.* **21**, 265 (1970).
3 E. W. Small and W. L. Peticolas, *Biopolymers* **10**, 1377 (1971).
4 S. C. Erfurth, E. J. Kiser and W. L. Peticolas, *Proc. Natl. Acad. Sci. USA* **16**, 938 (1972).
5 W. E. Small and W. L. Peticolas, *Biopolymers* **10**, 69 (1971).
6 G. J. Thomas Jr., G. D. Medeiros and K. A. Hartman, *Biochem. Biophys. Res. Comm.* **44**, 587 (1971).
7 N. Aylward and J. L. Koenig, *Macromolecules* **3**, 590 (1970).
8 T. Shimanouchi, M. Tsuboi and Y. Kyogoku, *Advan. Chem. Phys.* **7**, 435 (1964).
9 K. E. Brown, E. Kiser and W. L. Peticolas, *Biopolymers* **11**, 1209 (1972).
10 E. Kiser and W. L. Peticolas, (unpublished works).
11 E. W. Small, K. E. Brown and W. L. Peticolas, *Biopolymers* **11**, 1855 (1972).

35. THE RAMAN SPECTRA OF ORIENTED ISOTACTIC POLYPROPYLENE

R. T. Bailey, A. J. Hyde and J. J. Kim
Department of Pure and Applied Chemistry, University of Strathclyde, Glasgow, Scotland

INTRODUCTION

It is well known that the dynamical physical properties of polymers are influenced by the local environment of the polymer chains in the bulk solid. Factors such as degree of crystallinity, chain orientation and folding and packing of the polymer chains all have a marked effect on the dynamics of the polymer chain and thus on its physical characteristics. The vibrational Raman spectra of polymers can be used as a convenient molecular probe for observing these effects since the polymer molecule vibrations are strongly influenced by the intermolecular interactions in the solid. Isotactic polypropylene (IPP) was chosen as a model compound for this work since it is readily obtained in various degrees of crystallinity and of chain orientation. This chapter describes initial work on two kinds of material, highly oriented and crystalline material and randomly oriented crystalline IPP. One of the ultimate objectives of this work is the study of the interrelationships between crystallinity and chain orientation and their effect on polymer dynamics. In addition, Snyder[1] has recently worked out Raman scattering activities for uniaxially oriented molecules. This system presents the opportunity of evaluating the applicability of these calculations to oriented polymer systems, particularly those which are opaque, to visible radiation.

In IPP the repeating unit consists of three monomer units related to each other by a three fold helix axis. The normal vibrations can be treated under the factor group of the line group which is isomorphous to the point group $C(2\pi/3)$. For crystalline IPP the 77 modes are classified under factor group C_3 into 25 parallel modes (A) and 52 perpendicular modes (E). The A modes will be polarized and the E modes depolarized in the Raman spectrum.

The vibrational spectra of IPP have been extensively studied.[2–7] The normal modes of crystalline IPP have been calculated by Snyder and Schachtschneider[8] and Tadakoro *et al.*[5] No polarized Raman studies have been reported on crystalline IPP either random or oriented but Vasko and Koenig made depolarization studies on molten and smectic IPP.[7]

EXPERIMENTAL

The Raman spectra were obtained using a Raman system developed in our laboratory. This was based on a f/5 300 mm FL double monochromator equipped with 1800 g mm^{-1}

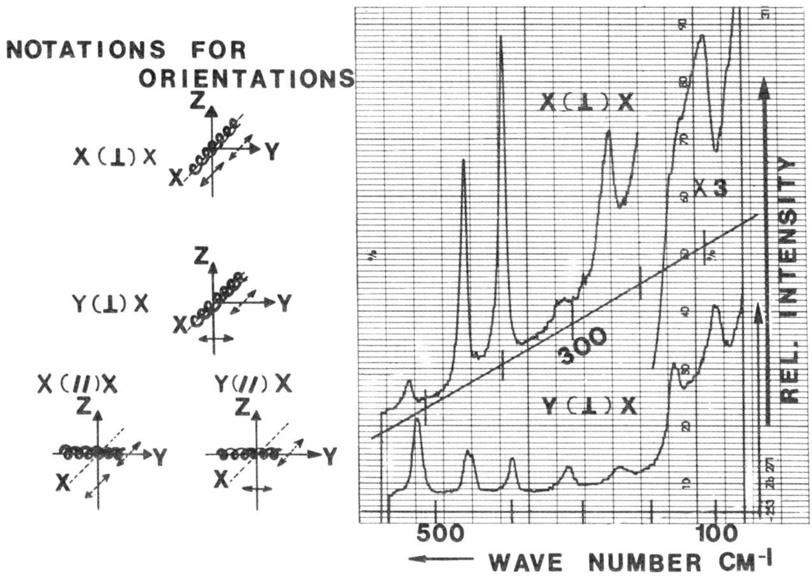

Fig. 1. Raman spectra of oriented polypropylene.

holographic gratings. A Brookdeal phase sensitive system with narrow band tracking filter and a photon counting system were used for detection. Excitation was accomplished with a coherent radiation argon-krypton mixed gas laser.

Most spectra of IPP were recorded using the photon counting system and spectral slit widths of 3–5 cm^{-1}. Wavelength accuracy was about ± 2 cm^{-1}. To date, work has been carried out on two samples of IPP, a highly crystalline and oriented sample obtained from I.C.I. and a highly crystalline sample of monoclinic IPP obtained from Shell. The oriented sample was in the form of a translucent rod, opaque to visible light. X-ray measurements established a crystallinity of greater than 80% and a high degree of orientation along the principal axis. The random IPP was in the form of small (2 mm X 2 mm) pellets again opaque to visible light. With the oriented sample the laser beam was positioned near the edge of the sample to minimize polarization scrambling by the translucent samples.

In Fig. 1 the Raman scattering geometrics used are defined. The laser radiation is directed along the Z axis with its electric vector oriented along the X or the Y axes. The scattered light is observed along the Y axis, at 90° to the laser direction, so it has components in the X and Z directions. The IPP sample was oriented with its unique axis parallel to the X or Y directions. The abbreviated rotation $F(f)F'$ was used to indicate the geometry of the Raman experiment. F denotes the polarization of the laser radiation and F' that of the scattered radiation. The orientation of IPP is identified by f. The sample can be oriented with its unique axis either parallel (||) or perpendicular (⊥) to the Y axis. Standard notation is used in designating the components of the derived tensor in the space fixed coordinate system. Thus for excitation polarized in the F direction, the intensity of the Raman scattered light polarized in the F' direction will be proportional to $\alpha^2_{FF'}$.

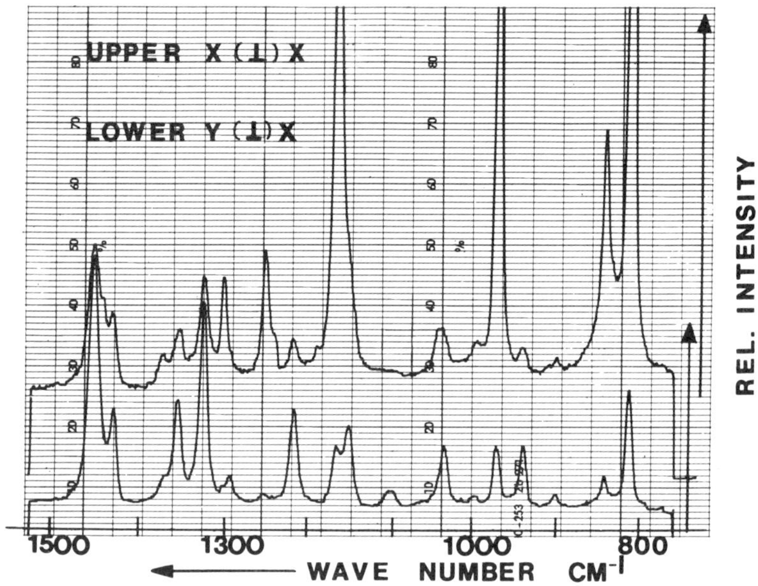

Fig. 2. Raman spectra of oriented polypropylene.

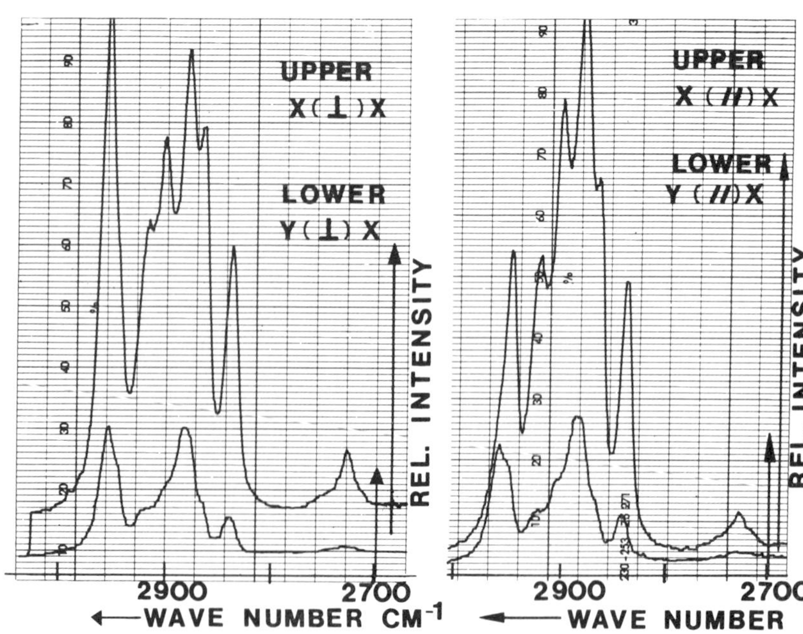

Fig. 3. Raman spectra of oriented polypropylene.

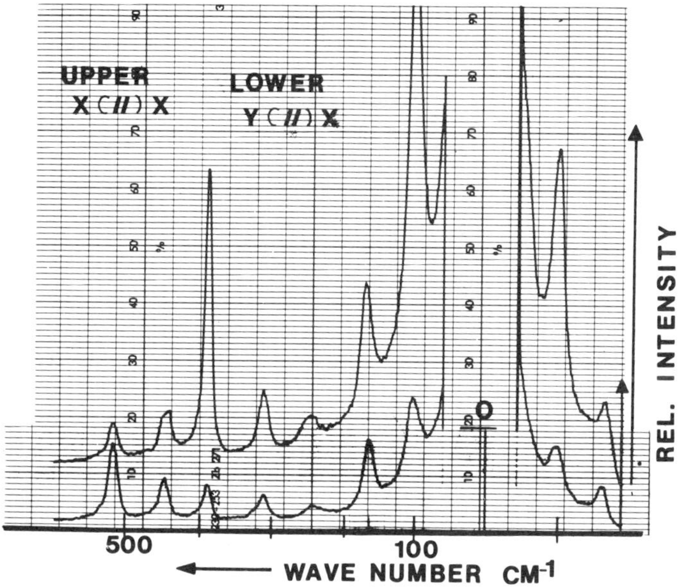

Fig. 4. Raman spectra of oriented polypropylene.

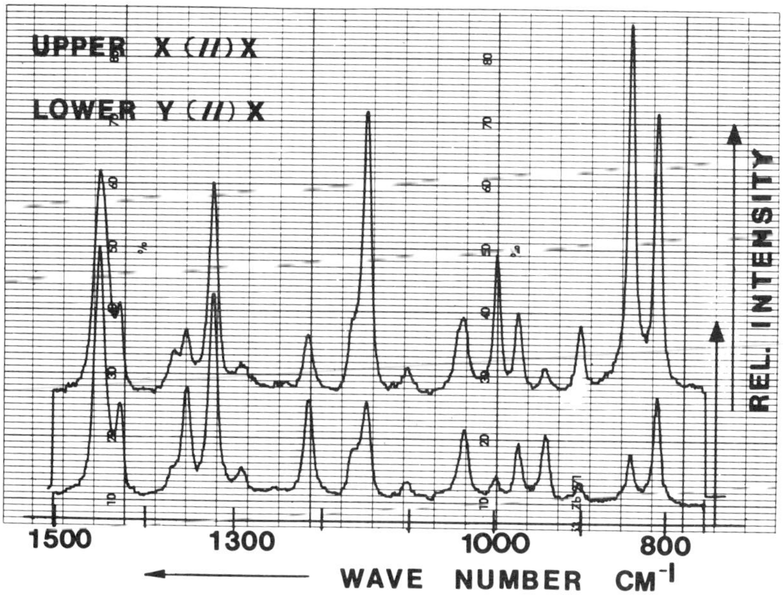

Fig. 5. Raman spectra of oriented polypropylene.

The Raman spectra are illustrated in Figs. 1–5. The geometry of the scattering experiment is shown on each trace. The Raman spectra of the random IPP sample were also obtained. These were completely independent of sample orientation. The observed frequencies for oriented and random IPP are shown in Table 1.

Table 1. Vibrational spectra of polypropylene.

Raman		Infrared			Present Work	
Frequency	Polarization	Frequency	Dichroism	Species	Raman	Species
		2965	$\parallel$		2957	*A, E*
2957	P	2953	$\perp$	*A*	2951	*A, E*
2920	P	2921	$\perp$	*A*	2923	*A*
2905	P				2905	*A*
2883	P	2877	$\perp$	*A*	2884	*A, E*
2871	P	2869	$\perp$	*A*	2869	*A*
2840	P	2840		*A*	2840	*A, E*
1457		1456	$\perp$	*A*	1458	*A, E*
		1450	$\parallel$	*A, E*	1449	*A*
1435		1434	$\perp$	*A, E*	1437	*A, E*
1371		1376	$\perp$	*A, E*	1377	*A, E*
1360		1357	$\perp$	*A, E*	1365	*A, E*
1330		1326	$\perp$	*A, E*	1334	*A, E*
1307	P	1305	$\perp$	*A*	1305	*A*
1296		1296	$\perp$	*E*	1295	*E*
1257	P	1255	$\parallel$	*A*	1256	*A*
1219		1218	$\perp$	*E*	1220	*E*
1167		1166	$\parallel$	*A*	1167	*A*
1152		1153	$\perp$	*E*	1151	*E*
1102		1101	$\perp$	*E*	1103	*E*
		1043	$\parallel$	*A*	1044	*A*
1034					1038	*E*
		1024		*E*		
996	P	998	$\parallel$	*A*	997	*A*
973		974	$\parallel$	*A, E*	973	*A, E*
942		940	$\perp$	*E*	942	*E*
900	P	900	$\perp, \parallel$	*A, E*	900	*A, E*
841	P	841	$\parallel$	*A*	841	*A*
806	P	806	$\perp$	*A, E*	809	*A, E*
530		528	$\perp$	*E*	528	*E*
458	P	456	$\perp$	*A*	456	*A*
		452	$\parallel$	*A, E*	451	*E*
398	P	396	$\parallel$	*A*	398	*A*
321		318	$\perp$	*E*	321	*E*
252	P	248	$\parallel$	*A*	252	*A*
171		168	$\perp$	*E*	171	*A, E*
					101	*A, E*

RESULTS AND DISCUSSION

For gases and liquids where the molecular orientation is perfectly random only two independent quantities, the invariants of the derived polarizability tensor can be determined. In the case of perfectly oriented crystals all six independent elements of the derived polarizability tensor can in principle be determined. Oriented polymers lie between these extremes. Where the polymer molecule is uniaxially oriented Snyder[1] has shown that the reconstructed scattering tensor has four independent elements by which we can differentiate Raman lines. Raman scattering activities for two cases of uniaxial orientation have been worked out. In the first case, the principal axis of the polymer is held in a fixed orientation whilst the other two axes are randomly oriented. In the second case the principal axis is confined to a plane while as before, the other two axes are randomly oriented. Expressions were tabulated for scattering activities in terms of the elements of the derived polarizability tensor for three orthogonal sample orientations and for both right angle and back viewing geometries. The scattering activities applicable in the present instance are those for a fixed orientation of the principal axis and right angle viewing. Normally, it is not possible to determine the absolute values of the polarizability elements and thus the Raman intensities. However, in many cases, the intensity changes brought about by variation of the sample orientation or laser polarization can be used to establish the symmetry of molecular vibrations. For example, using the geometry $Y(\perp)X$, only E class vibrations are excited while with geometry $X(\perp)X$, vibrations of A symmetry should be observed. Furthermore, by monitoring the intensities of vibrational transitions in appropriate orientations it is sometimes possible to deduce information on the nature and orientation of the normal mode involved.

ASSIGNMENTS

The most recent work on the vibrational Raman spectra of IPP was carried out by Vasko and Koenig.[7] Their results together with previous infrared work are summarized in Table 1. The most useful scattering geometries for identifying the species of the normal modes in IPP are illustrated in Figs. 1–3. Since, however, some scrambling of the radiation will occur and also the uniaxial orientation is not perfect, the scattering will not be 'pure' A or E polarization. The overlapping of A and E modes in several cases introduces further complications. However, in most cases the A and E modes are clearly identified. Assignments made on this basis are included in Table 1. In most cases, they agree well with previous results and will not be discussed further, but several points emerge which are worthy of further discussion. In Fig. 1, A modes only should appear with geometry $X(\perp)X$ while E modes should be preferentially excited with $Y(\perp)X$ geometry. A very low frequency band appears at 101 cm^{-1} which is predominantly A in character but has a component of E polarization. With the electric vector of the radiation along the chain axis this band shows fine structure on the high frequency side with a spacing of about 16 cm^{-1}. This fine structure disappears when the excitation is perpendicular to the chain axis (*cf.* Fig. 4). This suggests that this fine structure arises from the coupling of the acoustic longitudinal chain modes with the 101 cm^{-1} mode. These modes are of course preferentially excited when the laser polarization is along the chain axis. This band is currently under further investigation. The 101 cm^{-1} band, which consists mainly of chain bending motions,[5, 7] does not appear to have been observed previously. The 171 cm^{-1} frequency also has an A component in addition to its E class component. In the E polarization the band at 456 cm^{-1} is resolved into two components in agreement with the infrared measurements but one band appears to have A symmetry (456 cm^{-1}) and the other E symmetry (451 cm^{-1}). The assignments between 800 and 1500 cm^{-1} (Fig. 2) agree very closely with the infrared results illustrating the value of the technique. There is however some disagreement in the assignment of the bands at 1449 and 1458 cm^{-1}. The CH region is shown in Fig. 3. It is apparent that the *A* modes dominate the scattering both when the laser polarization is parallel and perpendicular to the unique axis of IPP. However, some bands have quite strong *E* components and it is possible to make definitive assignments for these modes. Previously only the *A* species vibrations had been assigned in this region. In spectra shown in Figs. 3, 4 and 5 the excitation geometries $X(\|)X$ and $Y(\|)X$ are used. The former selects *A* and *E* modes while the latter geometry selects *E* modes only. The intensities of the *A* modes in this geometry are generally lower than in the perpendicular geometry particularly the modes associated with the bands at 456, 809, 973 and 1167 cm^{-1}. This demonstrates that the polarizability derivatives along the chain axis are larger than those perpendicular to the axis for these modes. However, other modes such as those at 841 and 2884 cm^{-1} show the reverse behaviour.

In contrast, the spectra of the random sample of IPP show no orientation dependence. When however depolarization measurements are made in the usual fashion the same bands were found to be polarized as found in molten IPP. Further work is in progress.

REFERENCES

1. R. G. Snyder, *J. Mol. Spectry.* **37**, 353 (1971).
2. M. C. Tobin, *J. Phys. Chem.* **64**, 216 (1960).
3. J. P. Luongo, *J. Appl. Polym. Sci.* **9**, 302 (1960).
4. T. Miyazawa, *J. Polym. Sci., Part B* **2**, 847 (1964).
5. H. Tadakoro, M. Kobayashi, M. Ukita, K. Yasufuku, and S. Murahashi, *J. Chem. Phys.* **42**, 1432 (1965).
6. R. F. Schaufele, *J. Opt. Soc. Amer.* **57**, 105 (1967).
7. P. D. Vasko and J. L. Koenig, *Macromolecules* **3**, 597 (1970).
8. R. G. Snyder and J. H. Schachtschneider, *Spectrochim. Acta* **20**, 853 (1964).

36. RECENT DEVELOPMENTS AND ADVANCES IN THE INTERPRETATION OF RAMAN SPECTRA OF POLYMERS AS STRUCTURALLY IMPERFECT SYSTEMS

G. Zerbi

Istituto di Chimica delle Macromolecole del CNR, Milano, Italy

SUMMARY

The interpretation of the vibrational spectrum (infrared, Raman and inelastic neutron scattering) of polymers is generally carried out quantitatively on the basis of normal co-ordinate calculations on an ideal system as a single chain or, when feasible, as a three-dimensional ordered arrangement of chains. Several non-negligible absorptions in infrared or lines in the Raman however have been neglected or their interpretation proposed on a qualitative way.

A more quantitative approach to the problem has been attempted in the last few years in our laboratory in order to account for the dynamics of a partially or completely disordered model which approaches the reality of a partially crystalline or amorphous polymer. The mathematical methods have already been presented in the literature and a few applications have been discussed.

In this paper we present some recent results on polytetrafluoroethylene (Teflon) whose lattice dynamics and whose structure still remains an unsolved problem. The densities of vibrational states $g(\omega)$ for several disordered or ordered models were calculated and compared with the vibrational spectrum. A few Raman lines observed correspond to critical points in the density of states inactive for the ideal case but which become activated because of the disorder. Local and cooperative motions will be discussed together with the limitations of our approach.

SECTION VI
RESONANCE RAMAN EFFECT

37. RESONANCE RAMAN SPECTRA

H. J. Bernstein
Division of Chemistry, National Research Council of Canada, Ottawa, Canada

Irradiation of I_2 gas with various lines of an argon ion laser gives spectra of widely differing appearance. Using 5145 Å one obtains the well known resonance fluorescence spectrum[1] [Fig. 1(a)]. It is characterized by having sharp lines and rotational structure with intensity due to Franck Condon factors, and the band is depolarized. Irradiation into the continuum with 4880 Å produces a resonance Raman spectrum.[2] It is characterized by having broad features; increasing width with increasing vibrational quantum number; intensities monotonically decreasing with increasing vibrational quantum number; and the band is polarized [Fig. 1(b)]. If excitation is with 5017 Å [Fig. 1(c)] the Stokes lines are due to fluorescence (2ν being very weak), while the anti-Stokes lines are resonance Raman lines. It is clear that $I(\nu = -1)/I(\nu = +1)$, the anti-Stokes/Stokes intensity ratio, can be calculated from the Raman frequency of the observed anti-Stokes fundamental. The Stokes line intensity when compared then with the $\nu = 1$ fluorescence band is found to be only about one thousandth of the intensity of the fluorescence features. Thus the intensity from molecules giving resonance Raman spectra in the presence of fluorescence is usually obscured by the strong r.f. lines unless for example the fluorescence is quenched with about 25 atm of argon[3] or by the solvent when I_2 is dissolved in solution.[4] In fact, when excitation at all wavelengths through the absorption band of a liquid, solution or solid occurs, one observes a resonance Raman spectrum only because the fluorescence is quenched.

It is worth calling attention at this time to the various types of resonance Raman spectra which can be obtained. Most of the mercury arc Raman studies[5] were concerned with irradiation whose frequency was between the dissociation limits of the ground state and the minimum of the potential of the nearest excited state having its own vibrational energy levels [see Fig. 2(a)]. This is the case first treated by Placzek[6] and the intensity is related to the frequency of the exciting line ν_0 by

$$I \propto (\nu_0 - \nu)^4 \tag{1}$$

With a large value for $(\nu_0 - \nu_e)$ where ν_e is the frequency of the intermediate state, the intensity has this simple dependence on ν_0. As ν_0 approaches the frequency of the minimum in the excited state [Fig. 2(b)] the resonance denominator in the expression for the

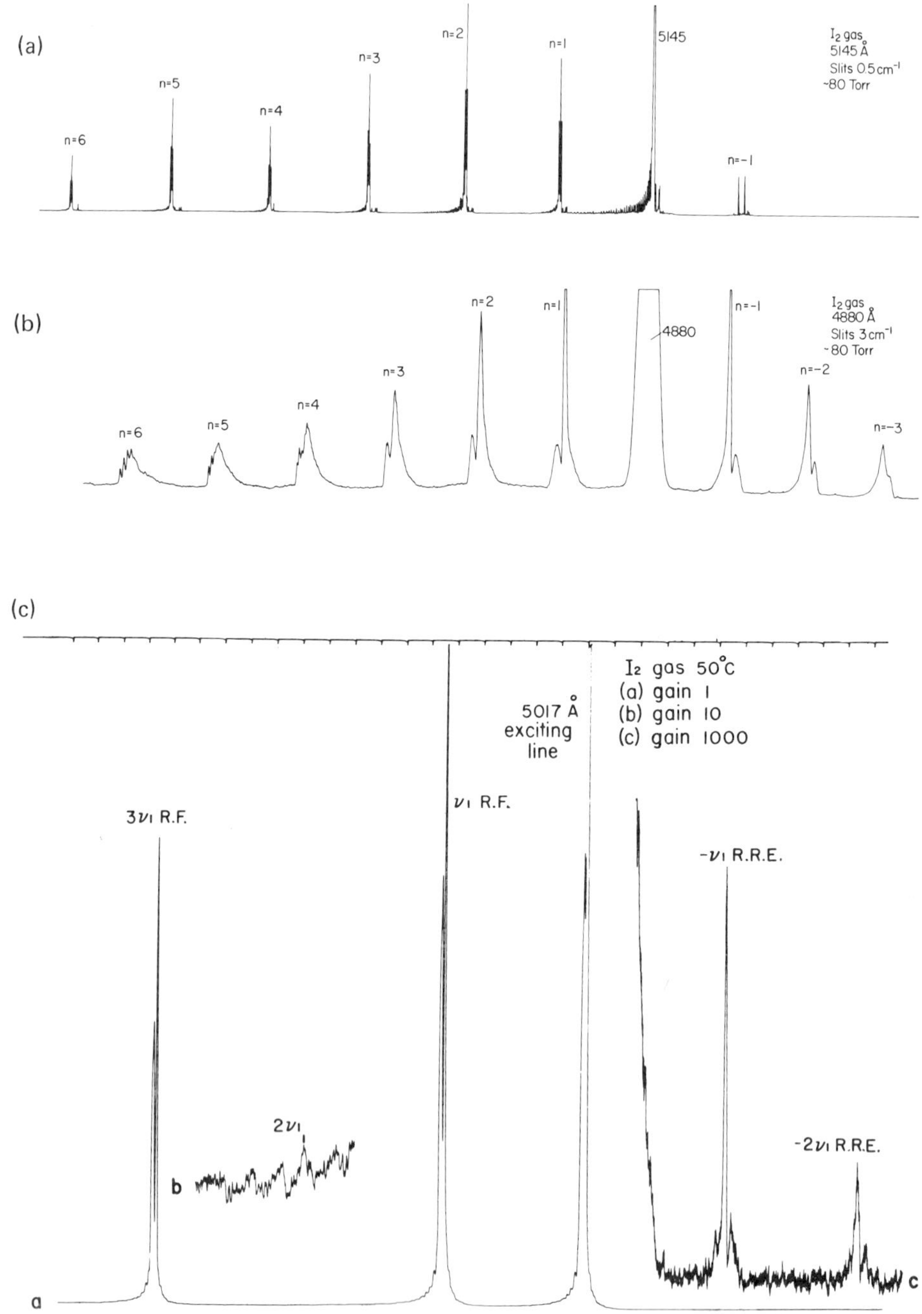

Fig. 1. (a) Fluorescence spectrum of I_2 vapour excited with 5145 Å. (b) Resonance Raman spectrum of I_2 vapour excited with 4880 Å. (c) Stokes fluorescence and anti-Stokes resonance Raman spectrum of I_2 vapour observed by excitation with 5017 Å.

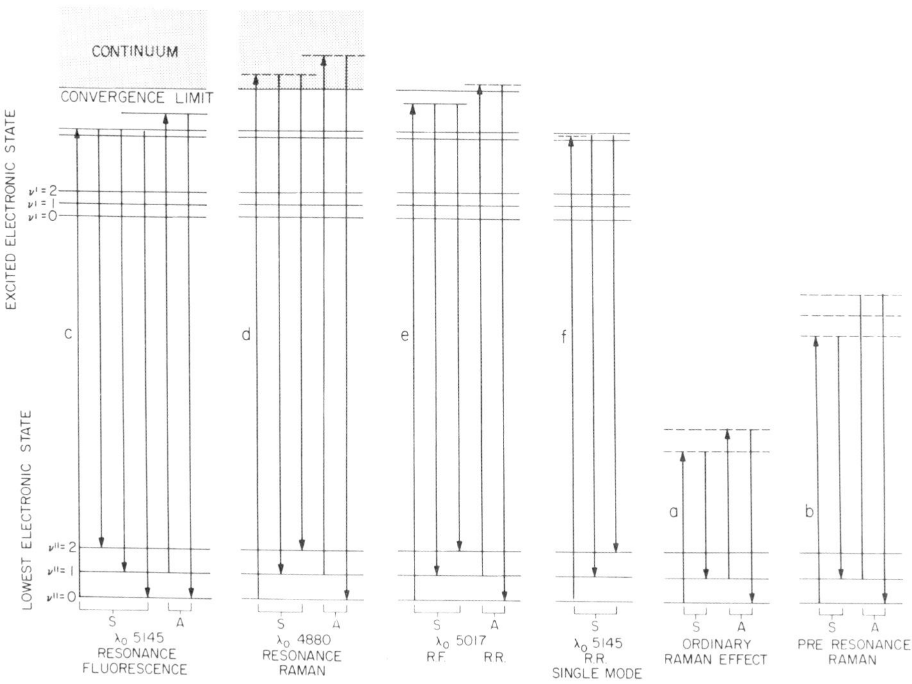

Fig. 2. (a) Energy level diagrams for Raman effects. Ordinary Raman effect; S = Stokes, A = anti-Stokes. (b) Pre-resonance Raman. (c) Resonance fluorescence, when exciting line is into the energy levels of the excited state. (d) Resonance Raman, when exciting line is in the continuum. (e) Where Stokes is fluorescence and anti-Stokes is resonance Raman. (f) Single mode excitation into the 'dark'.

intensity becomes smaller and a one or two state theory developed by Albrecht[7] is applicable.

$$I \propto [\nu^2(\nu_e^2 + \nu_0^2)/(\nu_e^2 - \nu_0^2)^2]^2 = F_A^2 \tag{2}$$

for a single state e, and

$$I \propto [2\nu^2(\nu_e\nu_s + \nu_0^2)/(\nu_e^2 - \nu_0^2)(\nu_s^2 - \nu_0^2)]^2 = F_B^2 \tag{3}$$

for two states e and s. Such a case has been observed for CrO_4^{2-} ion in solid or aqueous solution[8] as well as *p*-nitroaniline and β carotene.[9] For the case in Fig. 2(c), fluorescence is obtained and the Raman spectrum will be observed only when the quenching is nearly complete. For the case in Fig. 2(d), excitation is into the continuum and the resulting resonance Raman spectrum depends strongly on the electronic absorption coefficient at the frequency of the incident light. The theory for the dependence of the intensity as a function of the incident light frequency[10, 11] for the case shown in Fig. 2(d) has been given by Berjot, Jacon and Bernard.[12] The intensity dependence of the resonance Raman line on vibrational quantum number at fixed irradiation frequency[2] is yet to be explained. From Fig. 2(e) it is clear that the Stokes features are of the fluorescence type whereas the anti-Stokes features are resonance Raman. This is so, of course, because excitation from

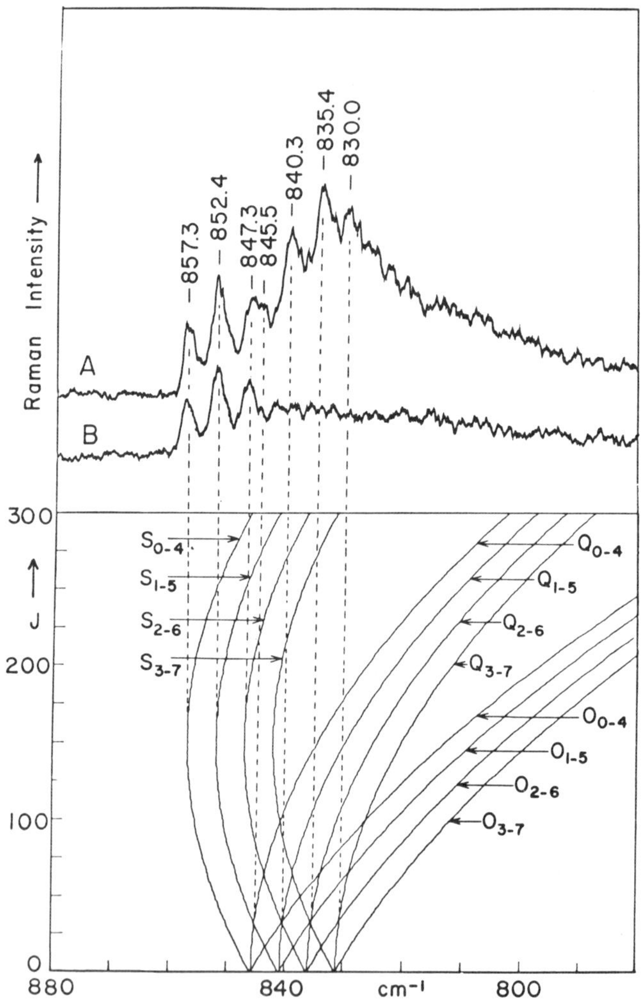

Fig. 3. Vibration–rotation spectrum of the fourth harmonic 4880 Å, slits 1.5 cm^{-1}. A is $I(\parallel)$, B is $(\perp)$, and computed Fortrat diagrams for S, O and Q branches.

the molecules in the first vibrational level is in the continuum from which resonance Raman spectra are observed. This was first observed and explained in Fig. 9 of ref. 2, the inference being that since there is an anti-Stokes resonance Raman effect there is also a Stokes resonance Raman effect.

In Fig. 2(f) one has the case where a single mode argon ion laser line at 5145 Å is used and it is narrow enough for its energy to lie in between the vibrational rotational levels. We are now off resonance fluorescence and would obtain the residual resonance Raman spectrum.

The theory for these cases when both the fluorescence and resonance Raman processes are taking place, i.e. when excitation is into the potential well of the excited state, has not yet been formulated.

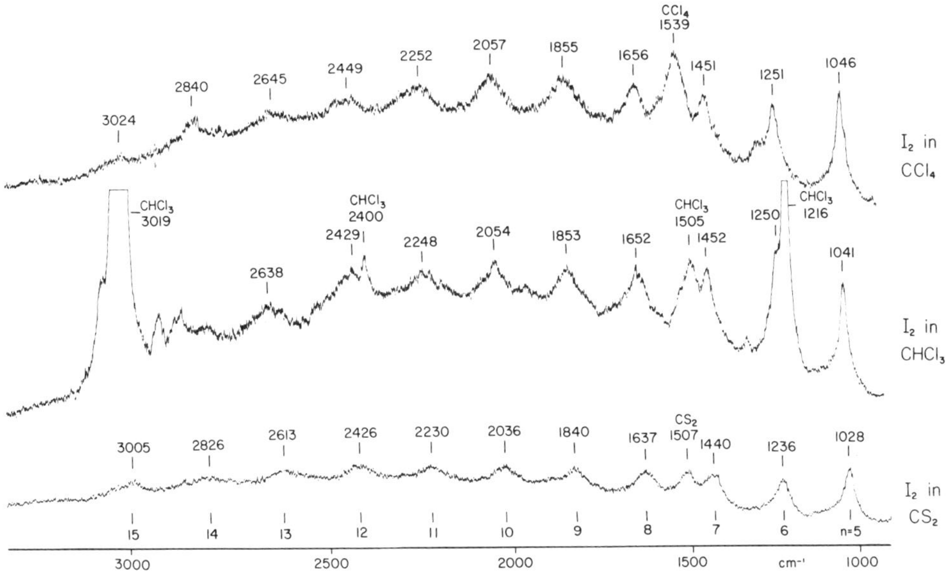

Fig. 4. I_2 (10^{-2} mole l^{-1}) in CCl_4, $CHCl_3$ and CS_2, 5145 Å, slits 10 cm^{-1}.

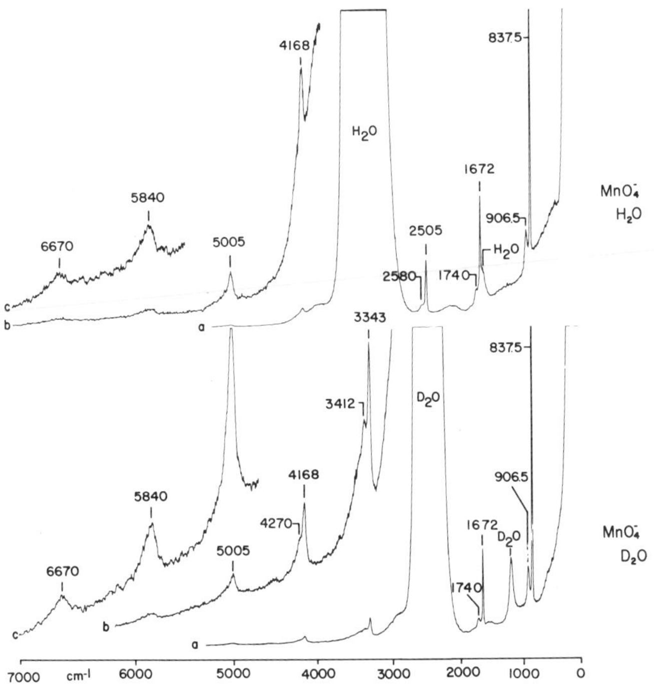

Fig. 5. $KMnO_4$ in H_2O and D_2O (1^{-3} M/l), 4880 Å, 7.5–10.0 cm^{-1} slits rotating sample.

Table 1. The relative intensities of the ν_1 vibration and its overtones of MnO_4^- in D_2O solution at different exciting wavelengths. ϵ_0 = extinction coefficient for the exciting line.

	Wavelength in Å							
	5145	5017	4965	4880	4765	4727	4658	4579
ϵ_0	1760	1470	1190	1010	570	510	350	220
ν_1	1.00[a]	0.65	0.62	0.69	0.34	0.32	0.27	0.21
$2\nu_1$	0.44	0.35	0.38	0.24	0.15	0.15	0.069	0.055
$4\nu_1$	0.16	0.20	0.16	0.10	0.063	–	–	–
$5\nu_1$	0.11	0.084	0.089	0.084	0.024	–	–	–
$6\nu_1$	0.094	–	0.039	0.050	0.034	–	–	–

[a] Normalized to ν_1 (5145 Å); the relative scattering power of the ν_1 fundamental of MnO_4^- to the bending mode of D_2O is 104:000

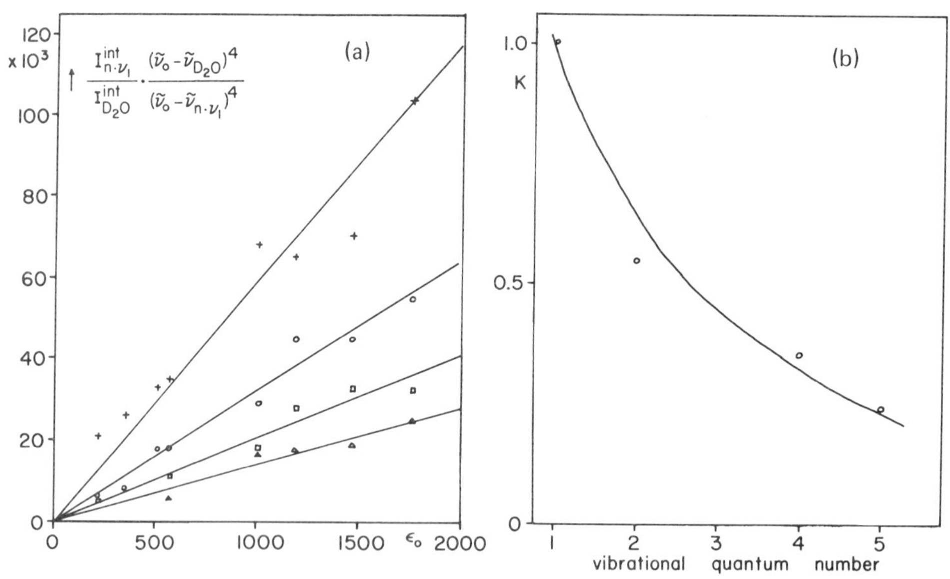

Fig. 6. (a) Relative intensity of overtone progression of ν_1 for MnO^{4-} in D_2O as a function of the extinction coefficients of the exciting line. + for ν_1, ○ for $2\nu_1$, □ for $4\nu_1$ and △ for $5\nu_1$. (b) The slopes of the lines in (a) versus n.

With slits of 1 cm^{-1} the broad features in Fig. 1(b) show fine structure consisting of S, Q and O branches so that the increased broadening with increasing quantum number can be explained[13] (see Fig. 3).

Up until this point we have been discussing the various cases which may be encountered in gases. In I_2 solutions, as previously mentioned, the fluorescence is quenched so that irradiation into any part of the absorption band produces a resonance Raman spectrum.[4]

In order to obtain spectra of absorbing substances in solution, liquid and solid phases one must avoid the local heating effect in the sample due to the laser otherwise the beam diffuses or the sample burns. This is achieved by rotating the sample about 2500

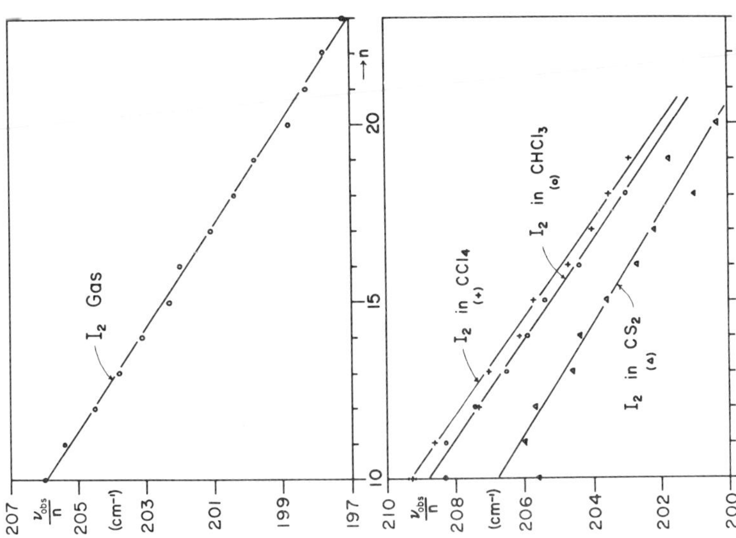

Fig. 7. Plot of $\Delta G/n$ versus n for I_2 in solutions.

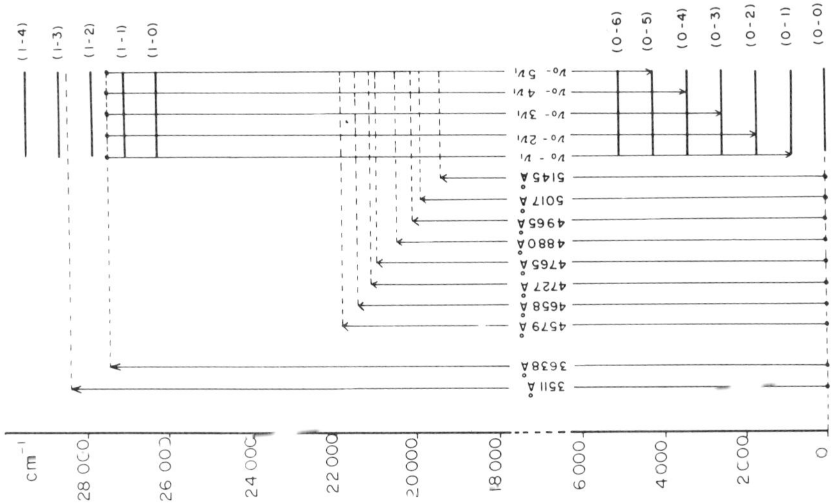

Fig. 8. Energy levels of K_2CrO_4 solid showing positions of exciting line.

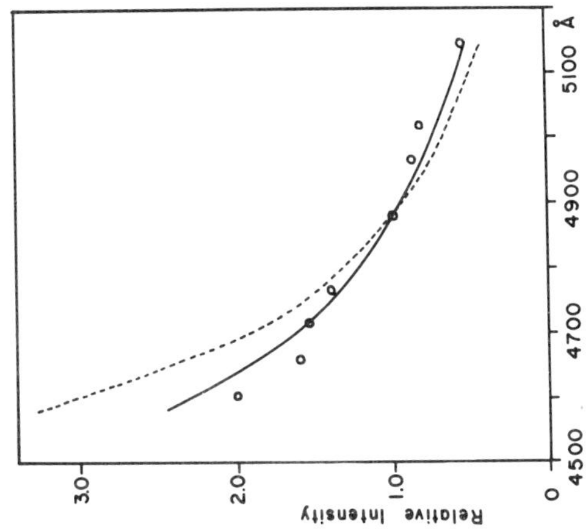

Fig. 9. One state calculated function - - - -.
Two state calculated function ——.

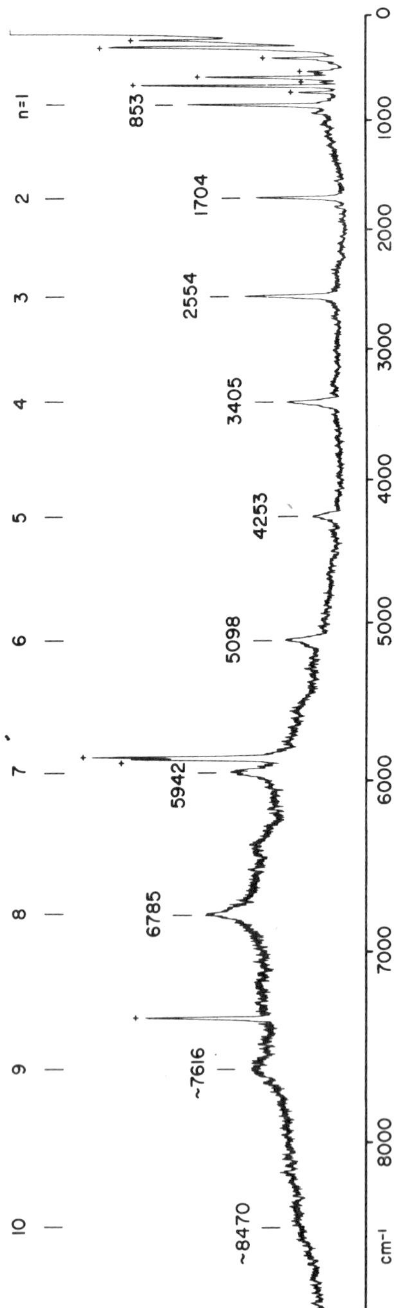

Fig. 10. Spectrum of solid K_2CrO_4 obtained for 3638 Å excitation (5 mW) and rotating samples. Slits 10 cm^{-1}; lines marked + are laser lines.

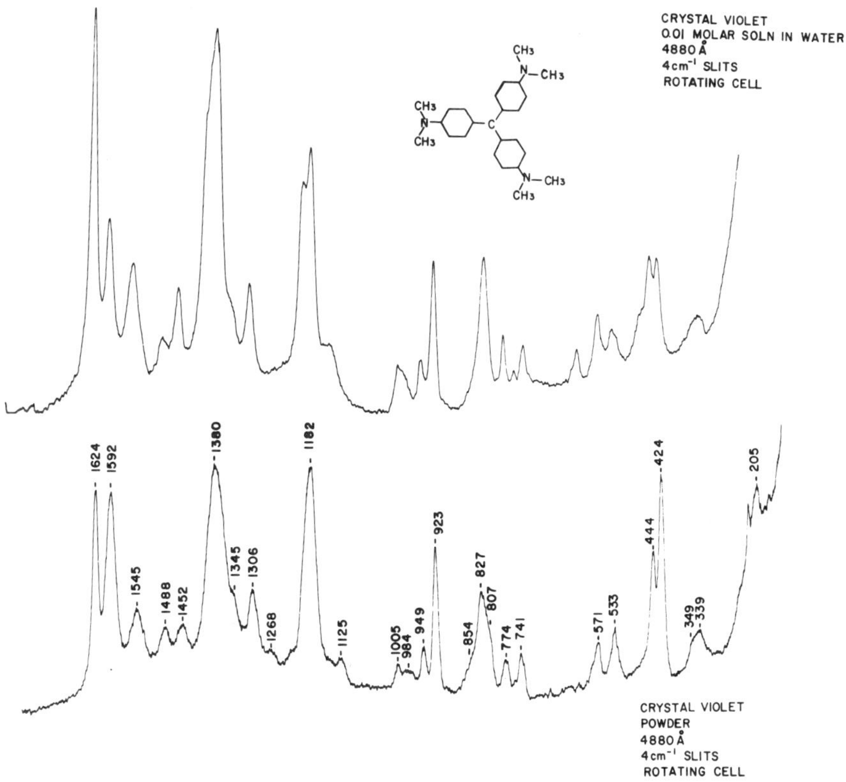

Fig. 11. Aqueous solution and solid spectrum of crystal violet.

r.p.m. so that a cool sample spot is always observed by the spectrometer. This has been used for solution[4b] and solids[14] where long progressions of overtones have been obtained for I_2 (Fig. 4) and permanganate ion in solution[16] (Fig. 5). The intensity of all lines in the progressions of $\nu_1(a)$ of the permanganate ion were measured (see Table 1) for several of the argon laser exciting lines and the scattering cross-section was found to be of the form ϵ_0/n where ϵ_0 is the extinction coefficient (see Fig. 6) at the laser frequency ν_0 and n is the vibrational quantum number.

From these progressions in solution it may be possible to obtain the effect of intermolecular interacting on the ground state properties of I_2 and the other molecules, since as Fig. 7 shows there is a frequency change with solvent.

For chromate ions in solution and solid[8,15] the argon lines fall in the region shown in Fig. 2(b) and are located on the energy level diagram in Fig. 8. The intensity calculated by the two state function F_B^2 is within experimental error of the observed intensity of the $\nu_1(a)$ line excited by the different Ar^+ lines (see Fig. 9). Using the ultraviolet line at 3638 Å produces the progression shown in Fig. 10 for the chromate ion in solids.

With laser irradiation and sample rotating it is possible to obtain Raman spectra from aqueous solutions of transition metal complexes, dyes, pigments, etc. In many cases spectra of the corresponding solids may also be obtained.

In Fig. 11 the resonance Raman spectrum of the dye crystal violet is shown. It is

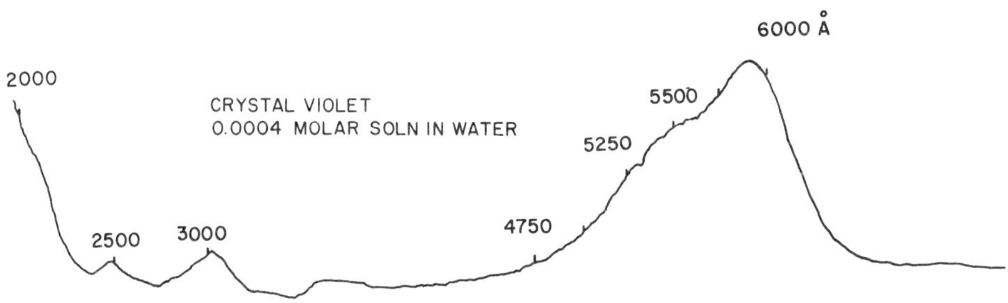

Fig. 12. Absorption spectrum of the same aqueous solution.

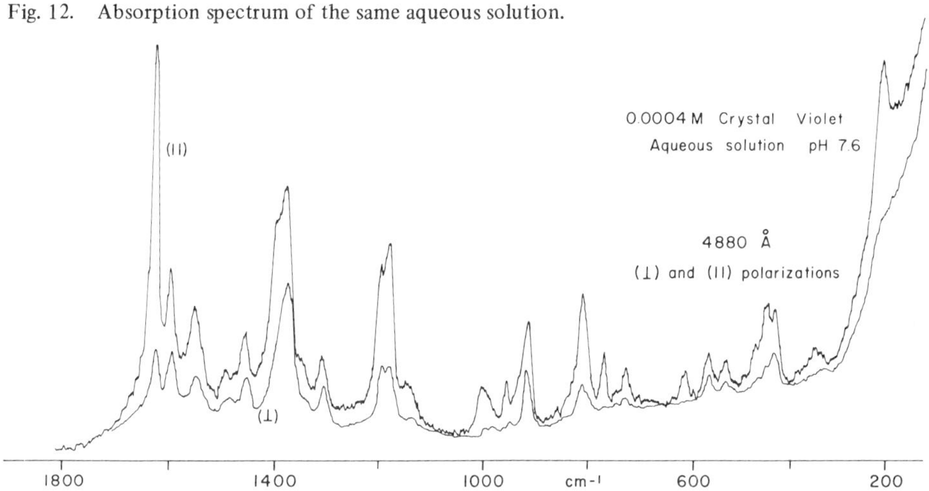

Fig. 13. $I(\|)$ and $I(\perp)$ for the same solution.

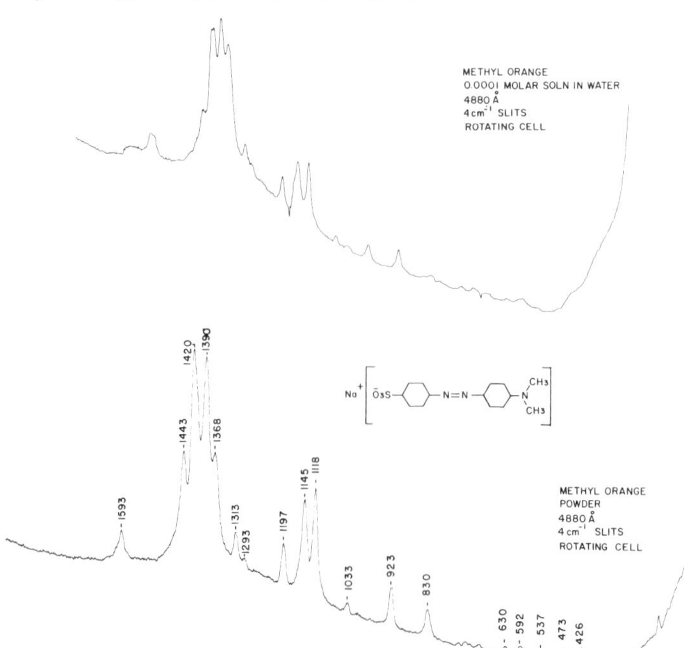

Fig. 14. Methyl orange in aqueous solution and powder.

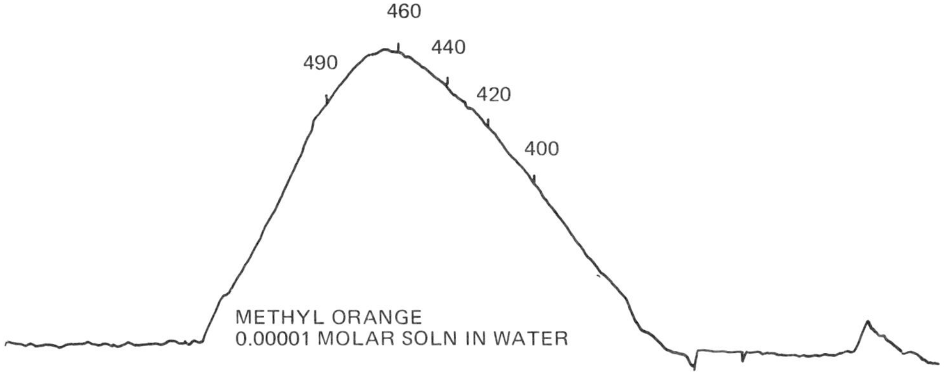

Fig. 15. Absorption spectrum of methyl orange.

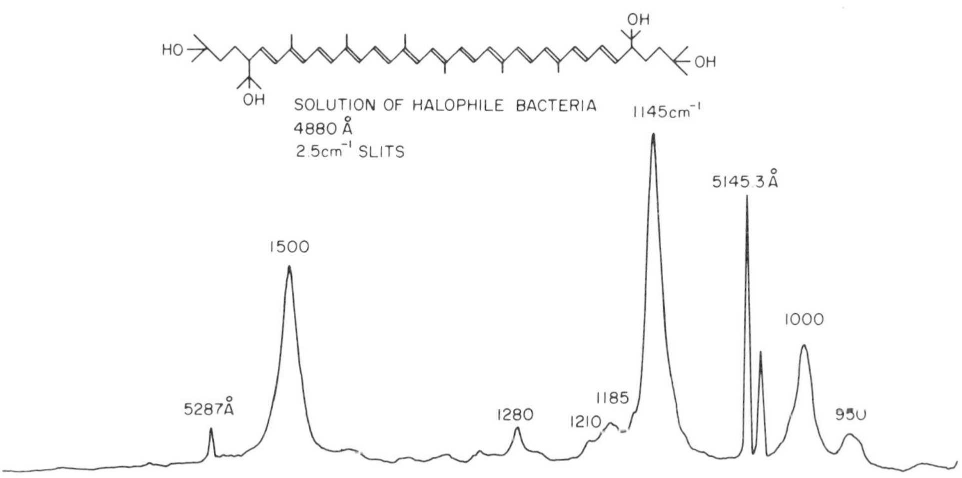

Fig. 16. Solution of bacteria sample ~ 0.001 ml.

clear that much more information is available than from the visible spectrum in Fig. 12. In Fig. 13 the polarized spectra are shown. From this ratio of $I(\|)/I(\perp)$ the depolarization ratio ρ_s is obtained. Under these experimental conditions no bands corresponding to the stretching vibration of C—H bonds were observed. This is not unexpected since the electronic system which is responsible for absorption is also responsible for the resonance Raman spectrum. Since hydrogens play little role in the mobile electron distribution we also expect that hydrogens play a minor role in the resonance Raman effect. The spectrum in Fig. 11 is then essentially from the skeleton.

In Fig. 14 we see the resonance Raman spectrum of methyl orange and in Fig. 15 its absorption spectrum. One use of these dyes is to act as a label in studies involving biologically important molecules. Such an interaction has been observed for methyl orange and bovine serum albumin.[17] These are under the conditions (such as low-concentration etc.) which are biologically relevant. Other studies are in progress using dye molecules to complex with biochemical systems, and as labels for immunological sera and carcinogens.

Figure 16 shows the resonance Raman spectrum of a cell membrane of bacteria[18] which has a faint ruby colour and a typical carotenoid-like spectrum.

It is clear that resonance Raman spectroscopy has much to offer and will truly come into its own when continuously tunable lasers are available.

ACKNOWLEDGEMENT

This chapter is based on the work of P. Carey, W. Kiefer and W. Holzer while post-doctoral Fellows at NRC, and my colleague W. F. Murphy.

REFERENCES

1 D. H. Rank and B. S. Rao, *J. Mol. Spectry.* **13**, 34 (1964).
2 W. Holzer, W. F. Murphy and H. J. Bernstein, *J. Chem. Phys.* **52**, 399 (1970).
3 I. R. Beattie, G. A. Ozin and R. O. Perry, *J. Chem. Soc. (A)* 2071 (1970).
4 (a) O. Mortensen, *J. Mol. Spectry.* **39**, 48 (1971); (b) W. Kiefer and H. J. Bernstein, *Appl. Spectry.* **25**, 500 (1971).
5 J. Behringer, *Raman Spectroscopy*, (Ed. H. Szymanski), Plenum Press, New York, 1967, has reviewed pre-resonance Raman effect. Shorygin and co-workers were first to observe overtone progressions. P. P. Shorygin and T. M. Ivanova, *Dokl. Akad. Nauk SSSR* **150**, 533 (1963); *Opt. i Spektroskopiya* **15**, 176 (1963); also T. M. Ivanova, L. A. Yanovskaya and P. P. Shorygin, *Opt. i Spektroskopiya* **18**, 206 (1965).
6 G. Placzek, *Handbuch der Radiologie*, Vol. 6, Part 2, Akademische Verlags Gesellschaft, Leipzig, 1934.
7 A. C. Albrecht and M. C. Hutley, *J. Chem. Phys.* **55**, 4438 (1971).
8 W. Kiefer and H. J. Bernstein, *J. Mol. Phys.* **23**, 835 (1972).
9 M. C. Hutley and D. J. Jacobs, *Chem. Phys. Letters* **6**, 269 (1970).
10 J. Behringer, *Z. Physik* **229**, 209 (1969).
11 O. Mortensen, *Mol. Phys.* **22**, 179 (1971).
12 (a) M. Jacon, M. Berjot and L. Bernard, *Compt. Rend.* **273B**, 596 (1971); **273B**, 956 (1971); (b) M. Berjot, M. Jacon and L. Bernard, *Opt. Commun.* **4**, 117 (1971); **4**, 246 (1971).
13 W. Kiefer and H. J. Bernstein, *J. Mol. Spectry.* **43**, 366 (1972).
14 W. Kiefer and H. J. Bernstein, *Appl. Spectry.* **25**, 500 (1971).
15 W. Kiefer and H. J. Bernstein, *Appl. Spectry.* **25**, 609 (1971).
16 W. Kiefer and H. J. Bernstein, *Chem. Phys. Letters* **8**, 381 (1971).
17 P. R. Carey, H. Schneider and H. J. Bernstein, *Biochem. Biophys. Res. Commun.* **47**, 588 (1972).
18 Courtesy of P. R. Carey.

38. RESONANCE RAMAN SPECTRA USING ULTRAVIOLET LASER EXCITATION

W. Kiefer
Sektion Physik der Universität München, München, Germany

and

H. J. Bernstein
Division of Chemistry, National Research Council of Canada, Ottawa, Canada

INTRODUCTION

Using the rotating sample technique for liquids[1] and for crystal powders[2] we have been able to investigate the resonance Raman effect (RRE) of the chromate (CrO_4^{2-}) and triiodide (I_3^-) ions. The resonance Raman spectra were excited with the ultraviolet laser lines 3638 and 3511 Å of an argon ion laser equipped with dielectric mirrors of high reflectivity in the ultraviolet.

CHROMATE ION

The electronic absorption spectrum of CrO_4^{2-} has a pronounced peak around 3700 Å (Fig. 1). It shows no vibrational fine structure in solution as is the case for the permanganate ion.[3] The vibrational pattern can be obtained in solid K_2CrO_4 at low temperature.[4, 5] In Fig. 1 the relative positions of the ultraviolet and visible lines of the argon ion laser are marked by dotted vertical lines. The two ultraviolet laser lines at 3638 Å and 3511 Å are close to the absorption peak and the visible lines have their positions on the long wavelength side of the absorption band. In this case one should expect RRE with ultraviolet excitation producing a progression of overtones, and a pre-RRE (with perhaps a weak overtone) when exciting with the visible laser lines.

Normal Raman spectra of K_2CrO_4 in aqueous solutions[6] and as single crystals[7, 8] have been reported by Stammreich *et al.*, Adams *et al.* and Carter *et al.* In aqueous solutions all Raman lines given by Stammreich *et al.* could be confirmed using the 8 visible argon ion laser lines. With the higher frequency laser lines at 4765–4579 Å a weak overtone of the ν_1 line at 848 cm^{-1} could also be detected.

RRE spectra in solutions of H_2O and D_2O and as pure solids were obtained using ultraviolet excitation at 3511 and 3638 Å. A highly polarized Raman line at 848 cm^{-1} (with 3638 Å excitation $\rho_s < 0.05$) is assigned to the totally symmetric vibration $\nu_1(A_1)$ and a weaker line at 888 cm^{-1} to the $\nu_3(F_2)$ antisymmetric stretching vibration in agreement with the assignment by Stammreich *et al.*[6] Four overtones of the symmetric stretching mode at 848 cm^{-1} could be observed with decreasing intensity, also a combination of $\nu_1 + \nu_3$ at about 1743 cm^{-1}. The RRE spectrum of the chromate ion in solution resembles that of the permanganate ion,[3] except the signal-to-noise ratio is much

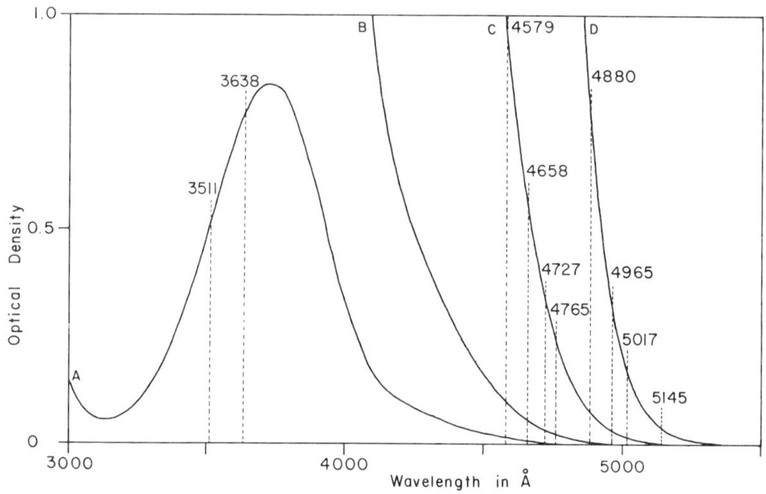

Fig. 1. Absorption spectrum of K_2CrO_4 in D_2O. Concentrations: (A) 2×10^{-4} mole/1; (B) 1×10^{-3} mole/1; (C) 1×10^{-2} mole/1; (D) 1×10^{-1} mole/1; pathlength 1 cm.

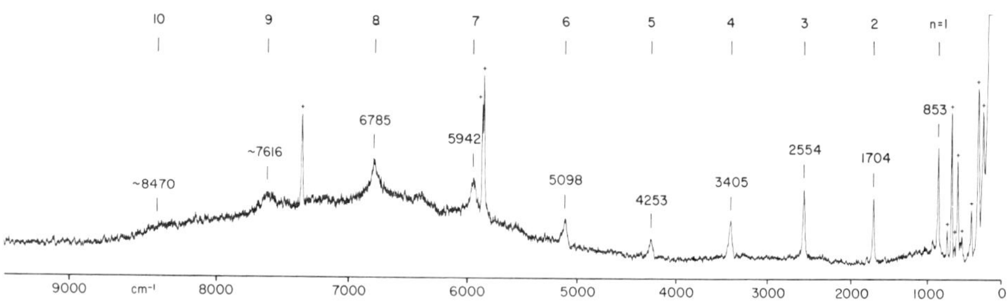

Fig. 2. The resonance Raman spectrum of solid K_2CrO_4, obtained with 3638 Å excitation and the rotating solid sample device [2]; n represents the vibrational quantum number; lines marked with a + sign are discharge lines from the argon lasers; $s = 200$ μs; T = 1 s; $V = 12.5$ Å/min, laser power ~ 5 mW.

lower due to the low laser power of about 5 to 10 mW in the ultraviolet laser lines and the use of a spectrometer which was designed for the visible region. Nevertheless, reasonable spectra were obtained, as can be seen in Fig. 2, where the RRE spectrum of solid K_2CrO_4 is shown. Lines marked with a + sign are argon discharge lines for the laser tube. No Raman lines with wavenumbers lower than 800 cm^{-1} could be observed because of the increasing transmission of the spike filter which was used for this laser line and which allows the appearance of many discharge lines in the low frequency spectrum of the solid. A progression of ten harmonics of the internal stretching mode at 853 cm^{-1} can be seen with decreasing intensity but increasing half bandwidth. The very broad band between 5000 and 8500 cm^{-1} is due to fluorescence. The wavenumbers, half bandwidths and the spectral slit widths and the assignment of the ν_1 progression in solid K_2CrO_4 are listed in Table 1.

The observation of high intensity overtones in the RRE allows one to determine

Table 1. The Raman spectrum and half bandwidths of the ν_1 overtone progression of solid K_2CrO_4.

Assignment	Observed Raman ν [cm^{-1}]	Half bandwidth $\Delta\nu_{1/2}$ [cm^{-1}]	Used slit width in [cm^{-1}]
ν_1	853	6	2.0
$2\nu_1$	1704	10	2.5
$3\nu_1$	2554	16	2.5
$4\nu_1$	3405	30	4.0
$5\nu_1$	4253	35	4.5
$6\nu_1$	5098	40	10.0
$7\nu_1$	5942	49	10.0
$8\nu_1$	6785	68	10.0
$9\nu_1$	7616	~ 100	20.0
$10\nu_1$	8470	–	–

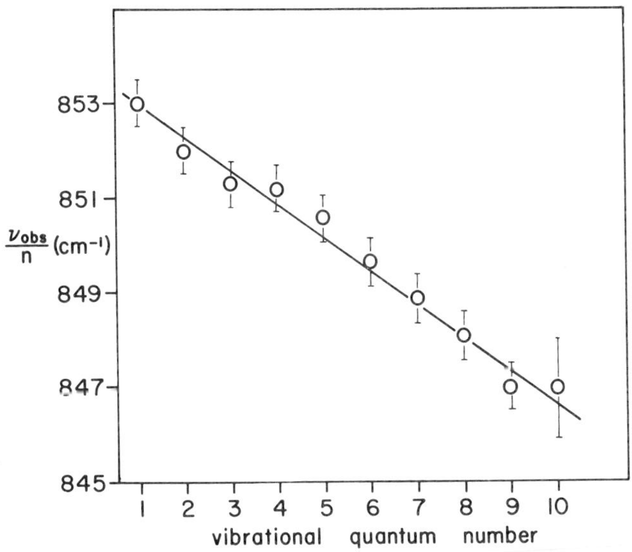

Fig. 3. Determination of the anharmonicity constant X_{11} in solid K_2CrO_4. ν_{obs}/n versus n.

anharmonicity constants of particular vibrations. The observed wavenumber $\nu(n)$ of any overtone of an anharmonic oscillator is given by[9]

$$\nu(n) = G(n) - G(0) = n\omega_1 - X_{11}(n^2 + n) + \ldots$$

where $G(n)$ is the term of the nth eigenvalue, n the vibrational quantum number, ω_1 is the harmonic frequency and X_{11} the anharmonicity constant. The slope of a plot of $\nu(n)/n$ versus n therefore gives X_{11}. Figure 3 shows a $\nu(n)/n$ versus n plot of the ν_1 vibration of the CrO_4^{2-} ion in solid potassium chromate. From this plot we have derived the following values for the CrO_4^{2-} ion

$$X_{11} = 0.71 \pm 0.1\ cm^{-1} \qquad \omega_1 = 854.4 \pm 0.5\ cm^{-1}$$

Table 2. Pre-RRE intensity results for CrO_4^{2-} in solution compared to theory.[10]

Exciting line in Å λ_0	Exciting line in cm^{-1} ν_0	$I_{rel}{}^a$ Experimental	$(F_B 2)rel^a$	$(F_A 2)rel^a$
5145	19,430	0.55	0.532	0.442
5017	19,930	0.84	0.711	0.643
4965	20,130	0.89	0.806	0.755
4880	20,490	1.000	1.000	1.000
4765	20,980	1.20	1.367	1.512
4727	21,150	1.35	1.526	1.751
4658	21,460	1.59	1.880	2.319
4579	21,830	2.00	2.423	3.280

[a] Scaled to unity at 4880 Å.

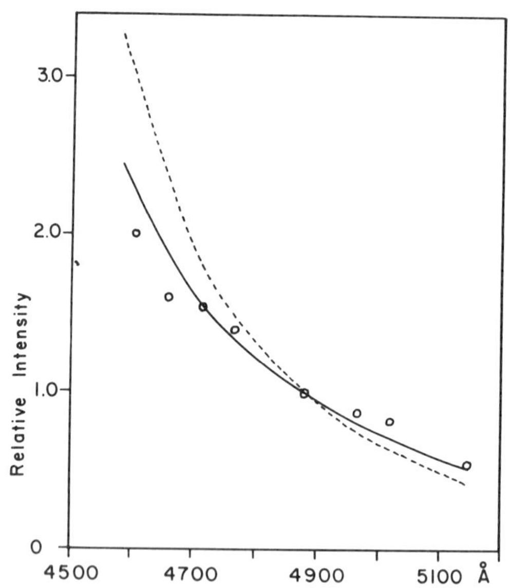

Fig. 4. Frequency factors for the ν_1 vibration of CrO_4^{2-} as a function of the exciting frequency with scaling to unity at 4880 Å. Experimental points are circled. —, $F_B{}^2$ ($\nu_e = 26\ 900\ cm^{-1}$, $\nu_s = 36\ 600\ cm^{-1}$); - - -, $F_A{}^2$ ($\nu_e = 26\ 900\ cm^{-1}$).

The positions of the 8 visible argon ion laser lines relative to the near ultraviolet band of the CrO_4^{2-} ion at about 3700 Å (see Fig. 1) allow an experimental study of pre-RRE intensities. Using these lines we have measured the relative intensity of the ν_1 fundamental as a function of the exciting frequency. As internal standard we have chosen the bending mode of D_2O at about 1200 cm^{-1}. In Table 2 the results normalized for 4880 Å excitation are in column 3.

Recently Albrecht and Hutley[10] have developed a quantum theory for the dependence of vibrational Raman intensity on the wavelength of incident light and applied the results to the pre-RRE of *p*-nitroaniline in benzene solution. They calculated

Table 3. Relative intensity of the first overtone of ν_1 of CrO_4^{2-} as a function of exciting frequency.

Exciting line in Å	$\frac{I_{2\nu_1}}{I_{\nu_1}}$	$\left[\frac{I_{2\nu_1}}{I_{\nu_1}}\right]_{rel}^{a}$
4765	0.025	1.00
4727	0.031	1.24
4658	0.035	1.40
4579	0.043	1.72
3638	~2/3	27.5

[a]Scaled to unity at 4765 Å.

two dimensionless frequency factors F_A^2 and F_B^2 which involve the exciting frequency

$$F_A^2 = [\nu^2(\nu_e^2 + \nu_0^2)/(\nu_e^2 - \nu_0^2)^2]^2$$

$$F_B^2 = [2\nu^2(\nu_e\nu_s + \nu_0^2)/(\nu_e^2 - \nu_0^2)(\nu_s^2 - \nu_0^2)]^2$$

where ν_0 is the exciting frequency, $\nu = \nu_0 - \Delta\nu$, $\Delta\nu$ being the Raman frequency. Subscripts *e* and *s* refer to electronic states with eigen-frequencies at ν_e and ν_s respectively. The *A* term describes the frequency dependence if only one electronic absorption band (ν_e) is taken into account, whereas the *B* term considers also a second electronic state which has its frequency position (ν_s) further in the ultraviolet.

In aqueous solution the chromate ion possesses two absorption bands with maxima at about 26900 cm^{-1} (ν_e) and 36600 cm^{-1} (ν_s); both absorption bands correspond to electronic transitions $^1A_1 \rightarrow {}^1T_2$.[5] Using these values for ν_e and ν_s we have calculated $F_A^2(\nu_0)$ and $F_B^2(\nu_0)$ for the 8 visible exciting frequencies of the argon ion laser. The results are in Table 2 columns 4 and 5. Figure 4 compares the experimental with the theoretical results. It is seen that the F_B^2 term represents the experimentally observed frequency dependence better than the F_A^2 term which contains only one absorption peak. The same observations have been made by Albrecht and Hutley for the *p*-nitroaniline example.[10]

In the RRE in contrast to the pre-RRE, the intensity of the overtones is remarkably high. It might be of interest to investigate whether there is any change of the intensity of the first overtone relative to the fundamental while approaching the strong electronic absorption peak from the long wavelength side. We could observe in D_2O solutions of K_2CrO_4 the appearance of the first overtone of the ν_1 line with the higher frequency laser lines. The intensities of the first overtones relative to the fundamental are presented in Table 3. The value is about 1/40 for 4765 Å excitation, but $I_{2\nu_1}/I_{\nu_1}$ should increase still more until the ratio of $I_{2\nu_1}/I_{\nu_1}$ becomes about 2/3 and the higher overtones $3\nu_1$, $4\nu_1$ and $5\nu_1$ are observed.

TRIIODIDE ION

During an extensive resonance Raman spectroscopic study of iodine in different organic solvents[11] we found an overtone progression of a fundamental located at round 110 cm^{-1} when excited by the ultraviolet argon ion laser lines. Further investigations[12] of

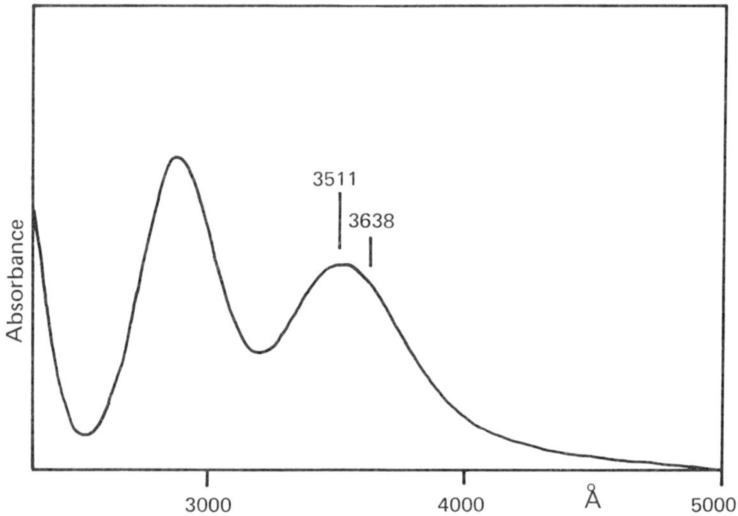

Fig. 5. Absorption spectrum of caesium triiodide in water.

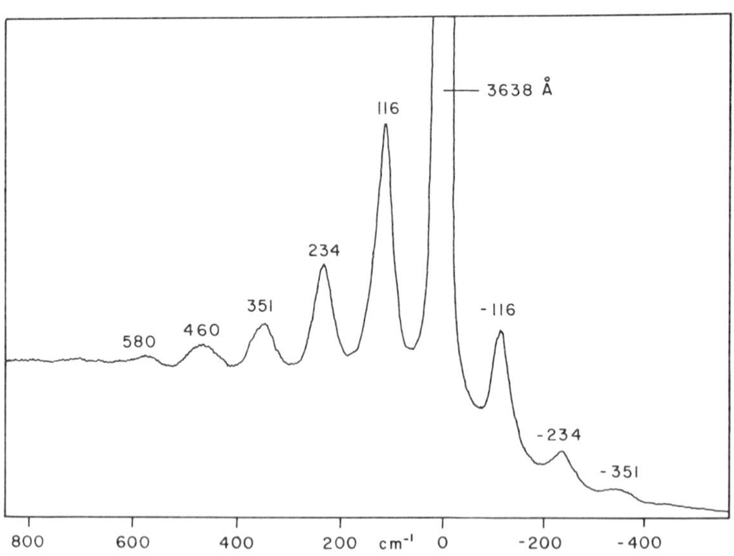

Fig. 6. The resonance spectrum of the I_3^- ion in water. Concentration: 2.5×10^{-4} mole CsI_3 per 1 H_2O. Exciting line: ν_0 = 3638 Å, *ca.* 10 mW; slit width s = 9 cm^{-1}; rotating cell.

caesium triiodide in aqueous and methanol solutions showed that this progression should be assigned to the totally symmetric stretching vibration of the I_3^- ion. This ion has two strong absorption peaks at 3500 and 2750 Å (Fig. 5). Again a high intensity overtone progression has been found in the Raman spectrum excited in the ultraviolet, as can be seen in Fig. 6. Table 4 summarizes the assignment of the observed Raman lines and gives their half bandwidth, depolarization ratio and relative intensity together with corres-

Table 4. The Raman spectrum, half bandwidth, depolarization ratio and relative intensity of the overtone progression of the I_3^- ion in water and methanol solutions.

Assignment	Water[a]				Methanol[b]		
	Wave-number (cm^{-1})	Half band width (cm^{-1})[c]	Depolar. ratio ρs	Relative intensity	Wave-number (cm^{-1})	Half band width (cm^{-1})[c]	Relative intensity
ν_1	116	35	0.33	1.00[d]	111	27	1.00[d]
$2\nu_1$	234	43	0.35	0.49	223	42	0.59
$3\nu_1$	351	55	0.34	0.30	342	46	0.33
$4\nu_1$	460	65	~0.35	~0.2	455	58	0.29
$5\nu_1$	580				572		
$6\nu_1$	~690				684		
$7\nu_1$					~809		
$8\nu_1$					~919		

[a] CsI_3 concentration 2.5 × 10^{-4} mole/l
[b] Iodine concentrations 10^{-3} to 10^{-4} mole/l
[c] Used slit width 9 cm^{-1}
[d] Normalized to 1.00 for ν_1

ponding values for methanol solutions. It is clear that the polarized Raman line at 116 cm^{-1} has to be assigned to the symmetric vibration ν_1 of the I_3^- ion and the Raman lines with higher wavenumbers to the harmonics of this vibration. The ν_1 progression shows a broadening of the band width at half height $\Delta\nu_{1/2}$ and also decreasing intensity with increasing vibrational quantum number. It should be mentioned that the overtone progression of the triiodide ion in methanol was obtained from a solution of methanol containing only iodine. The formation of the triiodide ion in this system can be explained by the reaction of the ionic structure of a charge-transfer complex $(MeOHI)^+I^-$ with iodine, which seems to be favoured at low iodine concentrations. Also a weak, polarized band was found at 143 cm^{-1} which is assigned to the asymmetric stretching vibration ν_3 of I_3^-.

CONCLUSION

We have shown that the symmetric stretching vibration of the CrO_4^{2-} and I_3^- ion produces a high intensity overtone progression when excited by a laser line which lies inside the absorption band. Furthermore, the resonance Raman effect can be used to confirm the presence of particular species in solutions in a convincing way.

REFERENCES

1 W. Kiefer and H. J. Bernstein, *Appl. Spectry.* **25**, 500 (1971).
2 W. Kiefer and H. J. Bernstein, *Appl. Spectry.* **25**, 609 (1971).
3 W. Kiefer and H. J. Bernstein, *Mol. Phys.* **23**, 835 (1972).
4 J. C. Duinker and C. J. Ballhausen, *Theoret. Chim. Acta* **12**, 325 (1968).
5 L. W. Johnson and S. P. McGlynn, *Chem. Phys. Letters* **7**, 618 (1970).
6 H. Stammreich, D. Bassi and O. Sala, *Spectrochim. Acta* **12**, 403 (1958).
7 D. M. Adams, M. A. Hooper and M. H. Lloyd, *J. Chem. Soc.* **27A**, 946 (1971).

8 R. L. Carter and C. E. Bricker, *Spectrochim. Acta* **27A**, 569 (1971).
9 G. Herzberg, *Molecular Spectra and Molecular Structure II. Infrared and Raman Spectra*, van Nostrand, Princeton, 1964, p. 205.
10 A. C. Albrecht and M. C. Hutley, *J. Chem. Phys.* **55**, 4438 (1971).
11 W. Kiefer and H. J. Bernstein, (to be published).
12 W. Kiefer and H. J. Bernstein, *Chem. Phys. Letters* (to be published).

39. THEORETICAL STUDY OF RESONANCE RAMAN SCATTERING BY THE METHODS OF QUANTUM ELECTRONICS

M. Jacon

Laboratoire de Recherches Optiques,
Faculté des Sciences, Reims, France

GENERAL INTRODUCTION

In theoretical studies of Raman scattering, two stages may be discerned. A first approach is given by studying the interaction of a molecule with the electromagnetic field, where the molecules of the scattering system can be considered to be independent of each other: this is the strict theory of the light–molecule interaction. This problem has been studied in 1934 by Placzek[1] and also by Weisskopf[2] who was especially interested in the problem of resonant scattering.

In the second stage, which is more complete, in addition to the light–molecule interaction, the interactions between the scattering molecule and the neighbouring molecules are taken into account. This allows, for instance, band profile analysis which nowadays has become more and more important.

In this work, we will study in a complete and straightforward manner, the light–molecule interaction, and we will also include the effects of the surrounding molecules in order to discuss some of our experimental results.

Let us consider first the light–molecule interaction. We will present in this paper the new results obtained with one of the modern methods of quantum electronics known as resolvant formalism.[3–6] For this purpose, we first give the fundamental equations of this formalism which will then be applied to the theory of resonance Raman scattering. We will not follow the usual separation preresonance and strict resonance Raman effect since this separation may not always be well defined. We prefer to proceed as follows: we present first the general expression of the scattering tensor which depends on the relative position of the frequency of excitation with respect to the absorption bands of the scattering molecule, and on the structure of the spectrum of the Hamiltonian of the molecule. This will allow us to study the following cases:

1. preresonance Raman effect together with resonance at excitation in the continuum;
2. resonance at excitation in a discrete level.

The particular case of excitation in a discrete level will be treated while including the interaction of the scattering molecule with the neighbouring molecules. We may thus follow in this theory the transition from resonance fluorescence to resonance Raman effect, while the excited states of the scattering molecule are more and more perturbed by the environment. A detailed experimental investigation has been carried out by Berjot[7] at the Laboratoire de Recherches Optiques de la Faculté des Sciences de Reims.

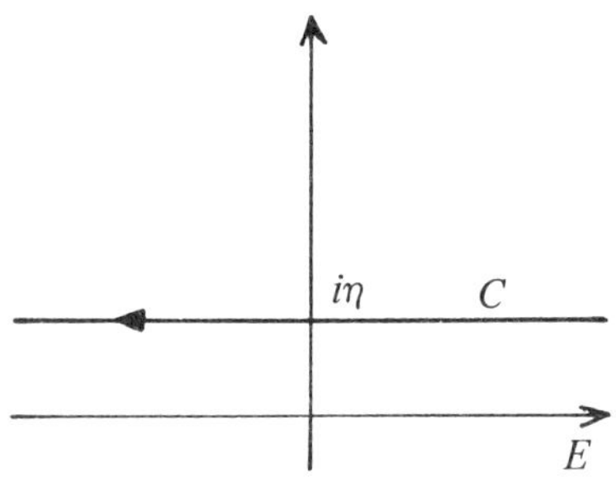

Fig. 1. The contour of integration for Eqn. (4).

THE LIGHT–MOLECULE INTERACTION

The calculation of the amplitudes of the transition probability using the resolvant formalism

The system studied consists of one molecule and the interacting electromagnetic field. Its Hamiltonian is $H = H_0 + V$, where H_0 is the free Hamiltonian and V the Hamiltonian of the interaction.

The system is at time $t = 0$ in the eigenstate $|1\rangle$ of the Hamiltonian H_0. Its state vector at time t is determined by means of the evolution operator $U(t, 0) = \exp(-iHt/\hbar)$ $(t \geqslant 0)$ by the relation

$$|\psi(t)\rangle = U(t, 0)|1\rangle \tag{1}$$

The amplitude of the transition probability at time t is given by

$$\langle 2|\psi(t)\rangle = \langle 2|U(t, 0)|1\rangle \tag{2}$$

This probability amplitude can be expressed[3–6, 8] using the Green's function (or resolvant operator)

$$G(z) = (z - H)^{-1} \tag{3}$$

by the following integral

$$\langle 2|U(t, 0)|1\rangle = \frac{1}{2\pi i}\int_C \exp(-izt/\hbar)\langle 2|G(z)|1\rangle\,dz \tag{4}$$

The contour C is infinitely close to the real axis (Fig. 1): $z = E + i\eta$, where E is the energy of the system. The calculation of this integral is simplified by introducing the projector P_1 on the initial state and the complementary projector $Q_1 = 1 - P_1$.

The probability amplitude of the transition into a state $|2\rangle \neq |1\rangle$ can be written as

$$\langle 2|U(t,0)|1\rangle = \frac{1}{2\pi i} \int_C \exp(-izt/\hbar) \langle 2|Q_1 G(z) P_1|1\rangle \, dz \tag{5}$$

It has been shown by Mower[5,6] that the two projections $P_1G(z)P_1$ and $Q_1G(z)P_1$ of the Green's function $G(z)$ are determined by the equations

$$P_1G(z)P_1 = P_1(z - H_0 - P_1R(z)P_1)^{-1} \tag{6}$$

and

$$Q_1G(z)P_1 = Q_1(z - Q_1HQ_1)^{-1} Q_1VP_1G(z)P_1 \tag{7}$$

with

$$R(z) = V + VQ_1(z - Q_1HQ_1)^{-1} Q_1V \tag{8}$$

It follows that

$$Q_1G(z)P_1 = Q_1(z - H_0)^{-1} Q_1R(z) P_1G(z)P_1 \tag{9}$$

and

$$R(z) = V + VQ_1(z - H_0)^{-1} Q_1R(z) \tag{10}$$

Equation (10) allows us to calculate $R(z)$ by successive approximations. On the other hand, the S-matrix which gives the scattering cross sections is completely determined[3,4] by the operator $R(z)$. One obtains

$$S = 1 - 2\pi i\delta(E_2 - E_1) R(E_1 + i0) \tag{11}$$

$$R(E_1 + i0) = \lim_{\eta \to 0^+} R(E_1 + i\eta)$$

E_1 and E_2 are the energies of the initial and final states respectively.

One of the advantages of the resolvant formalism is that the different processes of the interaction between light and molecule can be followed clearly in the calculations. The operator $P_1G(z)P_1$ determines the probability to find the system in the state $|1\rangle$ at time t: $P_1G(z)P_1$ therefore describes the propagation of the system in this state. This operator describes the natural width and the Lamb shift of the state $|1\rangle$. On the other hand, $Q_1G(z)P_1$ which determines the probability of transition from state $|1\rangle$ to state $|2\rangle \neq |1\rangle$ may be interpreted in the following way, when reading Eqn. (9) from right to left:

1. Evolution of the interacting system in the initial state $|1\rangle$ given by $P_1G(z)P_1$.
2. Transition by means of the operator $R(z)$.
3. Free evolution of the system in the final state $|2\rangle$ given by the operator $Q_1(z - H_0)^{-1} Q_1$; $(z - H_0)^{-1}$ corresponds indeed to the propagator of the free system.

The interaction processes can also be interpreted using the Feynman diagrams.[8,9]

General expression of the Raman scattering tensor

In a previous work, the resolvant formalism has enabled us to account for the natural line width of the Raman bands in the intensity calculations.[8] When this width is neglected, the formula for the scattering cross section calculated by Placzeck is found. This theory is based on the following approximations

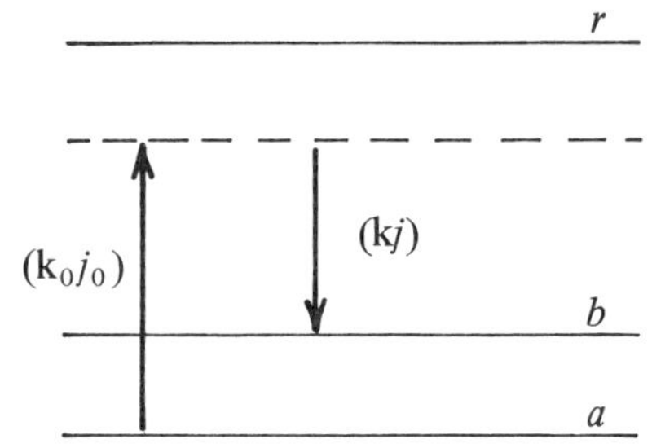

Fig. 2. Energy diagram for Raman scattering.

$$Q_1G(z)P_1 = Q_1(z-H_0)^{-1}\,Q_1R(z)P_1(z-H_0)^{-1}P_1 \tag{12}$$

and

$$R(z) \cong V + VQ_1(z-H_0)^{-1}Q_1V \tag{13}$$

Equation (13) is the second order approximation of the operator $R(z)$. The Raman intensity of the transition given in Fig. 2 is then completely determined by the following matrix element of $R(z)$

$$\langle b;\mathbf{k}j\,|R(E_1+i0)|\,a;\mathbf{k}_0j_0\rangle \tag{14}$$

$$= -\frac{1}{\hbar}\,\underset{r}{\mathsf{S}}\left[\frac{\langle b;\mathbf{k}j|V|r\rangle\langle r|V|a;\mathbf{k}_0j_0\rangle}{\omega_{ra}-\omega_0-i0}+\frac{\langle b;\mathbf{k}j|V|r;\mathbf{k}j,\mathbf{k}_0j_0\rangle\langle r;\mathbf{k}j,\mathbf{k}_0j_0|V|a;\mathbf{k}_0j_0\rangle}{\omega_{rb}+\omega_0-i0}\right]$$

(We have written the denominators of this formula by using the law of energy conservation: $E_1 = E_2$). We call this matrix element the scattering tensor. The symbol S_r denotes the summation over all the intermediate states of the molecule which may be of two types: discrete states and states belonging to a continuum.

Equation (14) is not valid in the case of resonance with a discrete state because one of the terms of the tensor becomes infinite. However, in the case when the frequency ω_0 coincides with an eigen-frequency of the continuous spectrum of the molecule, 'resonance catastrophy' does not occur and Eqn. (14) remains finite. This results[8] from the operator relation

$$\frac{1}{E-H_0+i0} = \mathcal{P}\,\frac{1}{E-H_0} - i\pi\delta(E-H_0) \tag{15}$$

from which the following expression of the scattering tensor can be obtained

$$\langle b;\mathbf{k}j|R(E_1+i0)|a;\mathbf{k}_0j_0\rangle = -\frac{i\pi}{\hbar}\left[\langle b;\mathbf{k}j|V|\alpha\rangle\langle\alpha|V|a;\mathbf{k}_0j_0\rangle\rho_\alpha(\omega')\right]_{\omega'=\omega_0} \tag{16}$$

$$+\,\mathcal{P}\int_{\text{continuum}}\frac{\langle b;\mathbf{k}j|V|\alpha\rangle\langle\alpha|V|a;\mathbf{k}_0j_0\rangle}{\hbar(\omega_0-\omega')}\,\rho_\alpha(\omega')\,d\omega'$$

[Continuation of Eqn. (16)]

$$+\int_{\text{continuum}} \frac{\langle b;\mathbf{k}j\,|V|\,\alpha;\mathbf{k}j,\mathbf{k}_0j_0\rangle\langle\alpha;\mathbf{k}j,\mathbf{k}_0j_0\,|V|\,\alpha;\mathbf{k}_0j_0\rangle}{E_b-E'-\hbar\omega_0}\rho_\alpha(\omega')\,d\omega'$$

$$-\sum_r\left(\frac{\langle b;\mathbf{k}j|\,V|r\rangle\langle r|\,V|a;\mathbf{k}_0j_0\rangle}{\hbar(\omega_{ra}-\omega_0)}+\frac{\langle b;\mathbf{k}j|\,V|r;\mathbf{k}j,k_0j_0\rangle\langle r;\mathbf{k}j,\mathbf{k}_0j_0\,|\,V|a;\mathbf{k}_0j_0\rangle}{\hbar(\omega_{rb}+\omega_0)}\right) \quad (16)$$

In this formula, we have noted by $E'(\alpha)=\hbar\omega'$ the energy of a state $|\alpha\rangle$ of the continuum, to be counted from the origin of the energy E_a of the initial state, and by $\rho_\alpha(E')$ the density of the states of the continuum. The first term of Eqn. (16) is the resonant term of the scattering amplitude and corresponds to a real intermediate transition ($\hbar\omega_0=\hbar\omega'$). The two integrals on the continuum and the summation Σ_r on the discrete states correspond to a virtual intermediate transition during which the energy is not conserved.[8]

The case of resonance with a discrete state will be treated now. Here, it is necessary to account for the natural width of the various intermediate states. The calculation has been worked out by Weisskopf.[2] The result may be obtained in a simpler way by the resolvant formalism.[10] For this calculation, the approximations (12) and (13) are not sufficient and we have to use the rigorous expression of $R(z)$ given in Eqn. (8). This gives

$$Q_1G(z)P_1 = Q_1(z-H_0)^{-1}Q_1V\sum_3[P_3(z-Q_1HQ_1)^{-1}P_3]VP_1(z-H_0)^{-1}P_1 \quad (17)$$

$\sum_3 P_3$ denotes the projector on the set of the possible intermediate states. We obtain then the expression of the scattering tensor

$$\langle b;\mathbf{k}j|R(E_1+i0)|a;\mathbf{k}_0j_0\rangle$$

$$=-\frac{1}{\hbar}\sum\left(\frac{\langle b;\mathbf{k}j|\,V|r\rangle\langle r|\,V|a;\mathbf{k}_0j_0\rangle}{\bar\omega_{ra}-\omega_0-i\Gamma_r/2\hbar}+\frac{\langle b;\mathbf{k}j|\,V|r;\mathbf{k}j,\mathbf{k}_0j_0\rangle\langle r;\mathbf{k}j,\mathbf{k}_0j_0\,|\,V|a;\mathbf{k}_0j_0\rangle}{\bar\omega_{rb}+\omega_0-i\Gamma_r/2\hbar}\right) \quad (18)$$

with $\hbar\bar\omega_{ra} = \bar E_r-E_a = E_r+\Delta_r-E_a.$

Γ_r is the natural width of the state r and Δ_r its energy displacement (Lamb shift).

In general, the expression for the Raman scattering tensor is very complicated because all the molecular states intervene, which is the intrinsic difficulty in Raman theory. Therefore, we will now present various methods to obtain simpler theoretical expressions which may be compared with experimental results.

The scattering tensor in the Born–Oppenheimer approximation of the matrix elements of the electric dipole operator: the study of Raman effect obtained by excitation in an absorption continuum

The straightforward way to obtain numerical results for the elements of the scattering tensor, either given by Eqn. (14) or by Eqn. (18), would be to introduce the Born–Oppenheimer approximation into the molecular wave functions and then to calculate explicitly the matrix elements of the dipole operator. Unfortunately, in general this is impossible since the excited states of the molecules are not well known. However, there are exceptions and the great interest in the resonance Raman spectra of the halogen molecules[11, 12] consists in the fact that here, theory can be tested in a system like the iodine molecule where the vibro-electronic structure is well known.

We give below the expression of the intensities of Raman bands obtained in ref. 13 when excitation takes place in the continuum.

$$\frac{dI_{ab}}{d\Omega} = \frac{I_0}{c^4}(\omega_0 + \omega_{ab})^4 \left| \frac{i\pi}{\hbar} [\langle b|\mathbf{D}.\mathbf{e}_{\mathbf{k}j}|\alpha\rangle\langle\alpha|\mathbf{D}.\mathbf{e}_{\mathbf{k}_0 j_0}|a\rangle\, \rho_\alpha(\omega')]_{\omega'=\omega_0} \right.$$
$$+ \mathcal{P} \int_{\text{continuum}} \frac{\langle b|\mathbf{D}.\mathbf{e}_{\mathbf{k}j}|\alpha\rangle\langle\alpha|\mathbf{D}.\mathbf{e}_{\mathbf{k}_0 j_0}|a\rangle}{\hbar(\omega' - \omega_0)}\rho_\alpha(\omega')\,d\omega' + \int_{\text{continuum}} \frac{\langle b|\mathbf{D}.\mathbf{e}_{\mathbf{k}_0 j_0}|\alpha\rangle\langle\alpha|\mathbf{D}.\mathbf{e}_{\mathbf{k}j}|a\rangle}{E' - E_b + \hbar\omega_0}\rho_\alpha(\omega')\,d\omega'$$
$$\left. + \sum_r \left(\frac{\langle b|\mathbf{D}.\mathbf{e}_{\mathbf{k}j}|r\rangle\langle r|\mathbf{D}.\mathbf{e}_{\mathbf{k}_0 j_0}|a\rangle}{\hbar(\omega_{ra} - \omega_0)} + \frac{\langle b|\mathbf{D}.\mathbf{e}_{\mathbf{k}_0 j_0}|r\rangle\langle r|\mathbf{D}.\mathbf{e}_{\mathbf{k}j}|a\rangle}{\hbar(\omega_{rb} + \omega_0)} \right) \right|^2 \tag{19}$$

This expression may be easily deduced from the Eqns. (11) and (16). The explicit calculation of the intensities in the Born–Oppenheimer approximation of the dipole matrix elements has been performed by Berjot[14] in the case of iodine and there is good agreement with the experimental results. In addition, these calculations have shown the following important fact: the resonant term of the scattering amplitude

$$A = \frac{i\pi}{\hbar} [\langle b|\mathbf{D}.\mathbf{e}_{\mathbf{k}j}|\alpha\rangle\langle\alpha|\mathbf{D}.\mathbf{e}_{\mathbf{k}_0 j_0}|a\rangle\, \rho_\alpha(\omega')]_{\omega'=\omega_0} \tag{20}$$

is not necessarily the most important. The other terms are by no means negligible and it may happen that the major contribution to intensity will arise from these terms. From this, it follows that the intensities of the Raman bands observed in resonance by excitation in a continuum do not only depend on the excited state whose energy coincides with the one of the incident photon, but on all the possible intermediate states. This result is characteristic of the resonance Raman effect, in contrast to resonance fluorescence where, in general, a single discrete excited state intervenes.

However, only in simple and well known systems is this kind of calculation of the intensities of the Raman bands possible. We shall therefore look for further approximations. The second method which we will discuss may also be applied to preresonance Raman effect as, under certain conditions, to resonance Raman effect in a continuum.

Operator expression of the scattering tensor in the Born–Oppenheimer approximation. The problem of the change of intensities as a function of excitation frequency

1. The hypothesis. For simplicity, we consider only one normal coordinate of vibration (diatomic molecule) and a single excited electronic state. First we give the conditions for the position of the exciting frequency:

(a) ω_0 should be sufficiently separated from the discrete eigen-frequencies of the molecule, which will allow the natural width Γ_r to be neglected.
(b) The energy $E_0 = \hbar\omega_0$ of the incident photon should obey the relation

$$|E_{e'e} - E_0| \gg \hbar\Omega \tag{21}$$

$|E_{e'e} - E_0|$ is the difference between the energy of the pure electronic transition of the molecule and the energy of the incident photon, and $\hbar\Omega$ is the energy of one quantum of the vibration of the molecule. The following calculations are thus valid for the preresonance Raman effect and also for the Raman effect excited in a continuum when the above conditions are obeyed: this is the case of the iodine molecule in the gas phase excited in the absorption continuum.[15]

As the natural width may be neglected, we may write the scattering tensor as follows:

$$\begin{aligned}&\langle b; \mathbf{k}j | R(E_1 + i0) | a; \mathbf{k}_0 j_0\rangle \\ &= -\frac{1}{\hbar} \underset{r}{\mathrm{S}} \left(\frac{\langle b; \mathbf{k}j | V | r\rangle\langle r | V | a; \mathbf{k}_0 j_0\rangle}{\omega_{ra} - \omega_0 - i0} + \frac{\langle b; \mathbf{k}j | V | r; \mathbf{k}j, \mathbf{k}_0 j_0\rangle\langle r; \mathbf{k}j, \mathbf{k}_0 j_0 | V | a; \mathbf{k}_0 j_0\rangle}{\omega_{rb} + \omega_0 - i0} \right)\end{aligned} \tag{22}$$

2. Operator expression of the scattering tensor in the Born–Oppenheimer approximation. In the Born–Oppenheimer approximation,[16] the Hamiltonian H_m of the molecule can be considered as being the sum of the electronic Hamiltonian $H_e(X)$ which depends partly on the coordinates X of the nuclei, and the Hamiltonian H_{en} of the nuclei for the electronic state e. We denote by e and e' the fundamental and excited electronic states. Using Eqn. (13), one may write the operator $R(z)$ as follows:

$$R(z) \cong V + VQ_1[z - H_e(X) - H_{en} - H_r]^{-1} Q_1 V \tag{23}$$

H_r is the Hamiltonian of the electromagnetic field.

The molecular eigenstates are written[16] as a tensor product

$$|ev\rangle = |e(X)\rangle \otimes |v\rangle \tag{24}$$

The notation $|e(X)\rangle$ shows explicitly the parametric dependence of the electronic state e with respect to the nuclear coordinates and $|v\rangle$ corresponds to the vibrational state of the nuclei.

The energy of the state $|ev\rangle$ is given by

$$E_{ev} = E_e(X = X_0) + \langle v | H_{en} | v\rangle \tag{25}$$

$E_e(X = X_0)$ is the energy of the electronic state e corresponding to the equilibrium configuration X_0 of the nuclei and $\langle v | H_{en} | v\rangle$ is the energy of vibration.

It results from the Born–Oppenheimer approximation that the dipole moments of the transition between two vibronic states $|ev\rangle$ and $|e'v'\rangle$ can be expressed as in Eqn. (26).

$$\langle ev|\mathbf{D}|e'v'\rangle = \langle v|\mathbf{D}_{ee'}|v'\rangle \tag{26}$$

where the electronic dipole moment of transition $\mathbf{D}_{ee'}$ is a function of the coordinates X of the nuclei.

It can be shown by means of Eqn. (23) that the expression of the Raman scattering tensor takes the following form

$$\langle \nu_2|T_{si}|\nu_1\rangle = \sum_{v'}\left(\frac{\langle \nu_2|D_{s,ee'}|\nu'\rangle\langle \nu'|D_{i;e'e}|\nu_1\rangle}{\langle v'|H_{e'n}|v'\rangle + E_1 - E_0} + \frac{\langle \nu_2|D_{i;ee'}|\nu'\rangle\langle \nu'|D_{s,e'e}|\nu_1\rangle}{\langle v'|H_{e'n}|v'\rangle + E_2 + E_0}\right) \tag{27}$$

with

$$\begin{aligned} E_1 &= E_{e'}(X = X_0') - E_e(X = X_0) - \langle \nu_1|H_{en}|\nu_1\rangle \\ E_2 &= E_{e'}(X = X_0') - E_e(X = X_0) - \langle \nu_2|H_{en}|\nu_2\rangle \end{aligned} \tag{28}$$

X_0' gives the equilibrium coordinates of the nuclei in the excited electronic state, D_{i} and D_{s} design the projections of the dipole moment on the polarization vectors of the incident and the scattered photon respectively. (In fact, the tensor in Eqn. (27) differs from the the tensor in Eqn. (22) by a constant, but we adopt this expression in order to enable a comparison with the theoretical expressions obtained by other authors).

Using the closure relation $\sum_{\nu'} |\nu'\rangle\langle \nu'| = 1$, the matrix element in Eqn. (27) can be written

$$\begin{aligned} \langle \nu_2|T_{si}|\nu_1\rangle &= \langle \nu_2|D_{s,ee'}(H_{e'n} + E_1 - E_0)^{-1}D_{i,ee'}|\nu_1\rangle \\ &+ \langle \nu_2|D_{i,ee'}(H_{e'n} + E_2 + E_0)^{-1}D_{s,e'e}|\nu_1\rangle \end{aligned} \tag{29}$$

which defines the tensor operator T in the space of the vibrational states of the nuclei.

3. Application of this result to the case of fundamentals and overtones. The expressions for the scattering tensor of the various Raman bands can be easily obtained using the two following developments:

(a) Expansion[3] of the dipole moments with respect to the creation and annihilation operators a^+ and a corresponding to the normal coordinate q.

$$\mathbf{D}_{ee'} = (\mathbf{D}_{ee'})_0 + \sqrt{\frac{\hbar}{2\Omega}}\left(\frac{\partial \mathbf{D}_{ee'}}{\partial q}\right)_0 (a + a^+) + \frac{\hbar}{4\Omega}\left(\frac{\partial^2 \mathbf{D}_{ee'}}{\partial q^2}\right)_0 (a + a^+)^2 \ldots \tag{30}$$

(b) Perturbation expansion[8] of the operator $(H_{e'n} + E)^{-1}$

$$(H_{e'n} + E)^{-1} = (H_{en} + E)^{-1} - (H_{en} + E)^{-1}\Delta H(H_{en} + E)^{-1} + \ldots \tag{31}$$

with

$$\Delta H = H_{e'n} - H_{en} \tag{32}$$

The convergence of this development requires Eqn. (21) to be fulfilled. On the other hand, ΔH will also be expanded in terms of a and a^+

$$\Delta H = (\Delta H)_0 + \sqrt{\frac{\hbar}{2\Omega}}\left(\frac{\partial \Delta H}{\partial q}\right)_0 (a + a^+) + \ldots \tag{33}$$

The evaluation of the scattering tensor of the fundamental bands at the lowest order approximation requires the first order expansion of Eqns. (30) and (31) which are then introduced into Eqn. (29). Thus, we obtain

$$\langle n+1 | T_x | n\rangle$$

$$= \sqrt{\frac{\hbar(n+1)}{2\Omega}}\left[\left(\frac{\partial}{\partial q}|D_{x,ee'}|^2\right)_0 \frac{2E_{e'e}}{E_{e'e}^2 - E_0^2} - 2|(D_{x,ee'})_0|^2 \left(\frac{\partial \Delta H}{\partial q}\right)_0 \frac{E_{e'e}^2 + E_0^2}{(E_{e'e}^2 - E_0^2)^2}\right] \tag{34}$$

In fact, this expression is similar to the one obtained by Shorygin.[17] Therefore, the theory developed here is thus able to reproduce, in a purely quantum mechanical way, the semi-classical theory of preresonance Raman effect.

In order to obtain the scattering tensor of the first overtone, the second order expansion of Eqns. (30) and (31) has to be used. This leads to the following result

$$\langle n+2 | T_x | n\rangle$$

$$= \frac{\hbar}{2\Omega}\sqrt{(n+1)(n+2)}\left\{\frac{E_{e'e}^2 + E_0^2}{(E_{e'e}^2 - E_0^2)^2}\left[-2\left(\frac{\partial}{\partial q}|D_{x,ee'}|^2\right)_0 \left(\frac{\partial \Delta H}{\partial q}\right)_0 - |(D_{x,ee'})_0|^2 \left(\frac{\partial^2 \Delta H}{\partial q^2}\right)_0\right]\right.$$

$$\left. + 2E_{e'e}\frac{E_{e'e}^2 + 3E_0^2}{(E_{e'e}^2 - E_0^2)^3}|(D_{x,ee'})_0|^2 \left(\frac{\partial \Delta H}{\partial q}\right)_0^2 + \frac{E_{e'e}}{E_{e'e}^2 - E_0^2}\left(\frac{\partial^2}{\partial q^2}|D_{x,ee'}|^2\right)_0\right\} \tag{35}$$

The terms of the tensor in Eqn. (35) which determine the variation of the intensity as a function of the exciting frequency are similar to the ones obtained by Kondilenko.[18] However, the other factors are different. Finally, we want to point out the fact that Eqn. (35) has been obtained in a relatively simple way. Thus, the calculation may be easily extended to higher order harmonics and combination bands.

4. Conclusion. The theory presented here holds in cases of preresonance Raman effect, and also in cases of resonance due to excitation in a continuum, if certain conditions which have been pointed out, are fulfilled. This theory confirms the preresonance theory of Shorygin and describes well the experimental results obtained in the case where excitation takes place in a continuum.[15]

In experimental work, one usually separates preresonance Raman effect from strict resonance Raman effect which is, in particular, characterized by a large number of overtones. However, with respect to the resonance frequency $\omega_{e'e} = E_{e'e/\hbar}$, the Raman effect observed in iodine in the gaseous state excited in the continuum of the absorption, acts theoretically in the same way as in preresonance Raman effect. For this reason, the two phenomena have not been discussed separately.

In order to explain the presence of a sequence of overtones, as observed in the iodine Raman spectrum, one has also to take into consideration that the terms of the scattering tensor which depend explicitly on the exciting frequency do not completely determine the intensities. The derivative $(\partial\Delta H/\partial q)_0$ which enters into the expression of the intensities for the fundamentals as square terms, and into the one for the first overtone by the fourth power, plays also an important role.[15]

Finally , the temperature dependence of scattering is also an important factor which can easily be accounted for in the theoretical expressions.

INTERACTION OF THE SCATTERING MOLECULE WITH THE NEIGHBOURING MOLECULES: CONTINUOUS TRANSITION FROM RESONANCE FLUORESCENCE TO RESONANCE RAMAN EFFECT

In the first part of this paper, we were only interested in the light–molecule interaction and we were able to determine in this case the scattering tensor for different conditions of optical excitation. Let us consider Eqn. (18) corresponding to optical excitation into or extremely close to a discrete level of a molecule: the damping terms which appear in the denominators are defined by the natural widths of the various states r. Obviously, such a theory cannot explain the phenomena arising from interactions between the scattering molecule and the neighbouring molecules, as for instance quenching of resonance fluorescence.

In this second part we will give an expression for the resonant scattering tensor, for the case of optical excitation very close to discrete levels of the molecule, which takes into account the perturbations by the neighbouring molecules. These calculations are necessary for the interpretation of the experiments on the continuous transition from resonance fluorescence to resonance Raman effect.[7]

The problem

We suppose that the experiment of resonant scattering is done on a discrete level of a molecule which is perturbed by other molecules (or atoms). In the special experiments which we performed,[7] an iodine molecule is perturbed by the atoms of rare gases. The total Hamiltonian of the system is as follows

$$\begin{aligned} H &= H_0 + H_1 \\ \text{where} \quad H_0 &= H_{\mathrm{m}} + H_{\mathrm{r}} + H_{\mathrm{p}} \\ H_1 &= V_{\mathrm{mr}} + V_{\mathrm{mp}} \end{aligned} \tag{36}$$

with the following definitions:

H_{m}: Hamiltonian of the scattering molecule
H_{r}: Hamiltonian of irradiation
H_{p}: Hamiltonian of the perturbers
V_{mr}: interaction of the scattering molecule with light
V_{mp}: interaction of the scattering molecule with perturbers.

The states of the non-interacting system, scattering molecule–light–perturbers, are described by the following state vectors:

$$|S\rangle = |s;\alpha\rangle = |m;\rho;\alpha\rangle = |m\rangle \otimes |\rho\rangle \otimes |\alpha\rangle \tag{37}$$

Here S describes the set of the quantum numbers which defines the total system

(S-system), s describes the quantum numbers defining the system light-scattering molecule (s-system).

$s = (m, \rho)$

m: molecular quantum numbers
ρ: irradiation quantum numbers
α: quantum numbers of the system of the perturbing atoms (p-system).

The equation for the evolution of the density matrices of the S- and s-systems

The evolution of the density matrix $\rho_{\mathrm{mrp}}(t)$ of the S-system is determined[19] by the following equation

$$\rho_{\mathrm{mrp}}(t) = \exp(-iLt/\hbar)\, \rho_{\mathrm{mrp}}(0) \tag{38}$$

L is a Liouville operator defined by

$$L = HI^* - IH^* \tag{39}$$

We use the notation of Baranger[20] to describe the elements of the matrices of the Liouville operators

$$\langle\langle S_1 S_1'^{+} | L | S_2 S_2'^{+} \rangle\rangle = \langle S_1 | H | S_1' \rangle\, \delta_{S_2 S_2'} - \delta_{S_1 S_1'} \langle S_2' | H | S_2 \rangle \tag{40}$$

We decompose $L = L_0 + L_1$, where L_0 belongs to the operator H_0 and L_1 to the operator H_1. Similarly, as has been done for the evolution operator $U(t, 0)$ of the state vector $|\Psi(t)\rangle$, we introduce[19] now the resolvant operator $G(z) = (z - L)^{-1}$ with respect to the Liouville operator L. The evolution Eqn. (38) can now be written as

$$\rho_{\mathrm{mrp}}(t) = \frac{1}{2\pi i} \int_c \exp(-izt/\hbar) \frac{1}{z - L}\, \mathrm{d}z\, \rho_{\mathrm{mrp}}(0) \tag{41}$$

On the other hand, the density matrix $\rho_{\mathrm{mrp}}(t)$ of the s-system is given[21] by

$$\rho_{\mathrm{mr}}(t) = \mathrm{Tr}_{\mathrm{p}}\, [\rho_{\mathrm{mrp}}(t)] \tag{42}$$

The symbol Tr_{p} describes the trace operation on the p-system. This last equation can be written, using Eqn. (41), in the following way

$$\rho_{\mathrm{mr}}(t) = \frac{1}{2\pi i} \int_c \exp(-izt/\hbar)\, \mathrm{Tr}_{\mathrm{p}}\left(\frac{1}{z - L} \rho_{\mathrm{p}}(0) \right) \mathrm{d}z\, \rho_{\mathrm{mr}}(0) \tag{43}$$

(We have here used the hypothesis that at time $t = 0$, there is no correlation between the s- and p-systems.)

The trace operation

In order to determine the density matrix $\rho_{\mathrm{mr}}(t)$, it is necessary to calculate

$\mathrm{Tr_p}\,[z-L)^{-1}\rho_\mathrm{p}(0)]$. This calculation can be done by introducing the Liouville operator T defined by the following relation

$$T(A\rho_\mathrm{p}) = \rho_\mathrm{p}\,\mathrm{Tr_p}\,(A\rho_\mathrm{p}) \tag{44}$$

It can easily be shown that this operator is idempotent. In addition, it has the following properties

$$\begin{aligned} &\text{(a)}\quad TAT\rho_\mathrm{p} = TA\rho_\mathrm{p} \\ &\text{(b)}\quad \mathrm{Tr_p}\,(TAT\rho_\mathrm{p}) = \mathrm{Tr_p}\,(A\rho_\mathrm{p}) \\ &\text{(c)}\quad (TAT)\,\rho_\mathrm{p} = \rho_\mathrm{p}\,[\mathrm{Tr_p}\,(TAT\rho_\mathrm{p})] \end{aligned} \tag{45}$$

Therefore, using (45), one may calculate

$$\mathrm{Tr_p}\,[G(z)\rho_\mathrm{p}(0)] = \mathrm{Tr_p}\,[(z-L)^{-1}\rho_\mathrm{p}(0)] = \mathrm{Tr_p}\,[TGT\rho_\mathrm{p}(0)] \tag{46}$$

First of all, the projection $TG(z)T$ of the resolvant $G(z)$ has to be determined. The calculation is formally identical to the one of the projection $P_1(z-H)^{-1}P_1$ given on p. 326. The result is

$$\begin{aligned} TG(z)T &= T[z-L_0-TR(z)T]^{-1} \\ R(z) &= L_1 + L_1S(z-SLS)^{-1}SL_1 \\ S &= 1-T \end{aligned} \tag{47}$$

(The operator $R(z)$ obtained here is the same as the operator $M_c(\omega)$ of Fano.[19])

The expression $TG(z)T$ can be developed into the following series

$$TG(z)T = T(z-L_0)^{-1}\sum_{n=0}^{\infty}\,[TR(z)T(z-L_0)^{-1}]^n \tag{48}$$

and, using the property (45c), we obtain as a final result

$$\mathrm{Tr_p}\,[G(z)\rho_\mathrm{p}(0)] = \frac{1}{z-L_{0s}-\mathrm{Tr_p}\,[R(z)\rho_\mathrm{p}(0)]} \tag{49}$$

L_{0s} describes the Liouville operator of the s-system without interaction. This expression shows that the 'effective Liouville interaction' for the s-system is $\mathrm{Tr_p}\,[R(z)\rho_\mathrm{p}(0)]$.

Introduction of the projector P_{11} on the initial state of the s-system

We are indeed interested in the initial state of the s-system in which the scattering molecule is found in state $|a\rangle$ together with an incident photon (k_0, j_0) of wave vector k_0 and polarization j_0. We designate this initial state by s_1. We have

$$\langle s'|\rho_\mathrm{mr}(0)|s\rangle = \delta_{s's_1}\,\delta_{ss_1}\,\langle s_1|\rho_\mathrm{mr}(0)|s_1\rangle \tag{50}$$

As in the first part of this paper, we introduce the Liouville projector P_{11} into this state

$$P_{11} = |s_1s_1^+\rangle\!\rangle\langle\!\langle s_1s_1^+| \tag{51}$$

Therefore, we have

$$\langle s'|P_{11}\rho_{mr}(0)|s\rangle = \delta_{s's_1}\delta_{ss_1}\langle s_1|\rho_{mr}(0)|s_1\rangle \tag{52}$$

The two projections of the density matrix of the s-system are given by

$$P_{11}\rho_{mr}(t) = \frac{1}{2\pi i}\int_c \exp(-izt/\hbar)\, P_{11}\, \mathrm{Tr}_p[G(z)\rho_p(0)]P_{11}\, dz\rho_{mr}(0) \tag{53}$$

and

$$Q_{11}\rho_{mr}(t) = \frac{1}{2\pi i}\int_c \exp(-izt/\hbar)\, Q_{11}\, \mathrm{Tr}_p[G(z)\rho_p(0)]P_{11}\, dz\rho_{mr}(0) \tag{54}$$

$$(Q_{11} = 1-P_{11})$$

with

$$\mathrm{Tr}_p[G(z)\rho_p(0)] = \frac{1}{z-L_{0s}-\mathrm{Tr}_p[R(z)\rho_p(0)]} \tag{55}$$

The formalism of the problem is thus equivalent to the one presented on p. 326 and one has to calculate the two projections $P_{11}\mathrm{Tr}_p(G\rho_p^0)P_{11}$ and $Q_{11}\mathrm{Tr}_p(G\rho_p^0)P_{11}$. $[\rho_p^0 = \rho_p(0)]$. These are given by

$$P_{11}\mathrm{Tr}_p(G\rho_p^0)P_{11} = P_{11}[z-L_{0s}-P_{11}R'(z)P_{11}]^{-1} \tag{56}$$

$$\begin{aligned} Q_{11}\mathrm{Tr}_p(G\rho_p^0)P_{11} = {} & Q_{11}[z-L_{0s}-Q_{11}\mathrm{Tr}_p(R\rho_p^0)Q_{11}]^{-1} \\ & \times Q_{11}\mathrm{Tr}_p(R\rho_p^0)P_{11}\mathrm{Tr}_p(G\rho_p^0)P_{11} \end{aligned} \tag{57}$$

with

$$R'(z) = \mathrm{Tr}_p(R\rho_p^0) + \mathrm{Tr}_p(R\rho_p^0)Q_{11}[z-L_{0s}-Q_{11}\mathrm{Tr}_p(R\rho_p^0)Q_{11}]^{-1}Q_{11}\mathrm{Tr}_p(R\rho_p^0) \tag{58}$$

Similarly, the following expression for $R'(z)$ can be given

$$R'(z) = \mathrm{Tr}_p(R\rho_p^0)\sum_{n=0}^{\infty}[(z-L_{0s})^{-1}Q_{11}\mathrm{Tr}_p(R\rho_p^0)]^n \tag{59}$$

This enables us to evaluate $R'(z)$ by successive approximations starting from $\mathrm{Tr}_p(R\rho_p^0)$.

By means of these fundamental expressions, we will now calculate the scattering cross section.

Calculation of the Raman scattering cross section

We denote

$$\bar{G}(z) = \mathrm{Tr}_p[G(z)\rho_p(0)] \tag{60}$$

$$\bar{L} = L_{0s} + \mathrm{Tr}_p[R(z)\rho_p(0)] = L_{0s} + \bar{R}(z) \tag{61}$$

In the same way as we have introduced a projector P_{11} on the initial state of the s-system, we shall now introduce a projector P_{22} on the final state in order to obtain an expression for the resolvant which is symmetric with respect to the two states. The transition probability is thus determined by $P_{22}\bar{G}(z)P_{11}$.

General expression for $P_{22}\bar{G}(z)P_{11}$. The following relation holds

$$P_{22}\bar{G}(z)P_{11} = P_{22}Q_{11}\bar{G}(z)P_{11} = P_{22}(z-Q_{11}\bar{L}Q_{11})^{-1}Q_{11}\bar{R}(z)P_{11}\bar{G}(z)P_{11} \tag{62}$$

with

$$P_{11}\bar{G}(z)P_{11} = P_{11}[z-L_{os}-P_{11}R'(z)P_{11}]^{-1} \tag{63}$$

In Eqn. (62), the operator $Q_{11}\bar{R}(z)P_{11}$ describes the transition of the scattering molecule from the initial state a to an excited state r, by means of the electromagnetic field. Therefore, we may write

$$Q_{11}\bar{R}(z)P_{11} \cong Q_{11}L_{1mr}P_{11} = Q_{22}L_{1mr}P_{11} \tag{64}$$

L_{1mr} is the Liouville operator associated with the interaction V_{mr}. The relation on the right has been obtained from the following properties

$$Q_{22} = Q_{11}+P_{11}-P_{22} \tag{65}$$

and

$$P_{11}L_{1mr}P_{11} = P_{22}L_{1mr}P_{11} = 0 \tag{66}$$

The last equation is fulfilled when V_{mr} is expressed in the dipole approximation, which we assume. It results from the structure of the matrix elements of L_{1mr}

$$\langle\langle 12^+|L_{1mr}|34^+\rangle\rangle = \langle 1|V_{mr}|3\rangle\delta_{24}-\delta_{13}\langle 4|V_{mr}|2\rangle \tag{67}$$

Thus, we have

$$P_{22}\bar{G}(z)P_{11} = P_{22}(z-Q_{11}\bar{L}Q_{11})^{-1}Q_{22}L_{1mr}P_{11}\bar{G}(z)P_{11} \tag{68}$$

We will now continue to develop the operator $P_{22}\bar{G}(z)P_{11}$.

Calculation of $P_{22}(z-Q_{11}\bar{L}Q_{11})^{-1}Q_{22}$. The principle of the calculation has already been given.[5, 6, 8] One obtains

$$P_{22}(z-Q_{11}\bar{L}Q_{11})^{-1}Q_{22} = P_{22}\bar{G}_{11}(z)Q_{22} \cong P_{22}L_{1mr}Q_{22}(z-Q_{11;22}\bar{L}Q_{11;22})^{-1}Q_{22} \tag{69}$$

(For the same reasons as above, we have written $P_{22}\bar{R}(z)Q_{22} \cong P_{22}L_{1mr}Q_{22}$.)

On the other hand, we have

$$Q_{11;22} = Q_{11}Q_{22} = Q_{22}Q_{11} = 1-P_{11}-P_{22} \tag{70}$$

$$P_{22}G_{11}(z)P_{22} = P_{22}[z-L_{os}-P_{22}\bar{R}_{11}(z)P_{22}]^{-1} \tag{71}$$

and

$$\bar{R}_{11}(z) = \bar{R}(z)+\bar{R}(z)Q_{22}(z-Q_{11;22}\bar{L}Q_{11;22})^{-1}Q_{22}\bar{R}(z) \tag{72}$$

This gives

$$P_{22}\bar{G}(z)P_{11} = P_{22}\bar{G}_{11}P_{22}L_{1mr}Q_{22}(z - Q_{11;\,22}\bar{L}Q_{11;\,22})^{-1}Q_{22}L_{1mr}P_{11}\bar{G}(z)P_{11} \tag{73}$$

In this expression, Q_{22} can be replaced by the projector on the set of the possible intermediate states. Taking into account Eqn. (66), this gives

$$P_{22}\bar{G}(z)P_{11} = P_{22}\bar{G}_{11}(z)P_{22}L_{1mr}P_{23}(z - Q_{11;\,22}\bar{L}Q_{11;\,22})^{-1}P_{13}\,L_{1mr}P_{11}\bar{G}(z)P_{11} \tag{74}$$

with

$$P_{23} = P_{32} = \sum_{\alpha}(P_{2\alpha} + P_{\alpha 2}) \quad \text{and} \quad P_{13} = P_{31} = \sum_{\beta}(P_{1\beta} + P_{\beta 1}) \tag{75}$$

In fact, it follows from Eqn. (67) that L_{1mr} couples only vectors with two equal indices.

Calculation of the operator $P_{23}(z - Q_{11;\,22}\bar{L}Q_{11;\,22})^{-1}P_{13}$. The calculation is identical with the one used to obtain Eqn. (69) and gives the following result

$$\begin{aligned} &P_{23}(z - Q_{11;\,22}\bar{L}Q_{11;\,22})^{-1}P_{13} = P_{23}\bar{G}_{11;\,22}(z)P_{13} \\ &\cong P_{23}(z - Q_{11;\,22;\,13}\bar{L}Q_{11;\,22;\,13})^{-1}Q_{13}L_{1mr}P_{13}\bar{G}_{11;\,22}(z)P_{13} \end{aligned} \tag{76}$$

In order to obtain an expression of fourth order in L_{1mr}, we apply once more the above process and 'cut off' the interaction at this order. Since. we want to determine the the transition probability from state $|1\rangle$ to state $|2\rangle$, the final projector should be P_{12}. We may thus write

$$\begin{aligned} &P_{23}(z - Q_{11;\,22}\bar{L}Q_{11;\,22})^{-1}P_{13} \\ &\simeq P_{23}(z - Q_{11;\,22;\,13}\bar{L}Q_{11;\,22;\,13})^{-1}P_{23}L_{1mr}P_{12}(z - L_{0s})^{-1}P_{12}L_{1mr}P_{13}\bar{G}_{11;\,22}(z)P_{13} \\ &= P_{23}\bar{G}_{11;\,22;\,13}P_{23}L_{1mr}P_{12}(z - L_{0s})^{-1}P_{12}L_{1mr}P_{13}\bar{G}_{11;\,22}(z)P_{13} \end{aligned} \tag{77}$$

One obtains finally the following expression of $P_{22}\bar{G}(z)P_{11}$

$$\begin{aligned} P_{22}\bar{G}(z)P_{11} = {} & P_{22}G_{11}(z)P_{22}L_{1mr}P_{23}\bar{G}_{11;\,22;\,13}(z)P_{23}L_{1mr}P_{21}(z - L_{0s})^{-1}P_{21} \\ & \times P_{21}L_{1mr}P_{31}\bar{G}_{11;\,22}(z)P_{31}L_{1mr}P_{11}\bar{G}(z)P_{11} \end{aligned} \tag{78}$$

In this formula, the various operators $\bar{R}_{11}, \bar{R}_{11;\,22}, \bar{R}_{11;\,22;\,13}$ are evaluated at the lowest order

$$\begin{aligned} P_{22}\bar{G}(z)P_{11} = {} & P_{22}[z - L_{0s} - P_{22}\bar{R}(z)P_{22}]^{-1}P_{22}L_{1mr}P_{23}[z - L_{0s} - P_{23}\bar{R}(z)P_{23}]^{-1}P_{23}L_{1mr}P_{21} \\ & \times P_{21}(z - L_{0s})^{-1}P_{21} \\ & \times P_{21}L_{1mr}P_{31}[z - L_{0s} - P_{31}\bar{R}(z)P_{31}]^{-1}P_{31}L_{1mr}P_{11}[z - L_{0s} - P_{11}\bar{R}(z)P_{11}]^{-1}P_{11} \end{aligned} \tag{79}$$

This expression of $P_{22}\bar{G}(z)P_{11}$ is especially remarkable because it enables us to visualize the diffusion process in the same way as has been done for the resolvant operator $(z - H)^{-1}$. The above expression is symmetric with respect to the two states $|1\rangle$ and $|2\rangle$ and will therefore give an expression for the transition probability in the form of the square modulus of a matrix element.

In particular, the evolution of the scattering molecule in the different excited states is described by the operators

$$P_{31}[z - L_{0s} - P_{31}\bar{R}(z)P_{31}]^{-1}P_{31} \tag{80}$$

and

$$P_{23}[z - L_{0s} - P_{23}\bar{R}(z)P_{23}]^{-1}P_{23} \tag{81}$$

These excited states evolve under the influence of the effective Liouville interaction $\mathrm{Tr}_{\mathrm{p}}\,[R(z)\rho_{\mathrm{p}}(0)]$.

Application for the calculation of the Raman scattering cross section. In this calculation, we shall draw attention to the widths of the excited molecular states. However, we neglect the width of the initial and final states of the transition. Indeed, we explain the experimental results of the continuous transition from resonance fluorescence to resonance Raman effect by taking into account the broadening of the excited states by collisions. This means especially that the following approximations are used

$$P_{22}\bar{G}_{11}(z)P_{22} \cong P_{22}(z - L_{0s})^{-1}P_{22} = z^{-1} \tag{82}$$

and

$$P_{11}\bar{G}(z)P_{11} \cong P_{11}(z - L_{0s})^{-1}P_{11} = z^{-1} \tag{83}$$

The evolution of the scattering molecule in the excited state is completely described by the Eqns. (80) and (81). The calculation of the matrix elements of these operators enables us to discuss the various phenomena due to the interaction of the scattering molecule with the perturbing atoms, particularly the quenching of resonance fluorescence.

The operators (80) and (81) are the inverse of non-diagonal operators. The problem of the calculation of the matrix elements of these operators consists of inverting a matrix of a finite number of elements and poses no fundamental difficulties. We limit ourselves to the approximation where the non-diagonal elements of these operators are neglected. This first approximation enables us to see the width of the excited states caused by collisions with the perturbing atoms[19] and describes the experiments of continuous transition from resonance fluorescence to resonance Raman effect, at least in a first approximation.

We use the following approximation of $P_{22}\bar{G}(z)P_{11}$

$$P_{22}\bar{G}(z)P_{11} = \frac{1}{z^2}\sum_{\alpha\beta}\Big\{P_{22}L_{1\mathrm{mr}}P_{2\alpha}[z - L_{0s} - P_{23}\,\bar{R}(z)P_{23}]^{-1}P_{2\alpha}L_{1\mathrm{mr}}P_{21}$$

$$\times P_{21}(z - L_{0s})^{-1}P_{21} \times P_{21}L_{1\mathrm{mr}}P_{\beta 1}[z - L_{0s} - P_{31}\bar{R}(z)P_{31}]^{-1}P_{\beta 1}L_{1\mathrm{mr}}P_{11}\Big\} \tag{84}$$

+ identical term where $P_{2\alpha}$ and $P_{\beta 1}$ are replaced by $P_{\alpha 2}$ and $P_{1\beta}$ respectively.

Calculation of the matrix element $\langle\langle 22^+|P_{22}\bar{G}(z)P_{11}|11^+\rangle\rangle$. The last step left is to write explicitly the matrix elements which appear in the general expression (84). It can be shown that they are in particular of the following type

$$M_{b\mathrm{k}j(r)^+} = \langle\langle b\mathrm{k}j(r)^+|P_2[z - L_{0s} - P_{23}\bar{R}(z)P_{23}]^{-1}P_{2\alpha}|b\mathrm{k}j(r)^+\rangle\rangle \tag{85}$$

which can be written, with a very good approximation

$$M_{bkj(r)^+} = \frac{1}{z - L_{0s} - \hbar ck + E_r + \Delta_r - i\Gamma_r/2} \tag{86}$$

when the widths of the states a and b are neglected.

In this expression, Γ_r represents the sum of the natural width Γ_r^n and the collisional width Γ_r^c. This width Γ_r^c is a function of the average scattering cross section of the collision between the scattering molecule and the perturbers and it is proportional[22] to the density n of the perturbers

$$\Gamma_r^c = \langle nv\sigma \rangle \tag{87}$$

After integration of Eqn. (84), we obtain finally the following expression of the transition probability per unit time

$$\frac{d}{dt}\langle b; \mathbf{k}j | \rho_{mr}(t) | b; \mathbf{k}j \rangle \cong \frac{2\pi}{\hbar}\,\delta(E_a + \hbar ck_0 - E_b - \hbar ck)$$
$$\times \left| \sum_r \left(\frac{\langle b; \mathbf{k}j | V | r \rangle \langle r | V | a; \mathbf{k}_0 j_0 \rangle}{\hbar(\omega_{ra} - \omega_0) - i\Gamma_r/2} + \text{anti-resonant term} \right) \right|^2 \langle a; \mathbf{k}_0 j_0 | \rho_{mr}(0) | a; \mathbf{k}_0 j_0 \rangle \tag{88}$$

with $\bar{\omega}_{ra} = (E_r + \Delta_r - E_a)/\hbar$ where Δ_r = energy displacement and $\Gamma_r = \Gamma_r^n + \Gamma_r^c$.

This expression for the matrix element $\rho_{mr}(t)$ generalizes the expression obtained for non-interacting molecules, for example by Weisskopf.[2] The limits of validity have been summarized in the text. The above expression shows clearly the influence of the collisions caused by the perturbing atoms on the various excited states and describes the phenomenon of quenching of the resonance fluorescence. This expression will enable use to interpret the experimental results[7] on the continuous transition from resonance fluorescence to resonance Raman effect.

GENERAL CONCLUSION

We have shown that the problems of the light–molecule interaction can be solved in a straightforward manner by the methods of quantum electronics. In particular, we have seen that the different stages of the interaction processes are well described, which enables the theoretical developments to be physically interpreted.

The general expressions of the Raman scattering tensor corresponding to the various conditions of optical excitation have been determined and we have also described different methods of approximation. One of these methods, which uses operatorial developments, reproduces in a purely quantum mechanical way, the semi-classical theory of Shorygin. On the other hand, it gives a physical interpretation of Raman scattering in terms of creation and annihilation of quanta of vibration. In the last section, the resolvant formalism has been applied to a theoretical study of the evolution of the density matrix of the light-scattering molecule system. We have thus shown the influence of the environment of the scattering molecule on the scattered spectrum. This result enables us to explain the experiments on the continuous transition from resonance fluorescence to resonance Raman effect.[7]

Thus, we think that the formalism we have presented in this work leads to a more rigorous theory which may be used as a basis for new developments in the study of Raman scattering in the determination of molecular structures and the study of molecular interactions.

REFERENCES

1 G. Placzek, *Handbuch der Radiologie* **6**, 209 (1934).
2 W. Weisskopf, *Ann. Physik* **9**, 23 (1931).
3 A. Messiah, *Mécanique Quantique*, Dunod, Paris, 1965.
4 M. L. Goldberger and K. M. Watson, *Collision Theory*, Wiley, New York, 1967.
5 L. Mower, *Phys. Rev.* **142**, 799 (1966).
6 L. Mower, *Phys. Rev.* **165**, 145 (1968).
7 M. Berjot and L. Bernard, *Advances in Raman Spectroscopy,* Vol. 1 (Ed. J. P. Mathieu), Heyden & Son, London, 1973, p. 343.
8 M. Jacon, R. Germinet, M. Berjot and L. Bernard, *J. Phys. Radium* **32**, 517 (1971).
9 R. Wallace, *Mol. Phys.* **11**, 457 (1966).
10 M. Jacon, Annales de l'Université et de l'Arers (to be published).
11 W. Holzer, W. F. Murphy and H. J. Bernstein, *J. Chem. Phys.* **52**, 399 (1970).
12 R. J. Gillespie and M. J. Morton, *J. Mol. Spectry.* **30**, 178 (1969).
13 M. Jacon, M. Berjot and L. Bernard, *Compt. Rend.* **273B**, 595 (1971).
14 M. Berjot, M. Jacon and L. Bernard, *Opt. Commun.* **4**, 3, 246 (1971).
15 M. Berjot, M. Jacon and L. Bernard, *Opt. Commun.* (to be published).
16 M. Born and K. Huang, *Dynamical Theory and Crystal Lattices*, Clarendon Press, Oxford, 1956.
17 *Raman Spectroscopy – Theory and Practice*, (Ed. M. A. Szymanski), Plenum Press, New York, 1967.
18 I. I. Kondilenko, *Opt. Spectry.* **13**, 369 (1962).
19 U. Fano, *Phys. Rev.* **131**, 259 (1963).
20 M. Baranger, *Phys. Rev.* **111**, 494 (1958).
21 U. Fano, *Rev. Mod. Phys.* **29**, 74 (1957).
22 M. Baranger, *Phys. Rev.* **112**, 855 (1958).

40. PASSAGE CONTINU DE LA FLUORESCENCE DE RESONANCE A L'EFFET RAMAN DE RESONANCE

M. Berjot et L. Bernard
Laboratoire de Recherches Optiques, Faculté des Sciences, Reims, France

ABSTRACT

The study of the scattering tensor in optical excitation of a molecule into a discrete state shows that under certain experimental conditions either resonance fluorescence may be obtained or a spectrum similar to the one of resonance Raman effect observed in the case of excitation in a continuum.

We present a series of spectra which illustrate the continuous transition of resonance fluorescence of iodine molecules excited by the argon ion laser line 5017 Å to resonance Raman effect, when the intermediate levels interacting in the scattering process are broadened by collisions.

The change in depolarization ratio and in line shape (from a fluorescence band to a resonance Raman band) is given as a function of the pressure of the perturbing gas.

INTRODUCTION

Nous nous proposons ici, de présenter les résultats expérimentaux qui, conjointement avec la théorie développée par Jacon[1] permettent d'apporter quelques précisions sur les relations entre la fluorescence de résonance et l'effet Raman de résonance; le problème de la distinction entre ces deux phénomènes a pris toute son importance depuis les travaux de Holzer, Murphy et Bernstein[2] ainsi que ceux de Gillespie et Morton.[3]

Si on excite de la vapeur d'iode ou de brome par une radiation de fréquence ω_0 portant la molécule sur un niveau α_0 du continuum d'absorption, on obtient un spectre qui a tout à fait l'aspect d'un spectre Raman. Par contre, si la fréquence excitatrice ω_0 coincide avec une fréquence propre discrète $\omega_{r_0 a}$ de la molécule, on observe, sous faible pression, un spectre de fluorescence de résonance. Dans les deux cas, il s'agit d'une diffusion résonnante au sens mathématique du terme puisque la fréquence excitatrice coincide à chaque fois avec une fréquence d'absorption de la molécule.

Nous rappelons dans le Tableau 1 les caractéristiques principales de ces deux phénomènes de résonance.

ASPECT THEORIQUE DE LA DIFFUSION RESONNANTE

Les différences expérimentales que l'on observe dans les deux cas que nous venons de signaler peuvent s'interpréter facilement si on examine la structure de la matrice S qui détermine les sections efficaces de diffusion.

Tableau 1. Différences expérimentales entre la fluorescence de résonance et l'effet Raman d'après Holzer *et al.*[2]

	Fluorescence de résonance	Effet Raman de résonance
Structure des bandes	Doublet constitué de raies très fines à basse pression	Branche Q relativement large et non structurée, encadrée par 2 ailes de rotation
Structure des harmoniques	Intensité des harmoniques variable; certains sont absents	La branche Q continue à s'élargir et l'intensité de crête diminue progressivement
Facteur de dépolarisation	Totalement dépolarisée à basse pression: $\rho_S = 0{,}75$	Toutes les bandes sans exception sont polarisées: $\rho_S < 0{,}45$
Comportement lorsque la pression croît	L'intensité décroît ('Quenching'). La structure de doublet se transforme en structure de multiplet	L'intensité s'accroît avec le nombre des centres diffusants. Le profil de raie est inchangé
Action d'un gaz étranger	Même comportement que précédemment	Aucune influence sensible

L'excitation optique s'effectue dans le continuum d'absorption de la molécule diffusante

ω_0 coïncide alors avec une fréquence d'absorption de ce continuum. Dans de telles conditions d'excitation, le phénomène le plus important est la *dissociation* de la molécule par absorption du photon incident. Cependant, on observe également un phénomène de diffusion qui présente toutes les caractéristiques de l'effet Raman (Fig. 1). Nous avons montré[4] que l'élément de la matrice S s'exprime rigoureusement de la manière suivante

$$\langle 2|S|1\rangle = \frac{2\pi i}{\hbar}\,\delta(\omega - \omega_R)\,[A + B + C] \tag{1}$$

$$A = \frac{i\pi}{\hbar}\,[\langle b, \mathrm{k}j|V|\alpha\rangle\langle\alpha|V|a, \mathrm{k}_0 j_0\rangle\,\rho_\alpha(\omega')]_{\omega'=\omega_0}$$

$$B = \int_{D_e/\hbar}^{\infty} \frac{\langle b, \mathrm{k}j|V|\alpha\rangle\langle\alpha|V|a, \mathrm{k}_0 j_0\rangle}{\hbar(\omega'_{ra} - \omega_0)}\,\rho_\alpha(\omega')\,\mathrm{d}\omega' + \text{termes antirésonnants}$$

$$C = \sum_r \left(\frac{\langle b, \mathrm{k}j|V|r\rangle\langle r|V|a, \mathrm{k}_0 j_0\rangle}{\hbar(\omega_{ra} - \omega_0)} + \text{termes antirésonnants}\right)$$

$E' = \hbar'\omega'$ désigne l'énergie d'un état du continuum.
$\omega_R = \omega_0 + (E_a - E_b)/\hbar$.

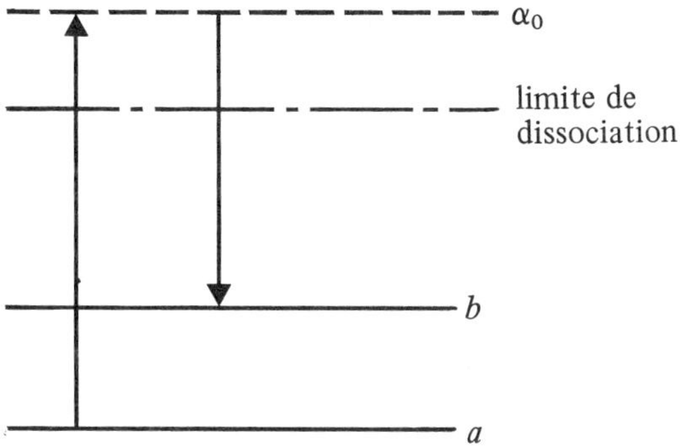

Fig. 1. Schéma d'excitation de la molécule dans le continuum d'absorption.

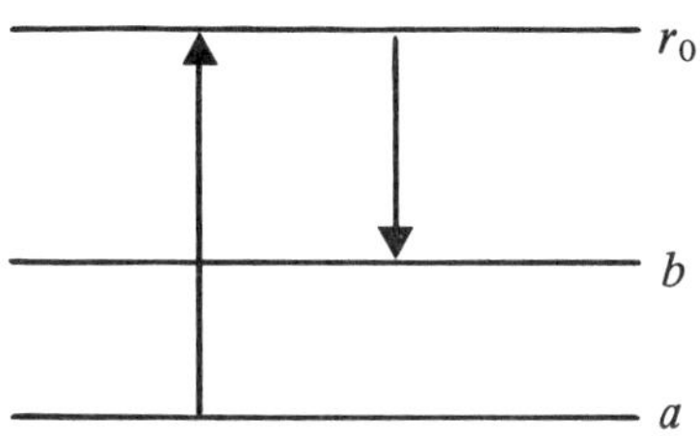

Fig. 2. Schéma d'excitation de la molécule sur un niveau discret.

A représente le terme résonnant de l'amplitude de diffusion et ne fait intervenir que le seul état α_0 du continuum en résonance avec le photon incident. B et C correspondent respectivement aux transitions *virtuelles*, qui s'effectuent par l'intermédiaire des autres états α du continuum et par les états discrets r.

Les termes A, B et C étant sensiblement du même ordre de grandeur ainsi que nous l'avons déjà montré,[5] on doit tenir compte de tous les états intermédiaires de la molécule. Le phénomène observé correspond à l'effet Raman de résonance.

La fréquence ω_0 du rayonnement incident coïncide avec une fréquence propre discrète ω_{r_0a} de la molécule

Dans ce cas (Fig. 2), l'élément de la matrice S qui détermine l'intensité des raies émises, s'écrit alors

$$\langle 2|S|1\rangle = \frac{2\pi i}{\hbar}\,\delta(\omega-\omega_R)\left(\frac{\langle b,\mathbf{k}j|V|r_0\rangle\langle r_0|V|a,\mathbf{k}_0j_0\rangle}{\hbar(\omega_{r_0a}-\omega_0)-\dfrac{i\Gamma_{r_0}}{2}}\right. \tag{2}$$

$$\left. + \sum_{r\neq r_0}\frac{\langle b,\mathbf{k}j|V|r\rangle\langle r|V|a,\mathbf{k}_0j_0\rangle}{\hbar(\omega_{ra}-\omega_0)-\dfrac{i\Gamma_r}{2}} + \text{termes antirésonnants}\right)$$

Nous avons mis en évidence dans cette formule le terme résonnant pour lequel

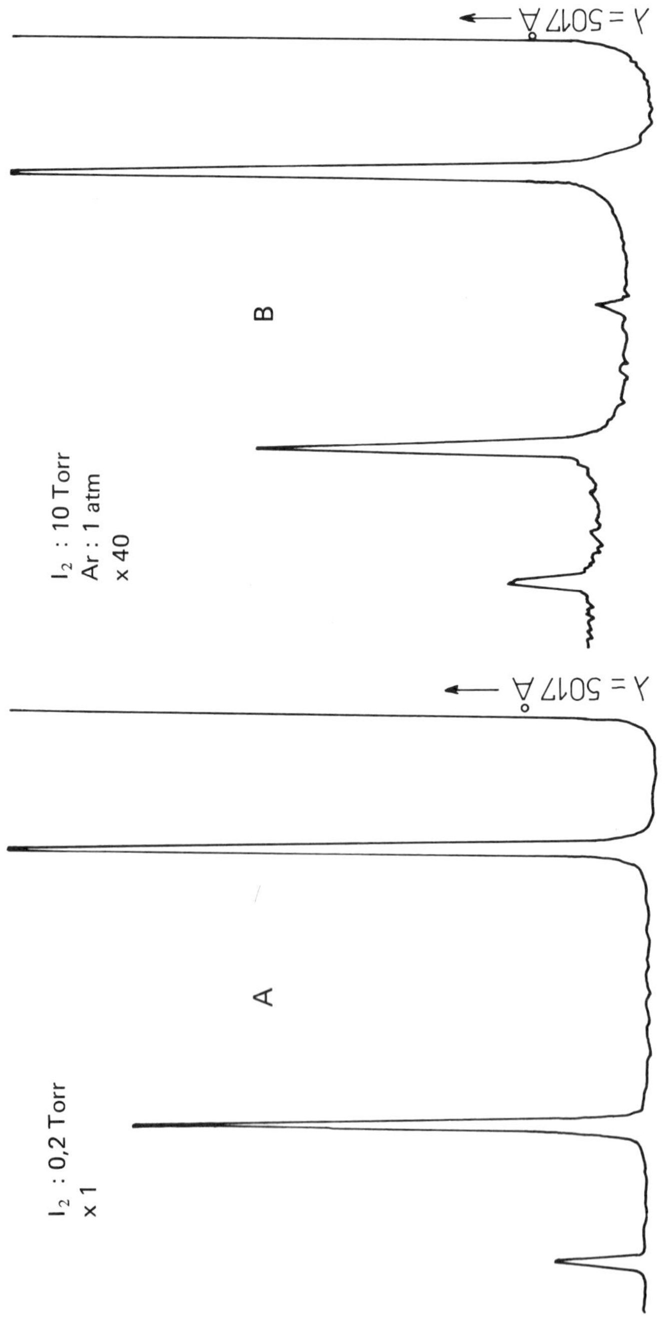
λ = 5017 Å
V''=1
V''=2
B
V''=3
I2 : 10 Torr
Ar : 1 atm
x 40
V''=4
λ = 5017 Å
V''=1
V''=2
A
V''=3
I2 : 0,2 Torr
x 1
V''=4

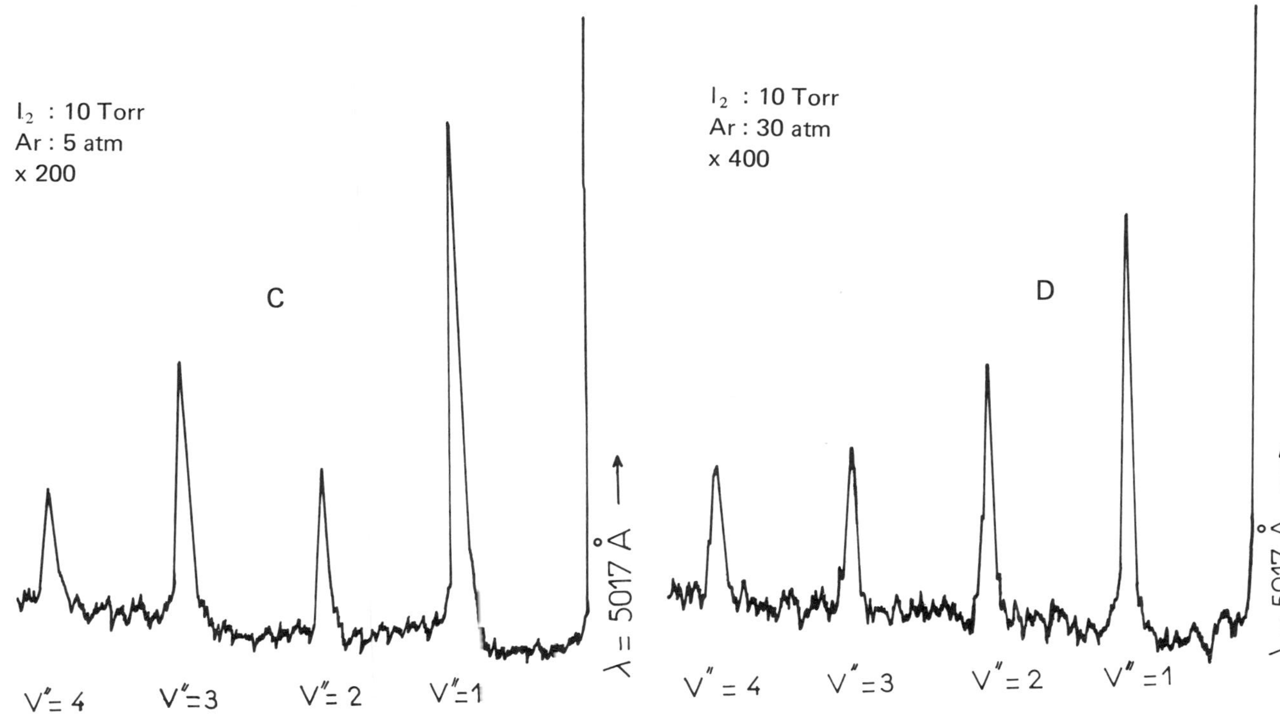

Fig. 3. Evolution progressive du spectre de fluorescence de résonance (A) de l'iode en presence d'argon, vers une structure de spectre Raman de résonance (D). I_2 : 0,2 torr – 10 torr; fentes 6 cm^{-1} ; constante de temps 4 s, 25 cm^{-1} /nm; excitation λ = 5017 Å (laser Ar^+).

$\omega_0 = \omega_{r_0 a}$. Les termes Γ_r représentent les largeurs des différents états r.

$\Gamma_r = \gamma_r + \Gamma_r^c$

γ_r est la largeur naturelle de l'état r.

Γ_r^c est la largeur par collisions de cet état.

A basse pression, les largeurs Γ_r étant faibles par rapport à l'écart entre les différents niveaux de vibration rotation de l'état électronique excité, le premier terme est largement prépondérant et les propriétés du spectre observé s'interprètent en ne tenant compte dans l'expression (2) que de ce seul terme résonnant qui correspond au passage *réel* de la molécule par l'état r_0, les autres termes qui correspondent à des transitions virtuelles étant négligeables.

Ce phénomène de diffusion, incontestablement caractérisé par l'existence du seul état intermédiaire discret r_0 de durée de vie moyenne $\tau = \hbar/\Gamma_r$, correspond à fluorescence de résonance.

PRINCIPE DU PASSAGE CONTINU DE LA FLUORESCENCE DE RESONANCE A L'EFFET RAMAN DE RESONANCE

Revenons maintenant à l'expression (2) et supposons que par une méthode appropriée, on augmente progressivement la largeur Γ_r des états intermédiaires r; dans ce cas, l'importance relative du terme résonnant de l'expression (2) par rapport aux autres termes diminue de plus en plus lorsque la largeur des états augmente. Ce résultat peut par exemple être obtenu par une augmentation de pression. Ce phénomène, pour des pressions de deux à trois atmosphères est bien connu: c'est le 'Quenching de la fluorescence'. Il est évident que dans ces conditions, pour décrire le processus de diffusion correspondant à des pressions élevées, on doit tenir compte de tous les termes de (2). On doit alors observer une modification continue du spectre de diffusion résonnante, le spectre à basse pression présentant les caractéristiques d'un spectre de fluorescence de résonance qui vont se modifier au fur et à mesure que la pression croît pour donner finalement un spectre Raman de résonance. Ces considérations nous ont d'ailleurs permis d'interpréter[6] les expériences de Beattie et de ses collaborateurs[7] qui avaient observé une diminution nette du facteur de dépolarisation ρ_S de la raie de fluorescence $v' = 43 \rightarrow v'' = 1$ de l'iode excitée par la raie $\lambda_0 = 5145$ Å lorsque de l'argon sous pression était introduit dans la cellule.

RESULTATS EXPERIMENTAUX

Nous allons montrer, qu'il est effectivement possible de passer d'une façon continue de la fluorescence de résonance à l'effet Raman de résonance lorsque la molécule est excitée optiquement sur un niveau discret, que l'on élargit progressivement.

L'excitation optique est réalisée par les raies d'un laser Ar^+ (5145 Å et 5017 Å dans le cas de l'iode, 5145 Å dans le cas du brome).

L'extinction de la fluorescence et l'élargissement des niveaux sont obtenus par introduction, dans la cellule, de gaz inertes (He, Ne, Ar, Kr et Xe) sous des pressions variant de 0–50 atmos.

Les Figures 3 (A, B, C et D) permettent de suivre par exemple dans le cas de la vapeur d'iode excitée par la raie $\lambda = 5017$ Å, l'évolution progressive du spectre de fluorescence de résonance, vers une structure de spectre Raman de résonance analogue à ceux obtenus par excitation dans le continuum d'absorption ($\lambda_0 < 4995$ Å). En particulier, on

Tableau 2. Variation du facteur de dépolarisation des raies de fluorescence de l'iode excitée par $\lambda_0 = 5017$ Å en présence d'argon.

Pression A_r en atmos	0	1	5	10	20	30
ρ_S pour $v'' = 1$	0,75 ± 0,02	0,76 ± 0,02	0,55 ± 0,05	0,46 ± 0,05	0,36 ± 0,05	0,35 ± 0,05
ρ_S pour $v'' = 2$	absent	–	0,36 ± 0,05	0,36 ± 0,05	0,35 ± 0,05	0,35 ± 0,05
ρ_S pour $v'' = 3$	0,75 ± 0,02	0,75 ± 0,02	0,60 ± 0,05	0,44 ± 0,05	0,35 ± 0,05	0,35 ± 0,05

constate sur ces figures que le premier harmonique ($v'' = 2$) qui est inobservable en fluorescence de résonance[8] (intégrales de recouvrement ayant une valeur quasi nulle) est présent dans le spectre Raman de résonance.

Le Tableau 2 permet également de suivre l'évolution du facteur de dépolarisation ρ_S des raies qui passe progressivement de la valeur 0,75 (fluorescence de résonance) à 0,35 environ (effet Raman de résonance) lorsque la pression d'argon varie de 0–30 atmos.

La Fig. 4 permet de suivre l'évolution de la structure d'un doublet de fluorescence vers une structure de raie Raman lorsque la pression augmente. On constate en particulier que la branche R du doublet de fluorescence (dont la fréquence coïncide rigoureusement avec celle de la branche Q en effet Raman) diminue beaucoup moins vite en intensité que la branche P. D'autre part, la raie de fluorescence de résonance, s'élargit progressivement lorsque la pression augmente; cet élargissement est d'autant plus sensible que la numéro d'ordre de l'harmonique est élevé.

Nous avons constaté, d'autre part que:

1. Dans le cas d'une excitation de la fluorescence de l'iode par la raie $\lambda_0 - 5017$ Å ($v' \simeq 62$), les différents gaz perturbateurs utilisés (He, Ne, Ar, Kr, et Xe) ont une 'efficacité' comparable pour réaliser ce passage continu d'un phénomène à l'autre.
2. Dans le cas, d'une excitation de l'iode par 5145 Å ($v' = 43$), les gaz perturbateurs sont par ordre d'efficacité croissante: He, Ne, Ar, Kr et Xe. Cependant, même avec le xenon, qui est le plus efficace, il est nécessaire d'élever la pression jusqu'à 50 atmos pour obtenir une véritable structure de spectre Raman alors qu'il suffit d'une pression de 25 atmos avec $\lambda_0 = 5017$ Å pour obtenir ce résultat.

INTERPRETATION DES RESULTATS

Le processus de 'Quenching' est induit par une interaction non symétrique de Van der Waals entre la molécule d'iode et le gaz perturbateur selon le schéma:

$$I_2^* + M \rightarrow 2I(^2P_{3/2}) + M \tag{3}$$

Des calculs effectués sur cette base par Selwyn[9] montrent que la probabilité de 'Quenching' par collisions est proportionnelle à

$$I_p \ \alpha \ \mu^{1/2}/R_c^3$$

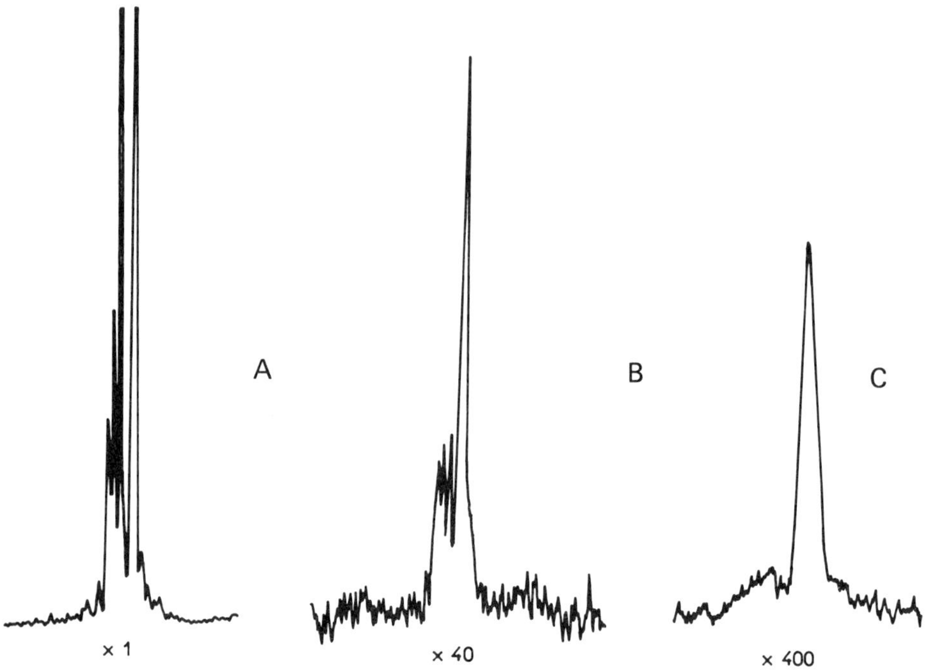

Fig. 4. Evolution de la structure de la bande $\nu'' = 1$ de la fluorescence de résonance de l'iode en présence d'argon. (A) I_2: 0,2 torr. (B) I_2: 0,2 torr, Ar: 1 atmos. (C) I_2: 10 torr, Ar: 30 atmos. Fentes: 1 cm^{-1} (A, B), fentes 4 cm^{-1} (C), constante de temps 1 s, 5 cm^{-1}/nm; Excitation $\lambda = 5017$ Å.

où I_p et α sont respectivement le potentiel d'ionisation et la polarisabilité du perturbateur.

μ est la masse réduite de collision.

R_c étant la distance minimum d'approche.

Steinfeld et Schweid [10] ont montré que si la probabilité de Quenching par collisions était bien proportionnelle au 'paramètre de Selwyn', pour $\nu' = 15$, $\nu' = 25$ et $\nu'' = 43$, il n'en était pas de même pour $\nu' > 50$.

En effet, dans ce cas, tous les gaz étudiés ont la même efficience de 'Quenching'. Il semble que ce phénomène puisse être attribué, comme l'a suggéré Callear,[11] à des 'relaxations par collisions' qui auraient lieu suivant le schéma:

$$I_2^* + M \longrightarrow MI + I^* (^2P_{1/2}) \quad (4)$$

et où la faible énergie supplémentaire nécessaire pour porter l'atome dans un état excité proviendrait de l'énergie de formation du complexe MI. Ce processus qui intervient en plus de la prédissociation induite par collisions (Eqn. 3) est vraisemblablement responsable de la valeur uniformément élevée des sections efficaces de collisions observées [10] et de l'efficacité comparable des différents gaz nobles pour réaliser le passage continu de la fluorescence de résonance à l'effet Raman de résonance dans le cas de l'excitation par $\lambda = 5017$ Å.

CONCLUSION

A partir de ces considérations, nous pouvons en conclure que la 'fluorescence de résonance' et 'l'effet Raman de résonance' ne sont que deux aspects différents d'un même phénomène qui est le 'processus de diffusion résonnante'; la fluorescence de résonance est obtenue lorsque les conditions expérimentales sont telles que dans l'amplitude de diffusion, le seul état intermédiaire r_0 en résonance avec le photon incident domine largement, tandis que l'effet Raman est obtenu lorsque plusieurs états intermédiaires apportent une contribution du même ordre de grandeur à l'amplitude de diffusion.

REFERENCES

1 M. Jacon, *Advances in Raman Spectroscopy*, Vol. 1 (Ed. J. P. Mathieu), Heyden & Son, London, 1973, p. 325.
2 W. Holzer, W. F. Murphy et H. J. Bernstein, *J. Chem. Phys.* **52**, 399 (1970).
3 R. I. Gillespie and M. J. Morton, *J. Mol. Spectry.* **30**, 178 (1969).
4 M. Jacon, M. Berjot et L. Bernard, *C. R. Acad. Sci. Paris* **273**, 595 (1971).
5 M. Berjot, M. Jacon et L. Bernard, *Opt. Commun.* **4**, 246 (1971).
6 M. Jacon, M. Berjot et L. Bernard, *C. R. Acad. Sci. Paris* **273**, 956 (1971).
7 I. R. Beattie, C. A. Ozin et R. O. Perry, *J. Chem. Phys.* **A12**, 2971 (1970).
8 D. Lobet, M. Berjot et L. Bernard, *C. R. Acad. Sci. Paris* **273**, 207 (1971).
9 J. Selwyn et J. I. Steinfeld, *Chem. Phys. Letters* **4**, 217 (1969).
10 J. I. Steinfeld et A. N. Schweid, *J. Chem. Phys.* **53**, 2304 (1970).
11 A. B. Callear et J. F. Wilson, *Trans. Faraday Soc.* **63**, 1358 (1967).

41. NEAR RESONANCE SCATTERING AND FLUORESCENCE

C. M. Penney
General Electric Company, Corporate Research and Development Schenectady, New York, USA

SUMMARY

If the wavelength of a light beam passing through a gas is brought near resonance with a strong absorption line, the resulting light scattering by the molecules can increase rapidly until, as the incident light wavelength enters the absorption region, this scattering often assumes such properties of fluorescence as an exponential decay in time and non-linear dependence on gas pressure (quenching). In this paper we present theoretical and experimental work regarding what happens, in the approach to an absorption region, to characteristics of the re-emitted radiation such as its time dependence, intensity, sensitivity to quenching, and spectral distribution. The theoretical work will be based largely on a classical model which is easily visualized yet contains the essential features of a relatively complicated quantum analysis. The experimental work concerns predominantly the gases NO_2 and I_2 and incident light in the visible. We will report quantitative measurements of the intensity of re-emitted light based on the spectral calibration set up in our laboratory for Raman scattering cross section measurements.[1]

Reference

1 C. M. Penney, L. M. Goldman and M. Lapp, *Nature* **235**, 110 (1972).

42. DOUBLE SPIN-FLIP RESONANT RAMAN EFFECT

J. F. Scott
Bell Telephone Laboratories, Inc., Quantum Electronics Research Department, Holmdel, New Jersey, USA
and
*Department of Physics and Astrophysics,**
University of Colorado, Boulder, Colorado, USA

INTRODUCTION

About five years ago it was observed by Mooradian and Wright[1, 2] and by Slusher, Patel and Fleury[3–5] that inelastic light scattering provided an excellent probe of the collective (plasmon) and noncollective electron states of narrow bandgap semiconductors in both zero field and in large magnetic fields.

More recently, Daman, Leite, Shah, Fleury and Scott have applied the Raman technique to systematically elucidate electron behaviour in the wide bandgap II–VI semiconductors CdS and ZnSe. We achieved the first plasmon scattering in the visible portion of the spectrum[6] and examined asymmetric plasmon lineshapes.[7] We have also studied the non-collective electron transitions in fields from zero to 105 kG.[7] Of special interest has been spin-flip scattering. This process is diagrammed schematically in Fig. 1. An electron in the valence band is excited optically to a $\downarrow$ state above the Fermi surface while another electron in $\uparrow$ state of the same Landau level drops down to fill the hole in the valence band. This process is allowed only because the valence band has mixed spin character $a|\uparrow\rangle + b|\downarrow\rangle$ due to spin-orbit coupling. That is, if there were no spin-orbit coupling and the valence band were pure $|\uparrow\rangle$, one of the two electron transitions shown in Fig. 1 would be spin forbidden ($\Delta S = 1$) and the process would have zero cross-section. It is of interest to point out that the gyromagnetic ratio (g-value) of the conduction electrons differs from 2 due to spin-orbit coupling also. In III–V semiconductors the formula below has been found[8] to provide a good approximation

$$g = 2\left\{1 - \left(\frac{m}{m^*} - 1\right)\left[\frac{\Delta}{3E_g + 2\Delta}\right]\right\} \tag{1}$$

where m^* is the effective mass; E_g, the bandgap; and Δ, the spin-orbit splitting of the valence band. Thus we see that for large spin-orbit coupling, spin-flip scattering will be intense, and g will be very different from $+2$. These considerations favour studies of narrow gap semiconductors such as InSb ($g = -55$); however, we have found that this

* Permanent address.

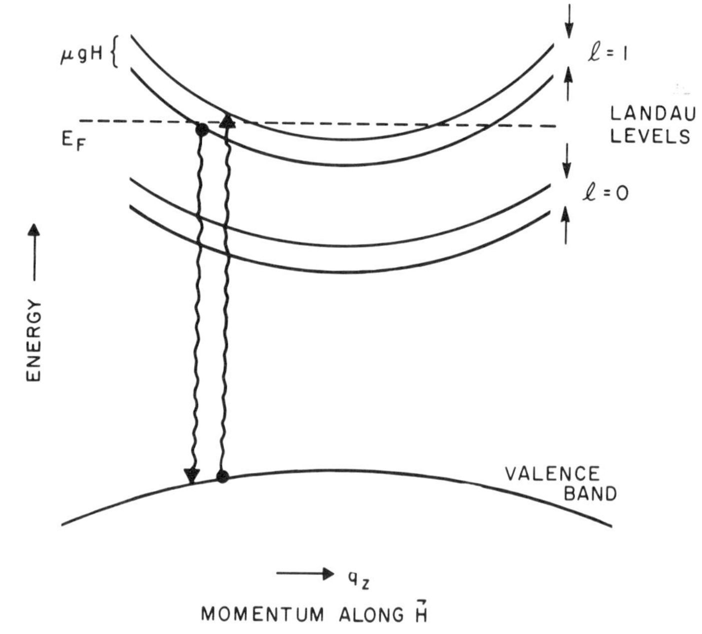

Fig. 1. Energy level diagram for spin-flip scattering in a degenerate semiconductor.

deterrent to visible spin-flip spectroscopy is more than compensated by the high quantum efficiency of S-20 photocathode detectors.

SPIN LIFETIMES AND LINEWIDTHS

When two electrons collide in a semiconductor, they normally do not flip either spin. Consequently, the spin lifetime T is many orders of magnitude longer than τ, the collision time. In commercially available samples of CdS or ZnSe the Hall mobility is typically 50–300 $cm^2/V\text{-}s$, which corresponds to a collision time of $(0.5\text{–}3.0) \times 10^{-13}$ s. Since the spin-flip energy

$$\omega_s = \mu g H \tag{2}$$

is only about 7 cm^{-1} or 2×10^{11} Hz even at $H = 100$ kG, such a collision time would render spin-flip lines unobservably broad if spin lifetimes were comparably short. Fortunately, they are not. The spin-lattice (longitudinal) relaxation time has been estimated by Yafet as[9]

$$\frac{T_1}{\tau} = (2 - g)^2 \frac{m^*}{m} \left(\frac{kT}{E_g}\right)^2 \tag{3}$$

This time T, is defined as the interval in which, on average, an electron spin will actually invert. It is typically 10^6 longer than τ. In addition to T_1 relaxation times, any discussion of spin linewidths must include transverse relaxation times, T_2. T_2 is a dephasing time, which measures the interval in which, on average, two electron spins in a semiconductor, precessing about an external field $\mathbf{H}_0$ will get out of phase by exactly one radian. Both T_1

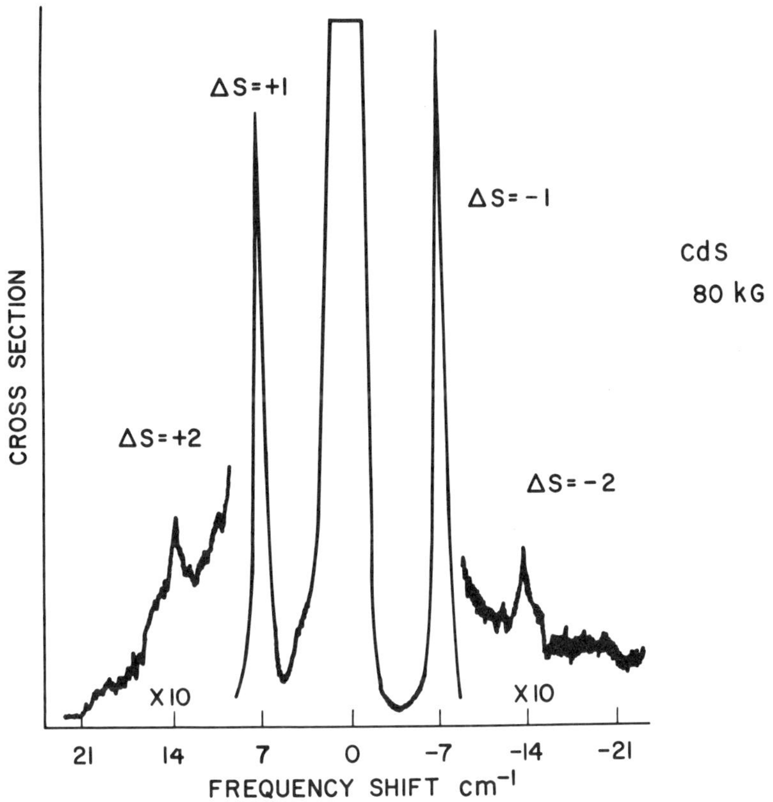

Fig. 2. Single and double spin-flip spectra in CdS.

and T_2 arise in the Bloch Equations describing spin dynamics in an external field

$$\frac{dM_z}{dt} = \alpha(\mathbf{M} \times \mathbf{H})_z + (M_0 - M_z)/T_1 \tag{4}$$

$$\frac{dM_x}{dt} = \alpha(\mathbf{M} \times \mathbf{H})_x - M_x/T_2 \tag{5}$$

$$\frac{dM_y}{dt} = \alpha(\mathbf{M} \times \mathbf{H})_y - M_y/T_2 \tag{6}$$

where $\hat{Z}$ is along $\mathbf{H}_0$ and α is proportional to g.

In an extensive study of the linewidths and lineshapes of spontaneous free-electron spin-flip lines in CdS and ZnSe we have found that T_2 is dominant; i.e. $T_1 \gg T_2$. Moreover,

$$\Gamma \simeq \frac{1}{T_2} = 2q^2 \frac{kT}{m^*} \tau_0 \tag{7}$$

an equation for spin diffusion contribution to linewidth due to Auyang[10] is obeyed [here

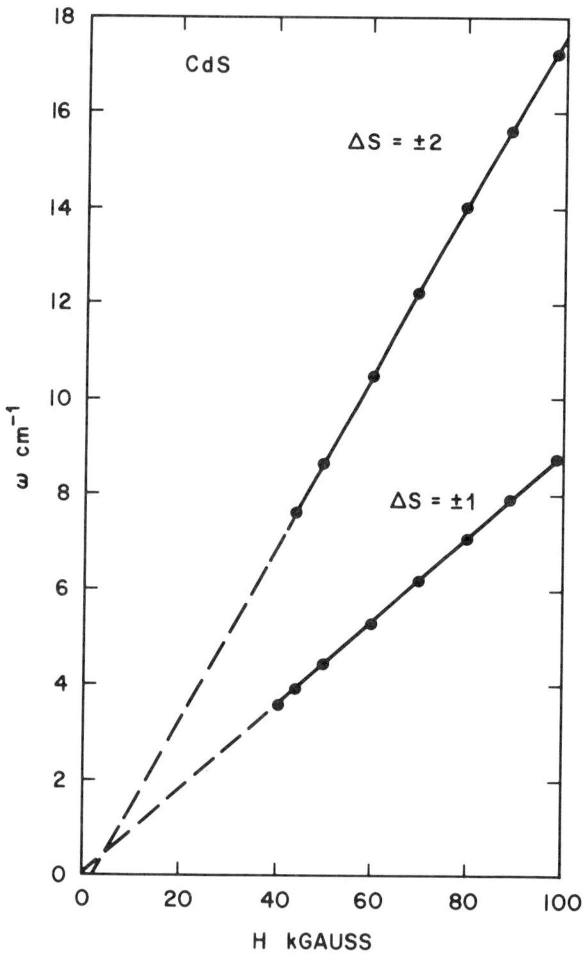

Fig. 3. ω versus H for $\Delta S = 1$ and $\Delta S = 2$ processes in CdS.

q is the momentum transfer in the scattering process; $q^2 = 2k_L^2(1 - \cos\theta)$, where $k_L = 2\pi n\omega_L$ is the laser photon momentum and θ is the scattering angle]. However, τ_0 is observed to be about twelve times less than τ, where τ is measured from plasmon linewidth or Hall mobility. No theories presently exist for τ_0, except that it was previously thought that $\tau_0 \approx \tau$ would hold. The experimental results are summarized in a recent paper by Scott, Fleury and Damen.[11]

BOUND ELECTRON SPIN-FLIP

When CdS or ZnSe is cooled below about 80 K, the spin-flip selection rules change. Whereas only α_{ij} with $i \neq j$ and $\hat{\mathbf{i}}$ or $\hat{\mathbf{j}} \parallel \mathbf{H}$ are observed above 80 K, all tensor components are strong at low temperatures. In addition, the q-dependence of the linewidth given in Eqn. (7) disappears. The line is very sharp for all scattering angles; apparently there is no spin diffusion – the carriers are no longer mobile.

We interpret these results as spin-flip scattering from electrons bound to shallow donors (impurities). In our samples the donors are In and/or Ga. Spin-flip scattering from electrons bound to donors was first studied by Thomas and Hopfield,[12] who employed, primarily, Br-doped samples.

The new observation in our bound-electron spin-flip spectra is shown in Fig. 2. Weak features near $\omega = 2\omega_s$ are observed, which we interpret as 'Double Spin-Flip' processes. Figure 3 shows that these features do indeed shift with magnetic field by an amount which is approximately twice that of the single spin-flip line ($g = 1.86 \pm 0.04$). In particular, the shape ω_s/H is exactly twice that of the single spin-flip, but there is a small, H-independent down-shift; $\omega(\Delta S = \pm 2)$ is 0.25 ± 0.05 cm^{-1} less than twice $\omega(\Delta S = 1)$.

Most of the principal characteristics of double spin-flip spectra have been reported in a letter[13] with Damen. We summarize them below:

1. The effect is observed only below 25 K.
2. Between 10 and 25 K the intensity is H-independent.
3. At 2 K the intensity varies roughly as $1/H$ for H between 40 and 100 kG.
4. The double spin-flip intensity is about 1.5% of the single spin-flip in nominally 'pure' CdS ($n \sim 10^{16}$ cm^{-3}) at 10 K and 95 kG for $\lambda = 4880$ Å.
5. The double spin-flip intensity becomes larger relative to the single spin-flip, when the donor concentration is increased, reaching 10% for 10^{18} In cm^{-3} at 10 K at 95 kG for $\lambda = 4880$ Å.

RESONANCE

Double spin-flip scattering is observed only very near resonance in CdS. It can be observed easily at 4880 Å, but not at all at 4695 Å, although the signal-to-noise for 4695 Å is 400:1. In other words, the double spin-flip scattering is more highly resonant than is single spin-flip scattering. This situation is analogous to multiple LO-phonon scattering near resonance, where the 2LO cross-section increases much faster with laser frequency than does the LO.[14]

THEORY

We interpret the double spin-flip scattering as the simultaneous flipping of two spins of electrons bound to nearby or adjacent donors. This interpretation has been formalized and made quantitative by Economou, Ruvalds and Ngai.[15] Both the small (0.25 cm^{-1}) shift in $\omega(\Delta S = 2)$ and the surprising large intensity of the $\Delta S = 2$ line are interpreted in terms of exchange interaction between nearby donors.

A more careful study of the double spin-flip lineshape will be required before we can tell whether the 0.25 cm^{-1} shift is a true binding energy (analogous to the two-phonon bound state in diamond)[16] or simply the effect of dispersion introduced by exchange in the otherwise dispersionless bound-electron spin level.

SUMMARY

In conclusion, we have presented data and analyses for the only $\Delta S = 2$ process known in any branch of spectroscopy.

REFERENCES

1 A. Mooradian and G. B. Wright, *Phys. Rev. Letters* **16**, 999 (1966).
2 A. Mooradian, *Phys. Rev. Letters* **20**, 1102 (1968).

3 R. E. Slusher, C. K. N. Patel and P. A. Fleury, *Phys. Rev. Letters* **17**, 1011 (1966).
4 C. K. N. Patel and R. E. Slusher, *Phys. Rev.* **167**, 413 (1968).
5 R. E. Slusher and C. K. N. Patel, *Phys. Rev.* **177**, 1200 (1969).
6 J. F. Scott, T. C. Damen, R. C. C. Leite and Jagdeep Shah, *Phys. Rev.* **B1**, 4330 (1970).
7 J. F. Scott, T. C. Damen, J. Ruvalds and A. Zawadowski, *Phys. Rev.* **B3**, 1295 (1971); P. A. Fleury and J. F. Scott, *Phys. Rev.* **B3**, 1979 (1971).
8 L. M. Roth, B. Lax and S. Zwerdling, *Phys. Rev.* **114**, 90 (1959).
9 Y. Yafet, *Solid State Phys.* **14**, 1 (1963).
10 Y. C. S. Auyang, *Thesis,* M.I.T., 1971.
11 J. F. Scott, P. A. Fleury and T. C. Damen, *Phys. Rev. B* (1972) (in press); see also *Bull. Amer. Phys. Soc.* **17**, 335 (1972); *J. Opt. Soc. Amer.* **62**, 730 (1972); and *J. Quant. Elec.* **8**, 531 (1972) for brief reports by the same authors.
12 D. G. Thomas and J. J. Hopfield, *Phys. Rev.* **175**, 1020 (1968).
13 J. F. Scott and T. C. Damen, *Phys. Rev. Letters* **29** (1972) (in press).
14 J. F. Scott, R. C. C. Leite and T. C. Damen, *Phys. Rev.* **188**, 1285 (1969).
15 E. N. Economou, J. Ruvalds and K. L. Ngai, *Phys. Rev. Letters* **29** (1972) (in press).
16 M. H. Cohen and J. Ruvalds, *Phys. Rev. Letters* **23**, 1378 (1969).

43. PRERESONANCE RAMAN EFFECTS OF SOME CHLOROETHYLENES

M. Ito, Y. Takagi, N. Mikami, Y. Udagawa and K. Kaya
Department of Chemistry, Tohoku University, Sendai, Japan

Raman intensity of molecular vibration is closely related with the electronic excited states of the molecule, vibronic coupling among the various electronic states and excitation frequency. Because of the complexity of the relationships among these quantities, theoretical study of Raman intensity is one of the most difficult problems to attack. Experimentally, the study of the dependence of Raman intensity on the wavelengths of incident light is a useful approach to find correlation between the Raman intensities and electronic excited states of a molecule. With this view, we have studied in a previous paper[1] the Raman spectra of the halogenomethanes excited by several lasers ranging from visible to ultraviolet wavelengths, and a marked intensity enhancement has been found for the Raman lines of particular vibrational modes when the exciting frequency approaches one of the electronic states of the molecule. It will be interesting to extend the study to molecules possessing π-conjugated system, for which fairly accurate data are available for the π-electronic states both from theory and experiment. In the present study, some derivatives of ethylene have been subjected to study as the most simple π-electron systems.

The Raman spectra of *trans*-dichloroethylene 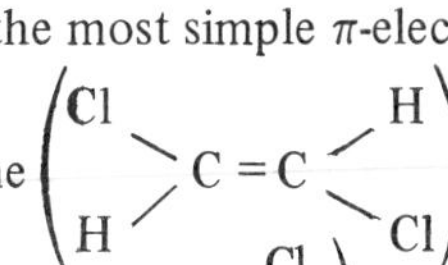 trichloroethylene 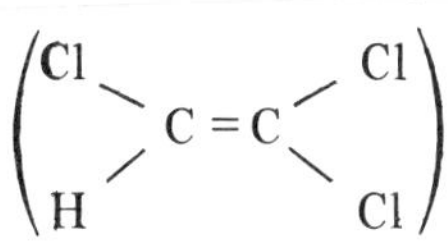 and tetrachloroethylene 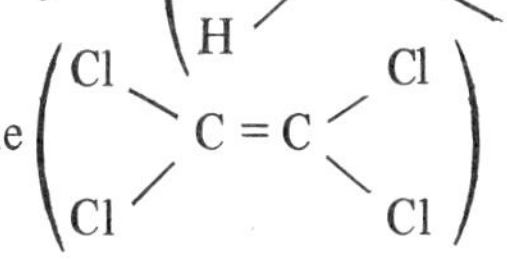 were measured with incident light of various wavelengths in the visible and ultraviolet range. The excitation wavelengths were 6328 Å (He–Ne laser), 5145 Å, 4880 Å, 4579 Å (argon ion laser) and 3371 Å (N_2 laser). The spectra were measured with a Nalumi grating double monochromator by a photon counting detection method for He–Ne and argon laser excitation and by a pulse-gated photon counting method for N_2 pulsed laser excitation.[1]

The Raman line of 990 cm^{-1} of benzene which was dissolved in each sample at a constant concentration was used as an internal standard of Raman intensity. Benzene is a good internal standard since the Raman line of 990 cm^{-1} is very strong compared with the other Raman lines of benzene and addition of a small amount of benzene does not greatly deteriorate the sample spectra except for narrow regions around 990 cm^{-1} and 3000 cm^{-1}.

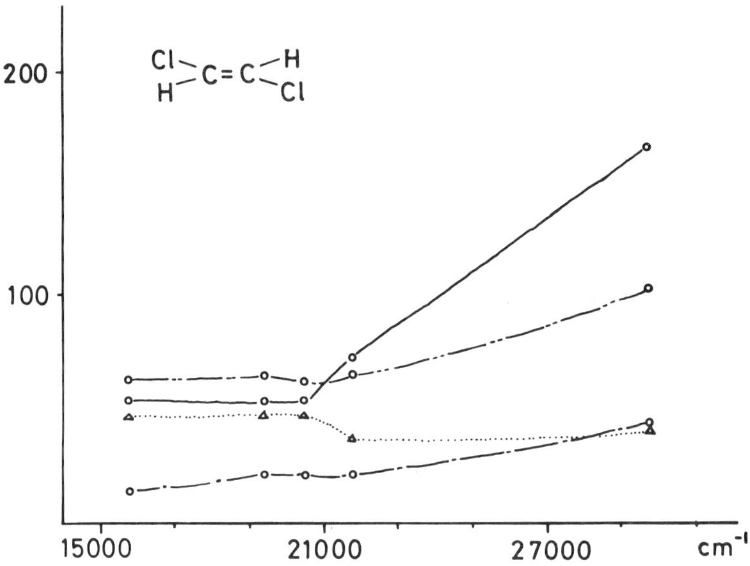

Fig. 1. Dependence of Raman intensities of 1,2-*trans*-dichloroethylene on the excitation wavelengths. –x–: 349 cm^{-1} line; –o–: 758 cm^{-1} line; –△–: 844 cm^{-1} line; –⊙–: 1270 cm^{-1} line; –●–: 1576 cm^{-1} line. The intensities were referred to that of 992 cm^{-1} of benzene used as an internal standard.

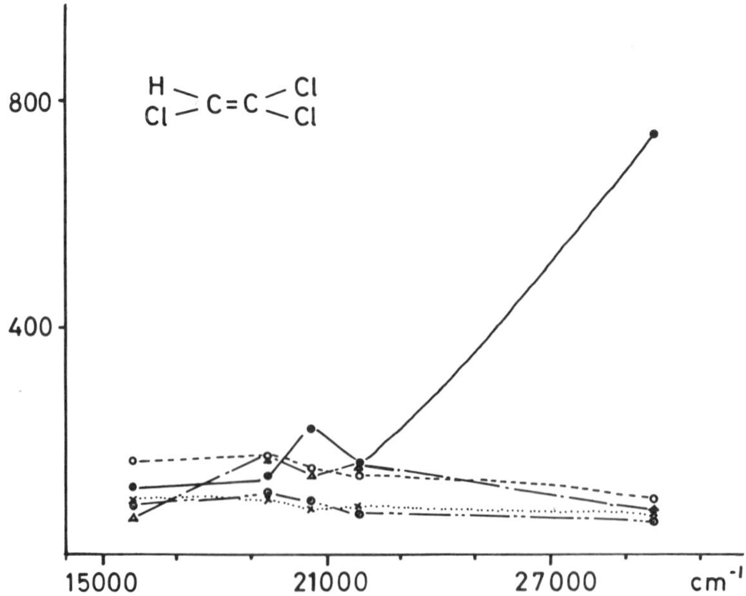

Fig. 2. Dependence of the Raman intensities of trichloroethylene on the excitation wavelengths. –x–: 172 cm^{-1} line; –⊙–: 274 cm^{-1} line; –△–: 380 cm^{-1} line; –○–: 628 cm^{-1} line; –●–: 1587 cm^{-1} line.

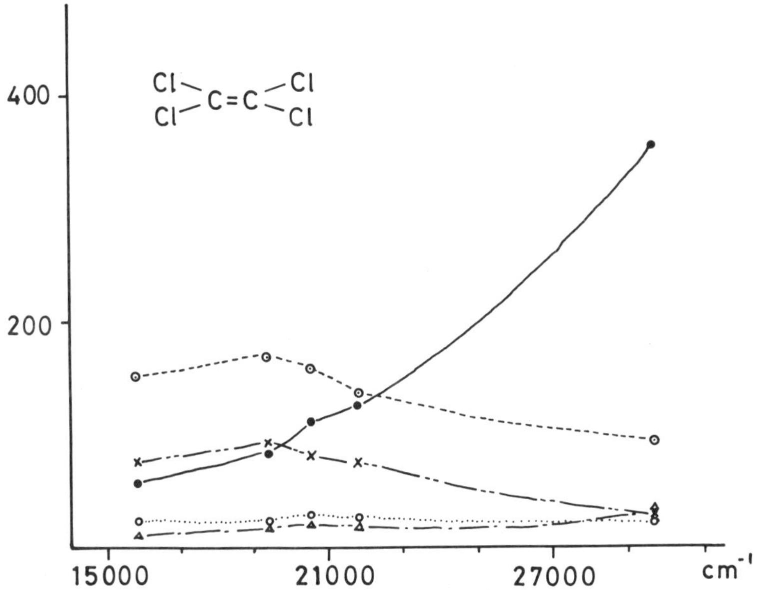

Fig. 3. Dependence of the Raman intensities of tetrachloroethylene on the excitation wavelengths. –x–: 237 cm^{-1} line; –o–: 340 cm^{-1} line; –⊙–: 450 cm^{-1} line; –△–: 514 cm^{-1} line; –•–: 1575 cm^{-1} line.

The intensities of the Raman lines were evaluated from the integrated areas of the individual Raman lines recorded in an expanded scale. The intensities were corrected for variation of sensitivity of a photomultiplier with wavelength.

In Figs. 1–3 are shown the relative intensities of the Raman lines of 1,2-*trans*-dichloroethylene, trichloroethylene and tetrachloroethylene, respectively, for various excitation wavelengths. The intensities are referred to that of the 990 cm^{-1} line of benzene dissolved in each sample at a constant concentration, which was taken to be 100. A typical example of the wavelength dependence of the Raman spectra is shown in Fig. 4 for tetrachloroethylene.

It is seen from the figures that the Raman lines of about 1580 cm^{-1} greatly increase their intensities when the frequency of exciting light increases. The intensity enhancements at 3371 Å relative to the intensity at 6328 Å are about 3.0, 5.5 and 6.0 for 1,2-*trans*-dichloroethylene, trichloroethylene and tetrachloroethylene, respectively. For the other Raman lines intensity variation is generally smaller than the above. The Raman line of 349 cm^{-1} of dichloroethylene exhibits a peculiar intensity change as seen in Fig. 1.

The Raman lines of ~ 1580 cm^{-1} showing the prominent intensity enhancement are obviously due to the stretching vibration of C = C double bond. It is concluded, therefore, that the Raman line of the C = C stretching vibration is preferentially subject to intensity enhancement and the effect increases successively with the number of chlorine atoms substituted. For comparison, we measured the vapour Raman spectra of the parent molecule, ethylene, excited by 4880 Å argon line and by 3371 Å nitrogen line. We found that the two spectra were practically identical with respect to the relative intensities and that there was no selective effect for a particular Raman line.

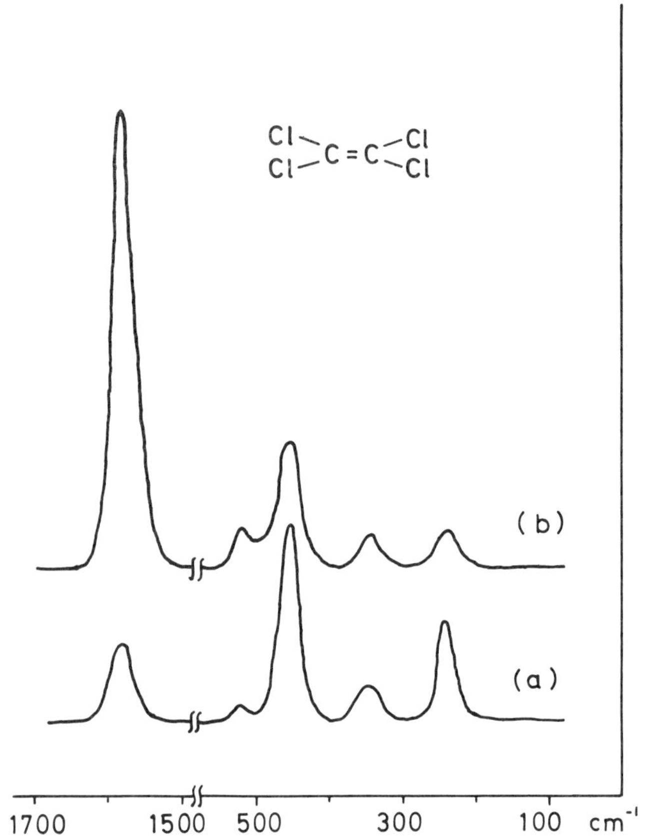

Fig. 4. Raman spectra of tetrachloroethylene (a) excited by 6328 Å and (b) excited by 3371 Å.

According to the Kramers–Heisenberg's dispersion equation,[2] all the electronic excited states of the molecules whose electronic transitions to the ground state are allowed are responsible for the Raman intensities. When the frequency of the exciting light approaches one of the electronic excited states, particular Raman lines are enhanced in intensity whose normal vibrations are closely associated with the electronic excited state concerned.[3] In the case of the ethylenes, the lowest excited π–π^* electronic state is one of the most eligible states. The maximum positions and intensities of the π–π^* electronic absorptions are 1621 Å, 1950 Å, 1960 Å and 1972 Å, for ethylene, 1,2-*trans*-dichloroethylene, trichloroethylene and tetrachloroethylene, respectively. The frequencies of the maximum positions decrease in the above order, and the order is in accordance with that of increasing effect of the intensity enhancement of the C = C stretching Raman line. Therefore, it may be qualitatively argued that the π–π^* electronic state is involved in the observed effect. According to the theory,[4] the Raman line due to the normal vibration whose coordinate origin is greatly displaced in the electronic excited state from that in the ground state preferentially appears in the Raman spectrum. Since the C = C bond length of the ethylenes is greatly altered in the π–π^* excited state, the Raman line of C = C stretching vibration is expected to be enhanced in intensity when the wavelength of the exciting light approaches the $\pi-\pi^*$ absorption. No appreciable effect on ethylene is ascribed to too far location of the $\pi-\pi^*$ absorption from the excitation wavelengths used.

The above interpretation explains very roughly the observed difference in preresonance effect between ethylene and the substituted ethylenes. However, it does not explain the difference among the substituted ethylenes, that is, the locations of the $\pi-\pi^*$ transitions are too slightly different to explain rather large differences in the preresonance effect among three molecules. Substitution of hydrogen atoms of ethylene by chlorine atoms introduces additional low-lying electronic excited states. They are $n-\pi^*$ and $n-\sigma^*$ states arising from excitation transfer of the non-bonding electrons localized on chlorine atoms to the anti-bonding π^* and σ^* orbitals. These states are not exactly located yet for the chlorine-substituted ethylenes. However, it is highly expected that the $n-\pi^*$ state and some of the $n-\sigma^*$ state lie at a lower energy than the $\pi-\pi^*$ state. Actually, there are broad and structureless absorptions extending towards the longer wavelength side of the $\pi-\pi^*$ absorption peaks in the chlorine-substituted ethylenes. Furthermore, the extension of the absorption towards the longer wavelength side seems to increase in the order of the number of chlorine atoms substituted. The electronic states corresponding to these absorptions, which are probably $n-\sigma^*$ absorptions, might participate in the preresonance effect of the C = C stretching Raman lines in addition to the contribution from the $\pi-\pi^*$ electronic states.

REFERENCES

1 K. Kaya, N. Mikami, Y. Udagawa and M. Ito, *Chem. Phys. Letters* **13**, 221 (1972).

2 G. P. Barnett and A. C. Albrecht in *Raman Spectroscopy, Vol. 2,* (Ed. H. A. Szymanski), Plenum Press, New York, 1970, p.207.

3 A. C. Albrecht, *J. Chem. Phys.* **33**, 156 (1960).

4 F. A. Savin, *Opt. Spectry.* **19**, 308 (1965).

44. RESONANCE RAMAN SPECTRA OF THE CATIONS I_2^+, Br_2^+ AND Te_4^{2+}

M. Booth and M. J. Morton
Department of Chemistry, McMaster University, Hamilton, Ontario, Canada
and
R. J. Gillespie
Laboratoire des Acides Mineraux, Université de Montpellier, France

INTRODUCTION

With the advent of highly intense and highly collimated laser beams as light sources the observation of resonance Raman spectra has become considerably simpler than it has been in the past and we have been able to obtain such resonance Raman spectra from several highly coloured polyatomic cations of the non-metallic elements.

THE I_2^+ CATION

The intensely blue coloured I_2^+ ion can be prepared by oxidizing iodine with $S_2O_6F_2$ in solution in fluorosulphuric acid according to the equation

$$2I_2 + S_2O_6F_2 \rightarrow 2I_2^+ + 2SO_3F^-$$

or by dissolving iodine in 65% oleum when oxidation occurs according to the equation

$$2I_2 + 5SO_3 + H_2S_4O_{13} \rightarrow 2I_2^+ + 2HS_4O_{13}^- + SO_2$$

These solutions of I_2^+ have a strong absorption band with a maximum at 640 nm.[1] Figure 1 shows the resonance Raman scattering obtained from a 10^{-2} m solution of I_2^+ in HSO_3F contained in a glass melting point tube of 1.2 mm I.D. and using the 6328 Å radiation from a He–Ne laser. Both the Stokes and anti-Stokes lines are clearly visible. The fundamental has a frequency of 238 cm^{-1} and five overtones may be observed in the Stokes lines. The height of these overtones diminishes with increasing frequency and at the same time they become broader (Table 1). We note that the overtones are of comparable intensity to the fundamental and only diminish slowly in intensity with increasing frequency. Similar strong overtones have been observed in the resonance Raman spectra of the gaseous halogens Cl_2, Br_2 and I_2.[3] At a concentration of 10^{-2} m no spectrum from the solvent is observed as the relatively weak non-resonance Raman scattering from the solvent molecules is strongly absorbed by the solution. On decreasing the concentration of the solution the intensity of the resonance spectrum from I_2^+ decreased and the intensity of the spectrum of the solvent increased (Fig. 2). On the other hand when the concen-

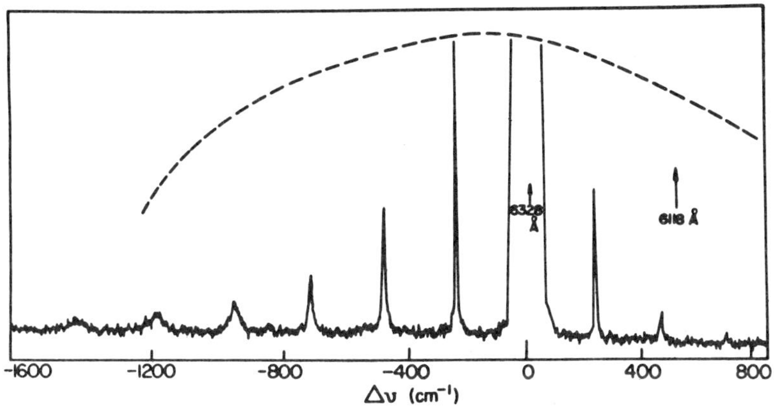

Fig. 1. Resonance Raman spectrum of a 10^{-2} M solution of I_2^+ in fluorosulphuric acid: (---), contour of visible absorption band.

Table 1. Frequency shifts, peak heights and widths of lines in the resonance Raman spectrum of I_2^+.

		Anti-Stokes			Stokes			
$\Delta\nu$ cm^{-1}		710	474	238	−238	−476	−716	−950
Relative peak heights	25°	1.7	8	50	100	42	19	8
	−80°	–	3	30	100	41	18	8
Half widths cm^{-1}		15	10	6	6	10	15	25
Relative intensities		–	5	30	100	68	45	33

tration of the solution is increased the solution becomes so absorbing that no spectrum was observed at all. For any given sample cell there is an optimum concentration at which a resonance Raman spectrum of maximum intensity will be observed.

Similar but weaker spectra were observed using the 6118 Å He–Ne radiation. Comparison of the intensity of the I_2^+ lines with that of the 400 cm^{-1} line of fluorosulphuric acid showed that the spectra were approximately twice as strong using 6328 Å radiation as with 6118 Å radiation (Table 2).

Depolarization measurements were made on the fundamental and the first and second overtones and on two anti-Stokes lines and a depolarization ratio of 0.3 was found in each case. For a $\pi \to \pi^*$ transition the value of the depolarization ratio is expected to be 1/3.[4]

The vibrational frequency of the I_2^+ ion is higher than that of the I_2 molecule (213 cm^{-1}) and this is consistent with the fact that I_2^+ contains one less electron in an anti-bonding orbital and has a formal bond order of 1.5.

THE Br_2^+ CATION

The Br_2^+ cation can be obtained by the oxidation of bromine with $S_2O_6F_2$

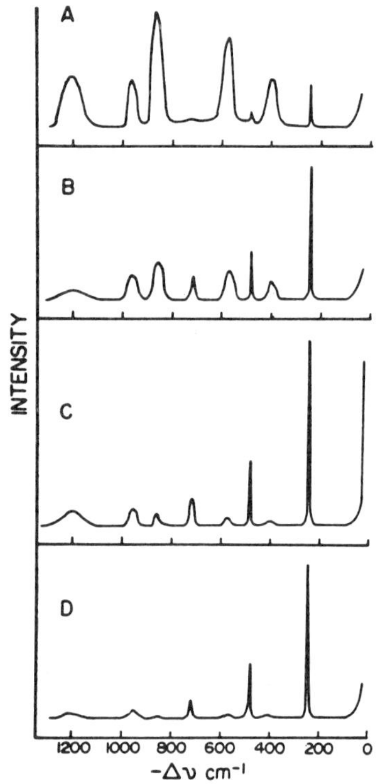

Fig. 2. Resonance Raman spectra of the I_2^+ ion in solution in fluorosulphuric acid: A, 10^{-4} M; B, 5×10^{-4} M; C, 2×10^{-3} M; D, 10^{-2} M.

Table 2. Raman intensity variations with laser line for a 5×10^{-4} m solution of I_2^+.

	6328 Å	6118 Å	Ratio
$\frac{\text{Intensity } 238 \text{ cm}^{-1}}{\text{Intensity } HSO_3F \text{ at } 400 \text{ cm}^{-1}}$	4.8	2.3	2.1
$\frac{\text{Intensity overtone at } 476 \text{ cm}^{-1}}{\text{Intensity } HSO_3F \text{ at } 400 \text{ cm}^{-1}}$	1.6	0.7	2.3

according to the equation

$$2Br_2 + S_2O_6F_2 \rightarrow Br_2^+ + 2SO_3F^-$$

However the Br_2^+ ion is more highly electrophilic than the I_2^+ ion as a consequence of its smaller size and more highly concentrated charge and it is not stable in HSO_3F but is completely disproportionated to Br_3^+ and $BrSO_3F$. The less basic superacid medium HSO_3F-SbF_5-SO_3 is needed to stabilize the Br_2^+ cation which gives a cherry red solution with an absorption maximum at 510 cm^{-1}. Even in this solution there is some disproportionation to Br_3^+ and $BrSO_3F$. Consequently the Raman spectra of these solutions (Fig. 3)

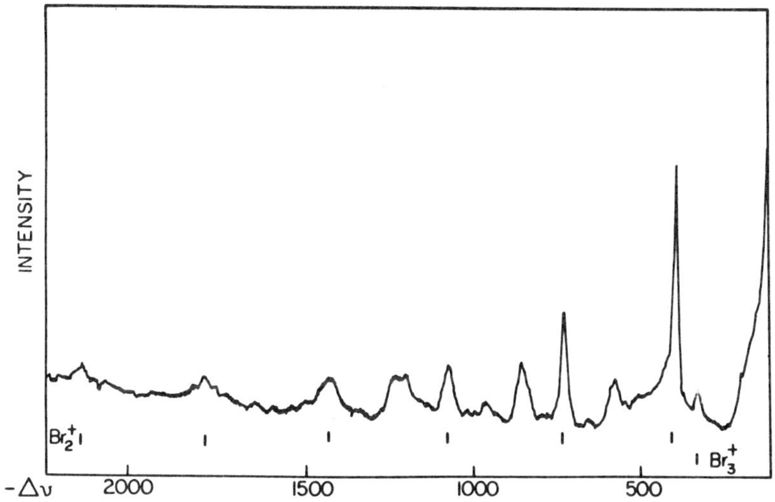

Fig. 3. Resonance Raman spectrum of a 10^{-2} M solution of the Br_2^+ cation in super acid.

Table 3. Frequency shifts, peak heights and widths of lines in the resonance Raman spectrum of a 10^{-1}m solution of Br_2^+ at 5145 Å.

$\Delta\nu$ cm^{-1}	360	716	1070	1430	1780	2130
Relative peak heights	100	50	20	10	6	4
Half widths cm^{-1}	10	22	30	50	>50	>50
Relative intensities	100	100	60	50	–	--

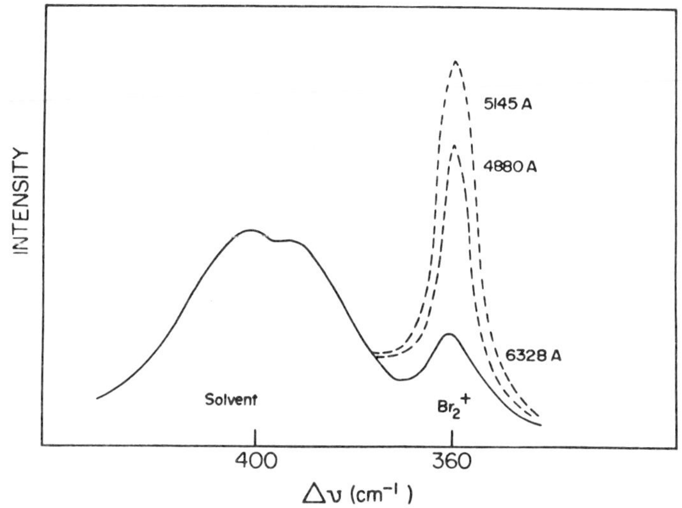

Fig. 4. Resonance Raman spectrum of the Br_2^+ cation: variation with exciting wavelength.

Table 4. Variation of depolarization ratio, and intensities of the Raman fundamental and overtone of the Br_2^+ cation with laser wavelength.

Laser wavelength Å	Area of fundamental[a]	Relative area of overtone[b]	ρ
6328	25	0.25	0.27
5145	100	1.0	0.30
5017	95	1.0	0.28
4965	80	1.0	0.25
4880	70	1.0	–
4765	65	1.0	0.20
4579	30	1.0	–

[a] Areas measured relative to that of fundamental at 5145 Å excitation.
[b] Peak area measured relative to that of the 400 cm^{-1} solvent band.

Table 5. Resonance Raman spectrum of Te_4^+ in concentrated H_2SO_4 using 5145 Å exciting radiation.

$\Delta\nu$ (cm^{-1})	Intensity	Polarization	Assignment
113	m	dp	$\nu_3 (B_{2g})$
219	s	p	$\nu_1(A_{1g}) + 2\nu_3$?
326	w	dp	$\nu_1 + \nu_3$ or $3\nu_3$
434	m	p?	$2\nu_1$, H_2SO_4
563	m	dp	H_2SO_4
650	w	p?	$3\nu_1$

Table 6. Variation of intensities of Raman lines with frequency of exciting radiation.

Å	ν_3(173 cm^{-1})	ν_1(219 cm^{-1})	$2\nu_1$	$3\nu_1$
4579	–	0.36	–	–
4765	0.10	0.59	0.18	0.11
4880	0.15	1.5	0.4	0.3
4965	0.23	1.2	0.3	0.2
5017	0.2	1.5	0.5	0.3
5145	0.3	1.7	0.3	0.4
6328	–	0.18	–	–

are somewhat complex; nevertheless the sharp intense resonance of the Br_2^+ cation at 360 cm^{-1} is easily observed and overtones at 716, 1070, 1430, 1780 and 2130 cm^{-1} may also be distinguished among the peaks due to the solvent and $BrSO_3F$, etc. As for I_2^+ the heights of the overtone peaks steadily decrease while their width increases but the decrease in the intensity of the overtones with increasing frequency is apparently even slower than with I_2^+ (Table 3). The intensity of the resonance Raman spectrum was found to be greatest using 5145 Å radiation and was diminished at both higher and lower wavelengths (Fig. 4).

Comparative intensity measurements were made with reference to the 400 cm^{-1} solvent peak as standard. The depolarization ratio has a value of 0.3 for the fundamental at 5145 Å but appears to decrease somewhat at both shorter and longer exciting wavelengths (Table 4). The relative area of the first overtone also appeared to decrease using longer wavelengths exciting radiation but apparently not with shorter wavelengths (Table 4).

THE Te_4^{2+} CATION

Tellurium is oxidized by concentrated sulphuric acid to a red solution that contains the square planar Te_4^{2+} cation.[6] The structure of this cation has been shown to be square planar by an X-ray crystallographic investigation of the salt $Te_4(AlCl_4)_2$.[7] Moreover the analagous Se_4^{2+} and S_4^{2+} cations have both been shown to have square planar structures and their vibrational spectra have been observed and assigned.[8,9]

Resonance Raman spectra have been obtained from solutions of Te_4^{2+} having concentrations of approximately 10^{-2} m and in capillary tubes having internal diameters of 0.3 to 1.2 mm. Te_4^{2+} has an absorption maximum at 510 nm and as expected the strongest resonance Raman spectra were observed using 5145 Å radiation. The bands observed in a typical spectrum up to 650 cm^{-1} are listed in Table 5. For a square planar molecule three Raman active bands are expected $\nu_1(A_{1g})$, $\nu_2(B_{1g})$ and $\nu_3(B_{2g})$ and of these ν_1 is observed strongly in the resonance Raman spectrum and ν_3 is observed very weakly. The first and second overtones of ν_1 are observed and either a combination band of ν_1 and ν_3 or the second overtone of ν_3 in which case the second overtone of ν_3 must be hidden by the first overtone of ν_1. Although the second overtone of ν_1 coincides with a sulphuric acid peak at 434 cm^{-1} its presence is clearly indicated by the considerable enhancement of the intensity of this peak when compared with the 563 cm^{-1} peak in the solvent alone.

Table 6 shows the relative intensities of ν_1 and its overtones and ν_3 at various exciting frequencies. In each case frequencies were measured with reference to the 560 cm^{-1} peak of the solvent. It may be seen that both ν_1 and ν_3 appear to reach a maximum intensity at around 5000 Å which is close to the maximum of the visible absorption band. The intensities of the overtones appear to diminish somewhat more rapidly than was observed for I_2^+ and Br_2^+. The depolarization ratio for the 213 cm^{-1} was found to be approximately 0.3 at all wavelengths.

REFERENCES

1 R. J. Gillespie and J. B. Milne, *Inorg. Chem.* **5**, 1236 (1966).

2 R. J. Gillespie and K. C. Malhotra, *Inorg. Chem.* **8**, 1751 (1969).

3 W. Holzer, W. F. Murphy and H. J. Bernstein, *J. Chem. Phys.* **52**, 399 (1970).

4 J. Behringer, *Raman Spectroscopy*, (Ed. H. A. Szymanski), Plenum Press, New York, 1967.

5 R. J. Gillespie and M. J. Morton, *Inorg. Chem.* **11**, 586 (1972).

6 J. Barr, R. J. Gillespie, G. P. Pez, P. K. Ummat and O. C. Vaidya, *Inorg. Chem.* **10**, 362 (1971).

7 T. W. Couch, D. A. Lokken and J. D. Corbett, *Inorg. Chem.* **11**, 357 (1972).

8 R. J. Gillespie, J. Passmore, P. K. Ummat and O. C. Vaidya, *Inorg. Chem.* **10**, 1327 (1971).

9 I. D. Brown, D. B. Crump and R. J. Gillespie, *Inorg. Chem.* **10**, 2319 (1971).

45. RESONANCE RAMAN SPECTRUM OF THE POLY(VINYL ALCOHOL)-IODINE COMPLEX

F. Inagaki, I. Harada, T. Shimanouchi and M. Tasumi
Department of Chemistry and Department of Biophysics and Biochemistry, University of Tokyo, Hongo, Tokyo, Japan

INTRODUCTION

It is attractive to apply the resonance Raman effect to the study of the molecular structure of trace amounts of coloured substances in solution. As the concentration of a coloured substance detectable by the resonance Raman effect is very low (*ca.* 10^{-4} mole/1) compared with the ordinary Raman effect or the infrared absorption, this method will find various applications not only in physical chemistry but also in analytical and biological chemistry.

One of us (MT) has studied the resonance Raman spectra observed from dilute aqueous solutions of the starch–iodine complex.[1] The observed Raman bands have been assigned to the vibrations of the I_3^- ions aligned linearly in the cylindrical cavity inside the helical starch molecule. Since poly(vinyl alcohol) (PVA) is known to form a blue complex with iodine just as starch does, we have taken up this complex in order to check whether it is structurally similar to the starch–iodine complex. The results show that the PVA–iodine complex gives a spectrum similar to that of the starch–iodine complex. Under certain conditions, however, a completely different spectrum is observed for the PVA–iodine complex. The details will be reported in this chapter.

EXPERIMENTAL

Two kinds of PVA sample were used. One (sample A) had an average degree of polymerization (DP) of 500 and the other (sample B) of 2000. Each sample solution of the PVA–iodine complex was prepared as follows: 1 ml of the mixed solution of iodine (0.0029 mol/l) and potassium iodide (0.12 mol/l) was added to 5 ml of 0.2 wt % aqueous solution of PVA and then diluted to 25 ml with boric acid solutions of various concentrations (0.05–0.30 mol/l). In Table 1, the numbering of sample solutions is given. The concentration of iodine in each sample was 1.1520×10^{-4} mol/l as I_2.

The Raman spectra were recorded using a He–Ne laser (632.8 nm) and an Ar^+ ion laser (488.0 nm) for excitation. When the He–Ne laser was used, a spectrometer was employed which consists of a Spex 1401 double monochromator, a photomultiplier, a phase-sensitive detector and a recorder. A Raman spectrometer constructed by Kawaguchi-Denki Co. and a JEOL JRS–Ul spectrometer were also used, both with an Ar^+ ion laser.

Table 1. The numbering of solutions.

PVA	H_3BO_3 concentration in mol/l					
	0.05	0.10	0.15	0.20	0.25	0.30
DP = 500	A–1	A–2	A–3	A–4	A–5	A–6
DP = 2000	B–1	B–2	B–3	B–4	B–5	B–6

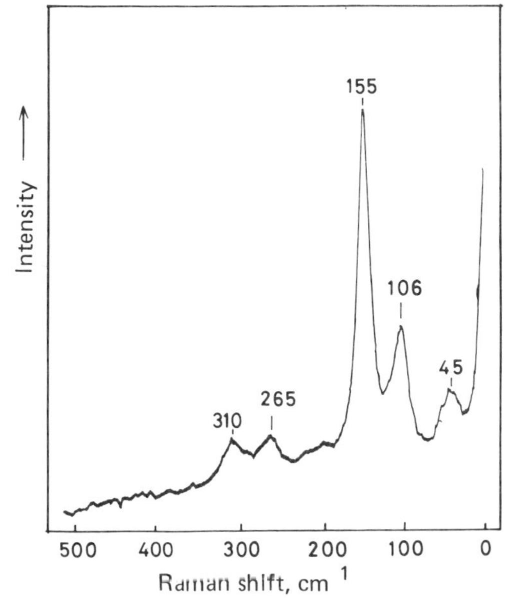

Fig. 1. The Raman spectrum of the solution B–3 taken with the 632.8 nm excitation.

The Raman scattering was observed at right angles to the exciting laser beam which radiates the sample very close to the quartz cell wall. Hitachi EPS–3 and 124 spectrophotometers were used to measure the absorption spectrum in the ultraviolet and visible regions.

RESULTS AND DISCUSSIONS

In Fig. 1 is shown an example of the Raman spectrum of the PVA–iodine complex obtained with the 632.8 nm line of the He–Ne laser. A close resemblance is found between this and the spectrum reported previously for the starch–iodine complex.[1] It seems reasonable, therefore, to presume that the structure of the PVA–iodine complex is similar to that of the starch–iodine complex. In other words, the present result is compatible with the model structure of the PVA–iodine complex proposed by Zwick[2] that a linear polyiodide chain is enclosed by a helical PVA molecule and the latter, in turn, is surrounded by borates which play the role of maintaining the PVA helix by forming the –C–O–B–O–C– bridge.

By analogy with the starch–iodine case the three bands at 45, 106 and 155 cm^{-1} in

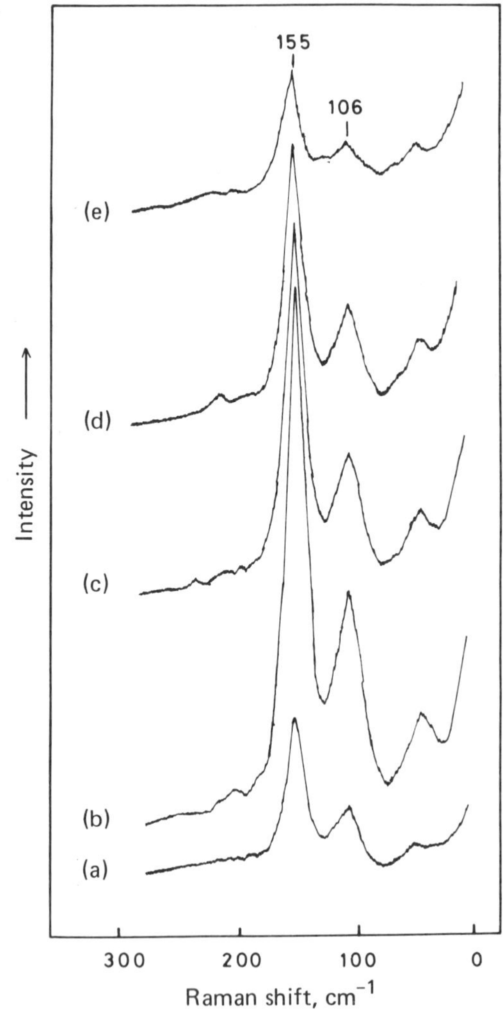

Fig. 2. The Raman spectra of sample B taken with the 632.8 nm excitation. Iodine concentrations are as I_2 : (a) 2.09 × 10^{-4} mol/l; (b) 1.05 × 10^{-4} mol/l; (c) 0.52 × 10^{-4} mol/l; (d) 0.26 × 10^{-4} mol/l; (e) 0.13 × 10^{-4} mol/l.

Fig. 1 may be assigned, respectively, to the bending, symmetric stretching, and antisymmetric stretching vibrations of the I_3^- ion. The other two at 265 and 310 cm^{-1} may be assigned to a combination tone (265 ~ 106 + 155) and an overtone (310 ~ 155 × 2).

In the following section we will discuss various spectral characteristics observed under different conditions.

Effect of concentration on the Raman intensities

In Fig. 2 is shown the dependence of the Raman intensities on iodine concentration observed for sample B. Apparent Raman intensity reaches its maximum at the iodine concentration of 1.0488 × 10^{-4} mole/l. Similar tendency was also observed in the solution of sample A.

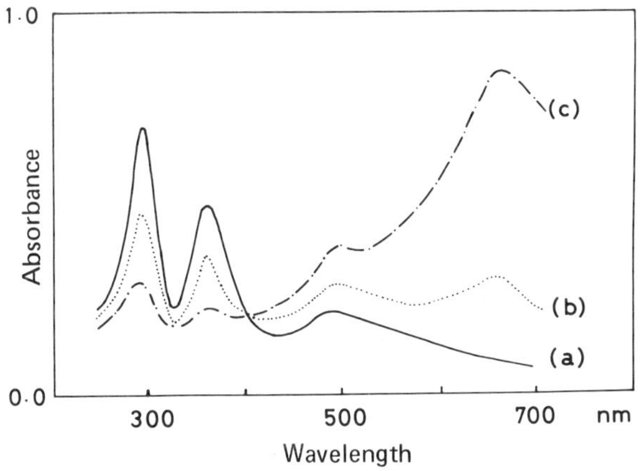

Fig. 3. The absorption spectrum of sample A in the ultraviolet and visible regions. The solutions used are: (a) A–1; (b) A–3; (c) A–5.

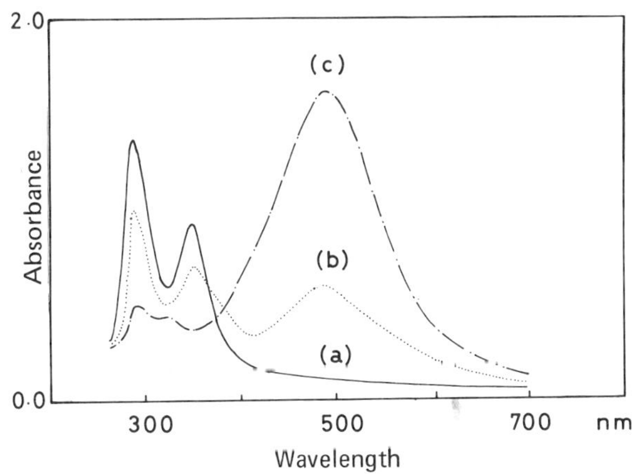

Fig. 4. The absorption spectrum of sample A in the ultraviolet and visible regions. PVA concentrations are: (a) 0.04 wt%; (b) 0.2 wt %; (c) 1.0 wt %. No boric acid was added.

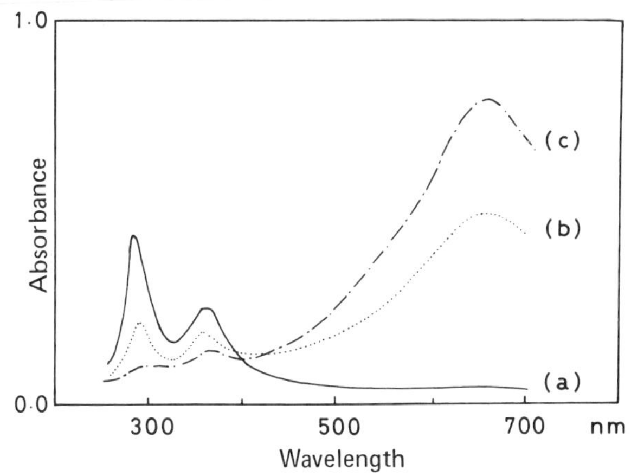

Fig. 5. The absorption spectrum of sample B in the ultraviolet and visible regions. Boric acid concentrations are: (a) B–1; (b) B–3; (c) B–5.

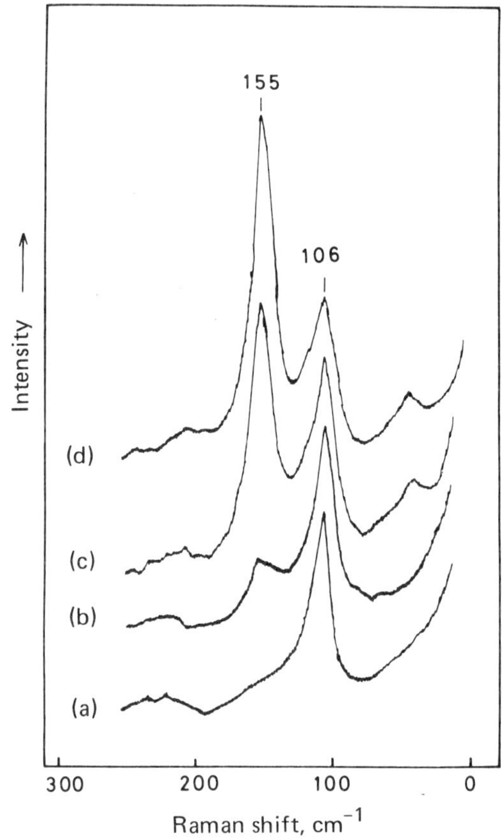

Fig. 6. The Raman spectra of the solution series A taken with the 632.8 nm excitation. (a) A–1; (b) A–2; (c) A–3; (d) A–5.

Correlation between the absorption and the Raman spectra

Ultraviolet and visible absorption spectra. The absorption spectrum of sample A in the ultraviolet and visible regions is shown in Fig. 3. The colour of the complex varied from red to blue as the concentration of boric acid was increased. The absorption bands at 290 nm and 350 nm are due to the I_3^- ion. The band at 490 nm is close to the absorption wavelength of the I_2 molecule. However, only a small amount of the I_2 molecule would remain in the solutions treated here owing to the presence of excess potassium iodide. As the extinction coefficient of the I_2 molecule is small ($\epsilon = 746$ l mole^{-1} cm^{-1}), it is unlikely that the 490 nm band in Fig. 3 arises from I_2. Therefore this band should be assigned to the PVA (sample A)–iodine complex. This is supported also by the observation that the intensity of this band increases with increasing PVA concentration, while, on the other hand, the intensities of the 290 nm and 350 nm bands of the I_3^- ion decrease as shown in Fig. 4. The band at 670 nm appears very strongly when the boric acid concentration is high.

In Fig. 5 are shown the absorption spectra of the solution series B. The colour of the solution B–1 (yellow) was exactly the same as that of the solution of iodine and

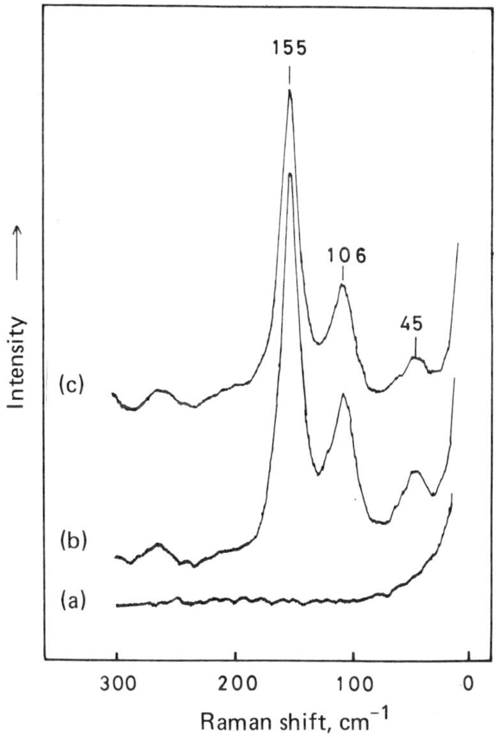

Fig. 7. The Raman spectra of the solution series B taken with the 632.8 nm excitation. (a) B–1; (b) B–3; (c) B–5.

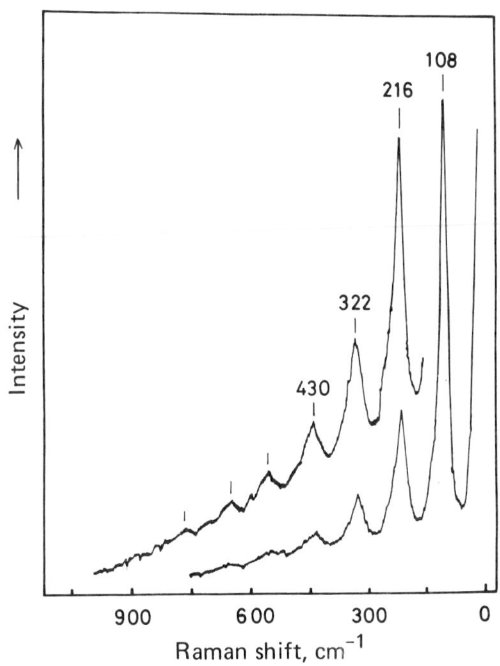

Fig. 8. The Raman spectrum of the solution A–4 with the 488.0 nm line excitation. The solution A–2 gives essentially the same spectrum.

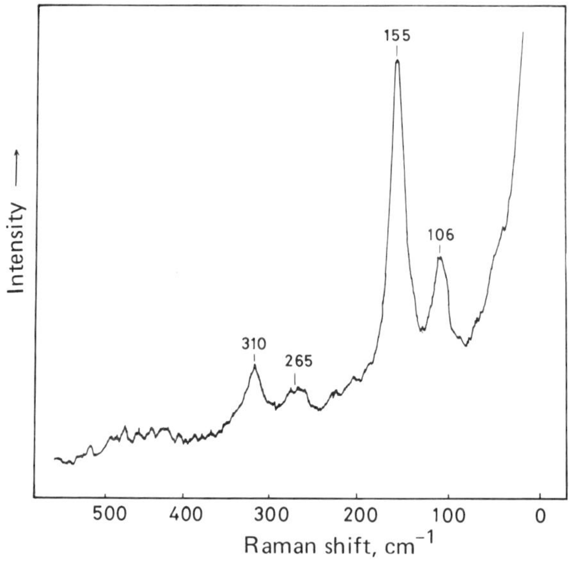

Fig. 9. The Raman spectrum of the solution B–3 taken with the 488.0 nm line excitation.

potassium iodide. This would mean that sample B does not make a complex with iodine at low boric acid concentration. At higher boric acid concentrations sample B also shows a strong band at 670 nm.

It is noted that the 490 nm band found in the spectra of sample A is not observed in any of the spectra of sample B. This difference gives rise to an important effect in the Raman spectra as will be discussed later.

The Raman spectra observed with 632.8 nm excitation. The Raman spectra of the solution series A and B are shown in Figs. 6 and 7, respectively, which were obtained with the 632.8 nm line of He–Ne laser. A marked change may be seen in the spectra of the solution series A. The Raman band at 155 cm^{-1} (the antisymmetric stretching mode) is not observed in the reddish solution A–1. When the colour of the complex becomes blue, this band is strongly enhanced. In contrast with this, the intensity of the 106 cm^{-1} band (symmetric stretching mode) remains nearly constant throughout the solution series A, independent of the colour of the complex.

No Raman bands are observed for the yellow solution B–1. However, as the colour becomes increasingly blue by the addition of the boric acid, strong Raman bands appear for the solutions B–3, B–4 and B–5.

The Raman spectra observed with the 488.0 nm excitation. In Fig. 8 is shown the spectrum of the solution A–4 obtained with the 488.0 nm line of the Ar^+ ion laser. It is noted that the Raman band at 106 cm^{-1} is strongly enhanced and overtones may be observed until the sixth. The Raman line at 155 cm^{-1} is observed only as a weak shoulder. This is in contrast with Fig. 1 in which the band at 155 cm^{-1} is much stronger than the band at 106 cm^{-1}.

For sample B, on the other hand, essentially the same spectra were observed as those obtained with He–Ne 632.8 nm line (Fig. 9). The different spectral behaviour be-

Table 2. Summary of the resonance effect.[a]

	PVA–iodine					
	A		B			
Exciting line	red	blue	yellow	blue	Starch–iodine[b]	Free I_3^-
632.8 nm	s	s,A	none	s,A	s,A	s
488.0 nm	S	S	none	s,A	s,A	s

[a] The capital letters S and A indicate that the intensities of the I_3^- *symmetric* and *antisymmetric* modes are, respectively, enhanced greatly. The small symbol s means that the intensity of the I_3^- *symmetric* stretching is enhanced, but not as greatly as the case indicated by the capital letter.
[b] Taken from ref. 1.

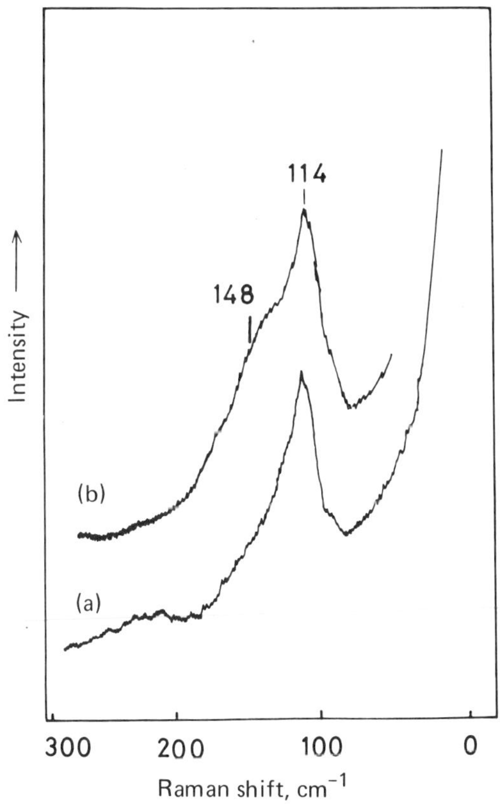

Fig. 10. The Raman spectra of aqueous solutions of potassium iodide. (a) excited with the 488.0 nm line; (b) excited with the 632.8 nm line.

tween sample A and sample B with respect to the 488.0 nm excitation seems to be associated with the fact that sample A has the 490 nm band which is resonant with the exciting line, whereas sample B has no such band.

The results obtained above are summarized in Table 2, together with those for the starch–iodine complex and the free I_3^- ion. The Raman spectra of the free I_3^- ion in aqueous solution are shown in Fig. 10. The measurements were made under much higher I_3^- concentrations than those used in the Raman measurements of the complexes. This indicates that the resonance effect is much weaker for the free I_3^- ion than for I_3^- in the complexes, reflecting the experimental conditions that the exciting wavelengths used are outside of the absorption bands of the free I_3^- ion found at 350 and 290 nm (preresonance case).

The following points in Table 2 are particularly noteworthy:

1. The intensity of the I_3^- antisymmetric stretching mode is strongly enhanced in the complex state under some conditions, whereas it is only weakly observed in the free state. This is undoubtedly due to the presence of the absorption band of the complex at 670 nm (PVA–iodine) or *ca.* 600 nm (starch–iodine).
2. The intensity of the I_3^- symmetric stretching mode is enhanced in the free I_3^- spectra. It is strongly enhanced also in the complex-state spectra, instead of the antisymmetric mode, if the exciting wavelength is resonant with the absorption band at 490 nm.

These results clearly show that the alternative enhancement of different Raman bands can be achieved by using the resonance effect. As far as we know, such a clear example of alternative enhancement has not been reported.

REFERENCES

1 M. Tasumi, *Chem. Letters* **1**, 75 (1972).
2 M. M. Zwick, *J. Appl. Polymer Sci.* **9**, 2393 (1965).

46. RAMAN SCATTERING OF LIGHT BY MOLECULAR IMPURITIES IN ALKALI HALIDES

K. K. Rebane, L. A. Rebane, T. J. Haldre
and A. A. Gorokhovski

Institute of Physics and Astronomy of the Estonian S.S.R. Academy of Sciences, Tartu, Estonian S.S.R., USSR

RAMAN SPECTRA OF MOLECULAR IONS NO_2^- AND NO_3^-

Single crystals of KCl, KBr, and KI doped with 1 mol % KNO_2^-, contained comparable concentrations (about $10^{19}\,cm^{-3}$) of NO_2^- and NO_3^- ions. Excitation was carried out with an argon laser in the [001] direction of the crystal. Scattered light was registered in the [010] direction. Spectra were measured at temperatures 300 and 10 K.

Excitation $\lambda = 4880$ Å (20486 cm^{-1}) for the NO_2^- ion is out of resonance with the absorption by impurity ions[1] and the main lattice.* The Raman spectrum contains maxima of all intramolecular (local) vibrations of NO_2^- and NO_3^- ions (see Table 1). Intensities of lines are in accordance with selectrion rules for free molecules.[3] The lines of totally symmetric vibrations ν_1 of NO_2^- (Fig. 1) and ν_1 of NO_3^- have the highest intensities. Lines ν_2 of NO_2^- and also ν_3 and ν_4 of NO_3^- have nearly the same intensities. Vibrations ν_3 and ν_4 of NO_3 which are twofold degenerated for the free molecule, remain unsplit in the Raman spectrum (they do not split in the infrared absorption spectrum either[4]), which gives evidence about a sufficiently high symmetry of the NO_3^- centre in the crystal, corresponding to the orientation of axis C_3 of the molecule along [111] directions of the crystal. Line ν_2 of the NO_3^- ion, forbidden for the free ion, appears in the spectrum with lowest intensity.

Line ν_3 of the NO_2^- ion reveals a striking dependence on temperature; at room temperature it appears in the spectrum as a weak broad 'step' but at the 10 K line ν_3 has the width and intensity comparable to those of line ν_3 of NO_3^- (Fig. 1).

On the basis of vibronic spectra[1] and also infrared absorption spectra[5] it is known that in KCl and KBr crystals the NO_2 ion has a one-dimensional (hindered) rotational degree of freedom around axis a, perpendicular to axis C_2 of the molecule and oriented along direction [100] in the crystal. Since the polarization tensor of vibration ν_3 of the free ion NO_2^- has no diagonal components, the free rotation of the molecule around axis a for vibrational–rotational transitions in the Raman spectrum leads to the selection rule for the rotational quantum number in the model of rigid rotator $\Delta K = \pm 2$.[3] This also means that pure vibrational transition is forbidden.

Considerable broadening of ν_3 lines observed in the spectrum with the rise of

* It should, however, be mentioned that the NO_2^- ion has a forbidden electronic transition $^3B_1 - {}^1A_1$ in this region, which was observed in nitrite salts.[2]

Table 1. The fundamental frequencies of NO_2^- and NO_3^- impurity ions from the Raman spectra.

	KCl $T = 10$ K cm^{-1}	KBr $T = 10$ K cm^{-1}	KI $T = 300$ K cm^{-1}
NO_2^-	1324	1312	1310
	800	800	787
	1288	1278	1257
NO_3^-	1060	1055	1050
	—	845	825
	1388	1380	1370
	737	720	690

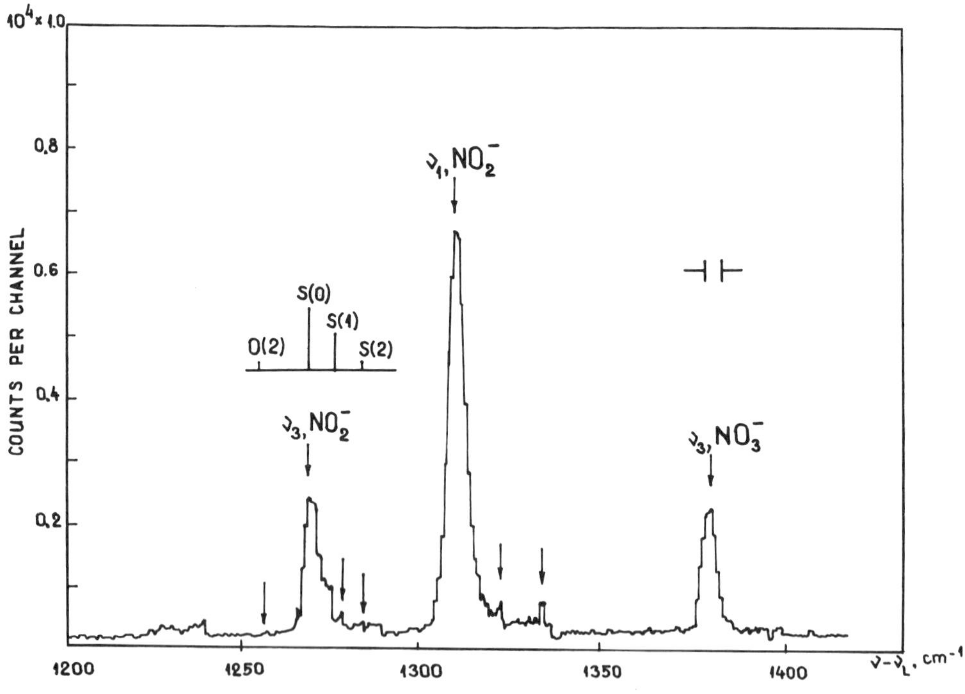

Fig. 1. Part of the Raman scattering spectrum of KBr crystal doped with 1 mol % KNO_2 at temperatures about 10 K, scanning 18 s per channel. Above the scheme of the rotational structure of ν_3 line of the NO_2^- ion is given.

temperature may be interpreted as a result of the rotation of NO_2^- in the crystal.[6] Making use of the scheme of rotational terms found on the basis of the rotational structure of the vibronic transitions in the absorption and luminescence spectra of the NO_2^- ion[7,8] (rotational terms in the ground electronic state are described by a model of rotator $F = AK^2$ where $A = 2\ cm^{-1}$), and also the above mentioned selection rule $\Delta K = \pm 2$, we obtain such a rotational structure of line in the Raman spectrum which is, on the whole, in agreement with the observed one (Fig. 1). The highest intensity transition is $S(0)$ transition, i.e. the transition $V = 0 \rightarrow V = 1$ and $K = 0 \rightarrow K = 2$ in which both vibrational and rotational quantum numbers are changed. According to the scheme of rotational levels this transition must be shifted by $+ 8\ cm^{-1}$ relative to pure vibrational transition $Q(0)$. At 10 K also, two following lines of the S-branch must be observable – transitions $S(1)$ and $S(2)$, whose intensities must decrease in accordance with the population of levels $K = 1$ and 2. The first maximum on the anti-Stokes side from transition $S(0)$ should be expected at a distance of $-16\ cm^{-1}$: that would be transition $O(2)$ which has a low intensity in accordance with the population of level K = 2. The shape of line ν_3 of the NO_2^- ion corresponds with the predicted one: line ν_3 has a gradual slope towards the Stokes side with steps at + 8 and 16 cm^{-1} which may be ascribed to transitions $S(1)$ and $S(2)$, and a sharp slope towards the anti-Stokes side.

For vibrations ν_1 and ν_2 of the NO_2^- ion transitions $Q(0)$ are evidently observed in the spectrum (for the free ion a transition with $\Delta K = 0$ is allowed). Line ν_1 (Fig. 1) has two additional weak maxima on the Stokes side, shifted from ν_1 by +13 and +25 cm^{-1}, which may be accounted for by the rotational structure.

The part of the spectrum lying in the region of 50–400 cm^{-1} on the long-wave side from the Rayleigh line contains several rather sharp maxima on a continuous background. This region may be interpreted as the phonon wing of the Rayleigh line of scattering. Its one phonon part is induced by impurity ions NO_2^-, NO_3^-, as well as by other defects. A number of maxima (at 90, 136, 150 and 173 cm^{-1}) are close to those observed and interpreted by Fitchen *et al.*[9] as being due to crystal vibrations near NO_2^-. In our case, when crystals contained comparable concentrations of NO_2^- and NO_3^- ions, the interpretation given above requires additional arguments. Besides, a band was observed at 250 cm^{-1} and a number of weak maxima in the region of 300 cm^{-1}.

The most reliable data about the local dynamics in the impurity centre under study must arise from the phonon wing adjacent to the Raman line from the intramolecular vibration of the given impurity. The intensity of scattering in the region of the phonon wing is expected to be about an order of magnitude lower than in the line itself. We observed quite a weak broad band at 100–200 cm^{-1} from line ν_2 of NO_2^-, which may be considered to be the phonon wing of line ν_2.

SECONDARY EMISSION SPECTRUM OF S_2^- CENTRES IN KI

Samples were prepared by heating KI monocrystals in sulphur vapour (up to 200 h) at a temperature close to the melting point of KI. As a result S_2^- and S_3^- ions diffused into the crystal.

Excitation by the 4880 Å line is resonant for S_2^- in KI. According to the data about vibronic spectra[10,11] the frequency of the pure electronic transition was estimated as 20020 cm^{-1} and the vibration frequency of S_2^- in the ground and excited electronic states is equal $\nu_0 = 600\ cm^{-1}$ and $\nu_0' = (350 \pm 50)\ cm^{-1}$, respectively. Thus, excitation leads S_2^- centres into the state which exceeds the vibrational level $V' = 1$ by about 130 cm^{-1}, i.e. the state where both local and crystal vibrations are excited. According to refs. 12, 13 and 14, in cases of resonant excitation the secondary emission spectrum of a rapidly relaxing impurity centre may be presented as a super-position of two spectra:

1. no-phonon (Rayleigh) line of scattering and its vibrational replicas, and
2. no-phonon (pure electronic) luminescence line and its vibrational replicas.

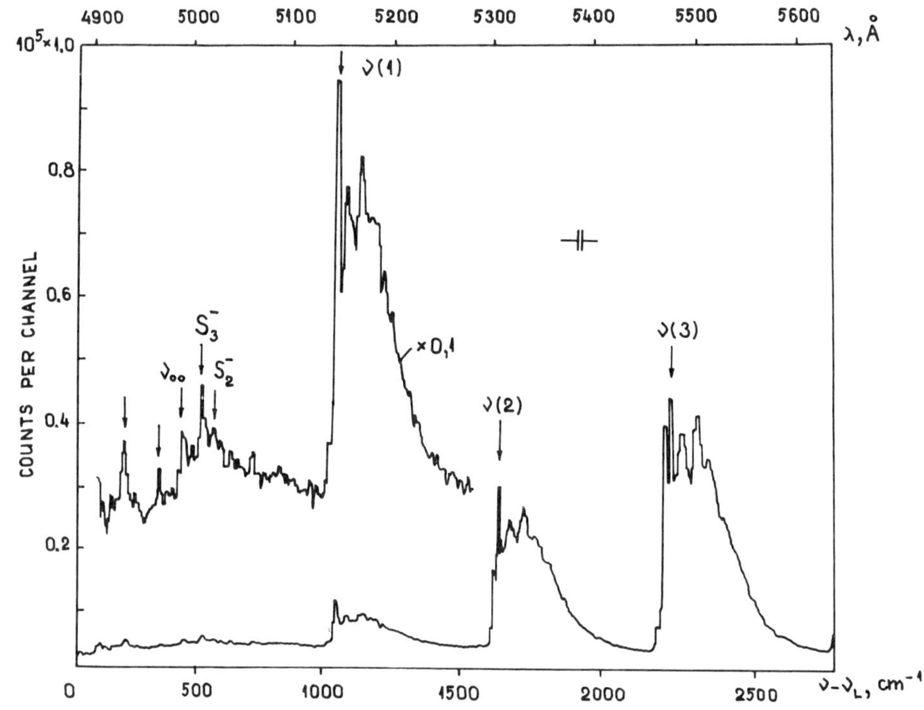

Fig. 2. Secondary emission spectrum of $KI-S_2^-$ crystal at a temperature about 10 K, scanning 3.6 s per channel. The lower scale is related to the excitation line ν_L.

The scattering spectrum (1) contains the Rayleigh line whose frequency coincides with that of exciting line ν_L, and all the Raman lines whose frequencies are shifted to various combinations of the vibrational frequencies of the crystal and impurity centre. The luminescence spectrum (2) contains, besides ordinary luminescence, also hot luminescence which is much weaker in rapidly relaxing centres than ordinary luminescence and is comparable to scattering by its intensity.

The general appearance of the secondary emission spectrum of the $KI-S_2^-$ crystal is given in Fig. 2. In accordance with the theory it contains both scattering and luminescence spectra.

Luminescence of S_2^- centres contains a pure electronic line $\nu_{00} = 20034 \pm 2$ cm^{-1} and its replicas by the local (intramolecular) vibration $\nu(V)$ ($V = 1, 2$, etc.). Because of big Stokes losses on the local vibration, which make eight vibrational quanta, the pure electronic line and its first vibrational replicas have not been registered earlier.[10] Each no-phonon line is accompanied by a phonon wing due to the excitation of crystal vibrations.

Comparison of the detailed structure of phonon wings in luminescence in cases of broad band[15] and laser excitation was of interest. The latter guaranteed polarized and monochromatic excitation leading the S_2^- centre into the state where $V' = 1$ and a very narrow region of lattice vibrations (near 130 cm^{-1}) is excited. In this situation differences depending on the method of excitation are possible. In cases of laser excitation, especially hot transitions in crystal vibrations could be possible. Comparison of phonon wings (Figs. 2 and 3) shows a general coincidence of the maxima of phonon wings (with the

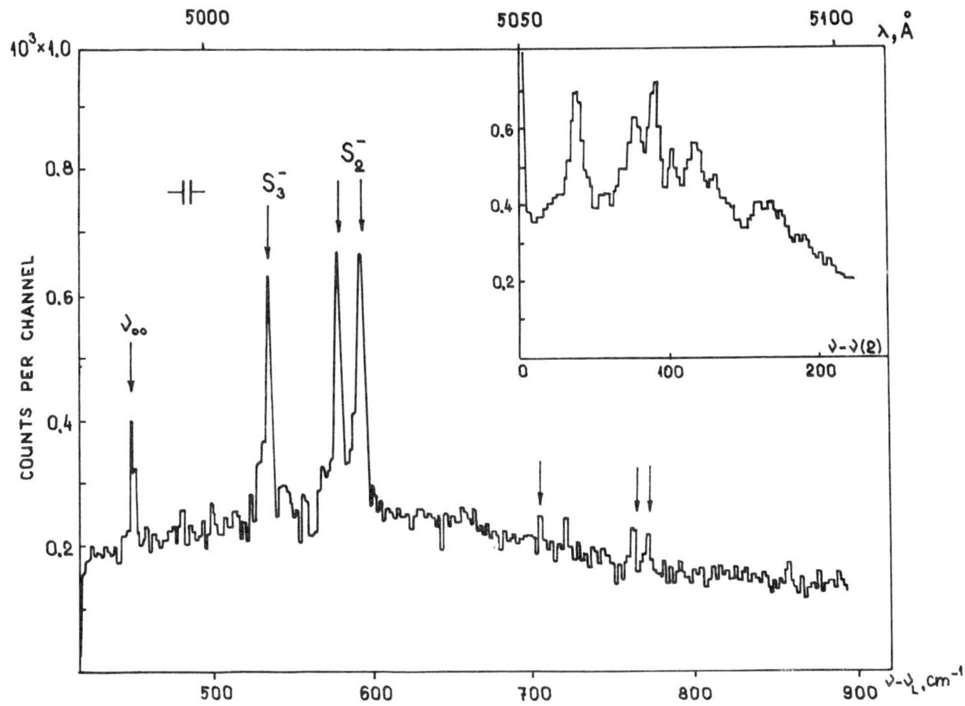

Fig. 3. Part of the spectrum of Fig. 2 containing the region of the zero vibronic group of luminescence and the scattering lines from the vibrations of S_2^- and S_3^- ions. The phonon wing in the luminescence spectrum of S_2^- in the case of wide band excitation[15] is given above.

exception of the peak at 123 cm^{-1} which is absent in cases of laser excitation). No indications about hot luminescence in crystal vibrations were observed.

We also searched for hot luminesence lines over the intramolecular vibration but they were not found. It is easy to understand that it is really difficult to discover hot luminescence in KI–S_2^-. Being based on the given vibronic spectra and taking into account that our excitation makes possible hot transitions from level $V' = 1$ only, one may expect the appearance of a series of hot transitions $1' \to V$. The lines of this series must be shifted from zero-phonon lines $0' \to V$ of the ordinary luminescence by about +350 cm^{-1}. From these only transition $1' \to 0$ falls into the short wavelength region from the ν_{00}, which is convenient for registration. But this transition has, firstly, a very low relative probability (very small Franck–Condon factor), and, secondly, falls on the scattering region (ν_L − 100) cm^{-1}, i.e. in the intensive region of scattering by crystal vibrations. As a result of big Stokes losses on the intramolecular vibration the rest of the first terms of the hot series are also weak. Hot lines themselves fall on the regions of the phonon wings of corresponding ordinary luminescence vibronic bands. That is why S_2^- centres are inconvenient objects for detecting hot luminescence.

Raman scattering lines from intramolecular vibrations overlap with the phonon wing of pure electronic luminescence lines (Fig. 2). Here scattering lines of S_2^- and S_3^- ions can be seen. In Fig. 3 this region, beginning at ν_{00}, is presented separately. The line at 531 cm^{-1} belongs to the totally symmetric vibration of the S_3^- ion.[16] Scattering on the vibrations of the S_2^- ion gives $\nu = 595\ cm^{-1}$ in accordance with luminescence data. Another narrow line at $579\ cm^{-1}$ probably belongs to the S_2^- ion too.

At room temperature luminescence of S_2^- centres is quenched and only scattering remains. Therefore the second and third vibrational recurrences also become visible: at $1185 \pm 5 \simeq 2 \times 595\ cm^{-1}$ and $1783 \pm 5 \simeq 3 \times 595\ cm^{-1}$ for S_2^-, and at 1093 and 1634 cm^{-1} for S_3^-.

In the low frequency region of the scattering spectrum the phonon wing of the Rayleigh scattering line is clearly observable — it is scattering from crystal vibrations induced by S_2^- and S_3^- impurities. Here the maximum emerges at 85 cm^{-1} which is close to the frequency of the gap mode in the phonon wing of KI–S_2^- luminescence. In the region of two-phonon transitions there are also a number of wide maxima at 170, 183, 215 and 270 cm^{-1}, which are not yet interpreted.

REFERENCES

1 R. Avarmaa, *Izv. Akad. Nauk Est. SSR: Ser. Tekhn. Fiz. Mat. Nauk* **17**, 78 (1968).

2 R. M. Hochstrasser and A. P. Marchetti, *J. Chem. Phys.* **50**, 1727 (1969).

3 M. M. Sushchinskii, *Spektry Kombinationnogo Rassejanija Molekul i Kristallov*, Moscow, 1969.

4 T. Mauring, *Izv. Akad. Nauk Est. SSR: Ser. Tekhn. Fiz. Mat. Nauk* **18**, 105 (1969).

5 V. Narayanamurti, W. D. Seward and R. O. Pohl, *Phys. Rev.* **148**, 481 (1966).

6 R. Callender and P. S. Pershan, *Phys. Rev.* **A2**, 672 (1970).

7 R. Avarmaa and L. Rebane, *Phys. Status Solidi* **35**, 107 (1969); *Izv. Akad. Nauk Est. SSR: Ser. Tekhn. Fiz. Mat. Nauk* **18**, 117 (1969).

8 L. A. Rebane, *Czech. J. Phys.* **B20**, 608 (1970).

9 I. W. Shephad, A. R. Evans and D. B. Fitchen, *Phys. Rev. Letters* **27A**, 171 (1968); A. R. Evans and D. B. Fitchen, Report N 1326, 1970.

10 L. Rebane, *Izv. Akad. Nauk Est. SSR: Ser. Tekhn. Fiz. Mat. Nauk* **17**, 72 (1968).

11 I. Sildos and L. Rebane, *Izv. Akad. Nauk Est. SSR: Ser. Tekhn. Fiz. Mat. Nauk* **20**, 354 (1971).

12 K. Rebane, V. Hizhnyakov and I. Tehver, *Izv. Akad. Nauk Est. SSR: Ser. Tekhn. Fiz. Mat. Nauk* **16**, 207 (1967).

13 V. V. Hizhnyakov. K. K. Rebane and I. J. Tehver, in *Light Scattering Spectra of Solids* (Ed. G. B. Wright), Proceedings of the International Conference held at New York University, September 1968, Springer-Verlag, 1968, p. 513.

14 K. K. Rebane, *Vtoritchnoje svetchenije primesnogo tsentra kristalla,* Lecture held at Summer School on luminescence, Baical Buchta Pestchanaja, July 1969, Estonian SSR Academy of Sciences, Tartu, 1970.

15 L. A. Rebane, *Opt. i Spektroskopiya* **31**, 230 (1971).

16 W. Holzer, F. Murphy and H. J. Berstein, *J. Mol. Spectry.* **32**, 13 (1969).

47. SPECTRE RAMAN D'UN MONOCRISTAL DE FLUORINE QUADRATIQUE DOPEE A L'EUROPIUM Eu^{3+}

G. Lucazeau

Laboratoire de Chimie Physique, C.N.R.S., Thiais, France

et

J. A. Koningstein

Department of Chemistry, Carlton University, Ottawa, Canada

ABSTRACT

The polarized Raman spectra of a tetragonal single crystal of CaF_2 doped with Eu^{3+} –(Eu^{3+})/(Ca^{2+}) = 0.005 is reported. The main features of the spectrum are on one hand, the $^7F_0 \rightarrow {}^7F_1$ and $^7F'_0 \rightarrow {}^7F_2$ electronic transitions of the Eu^{3+} ions located at sites of C_2 symmetry and on the other hand transitions which involve their motion. Some of the electronic transitions are related to an anti-symmetric scattering tensor, and this has been used in the assignment of bands in the Raman spectra. The comparison of fluorescence and Raman data have been made.

INTRODUCTION

Le spectre Raman du premier ordre de la fluorine (cubique) est caractérisé par une bande située à 322 cm^{-1} qui correspond au phonon de symétrie T_{2g}, seul actif en diffusion Raman.[1] A notre connaissance, il n'existe pas de spectre Raman du second ordre de CaF_2 pure; seules des études théoriques donnent les courbes de densité de phonons.[2-4]

Cependant, il existe de nombreuses études Raman de cristaux de fluorine, dopés[5-9] avec des impuretés; dans certains travaux l'effet Raman du second ordre a été invoqué à tort. En effet, la présence de bandes de fluorescence peut rendre difficile l'interprétation de ces spectres.[9] Les spectres Raman de Ce^{2+}: CaF_2[5] et de Eu^{2+} : CaF_2[6] sont très semblables et ont été interprétés: le premier sur la base d'un spectre de phonons induit par les impuretés Ce^{2+}, le second en termes de phonons associés à des cages EuF_8.

Nous discutons ici, le spectre Raman d'un monocristal de Eu^{3+} : CaF_2, le diagramme des niveaux électroniques de l'ion europium est bien connu,* et de plus cet ion est un excellent candidat à l'effet Raman électronique.[10] Certaines transitions sont reliées à des tenseurs antisymétriques, ce qui facilite leur attribution. Le cristal Eu^{3+} : CaF_2 étudié est de symétrie quadratique (les ions Eu^{3+}, même dans une proportion aussi faible (Eu^{3+}/Ca^{2+} = 0,005) ont pu provoquer une déformation du cristal).

Les données Raman relatives à tous les ions du réseau ont été utilisées dans l'attribution des spectres Raman et de fluorescence.

---

* La configuration électronique est: Xe $4f^6\,5s^2\,5p^6$. L'état de plus haute multiplicité est un état ^7F et J peut donc prendre 7 valeurs (0 à 6). L'état fondamental est $^7F_0$, le premier état excité $^7F_1$ suivant la symétrie de l'environnement peut, au maximum éclater en 3 niveaux, correspondants aux 3 valeurs possibles (+1, 0, −1) de J_z. Le second état excité est $^7F_2$, pour lequel J_z peut prendre 5 valeurs. Le deuxième terme est situé 17000 cm^{-1} au dessus du niveau fondamental et est $^5D_0$.

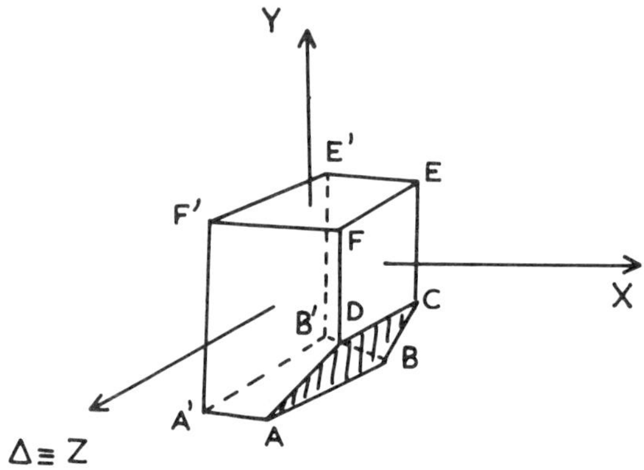

Fig. 1. Monocristal de Eu^{3+} CaF_2 utilisé dans nos expériences A'F': 4 mm; E'F': 3 mm. ABCD fait un angle de 45° avec CDEF. Δ est l'axe optique parallèle à l'axe cristallographique z d'ordre 4.

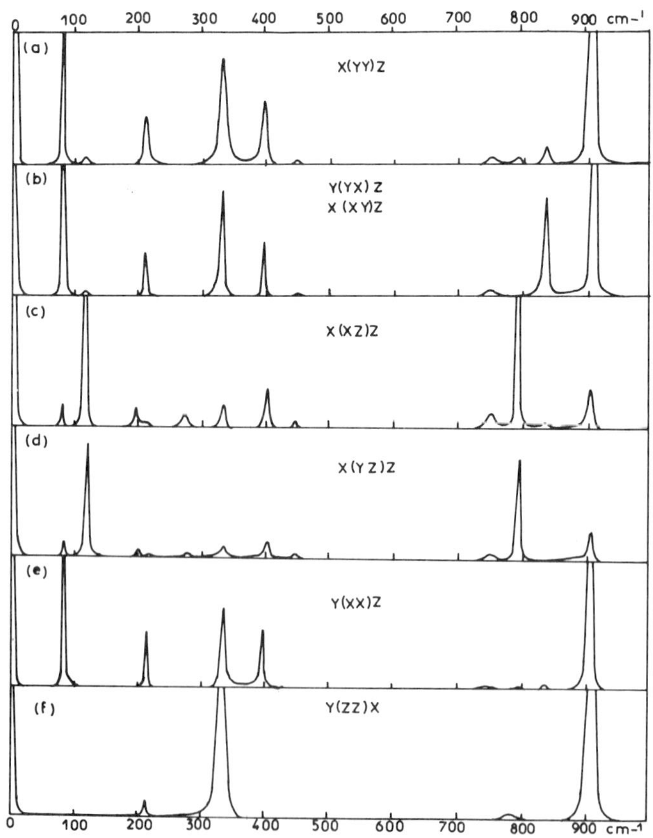

Fig. 2. Spectres Raman polarisés de Eu^{3+} CaF_2 à 300 K. $x(xz)z$ signifie que la lumière incidente se propage selon x et est polarisée suivant z tandis que la lumière diffusée se propage suivant z et est analysée suivant x. Cette notation donne directement l'élément de tenseur xz.

Tableau 1. Position des bandes des spectres Raman et de fluorescence de Eu^{3+}: CaF_2.

Raman	Tenseur	Symétrie	Fluorescence	Attribution
83 cm^{-1}	$\alpha_{xx}-\alpha_{yy}, \alpha_{xy}$	$B_{1g}+B_{2g}$	88 cm^{-1}	phonon (transl. RE)
117 cm^{-1}	$(\alpha_{xz}, \alpha_{yz})$	E_g	117 cm^{-1}	phonon (transl. RE)
199 cm^{-1}	$(\alpha_{xz}, \alpha_{yz})$	E_g	190 cm^{-1}	phonon (transl. RE)
214 cm^{-1}	$\alpha_{xx}+\alpha_{yy}, \alpha_{zz}$ α_{xy} $(\alpha_{xz}, \alpha_{yz})$	A_{1g} $+B_{2g}$ $+E_g$	211 cm^{-1}(w)	phonon
238[a] cm^{-1} (vw)	–	–		phonon
276 cm^{-1}	$(\alpha_{xz}, \alpha_{yz})$	E_g		$^7F_0-{}^7F_1$
313 cm^{-1} (vw)	–	–	308 cm^{-1}(st)	phonon
336 cm^{-1}	$\alpha_{xx}+\alpha_{yy}, \alpha_{zz}$ $\alpha_{xx}-\alpha_{yy}, \alpha_{xy}$ $(\alpha_{xz}, \alpha_{yz})$	A_{1g} $+B_{1g}+B_{2g}$ $+E_g$	331 cm^{-1}(st)	phonon
376[b] cm^{-1}	–	–		$^7F_1-{}^7F_2$
400 cm^{-1}	$\alpha_{xx}-\alpha_{yy}$ $\alpha_{xy}-\alpha_{yx}$	B_{1g} $+A_{2g}$	400 (br)	$^7F_0-{}^7F_1$
406 cm^{-1}	$(\alpha_{xz}-\alpha_{yz})$	E_g		$^7F_0-{}^7F_1$
450 cm^{-1} (vw)	–	–	460 cm^{-1}(st)	phonon (CaF_2)
483[b] cm^{-1} (w)	–	–		$^7F_1-{}^7F_2$
573[b] cm^{-1} (w)	–	–		$^7F_1-{}^7F_2$
620 cm^{-1} (w br)	–	–		phonon (CaF_2)
755 cm^{-1}	$\alpha_{xx}-\alpha_{yy}$ or $\alpha_{xx}+\alpha_{yy}, \alpha_{zz}$ α_{xy} $(\alpha_{xz}, \alpha_{yz})$	A_{1g} or B_{1g} $+B_{2g}$ $+E_g$	755 cm^{-1}(v st)	$^7F_0-{}^7F_2$
797 cm^{-1}	$(\alpha_{xz}, \alpha_{yz})$	E_g	797 cm^{-1}	$^7F_0-{}^7F_2$
839 cm^{-1}	$\alpha_{xx}-\alpha_{yy}$ α_{xy}	B_{1g} $+B_{2g}$	840 cm^{-1}	$^7F_0-{}^7F_2$
910 cm^{-1}	$\alpha_{xx}+\alpha_{yy}, \alpha_{zz}$ $\alpha_{xx}-\alpha_{yy}$, α_{xy}, $(\alpha_{xz}, \alpha_{yz})$	A_{1g} $+B_{1g}$ $+B_{2g}$ $+F_g$	906 cm^{-1}	$^7F_0-{}^7F_2$
990 cm^{-1} (vw)	–	–	989 cm^{-1}	$^7F_0-{}^7F_2$

[a] Bandes observées à 20 K (Raman)
[b] Bandes disparaissant à 20 K (Raman)

PARTIE EXPERIMENTALE

Le monocristal de Eu^{3+} : CaF_2 est de source commerciale. Sa pureté a été vérifée par microanalyse et la concentration en Eu^{3+} (0,027 en poids) a été déterminée par analyse par activation. Le cristal a été taillé et orienté suivant ses axes cristallographiques (Fig. 1). Des mesures faites par rayons X ont montré l'existence d'un axe C_4 unique, perpendiculaire à deux axes C_2. L'observation au microscope a confirmé que le cristal était uniaxe et très faiblement biréfringent. Les spectres Raman et de fluorescence de ce cristal ont été enregistrés à 20 K et 300 K en utilisant successivement les raies 4579,

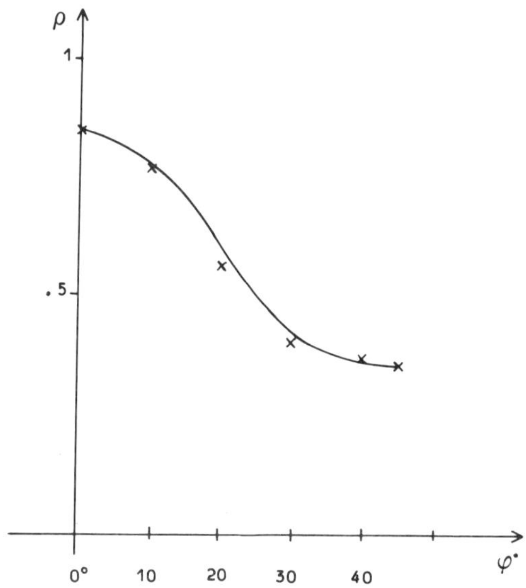

Fig. 3. Rapport d'intensité des bandes à 214 et 400 cm^{-1} reliées à l'élément α_{xz} en fonction de l'angle de rotation autour de l'axe z du cristal.

4765, 4880 et 5145 Å d'un laser à argon ionisé et la raie 6880 Å d'un laser à hélium-néon; ceci nous a permis d'être certains que les raies Raman mesurées n'étaient pas dues à la fluorescence. Les spectres ont été enregistrés sur un spectromètre Coderg Ph1. La largeur spectrale était de 2 cm^{-1} dans la plupart des spectres.

Nous avons toujours aligné les directions de propagation et de polarisation de la lumière incidente ou diffusée avec les axes cristallographiques. Lorsque nous avons dû tourner le cristal autour d'un axe, nous avons pris soin de le choisir comme axe optique et de faire propager la lumière diffusée le long de cet axe. Cependant, nos résultats suggèrent que l'effet de la biréfringenece n'affecte pas nos mesures d'intensité Raman même lorsqu'on s'écarte des conditions mentionnées ci-dessus. Les spectres Raman (Fig. 2) donnent tous les éléments de la matrice de diffusion caractérisant chaque bande.

Les fréquences des bandes Raman et de fluorescence sont rassemblées dans le Tableau 1.

Les transitions Raman électroniques de Eu^{3+} sont attendues dans les domaines 260–500 cm^{-1} ($^7F_0 \rightarrow {}^7F_1$) et 700–1000 cm^{-1} ($^7F_0 \rightarrow {}^7F_2$). Les transitions du premier multiplet sont reliées à des tenseurs antisymétriques.[11] Le type de tenseur xy–yx peut être mis en évidence en tournant le cristal autour de l'axe z: l'intensité diffusée correspondant au montage expérimental $X(XY)Z$ est indépendante de l'angle de rotation φz. Au contraire, dans le cas d'un tenseur symétrique $xy + yx$ l'intensité devient nulle pour $\varphi = 45°$ (Fig. 3). De même un tenseur $xx + yy$ peut être distingué de $xx - yy$ par rotation du cristal autour de son axe z . Dans ces mesures, la lumière incidente et diffusée se propageait respectivement le long des axes x et z. Les faces ABCD et CDEF (Fig. 1) ont été choisies pour les mesures avec $\varphi = 0°$ et $\varphi = 45°$, la face A′B′E′F′ a servi aux mesures intermédiaires $\varphi' = 10, 20, 30$ et 40°. La réfraction du faisceau incident n'a pas introduit d'erreur importante car les résultats obtenus avec $\varphi' = 40$ et $\varphi = 45°$ sont très voisins.

SPECTRE DE PHONON DE $Eu^{3+}:CaF_2$

En comparant le spectre Raman de CaF_2 et de $Eu^{3+}:CaF_2$, il est raisonnable de faire correspondre la bande qui apparait à 336 cm^{-1} pour $Eu^{3+}CaF_2$ au phonon de CaF_2 dont la fréquence Raman est 322 cm^{-1}; cette bande étant T_{2g}, le tenseur de diffusion correspondant ne possède pas d'éléments diagonaux. Au contraire, dans le tenseur caractérisant la bande à 336 cm^{-1}, entrent des éléments diagonaux et non diagonaux. Ce qui confirme bien que le cristal n'est plus cubique.

Les tenseurs observés (Tableau 1) peuvent être divisés en trois groupes. Au premier appartiennent des tenseurs de la forme (xz, yz) caractérisant les bandes Raman à 117, 199 et 797 cm^{-1}. Au second appartiennent des tenseurs de la forme $xx - yy$, xy et caractérisant les bandes à 83, 400 et 839 cm^{-1}. Enfin, les autres bandes sont reliées à des tenseurs plus compliqués, combinaisons des types précédents. L'observation de tenseurs du type $xx - yy$, xy c'est à dire dans lesquels $xy \neq xx$ nous permet d'en conclure que ce cristal uniaxe appartient à un groupe d'espace quadratique et d'éliminer les groupes trigonaux et hexagonaux.[12]

Le groupe d'espace de CaF_2 est O_h^5, et les ions Ca^{++} et F^- occupent respectivement des sites de symétrie O_h et T_d. Pour $k = 0$, le phonon de symétrie T_{2g} actif en Raman (322 cm^{-1}) implique essentiellement le mouvement des ions F^-, l'un contre l'autre, les ions Ca^{++} restant au repos.[3] Les phonons T_{1u}, TO et LO (inactifs en Raman) impliquant le mouvement des ions Ca^{++} contre le réseau des F^- ont respectivement pour fréquence 254 et 463 cm^{-1}.[13] Pour $Eu^{3+}:CaF_2$, nous associons la bande à 336 cm^{-1} au phonon mettant en jeu les déplacement des atomes de fluor l'un contre l'autre, cette bande est plus large, plus intense que dans la fluorine, et si nous supposons que la levée de dégénerescence ne contribue qu'à son élargissement, ceci peut expliquer la présence des composantes xx, yy, zz, xy, xz, yz dans la matrice de diffusion. Le tenseur de ce phonon peut être correctement prévu si la symétrie de site des ions F^- est donnée par le groupe C_2 et si le groupe facteur est C_4, D_4, C_{4v}, C_{4h} ou D_{4h}, pour ce dernier, un site C_{2v} étant acceptable. De très faibles bandes dans le spectre Raman $Eu^{3+}:CaF_2$ sont observées à 238 et 450 cm^{-1}, et on peut attribuer ces bandes aux phonons correspondants aux bandes à 254 et 463 cm^{-1} dans CaF_2; leur activité Raman doit être induite par les 'impuretés' Eu^{3+} plutôt que due à un changement radical de symétrie de site, qui, on l'a vu pour les ions F^- autorise l'apparition de composantes xy, xz très intenses. Par conséquent, si nous imposons aux phonons correspondants qui mettent en jeu les mouvements des ions Ca^{2+}, de rester inactif en Raman, il est nécessaire que T_x, T_y, T_z de Ca^{2+} appartiennent à des représentations 'ungerade' du groupe de site C_{4h}, ce qui autorise comme groupe facteur le groupe D_{4h}. Dans le Tableau 1 et Fig. 4 on peut constater la correspondance entre le spectre Raman et le spectre de fluorescence et on peut en déduire que le spectre de fluorescence est en grande partie dû à des transitions vibroniques. Le spectre Raman de $Eu^{3+}:CaF_2$ est cependant très différent de celui de $Ce^{2+}:CaF_2$ (cubique) ou $Eu^{2+}:CaF_2$; dans ces derniers, les bandes sont beaucoup plus larges. Les bandes observées entre 80 et 225 cm^{-1} sont attribuées à des vibrations impliquant les déplacements des ions Eu^{3+}. En effet, dans des grenats de terres rares[14] les ions de terres rares sont situés à des sites D_2 et pour $k = 0$, les phonons correspondant à leur mouvement sont situés à fréquence inférieure à 250 cm^{-1}

Etant donné la complète levée de dégénerescence des niveaux électroniques, la symétrie de site doit être basse. Les translations de Eu^{3+} ne peuvent engendrer deux tenseurs de la forme zx, zy et un de la forme xx, yy, zz, xy qu'en admettant une symétrie de site C_2. Si les ions Eu^{3+} substituent les ions Ca^{2+}, l'axe C_2 de leur site coincide avec un

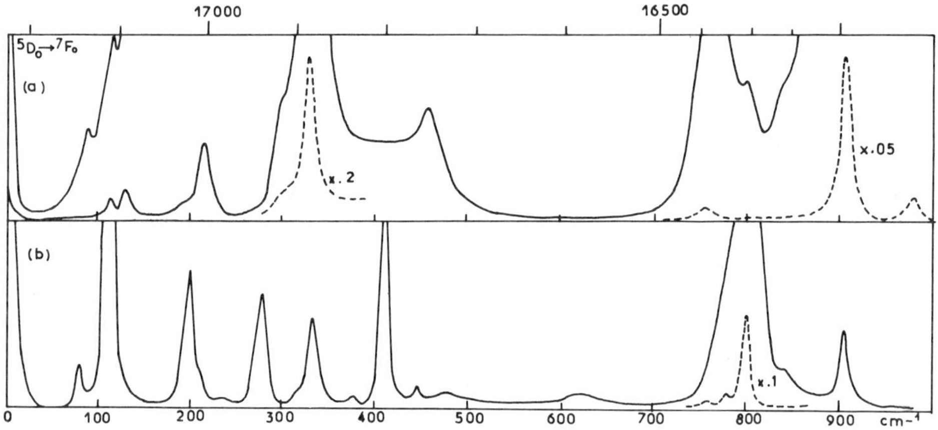

Fig. 4. (a) Spectre de fluorescence de Eu^{3+} CaF_2 entre 16200 et 17200 cm^{-1} à 300 K, excité avec la raie 4765 Å d'un laser à argon. (b) Spectre Raman de Eu^{3+} CaF_2 à 300 K, la largeur spectrale étant de 6 cm^{-1}. Les polariseur et analyseur étaient en position horizontale et le cristal était volontairement mal orienté, pour faire apparaitre le maximum de bandes.

axe cristallographique, et les tenseurs attendus sont ceux observés. Cependant, si les ions Eu^{3+} sont distribués sur différents sites, C_2 du cristal D_{4h}, on attend un tenseur du type $A_{1g} + B_{1g} + B_{2g}$ $(xx + yy, zz, xx - yy, xy)$ et deux du type E_g (yz, zx) qui sont également en accord, à la précision des mesures près, avec nos résultats (Tableau 1).

Pour conclure ce paragraphe, nous pensons que le spectre Raman de ce composé peut être interprété sur la base d'une analyse de groupe facteur, comme si, bien que les impuretés détruisent la symétrie de translation, le groupe facteur était D_{4h}, les groupes de site des ions F^- et Ca^{++} étaient respectivement C_2 et C_{4h}, et le groupe de site de Eu^{3+} était C_2 (les charges de compensation peuvent contribuer à la symétrie de site). De plus, on peut remarquer qu'il n'est pas nécessaire de faire appel à des phénomènes de second ordre pour interpréter nos spectres: toutes les bandes Raman ont une demie largeur comprise entre 2 et 8 cm^{-1} et ne peuvent pas être correlées aux courbes de densité d'états de CaF_2.[3] Enfin, toutes les bandes sont intenses (par exemple la bande à 336 cm^{-1} est environ 10 fois plus forte que dans la fluorine) et ceci doit être due aux ions Eu^{3+} qui sont à l'origine de l'effet Raman de résonance.

NIVEAUX ELECTRONIQUES DE Eu^{3+} : CaF_2

L'étude des transitions Raman électroniques de Eu^{3+} dans des grenats[10] a révélé la présence de tenseurs antisymétriques associés aux transitions $^7F_0 \rightarrow {}^7F_1$. L'attribution des bandes dans Eu^{3+}:CaF_2 comprises entre 260 et 425 cm^{-1} d'une part et entre 700 et 1000 cm^{-1}, d'autre part ainsi que l'observation d'un tenseur antisymétrique pour la bande à 400 cm^{-1} est en accord avec les résultats obtenus sur EuGaG.

En considérant que Eu^{3+} occupe un site de symétrie C_2 les niveaux $J = 1$ et $J = 2$ éclatent respectivement en trois $(A + 2B)$ et cinq $(3A + 2B)$ niveaux. Les tenseurs liés aux transitions Raman électroniques $J = 0 \rightarrow J = 1$ sont symétriques, trois sont du type $xx, yy, zz, xy, xx\text{-}yy$ et deux sont du type yz, xz; les tenseurs liés aux transitions $J = 0 \rightarrow J = 1$ sont attendus antisymétriques un de la forme $xx, yy, zz, xy, xy\text{-}yx$ et deux du type $yz, xz, xz\text{-}zx$. En corrélant le groupe de site au groupe D_{4h}, les tenseurs attendus sont ceux donnés dans

Tableau 2. Niveaux électroniques de basse énergie de $Eu^{3+}:CaF_2$ et leur attribution sur la base d'une symétrie de site C_2.

Raman (cm^{-1})	Fluorescence [a]	Tenseurs observés	Tenseurs prévus	Symétrie
0	0	–	$\alpha_{xx}+\alpha_{yy}, \alpha_{zz}$ (Rayleigh)	$A\ (^7F_0)$
276	–	$(\alpha_{xz}, \alpha_{yz})$	$(\alpha_{xz}-\alpha_{zx}, \alpha_{yz}-\alpha_{zy})$	$B\ (^7F_1)$
400	400	$\alpha_{xx}-\alpha_{yy}$	$\alpha_{xx}-\alpha_{yy}$	$A\ (^7F_1)$
		$\alpha_{xy}-\alpha_{yx}$	$\alpha_{xy}-\alpha_{yx}$	
406		$(\alpha_{xz}, \alpha_{yz})$	$(\alpha_{xz}-\alpha_{zx}, \alpha_{yz}-\alpha_{zy})$	$B\ (^7F_1)$
755	755	$\alpha_{xx}+\alpha_{yy}, \alpha_{zz}$	$\alpha_{xx}+\alpha_{yy}, \alpha_{zz}$	$A\ (^7F_2)$
		α_{xy}	α_{xy}	
		$(\alpha_{xz}, \alpha_{yz})$	$(\alpha_{xz}, \alpha_{yz})$	
		–	$\alpha_{xx}-\alpha_{yy}$	
797	790	$(\alpha_{xz}, \alpha_{yz})$	$(\alpha_{xz}, \alpha_{yz})$	$B\ (^7F_2)$
839	–	$\alpha_{xx}-\alpha_{yy}$	$\alpha_{xx}-\alpha_{yy}$	$A\ (^7F_2)$
		α_{xy}	α_{xy}	
			$(\alpha_{xz}, \alpha_{yz})$	
			$\alpha_{xx}+\alpha_{yy}, \alpha_{zz}$	
910	910	$\alpha_{xx}+\alpha_{yy}, \alpha_{zz}$	$\alpha_{xx}+\alpha_{yy}, \alpha_{zz}$	$A\ (^7F_2)$
		$\alpha_{xy}, \alpha_{xx}-\alpha_{yy}$	$\alpha_{xy}, \alpha_{xx}-\alpha_{yy}$	
		$(\alpha_{xz}, \alpha_{yz})$	$(\alpha_{xz}, \alpha_{yz})$	
990	989	–	$(\alpha_{xz}, \alpha_{yz})$	$B\ (^7F_2)$

[a] Les transitions de fluorescence ont pour origine le niveau $^5D_0$ de Eu^{3+} CaF_2 situé à 17227 cm^{-1} au dessus de $^7F_0$.

le Tableau 2. Ce tableau donne la position des transitions Raman électroniques et les tenseurs observés et prévus pour une symétrie de site C_2 pour Eu^{3+}.

L'attribution des bandes Raman vibroniques et électroniques dues aux ions Eu^{3+} est le résultat de la comparaison des tenseurs attendus pour des sites de différentes symétries. A partir de la symétrie des niveaux, on peut calculer certains paramètres du champ cristallin et on a trouvé $A_2^0 = +13\ cm^{-1}$ et $A_2^2/A_2^0 = 26\ cm^{-1}$, et les valeurs des paramètres d'ordre quatre restant inférieures à celles de EuGaG.[10]

Enfin, on peut obtenir confirmation du groupe de site de Eu^{3+} en comparant les bandes Raman à 755 et 910 cm^{-1} avec les bandes de fluorescence correspondantes. En effet, les niveaux $^7F_2$ impliqués dans ces transitions doivent appartenir à la représentation totalement symétrique de tout sous groupe de D_{4h}, pour expliquer la présence de *xx, yy, zz*. Par ailleurs, D_0 appartenant également à cette représentation (car $J = 0$), une composante du moment dipolaire doit aussi appartenir à cette représentation et parmi le sous groupe de D_{4h}, seuls C_4, C_{4v}, C_{2v}, C_2, C_i ou C_s sont possibles: C_i conduirait à rendre interdites en Raman les transitions vibrationnelles des Eu^{3+} et les groupes C_{4v}, C_4, C_{2v} et C_s conduisent à des tenseurs très différents de ceux observés.

Avant de conclure, nous voulons faire quelques remarques:

1. Le cristal de MnF_2 appartient au groupe D_{4h} et présente deux bandes dans son spectre Raman: à 247 cm^{-1} (E_g) et 341 cm^{-1} (A_{1g}).[15] Dans nos spectres, la bande à 214 cm^{-1} possède le même caractère que celle à 336 cm^{-1} ($A_{1g}+B_{1g}+B_{2g}+E_g$) et ne doit donc

pas avoir la même origine que le phonon E_g de MnF_2. Elle pourrait être due à une transition vibrationnelle impliquant les charges de compensation au voisinage de Eu^{3+}, rendues intenses par un phénomène de résonance.

2. Le spectre donné par $x(xz)z$ ne change pas lorsqu'on tourne le cristal autour de son axe z et ceci n'est pas surprenant si on considère les propriétés de transformation des tenseurs observés: dans nos spectres lorsqu'on applique une rotation de 45° autour de z. Les tenseurs du premiers groupe xz, yz se transforment en $xz\sqrt{2}$. Dans ceux du second groupe xx-yy, xy, les éléments diagonaux s'échangent avec les non diagonaux et par conséquent xz reste nul. Cependant xy change et la variation d'intensité dépend du rapport xx/xy; pour la bande 839 cm^{-1}, il est égal à 0.1 et ceci explique son abaissement d'intensité pour une rotation de 45°. Les tenseurs du troisième groupe étant des combinaisons des deux premiers xz restera également nul.

CONCLUSIONS

La technique de la diffusion Raman a été employée pour déterminer la symétrie de site des ions ainsi que le groupe facteur d'un cristal de structure inconnue $Eu^{3+}:CaF_2$. Les transitions Raman électronique et vibrationelle ainsi que les données de fluorescence ont été analysées dans cette étude. L'existence de tenseur de diffusion antisymétrique nous a aidé à attribuer les transitions électroniques dans nos spectres Raman. Enfin, on peut constater qu'il n'est pas possible d'observer les niveaux $J = 1$ dans le spectre de fluorescence, et ceci demontre l'importance de la spectroscopie Raman électronique.

REMERCIEMENTS

Nous tenons à remercier Madame Lenain (Département de Recherches Physiques, Paris VI) qui a orienté et poli le cristal, ainsi que Monsieur Leroy qui a enregistré les spectres à basse températures. Nous sommes également reconnaissants aux Drs. Krauzman et Guillermet pour les discussions fructueuses que nous avons eues avec eux.

REFERENCES

1 D. C. Press, *Proc. Indian Acad. Sci., Sect. A* **31**, 56 (1950).
2 M. M. Elcombe et A. N. Pryor, *J. Phys. Chem.; Solid State Phys.* **3**, 492 (1970).
3 S. Genesan et E. Burstein, *J. Phys. Radium* **26**, 645 (1965).
4 L. E. Gurevich, I. P. Ipatova et A. A. Klotchichin, *J. Phys. Radium* **26**, 745 (1965).
5 A. Kiel et J. F. Scott, *Phys. Rev. 2B* **6**, 2033 (1970).
6 Z. Kam et E. Cohen, *Light Scattering in Solids,* (Ed. M. Balkanski), Flammarion Sciences, France, 1971, p. 253.
7 J. P. Russell, *Proc. Phys. Soc.* **85**, 194 (1965).
8 R. S. Krishnan et N. Krishnamurthy, *J. Phys. Radium* **26**, 633 (1965).
9 A. R. Gee, D. C. O'Shea et H. Z. Cummins, *Solid State Commun.* (1966).
10 I. Dabrowski, P. Grunberg et J. A. Koningstein, *J. Chem. Phys.* **56**, 1264 (1972).
11 O. S. Mortensen et J. A. Koningstein, *J. Chem. Phys.* **48**, 3971 (1968).
12 E. B. Wilson, J. C. Decius et P. C. Cross, *Molecular Vibrations*, McGraw-Hill, New York, 1955.
13 W. Kaiser, W. G. Spitzer, R. H. Kaiser et H. E. Howarth, *Phys. Rev.* **127**, 1950 (1962).
14 G. Mace, G. Schaack, Toa-Ning Ng et J. A. Koningstein, *Z. Physik.* **230**, 391 (1970).
15 S. P. S. Porto, P. A. Fleury et T. C. Damen, *Phys. Rev.* **154**, 522 (1967).

48. INTENSITY AND LINE SHAPE STUDY OF RESONANCE RAMAN EFFECT OF S_2^- AND S_3^- DOPED IN A POTASSIUM IODIDE SINGLE CRYSTAL

W. Holzer, S. Racine and J. Cipriani
Département des Recherches Physiques, Laboratoire associé au C.N.R.S., Université de Paris VI, France

INTRODUCTION

Recently the Raman spectrum of S_2^- and S_3^- doped in alkali halide single crystals has been studied [1] and the resonance Raman effect has been observed. A first quantitative study on the dependence of the scattering cross section on the exciting frequency has been performed [2] on the system NaCl : S_3^-. The system S_2^- in alkali halide crystals has also been studied by EPR,[3] by emission spectroscopy[4,5] and by mechanical stress experiments;[6] the S_3^- ion has been studied by EPR [7] and ENDOR techniques.[8] Negative ions doped in alkali halide crystals are very convenient and interesting systems to investigate, since simple molecules, which are otherwise unstable, may be isolated and studied in an ordered matrix from liquid helium temperature to room temperature. On the other hand, the impurity-host interactions will provide additional information on the well known properties of the host lattice.

The occurrence of a strong resonance Raman effect for S_2^- and S_3^- in KI, when excited by argon ion laser irradiation, provides a very sensitive method by which to study the dynamics of these molecules in the lattice at very low ion concentration.

Since the molecules are simple ones, it is hoped that quantitative results on the resonance Raman effect, for instance the dependence of the scattering cross section on the exciting frequency, may provide suitable information for a theoretical approach.[9] The laser lines 4579 Å to 5145 Å are indeed well situated in the intermediate region between the absorption bands[1,6] of S_2^- and S_3^- in KI, centred at ~ 4000 Å and ~ 6000 Å respectively. We will here present some first results on the Raman spectrum of the local vibration of these molecules and the impurity-induced first order Raman spectrum close to the exciting line obtained at temperatures from 20–300 K; the emission spectrum will also be discussed.

EXPERIMENTAL

The spectra have been obtained with an argon ion laser (165 Spectra Physics) and a double monochromator (Coderg PH1) with photon counting. The cryostat is a modified Coderg conduction cooling cryostat; the lowest temperature obtained has been 15 K, measured directly on the crystal, with the laser beam present. The crystal has been supplied by Dr. J. Morton, NRC, Canada and prepared as indicated in ref. 3. The crystal

Table 1. Ratio of the scattering cross sections (*Y(XX)Z*) of S_3^- and S_2^- in KI as a function of excitation wavelength (Error ± 10%).

Excitation (Å)	4579	4727	4765	4880	4965	5017	5145
S_3^-/S_2^-	0.11	0.22	0.24	0.35	0.42	0.55	1.00

is cleaved perpendicular to the (100) (010) and (001) axis; we have taken spectra in the *Y(XX)Z* and *Y(XY)Z* geometry. The focused laser beam has been passed as close as possible to the crystal surface in order to avoid absorption losses.

RESULTS AT ROOM TEMPERATURE

Raman frequencies

The Raman frequencies of the peaks of the fundamental band of S_2^- and S_3^- doped in KI have been measured in the *Y(XX)Z* geometry by using several laser lines (laser power < 200 MW) to be 597.5 ± 1 cm^{-1} and 548 ± 1 cm^{-1} respectively. The contribution of hot bands may easily shift the frequency to lower values at elevated laser power.

The S_2^- band consists of two components; the *Y(XX)Z* and *Y(XY)Z* spectra have different shapes and the splitting in the overtones increases from 9 cm^{-1} in the first to 14 cm^{-1} in the second overtone. The isotope component due to $S^{32}S^{34-}$ is, however, not discernible at room temperature.

Intensity measurements

The ratio of the integrated intensities of S_2^- in the *Y(XY)Z* and *Y(XX)Z* geometry (which we call the *Y/X* ratio) has been measured to be 0.47 ± 0.02 for excitation from 4579 Å to 4145 Å. This requires the induced dipole moment to be parallel to the (110) direction (*Y/X* ratio = 0.50), the same direction as the molecular axis of S_2^- in KI.[3] The *Y/X* ratio of S_3^- increases from 0.50 ± 0.02 at 4765 Å to 0.58 ± 0.02 at 5145 Å.

The different resonance behaviour of S_2^- and S_3^- due to the different position of the absorption bands is well reflected in the ratio of the integrated intensities, measured at the same crystal spot using all available laser lines. Since the bands are only separated by 50 cm^{-1} no corrections have to be applied. The results, normalized to the 5145 excitation and obtained for *Y(XX)Z* geometry, are given in Table 1.

We studied the low frequency spectrum of the sulphur-doped KI crystal very carefully, but at room temperature we could neither detect any change in shape while exciting from 4579 Å to 5145 Å, nor any differences from the spectrum given for pure KI.[10] We think it therefore justified to use the integrated intensities of some characteristic bands of the low frequency spectrum as internal standard for the intensities of S_2^- and S_3^-. The difference in absorption of the scattered light by the very thin crystal layer over a frequency range of 500 cm^{-1} may be neglected. The ν^4 dependence is also taken into account by this method. The scattering cross sections, measured relative to the internal standard and normalized to the results at 4579 Å for S_2^- and at 5145 Å for S_3^-, is given in Fig. 1. The increase of the scattering cross section of S_2^-, when excitation approaches the absorption band from the lower energy side, is more pronounced than the one of S_3^- when the absorption band is approached from the higher energy side.

The intensity ratios of the first overtone to the fundamental band of S_2^- and S_3^- follow about the same dependence as given for the scattering cross section in Fig. 1.

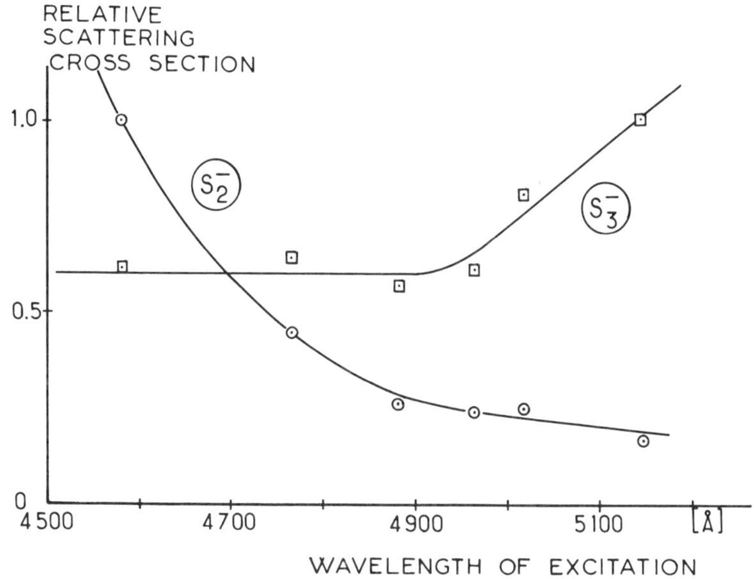

Fig. 1. Relative scattering cross sections ($Y(XX)Z$) at 300 K of the Raman bands of S_2^- and S_3^- doped in KI as a function of wavelength of excitation.

LOW TEMPERATURE RESULTS

Frequencies and identification of the spectra

We have observed dramatic changes of the spectrum of S_2^- and S_3^- at temperatures below 30 K. We show in Fig. 2 the spectrum of S_3^-; at 20 K the frequency is suddenly shifted by -10 cm^{-1} to 538 cm^{-1}. The transition region has not been studied in detail.

The S_2^- band at 25 K given in Fig. 3 is split into two components with the main peaks at 598 cm^{-1} and 582.5 cm^{-1}. We do not have any definite explanation for the band designated by $^*S_2^-$ at 582.5 cm^{-1}; however we have solved the following problems:

1. The bands at 598 cm^{-1} and 582.5 cm^{-1} are certainly S_2^- bands since the calculated frequency shift ($-$1.50%) and intensity (9%) of the $S^{32}S^{34-}$ isotope bands are in perfect agreement with the bands observed in the fundamental spectrum (see Fig. 3), and the overtone spectrum.
2. The bands belong to S_2^- molecules with two different vibrational constants. From the fundamental frequencies 598 ± 1, 582.5 ± 1 cm^{-1} and the overtone frequencies 1191 ± 1, 1159 ± 1 cm^{-1} and 1783, 1730 cm^{-1}, we determine $\omega_e = 602 \pm 1$ cm^{-1} and $x_e\omega_e =$ 2 cm^{-1} for S_2^- and $\omega_e = 587 \pm 1$ cm^{-1} and $x_e\omega_e = 2$ cm^{-1} for $^*S_2^-$.
3. The Y/X ratio of the fundamental and first overtone band is 0.49 ± 0.02 for the S_2^- and the $^*S_2^-$ band as well, indicating the same direction (110) for the induced dipole and also the same orientation of the molecules in the lattice.
4. The $^*S_2^-$ band at 582.5 cm^{-1} becomes visible at 40 K and its intensity relative to the S_2^- band increases up to 0.5 at 20 K. At lower temperatures the S_2^- band at 598 cm^{-1} decreases (presumably due to some coupling effects with the lattice) and the ratio may be >1.
5. The band at 582.5 cm^{-1} is considered to be due to a perturbed electronic ground state,

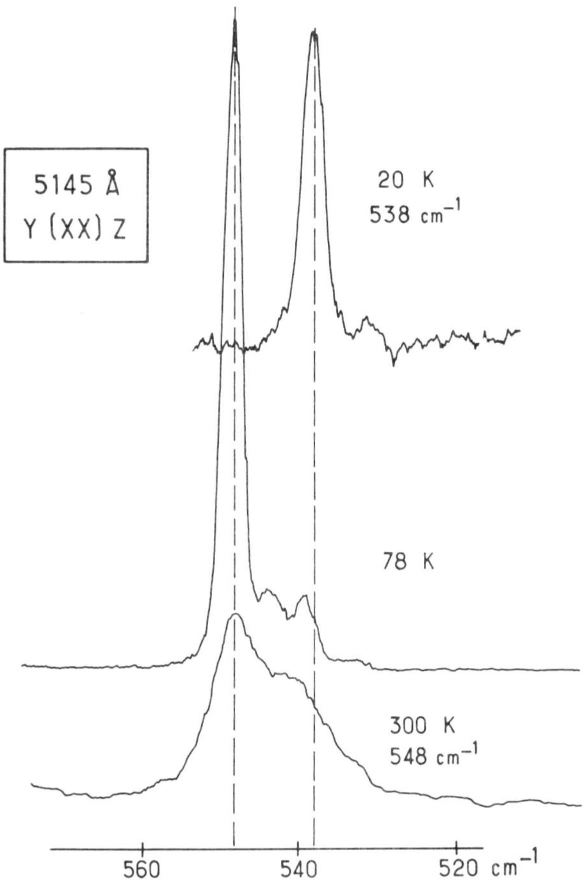

Fig. 2. Evolution of the shape of the Raman band of S_3^- with temperature (the intensity scale is not the same for the three spectra).

but not due to molecules pumped into an excited state by the laser irradiation. To these excited molecules we would attribute a sharp line of 362 ± 1 cm^{-1} appearing at 20–25 K using 4579 Å excitation (the line is also present at 4765 Å and 4880 Å, however not at 5145 Å excitation).

Frequencies and intensities of the impurity induced spectrum of KI

These experiments have been performed in a part of the crystal where S_3^- can be considered to be absent. The first order Raman spectrum of KI, induced by S_2^- is given in Fig. 4. The *Y(XX)Z* spectrum ($A_{1g} + 4E_g$) shows a sharp line at 87.5 cm^{-1} on top of a broad step and the *Y(XY)Z* spectrum (T_{2g}) shows a slightly broader strong line at 49 cm^{-1}.

The two spectra are similar to the ones observed[11] in the system KI : NO_2^-, the intensity distribution is about the same, but the T_{2g} mode is shifted to 57 cm^{-1} and a wing appears at its low frequency side. The spectra obtained in the system KI : Tl^+ [12] are quite different from our spectra.

We have studied the relative integrated intensities of the above discussed first order

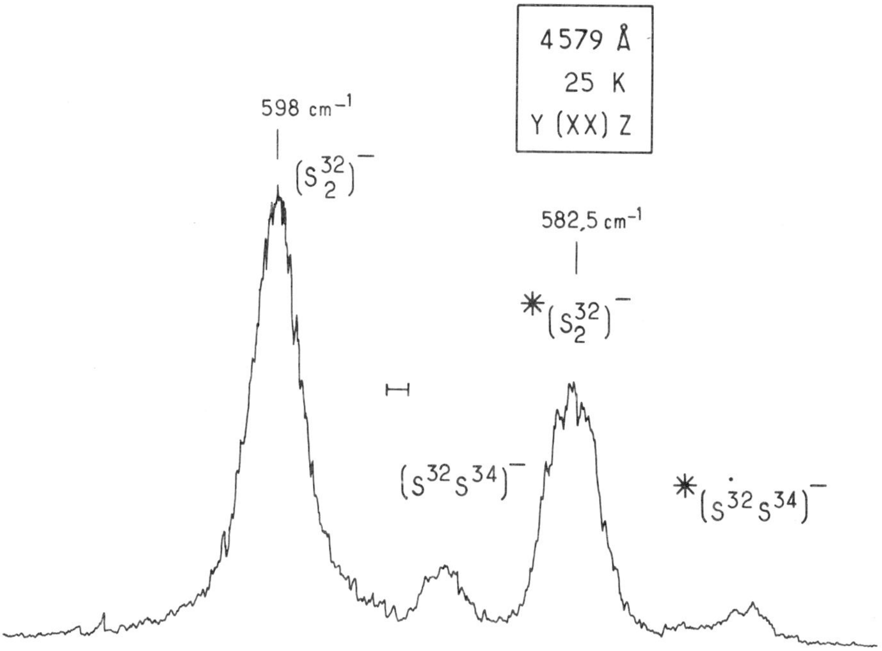

Fig. 3. Components of the S_2^- Raman band at 25 K.

Raman spectrum and the second order Raman spectrum of KI as a function of excitation frequency at 20 K. The second order Raman spectrum is much weaker than the one at room temperature and contains bands at 180 cm^{-1} and 212, 223, 233 cm^{-1}. The results are given in Table 2. It can be seen that there are no strong resonance effects present in the impurity-induced first order Raman spectrum of KI. A low frequency electronic Raman band due to the transition between the two $^2\Pi_g$ levels of S_2^- has not been observed.

Intensity measurements on the S_3^- and S_2^- bands

We have measured the temperature dependence of the S_2^- band at 598 cm^{-1} and the S_3^- band at 548 cm^{-1} when the temperature was changed from 300 K to 30 K, using 5145 Å excitation. The intensity of the S_2^- band stays constant within 15%, while the S_3^- intensity increases by a factor of 5.2.

The experiment on the dependence of the scattering cross section on the excitation frequency has been repeated at 25 K for the bands of S_2^- at 598 cm^{-1}, $^*S_2^-$ at 582.5 cm^{-1} and S_3^- at 538 cm^{-1}. We have used the intensity of the bands of the second order phonon spectrum of KI at 180 cm^{-1} and 212, 223, 233 cm^{-1} (see Table 2) as internal intensity standard. The scattering cross sections, normalized to the results at 4579 Å for S_2^- and at 5145 Å for S_3^- are given in Fig. 5. The trend of the S_2^- and S_3^- intensities is the same as at room temperature but the slope of the S_3^- plot is steeper. It is interesting to note that the results for S_3^- in KI at 25 K are almost identical with the one of S_3^- in NaCl at 300 K,[2] where the S_3^- ion is rigorously fixed in a bent three ion vacancy.

The curve for $^*S_2^-$ differs considerably from the curve for S_2^- scaled down by a constant factor. The ratio changes from 0.33 at 4579 Å to 0.55 at 5145 Å with an obvious

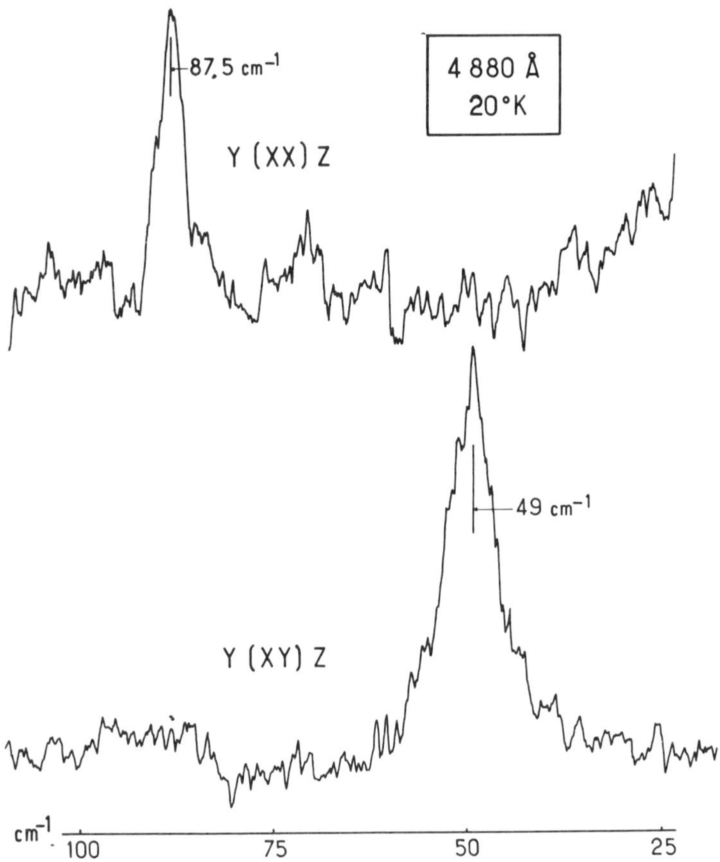

Fig. 4. Impurity-induced first order Raman spectrum of $KI:S_2^-$ at 20 K (fast scan).

Table 2. Relative scattering cross sections of the low frequency bands of $KI:S_2^-$ observed at several laser excitations.

Excitation	$Y(XY)Z$	$Y(XX)Z$				
(Å)	49 cm^{-1}	87.5 cm$^{-1\,a}$	180	212	223	233 cm^{-1}
4579	0.55	0.26	0.38		1	
4765	0.52	0.24	0.41		1	
4880	0.52	0.21	0.53		1	
4965	0.59	0.20	–		1	
5017	0.53	0.20	–		1	
5145	0.44	0.20	0.46		1	

[a] The broad step is not included.

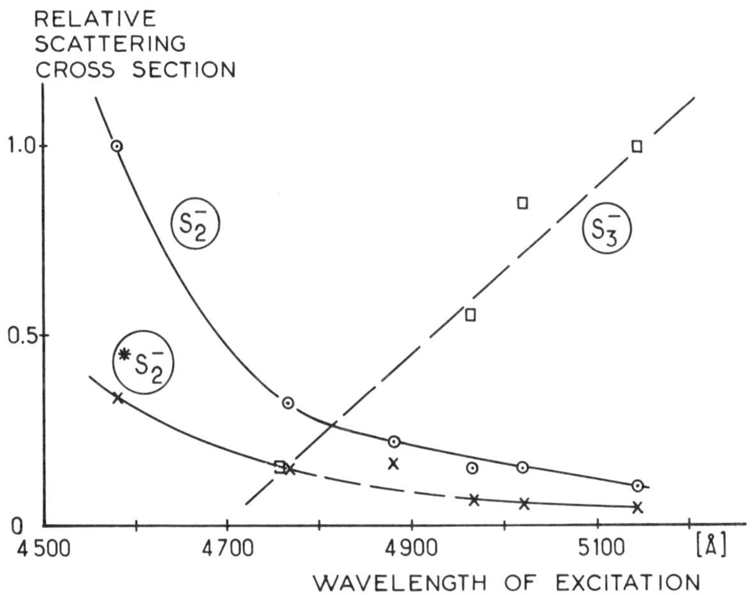

Fig. 5. Relative scattering cross sections ($Y(XX)Z$) at 20 K of the Raman bands of S_2^-, $^*S_2^-$ (582.5 cm^{-1}), and S_3^- doped in KI as a function of wavelength of excitation.

jump to 0.70 at 4880 Å.

The intensity of the line at 362 cm^{-1}, which may be due to an excited state of S_2^-, has been measured as well; it decreases from 1.0 at 4579 Å, 0.57 at 4765 Å, 0.35 at 4880 Å to < 0.04 at 5145 Å.

Laser-induced emission spectrum

We have observed with all laser lines the vibrational components $0 \rightarrow \nu'', \nu'' = 0, 1, \ldots 15$ of the emission from an excited electronic state of S_2^-. We could, however, not detect any hot luminescence.[13] The emission at low ν'' is most intense at 20 K; this may explain why only bands of $\nu'' > 4$ have been observed at 4 K.[4] The position of the non-phonon peak of the 0–0 transition was determined directly to be 20026 ± 1 cm^{-1}, (with reference to the 49 and 87.5 cm^{-1} first order line of KI, excited at 4965 Å) to be compared with 20035 ± 3 cm^{-1} of ref. 4. The bands 0–4 to 0–0 are peculiar, since a second non-phonon line of S_2^{32-} appears ~ 20 cm^{-1} lower in energy which takes over all the intensity at the 0–0 and 0–1 transitions at 20 K. The first sequences are: --- 20026; --- 19429; 18858–18837; 18266–18248; 17682–17665 and 17105 --- cm^{-1}. Thus the 0–0 transition observed by us at 20 K does not belong to the system given in ref. 4. The observed splitting is the only evidence of the splitting of the ground state $^2\Pi_g$ of S_2^-, since electronic Raman effect has not been observed. The vibrational constants deduced from our emission data for S_2^- agree well with the ones obtained by Raman spectroscopy.

ACKNOWLEDGEMENTS

The authors thank Dr. H. J. Bernstein and Dr. J. R. Morton for providing the doped KI crystal.

REFERENCES

1 W. Holzer, W. F. Murphy and H. J. Bernstein, *J. Mol. Spectry.* **32**, 13 (1969).
2 W. Holzer, W. F. Murphy and H. J. Bernstein, *Chem. Phys. Letters* **4**, 641 (1970).
3 L. E. Vanotti and J. R. Morton, *Phys. Rev.* **161**, 282 (1967).
4 J. Rolfe, *J. Chem. Phys.* **49**, 4193 (1968).
5 L. A. Rebane, *Opt. Spectry.* **31**, 123 (1971).
6 K. Bachmann, W. Känzig, H. R. Zeller and A. Zimmermann, *Physik der Kondensierten Materie* 7, 360 (1968).
7 J. Schneider, B. Dischler and A. Räuber, *Phys. Status Solidi* **13**, 141 (1966).
8 J. Suwalski and H. Seidel, *Phys. Status Solidi* **13**, 159 (1966).
9 M. Jacon, *Advances in Raman Spectroscopy,* Vol. 1 (Ed. J. P. Mathieu), Heyden & Son, London, 1973, p. 325.
10 M. Krauzman, Thèse d'Etat, Faculté des Sciences, Paris, 1969; *C. R. Acad. Sci.* **265B**, 689 (1967).
11 A. R. Evans and D. B. Fitchen, *Phys. Rev.* **B2**, 1074 (1970).
12 R. T. Harley, J. B. Page and C. T. Walker, *Phys. Rev.* **B3**, 1365 (1971).
13 K. K. Rebane, R. A. Avarmaa, L. A. Rebane and P. A. Saari, *Light Scattering in Solids* (Ed. M. Balkanski), Flammarion, Paris, 1971, p. 72.

49. RESONANCE RAMAN EFFECT OF I_3^- ION BY ULTRAVIOLET LASER EXCITATION

M. Ito,* K. Kaya, N. Mikami and Y. Udagawa
Department of Chemistry, Tohoku University, Sendai, Japan

INTRODUCTION

Resonance Raman effect (RRE) has attracted great interest because of its importance in elucidating the coupling mechanism between the intermediate electronic states and the molecular vibrations which appear in Raman scattering. To elucidate the coupling mechanism, the observation of the RRE of the simple molecules whose electronic and vibrational states are well established is most desirable.

Recent development of laser Raman spectroscopy made it feasible to observe RRE and some interesting results were reported for several diatomic molecules.[1-3] However, in the case of the diatomic molecule, we have only one vibrational mode. It is trivial therefore to discuss which mode of vibration is subjected to the great resonance effect in the resonant condition. The triatomic molecule has three different vibrational modes and it is the most simple molecule which is able to answer the above question.

In the present study, we observed the RRE of triiodide ion (I_3^-) excited by the nitrogen pulsed laser whose wavelength is 3371 Å. The results obtained were discussed in relation to the theory recently developed by Nafie *et al.*[4]

EXPERIMENTAL

Three kinds of lasers were used as the exciting light sources for the Raman scattering measurements. They were JEOL JLG–H–08M He–Ne laser, a CRL 52A Ar ion laser and a nitrogen pulsed laser constructed in our laboratory. Of these, He–Ne and Ar ion lasers are continuous working lasers and a JEOL photon counting apparatus was used for recording the Raman spectra. The Raman measurement by the use of the pulsed nitrogen laser was made with a pulse-gated photon counting device.[5] The spectrometer used was a Nalumi grating double monochromator of a linear dispersion 11 Å mm^{-1}.

The I_3^- ion was prepared by dissolving iodine into saturated methanol solution of potassium iodide.† Iodine in methanol has the absorption maximum at 440 nm and it gives a Raman line at 201 cm^{-1} which is ascribed to the stretching vibration of the iodine

---

* Presented by M. I.

† I_3^- ion is also produced by dissolving equivalent molar weights of potassium iodide and iodine into water. The electronic absorption spectrum and the Raman spectrum of the ion in water are almost the same as those in methanol.

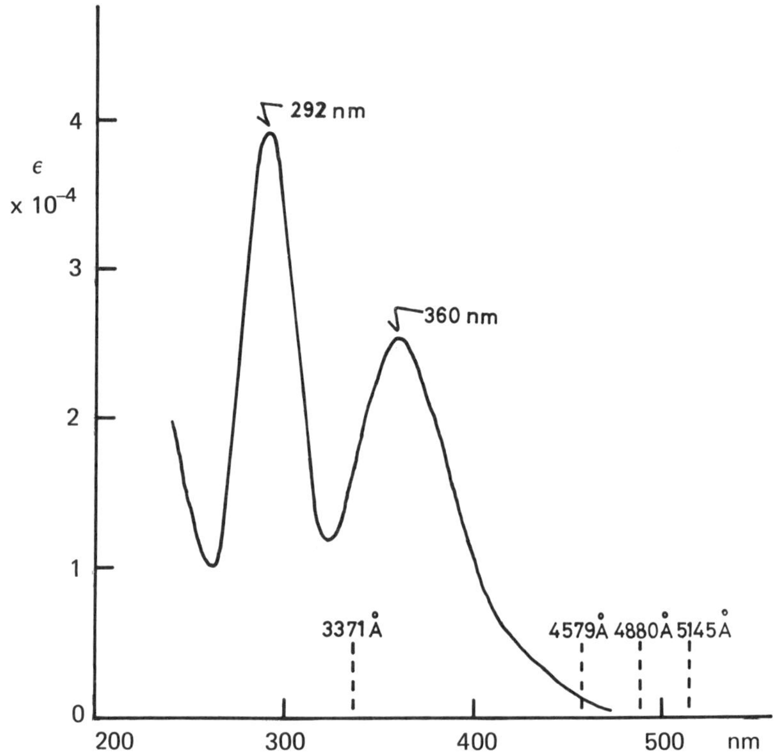

Fig. 1. The ultraviolet absorption spectrum of the I_3^- ion in methanol.

molecule forming electron-donor-acceptor complex with methanol.[6] As soon as potassium iodide is added into the above solution, the 440 nm band disappears and the absorption bands at 360 and 290 nm characteristic to I_3^- ion appear. The Raman lines are observed in this case clearly at 112 and 143 cm^{-1} with no trace of the 201 cm^{-1} line. The concentration of I_3^- ion in the solution was estimated from the observed absorbance and the reported absorption coefficient of the ultraviolet absorption.[7] Figure 1 represents the electronic absorption spectrum of I_3^- ion in methanol.

Commercially available G.R. grade potassium iodide was used without further purification and G.R. grade iodine was purified by sublimation. Methanol used as the solvent was purified by repeated distillations.

RESULTS AND DISCUSSION

The I_3^- ion has two strong absorption bands in the ultraviolet region whose peaks are at 360 and 290 nm (the molar absorption coefficients are 25,500 and 38,800, respectively) as seen in Fig. 1. The electronic absorption of I_3^- ion has been studied by Popov and Swensen[7] and by Ginn and Wood.[8] Pimentel[9] has discussed the nature of the chemical bond of I_3^- ion and concluded that the bond has mostly *p* character. From these studies, it is evident that the 360 nm band of I_3^- ion is due to a π–σ^* transition whose polarization direction is perpendicular to the molecular axis (where π represents the non-bonding $5p\pi$ orbital and σ^* the antibonding $5p\sigma$ orbital of the molecule). Apparently the excitation

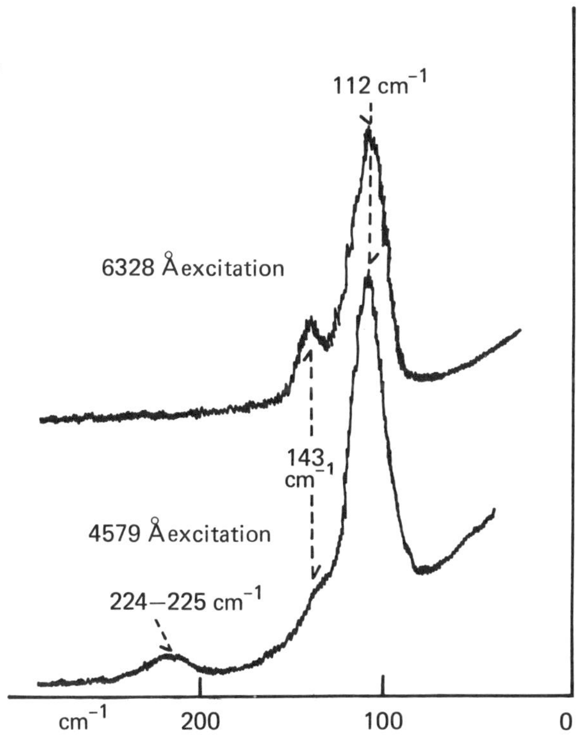

Fig. 2. Raman spectra of the I_3^- ion. Upper curve is the spectrum excited by 6328 Å He–Ne laser and the lower one is the spectrum excited by 4579 Å Ar ion laser.

wavelength of N_2 laser (3371 Å) falls within this absorption region, while the excitation wavelengths of Ar laser (5145 and 4880 Å) and He–Ne laser (6328 Å) are out of the absorption band (the wavelength 4579 Å of Ar laser is situated at the longer wavelength edge of the absorption). The upper curve of Fig. 2 is the Raman spectrum of I_3^- ion in methanol (5×10^{-2} m l^{-1}) excited by He–Ne 6328 Å line. The Raman lines at 112 and 143 cm^{-1} have been assigned by Maki and Forneris[10] to the symmetric and antisymmetric stretching vibrations, respectively. The bending vibration is extremely weak in the Raman spectrum, but it is reported to appear in the infrared spectrum at 69 cm^{-1}. Changing the excitation wavelength to shorter wavelength, the intensity of the 112 cm^{-1} Raman line tends to increase compared to that of the 143 cm^{-1} line. The relative intensity ratio of the two lines is: 3.9, 5.4, 7.0 and 10.3 for 6328, 5145, 4880 and 4579 Å excitations, respectively. Moreover, the second overtone of the 112 cm^{-1} band could be observed by the 4579 Å excitation as seen in the lower curve of Fig. 2. From the above results, it is concluded that the symmetric stretching vibration undergoes selective enhancement at the preresonant condition.

Figure 3 is the Raman spectrum of the I_3^- ion in methanol solution (10^{-5} m l^{-1}) excited by 3371 Å N_2 laser. It exhibits a feature characteristic to the RRE, that is, the overtones of the symmetric stretching vibration of 112 cm^{-1} appear clearly up to the

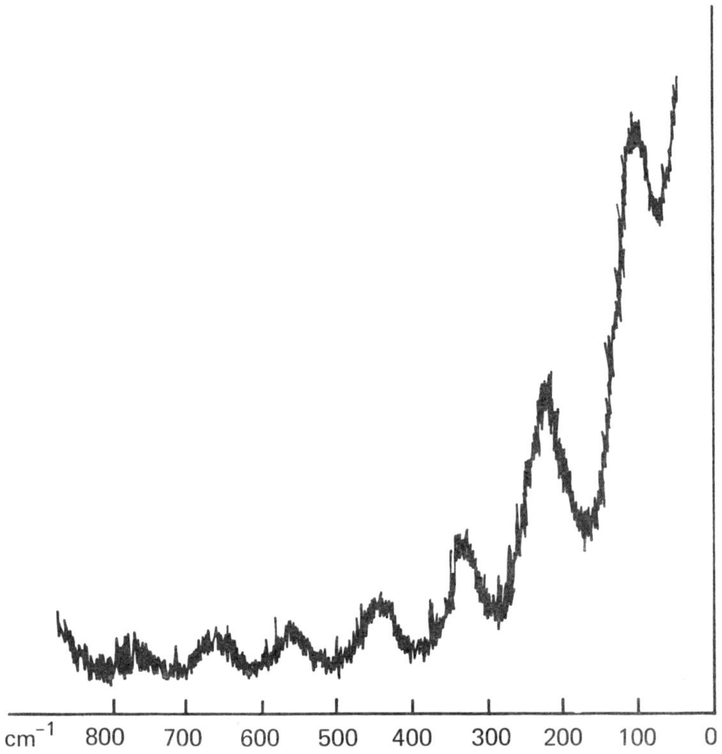

Fig. 3. Resonance Raman spectrum of the I_3^- ion excited by 3371 Å N_2 laser.

seventh orders.*,† The result indicates the apparent breakdown of the harmonic oscillator selection rule of the vibrational Raman scattering as well as the enormous intensity enhancement of the particular vibrational Raman line in resonant condition. We want to emphasize that the particular vibration being subjected to the RRE is only the totally symmetric stretching vibration.

Recently, Nafie, Stein and Peticolas[4] presented a new formulation for the vibrational Raman scattering by the use of time-ordering perturbation theory and obtained partial success in explanation of the RRE of iodine vapour reported by Holtzer *et al.*[2] Applying their theory to polyatomic molecules, appearance of higher overtones is limited only for totally symmetric vibrations irrespective of the nature of the electronically excited state concerned. Our result agrees completely with the prediction and gives support to the theory developed by Nafie *et al.*[4]

Very recently, Tasumi[11] has reported the RRE of starch-iodine complex, in which

* On the anti-Stokes side, the overtones up to the third orders were observed.

† Because of the low signal-to-noise ratio and the inherent wide width of the N_2 laser line (10 cm^{-1}), it is difficult to decide the exact wave numbers of the fundamental and overtones of the 112 cm^{-1} Raman line. The estimated wave numbers for which the experimental errors were taken into account are as follows
103–116 cm^{-1} (n = 1), 224–241 cm^{-1} (n = 2), 336–353 cm^{-1} (n = 3), 439–460 cm^{-1} (n = 4), 567–583 cm^{-1} (n = 5), 676–701 cm^{-1} (n = 6), 768–793 cm^{-1} (n = 7).

iodine is regarded to exist in a form of I_3^- ion in the helix of starch. He argued intensity enhancement of the antisymmetric stretching vibration of the I_3^- ion in starch in resonant condition. This is quite contrary to our result. However, judged from the electronic absorption spectra given in his paper, the Raman spectrum he obtained is probably due to the I_2 molecule which is strongly perturbed by the complex formation. Thus, his result seems not to be taken as evidence against the theory, although further examination is needed.

ACKNOWLEDGEMENT

We are grateful to Dr. Tasumi of Tokyo University for his kind suggestions and discussions.

REFERENCES

1 R. J. Gillespie and M. J. Morton, *J. Mol. Spectry.* **30**, 178 (1969).
2 W. Holtzer, W. F. Murphy and H. J. Bernstein, *J. Chem. Phys.* **52**, 399 (1970).
3 O. S. Mortensen, *J. Mol. Spectry.* **39**, 48 (1971).
4 L. A. Nafie, P. Stein and W. L. Peticolas, *Chem. Phys. Letters* **12**, 131 (1971).
5 K. Kaya, N. Mikami, Y. Udagawa and M. Ito, *Chem. Phys. Letters* **13**, 221 (1972).
6 P. Klaboe, *J. Amer. Chem. Soc.* **89**, 3667 (1967).
7 A. I. Popov and R. F. Swensen, *J. Amer. Chem. Soc.* **77**, 3724 (1955).
8 G. W. Ginn and J. L. Wood, *Trans. Faraday Soc.* **62**, 777 (1966).
9 G. C. Pimentel, *J. Chem. Phys.* **19**, 446 (1951).
10 A. G. Maki and R. Forneris, *Spectrochim. Acta* **23A**, 867 (1967).
11 M. Tasumi, *Chem. Letters* 75 (1972).

SECTION VII
BAND PROFILES IN GASES & LIQUIDS

50. SPECTRES INFRAROUGE ET RAMAN DES LIQUIDES

S. Bratos
*Laboratoire de Physique Théorique des Liquides,
Université de Paris VI, France*

SUMMARY

Les données expérimentales essentielles sont tout d'abord présentées et analysées en vue de mettre en évidence les mécanismes produisant les profils des bandes infrarouge et Raman des liquides. Les mouvements de vibration et de rotation jouent un rôle prépondérant dans ce contexte. Les éléments de base de la théorie des profils des bandes sont ensui discutés, le cas des solutions inertes des molécules diatomiques étant examiné en plus de détail. Les contributions apportées par la spectroscopie infrarouge et celles apportées par la spectroscopie Raman sont comparées. On montre, en particulier, que la spectroscopie Raman permet une séparation des effets de relaxation vibrationnelle et rotationnelle. Cette possibilité n'existe pas dans la spectroscopie infrarouge.

51. USE OF LIGHT-DIFFUSION SPECTROSCOPY TO STUDY MOLECULAR MOTIONS IN LIQUID PHASE

H. J. Bernstein and W. Murphy
Division of Chemistry, National Research Council of Canada, Ottawa, Canada
and
J. P. Perchard
Laboratoire de Recherches de Spectrochimie Moléculaire, Faculté des Sciences de Paris, France

As an example of the use of Raman spectroscopy for the study of molecular motions in the liquid phase, we have chosen hydrochloric and hydrobromic acids and their deuterated homologues in the pure liquid state or dissolved in some solvents. Consequently, we have divided this chapter into two parts: the first devoted to results for pure liquids and the second to those for solutions.

For pure liquids, band profiles for both isotropic and anisotropic scattering have been studied by three methods:

1. Band moment analysis of anisotropic components, to obtain the torque acting on a rotating molecule.
2. Fourier transform analysis to obtain correlation functions and correlation times.
3. Comparison of Raman and Rayleigh wings of the anisotropic scattering, to discuss the origin of these wings.

Only the first two points will be discussed in this chapter.

We should like first to emphasize the difficulties of quantitative intensity measurements of depolarized bands, the wings of which are spread over nearly 1000 wavenumbers. For these molecules, Fig. 1 shows the band profile for HBr and DBr molecules: for HBr the band is centred at about 2500 cm^{-1}; with wings spread from 2000 to 3000 cm^{-1}.

In Table 1, we have gathered the mean values of the second and fourth moments; the second moments are compared with theoretical values. It appears that the experimental values are reproducibly lower than the predicted ones by about 20%. From these values mean square torques are obtained. As these quantities should be isotopically independent, the discrepancy between the results for HCl/DCl and HBr/DBr gives an idea of the accuracy of such a determination.

We next discuss the Fourier transform analysis. According to the Bratos theory, the Fourier transform of the anisotropic component gives the product of the correlation function $G_v(t).G_R(t)$ while the Fourier transform of the isotropic component gives $G_v(t)$ only. Thus, from their ratio, $G_R(t)$ alone can be obtained. As an example of such a determination we show in Fig. 2 the result for HCl. There is one way to check the validity of this method, which is to determine the correlation time from the slope of the diffusional

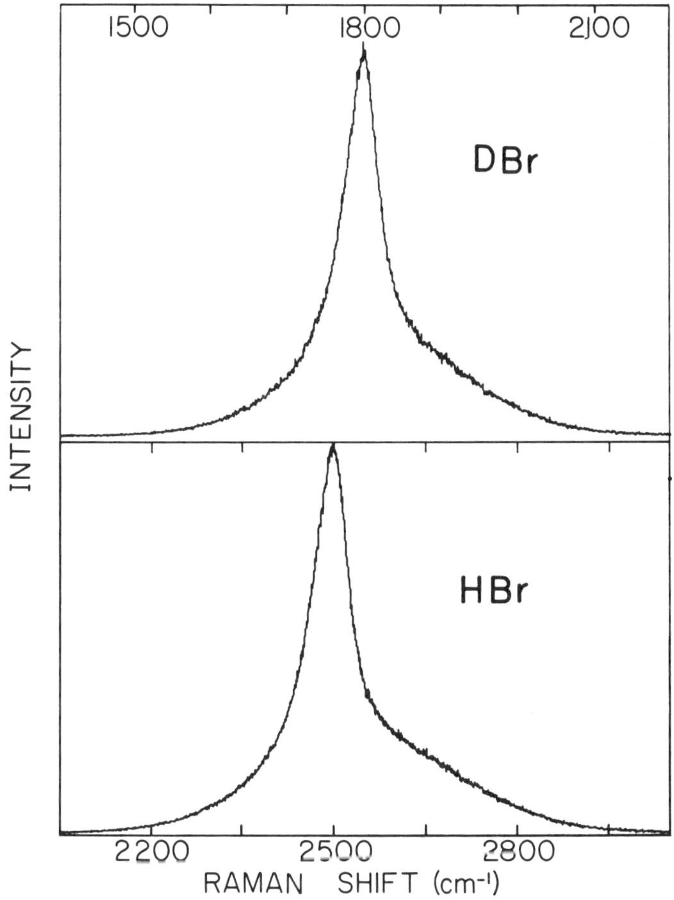

Fig. 1. HBr and DBr perpendicular components. Slitwidths: 2.5 to 3 cm^{-1}.

Table 1. Band moment calculations and determination of the mean square torques.

		M(2) (10^3 cm^{-2})	M(4) (10^9 cm^{-4})	$\langle\tau^2\rangle$ (10^4 cm^{-2})
HCl	Theoretical	23.7		
	Raman	20.8	1.80	53.4
	Rayleigh	16.7	1.44	49.9
DCl	Theoretical	12.5		
	Raman	10.2	0.448	47.9
HBr	Theoretical	19.0		
	Raman	16.6	1.18	52.1
	Rayleigh	20.4	1.65	62.7
DBr	Theoretical	9.92		
	Raman	8.17	0.293	55.2
	Rayleigh	8.16	0.296	56.1

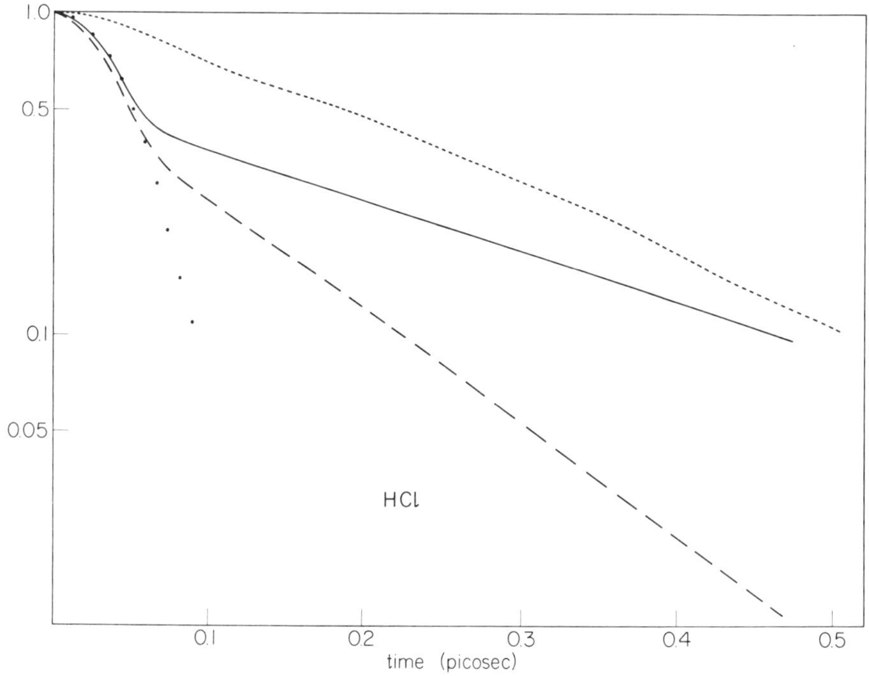

Fig. 2. HCl correlation functions: ------ $G_v(t)$; —— $G_R(t)$; – – – $G_v(t) \times G_R(t)$; ..... correlation function for HCl gas.

Table 2. Correlation times ($\tau \times 10^{13}$ s) from Raman and Rayleigh results.

	From $G_v \times G_R$ [a]	From G_v [b]	From G_R [c]	Rayleigh
HCl	1.24	2.19	2.77	2.40
DCl	1.53	3.35	2.75	
HBr	1.64	3.31	3.28	2.90
DBr	2.02	5.05	3.32	2.88

[a] From perpendicular component.
[b] From parallel component.
[c] After elimination of G_v from $G_v \times G_R$.

part of the curve. The results are shown in Table 2. For the four molecules we give the correlation time obtained from the isotropic compound, from the anisotropic component, and finally the rotational correlation time obtained from $G_R(t)$. As predicted by the theory, these last values are nearly isotopically independent, and are close to the values obtained from the Rayleigh band analysis.

We shall now discuss some of our results on solutions of the same compounds dissolved in three kinds of solvents: an active solvent (SO_2), inert solvents (SF_6 and C_2F_6) and an intermediate solvent (CCl_4). As for pure liquids, a band moment and a Fourier transform analysis were carried out but only the band moment analysis will be discussed here. Figure 3 shows the band profile of anisotropic components for the HCl molecule. In

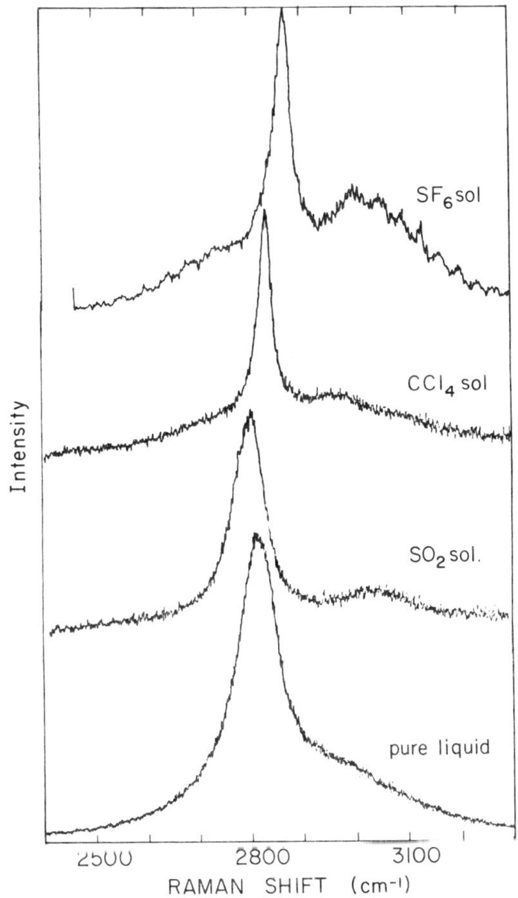

Fig. 3. Comparison between perpendicular components of HCl liquid and HCl dissolved in three solvents. Slitwidth of 1.6 cm^{-1} (pure liquid), 4 cm^{-1} (SO_2 sol), 3.6 cm^{-1} (CCl_4 and SF_6 solutions).

spite of the low intensity of the scattered light, it was possible to observe an *OQS* type structure, with a solvent-dependent separation between *O* and *S* or *Q* and *S* maxima: the more active the solvent, the greater the separation. In the case of the solutions in inert solvents, it is possible to observe several maxima on the *O* and *S* branches, which correspond to the rotational lines in the gas phase. A discrete rotational spectrum was also observed near the exciting line for solutions in the same solvents. Figure 4 shows such spectra in the pure rotation and vibration–rotation region. From the frequencies of these maxima it is concluded that rotational parameters do not change appreciably between the gas and solution.

Table 3 gathers the results of band moment analysis: second and fourth moments, then mean square torque acting on the molecule are obtained. It is seen that these last values are not strongly dependent on isotropic substitution. They are also in rather good agreement with recent infrared results.

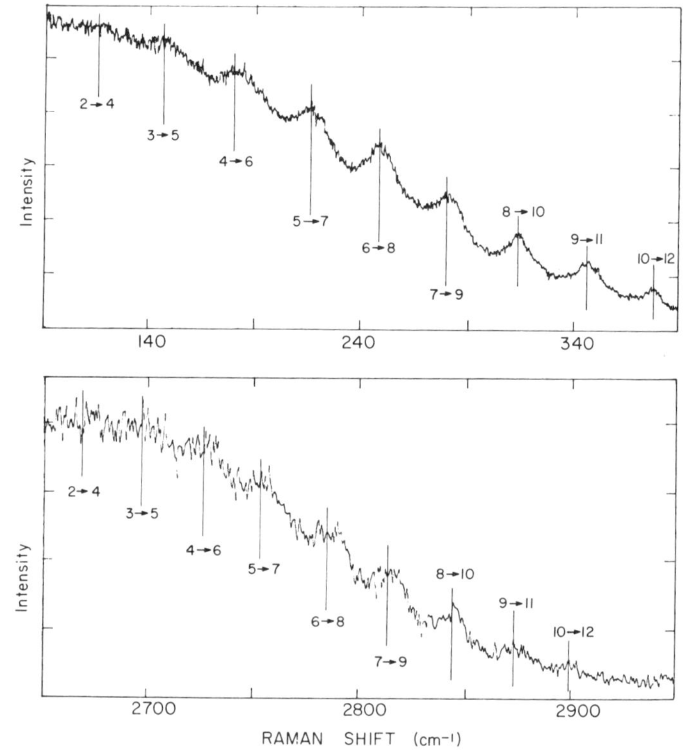

Fig. 4. The anisotropic component of the S-branches of rotation and vibration–rotation spectra for HBr in SF_6. Slitwidth of 2 cm^{-1} (rotation) and 3.5 cm^{-1} (vibration–rotation spectrum).

Table 3. Band moment calculations and determination of mean square torque.

Compound	Solvent	M(2) (10^3 cm^{-2})	M(4) (10^9 cm^{-4})	$\langle\tau^2\rangle$ (10^4 cm^{-2})
HCl	SF_6	24.7	1.68	4.7
M(2) theor.	CCl_4	21.1	1.52	26.0
= 23.7 × 10^3	SO_2	19.3	1.52	41.1
DCl	SF_6	13.3	4.85	4.2
M(2) theor.	CCl_4	10.4	3.73	23.3
= 12.5 × 10^3	SO_2	9.4	0.411	50.0
HBr M(2) theor.	SF_6	18.3	0.913	2.3
= 19.1 × 10^3	C_2F_6 288 K	19.2	0.984	0
DBr M(2) theor.	SF_6	8.0	0.180	4.3
= 9.9 × 10^3	C_2F_6 288 K	9.3	0.244	6.4

To conclude, it should be said that the complete study (*Mol. Phys.* **23**, 499 (1972)) is a demonstration that Raman spectroscopy is a powerful tool in the study of molecular motions in liquids. The possibility of separately analysing the vibrational and rotational relaxation mechanisms makes it especially useful.

52. RAMAN SCATTERING: VIBRATIONAL AND REORIENTATIONAL MOTIONS IN LIQUID METHYL IODIDE

M. Constant

Laboratoire de Spectroscopie Raman, Université des Sciences et Techniques de Lille, Villeneuve d'Ascq, France

and

R. Fauquembergue

Centre de Recherche sur les Propriétés Hyperfréquences des Milieux Condensés, Université des Sciences et Techniques de Lille, Villeneuve d'Ascq, France

INTRODUCTION

The width and profile of vibrational bands in liquids are produced by both vibrational and reorientational relaxation mechanisms. Infrared spectroscopy is not capable of distinguishing the two contributions to the bands from these two motions. Using Raman scattering, however, a measurement of the bands profiles and the analysis of these bands by polarization measurements, leads directly to the two contributions.

The theoretical work of Bratos and Marechal on infrared spectroscopy [1] and later on Raman spectroscopy [2] was the basic inspiration to us in 1970 in developing an experimental method of studying molecular motions of symmetric tops in liquids.[3, 4, 11] This method is adapted to Raman spectroscopy and it allows the separation of the vibrational relaxation due to internal motions from the orientational effects. Recently, Bartoli and Litovitz [5, 6] have given in 1972 a similar method of obtaining the correlation times. This was done for the reorientational motions of a dozen symmetric tops. In this chapter the results obtained with a method of the same type in the study of vibrational and orientational correlation functions are presented. The vibrational bands C–I of the CH_3I molecule in solution in hexane and in carbon tetrachloride are studied. The comparison of these results to those obtained from far infrared absorption spectra is made.

Experimentally the two components I_{pol} and I_{depol} are measured by means of a Coderg spectrometer. These components have been defined by Gordon.[7] The first component takes into account only the vibrational relaxation by measurements of ϕ_v, whereas the second component depends on both vibrational and reorientational motions.[2] Bratos and Marechal have shown that

$$I_{pol} = k \int_{-\infty}^{\infty} \phi_v \exp(-i\omega t)\, dt \tag{1}$$

$$I_{depol} = k' \int_{-\infty}^{+\infty} \phi_v \phi_{2R} \exp(-i\omega t)\, dt \tag{2}$$

where

$$\phi_{2R} = \frac{3}{2} \langle \left\{ [\mathbf{u}(0)\mathbf{u}(t)]^2 - \frac{1}{3} \right\} \rangle$$

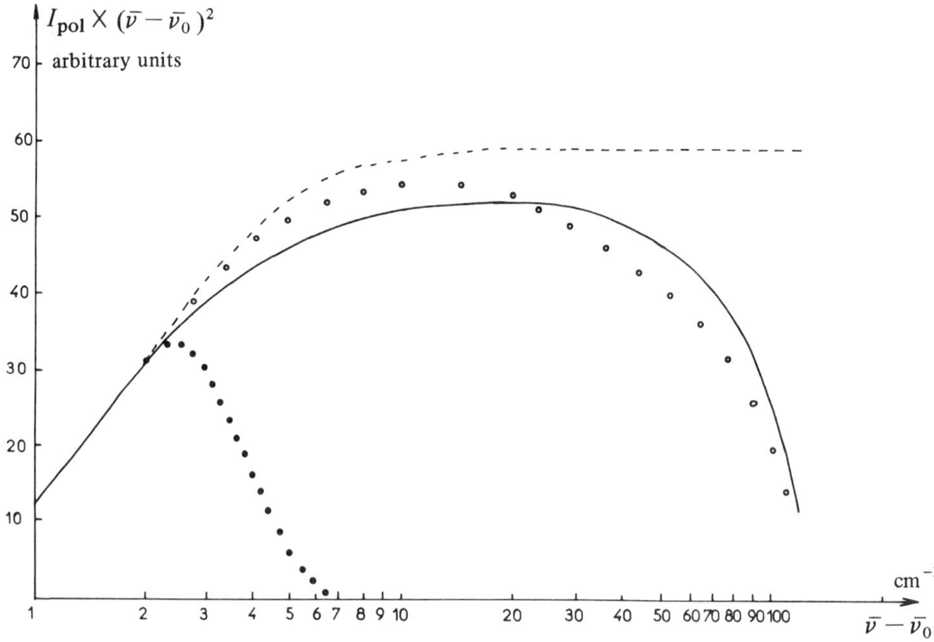

Fig. 1. Polarized spectra of 525 cm^{-1} line of methyl iodide at 25° C. —— experimental spectra; ○ ○ ○ ○ calculated spectra with the use of Eqn. 7; - - - - lorentzian spectra; ● ● ● ● gaussian spectra.

k, k' are constants, $\mathbf{u}(t)$ is a unit vector along the direction of the dipole moment of the molecule and the symbol ⟨ ⟩ represents an ensemble average value.

VIBRATIONAL RELAXATION RESULTS

The vibrational relaxation spectra (I_{pol}) of the symmetric vibrations C–I of methyl iodide is shown in Fig. 1. The vibrational relaxation function, obtained from Fourier transform of the isotropic component (I_{pol}) of the Raman band taking into account the instrumental profiles is plotted in Fig. 2 for methyl iodide, and methyl iodide in solution in hexane and carbon tetrachloride.

The spectra and the time dependence of the vibrational relaxation function can be divided into different portions. In the low frequency region the bands are nearly lorentzian and describe very accurately the exponential long-time portion of correlation functions. After reaching a plateau corresponding to a lorentzian form in the range of 10–40 cm^{-1} the factor $I_{pol} \times \omega^2$ decreases very rapidly with increasing frequency. This portion corresponds to zero first derivative of the correlation function, at the origin.

In order to interpret the present results it is necessary to recall a theoretical result[2] defining the vibration correlation function

$$\phi_v = < \exp\left(-i\int_0^t \Omega_v(t')\mathrm{d}t'\right) > \text{ and } \Omega_V^2 = \overline{\Omega_v^2} \qquad (3)$$

where Ω_v represents the instantaneous deviation of the vibrational transition with respect

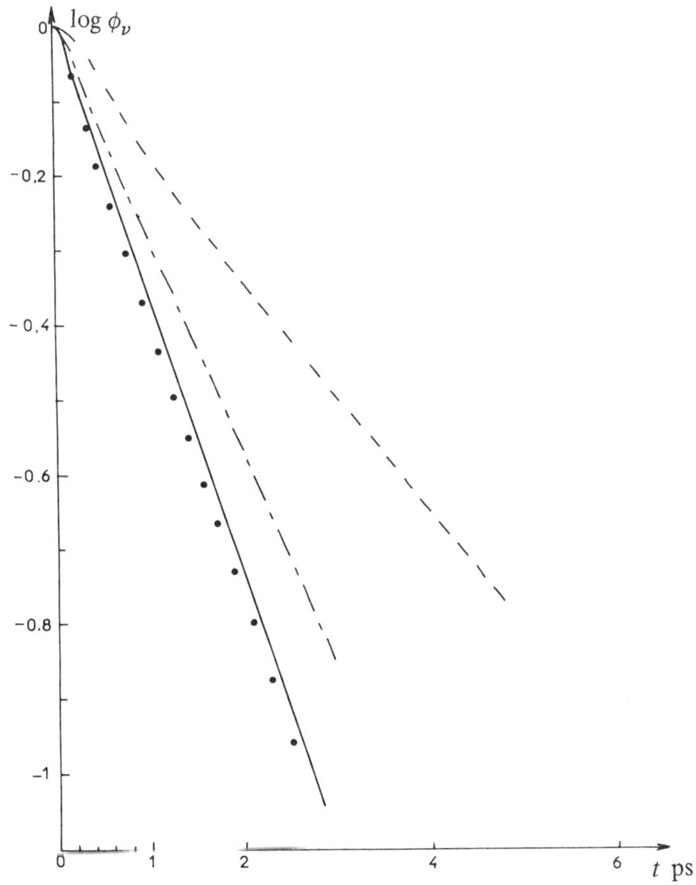

Fig. 2 Logarithm vibrational correlation function ϕ_v : (——) CH_3I; (- - - -) 20 mole percent CH_3I-hexane solution; (– · – ·) 20 mole percent CH_3I-CCl_4 solution; (• • • •) calculated values with the use of Eqn. 6.

to the average value and the symbol ⟨ ⟩ means an ensemble average value. For the study of the evolution of the function we must introduce the correlation time $\tau_{\Omega V}$ of the stochastic function $\Omega_v(t)$.

If the modulation of the frequency is slow i.e. $\tau_{\Omega V} . \Omega_v \gg 1$ then the distribution of the frequencies only determine the time dependence of the correlation function. Particularly, in the case where an active molecule is perturbed by a great number of neighbours the ϕ_v and I_{pol} have a gaussian shape. In this case the width $\Delta\nu$ of the polarized component is given by

$$\Delta\nu = 2.3 . \nu_V \qquad (4)$$

and $$\nu_V = \Omega_V/2\pi$$

If the modulation of the frequency is rapid i.e. $\tau_{\Omega V} . \Omega_V \ll 1$, then ϕ_v has an

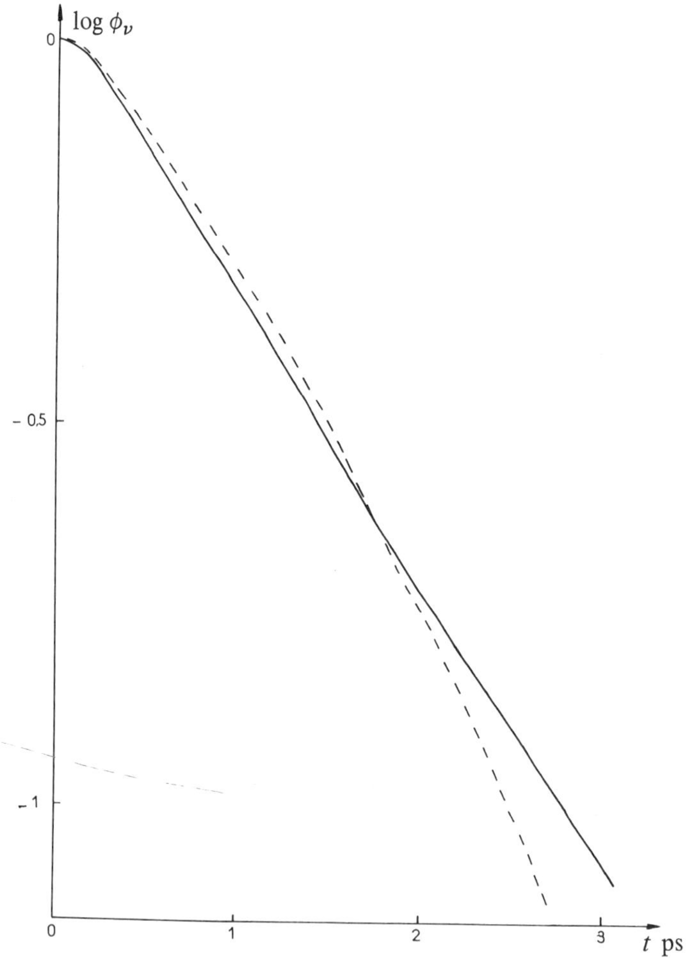

Fig. 3. Temperature dependence of vibrational correlation function for CH_3I. (——) function at 25° C; (- - - -) function at – 30° C.

exponential form[8] for the long time and we have

$$\Delta\nu = 2\,\nu_V\,(\Omega_V\,\tau_{\Omega V}) = 1/(\pi\,\tau_v) \tag{5}$$

In this case, in assuming an exponential form for the autocorrelation function of ϕ_v and a rapid modulation we have[9]

$$\phi_v = \exp\,(-\,t/\tau_v)\,\exp\,(\tau_{\Omega V}/\tau_v)\exp\,[(-\,\tau_{\Omega v}/\tau_v)\,\exp\,(-\,t/\tau_{\Omega V})] \tag{6}$$

and

$$I_{\text{pol}} = \frac{I_{\text{max}}}{(1+\omega^2\tau_v^2)(1+\omega^2\tau_{\Omega V}^2)} \tag{7}$$

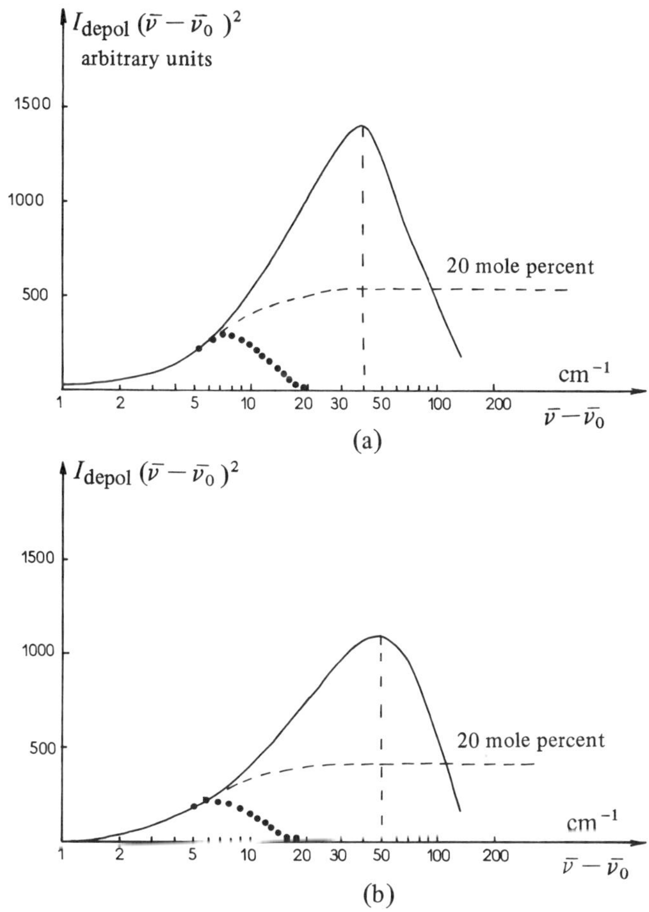

Fig. 4. Depolarized spectra of 525 cm^{-1} line of methyl iodide in solution in hexane (a) and carbon tetrachloride (b) at 25°C. (——) experimental spectra; (- - - -) lorentzian spectra; (● ● ● ●) gaussian spectra.

From Eqns. (4) and (5) and our experimental results, it is possible to determine the characteristic mechanisms of the vibrational motions. It is sufficient to compare the experimental widths $\Delta\nu$ to the value ν_V defined from the spectra by the experimental second order moment. For CH_3I, at 25°C, it is found that the ratio $\Delta\nu/\nu_V$ is 0.42, which is lower than the value 2.3, which corresponds to a gaussian form.

It was therefore concluded that a fairly rapid mechanism of modulation must probably define the vibrational relaxation in the present case. Under these conditions it seemed interesting to verify experimentally the relations giving the expressions for ϕ_v and for I_{pol}. Figures 1 and 2 show the comparison of the theoretical and experimental curves and it is seen that the agreement is satisfactory.

The study of the spectra as a function of temperature shown in Fig. 3 reveals that the function ϕ_v at low temperatures takes a gaussian form. It therefore seems possible to

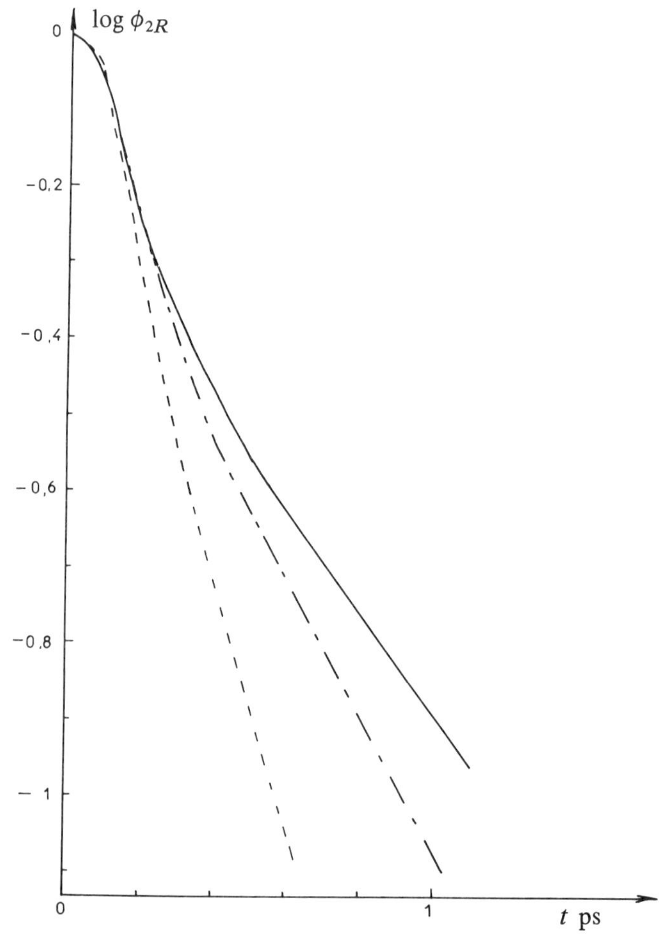

Fig. 5. Logarithm reorientational correlation ϕ_{2R} functions. (——) CH_3I; (- - - -) 20 mole percent CH_3I-hexane solution; (- — - —) 20 moles percent CH_3I-CCl_4 solution.

pass successively from a rapid modulation mechanism to a slow modulation mechanism just by lowering the temperature.

REORIENTATIONAL RELAXATION RESULTS

It is also possible to obtain, from the depolarized component I_{depol}, with the help of ϕ_v obtained by the method presented above, the correlation function ϕ_{2R} which characterizes the reorientational motions in liquids. The orientational relaxation spectra of the symmetric vibration C–I of methyl iodide in solution in hexane and in carbon tetrachloride are shown in Fig. 4.

The orientational correlation functions obtained are represented in Fig. 5. These functions have zero first derivative at the origin, and after a rapid decrease in the range of $0.2–0.4 \times 10^{-12}$s, they take an exponential form. It is interesting to compare these results with those of absorption spectra. In the far infrared spectra it is possible to relate the

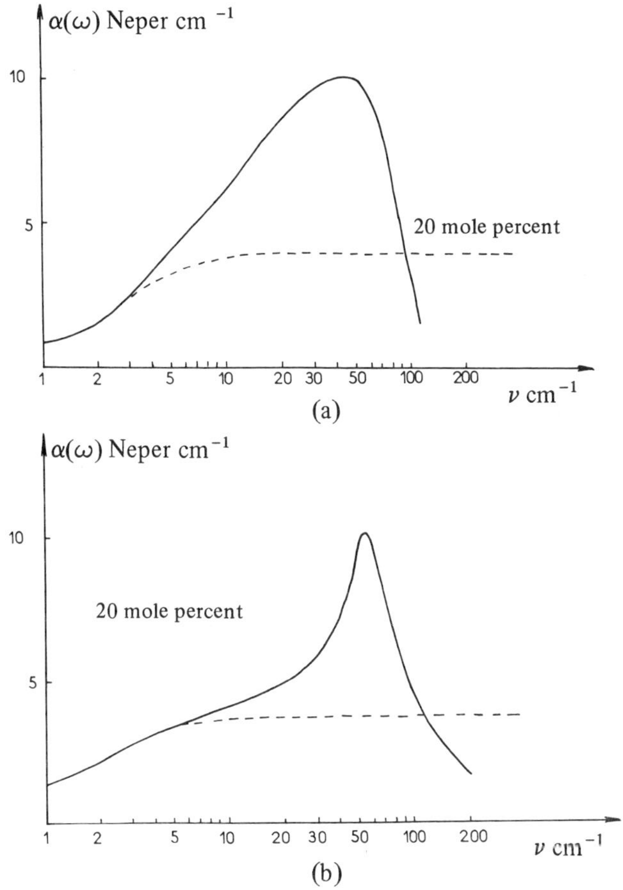

Fig. 6. Far infrared spectra of methyl iodide in solution in hexane (a) and in carbon tetrachloride (b), (——) experimental data; (- - - -) represents Debye plateau.

absorption coefficient $\alpha(\omega)$ to the reorientational correlation function $\phi_{1R} = \langle \mathbf{u}(0)\mathbf{u}(t)\rangle$. We have

$$\alpha(\omega) = k''\,\omega^2 \int_{-\infty}^{\infty} \phi_{1R} \exp(-i\omega t)\,\mathrm{d}t \tag{8}$$

It is to be noticed here that the above relation (8) is true only if a correlation between the different dipole moments does not exist. It is therefore necessary to work with dilute solutions.

Equations (2) and (8) show that it is interesting to compare the factor $I_{\mathrm{depol}} \times \omega^2$ (Fig. 4) to the absorption coefficient $\alpha(\omega)$ (Fig. 6). Experimentally, $I_{\mathrm{depol}}(\omega)$ and $\alpha(\omega)$ exhibit the same qualitative features. The functions ϕ_{1R} obtained from far infrared absorption spectra are given in Figs. 7 and 8. These functions show a time dependence which is analogous to the functions ϕ_{2R}. In several cases we have calculated the ratio τ_{1R}/τ_{2R} of the two correlation times defined by the following relations

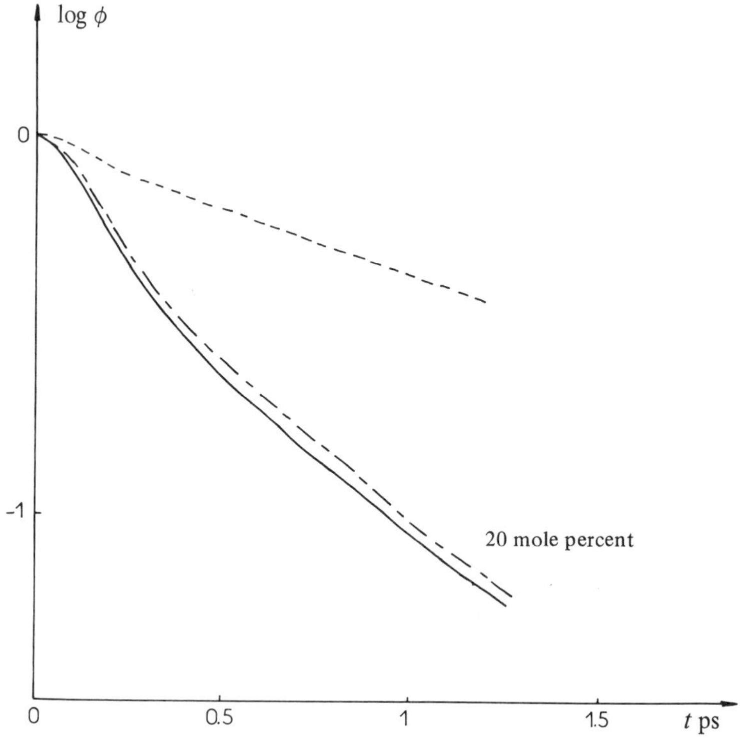

Fig. 7. Logarithm reorientational correlation function for methyl iodide in solution in carbon tetrachloride. (——) experimental ϕ_{2R} function; (- - - -) experimental ϕ_{1R} function; (– – – -) calculated ϕ_{2R} function with the use of Harp and Berne theory.

$$\tau_{2R} = \int_0^\infty \phi_{2R} \, dt \tag{9}$$

$$\tau_{1R} = \int_0^\infty \phi_{1R} \, dt \tag{10}$$

It is found that the value of this ratio is between 2.3 and 3. Thus, to a first approximation, the reorientational motions are taking place following a rotational diffusion process.

A more quantitative comparison of the ϕ_{1R} and ϕ_{2R} functions can be made by applying the theory given by Harp and Berne.[10] These authors have assumed that the entropy due to the reorientational degrees of freedom, was maximum in the liquids. On this basis they have shown that ϕ_{1R} and ϕ_{2R} were related by the following relations

$$\phi_{1R} = -\frac{1}{\beta} + \coth \beta \tag{11}$$

$$\phi_{2R} = 1 + \frac{3}{\beta}\left[\frac{1}{\beta} - \coth \beta\right] \tag{12}$$

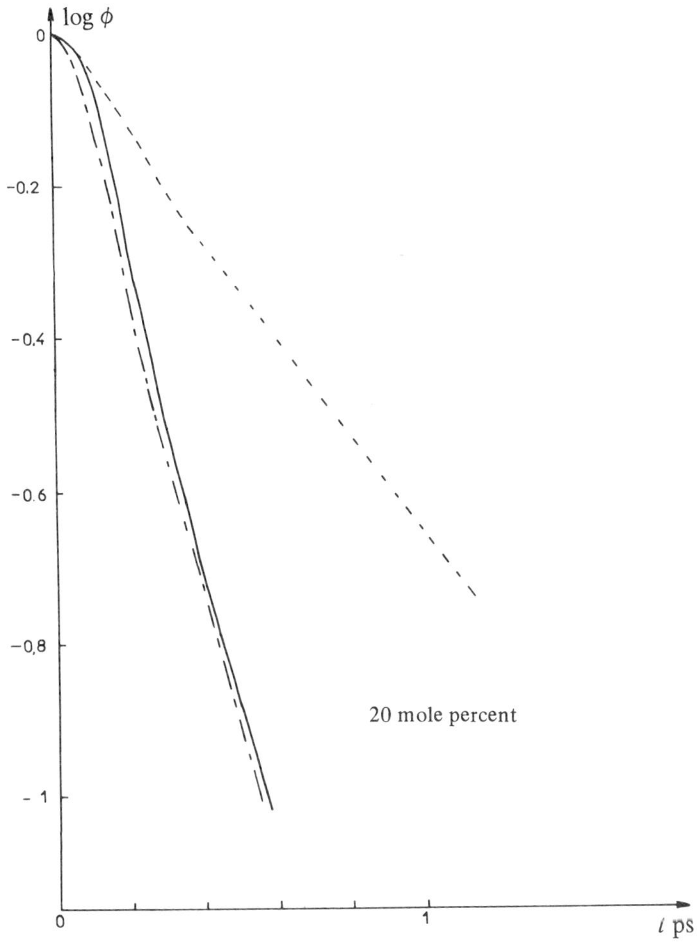

Fig. 8. Logarithm reorientational correlation function for methyl iodide in solution in hexane. (——) experimental ϕ_{2R} function; (- - - -) experimental ϕ_{1R} function; (– - – -) calculated ϕ_{2R} function with the use of Harp and Berne theory.

This theory gives good agreement in the case of the CH_3I molecule in solution in hexane and in carbon tetrachloride. This is shown in Figs. 7 and 8, where the functions ϕ_{1R} and ϕ_{2R} obtained from the experiments, and also the function ϕ_{2R} calculated from ϕ_{1R} by using the results of Harp and Berne, are presented.

In conclusion, it has been shown how, using Raman vibrational scattering it is possible to study vibrational and reorientational motions of symmetric tops.

REFERENCES

1 S. Bratos, S. Rios and Y. Guissani, *J. Chem. Phys.* **52**, 439 (1970).
2 S. Bratos and E. Marechal, *Phys. Rev.* **4**, 1078 (1971).
3 M. Constant, *Diplôme études approfondis*, Lille, September 1969.

4 M. Constant, M. Delhaye and R. Fauquembergue, *C.R. Acad. Sci. Paris* **271B**, 1117 (1970).
5 F. S. Bartoli and T. A. Litovitz, *J. Chem. Phys.* **56**, 413 (1972).
6 F. S. Bartoli and T. A. Litovitz, *J. Chem. Phys.* **56**, 404 (1972).
7 R. G. Gordon, *Advances in Magnetic Resonance*, Vol. 3, Academic Press, New York, 1968, p. 11.
8 R. Kubo, *Lectures in theoretical Physics*, Interscience, New York, 1959.
9 Y. Leroy, E. Constant and P. Desplanques, *J. Chim. Phys.* **64**, 1499 (1967).
10 G. D. Harp and B. J. Berne, *Phys. Rev.* **A2**, 975 (1970).
11 M. Constant and R. Fauquembergue, *C.R. Acad. Sci. Paris* **272B**, 1293 (1971).

53. STRUCTURE ROTATIONNELLE DE LA DIFFUSION RAMAN DE MOLECULES LEGERES DISSOUTES EN MILIEU LIQUIDE

Y. Le Duff, W. Holzer et R. Ouillon

Département des Recherches Physiques, Laboratoire associé au C.N.R.S., Université de Paris VI, France

ABSTRACT

We have studied Raman scattering of light diatomic molecules dissolved at low concentration in liquids. For H_2 and D_2 in SF_6 and CCl_4 we present precise measurements on the linewidth, the position and the intensity of the bands $S_0(J)$, $Q_1(J)$ and $S_1(J)$. The vibrational and rotational contribution to the band shape will be discussed. For HF, we analyse the spectra of HF in SF_6.

INTRODUCTION

Depuis quelques années l'augmentation de puissance des lasers permet d'étudier avec une précision satisfaisante la diffusion Raman de molécules en solution très diluée. Le cas d'une molécule diluée dans un solvant liquide représente une situation plus favorable que celui d'un liquide pur pour aborder théoriquement le problème des forces et de la dynamique moléculaires en milieu liquide.[1]

Notre travail concerne des molécules diatomiques légères (H_2, D_2, HF) dissoutes dans des solvants inertes. Dans ces molécules l'éloignement relatif exceptionnel des niveaux de rotation permet une étude quantitative de la structure rotationnelle en phase condensée.

Nous avons déjà montré par spectroscopie Raman, que la rotation des molécules H_2 et D_2 dissoutes dans SF_6 liquide peut être considérée comme quasi libre.[2] La structure rotationnelle est encore visible dans les spectres Raman de HCl et HBr dissous.[3] Ici nous présentons pour H_2 et D_2 en solution dans SF_6 et CCl_4 des mesures précises de la largeur, la position et les intensités de plusieurs raies $S_0(J)$ de la rotation pure et des composantes rotationnelles $Q_1(J)$ et $S_1(J)$ de la bande de vibration. La forme des raies est également discutée. Nous considérons ensuite la diffusion de HF dissous dans SF_6.

APPAREILLAGE

Les spectres ont été obtenus avec la raie 5145 Å d'un laser à argon ionisé (Spectra Physics, modèle 165) un double monochromateur (Coderg PH1), un photomultiplicateur (EMI 9558) refroidi et un compteur de photon. Nous utilisons d'autre part une cuve à pression avec des fenêtres en silice ou en saphir (HF). La concentration en H_2 et D_2 était d'environ 1 mole pour 50 de solvant. Les différentes polarisations ont été étudiées[4] en plaçant dans le faisceau diffusé un polariseur.

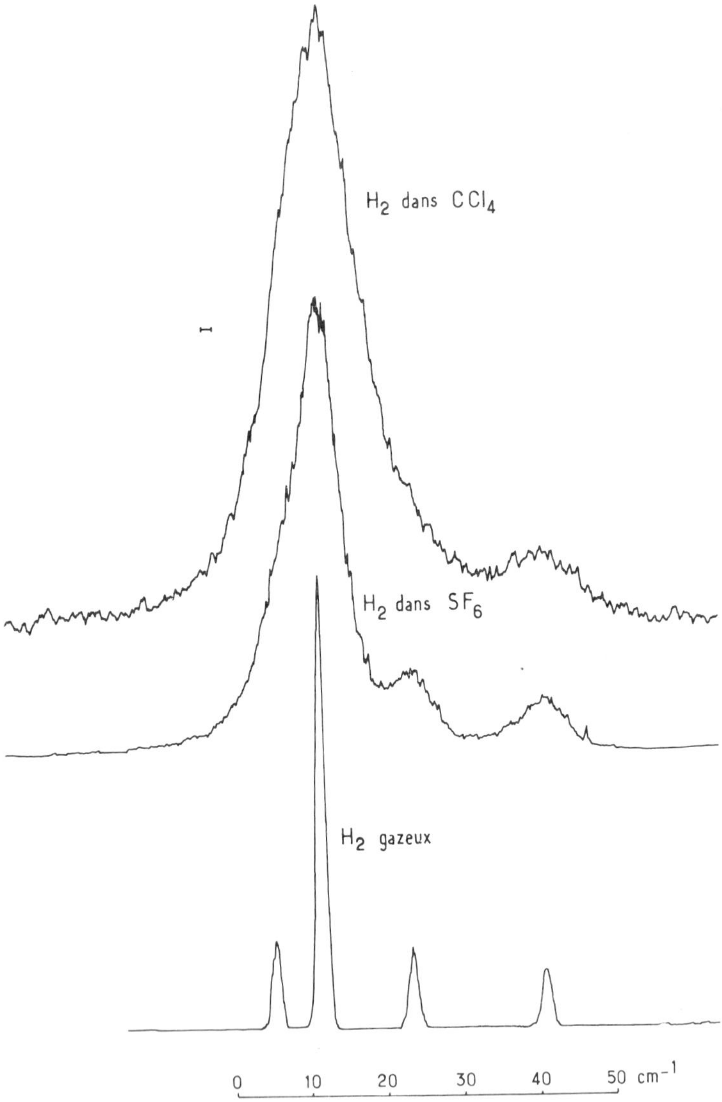

Fig. 1. Branche Q_1 (J) à température ambiante pour H_2/CCl_4, H_2/SF_6 et H_2 gazeux à 1 atm. La largeur de la fente, identique pour les trois spectres, est indiquée.

RESULTATS SUR H_2 et D_2

Bien que les spectres Raman de la molécule H_2 présentent des caractères de rotation libre aussi bien dans SF_6 que CCl_4 comme le montre qualitativement la Fig. 1 nos résultats montrent une action très différente de ces solvants sur la diffusion Raman.

Largeur

Le Tableau 1 présente nos mesures relatives aux largeurs à mi-hauteur des raies de rotation pure et de vibration. Elles peuvent être comparées avec des résultats obtenus à

Tableau 1. Largeur à mi-hauteur des raies Raman de H_2 et D_2 (en cm^{-1} et à température ambiante). La correction relative à la fonction d'appareil a été soustraite dans nos mesures.

	H_2 gazeux 2000 atm (ref. 5)	H_2 + A gazeux à 2000 atm (ref. 5)	H_2/SF_6	H_2/CCl_4	D_2/SF_6
$S_0(0)$	2,4		2,6±0,2		3,3±0,3
$S_0(1)$	3,6	2,1	2,6±0,2	25±4	2,5±0,2
$S_0(2)$			2,0±0,2		1,9±0,2
$S_0(3)$			1,9±0,2	8,6±1	
$Q_1(1)$ polar$_{//}$	1,4	5	7,0±0,5	10,5±1,5	5±1
$Q_1(1)$ polar$_{\perp}$			7±1	15,5±2	
$S_1(1)$			8,5±1	36±4	6±1

haute pression en phase gazeuse.[5] Dans le cas de la vibration pure, les composantes perpendiculaire (anisotrope) et parallèle (essentiellement isotrope) ont été étudiées séparément. Les fentes spectrales utilisées étaient au moins trois fois plus fines que les raies des spectres enregistrés. Les profils ont pu être déconvolués avec une précision suffisante en assimilant la fonction d'appareil à une gaussienne.

Il est montré (voir Tableau 1) que l'élargissement des raies de rotation $S_0(J)$ est une fonction de nombre quantique J. Une telle dépendance est observée pour la première fois dans la diffusion Raman en milieu liquide, donc pour des densités relativement élevées. Elle avait été prévue théoriquement[6] pour des intéractions quadripolaires dans des gaz à pression moyenne.

La comparaison des élargissements pour les raies $S_0(J)$ et $Q_1(J)$ notamment dans H_2/CCl_4 indique que la perturbation du mouvement de rotation de H_2 par le solvant se manifeste différemment pour les raies de rotation pure et de vibration pure. De même les composantes perpendiculaires (anisotropes) des raies $Q_1(J)$ sont plus élargies que les composantes parallèles (essentiellement isotropes). Des effets semblables existent également dans le modèle gazeux de Van Kranendonk[6] où il est montré que le changement de phase (phase shift) induit par les collisions n'influe pas sur la composante isotrope de la vibration. Cependant l'élargissement des raies $S_0(J)$ correspondant aux transitions entre niveaux de faible énergie (*cf.* $S_0(1)$ de H_2/CCl_4) semble être une caractéristique des intéractions fortes de la phase liquide. Ceci montre qu'il n'est en général pas possible de parler d'une largeur rotationnelle unique dans les liquides. Notons que la largeur de la raie $S_1(1)$ de vibration-rotation (dans chacun des solvants) représente approximativement la superposition d'une certaine largeur rotationnelle donnée par $S_0(1)$ et d'une largeur vibrationnelle dont la composante isotrope de $Q_1(1)$ représente une borne supérieure.

Position

Les déplacements liquide-gaz ont été mesurés pour chaque raie; ils sont notés dans le Tableau 2. Le déplacement de la vibration dans H_2/CCl_4 est plus de cinq fois supérieur

Tableau 2. Déplacement Raman gaz-liquide $\Delta\nu$ en cm^{-1} pour H_2 et D_2 dissous dans SF_6 et CCl_4 (précision 1 cm^{-1}).

	H_2/SF_6	H_2/CCl_4	D_2/SF_6
$S_0(1)$	1	1±2	<1
$S_0(3)$	1	4±2	<1
$Q(1)$	3	20,7±0,7	
$Q(3)$	3	21±2	
$Q(4)$			3
$S_1(1)$	4	21±2	3

à celui observé dans H_2/SF_6. Cet effet reflète la plus grande perturbation par CCl_4 qui s'est déjà manifestée dans le Tableau 1.

Pour les raies de rotation le déplacement est du même ordre de grandeur que notre précision (sauf pour $S_0(3)$ dans H_2/CCl_4).

On vérifie également que l'intéraction vibration-rotation n'est pas modifiée, même dans CCl_4 comme l'indique la conservation de la distance spectrale entre $Q_1(1)$ et $Q_1(3)$ (Fig. 1).

Intensité

Nous avons comparé pour H_2 et D_2 les intensités intégrées des raies Raman de H_2 et D_2 en solution dans SF_6 avec les intensités correspondantes pour ces molécules à l'état gazeux (1 atm). Ces mesures portent dans le cas de H_2 sur $S_0(0)$, $S_0(1)$, $S_0(2)$, $S_0(3)$, la branche Q_1 et $S_1(0)$, $S_1(1)$, dans le cas de D_2 sur $S_0(0)$, $S_0(1)$, $S_0(2)$, la branche Q_1 et $S_1(0)$, $S_1(1)$, $S_1(2)$. Il apparaît que l'intensité des raies de H_2/SF_6 relativement aux raies de H_2 gazeux est conservée avec une incertitude de 10% aussi bien pour la rotation pure que pour la vibration. Le même résultat est valable pour D_2/SF_6 et D_2.

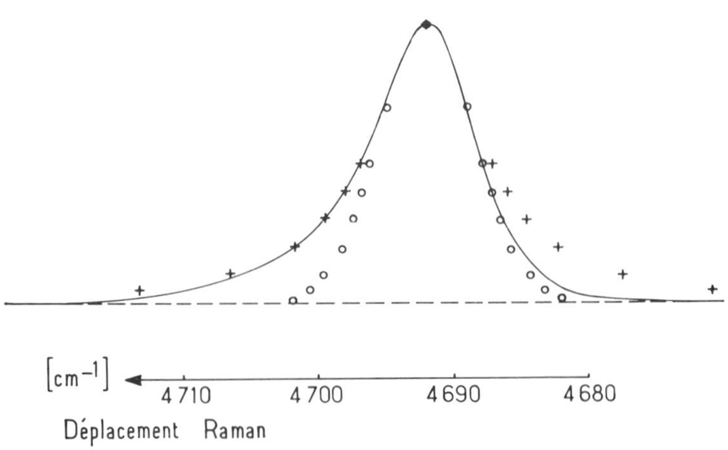

Fig. 2. Profil expérimental de la raie de vibration-rotation $S_1(1)$ pour H_2 dans SF_6 comparé avec un profil lorentzien (+) et un profil gaussien (○).

Forme

Le profil des raies de rotation pure de H_2 et D_2 dissous est symétrique, assez proche d'une lorentzienne pour H_2/SF_6 et D_2/SF_6, mais avec moins de pied pour H_2/CCl_4. Par contre les raies de vibration $Q_1(J)$ et $S_1(J)$ de H_2 et D_2 dans SF_6 sont nettement dissymétriques: le pied est plus étalé du côté opposé à l'excitatrice. Nous montrons cette dissymétrie dans la raie bien isolée $S_1(1)$ de H_2/SF_6 (Fig. 2). La décomposition de la branche Q_1 à l'aide d'une méthode de simulation[7] est en cours.

On peut suggérer, que la dissymétrie observée reflète la distribution statistique de la perturbation de la fréquence de vibration pour les différentes molécules H_2 dans le solvant, les effets rotationnels conduisant plutôt à un profil symétrique. Les raies de vibration de H_2 dans CCl_4 ne présentent pas la dissymétrie observée dans SF_6.

Facteur de dépolarisation

L'intéraction du solvant sur les molécules dissoutes se manifeste également sur le facteur de dépolarisation ρ_s^Q de la branche de vibration.

Des mesures très précises[4] de ρ_s^Q pour la bande $Q_1(J), J = 0, 1, 2, 3$ donnent $\rho_s^Q =$ 0,016 ± 0,001 pour H_2 gazeux (1 atm). Des mesures semblables en solution conduisent à

$$\rho_s^Q = 0{,}019 \pm 0{,}003 \text{ dans } H_2/SF_6$$

$$\rho_s^Q = 0{,}030 \pm 0{,}002 \text{ dans } H_2/CCl_4$$

Ces valeurs mettent en évidence une modification de l'anisotropie par le solvant; il semble aussi que la partie anisotrope de la raie $Q_1(0)$ qui est rigoureusement absente[4] dans le spectre gazeux, apparait.

RESULTATS SUR HF

Les résultats préliminaires relatifs à HF donnés à la conférence doivent être réexaminés.

CONCLUSION

Nous avons montré l'existence d'une structure rotationnelle bien résolue dans les liquides. Son étude quantitative indique que les niveaux d'énergie la plus faible sont les plus perturbés. Dans le cas d'une intéraction du solvant relativement élevée les modifications des bandes de rotation pure et de vibration pure sont dues à des effets différents.

Ces résultats devraient apporter à la spectroscopie Raman un rôle prépondérant dans l'étude de la dynamique des molécules légères en milieu liquide.

REFERENCES

1 R. G. Gordon, *Advances in Magnetic Resonance,* Vol 3, Academic Press, New York, 1968; S. Bratos and E. Marechal, *Phys. Rev.* **4**, 1078 (1971).

2 W. Holzer, Y. Le Duff et R. Ouillon, *C. R. Acad. Sci. Paris* **273B**, 313 (1971).

3 J. P. Perchard, W. F. Murphy et J. Bernstein, *Chem. Phys. Letters* **8**, 559 (1971).

4 W. Holzer et Y. Le Duff, *Advances in Raman Spectroscopy*, Vol. 1 (Ed. J. P. Mathieu), Heyden & Son, London, 1973, p. 109.

5 A. D. May, V. Degen, J. C. Stryland et H. L. Welsh, *Can. J. Phys.* **39**, 1769 (1961).

6 J. Van Kranendonk, *Can. J. Phys.* **41**, 433 (1963).

7 K. Altmann et G. Strey, *Z. Naturforsch.* **27**, 65 (1972).

54. ETUDE DES RAIES DE DIFFUSION RAMAN DU DIOXYDE DE SOUFRE DISSOUS DANS DES LIQUIDES

R. Ouillon et Y. Le Duff

Département des Recherches Physiques, Laboratoire associé au C.N.R.S., Université de Paris VI, France

ABSTRACT

We have analysed the half width of the Raman bands of sulphur dioxide dissolved in various solvents at room temperature. The temperature dependence has been studied in detail: the half width of ν_1 (measured by an interferometric method) decreases with temperature and the half width of ν_3 increases. A comparison with the results of Rakov is presented.

INTRODUCTION

Les profils de raies Raman de molécules dissoutes sont une source d'informations intéressantes pour la connaissance des forces et de la dynamique moléculaire[1] en milieu liquide. D'autre part l'étude de molécules diluées permet d'observer la réponse d'un système unique en présence de solvants différents.

Dans ce cadre nous avons étudié la largeur des bandes de diffusion du dioxyde de soufre dissous à faible concentration dans plusieurs liquides. SO_2 est une molécule triatomique non linéaire appartenant au groupe C_{2v}. Elle présente trois vibrations fondamentales toutes actives en Raman (Fig. 1).

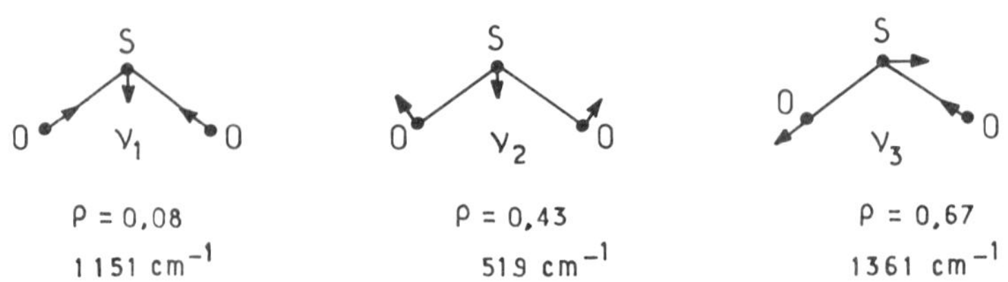

Fig. 1. Formes des vibrations de SO_2 et facteur de dépolarisation ρ_s (en lumière polarisée) mesuré pour SO_2 liquide.[2]

TECHNIQUE DE MESURE

Pour cette étude nous avons utilisé:

1. Pour les bandes larges (> 1 cm^{-1}); un laser spectra-Physics à argon ionisé, (modèle 165), et un double monochromateur Raman Coderg PH_1 (résolution maximale 0,6 cm^{-1} à $\lambda = 5000$ Å).

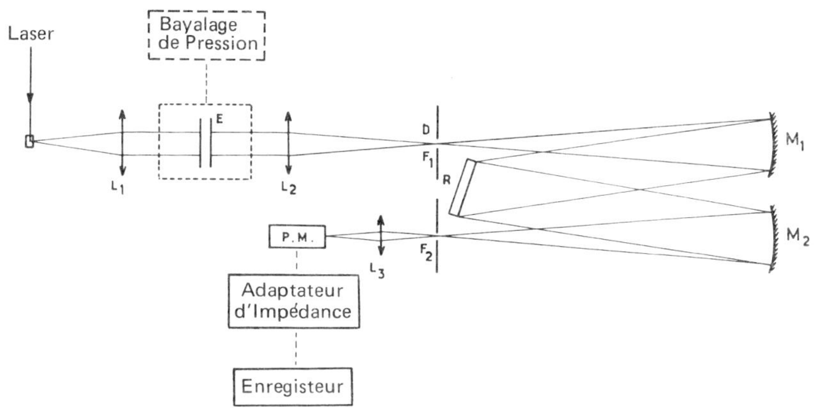

Fig. 2. Schéma du montage haute résolution.

Tableau 1. Largeur des raies Raman de SO_2 dissous dans des solvants liquides (à température ambiante).

	ν_1		ν_2		ν_3	
Solvants	position (cm^{-1})	$\Delta\nu_{1/2}$ (cm^{-1})	position (cm^{-1})	$\Delta\nu_{1/2}$ (cm^{-1})	position (cm^{-1})	$\Delta\nu_{1/2}$ (cm^{-1})
SO_2 gazeux	1151,8		519		1361	
$ClCF_2\ C\ Cl_2\ F$					1349 ±2,5	48 ±2
CS_2	1142,9 ±0,5	0,45 ±0,05			1340 ±2	41 ±2
CCl_4	1145,5 ±0,5	0,50 ±0,05			1343 ±1,5	38,5 ±2
$CHCl_3$	1145,9 ±0,5	0,8 ±0,2	521,5 ±1,5	14,5 ±2	1339 ±2	25,3 ±3
C_6H_6	1143,9 ±0,5	1,3 ±0,2	521 ±1,5	8 ±1	1335 ±1,5	20,3 ±2
CH_3COCH_3	1149,2 ±0,5	3,2 ±0,3			1331,5 ±1,5	27 ±3
C_2H_5OH	1149,1 ±0,5	4,4 ±0,3			1333 ±1,5	32 ±2,5

2. Pour les raies fines (< 0,5 cm^{-1}); un monochromateur à réseau associé à interféromètre Fabry-Pérot (Fig. 2). Le spectromètre est ouvert à $F/10$. Le diaphragme explorateur D de l'interféromètre est confondu avec la fente d'entrée F_1 du monochromateur.

Pour les enregistrements de la raie ν_3 effectués avec le spectromètre Coderg la largeur de la fonction d'appareil était au moins cinq fois plus étroite que celle de la raie. Les largeurs vraies ont pu être calculées par déconvolution à partir des profils expérimen-

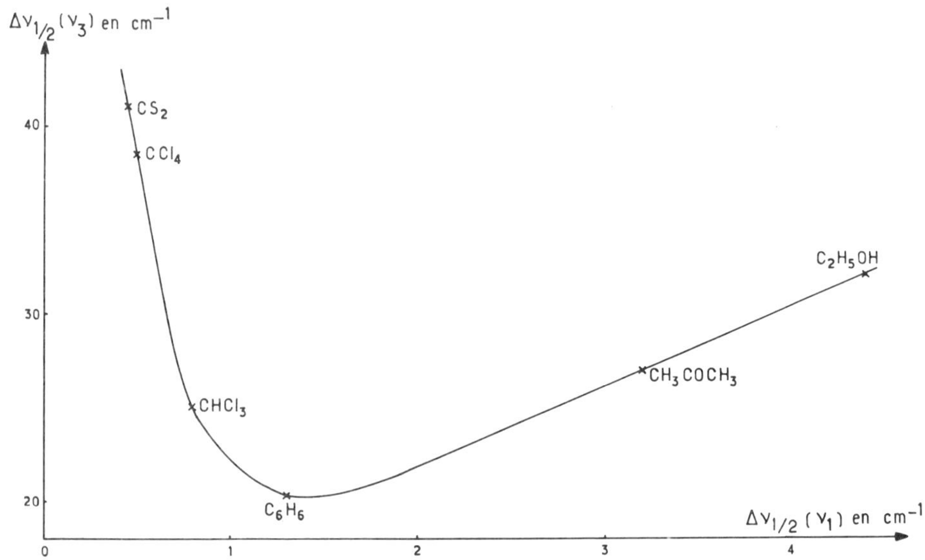

Fig. 3. Largeur comparée des bandes de diffusion Raman ν_1 et ν_3 de SO_2 dans différents liquides à température ambiante.

taux en assimilant ceux-ci à des lorentziennes et les fonctions d'appareil du spectromètre à des courbes de Gauss. On a procédé d'une manière analogue pour obtenir les largeurs de ν_1 à partir des profils expérimentaux.

La concentration des solutions étudiées était de l'ordre de cinq molécules de SO_2 pour cent molécules de solvant. Une étude de l'évolution de la largeur de ν_3 de SO_2 en solution dans CCl_4 (à $-2°C$) et dans CS_2 (à $-51°C$), en fonction de la concentration, a montré que l'influence de celle-ci était négligeable dans ce domaine de concentration. La précision sur la température mesurée avec un thermocouple est d'environ 3°.

Le dioxyde de soufre était un produit Matheson à 99,98% de pureté.

ETUDE DE LA LARGEUR DES RAIES DE DIFFUSION RAMAN DE SO_2 A TEMPERATURE AMBIANTE[3]

Les résultats expérimentaux sont résumés dans le Tableau 1. La raie ν_1, presque totalement polarisée, ($\rho_s = 0{,}08$), n'est pratiquement pas sensible aux mouvements de rotation de la molécule. Sa largeur représente donc un critère d'intéraction SO_2-solvant. Par contre la largeur de ν_3, raie dépolarisée, ($\rho_s = 0{,}67$), semble très sensible aux effets orientationnels. C'est ce que nous montrons sur la Fig. 3; dans une première région à faible intéraction $[\Delta\nu_{1/2}(\nu_1) < 1\ \text{cm}^{-1}]$ l'augmentation de l'intéraction diminue la largeur de ν_3; cela correspond à une contribution prépondérante des effets de rotation à la variation de $\Delta\nu_{1/2}(\nu_3)$; dans une deuxième région à forte intéraction $[\Delta\nu_{1/2}(\nu_1) > 1\ \text{cm}^{-1}]$ $\Delta\nu_{1/2}(\nu_3)$ évolue dans le même sens que $\Delta\nu_{1/2}(\nu_1)$; dans ces solvants, les mouvements de rotation de la molécule SO_2 sont très réduits et les intéractions moléculaires déterminent la variation de $\Delta\nu_{1/2}(\nu_3)$.

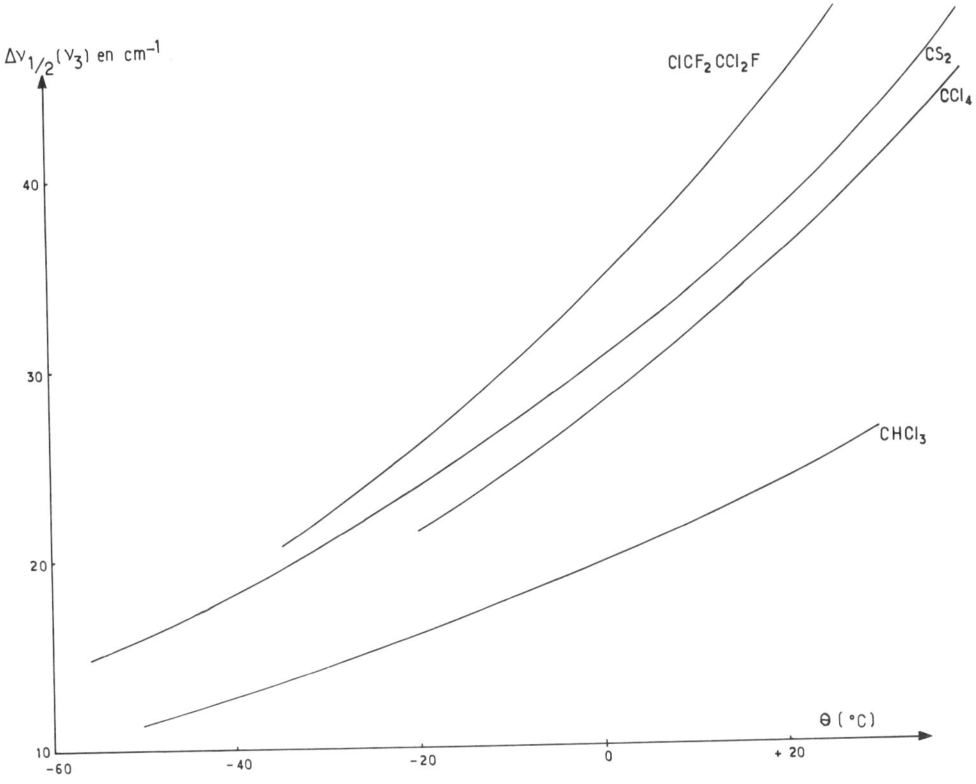

Fig. 4. Largeur à mi-hauteur de la bande Raman ν_3 et SO_2 dans différents solvants en fonction de la température.

ETUDE DES LARGEURS DE ν_3 ET ν_1 EN FONCTION DE LA TEMPERATURE

Raie ν_3

En analysant la largeur de ν_3 nous venons de voir qu'au moins dans les solvants peu actifs subsiste une certaine liberté de rotation de la molécule de SO_2. Rakov[4] a proposé pour la largeur d'une bande de diffusion Raman la loi de variation

$$\delta(T) = \delta_0 + A \exp(-U/kT)$$

où le second terme caractérise la largeur de la bande d'origine rotationnelle, U étant la barrière de potentiel qui s'oppose à la rotation de la molécule et où δ_0 est la largeur résiduelle dont l'origine s'explique par toutes les autres causes d'élargissement. Cette relation contient deux hypothèses qui méritent notre attention:

1. Elle suppose d'abord que les deux types d'élargissement de la bande sont relatifs à des profils lorentziens pour pouvoir additionner les deux demi-largeurs correspondantes.
2. D'autre part la largeur résiduelle δ_0 est supposée indépendante de la température. Nous verrons ci-après que cette condition n'est pas rigoureusement vérifiée pour la demi-largeur

Tableau 2. Valeurs de δ_0 et de U relatives à ν_3 de SO_2 en solution, obtenues par résolution graphique.

	SO_2/génétron 113	SO_2/CS_2	SO_2/CCl_4	$SO_2/CHCl_3$
δ_0 en cm^{-1}	4 ± 4	6,5 ± 3,5	5 ± 5	7,5 ± 2,5
U en Kcal/mole	2,3 ± 0,4	2,3 ± 0,5	2,6 ± 0,5	3,1 ± 0,9

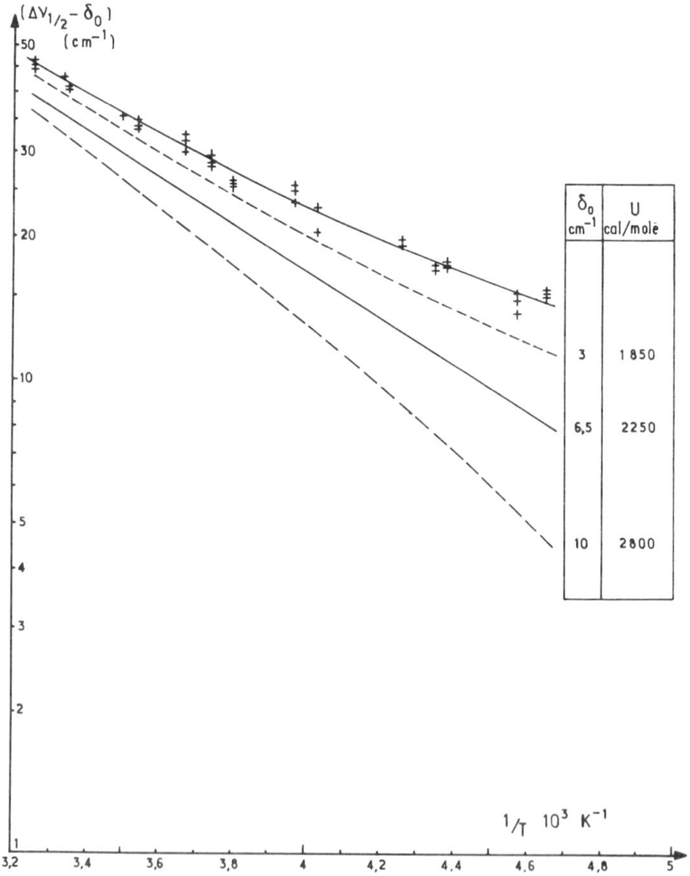

Fig. 5. Diagramme semi-logarithmique de $(\Delta\nu_{1/2} - \delta_0)$ en fonction de $1/T$ pour la bande ν_3 de SO_2 en solution dans CS_2.

de ν_1 (assimilable à δ_0 dans ce cas). Cependant la détermination de la largeur résiduelle de la bande à partir de cette relation est assez justifiée dans notre cas particulier car δ_0 est petit et sa variation avec la température est relativement faible comme nous le verrons dans le cas de ν_1.

La Fig. 4 montre l'évolution de la largeur de la bande Raman ν_3 de SO_2 dans

Tableau 3. Largeur résiduelle des raies ν_1 et ν_3 de SO_2 en solution dans CCl_4 et dans CS_2 obtenue en Raman et en infrarouge.

	δ_0 relatif à ν_1		δ_0 relatif à ν_3	
	Raman	Infrarouge[5]	Raman	Infrarouge[5]
CCl_4	0,45 ± 0,05	0	5 ± 5	0
CS_2	0,50 ± 0,05	0	6,5 ± 3,5	3

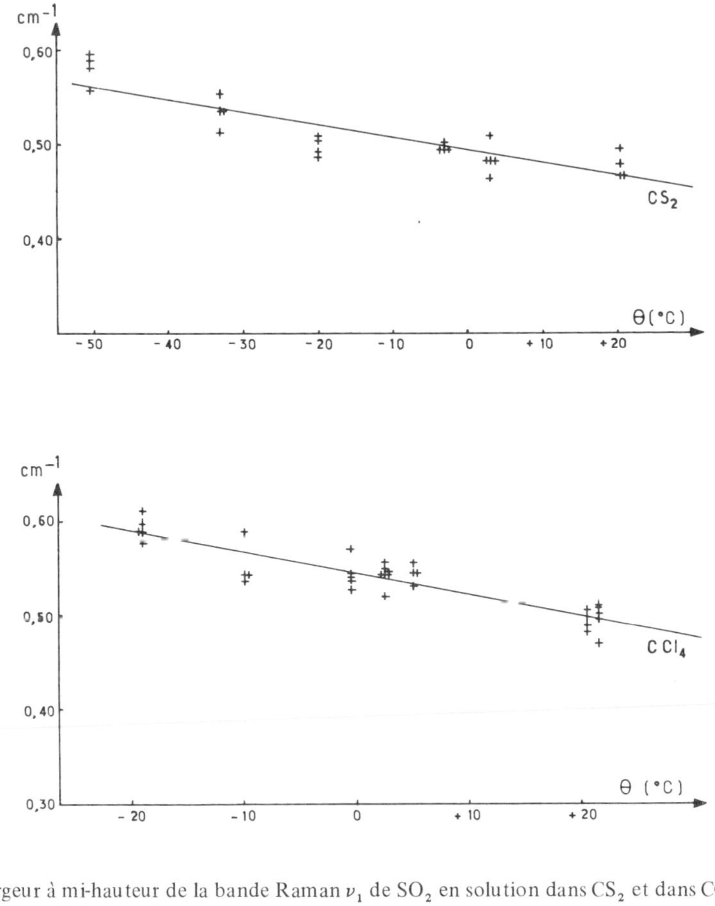

g. 6. Largeur à mi-hauteur de la bande Raman ν_1 de SO_2 en solution dans CS_2 et dans CCl_4 en fonction de la température.

différents solvants en fonction de la température. La dispersion de ces mesures, due à la très faible intensité de cette bande rend imprécise les valeurs de δ_0 et U que le Tableau 2 résume et qui sont déterminées à partir de la résolution graphique de la relation de Rakov comme l'indique la Fig. 5.

On constate sur ce tableau que le sulfure de carbone, le tétrachlorure de carbone et le génétron 113 ont des influences analogues sur le mouvement de réorientation de la molé-

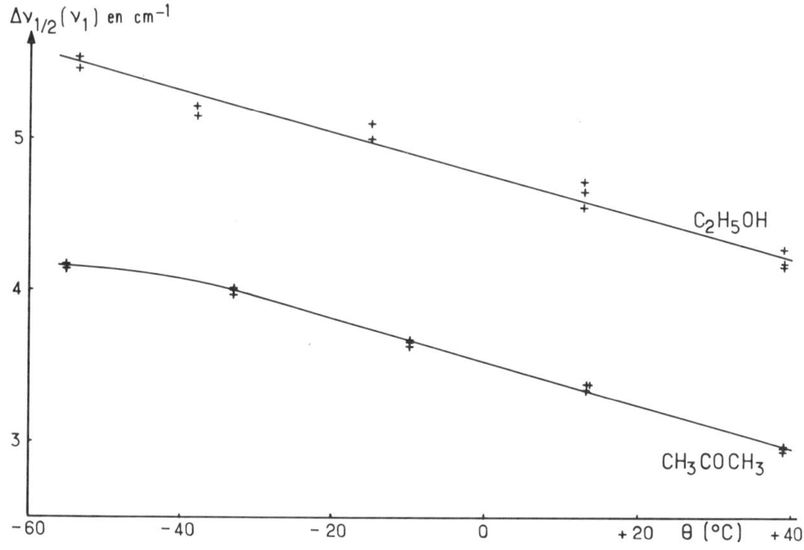

Fig. 7. Largeur à mi-hauteur de la bande Raman ν_1 de SO_2 en solution dans C_2H_5OH et dans CH_3COCH_3 en fonction de la température.

cule de SO_2. En effet, les valeurs de la barrière de potentiel U et de la largeur résiduelle δ_0 sont du même ordre de grandeur pour ces trois solvants. Pour le chloroforme la barrière de potentiel U plus haute caractérise une rotation plus génée. La valeur de la largeur résiduelle semble également supérieure aux précédentes ce qui signifie que les interactions intermoléculaires responsables de cette largeur sont plus fortes.

En absorption infrarouge Vincent et ses collaborateurs[5] ont étudié la largeur des bandes de SO_2 en solution et leur évolution en fonction de la température à partir de la relation de Rakov.[4] Pour les molécules diatomiques, et dans le cadre de sa théorie, Bratos[1] a montré que les profils isotropes d'une bande de diffusion Raman a même forme et donc même largeur que le profil infrarouge dû à la relaxation vibrationnelle pure. Dans la mesure où les profils résiduels infrarouge et Raman ont cette origine et si on peut extrapoler à la molécule de SO_2 les résultats précédents on peut comparer les largeurs résiduelles obtenues pour ν_1 et ν_3 dans les deux solvants communs (Tableau 3).

Les résultats obtenus par les deux techniques sont en bon accord malgré l'imprécision des valeurs de δ_0 relatives à ν_3 que nous avons obtenues. La diffusion Raman donne des résultats plus précis et surtout plus directs en ce qui concerne la largeur résiduelle de ν_1.

Raie ν_1

L'évolution de la largeur de ν_1 de SO_2 en fonction de la température en solution dans des solvants inertes (CS_2, CCl_4) et dans des solvants polaires (C_2H_5OH, CH_3COCH_3) est représentée sur les Figs. 6 et 7. On constate dans tous les cas un léger rétrécissement de la bande lorsque la température croît.

L'affinement d'une raie de diffusion quand on élève la température est un effet peu fréquent dans la spectroscopie Raman des liquides. Nous pouvons l'expliquer du moins en partie de la façon suivante: Quand on chauffe le liquide la densité diminue, la fonction de distribution radiale se modifie, il en résulte une diminution du nombre de plus proches voisins pour la molécule de SO_2 et donc un affinement des niveaux de vibration. Cet effet a déjà été signalé dans le cas de liquides polaires;[6] pour des molécules diatomiques dans des solvants inertes il a récemment été démontré théoriquement.[7]

REFERENCES

1 R. G. Gordon, *Advances in Magnetic Resonance*, Vol. 3, Academic Press, New York, 1968; S. Bratos et E. Marechal, *Phys. Rev.* **A4**, 1078 (1971).
2 M. Magat, *Tables annuelles de constantes*, XII, Hermann, Paris.
3 Y. Le Duff et R. Ouillon, *Compt. Rend.* **272B**, 757 (1971).
4 A. V. Rakov, *Proc. P. N. Lebedev Phys. Inst.* **27**, 109 (1965).
5 J. C. Briquet, *Mémoire de D. E. A. Paris*, 1971.
6 I. I. Kondilenko, V. E. Pogorelov et Khunkhue, *Opt. Spectry.* **28**, 367 (1970).
7 R. Levant, *Thèse de 3ème Cycle, Paris*, 1972.

55. APPLICATION DE LA SPECTROMETRIE RAMAN A L'ETUDE DES FLUCTUATIONS ORIENTATIONNELLES A L'ETAT LIQUIDE

P.-B. Caloine, J. Devaure, M. Perrot et J. Lascombe
Laboratoire de Spectroscopie Infrarouge,
Talence, France

SUMMARY

La spectrométrie Raman se prête bien à l'étude précise des fluctuations orientationnelles à l'état liquide si l'on considère des molécules pour lesquelles le couplage vibration–rotation est négligeable. En effet, dans ces conditions, la diffusion Raman de trace traduit seulement les fluctuations d'états vibrationnels tandis que la diffusion anisotrope est influencée par les fluctuations vibrationnelles et orientationnelles. Bratos et Marechal[1] ont montré que la déconvolution de la bande anisotrope par la bande isotrope permettait d'obtenir la fonction de corrélation orientationnelle

$$G_R(t) = \langle \tfrac{3}{2} \cos^2 \theta - \tfrac{1}{2} \rangle$$

Cette théorie a été appliquée à la triméthylamine à l'état liquide pour diverses températures et en solution dans divers solvants. Les fonctions de corrélation et les temps de corrélation seront discutés.

Pour les molécules plus petites, les effets d'interaction vibration–rotation peuvent ne plus être négligeables et entraîner un couplage des fluctuations vibrationnelles et orientationnelles; s'il en est ainsi la séparation devient difficile. Dans le but de préciser l'influence de cette interaction sur la partie isotrope des profils Raman, on a étudié les bandes de HCl à l'état gazeux et en solution. On conclut que pour les solutions, l'effet de l'interaction vibration–rotation est probablement négligeable et en tout cas beaucoup plus faible qu'à l'état gazeux, par suite d'un 'rétrécissement dynamique' de la bande.[2]

References

1 S. Bratos et E. Marechal, *Phys. Rev.* **A4**, 1078 (1971).
2 A. D. May, J. C. Stryland et G. Varghese, *Can. J. Phys.* **48**, 2331 (1970).

56. RAMAN LINESHAPES IN LIQUID CH_3I AND CD_3I*

H. Goldberg† and P. S. Pershan§

Gordon McKay Laboratory, Harvard University, Cambridge, Massachusetts, USA

INTRODUCTION

The Raman bands of a liquid can, in principle, provide much information on the vibrational and rotational motions of molecules in the liquid phase. However, a vital step in the analysis of the data is the separation of the vibrational and rotational effects. Experimentalists have, until recently, tended to skirt the problem by considering one or the other effect to be negligible. Unfortunately, in many cases the effects are of comparable magnitude. If one assumes that the vibrations and rotations are independent of each other then, at least for certain bands, an exact separation can be made by using the polarized and depolarized components of the Raman bands.[1,2] Virtually all of the analyses of the Raman and infrared data to date, both theoretical and experimental, have assumed the independence of the two effects in spite of the fact that it has long been recognized that rotation-vibration coupling plays a significant role in determining the line-shapes of bands in low density gases.[3,4] In this chapter we discuss the bands of CH_3I and CD_3I, emphasizing that important conclusions can be drawn about the rotation-vibration coupling as well as about the general nature of the molecular dynamics without relying on a detailed model of the rotation and vibration.

EXPERIMENTAL

The point group symmetry of CH_3I is C_{3v}. There are six vibrational modes, three totally symmetric (A_1), three non-totally symmetric, doubly degenerate (E). All modes are both Raman and infrared active. In Fig. 1 we show the complete Raman spectrum of methyl iodide. Figure 2 shows the analysis of the ν_3 (C–I stretch) band to illustrate several important features. Doge [5] and Bulanin [6] have suggested that the asymmetry of the line is due to the presence of hot bands. In order to separate the fundamental from the hot band, we assume that the lineshape due to the hot band is the same (but at this stage

---

* This work was supported in part by the Joint Services Electronics Program (U.S. Army, U.S. Navy and U.S. Air Force) under Contract No. N00014-67-A-0298-0006 and by the Division of Engineering and Applied Physics, Harvard University.

† Present address: Becton Center, Yale University, New Haven, Connecticut.

§ During the final preparation of this manuscript, partial support was provided by the National Science Foundation, Senior Postdoctoral Fellowship No. 51012 and the Université Paris-Sud, Laboratoire de Physique des Solides.

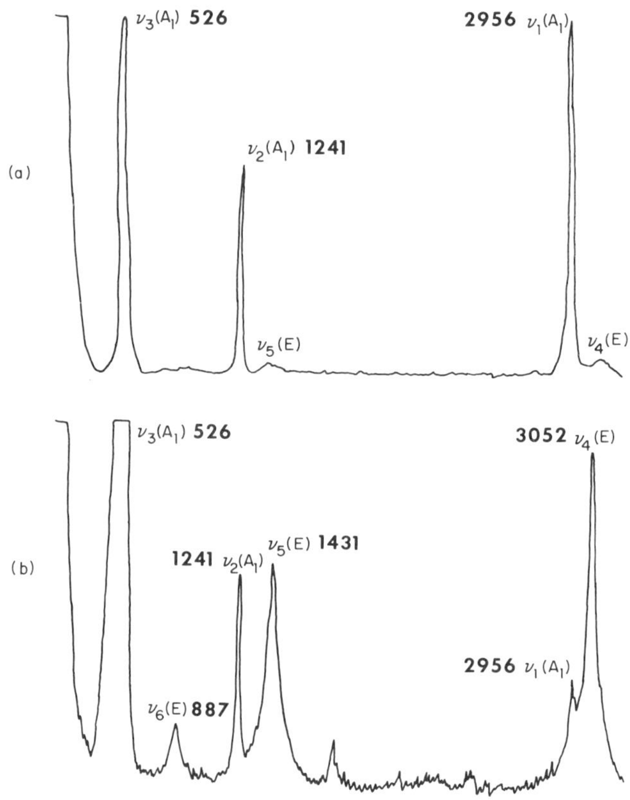

Fig. 1. Raman spectrum of pure liquid CH_3I. (a) polarized. (b) depolarized.

still unspecified) as that of the fundamental, and we numerically fit the observed spectrum $S^{obs}(\omega)$ by a calculated spectrum $S^{calc}(\omega)$ given by

$$S^{calc}(\omega) = S(\omega) + fS(\omega - \Delta\omega)$$

Here $S(\omega)$ represents the fundamental lineshape, $\Delta\omega$ the shift of the hot band from the fundamental and f the relative intensity of the hot band. The values of f obtained correspond well to the theoretical value.[7] The deviation of $S^{calc}(\omega)$ from $S^{obs}(\omega)$ in the wing of the polarized band occurs near $2\Delta\omega$, where the next hot band is expected.

Having removed the hot band, the procedures discussed in refs. 1 and 2 were followed to obtain the rotational and vibrational components and the corresponding correlation function of the A_1 bands. The vibrational lines were found to be quite well described by Lorentzians. The half widths at half maxima (HWHM) are given in Table 1 and are correct to about ± 15%. Note that the widths of the CD_3I bands are narrower than the corresponding bands of CH_3I. Zubova *et al.*[8] have observed the same phenomenon in the pair of molecules $CHCl_3$ and $CDCl_3$. Since one assumes that apart from the small mass difference the molecules are identical, the origin of the sizeable differences in the linewidths remains a mystery. The fact that the linewidths in the solutions are narrower

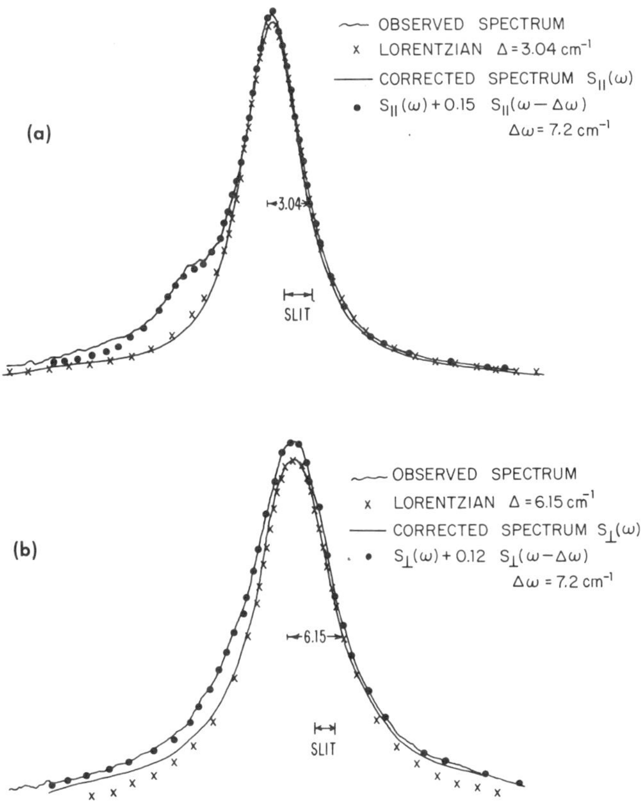

Fig. 2. Analysis of C–I stretch (ν_3, A_1). (a) polarized. (b) depolarized.

Table 1. Vibrational half widths (HWHM) of A_1 bands (cm^{-1}).

	present	CH_3I ref. 21	ref.22	CD_3I	1% CH_3I in CS_2	1% CD_3I in CS_2
ν_1	2.2	–	–	1.2	2.0	0.9
ν_2	2.2	–	2.6	1.9	1.5	0.8
ν_3	2.2	2.1	2.6	1.8	1.5	1.2

than those in the pure liquid follows from the fact that hydrogen bonding effects and 'resonant' coupling of equivalent vibrations on neighbouring CH_3I molecules are essentially eliminated in the solution as sources of broadening. Because of the absence of any good theoretical model for the vibrations, we turn to a discussion of the rotational phenomena.

The rotational lineshapes fit a Lorentzian in the centre of the band but tend to fall off more slowly than $1/\omega^2$ in the wings. The HWHM of the rotational lineshapes are listed in Table 2. From the table it appears that, within experimental uncertainty, the ν_2 width equals the ν_3 width in each of the systems where sufficient data are available. The two ν_1 widths on the other hand seem to be higher than the corresponding ν_2 and ν_3

Table 2. Rotational half widths (HWHM) of A_1 bands (cm^{-1}).

	CH_3I present	CH_3I ref. 21	ref. 22	CD_3I	1% CH_3I in CS_2	1% CD_3I in CS_2
ν_1	7.0	–	–	4.4	–	–
ν_2	4.8	–	3.8	3.2	–	7.5
ν_3	3.8	3.0	3.5	3.5	5.6	6.5

widths. Bulanin and Tonkov[6] have measured the ν_1 and ν_2 infrared bands of CH_3I in CCl_4 and have found that the former is broader than the latter by about 15%. The significance of the different widths is the following. It can be shown that if there is no rotation-vibration coupling then the rotational lineshapes (both infrared and Raman) of the A_1 bands must be identical to one another.[1] The fact that there is a discrepancy in the widths therefore implies that rotation-vibration coupling is present. This conclusion is somewhat surprising for the following reason. It has long been known that the Coriolis force can introduce strong coupling effects into the E modes because these vibrations are doubly degenerate. On the other hand, one expects that for the A_1 modes, being non-degenerate, the Coriolis perturbation should appear as a small higher order correction to the band shape. This in fact is the case, except in the particular situation where a weak A_1 band is near a strong E band. In this instance a very strong perturbation in the band shape can occur via a Coriolis coupling of the A_1 mode to the E mode.[9–11] (di Lauro and Mills[9] show a particularly dramatic example of the opposite situation in CH_3F, where a weak E band is greatly modified by a neighbouring strong A_1 band.)

As Fig. 1 indicates, the ν_1 bands of CH_3I and CD_3I, with their small depolarized components and proximity to the relatively intense ν_4 band, fulfill the conditions for a significant A_1–E type Coriolis coupling. The ν_2 and ν_3 bands on the other hand do not satisfy the conditions and so we do not expect this type of rotation-vibration coupling to be present. Clearly then one should look to these latter bands for further information on the molecular dynamics.

In particular we note that the rotational lines are symmetrical, i.e. have no Boltzmann asymmetry. This implies that a classical model should be adequate to describe the rotational motions of the molecules.[12] The rotation in this case is about an axis perpendicular to the symmetry axis because the A_1 bands are sensitive only to this motion.[1, 2] Previous studies[13, 14] of CH_3I and similar molecules have proposed that the simple rotational diffusion model[15] is the appropriate one for the description of the axial tumbling motion. This model predicts that the Raman rotational line is Lorentzian with a half width given by $6D_\perp$ where $D_\perp$ is the tumbling rotation diffusion constant. On the basis of the theory, one can also calculate values of $D_\perp$ from infrared, nuclear relaxation and microwave data, etc. In Table 3, we list a number of values of $D_\perp$ for pure methyl iodide obtained from various experiments [where $D_\perp(s^{-1}) = 2\pi c D\,(cm^{-1})$]. In the table ν_0 refers to pure rotation as opposed to the vibration-rotation bands. One measures ν_0 in experiments such as Rayleigh scattering, microwave and infrared absorption, and nuclear magnetic relaxation. The theoretical value given in the last column is based on a modified Debye–Stokes hydrodynamical model and was computed from the formula given in ref. 16. The general consistency of these various values of $D_\perp$ appears to substantiate both the use of the rotational diffusion model and the fact that for the ν_2 and ν_3 modes, rotation-

Table 3. Values of $D_\perp$ for CH_3I ($\times 10^{11}\,s^{-1}$).

	present	ref.21	ref.22	ref.23	ref.24	theory[16]
ν_0	–	1.0^a	1.6^b	–	$1.2–3.5^c$	
ν_3	1.2	1.1^d	1.0^d	1.4^e	–	1.2
ν_2	1.5	1.2^d	–	–	–	
ν_1	2.2	–	–	–	–	

^{a} Rayleigh
^{b} Far infrared
^{c} Dipolar relaxation
^{d} Raman
^{e} Near infrared

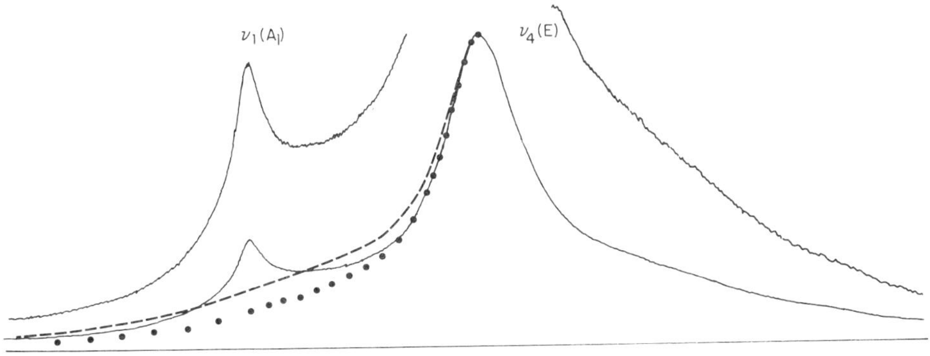

Fig. 3. Depolarized Raman spectrum of $\nu_4(E)$ band. – – – Stokes side of band reflected at band centre. • • • reflection of Stokes side multiplied by Boltzmann factor.

vibration coupling is absent. Further, the deviation of $D_\perp$ for the ν_1 mode again suggests that coupling is significant in this case.

We turn now to the E bands. Figure 3 shows the depolarized ν_4 band of CH_3I along with the depolarized component of the $\nu_1(A_1)$ band. To test for Boltzmann asymmetry we show superimposed on the anti-Stokes (left) side of the band the reflection of the Stokes side as well as the result of multiplying the Stokes side by the Boltzmann factor. In spite of the presence of the ν_1 band, it is readily apparent that the band exhibits Boltzmann asymmetry. We may immediately conclude that a purely classical model of the molecular dynamics will be inadequate to describe the E bands in contrast to the case of the A_1 bands. It may be shown[1,2] that the E bands reflect both the axial tumbling rotation and the spinning of the hydrogen atoms about the symmetry axis. Since we have already established that the former motion can be explained classically, it follows that the 'methyl group spinning' shows quantum behaviour. Physically this is reasonable because the moment of inertia $I_\parallel$ about the symmetry axis is approximately 20 times smaller than that about the perpendicular axis,[17] and hence the free rotational velocity (which is inversely proportional to $\sqrt{I_\parallel}$) for spinning is nearly 5 times that for tumbling. For a given collisional frequency, then, it is clear that a molecule will 'spin' through a much larger angle before interruption than it will 'tumble' and hence the former motion will look more like that of a free rotor than will the latter. We have observed that there is little difference in the widths of the E bands of the pure liquid when compared to those in the solutions. This is consistent with a relatively unhindered rotation independent of the environment. While other workers[13, 18] have concluded that the spinning rotation of methyl iodide and similar molecules is quasi-free, we believe that the examination for Boltzmann asymmetry discussed here is the simplest and most direct method for arriving at that conclusion. Note that the smaller breadth of Raman bands as compared to the corresponding infrared bands greatly facilitates the observation of Boltzmann asymmetry.

Because the point group of CH_3I is C_{3v}, it can be shown[1] that one cannot simply compare the rotational widths of the Raman E bands in the same manner as the Raman A_1 bands in order to test for rotation-vibration coupling (even if one could separate out the vibration widths). However, the infrared E bands can be shown to behave like the Raman A_1 band in that the absence of rotation-vibration coupling implies that for all of the E bands rotational lineshapes must be identical. Bulanin and Tonkov[6] have measured the infrared E bands of CH_3I and CD_3I in solutions with CS_2 and CCl_4. If one examines their data, it is clear that the half widths of these bands are not equal, and further that the differences are on the order of 10–20 cm^{-1}, which are too large to be reasonably attributed to vibrational effects. Hence we must conclude that in fact rotation-vibration coupling is present. This conclusion is consistent with the quasi-free nature of the spinning rotation in that it is found that there is an appreciable Coriolis effect in gaseous CH_3I. Jones and Thompson[19] find $+ 0.059, - 0.240$ and 0.206 for the values of the Coriolis coupling constant ζ for the ν_4, ν_5, and ν_6 bands respectively, where $|\zeta| \leqslant 1$ and $\zeta = 0$ implies no rotation-vibration interaction. We note that the ν_4 band is likely to be the one most amenable to quantitative study of the rotational dynamics since its small value of ζ indicates that at least in a first approximation we can ignore the Coriolis coupling.

Finally we mention the fact that the depolarization ratio, ρ, of an E band may contain some useful information on interactions. Ideally $\rho = 0.75$. Deviation from this value indicates that some disturbance of the symmetry has occurred, for example, by coupling to an A_1 mode because of a distortion of the molecule in the liquid. We have measured the depolarization ratio of the ν_5 band in pure CH_3I and have found that $\rho = 0.72 \pm 0.01$. Hence there is evidence for some small symmetry perturbation.[20]

CONCLUSIONS

We have found that a comparison of the bands of the same symmetry reveals the significance of rotation-vibration coupling in the different Raman bands of liquid methyl iodide and methyl-d_3 iodide. Further, we have been able to identify the bands which appear to be essentially free of such coupling and thus the best for detailed analysis. Additionally we have pointed out that the presence or absence of Boltzmann asymmetry in the lineshapes is a simple indicator as to the importance of quantum effects in the molecular dynamics.

REFERENCES

1 H. S. Goldberg, *Thesis*, Department of Engineering and Applied Physics, Harvard University, (unpublished), 1971.

2 F. J. Bartoli and T. A. Litovitz, *J. Chem. Phys.* **56**, 404 (1972).

3 G. Herzberg, *Molecular Spectra and Molecular Structure, Vol. II, Infrared and Raman Spectra of Polyatomic Molecules*, Van Nostrand, Princeton, 1945.

4 W. F. Edgell and R. E. Moynihan, *J. Chem. Phys.* **45**, 1205 (1966).

5 G. Doge, *Z. Naturforsch.* **23a**, 1405 (1968).
6 M. O. Bulanin and M. V. Tonkov, *Opt. Spektrosk.* **16**, 429 (1964); [*Opt. Spectry.* **16**, 234 (1964)].
7 G. Placzek, *The Rayleigh Raman Scattering*, UCRL Trans. 526(L), Clearinghouse for Federal and Scientific Information, Dept. of Commerce, 1959.
8 N. V. Zubova, N. E. Shalomeyeva, V. S. Gorelik and M. M. Sushchinskii, *Opt. Spektrosk.* **27**, 935 (1969); [*Opt. Spectry.* **27**, 508 (1969)].
9 C. di Lauro and I. M. Mills, *J. Mol. Spectry.* **21**, 386 (1966).
10 R. L. Dilling and P. M. Parker, *J. Mol. Spectry.* **22**, 178 (1967).
11 W. E. Blass, *J. Mol. Spectry.* **31**, 196 (1969).
12 R. G. Gordon, *Advan. Magn. Res.* **3**, 1 (1968).
13 T. T. Bopp, *J. Chem. Phys.* **47**, 3621 (1967).
14 J. Jonas and T. M. diGennaro, *J. Chem. Phys.* **50**, 2392 (1969).
15 E. N. Ivanov, *Sov. Phys. JETP* **18**, 1041 (1964).
16 D. E. Woessner, B. S. Snowden, Jr., and E. T. Strom, *Mol. Phys.* **14**, 265 (1968).
17 J. R. Lyerla, Jr., D. M. Grant and C. H. Wang, *J. Chem. Phys.* **55**, 4676 (1971).
18 W. B. Moniz, W. A. Steele and J. A. Dixon, *J. Chem. Phys.* **38**, 2418 (1963).
19 E. W. Jones and H. W. Thompson, *Proc. Roy. Soc.* **A288**, 50 (1965).
20 E. Priestly, Sarnoff Research Center, RCA, Princeton, N.J., (private communication).
21 F. J. Bartoli and T. A. Litovitz, *J. Chem. Phys.* **56**, 413 (1972).
22 M. Constant and M. Fauquembergue, *Compt. Rend.* **272**, B1293 (1971).
23 C. E. Favelukes, A. A. Clifford and B. Crawford, Jr., *J. Phys. Chem.* **72**, 982 (1968).
24 J. R. Lyerla, Jr., D. M. Grant and R. K. Harris, *J. Chem. Phys.* **75**, 585 (1971).

57. INTERMOLECULAR ENERGY TRANSFER IN LIQUID BENZENE: RAMAN LINE SHAPES, LINE WIDTH MEASUREMENTS AND VIBRATIONAL RELAXATION TIMES

J. E. Griffiths
Bell Laboratories, Murray Hill, New Jersey, USA

INTRODUCTION

In spontaneous Raman scattering in liquids, the important directly measureable parameters include the Raman displacement in cm^{-1}, the total and peak radiant intensities I_i and I_p respectively, the depolarization ratio ρ_s, the halfwidth of the band $\nu_{1/2}$(FWHH) and the detailed shape of the band profile. The present work focuses attention upon the line width parameter of the totally symmetric $\nu_2(a_{1g})$ fundamental of benzene in the liquid state from which evaluation of the mean lifetime τ_v of molecules in the upper state of that transition is possible. Furthermore, it is possible to influence the magnitude of the line width and hence the lifetime of the excited state in order to obtain additional insight into the dynamics and mechanisms of intermolecular interactions in the liquid state.

VIBRATIONAL RELAXATION

In a system of polyatomic molecules of the same kind in the gas phase at low pressure, rapid *intra*molecular vibration to vibration energy transfer maintains a continuous equilibrium of vibrational energy among the fundamental modes. The total vibrational energy ultimately suffers a vibration to translation relaxation process generally via the lowest energy vibrational mode. Ultrasonic dispersion techniques are commonly used to study this phenomenon. At increased pressures in the gas phase or in the liquid state, however, the time between collisions becomes extremely short and intermolecular vibration to vibration energy transfer processes may dominate the energy cascade.

Suppose that in a hypothetical collision, an energy level (E_i) of molecule A is close in energy to one labelled E_j in molecule B. Then, the cross section for energy transfer σ_{ij} in this near resonance situation will be considerably higher than if the levels were far apart. A consequence of such an energy transfer is that the mean lifetime τ_i of molecule A in its particular energy level E_i will be shorter than it would be if $E_i - E_j$ were large.

One way of obtaining further information about this phenomenon and the lifetime is through the line width parameter $\nu_{1/2}$ which is related to τ by the equation

$$\tau_i = (\pi c \nu_{1/2})^{-1}$$

Table 1. Properties of C_6H_6 and C_6D_6.

Property	C_6H_6	C_6D_6
Line width of ν_2	2.24 cm^{-1}	1.62 cm^{-1}
Energy mismatch $\nu_2-\nu_7$	6 cm^{-1}	115 cm^{-1}
$\nu_2-\nu_{19}$	11 cm^{-1}	150 cm^{-1}
$\nu_2-\nu_6$	− 18 cm^{-1}	− 20 cm^{-1}
$\nu_2-\nu_{17}$	− 186 cm^{-1}	80 cm^{-1}
Density ($\rho(H_6)-\rho(D_6)$)	0	0
Refractive index ($n(H_6)-n(D_6)$)	0	0
Chemical difference	0	0
Symmetry or normal coordinate difference	0	0

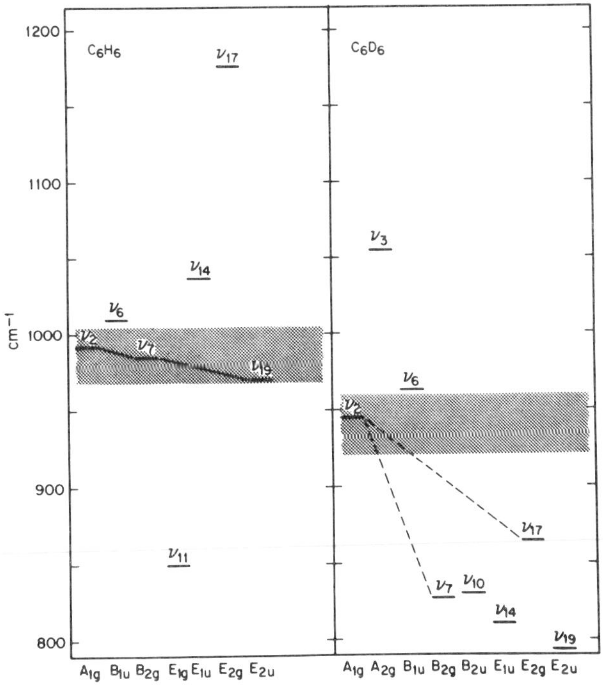

Fig. 1. Partial vibrational energy level diagram for C_6H_6 and C_6D_6.

This equation is valid if the overall line shape is Lorentzian and this will be confirmed later for the system studied.

Benzene was selected as our model system for a number of reasons. These are listed in Table 1. Note the large difference in the values for the line widths.

In Fig. 1, a partial energy level diagram is reproduced to show the relative positions of each level and clearly shows the degree of energy mismatch.

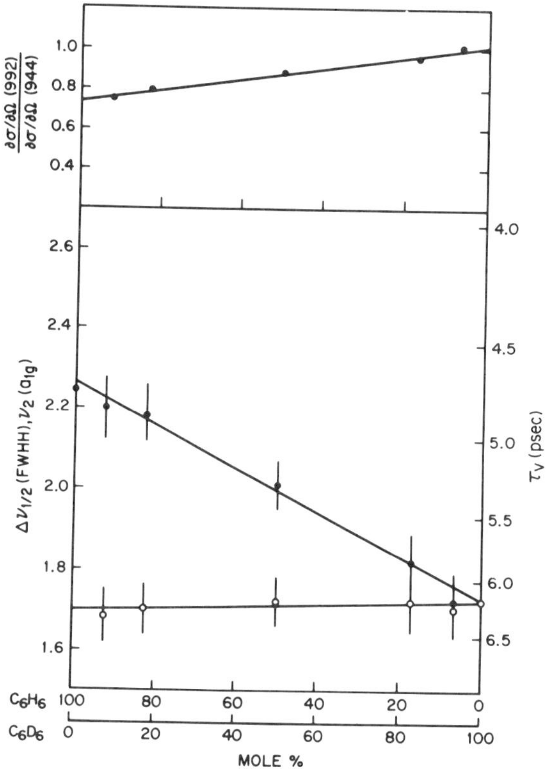

Fig. 2. Variation in line width and vibrational state lifetime for the $\nu_2\,(a_{1g})$ fundamentals with concentration in $C_6H_6-C_6D_6$ mixtures (lower graph). Variation in total differential cross section ratio for the 992 and 945 cm^{-1} bands with concentration in $C_6H_6-C_6D_6$ mixtures (upper graph).

The line widths of ν_2 in C_6H_6 and C_6D_6 are clearly quite different and suggest that a line broadening or line narrowing mechanism is operating. Since chemical and physical properties are similar for the two molecular species and since the energy level diagram suggests that an intermolecular energy transfer occurs in C_6H_6 and not in C_6D_6, we can seek to find further evidence by probing the line width parameter.

Intermolecular energy transfer will shorten the lifetime of the molecules in the ν_2' state and hence will broaden the line. At the other extreme at infinite dilution of light benzene in heavy benzene no $C_6H_6-C_6H_6$ collisions will occur and it follows that

$$\tau(\nu_2', C_6H_6) = \tau(\nu_2', C_6D_6)$$

Furthermore, at intermediate concentrations of C_6H_6 in C_6D_6, where less $C_6H_6-C_6H_6$ collisions occur than in pure C_6H_6, the line width parameter should be smaller than in pure C_6H_6 but larger than that of C_6D_6.

If intermolecular collisions between C_6H_6 and C_6D_6 are important in energy transfer, then the line width $\nu_{1/2}$ for C_6D_6 should increase with increased $C_6H_6-C_6D_6$ collisions. If energy transfer during $C_6H_6-C_6D_6$ collisions are unimportant because the energy mismatch term ($\sim$ 46 cm^{-1}) is too large, the line width of C_6D_6 will be concentration independent.

Table 2. Intermolecular energy transfer processes.[a]

1.	$C_6H_6(\nu_2') + C_6H_6(\nu_7) \rightarrow C_6H_6(\nu_2) + C_6H_6(\nu_7') + \Delta E_{2,7} = 6\ \mathrm{cm}^{-1}$
2.	$C_6H_6(\nu_2') + C_6H_6(\nu_{19}) \rightarrow C_6H_6(\nu_2) + C_6H_6(\nu_{19}') + \Delta E_{2,19} = 11\ \mathrm{cm}^{-1}$

[a] Primes denote the upper state ($\nu = 1$) and no prime above the ν denotes the lower state ($\nu = 0$).

Our data are shown in Fig. 2 and the observations are tabulated below:

1. Line width of C_6H_6 decreases with decreasing concentration.
2. Line width of C_6D_6 is concentration independent.
3. At zero C_6H_6 concentration, the line widths of C_6H_6 and C_6D_6 are equal.

The conclusions reached from these observations are that:

1. Benzene (C_6H_6) suffers a strong intermolecular energy transfer process or processes in the liquid state (Table 2).
2. C_6H_6–C_6D_6 and C_6D_6–C_6D_6 collisions are unimportant in intermolecular energy transfer processes.
3. An energy mismatch of $< 30\ \mathrm{cm}^{-1}$ is necessary for intermolecular energy transfer to be important.
4. The energy mismatch term must be positive for intermolecular energy transfer processes to be important.

Although we have demonstrated this principle for the ideal case in mixtures of C_6H_6 and C_6D_6, the phenomenon is expected to be a general one. In fact we have examined other systems using energy levels greater than the 991 cm^{-1} line of C_6H_6 and used the latter level as a sink to receive energy. The process was so efficient in near resonant cases that the line width of C_6H_6 was increased with dilution instead of being decreased.

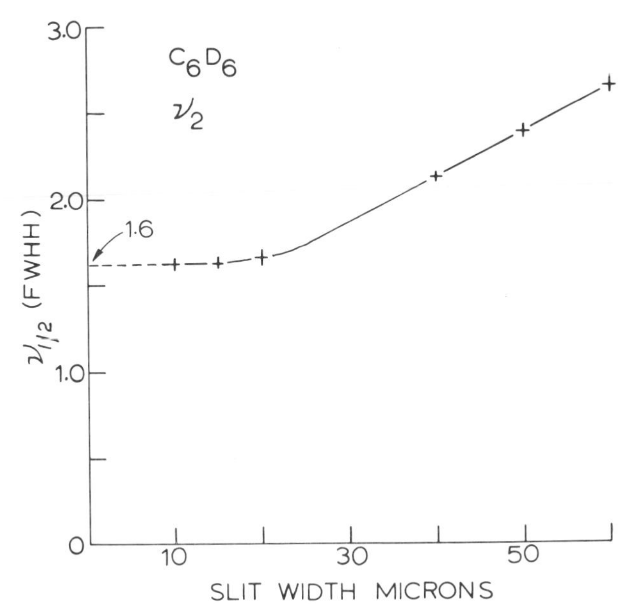

Fig. 3. Determination of line width (FWHH) of $\nu_2\,(a_{1g})$ in C_6D_6.

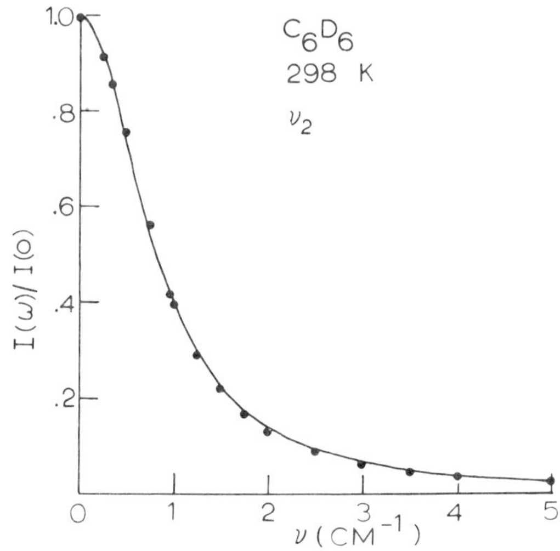

Fig. 4. Lorentz line shape of $\nu_2\,(a_{1g})$ in C_6D_6. Solid line is a calculated Lorentz function for $\nu_{1/2}$ = 0.8 cm^{-1} (HWHH) and circles are experimental points.

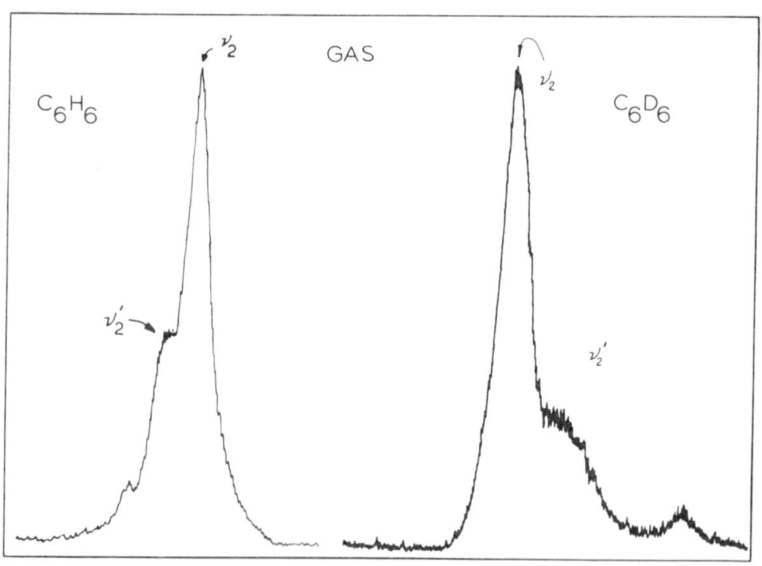

Fig. 5. Gas phase Raman spectra of $\nu_2\,(a_{1g})$ for C_6H_6 and C_6D_6 at $\rho \sim 30$ torr.

Furthermore, selection rules based upon symmetry do not appear to apply, at least in the systems studied so far.

EXPERIMENTAL DETAILS

In concluding it is important to mention briefly a few experimental details which have not yet been covered. These are briefly as follows:

1. Line widths were measured directly as a function of mechanical slit widths. An example is shown in Fig. 3.
2. Line shapes were shown to be Lorentzian at narrow slit widths where the slit function does not contribute significantly to the line profile (Fig. 4).
3. 90° scattering was used in all cases.
4. Excitation wavelengths were 4880 and 5145 Å.
5. Line shape of the ν_2 fundamentals in C_6H_6 and C_6D_6 are significantly different (Fig. 5). Anharmonicity corrections are approximately equal in magnitude but opposite in sign and reflect a fundamental difference in the shape of the vibrational potential well. Assignments of the hot bands were confirmed by the direct measurement of ν_2 ($v = 0 \rightarrow 1, 1 \rightarrow 2, 0 \rightarrow 2$). Figure 5 is for the gas phase and similar shapes are observed for the liquid phase although resolution of the hot bands is more clearly defined for the gas.

58. ROTATIONAL RAMAN BANDS IN SIMPLE FLUIDS

P. van Konynenberg and W. A. Steele
Department of Chemistry, Pennsylvania State University, University Park, Pennsylvania, USA

EXPERIMENTAL

Measurements of the Raman scattering from several molecular liquids were made using a 100 mW He–Ne laser polarized perpendicular to the scattering plane.[1] The laser light was focused into the sample and light scattered at a right angle to the incident beam was passed through an analyser oriented to transmit the depolarized component. The light was analysed with a Spex Model 1400 double monochromator operating in second order with gratings blazed for 10000 Å and the signal was detected with a cooled FW130 photomultiplier with S-29 cathode and photon-counting electronics.

A low temperature cell was machined from brass and was wrapped with cooling coils through which cold nitrogen (gas or liquid) could be circulated for attaining low temperatures. The cell was enclosed in a vacuum jacket designed to allow for independent alignment of the cell with the focused laser beam. Several features were incorporated into the cell design to reduce stray reflected light: the focused incident light passed through two small diaphragms placed between the entrance window and the scattering volume; after passing through the scattering volume, the beam subsequently passed through two more small diaphragms and was trapped in a tipped cone. Another large tipped cone was present opposite the viewing window. The entire interior was sandblasted and electrolytically finished flat black.

Reagent grade gas samples were first condensed into a small storage cylinder and then slowly distilled through a filter (Millipore, 0.32 μm) into the sample cell at a high vapour pressure. The first few samples were suddenly released into the atmosphere until the cell was free of dust. Spectra were obtained at slit widths corresponding to 1.5–2.0 cm^{-1} resolution at a scanning rate of 2 cm^{-1} per min with a time constant of 3 s. The laser power was continuously monitored and recorded during each spectrum; no power corrections were necessary. Because the liquid spectra were so broad, no slit width corrections were necessary.

The measured spectra consisted of smooth bands which appear to approach zero frequency shift with zero slope. However, a small sharp peak of width equal to the spectral slit width centred at zero frequency shift was always superimposed on the rotational spectra; despite careful considerations of scattering geometry and the use of

Table 1. Spectral moments.

System		$M(0)$ (cm^{-1}) Stokes	anti-Stokes	$M(1)$ (cm^{-1}) Stokes	anti-Stokes	$M(2)$ (cm^{-1}) Stokes	anti-Stokes
C_2H_6	180 K	38.8	32.5	19.0	−14.9	1184	1090
	225 K	41.5	33.9	18.9	−13.4	1153	737
	250 K	42.3	34.2	18.6	−13.3	1203	730
C_2H_4	170 K	22.4	17.9	15.8	− 8.0	960	675
	200 K	35.6	27.9	19.3	−12.5	1269	711
	230 K						
	270 K	52.1	42.1	22.3	−15.8	1558	990
CO_2	236 K	13.5	11.7	10.6	− 7.1	546	276
	258 K	18.9	17.1	11.5	− 9.2	622	431
	281 K	23.1	21.1	12.7	−10.2	705	489

long focal length optics, this feature was always present and was undoubtedly due to stray light. In the calculation of correlation functions this peak was eliminated by smooth extrapolation.

The systems studied are listed in Table 1 together with several spectral moments calculated separately for the Stokes and anti-Stokes branches of the experimental curves. In all cases, the scattered intensity at zero shift was normalized to unity. Consequently, the zero'th moment $M(0)$ is

$$M(0) = M_S(0) + M_A(0) = \int_{\text{band}} I(\omega)\, d\omega \tag{1}$$

where $I(\omega)$ denotes the spectral intensity at a frequency shift of ω wave numbers, and $M_S(0)$ and $M_A(0)$ denote the Stokes and anti-Stokes parts of the moment. The nth higher moment is given by

$$M(n) = M_S(n) + M_A(n) = \frac{1}{M(0)} \int_{\text{band}} \omega^n I(\omega) d\omega \tag{2}$$

The results listed in Table 1 clearly indicate that all spectra measured in this work are asymmetric; that is, the Stokes and anti-Stokes branches differ appreciably in width and shape and consequently give different moments.

In order to interpret these spectra, it is convenient to present them as Fourier transforms. As is well known, time-correlation functions[2] are obtained in this way which contain information concerning the time-dependence of the orientation of the axis of optical anisotropy of the molecule responsible for the scattering. However, the simple Fourier transform valid for ensembles of classical rotors does not give the desired correlation function when the quantum effects that cause asymmetry in the Raman bands are present. Several of the quantum time-correlation functions that have been suggested[3] can be calculated from an equation of the form

$$C_2(t^*) = \frac{1}{M(0)} \int_{\text{band}} f(\hbar\omega/kT)\, I(\omega^*) \exp(i\omega^* t^*)\, d\omega^* \tag{3}$$

where t^*, ω^* denote the time and angular frequency shift in reduced units obtained by

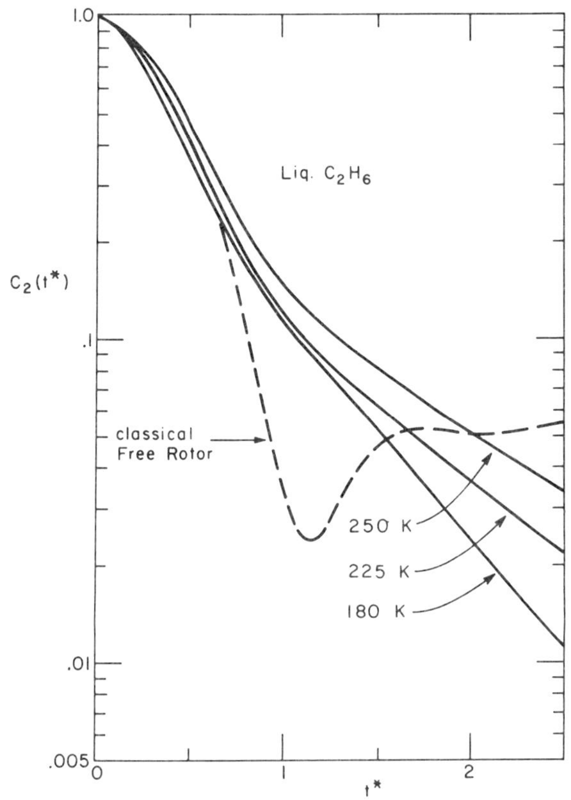

Fig. 1. Correlation functions for liquid ethane; the dashed line is the theoretical curve for freely rotating ethane molecules.

multiplying or dividing the unreduced variables by $\sqrt{kT/I}$, respectively (I is here the principal moment of inertia of the molecule). The quantum correction factors and the correlation functions that result include

the symmetrized correlation function $$f(\hbar\omega/kT) = \tfrac{1}{2}\,(1 + \exp(-\hbar\omega/kT)) \tag{4}$$

the canonical correlation function $$f(\hbar\omega/kT) = (kT/\hbar\omega)\,(1 - \exp(-\hbar\omega/kT)) \tag{5}$$

the complex-time correlation function $$f(\hbar\omega/kT) = \exp(-\hbar\omega/2kT) \tag{6}$$

It is easy to show that these correlation functions are identical to order $(\hbar\omega/kT)^2$; furthermore the symmetrized and the canonical correlation functions can be shown to be real even functions of time to all orders of $\hbar\omega/kT$. This means that the sine transform part of Eqn. (3) must be identically zero when accurate spectra are transformed using the weights given by Eqns. (4) or (5). The complex-time correlation function differs from the symmetrized or the canonical correlation functions only if quantum effects of the order of $(\hbar\omega/kT)^2$ or higher are important. Each of the three correlation functions was

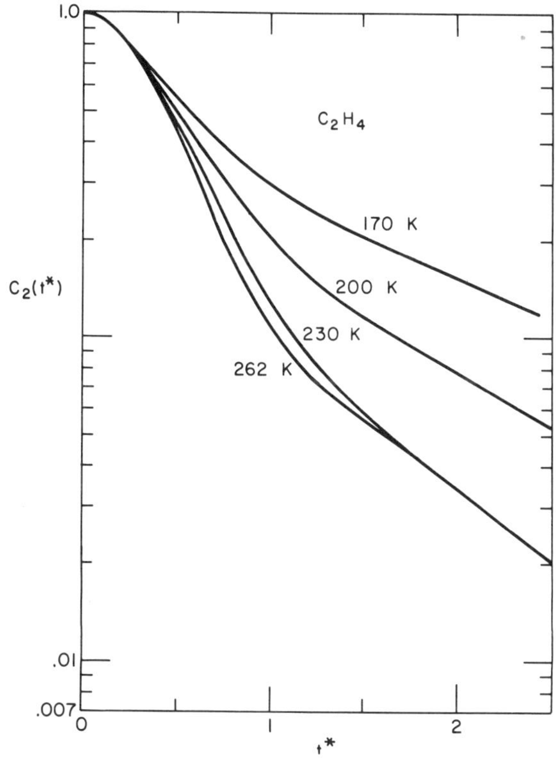

Fig. 2. Correlation functions calculated from the depolarized Raman scattering from liquid ethylene.

calculated by transforming the experimental spectra using Eqn. (3). It was found that all three were identical for all systems studied, within round-off errors of the transform calculation; furthermore, the sine transforms were all zero, to the same accuracy. Consequently, the cosine transforms presented in Figs. 1–3 represent quantum correlation functions which are related to molecular motions by

$$C_{\text{rot}}(t^*) = < 3/2 \cos^2 \delta\theta(t^*) - 1/2 > \tag{7}$$

where the brackets denote an ensemble average and $\delta\theta(t^*)$ is the reorientation angle of the optical axis in time t^*. In actual fact, the experimental correlation functions contain contributions due to collision-induced effects as well as rotational contribution; that is, time-dependent optical anisotropies due to molecular distortions while undergoing collision are superimposed upon the permanent anisotropy and give rise to Raman intensity. One can approximate the total time-correlation function by[1]

$$C_2(t^*) = f_{\text{rot}} C_{\text{rot}}(t^*) + f_{\text{coll}} C_{\text{coll}}(t^*) \tag{8}$$

where f_{rot} and f_{coll} denote the fractions of the total intensity due to the two factors, and $C_{\text{coll}}(t^*)$ is the normalized time-correlation of the collision-induced anisotropy. From

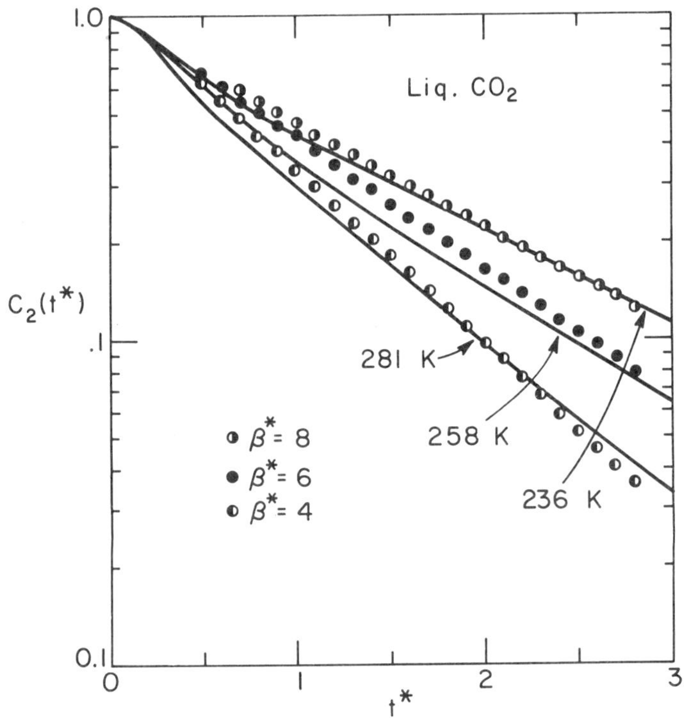

Fig. 3. Correlation functions for liquid carbon dioxide; the points show theoretical values calculated from the J-diffusion model.

other studies,[1] it has been deduced that f_{coll} is fairly small – roughly 0.1 to 0.2 for many molecules; furthermore, the time scale for the decay of these correlations is short compared with the time scale of the rotational correlations. Thus, if one focuses attention on the long-time decay of $C_2(t^*)$, one can be reasonably confident that this part of the curve is due to molecular reorientation. Indeed, the effects of collision-induced scattering can be seen most clearly in the values of the reduced spectral second moments $M^*(2)$ which are calculated from the values given in Table 1 by summing the Stokes and anti-Stokes parts and multiplying by $4\pi^2 c^2 I/kT$. For a classical ensemble of symmetric tops, $M^*(2) = 6$; when small quantum effects are present, one has[4]

$$M_{rot}^*(2) = 6[1 + 4\hbar^2/3IkT] \tag{9}$$

However, according to the approximation of Eqn. (8), one has

$$M^*(2) = f_{rot}M_{rot}^*(2) + f_{coll}M_{coll}^*(2) \tag{10}$$

where $M_{coll}^*(2)$ is generally greater than 6. Table 2 shows the experimental reduced second moments together with the reduced rotational moments calculated from Eqn. (9). It is evident that the collision-induced part of the spectrum causes $M^*(2)$ to increase by roughly 20% from the values for $M_{rot}^*(2)$, with the exception of the data for liquid ethane. It is possible that the disagreement between experimental and theory for this

Table 2. Reduced correlation times and second moments.

System	$M^*(2)$ Expt.	Theory rot.	τ_c^* Exponential slope	Zero'th moment	$\sqrt{I/kT}$ (ps)
C_2H_6, 180 K	13.71	6.08	0.64	0.54	0.412
225 K	9.12	6.07	0.94	0.60	0.369
250 K	8.40	6.06	1.17	0.67	0.350
C_2H_4 170 K	7.65	6.12	1.89	1.14	0.363
200 K	7.87	6.10	1.37	0.79	0.335
230 K		6.09	1.00	0.62	0.312
262 K	7.73	6.08	1.00	0.60	0.292
CO_2 236 K	6.41	6.04	1.46	1.41	0.469
258 K	7.51	6.03	1.20	1.03	0.448
281 K	7.82	6.03	0.94	0.88	0.430

system is due to errors in the measurements; the anomalous temperature dependence of the ethane correlation times relative to those for the other molecules in Table 2 is possibly also not a real effect. The intensities of the ethane spectra were appreciably less than any other substance studied due to the smaller optical anisotropy for ethane, and could well give rise to sufficient uncertainty, especially at large frequency shifts, to cause appreciable errors in the second moments of the spectra as well as in the zero'th moment of the correlation function. (Note that this reduced zero'th moment τ_c^* is equal to $I(0)/M^*(0)$, or $1/M^*(0)$ for the normalization used in this work.) Alternatively, the collision-induced scattering may be so large relative to the orientational part for ethane that it becomes the dominant factor in determining the time scale of the decay of the correlations.

THEORETICAL

In this section, the experimental curves will be compared with the predictions of several of the current theories of molecular reorientation in fluids. In order to do this, it will be assumed that quantum effects in excess of those accounted for in taking the transforms can be neglected. Thus, it is the general behaviour of the time-correlation functions of Figs. 1–3, particularly at $t^* > 0.5$, that is of interest. In addition, it will be assumed that the total correlation functions are indistinguishable from the rotational parts at the times of interest; that is, collision–induced effects can be neglected. Of course, the second moments listed in Table 2 indicate that this assumption results in appreciable error at short times; the detailed forms of the correct rotational correlation function may thus differ slightly from the curves in Figs. 1–3.

Theories of rotation in dense fluids can be divided into two general classes; random walk or jump models, in which reorientation is a Markov process; and dynamically coherent rotation that is perturbed or interrupted by the fluctuating intermolecular torques. Generally speaking, the random walk models give rise to correlation functions that are sums of exponential functions of time; for many of the simpler models, these sums contain only a single term

$$C_2(t^*) = \exp(-k\beta^* t^*) \tag{11}$$

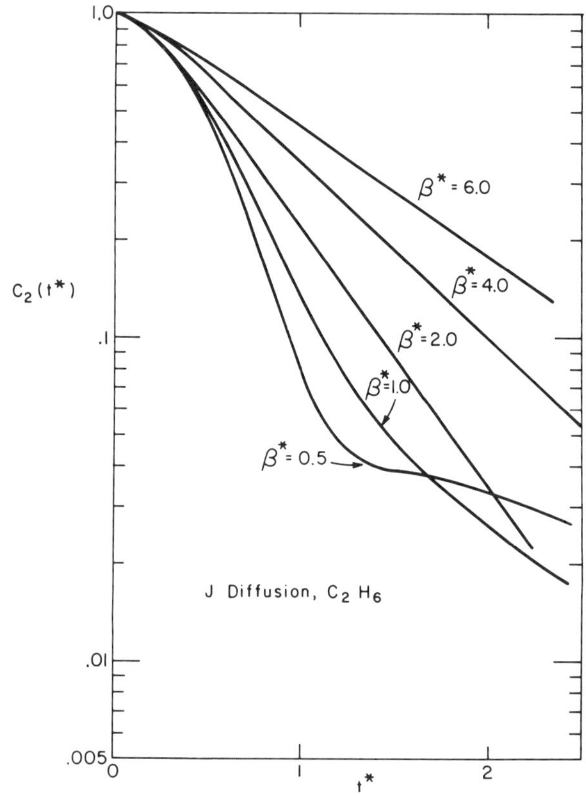

Fig. 4. Theoretical rotational correlation functions for liquid ethane calculated from the J-diffusion model.

where β^* is the reduced jump frequency and k is a constant that depends upon the detailed model. In Ivanov's 'large jump' model, $k = 1$;[5] in Ben Reuven's 'random jump' model, $k = 4/5$;[6] in the rotational diffusion model, k is equal to the mean square angular displacement in a jump. All these theories are based on the assumption that the reorientation occurs as a series of jumps, with the time between jumps being quite long compared with the time needed to make a jump. Since the time to make a jump through a finite angle of moderate size is of the same order of magnitude as that required for a freely rotating molecule to turn through the same angle, one estimates that the reduced jump durations will be of the order of $<\epsilon^2>^{-1/2}$, the inverse root mean square jump angle, and that β^* must then be large compared with $<\epsilon>^{1/2}$ if a real system is to be accurately represented by one of these Markoffian models. However, the theory is related to experiment by

$$k\beta^* = 1/\tau_c^* \tag{12}$$

and the τ_c^* values listed in Table 2 are of the order of unity. Thus, the experimental β^* obtained using one of the large jump models will be of the order of unity, which is not large compared with the r.m.s. jump angle unless one allows unphysically large jumps. The

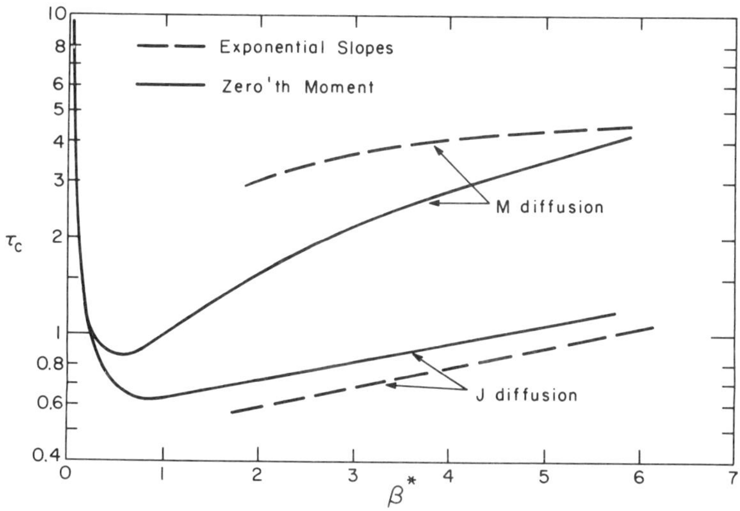

Fig. 5. Reduced correlation times calculated from the M- and J-diffusion models for liquid ethane.

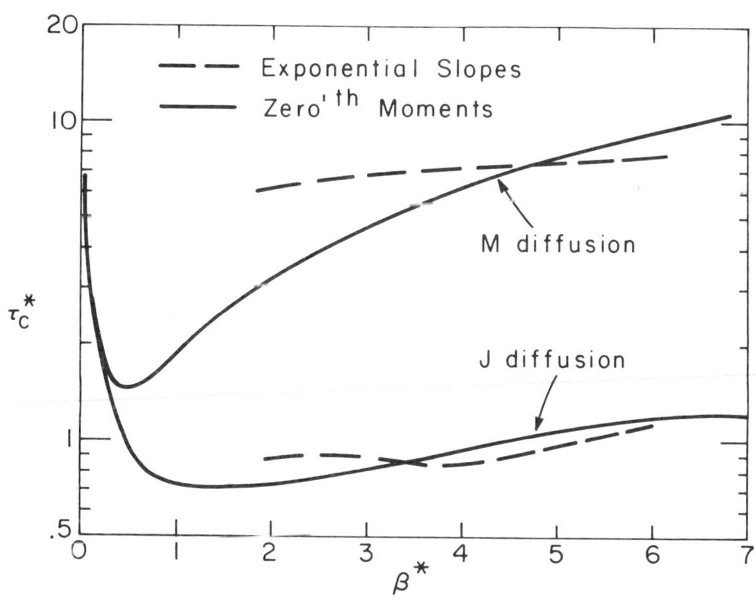

Fig. 6. Reduced correlation times calculated from the M- and J-diffusion models for a linear molecule.

rotational diffusion model can be rejected by a similar argument; in this case, $\beta^* = 1/\tau_c^* < \epsilon^2 >$, and is thus not small compared with $1/< \epsilon^2 >^{1/2}$ for infinitesimal jumps and the experimental correlation times.

Of course, the jump models considered are also defective in that they give correlation functions that disagree with most of the curves presented in Figs. 1–3. The models do not reproduce the Gaussian decays at short times; even at longer times where the experimental curves decay exponentially, the theory requires that the exponential decays extrapolate back to $C_2(0) = 1$, in disagreement with the data. It is perhaps unrealistic to expect that jump models will be valid when applied to fluids such as these. The temperatures are relatively high, approaching the critical temperature in some cases; furthermore, the angle-dependent parts of the intermolecular forces are not really large enough for these non-polar, symmetric molecules to cause the molecules to be confined in fixed orientations for long periods of time in the liquid.

A more attractive type of model for these systems is based on the idea that it is the angular momentum vector **L** that undergoes a random walk rather than the orientation itself.[7,8] In this case, a molecule rotates freely between 'collisions'; a 'collision' alters the angular momentum, but leaves the orientation unchanged; subsequently, the molecule again freely rotates (but with a different angular momentum) until the next 'collision' occurs. Two possibilities have been suggested for the effect of a 'collision' upon angular momentum: the M-diffusion model in which the orientation of **L** is randomized but the magnitude is unchanged; and the J-diffusion model, in which both the orientation and the magnitude of **L** are randomized. (Randomization here means a Maxwellian distribution for the magnitude of **L** and for the body-fixed L_z, in case of a symmetric top:) These models have the disadvantage that they do not lead to analytical expressions for the correlation functions of spectra; however, the computer calculations required are reasonably straightforward and lead to results such as those shown in Fig. 3–6. The theoretical correlation functions for reduced collision frequency β^* greater than one have the same form as the experimental curves – Gaussian decays at short times and exponential decays at long times. However, a more detailed comparison of theory and experiment is not completely successful. In the first place, the correlation times calculated for the M-diffusion model and plotted in Figs. 5 and 6 are always larger than the experimental values. Thus, it appears that there is no value of β^* that will bring this model into correspondence with any of the experimental systems. The situation is somewhat better for the J-diffusion model, as is shown in Fig. 3. Here, the theoretical curve for $\beta^* = 8$ agrees well with the data for $T = 238$ K ; a curve with $\beta^* = 5.6$ is close to the data for 258 K ; and $\beta^* = 4$ seems to give moderate agreement at 281 K.

Theoretical curves were also evaluated fro the ethane molecule, and results for the J-diffusion model are shown in Fig. 4. Agreement with experiment is noticeably poorer in this case than for CO_2. About all one can say is that the best agreement is obtained when β^* is somewhere between 2 and 5. It is not clear whether this is due to neglect of collision-induced effects or to errors in the experiments. As noted previously, the second moments for ethane are definitely larger than they should be, indicating perhaps that the experimental curves are decaying too rapidly at short times; nevertheless, the comparison with the free rotor curve shown in Fig. 1 indicates that the correlation functions decay at least more slowly than the free rotor at $t^* > 0.5$. On the other hand, the J-diffusion model implies that randomization of the body-fixed L_z occurs with the same probability as that of L_x and L_y; inasmuch as L_z is a symmetry axis, one might well expect that the torques which change this component of **L** will be weaker than those that alter L_x and L_y. It would be desirable to explore a symmetric top model that allows for different 'collision' frequencies for the different component of the body-fixed **L**.

Ethylene is of course slightly asymmetric top; since the dynamics of an ensemble of freely rotating asymmetric tops has not yet been worked out, it is not possible to perform calculations using the M- or J-diffusion models for this molecule. Computations based on a symmetric top approximation to ethylene are now in progress.

REFERENCES

1 P. van Konyenenberg and W. A. Steele, *J. Chem. Phys.* **56**, 4776 (1972).
2 R. G. Gordon, *Advan. Magn. Res.* **3**, 1 (1969); W. A. Steele, in *Transport Phenomena in Fluids* (Ed. H. J. M. Hanley), Dekker, New York, 1969.
3 J. T. Hynes, Ph.D. Thesis, Princeton Univ., 1969.
4 R. Gordon, *J. Chem. Phys.* **40**, 1973 (1964).
5 E. N. Ivanov, *Sov. Phys. JETP* **18**, 1041 (1964).
6 A. Ben Reuven and E. Zamir, *J. Chem. Phys.* **55**, 475 (1971).
7 R. G. Gordon, *J. Chem. Phys.* **44**, 1830 (1966).
8 A. St. Pierre and W. A. Steele, *J. Chem. Phys.* (in press).

59. INDUCED ANISOTROPY AND LIGHT SCATTERING IN LIQUIDS

H. E. Howard-Lock and R. S. Taylor
Department of Physics, University of Toronto, Ontario, Canada

INTRODUCTION

In liquids with molecules having non-spherical polarizability, the spectrum of scattered light contains the usual Rayleigh–Brillouin triplet, and in addition, a depolarized component known as the Rayleigh 'wing'. This depolarized component is centred at zero frequency and is often broad (sometimes extending to $> 100\ cm^{-1}$); it is believed to originate from molecular rotational motions in the liquid. These spectra for anisotropic molecules were first observed in 1928 by Raman and Krishnan,[1] and their intensities, line shapes and line widths have been studied by numerous authors.[2]

More recently, Stegeman and Stoicheff[3] showed that liquids composed of anisotropic molecules exhibit a narrow depolarized doublet superimposed on the depolarized Rayleigh background; this doublet is believed to originate from shear waves in liquids. In liquids composed of spherical-top molecules the depolarized Rayleigh wing was also observed,[1] but not the doublet. In the last three years, several theoretical and experimental studies on the wing have been made.

Levine and Birnbaum[4] have discussed the scattering of light arising from the change in anisotropic polarizability produced in pairs of atoms or molecules during collision and have derived expressions for the intensity of the spectrum and the degree of depolarization. McTague and Birnbaum[5] have observed depolarized spectra of gaseous argon and krypton at high pressure, in general agreement with this theory, and they attribute the spectra to colliding pairs. Also, Thibeau and Oksengorn[6] have developed a theory based on the fluctuations of the internal field at each atom, which explains their observations on the depolarization of Rayleigh scattering in compressed argon.

Gornall, Howard-Lock and Stoicheff[7] have observed depolarized Rayleigh spectra from liquids Xe, CCl_4, $SiCl_4$, $TiCl_4$, $SnCl_4$, $SnBr_4$, $C(CH_3)_4$ and $Si(CH_3)_4$. They used the equation derived by Levine and Birnbaum for collision-induced light scattering in rare gases, which is given approximately by

$$I(\nu) \propto \sqrt{\nu}\ \exp(-\nu/\nu_0) \tag{1}$$

to analyse their observed spectra for liquids. Here, ν is the frequency shift, and ν_0 is a parameter given by

$$\nu_0^{-1} = 2\pi\gamma_2^{-1}\left(\frac{\mu}{kT}\right)^{1/2} \tag{2}$$

where γ_2 is a constant, k is a Boltzmann's constant, μ is the reduced mass of a colliding pair and T is the absolute temperature. On a graph of $\log_{10}(I/\sqrt{\nu})$ versus ν, the experimental points fall on a straight line over most of the frequency range of the wing. The slopes of these lines are found to be roughly proportional to $(\mu/kT)^{1/2}$; the constant of proportionality, γ_2^{-1}, which is a measure of the distance between two colliding molecules for maximum induced anisotropy, was determined to be 0.6 Å for the molecular liquids. They concluded that the existence and shape of the depolarized Rayleigh spectra for liquids composed of isotropic molecules can be explained by a mechanism similar to that producing collision-induced light scattering in rare gases; and that induced anisotropy is a general phenomenon which should be exhibited by all liquids.

Howard-Lock also measured depolarized spectra for several anisotropic molecules including CS_2, C_6H_6, C_6D_6, C_6F_6 and $C_6H_5NO_2$. Preliminary analyses showed that the line shape followed a distribution similar to Eqn. (1) although the frequency weighting factor was raised to a higher power.

Volterra, Bucaro and Litovitz[8] have discussed the mechanism for depolarized light scattering from gaseous argon; they refer to Thibeau and Oksengorn's mechanism as a dipole-induced-dipole and Levine and Birnbaum's as electron overlap. At high gas densities, the dipole-induced-dipole mechanism is suppressed, and short ranged electron overlap and molecular frame distortion effects predominate. They assume a zero impact parameter for liquids, a Leonard-Jones potential, a $1/r^m$ form for the change in polarizability between two atoms, where m is the value appropriate to whether electron overlap or frame distortion processes predominate. Evaluating the Fourier transform of the energy spectrum numerically, Bucaro and Litovitz[9,10] found an intensity distribution

$$I \propto \nu^{2(m-7)/7} \exp(-\nu/\nu_0) \tag{3}$$

where ν_0 is related to the Leonard-Jones parameters and to the factor $(\mu/kT)^{1/2}$. For atomic liquids, the change in polarizability for an electron overlap effect is given by $\Delta\alpha \propto 1/r^9$, implying that $m = 9$. The resulting intensity distribution

$$I = I_0\nu^{4/7} \exp(-\nu/\nu_0) \tag{4}$$

is nearly identical to the expression found by Levine and Birnbaum and used by Howard-Lock *et al.*, although their choice of the factor $\sqrt{\nu}$ was a result of the mathematics of the binary collision theory. Consequently, on the basis of lineshape alone, it is not evident which mechanism is prevalent.

For isotropic (e.g. CCl_4) and anisotropic molecules (e.g. C_6H_6), Bucaro and Litovitz assumed a frame distortion mechanism in which the change of polarizability depended upon the first derivative of the repulsion term of the Leonard-Jones potential, giving $m = 13$ and $\nu^{12/7}$ weighting factor to the exponential term. With the 12/7 power, they obtained good fits for isotropic and moderately anisotropic molecules.

EXPERIMENTAL

We have observed the depolarized Rayleigh spectra of several liquids including both isotropic and anisotropic molecules. A He–Ne laser provided plane-polarized radiation of 15 mW power at 6328 Å. Reagent-grade liquids were filtered or vacuum distilled into glass Raman cells designed to minimize stray light. The cells were maintained at a temperature of 295 K. Depolarized radiation, I_{VH}, scattered at an angle of 90° was selected by a polarizer placed between the sample and spectrometer. The scattered light was analysed with a Coderg PH0 double monochromator, the total instrumental width (laser plus spectrometer) being 1.4 cm^{-1} for a slit width of 1 cm^{-1}. A photomultiplier tube (EMI 9558) with a dark current of 3×10^{-10} A was used for detection. This was followed by d.c. detection equipment and a chart recorder.

Table 1. Integrated intensities and polarization ratios.

	$I_{VV}^{(a)}$	$I_{VH}^{(a)}$	$I_{VH}^{(b)}$	$\rho^{(a)}$	$\rho^{(b)}$
Isotropic molecules					
CCl_4	92	1.2	0.8	0.03	0.70
$SiCl_4$	280	1.7	0.42	0.006	0.70
$GeCl_4$	97	1.2	0.77	0.03	0.70
Anisotropic molecules					
C_6H_{12}	75	1.4	1	0.02	0.72
C_6H_6	170	26	15	0.15	0.74
C_6F_6	140	49	8	0.36	0.70
$C_{14}H_{14}O_2$	140	70	9	0.52	0.71
$C_6H_5NO_2$	320	170	18	0.52	0.71

RESULTS AND DISCUSSION

Depolarized spectra having a continuous distribution of intensity out to 50–100 cm^{-1}, with a maximum at the exciting frequency, were observed for all the liquids studied. The spectra were all of very low intensity. The relative intensities of the depolarized spectra were found to be: $CCl_4 = 1$; $C_6H_6 = 3$; $C_6F_6 = 6$; $SiCl_4$, $TiCl_4$, $SnCl_4 \sim 1.2$. For comparison, the Rayleigh wing of $C_6H_5NO_2$ has an intensity of 20 on this scale and extends to 250 cm^{-1}. The intensities and polarization ratios were measured and the line shapes were analysed by a least squares curve fitting technique.

Integrated intensities of the Stoke's side were measured over two frequency ranges: (a) from 0 cm^{-1} out to where the spectrum and the background merge ($I_{VV}^{(a)}$ and $I_{VH}^{(a)}$) and (b) from 5 cm^{-1} out to the same point ($I_{VV}^{(b)}$ and $I_{VH}^{(b)}$). Measurement (b) is made in order to exclude the Rayleigh–Brillouin triplet, which in many cases accounts for 50% or more of the intensity of the I_{VV} spectrum. The polarization ratios calculated from the ratio I_{VH}/I_{VV} for ranges (a) and (b) are labelled $\rho^{(a)}$ and $\rho^{(b)}$ respectively. With the exclusion of the central part of the spectrum, the measured $\rho^{(b)}$ approaches the theoretical value of $\rho = 0.75$ for polarized incident radiation. The intensities and polarization ratios are listed in Table 1.

The spectra of all the liquids, including the isotropic molecular ones, show an increased intensity in the low frequency ($< 10\ cm^{-1}$) region. Lewis and Van Kranendonk[11] have given a theoretical explanation for this effect in rare gases. Radiation processes during successive binary collisions are correlated giving strong interference effects. Destructive interference produces a dip at zero frequency shift in the induced infrared absorption spectrum, and constructive interference produces an additional peak in the collision-induced light scattering spectrum. The theory holds for densities such that the time between collisions is much greater that the duration of the collision: the peak is ~ 65% Lorentzian in nature and is centred at zero frequency; it has a peak intensity ~ 1/4 of the peak intensity of the collision-induced line, and has a line width which is inversely proportional to the time between collisions. If this theory of intercollisional interference effects can be applied to the liquid state, we would expect such a peak to occur in all liquid spectra, although for anisotropic molecules it could be much weaker than the narrow Lorentzian spectrum produced by orientation fluctuations.

On the assumption that both collision-induced light scattering and intercollisional interference effects contribute to the spectrum for isotropic molecules, we constructed a composite theoretical line shape having the form

$$I(\nu) = \frac{I_R}{1 + \nu^2 \omega_R^{-2}} + I_C \nu^n \exp(-\nu/\nu_0) \tag{5}$$

Here, I is the observed intensity, I_R is the intensity of the relaxational component of half-width ω_R at frequency shift ν, and the second term gives the intensity of the collisional component and has the same form as Eqn. (3) with $n = 2(m-7)/7$. The same line shape can be applied to the spectrum of anisotropic molecules if it is assumed that I_R includes the more intense component caused by orientation scattering as well as the intercollisional term. The line shape can then be specified by three parameters ω_R, ν_0 and I_R/I_C.

We decomposed the spectral line shapes of molecular liquids into two components consisting of a narrow Lorentzian relaxational line I_R and a broad collision-induced line I_C, [Eqn. (5)] and we determined values for the parameters ω, ν_0, and m for these lines. The isotropic molecules were best fitted by $m = 9$; that is, a $\nu^{4/7}$ weighting factor in Eqns. (3) and (5). The anisotropic molecules were best fitted with $m = 13$; that is $\nu^{12/7}$. The former value is that of 'electron-overlap' and the latter implies a 'frame distortion' mechanism. The intensity ratio of the relaxational to the collisional components is approximately 1.3 for the isotropic molecules. As the isotropic polarizability is perturbed, this intensity ratio jumps to 6 (e.g. benzene); for extremely anisotropic molecules the ratio approaches 10 (e.g. benzyl benzoate). The half-width at half maximum of the relaxational component is relatively insensitive to the anisotropy, while the ratio of integrated intensities increased fourfold for the very anisotropic molecules. This last result is in agreement with the assumption that the relaxational peak contains a fairly strong contribution from 'orientation scattering' – for nonisotropic molecules the intensity of such a component is expected to be proportional to the optical anisotropy. Table 2 is a comparison of the measurements for CCl_4 and $C_6H_5NO_2$.

The ratio of the integrated intensity of the depolarized Rayleigh line from 5 cm^{-1} out, relative to the total integrated intensity of the 459 cm^{-1} band of CCl_4 is denoted by $I_{VH}{}^{(b)} : I_{CCl_4}{}^{(459)}$. An estimate of the half-width of the observed spectrum was made by measuring the intensity at 5 cm^{-1}, I_5, and the frequency ν, at which the intensity was one-half of I_5 : the quantity $\nu - 5$ is denoted by $\delta(5)$. After the depolarized spectrum was

Table 2. Comparison of depolarized Rayleigh scattering data for CCl_4 and $C_6H_5NO_2$.

	CCl_4	$C_6H_5NO_2$
$I_{VH}^{(b)} : I_{CCl_4}^{(459)}$	0.8 ± 0.05	20 ± 1
$\rho^{(a)}$	0.03 ± 0.001	0.5 ± 0.06
$\rho^{(b)}$	0.70	0.71
$\delta(5)$ in cm^{-1}	11 ± 0.2	5 ± 0.2
I_R/I_C at 5 cm^{-1}	1.3 ± 0.2	9 ± 1
ω_R in cm^{-1}	4 ± 0.5	4 ± 0.5
ν_0 in cm^{-1} for $m = 9, \nu^{4/7}$	14.9 ± 0.4	29 ± 0.6
ν_0 in cm^{-1} for $m = 13, \nu^{12/7}$	10.6 ± 0.3	19.4 ± 0.3

Table 3. Depolarized Rayleigh line parameters.

	I_R/I_C	ν_0 in cm^{-1}	ω_R in cm^{-1}	$\tau_C \times 10^{-12}$ s	$\tau_R \times 10^{-12}$ s
CCl_4	1.1	14.9	4.2	0.36	1.26
	2.0	9.2	6.0	0.58	0.88
$SiCl_4$	1.5	14.3	5.0	0.37	1.06
	2.7	9.2	6.0	0.58	0.88
$TiCl_4$	1.3	14.0	5.8	0.38	0.91
	3.0	9.5	7.5	0.56	0.71
$GeCl_4$	1.7	12.8	5.0	0.42	1.06
	3.3	8.7	6.5	0.61	0.81

decomposed into the two components I_R and I_C the following parameters were measured: the ratio of the intensities of the two components at 5 cm^{-1}; I_R/I_C at 5 cm^{-1}; the half-width at half maximum of the relaxational component, ω_R; and the collisional parameter ν_0. These values are listed in Table 2 for CCl_4 and $C_6H_5NO_2$ and in Table 3 for some isotropic liquids. There are two values quoted for the tetrahedral molecules; the upper one is based on a line shape with $\nu^{4/7}$, the lower one on $\nu^{12/7}$. The correlation times τ_C and τ_R, are also listed in units of 10^{-12} s for the collisional and relaxational components. τ_C represents the duration of the collision and is the order of 10^{-13} s; τ_R represents the time between collisions and is the order of 10^{-12} s.

REFERENCES

1. C. V. Raman and K. S. Krishnan, *Nature* **122**, 882 (1928).
2. S. Bhagavantam, *Scattering of Light and the Raman Effect*, Chemical Publications, New York, 1942; I. L. Fabelinski, *Molecular Scattering of Light*, Plenum Press, New York, 1968; V. S. Starunov, *Dokl. Akad. Nauk SSSR* **153**, 1055 (1963), [*Sov. Phys. Doklady* 8, 1205 (1964)]; and *Opt. i Spektroskopiya* **18**, 300 (1965) [*Opt. Spectry, (USSR)* **18**, 165 (1965)]; S. L. Shapiro and H. P. Broida, *Phys. Rev.* **154**, 129 (1967); H. C. Craddock, D. A. Jackson and J. G. Powles, *Mol. Phys.* **14**, 373 (1968); B. Simic-Glavaski, D. A. Jackson and J. G. Powles, *Phys. Letters* **34A**, 255 (1971).

3 G. I. A. Stegeman and B. P. Stoicheff, *Phys. Rev. Letters* **21**, 202 (1968).
4 H. B. Levine and G. Birnbaum, *Phys. Rev. Letters* **20**, 439 (1968).
5 J. P. McTague and G. Birnbaum, *Phys. Rev. Letters* **21**, 661 (1968).
6 M. Thibeau and B. Oksengorn, *Mol. Phys.* **15**, 579 (1968); *J. Phys. (Paris)* **30**, 47 (1969).
7 W. S. Gornall, H. E. Howard-Lock and B. P. Stoicheff, *Phys. Rev.* **A1**, 1288 (1970).
8 V. Volterra, J. A. Bucaro and T. A. Litovitz, *Phys. Rev. Letters* **26**, 55 (1971).
9 J. A. Bucaro and T. A. Litovitz, *J. Chem. Phys.* **54**, 3846 (1971).
10 J. A. Bucaro and T. A. Litovitz, *J. Chem. Phys.* **55**, 3585 (1971).
11 J. C. Lewis and J. Van Kranendonk, *Phys. Rev. Letters* **24**, 802 (1970).

60. THE SPECTRAL PROFILE OF COLLISION-INDUCED LIGHT SCATTERING FROM SOME MOLECULAR LIQUIDS

G. C. Tabisz, W. R. Wall, D. P. Shelton and J. Ho
Department of Physics, University of Manitoba,
Winnipeg, Manitoba, Canada

INTRODUCTION

Collision-induced light scattering (CIS) is a newly identified phenomenon which has increased the scope of the information obtainable about dense media by scattering methods.[1-4] While free, optically isotopic atoms or molecules do not depolarize light, an anisotropy can be induced in the total polarizability of a system of interacting atoms which produces a broad depolarized wing centred about the polarized Rayleigh line. Three models for the induction mechanism have been proposed based on studies of liquids and compressed gases.

(i) Dipole-induced dipole (DID).[1,2,5] The anisotropy results from the interaction between electric dipoles induced in the atoms by the incident light. The total field acting on a given atom is the sum of this external light field and the internal field due to the induced dipoles in its neighbours. Orientational fluctuations in the net induced moment due to the atomic motion yield the depolarized scattering. This is a long-range interaction giving an effective pair polarizability which varies as r^{-3} where r is the distance between the atoms in a collision pair.

(ii) Electron Overlap (EO).[3,4] The distortion of the electron clouds during close encounters is responsible for the induced polarizability. Originally, the form $r^2 \exp(-ar^2)$ was proposed but recently,[6] an r^{-9} dependence has been derived in the case of argon.

(iii) Frame Distortion.[7] Bucaro and Litovitz[8,9] have made an extensive study of CIS in atomic and molecular liquids and have concluded that frame distortion during close collisions which gives a pair polarizability varying as r^{-13} is the principal induction mechanism in liquids composed of isotropic and non-isotropic molecules.

With the use of several simplifying assumptions, principally binary collisions with zero impact parameter, Bucaro and Litovitz were able to derive an expression for the intensity I in the wings as a function of frequency ω

$$I(\omega) \propto \omega^{2[(m-7)/7]} \exp(-\omega/\omega_0) \qquad (1)$$

where the induced incremental polarizability is given by

$$\Delta\alpha(r) \propto r^{-m} \tag{2}$$

The constant ω_0 is inversely proportional to a characteristic time associated with the duration of the interaction giving rise to $\Delta\alpha$.

In the analysis of their spectra, they assumed a value for m and determined the corresponding ω_0 from their data. As there is evidence from experiments on compressed gases[1,2] that the DID contribution diminishes at high densities, they assumed that, in the case of atomic liquids, the electron overlap effect predominates. They found that $m = 9$ in the case of argon and xenon and $m = 13$ in the case of molecular liquids gave values of ω_0 in agreement with their calculations. This work is an excellent attempt to categorize CIS in liquids while providing insight into the interaction mechanisms responsible for the scattering. Primarily, it shows that binary interactions remain important even in the liquid phase.

Gersten[10,11] has pointed out that in the spectra of compressed rare gases, different parts of the profile are affected by different details of the interaction.

Encouraged by the success of Bucaro and Litovitz but mindful that the complications envisaged by Gersten should apply to liquids as well as gases, we have repeated observations of the CIS spectra in pure CCl_4 and C_6H_{12} with the goal of further investigating the form of spectral profile.

EXPERIMENTAL METHOD

The experimental set-up is typical of laser Raman scattering studies. Light from an argon-ion laser (CRL No. 52) operating at about 3/4 W at 4880 Å is passed through a half-wave plate and prism polarizer before being focused in the sample cell. The light scattered at 90° is collected by an $f/2$ system and brought to a double scanning monochromator (Jarrell–Ash No. 25–100) having a spectral slit width of 1 cm^{-1}. The detector is a cooled EMI 6256B photomultiplier tube used in the photon counting mode. With the discriminator levels set for an optimum signal-to-noise ratio at low light intensities, the dark count is 1/3 c.p.s. The output is stored in a multichannel analyser (Victoreen No. ST400M) whose channel advance is controlled by the stepping drive for the monochromator gratings. The geometry used is such that if the incident light is considered to travel in the x-direction, polarized in the z-direction, then observations are made in the z-direction. All experiments are performed at 295 K.

ANALYSIS OF RESULTS

If a frame distortion origin for the induced polarizability is assumed, Eqn. (1) leads to a linear relationship between $\ln\,(I(\omega)/\omega^{12/7})$ and ω with a slope $(-1/\omega_0)$. When this test is applied to the data, it is found that $\omega_0 = 11.9$ cm^{-1} for CCl_4 and 15.0 cm^{-1} for C_6H_{12} in excellent agreement with Bucaro and Litovitz (11.9 and 14.8 cm^{-1} respectively). The frequency range over which this linear relation holds is 30 to 80 cm^{-1} for CCl_4 and 30 to 120 cm^{-1} for C_6H_{12} (Fig. 1).

Frame distortion may not, however, be the only interaction mechanism contributing to the scattering. It was noted that much of the high frequency tail of the CCl_4 could not be accounted for by the above method. When $m = 9$ was tried, a value consistent with an electron overlap interaction (at least for rare gas atoms), Eqn. (1) held over an increased frequency range, 20 cm^{-1} to 100 cm^{-1}. Moreover the quality of the fit was much improved

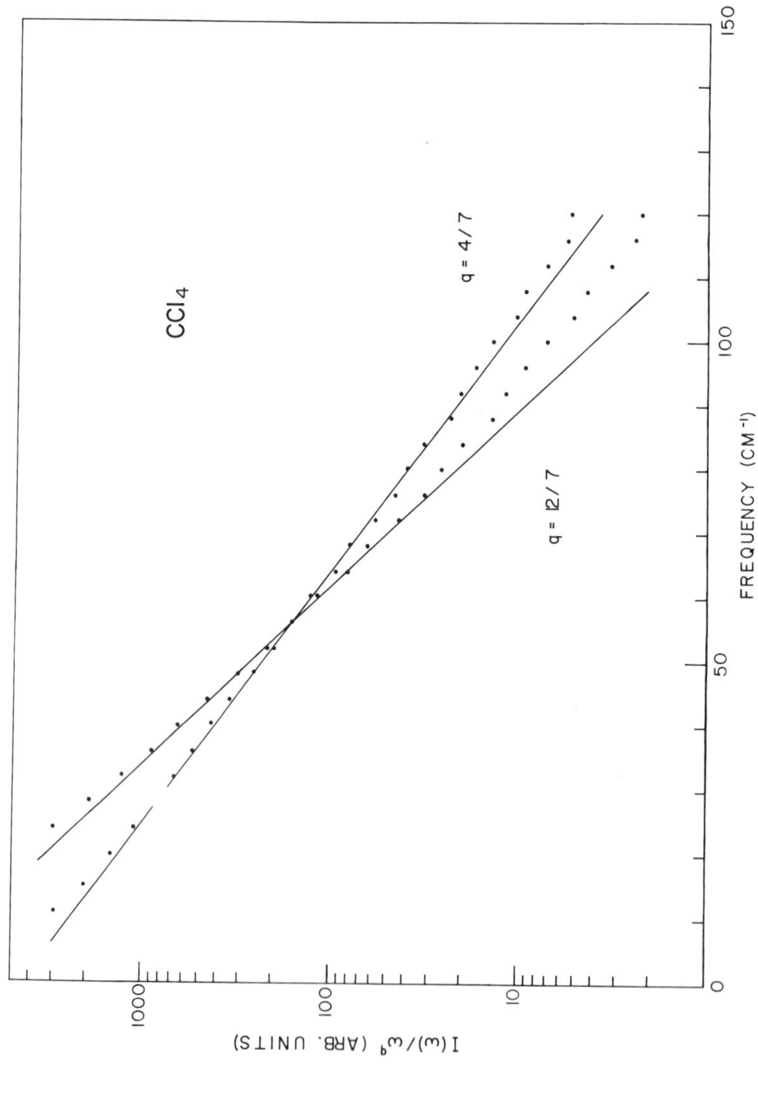

Fig. 1. $I(\omega)/\omega^q$ for the observed spectrum of CCl_4 at 295 K where q is 12/7 and 4/7 corresponding to $\Delta\alpha(r)$ varying as r^{-13} and r^{-9} respectively. Data points are shown only every 4 cm^{-1}.

Table 1. Parameters obtained from the least squares fit with Eqn. (4).

	Range (cm^{-1})	q	Effective m	ω_0 (cm^{-1})	S^2
CCl_4	16 → 100	0.36 ± 0.05	8.3 ± 0.2	18.0 ± 0.46	2.16
	30 → 80	0.74 ± 0.15	9.6 ± 0.5	15.2 ± 0.75	1.55
	40 → 100	0.11 ± 0.3	7.3 ± 1.1	19.2 ± 1.9	1.50
C_6H_{12}	16 → 100	0.28 ± 0.08	7.9 ± 0.28	23.1 ± 1.0	3.06
	30 → 80	0.71 ± 0.20	9.4 ± 0.68	20.3 ± 1.7	1.17
	40 → 100	1.73 ± 0.27	13.0 ± 0.98	15.0 ± 0.97	1.10
	40 → 120	1.68 ± 0.24	12.9 ± 0.84	15.2 ± 0.91	1.08

over that with $m = 13$. This result led us to perform the analysis based on the following considerations.

Most authors who have proposed mechanisms for collision-induced scattering conclude that the high frequency profile is described by an expression

$$I(\omega) = f(\omega) \exp(-\omega/\omega_0) \tag{3}$$

In the original model of Levine and Birnbaum, $f(\omega)$ is approximated by $\omega^{1/2}$; the more general function $\omega^{2[(m-7)/7]}$ shows clearly that the power of ω is a function of the induction process. Gersten[11] has emphasized that the exponential factor in Eqn. (3) is model independent (although the value of ω_0 is not). The accurate determination of $f(\omega)$ should lead to meaningful information not available through the exponential. Because of the presence of this strong exponential factor, the exact form of $f(\omega)$ is difficult to extract from the experimental data.

In an effort to acquire some estimate of $f(\omega)$ directly from the observed spectra, we have assumed that the high frequency profile can be represented by

$$I(\omega) \propto \omega^q \exp(-\omega/\omega_0) \tag{4}$$

where both q and ω_0 were taken as variable parameters and made a least squares fit to the spectra. The results given in Table 1 are averages over several spectra. The errors quoted represent a 99% confidence interval. For C_6H_{12}, the fits in the range 40 cm^{-1} to 120 cm^{-1} are the more significant as they are over a region sufficiently removed from the laser line to be free of contributions due to the permanent anisotropy of the molecule. The 'effective m' is determined by setting

$$q = (2/7)(m-7) \tag{5}$$

S^2 is χ^2 divided by the number of degrees of freedom.

DISCUSSION

The most striking result in Table 1 is the fact that for both liquids, the values of q and ω_0 depend on the frequency region over which the fit is made. For C_6H_{12} the result in the range 40 to 120 cm^{-1} agrees with that of Bucaro and Litovitz. On the other hand, for CCl_4, as more of the tail is included in the fit, the effective m decreases and ω_0 increases, nowhere approximating $m = 13$ or $\omega_0 = 11.9\ cm^{-1}$.

The failure of Eqn. (1) to hold over the entire tail with one set of values for q and ω_0 indicates that more than one type of interaction is operative in the scattering. For C_6H_{12}, short-range effects seem to be important with frame distortion especially dominant beyond 40 cm^{-1}. The theory of Bucaro and Litovitz holds well in this case: basically a single short-range interaction. For CCl_4, both short- and long-range interactions contribute. Here the theory of Bucaro and Litovitz is not entirely applicable as their calculations of ω_0 are valid only near the turning point. Judging from the effective m values, the short-range part includes both electron overlap and frame distortion. The long-range interaction could be of the dipole-induced dipole type.

It should be recalled that because of symmetry effects, the intensity of the scattering through DID at liquid densities is much less than that predicted from an extrapolation of the gas results using a quadratic dependence on density; however it should not necessarily vanish. In liquids the DID interaction would be effective through rapid local fluctuations in the positions of molecules, not describable by the trajectory dynamics used to derive Eqn. (1). Thus the spectral profile due to DID would broaden as the density increases and it would be expected to contribute to the far wings of spectrum.

The ω_0 found for $40 < \omega < 100$ corresponds to a time of 2.8×10^{-13} s which is of the order of magnitude of the time between collisions in a cell model for liquid CCl_4;[12] the time between collisions would be the time characterizing these fluctuations in positions.

Table 2. Parameters obtained from the least squares fit with Eqn. (6).

	Range (cm^{-1})	B	C	S^2
CCl_4	16 → 100	0.50 ± 0.03	−0.59 ± 0.20	2.85
	30 → 80	0.76 ± 0.07	1.15 ± 0.46	1.56
	40 → 100	0.76 ± 0.07	0.10 ± 1.22	1.44
C_6H_{12}	16 → 100	0.37	−1.24	6.67
	30 → 80	0.56 ± 0.20	0.42 ± 1.3	1.34
	40 → 100	1.04 ± 0.16	3.61 ± 1.2	1.07
	40 → 120	0.99 ± 0.12	3.33 ± 0.87	0.98

Further analysis

The expression developed by Shin[13]

$$I(\omega) \propto \omega^A \exp(-B\omega^{12/19} + C\omega^{6/19}) \tag{6}$$

was also fitted to the data with A fixed for $m = 13$. The results are given in Table 2.

For CCl_4 beyond 30 cm^{-1}, the B is in excellent agreement with predictions (0.77) and while C is uncertain, the expected value (0.76) is bracketed. With C_6H_{12}, B and C are much larger than the model suggests (0.66 and 0.69). The quality of the fit for both liquids is comparable to that using Eqn. (4).

This result leads to the conclusion that frame distortion is important for CCl_4 in the far tail. Since Eqn. (6) is a more exact formulation within the binary collision model than is Eqn. (1), it is more sensitive to the assumptions made. The neglect of the angular dependence of intermolecular forces in determining the collision trajectory may account for the discrepancy in the case of the non-spherical molecule C_6H_{12}.

While both the Bucaro and Litovitz and the Shin expressions can mathematically describe the observed spectra, the parameters required are not always in agreement with calculations and conflicting conclusions concerning the interaction arise. The question of the line shape in collision-induced scattering is still a problem.

REFERENCES

1. M. Thibeau, B. Oksengorn and B. Vodar, *J. Phys. (Paris)* **29**, 287 (1968).
2. M. Thibeau, G. C. Tabisz, B. Oksengorn and B. Vodar, *J. Quant. Spectry. Radiative Transfer* **10**, 839 (1970).
3. H. B. Levine and G. Birnbaum, *Phys. Rev. Letters* **20**, 439 (1968).
4. J. P. McTague and G. Birnbaum, *Phys. Rev. Letters* **21**, 661 (1968).
5. P. Lallemand, *Phys. Rev. Letters* **25**, 1079 (1970); *J. Phys. (Paris)* **32**, 10 (1971).
6. M. R. Vukcevich, private communication quoted in ref. 8.
7. H. S. Gabelnick and H. L. Strauss, *J. Chem. Phys.* **49**, 2334 (1968).
8. J. A. Bucaro and T. A. Litovitz, *J. Chem. Phys.* **54**, 3846 (1971).
9. J. A. Bucaro and T. A. Litovitz, *J. Chem. Phys.* **55**, 3585 (1971).
10. J. I. Gersten, R. E. Slusher and C. M. Surko, *Phys. Rev. Letters* **25**, 1739 (1970).
11. J. I. Gersten, *Phys. Rev.* **A4**, 98 (1971).
12. F. J. Bartoli and T. A. Litovitz, *J. Chem. Phys.* **56**, 413 (1972).
13. H. K. Shin, *J. Chem. Phys.* **56**, 2617 (1972).

61. MOLECULAR REORIENTATIONS IN LIQUID BENZENE: RAMAN LINE SHAPES AND ^{2}D NMR RELAXATION

J. E. Griffiths and K. T. Gillen
Bell Laboratories, Murray Hill, New Jersey, USA

INTRODUCTION

The most accessible experimental quantity in studies of rotational molecular motion in liquids is the rotational correlation time τ. This is the zero frequency Fourier transform of the orientational time correlation function. Values of τ can be estimated from a variety of experimental techniques, including NMR relaxation and Raman line shape analyses. Experimentally determined values of τ can be compared with values predicted from various theoretical models with consequent insight into the microdynamics of molecular motion in liquids.

One of the most important liquids for comparisons of this sort is benzene because it is both a species of high symmetry and can serve as the model compound for aromatic systems. Furthermore, substantial accurate data on the physical properties of benzene are available, data which are often essential for testing theoretical models. In discussing the reorientational motions of benzene most authors assume isotropic reorientation either *explicitly* or *implicitly*. Reorientations of benzene, like all symmetric top molecules, however, are characterized by two rotational diffusion constants $D_\perp$ and $D_\parallel$. These describe reorientations of the symmetry axis (the so-called flipping or tumbling motion) and reorientation around the symmetry axis respectively. For symmetric top molecules in which $D_\perp$ and $D_\parallel$ have been experimentally determined, it has been found that the temperature and magnitudes of $D_\perp$ and $D_\parallel$ are usually quite different. Accordingly, one might expect reorientations of benzene to be anisotropic and that all comparisons of experiment with theory based on isotropic reorientation may be of questionable value.

RAMAN LINE SHAPE

In this work we combine data from isotropic and anisotropic Raman line shapes and ^{2}D NMR relaxation times to evaluate the anisotropy of the reorientations of C_6D_6.

Analysis of the Raman line shape of the $\nu_2\,(a_{1g})$ fundamental in C_6D_6 yields information on $D_\perp$. The theory has recently been summarized and only needs a cursory introduction here.

All experiments were done using 90° scattering and linearly polarized incident light. With $I_\parallel(\omega)$ and $I_\perp(\omega)$ representing the strong and weak components of the scattered light, one can write

$$I_{\parallel}(\omega) = I_{\text{isot}}(\omega) + \tfrac{4}{3} I_{\text{anis}}(\omega) \tag{1}$$

$$I_{\perp}(\omega) = I_{\text{anis}}(\omega) \tag{2}$$

$$\rho_s = I_{\perp}(\omega)/I_{\parallel}(\omega) \tag{3}$$

where ρ_s is the depolarization ratio. $I_{\text{isot}}(\omega)$ represents the intrinsic vibrational line shape whereas $I_{\text{anis}}(\omega)$ is a convolution of the vibrational line shape and the orientational spectrum. The problem is to separate the two accurately.

$I_{\text{isot}}(\omega)$ can be extracted directly from measurements of $I_{\parallel}(\omega)$ and $I_{\perp}(\omega)$. Generally, however, the measured spectrum is a convolution of the true spectrum and the Gaussian slit function $S(\omega)$. One can write

$$I'_{\text{anis}}(\omega) = \int_{-\infty}^{\infty} I_{\text{or}}(\omega) \cdot I_{\text{isot}}(\omega) \cdot S(\omega) d\omega \tag{4}$$

and

$$I'_{\text{isot}}(\omega) = \int_{-\infty}^{\infty} I_{\text{isot}}(\omega) \cdot S(\omega) d\omega \tag{5}$$

where the primed and unprimed functions denote the measured and true shapes respectively. Superficially it would appear that the order of convolution is not important and hence that the slit function would cancel. This is not the case except when the line widths are very similar in magnitude. To obviate the large errors associated with this assumption, true line widths were determined directly as a function of slit width and a plot is shown in Fig. 1(D). For the 945 cm^{-1} line of benzene, $I_{\parallel}(\omega)$ can be determined directly using slit settings of 15 μm or less. In order to obtain $I_{\text{isot}}(\omega)$, $I_{\perp}(\omega)$ must be evaluated at the same slit width. The semi-width of $I_{\text{isot}}(\omega)$ was found to be 0.81 ± 0.01 cm^{-1} and was temperature independent. To determine $I(\omega)$ with a higher degree of confidence, however, the fact that the true line width is substantially larger than that of $I_{\parallel}(\omega)$ allows wider slits to be used with an attendant gain in S/N but without suffering a loss in accuracy in measuring the line width parameter. Examples of measured $I_{\parallel}(\omega)$ and $I_{\perp}(\omega)$ are shown in Fig. 1(A) and (B). The magnitudes of the count rate differences noted for $I_{\parallel}(\omega)$ and $I_{\perp}(\omega)$ make it mandatory that other errors be as small as possible and this is straightforward using previously established methods. Errors arising from monochromator polarization response, beam convergence, light collection angle, analyser light leak, exciting light polarization ellipticity, laser power fluctuations and stray light have all been minimized.

Implicit in the treatment of data is the assumption that $I_{\text{isot}}(\omega)$ and $I_{\text{anis}}(\omega)$ are Lorentzian and that vibrational and reorientational motions are not coupled. A test of this assumption at 25°C is shown in Fig. 1(C) in which $I_{\text{anis}}(\omega)$ and $I_{\text{isot}}(\omega)$, normalized to unit height, are displayed (points). The solid lines are calculated Lorentz functions based on the observed line widths. Because $I_{\text{isot}}(\omega)$ and $I_{\text{anis}}(\omega)$ are each Lorentzian and because $I_{\text{anis}}(\omega)$ is a convolution of $I_{\text{isot}}(\omega)$ with $I_{\text{or}}(\omega)$, the latter is also Lorentzian. Accordingly

$$\omega_{1/2}(\text{anis}) = \omega_{1/2}(\text{isot}) + \omega(\text{or}) \tag{6}$$

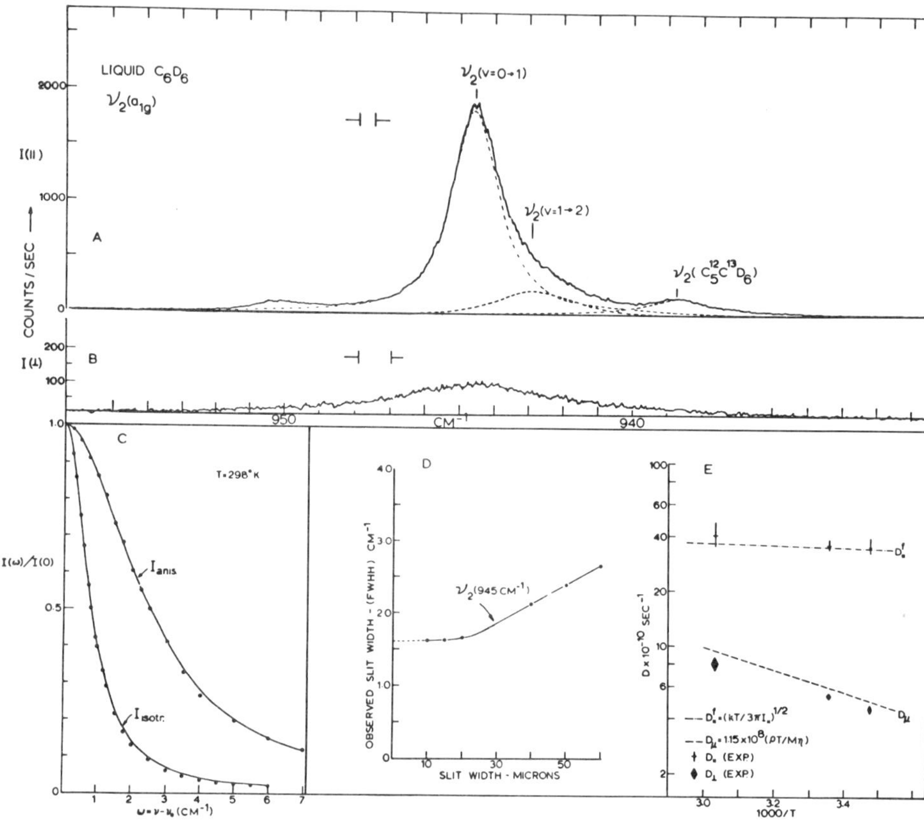

Fig. 1. (A) Polarized Raman spectrum, $I_{\|}(\omega)$, of $\nu_2\,(a_{1g})$ of C_6D_6 (ρ = 600 mW, λ_0 = 4880 Å, slit = 15 μm, slit height = 6 mm, time constant = 1 s, scan rate = 5 cm^{-1}/min). (B) Depolarized spectrum $I_{\perp}(\omega)$ (slit = 30 μm, time constant = 10 s, scan rate 5 cm^{-1}/min). (C) Normalized line shapes of $I_{isot}(\omega)$ and $I_{anis}(\omega)$ of the $\nu_2\,(a_{1g})$ band. (D) Line width (FWHH) of $\nu_2\,(a_{1g})$ of C_6D_6. (E) Experimental values of $D_{\perp}$ and $D_{\|}$ from the present work compared with values of D_{μ} and $D_{\|}^{f}$ calculated from hydrodynamic theory and SDFR theory respectively.

where the $\omega_{1/2}$ terms are semi-half widths. Furthermore, the perpendicular diffusion constant is given by

$$6D_{\perp} \;=\; 2\pi C\;\omega(\mathrm{or}) \;=\; \tau(\mathrm{or})^{-1} \tag{7}$$

where τ(or) is the reorientational correlation time.

From ν_2(945 cm^{-1}) of C_6D_6 we obtain τ_{or} = 3.8 ± 0.2 ps, 3.1 ± 0.15 ps, and 2.1 ± 0.15 ps at 14.5°C, 25°C and 56.5°C respectively. Values of $D_{\perp}$, calculated from Eqn. (7), are plotted versus $1000/T$ in Fig. 1(E). The activation energy for reorientation, determined from these points, is 2.7 ± 0.3 kcal/mole. For C_6H_6 at 25°C, analysis of ν_2(992 cm^{-1}) leads to τ_{or} = 2.8 ps, approximately 10% lower than that of C_6D_6. In view of the differences in their moments of inertia, this isotope effect is not surprising.

Attempts to obtain reliable values of $D_{\|}$ from analysis of degenerate Raman active bands have not enjoyed overwhelming success. Rakov[1] obtained $\omega_{\parallel}(\text{or}) = 1.8\ \text{cm}^{-1}$ for the 606 cm^{-1} line in C_6H_6 from which $\tau_{\parallel}(\text{or}) = (2D_{\perp} + 4D_{\parallel})^{-1}$ and $D_{\parallel} = 5.6 \times 10^{10}\ \text{s}^{-1}$. We disagree with the validity of the method used to obtain $\omega_{\parallel}(\text{or})$ and with the conclusion that isotropic reorientation is indicated in benzene. In subsequent discussion we show that reorientational motions in liquid benzene[2] in fact are highly anisotropic.

²D NMR RELAXATION

Values of T_{1D}, determined from $-11°C$ (supercooled liquid) to 73°C, have an Arrhenius temperature dependence given by

$$T_{1D} = 32.6 \exp(-1.86\ \text{kcal/mole}/RT) \tag{8}$$

and the results are in excellent agreement with earlier measurements.[3-5]

The ²D relaxation time T_{1D} is dominated by the quadrupole relaxation mechanism involving the interaction between the deuterium quadrupole moment eQ and its electric field gradient eq. Assuming the anisotropy of the deuterium electric field gradient ϵ is zero, T_{1D} can be expressed as

$$T_{1D}^{-1} = \tfrac{3}{2}\pi^2 (e^2qQ/h)^2 \tau_D \tag{9}$$

where e^2qQ/h is the deuterium quadrupole coupling constant. τ_D represents the reorientational correlation time for the principal axis of the electric field gradient tensor at the deuterium nucleus. This principal axis lies along the C–D bond direction and thus τ_D is related to the components of the rotational diffusion tensor by

$$\tau_D = 1/(24D_{\perp}) + 0.75/(2D_{\perp} + 4D_{\parallel}) \tag{10}$$

For C_6D_6, values of ²D e^2qQ/H have been determined as 193 ± 3 kHz (polycrystalline),[6] 186.6 ± 1.6 kHz (single crystal)[7] and 194 ± 4 kHz,[8] 196.5 ± 1.3 kHz[9] (C_6D_6 dissolved in a liquid crystal using NMR methods).

We will use the average value of 192.5 kHz for e^2qQ/h. By combining the last three equations, one obtains

$$1/(24D_{\perp}) + 0.75/(2D_{\perp} + 4D_{\parallel}) = 5.6 \times 10^{-14} \times \exp(1.86\ \text{kcal/mole}/RT) \tag{11}$$

Values of $D_{\perp}$ from the Raman results now can be used to evaluate $D_{\parallel}$ and the points appear in Fig. 1(E). If $\epsilon \neq 0$, these values of $D_{\parallel}$ would be slightly smaller.

It is obvious that the reorientational motion of benzene is highly anisotropic with $D_{\parallel}$ about 5 to 8 times larger than $D_{\perp}$ in marked contrast to earlier conclusions.

DISCUSSION

Experimental values of rotational diffusion constants D_i or reorientation correlation times τ_i are normally compared with values predicted by various theoretical models. Since most models offer adequate opportunities for adjustment of the magnitudes of the theoretical correlation times, the most significant and demanding test of a model is whether it predicts correctly the temperature dependence of the experimentally determined correlation times. Two of the most attractive models are the hydrodynamic and the slightly damped free rotor (SDFR) models.[10,11] These models, representing opposite microdynamic extremes, predict quite different values for the τ_i on temperature. Consequently, a comparison of experimental and theoretical correlation times and their temperature dependences represents a significant test of theory and can lead to a deeper understanding of the microdynamics of reorientations in the liquid state.

The hydrodynamic models explicitly assume small angle Brownian rotational diffusion and include an intimate connection between rotational and translational motions. The rotational diffusion constant D_h at temperature T is related to the viscosity η and the density ρ by

$$D_h = C_h T\rho/\eta \tag{12}$$

Using available η and ρ data, an Arrhenius activation energy $E_a = 2.8 \pm 0.1$ kcal/mole is predicted for the rotational diffusion constant of benzene and compared favourably with the value of 2.7 ± 0.3 kcal/mole determined for $D_\perp$ from the Raman results. Similar conclusions were reached for a number of other symmetric top molecules. The previous study also showed that the magnitude of $D_\perp$ can be reasonably well estimated using the microviscosity version of hydrodynamic theory. Such an approximation ($h = \mu$) gives

$$C_h = C_\mu = 1.5 \times 10^8/M \tag{13}$$

where M is the molecular weight. For C_6D_6, values of D_μ are represented by a dotted line in Fig. 1(E). The excellent agreement between $D_\perp$ (Raman) and D_μ with respect to magnitude and temperature dependence is obvious, and in an earlier paper[12] BCl_3 was the only exception noted to the correspondence of $D_\perp$ and D_μ. More accurate recent data for $\eta(T)$ indicate that BCl_3 is no longer an exception. Therefore, the results for C_6D_6 and BCl_3, when combined with the earlier results, suggest that the effect of a permanent electric dipole is unimportant in the relationship found between $D_\perp$ and η. In fact, since pseudo-spherical symmetric top molecules like ClO_3F ($\mu = 0.02$ D) and CCl_3F ($\mu = 0.46$ D) appear to be the only symmetric top molecules where $D_\perp$ is unrelated to η, we conclude that steric interactions are important in the coupling of rotational and translational motions assumed in the hydrodynamic theory. A similar conclusion was recently reached by Assink *et al.*[13]

At the opposite theoretical extreme is the slightly damped free rotor (SDFR) model in which molecular reorientations approach those in the free dilute gas phase. Large angle rotational jumps are assumed and one obtains

$$\tau_i = (\pi I_i/12\, kT)^{1/2} \tag{14}$$

for the rotational correlation time τ_i, where I_i is the appropriate moment of inertia. The SDFR model predicts an activation energy E_a for D_i of $RT/2$ (~ 0.3 kcal/mole in the room temperature region) which for C_6D_6 is in marked contrast to the value of 2.8 kcal/mole predicted by hydrodynamic theory. Since the SDFR model is predicted to apply for a reorientation whose intermolecular potential has little dependence on angle, one might expect the highly symmetric six-fold reorientation of C_6D_6 to be described by such a model. The fact that reorientation around this axis proceeds rapidly even in solid benzene[14] supports this view. Using Eqn. (14) and $I_{\|}$, values of $D_{\|}^f$ for the SDFR model have been calculated and the results are shown in Fig. 1(E). The curve is in excellent agreement with the experimental values of $D_{\|}$. It should be noted that Eqn. (10) was derived for small angle rotational diffusion, so that the experimental values of $D_{\|}$(exp) can only be considered semi-quantitative. However, by utilizing the approximate relationship in Eqn. (2.38) of Steele[10b] and the experimental values of $D_{\perp}$, it can be shown that the SDFR model for $\tau_{\|}$ of C_6D_6 is reasonably consistent with the experimental ^2D relaxation times.

Thus by combining data from Raman scattering and NMR relaxation, the reorientation motion of benzene in the liquid phase is found to be quite anisotropic. The strongly hindered perpendicular reorientation proceeds by small angle rotational diffusion and $D_{\perp}$ can be estimated from hydrodynamic theory. The symmetric parallel reorientation appears to be relatively free and hence proceeds with large angular jumps. Previous comparisons with theoretical models which utilized ^2D relaxation results alone are obviously of limited value since the ^2D correlation times come from a combination of two motions of quite different magnitudes and microdynamical character.

REFERENCES

1 A. V. Rakov, *Tr. Fiz. Inst. Akad. Nauk. SSSR* **27**, 111 (1964); *Opt. i Spectroskopiya* 7, 202 (1959) (*Opt. Spectry.* 7, 128 (1959)).

2 F. J. Bartoli and T. A. Litowitz, *J. Chem. Phys.* **56**, 404, 413 (1972).

3 D. E. Woessner, *J. Chem. Phys.* **40**, 2341 (1964).

4 G. Bonera and A. Rigamonti, *J. Chem. Phys.* **42**, 175 (1965).

5 J. G. Powles, M. Rhodes and J. H. Strange, *Mol. Phys.* **11**, 515 (1966).

6 J. Powell, W. Phillips, L. Melby and M. Panar, *J. Chem. Phys.* **43**, 3442 (1965).

7 P. Pyykko, *Ann. Univ. Turku., Ser. A* **88**, 93 (1966).

8 W. V. Caspary, F. Millett, M. Reichback and B. P. Dailey, *J. Chem. Phys.* **51**, 623 (1969).

9 P. Diehl and C. L. Khetrapal, *Can. J. Chem.* **47**, 1411 (1969).

10 (a) W. A. Steele, *J. Chem. Phys.* **38**, 2404 (1963); (b) W. A. Steele, *J. Chem. Phys.* **38**, 2410 (1963); (c) W. B. Moniz, W. A. Steele and J. A. Dixon, *J. Chem. Phys.* **38**, 2418 (1963).

11 P. W. Atkins, *Mol. Phys.* **17**, 321 (1969).

12 K. T. Gillen and J. H. Noggle, *J. Chem. Phys.* **53**, 801 (1970).

13 R. A. Assink, J. DeZwann and J. Jonas, *J. Chem. Phys.* **56**, 4975 (1972).

14 E. R. Andrew and R. G. Eades, *Proc. Roy. Soc. (London)* **A218**, 537 (1953).

62. CONTOURS DES BANDES DANS LES SPECTRES RAMAN DE GAZ

R. Gaufrès et S. Sportouch
Laboratoire de Spectroscopie Moléculaire, Université des Sciences et Techniques du Languedoc, Montpellier, France

ABSTRACT

In the past twenty years, a lot of work has been done on high resolution Raman spectroscopy, in order to obtain the pure rotation spectrum or the vibration–rotation spectrum, and hence the rotational constants of the molecule. But this work has generally concerned light molecules. Up to this date, only unresolved contours can be recorded for heavier molecules.

As early as 1931, Placzek and Teller gave the analytical expression of the rotational wings (*OP* and *RS*) for a symmetric top and for totally symmetric vibrations. This subject was then neglected due to lack of experimental support. Nowadays, with the commercial devices available, it is as easy to record contours for rotational wings, which are generally of poor intensity, as it is for *Q*-branches in the case of totally symmetric vibrations, and a few laboratories have recently contributed to this topic.

A review of the subject will be given, and the following sections will be emphasized:

1. *Q*-branch contours for totally symmetric vibrations and for linear molecules, symmetric and spherical tops.
2. *OP–RS* branches contours for degenerate vibrations and for spherical and symmetric tops.

The possibility of determining rotational constants and/or rotational temperatures will be discussed.

INTRODUCTION GENERALE

Détermination de constantes moléculaires par spectroscopie Raman

Depuis le travail fondamental de Placzek et Teller,[1] on connaît la structure de rotation des bandes de vibration–rotation dans le spectre de diffusion. La résolution expérimentale de cette structure permet d'atteindre les constantes moléculaires de rotation et d'interaction de vibration–rotation. Mais d'autres spectroscopies, comme l'absorption ou l'émission électroniques, conduisent à des information de même nature, et on peut se poser la question de l'intérêt de la détermination des constantes moléculaires (B_e, α_e, D_e, ζ etc.) par spectroscopie Raman alors qu'elle demeure sans doute encore la plus délicate des techniques que nous venons de citer. Toutefois, dans certains cas que nous signalerons dans cet exposé, la spectroscopie Raman peut apporter directement des informations difficiles ou pratiquement impossibles à atteindre par les autres spectroscopies.

Comme dans toute spectroscopie, meilleure est la résolution, plus nombreux et plus précis sont les résultats tirés de l'analyse du spectre. Ainsi, les spectres Raman de vibration–rotation et de rotation pure ont été assez largement étudiés en haute résolution sur des molécules légères, surtout par l'école canadienne. Stoicheff[2] a fait une mise au point des travaux faits dans ce domaine, des origines à 1958. Pour les molécules diatomiques homonucléaires, qui n'ont pas de spectre d'absorption infrarouge, les valeurs obtenues pour $B_e, \alpha_e, D_e, \beta_e$ peuvent être aussi précises que celles que donnent les spectres électroniques, et même davantage dans certains cas.

Dès que l'on envisage des molécules triatomiques, l'absorption infrarouge permet, en principe, d'atteindre toutes les constantes moléculaires de vibration, de rotation et d'interaction de vibration–rotation, par l'analyse en haute résolution des bandes fondamentales actives, et de bandes de combinaison judicieusement choisies. Les constantes relatives à des modes inactifs en absorption peuvent alors être obtenues par différence, mais lorsque ces modes sont actifs en diffusion, la spectroscopie Raman permet d'atteindre directement les constantes correspondantes. Ainsi, la constante α relative à la ν_1 de CO_2 et de CS_2 (B_{100}–B_{000}) a été mesurée par analyse de la structure de rotation des bandes Raman correspondantes.[3]

Avec les molécules du type toupie symétrique, apparait une nouvelle sorte de constante, les coefficients ζ de Coriolis, relatifs à chacun des modes dégénérés. Il est interéssant de connaître les ζ car ils dépendent, dans l'approximation d'ordre zéro, du potentiel harmonique de vibration, et peuvent constituer des données indépendantes dans le calcul du champ de force moléculaire.[4] Les coefficients ζ de Coriolis sont reliés entre eux au sein d'un même type de symétrie par la règle de la somme,[5] mais sont indépendants d'un type de symétrie à un autre, de telle sorte que, pour les modes actifs en diffusion seulement, le spectre Raman permet seul d'atteindre ces coefficients.

Intérêt et limite des contours non résolus

Le nombre des molécules pour lesquelles on peut espérer obtenir la résolution du spectre de rotation pure, et surtout du spectre de vibration–rotation, semble encore restreint. A résolution égale, la limite des possibilités est fixée non seulement par la valeur des moments principaux d'inertie, mais aussi par le type de molécule et par le libéralisme des règles de sélection. Si pour les molécules linéaires et les toupies sphériques, l'expression de l'énergie de rotation ne contient que le nombre quantique J, pour les toupies symétriques l'existence du nombre quantique K a pour effet de multiplier les termes spectraux et il s'ensuit que le nombre de transitions distinctes, lorsque $\Delta K = \pm 1$ ou ± 2, augmente considérablement. C'est pourquoi, dans la grande majorité des cas, on ne peut atteindre qu'un contour de bande.

L'intérêt de l'étude de ces contours est divers. Dans les cas simples leur analyse conduit à la détermination de constantes moléculaires. Mais les valeurs ainsi obtenues seront moins précises que celles que donne la haute résolution; elles seront toujours entâchées du doute qui demeure sur les résultats tirés d'un effet global, au prix d'hypothèses simplificatrices quelquefois invérifiables. Toutefois, utilisés avec prudence, les résultats tirés de l'analyse des contours pourraient permettre des progrès dans l'étude des molécules de masse moyenne.

Un autre aspect de l'étude des contours est le suivant: lorsque l'on connait les constantes moléculaires par ailleurs, on peut calculer *a priori* le contour d'une bande et l'utiliser dans l'analyse d'un tracé expérimental; on pourra par exemple séparer la contribution de la fondamentale de celle des bandes chaudes, déterminer une température de rotation, comme cela a été fait dans des flammes[6] soit encore vérifier le type de symétrie de la vibration à laquelle est due la bande considérée, lorsqu'il y a ambigüité.

Généralités sur les contours de bandes dans le spectre Raman

Dès 1931, Placzek et Teller[1] ont donné l'expression analytique des contours des ailes de rotation (*OP* et *RS*) des bandes des vibrations totalement symétriques et de la raie Rayleigh (rotation pure) et cela pour les toupies symétriques. Ces prévisions théoriques n'ont pas eu, pendant longtemps, d'application expérimentale, parce que les spectres Raman de gaz étaient difficiles à observer dans l'époque 'pré-laser', et aussi parce que l'enregistrement photographique se prête mal aux mesures d'intensité. Le premier travail expérimental sur les contours de bande dans le spectre Raman semble être celui de Mathai, Shepherd et Welsh[7] où l'aspect des bandes dégénérées du cyclopropane est comparé aux contours théoriques, pour les bandes de type E' et celles du type E''; les valeurs des ζ ont été ainsi déduites des contours expérimentaux.

Depuis 1970, des études plus systématiques ont été entreprises, que nous exposons dans ce qui suit. On peut distinguer, tant pour des raisons expérimentales que théoriques, d'une part l'étude des branches Q des bandes dues à des vibrations totalement symétriques, d'autre part celle des ailes de rotation, notamment pour les vibrations dégénérées.

Les contours des branches Q de bandes de vibrations symétriques (auxquels conviendraient mieux le terme de 'profils') sont relativement intenses en général. Leur étude peut être faite avec plus de rigueur car, en raison de la règle de sélection $\Delta J = 0$ (et $\Delta K = 0$ pour les toupies symétriques) les termes contenant les constantes D_J, D_K et D_{JK} dans l'expression des fréquences s'annulent.

Le calcul rigoureux des contours des ailes de rotation (branches O, P, R et S) fait en effet intervenir les constantes de distorsion centrifuge, lesquelles sont rarement connues et on pourra être amené à les négliger. En outre, l'intensité de ces ailes est généralement faible; elles sont enregistrées avec une résolution moins bonne que les branches Q intenses des vibrations symétriques. C'est pourquoi les auteurs ne s'intéressent le plus souvent qu'à la séparation $\Delta\tilde{\nu}_{OP-RS}$ des maximums de ces ailes. Notons qu'il peut être inexact de parler d'ailes de rotation *OP* et *RS*, car, dans le cas général, la branche Q est étalée dans une grande partie de la bande.

Il existe deux modes de calcul des contours des bandes non résolues. Il est toujours possible d'effectuer une *simulation de bande* par ordinateur, en sommant les intensités des raies individuelles, opération dont le principe est connu depuis longtemps.[8] Dans ce cas, on peut faire intervenir dans le calcul toutes les constantes moléculaires que l'on désire et simuler la déformation du contour par la fonction d'appareil.

On peut aussi chercher à établir des *expressions analytiques* des contours, comme l'ont fait Placzek et Teller.[1] Ces expressions sont intéressantes dans les cas simples car elles permettent la discussion de l'allure générale d'un contour ou d'un profil, en fonction des valeurs relatives des paramètres, lorsque différents cas de figure sont possibles. L'utilisation de ces expressions pourra faire appel en définitive à l'ordinateur, mais les calculs seront très abrégés par rapport à la simulation par addition de raies individuelles et ceci est intéressant lorsque l'on recherche par tâtonnements des paramètres conduisant au meilleur accord avec le tracé expérimental. Toutefois, l'établissement de ces expressions analytiques nécessite en général des approximations et elles correspondent donc à des modèles simplifiés, ce qu'il ne faudra pas perdre de vue.

Dans ce qui suit, nous traiterons des profils des branches Q des bandes dues à des vibrations totalement symétriques, puis des contours des bandes dues à des vibrations dégénérées.

PROFILS DES BRANCHES Q

Molécules linéaires

Nous donnerons avec quelque détail le calcul de l'expression analytique du profil dans ce cas qui est le plus simple. Le principe de ce calcul pourra s'appliquer à des cas plus compliqués.

Les fréquences Raman absolues des raies individuelles d'une branche Q, dans le spectre Stokes, sont données, pour la molécule linéaire, par

$$\tilde{\nu}_{\text{Ram}} = \tilde{\nu}_{\text{vib}} + \alpha J(J+1)$$

où $\tilde{\nu}_{\text{vib}}$ est la fréquence absolue de l'origine de la bande. Négligeant, de façon générale, les entiers petits devant J, ce qui suppose $Bhc \ll kT$, nous écrirons

$$\tilde{\nu}_{\text{Ram}} = \tilde{\nu}_{\text{vib}} + \alpha J^2 \tag{1}$$

Nous utiliserons, ici et dans tout ce qui suit, les fréquences rapportées à l'origine de la bande

$$\Delta\tilde{\nu} = \tilde{\nu}_{\text{Ram}} - \tilde{\nu}_{\text{vib}}$$

qu'il ne faut pas confondre avec les $\Delta\tilde{\nu}$ Raman usuels. Dans le spectre Stokes, les uns varient dans le sens opposé des autres, et dans le même sens dans le spectre anti-Stokes. Dans le cas présent

$$\Delta\tilde{\nu} = \alpha J^2 \tag{2}$$

Dans la diffusion de trace, les intensités sont données rigoureusement par

$$I(J) \propto (2J+1) \exp - [Bhc\, J(J+1)/kT]$$

soit, en effectuant les simplifications admises

$$I(J) \propto J \exp - [Bhc\, J^2/kT] \tag{3}$$

Dans la diffusion d'anisotropie, l'intensité des raies est donnée par

$$I(J) \propto (2J+1)\frac{J(J+1)}{(2J-1)(2J+3)} \exp - [Bhc\, J(J+1)/kT]$$

qui tend aussi vers la forme (3) pour J grand. Nous pouvons donc en déduire que, dans la limite de validité des approximations qui ont été faites, le résultat sera valable aussi bien pour $I_\perp$ que pour $I_\parallel$.

Pour passer de $I(J)$ à $I(\Delta\tilde{\nu})$, on écrira

$$I(\Delta\tilde{\nu}) = I(J)\left|\frac{\mathrm{d}J}{\mathrm{d}(\Delta\tilde{\nu})}\right| \tag{4}$$

où $|dJ/d(\Delta\tilde{\nu})|$ est la densité de raies, et on remplace J par son expression en fonction de $\Delta\tilde{\nu}$, tirée de (2)

$$J = (\Delta\tilde{\nu}/\alpha)^{1/2}$$

$$\left|\frac{dJ}{d(\Delta\tilde{\nu})}\right| = \frac{1}{2(\alpha\,\Delta\tilde{\nu})^{1/2}}$$

d'où

$$I(\Delta\tilde{\nu}) \propto \frac{1}{2|\alpha|} \exp - [a\,\Delta\tilde{\nu}] \tag{5}$$

où $a = Bhc/\alpha kT$.

La nécessité de faire intervenir les valeurs absolues de certaines grandeurs est due au fait que, si J est toujours positif, α et $\Delta\tilde{\nu}$ sont soit simultanément négatifs, soit positifs. L'expression (5) n'est à considérer que pour $a\,\Delta\tilde{\nu} > 0$.

Cette expression donne la distribution naturelle d'intensité, mais en fait le profil observé, ou distribution apparente, sera le produit de convolution de (5) par la fonction d'appareil. En admettant une fonction gaussienne, de forme $\exp - [b(\tilde{\nu}_{app} - \tilde{\nu})^2]$, on obtient l'expression

$$I(\Delta\tilde{\nu}_{app}) \propto \exp - \left[a\,\Delta\tilde{\nu}_{app} - \frac{a^2}{4b}\right] \operatorname{erf} c \left[\frac{a}{2b^{1/2}} - b^{1/2}\Delta\tilde{\nu}_{app}\right] \tag{6}$$

où a et b sont des paramètres définis plus haut, et $\Delta\tilde{\nu}_{app}$ est la fréquence affichée par le spectromètre, rapportée à l'origine de la bande. Cette forme mathématique donne une excellente concordance avec les profils expérimentaux,[9] ce qui a pu être vérifié sur une molécule dont les constantes sont très bien connues par ailleurs, comme SCO,[10] et cela à plusieurs températures. La comparaison des profils observés et calculés peut être faite en considérant les caractéristiques $\Delta\tilde{\nu}_{1/2}$, $\Delta\tilde{\nu}_{1/4}$ et $\Delta\tilde{\nu}_{1/8}$, largeurs à demi, quart et huitième de hauteur, dont deux sont indépendantes.

Une application possible de cette analyse est la détermination de la constante α pour la vibration considérée, lorsque B est connue. Ainsi, pour la molécule Cl_2, la valeur de 0,0015 cm^{-1} a pu être avancée,[9] alors que les valeurs les plus récentes, obtenues par absorption électronique, sont 0,00153 cm^{-1} [11] et 0,00149 cm^{-1}.[12]

La connaissance précise du profil d'une branche Q permet, en outre, l'analyse d'un système constitué par la fondamentale et, par exemple, un ensemble de bandes chaudes et les bandes dues aux variétés isotopiques. La région de la ν_1 de S_2C a été analysée[13] en partant de la fondamentale $(1, 0^\circ, 0) \longleftarrow (0, 0^\circ, 0)$ de $^{32}S_2C$, sur laquelle la valeur de α a été trouvée égale à $175 \times 10^{-6}\,cm^{-1}$, valeur légèrement supérieure à celle qu'avait trouvée Stoicheff[3] par analyse en haute résolution des branches O et S ($155 \times 10^{-6}\,cm^{-1}$). A partir du contour de la fondamentale, il est possible de reconstituer l'ensemble du spectre de la région en ajoutant, aux fréquences convenables, toutes les branches Q des transitions satellites (bandes chaudes et spectre des variétés isotopiques), de même largeur que la fondamentale, mais d'intensités différentes. Dans un premier temps, on n'a pris en considération que les fréquences repérées et attribuées par Stoicheff; la coïncidence du tracé

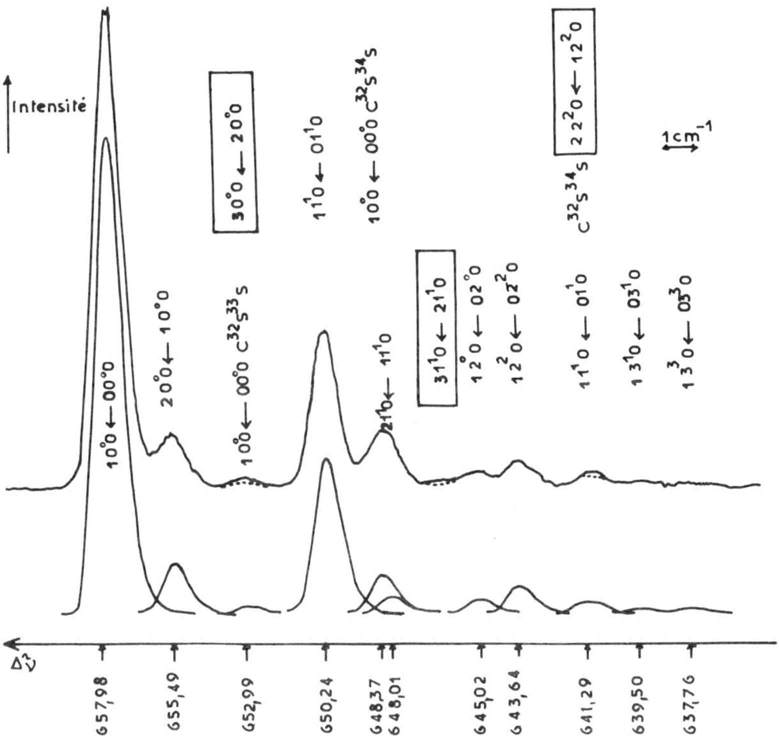

Fig. 1. Région de la ν_1 de S_2C. Etude à 320 K. Comparaison des contours calculé et observé. La somme des branches Q individuelles, représentées en bas, coincide très exactement avec le contour expérimental, représenté au-dessus, sauf en trois régions, où elle est représentée en pointillés. Ces décrochements correspondent à des transitions indiquées dans les cadres.

expérimental et du profil calculé est excellente, sauf en trois endroits où l'intensité est supérieure à celle que l'on a calculée (Fig. 1). En élevant la température, on a constaté que les trois décrochements deviennent plus importants (Fig. 2), et on peut les attribuer aux bandes chaudes suivantes

$$(3, 0^0, 0) \longleftarrow (2, 0^0, 0)$$

$$(3, 1^1, 0) \longleftarrow (2, 1^1, 0)$$

$$(2, 2^2, 0) \longleftarrow (1, 2^2, 0)$$

leurs fréquences et leurs intensités étant convenables. Ces bandes chaudes coïncident à peu près exactement avec d'autres branches Q, et elles n'avaient pu être observées sur la plaque photographique, même en haute résolution. Il y a donc là une possibilité de mettre en évidence des bandes faibles dans un massif.

Lapp, Goldman et Penney[6] ont effectué l'analyse des systèmes de bandes chaudes de O_2 et de N_2 dans des spectres Raman de flammes. Ils ont pu déterminer la température de vibration et une température moyenne de rotation; on peut remarquer en effet que, si les intensités relatives des bandes chaudes permettent de vérifier si les peuplements des

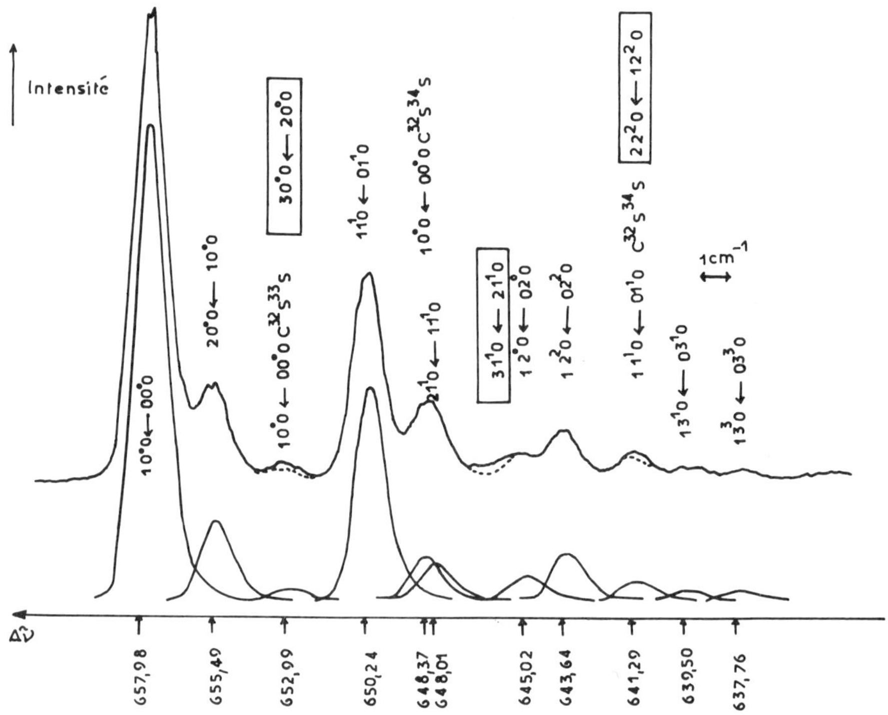

Fig. 2. Région de la ν_1 de S_2C. Même étude que la Fig. 1, à la température de 380 K. Les bandes chaudes apparaissent plus intenses qu'à 320 K.

niveaux de vibration correspondent à l'équilibre thermodynamique, le profil des branches Q ne permet d'atteindre qu'une température moyenne de rotation, sans que l'on puisse savoir si le système de rotateurs est en équilibre.

Toupies sphériques et symétriques

Les expressions analytiques des distributions naturelle et apparente d'intensité dans les branches Q des toupies sphériques sont aisément établies, par un calcul analogue à celui qui a été développé pour les molécules linéaires. Les fréquences seront encore données par (1), mais les intensités par

$$I(J) \propto (2J+1)^2 \exp - [Bhc\,J(J+1)/kT]$$

soit

$$I(J) \propto J^2 \exp - [Bhc\,J^2/kT] \tag{7}$$

et la distribution continue naturelle sera

$$I(\Delta\tilde{\nu}) \propto \frac{1}{2\,|\alpha|^{3/2}}\,\Delta\tilde{\nu}^{\,1/2} \exp - [a\,\Delta\tilde{\nu}] \tag{8}$$

expression différente de (5) par le terme $\Delta\tilde{\nu}^{\,1/2}$ essentiellement.

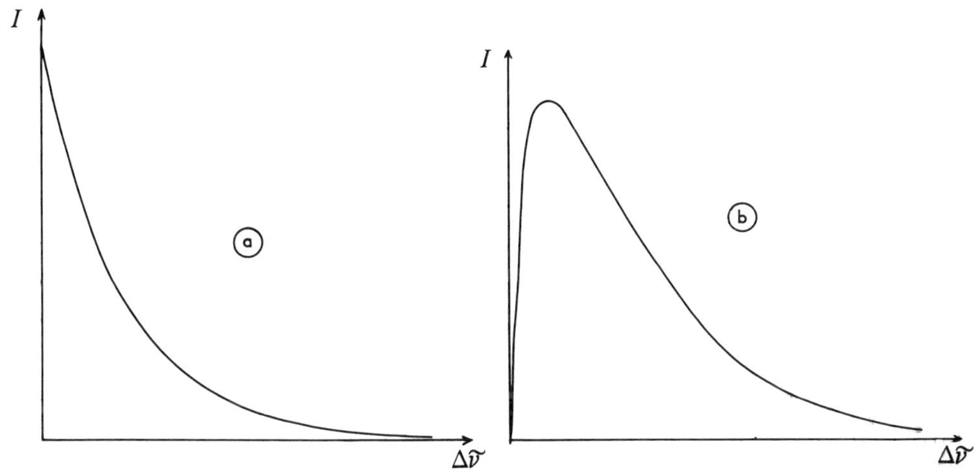

Fig. 3. Comparaison de contours naturels de branches Q de vibrations symétriques: (a) pour une molécule linéaire; (b) pour une toupie sphérique.

La convolution par une fonction gaussienne de constante b donne

$$I(\Delta\tilde{\nu}_{\text{app}}) \propto \exp - [b\,\Delta\tilde{\nu}_{\text{app}}] \int_0^\infty x^{1/2} \exp - [x^2 + Ax]\,\mathrm{d}x \qquad (9)$$

où x a remplacé dans le calcul $b^{-1/2}\,\Delta\tilde{\nu}$ et $A = b^{1/2}(a - 2b\,\Delta\tilde{\nu}_{\text{app}})$.

Dans la pratique, il est commode de dresser une table des valeurs de l'intégrale qui apparait dans (9), lesquelles ne dépendent bien sur que de A.

Sur la Fig. 3 ont été juxtaposés les profils naturels pour une molécule linéaire et pour une toupie sphérique. On peut remarquer que si le maximum du profil naturel coïncide avec l'origine de la bande pour la molécule linéaire, il n'en est plus de même pour la toupie sphérique. Sur les contours apparents, il n'y a coïncidence dans aucun cas (à moins que $\alpha = 0$) entre le maximum et l'origine de la bande. Pour la molécule linéaire, le décalage a été donné [13] en fonction des constantes a et b de (6).

Pour les toupies symétriques, les calculs sont plus compliqués. Dans la branche Q, il y a autant de raies individuelles que de couples de valeurs de J et de K et on peut la considérer comme la superposition d'une infinité de sous-branches ${}^{Q}Q_K$. En outre, l'intensité des raies dans la diffusion de trace ne dépend que de J alors que la diffusion d'anisotropie est fonction de J et de K,[1,14] si on met à part le facteur de Boltzmann, le même dans les deux cas. On doit donc traiter séparément les deux termes et les résultats qui vont suivre sont relatifs à la diffusion de trace, qu'expérimentalement on peut obtenir par la différence $I_{\parallel} - 4/3\,I_{\perp}$.

Nous traiterons le cas de la toupie aplatie, pour laquelle les fréquences sont données, pour J grand, par

$$\tilde{\nu}_{\text{Ram}} = \tilde{\nu}_{\text{vib}} + \alpha^B J^2 + (\alpha^C - \alpha^B) K^2 \qquad (10)$$

où α^B et α^C sont les constantes d'interaction de vibration–rotation, relatives aux deux constantes de rotation B et C respectivement.

Dans la diffusion de trace, on peut écrire, pour les intensités

$$I(J, K) \propto J \exp - [(BJ^2 + [C-B]K^2)hc/kT] \tag{11}$$

Pour une sous-branche prise isolément, pour laquelle $|K| = K_i$ et $J = K_i, K_i + 1, K_i + 2$, etc., la distribution naturelle d'intensité pourra être calculée selon la méthode décrite plus haut et on obtient

$$I(\Delta\tilde{\nu})_{K_i} \propto \frac{1}{2|\alpha^B|} \exp - \left[\frac{\sigma}{\alpha^B} \Delta\tilde{\nu}\right] \exp - [\sigma(\gamma - \gamma')K_i^2] \tag{12}$$

où

$$\gamma = \frac{C}{B} - 1, \quad \gamma' = \frac{\alpha^C}{\alpha^B} - 1 \quad \text{et} \quad \sigma = Bhc/kT$$

La première exponentielle est commune à toutes les sous-branches, alors que la seconde est caractéristique d'un sous-branche. A chaque fréquence, l'intensité totale sera la somme des intensités des sous-branches existant à la fréquence considérée, c'est à dire

$$I(\Delta\tilde{\nu}) \propto \frac{1}{2|\alpha^B|} \exp - \left[\frac{\sigma}{\alpha^B} \Delta\tilde{\nu}\right] \sum_K \exp - [\sigma(\gamma - \gamma')K^2] \tag{13}$$

où les sommations sur les K doit s'étendre aux valeurs de K qui correspondent à des sous-branches existant au $\Delta\tilde{\nu}$ considéré. Différents cas de figure doivent être considérés.

Une sous-branche caractérisée par $|K| = K_i$ débute à $J = K_i$, soit à $\Delta\tilde{\nu} = \alpha^C K_i^2$, et selon (10), elle se développe vers les $\Delta\tilde{\nu}$ croissants si $\alpha^B > 0$, vers les $\Delta\tilde{\nu}$ décroissants si $\alpha^B < 0$. Quatre cas de figure sont donc possibles (Fig. 4) selon les signes respectifs de α^B et de α^C. En fait, ils sont deux à deux symétriques par rapport à l'origine de la bande et se ramènent à deux cas fondamentaux qui correspondent à α^B et α^C de même signe ou de signes opposés. Nous traiterons donc que le cas où α^B est positif; on pourra étendre les résultats au cas où α^B est négatif en prenant dans la discussion, pour α^C et le domaine de $\Delta\tilde{\nu}$, les signes opposés à ceux que l'on envisage pour α^B positif.

Donc, si α^B est positif, selon (12) toutes les sous-branches sont des exponentielles décroissantes dans le sens des $\Delta\tilde{\nu}$ croissants. Si α^C est positif, la première ($K_i = 0$) débute à $\Delta\tilde{\nu} = 0$, la seconde ($K_i = 1$) à $\Delta\tilde{\nu} = \alpha^C$, la troisième à $\Delta\tilde{\nu} = 4\alpha^C$, c'est à dire à des $\Delta\tilde{\nu}$ positifs et il n'existe aucune sous-branche dans le domaine négatif des $\Delta\tilde{\nu}$.

Si α^C est négatif, les sous-branches se développent toujours vers les $\Delta\tilde{\nu}$ croissants selon (10), leur allure de variation, donnée par (12) et ne dépendant que du signe de α^B, est toujours la même, mais la première débute à $\Delta\tilde{\nu} = 0$, la seconde à $\Delta\tilde{\nu} = \alpha^C$, qui est négatif, la troisième à $\Delta\tilde{\nu} = 4\alpha^C$, de telle sorte que, dans le domaine des $\Delta\tilde{\nu}$ positifs, toutes les sous-branches existent, et qu'elles disparaissent une à une dans le domaine des $\Delta\tilde{\nu}$ négatifs; pour K_i infini, la sous-branche s'étend théoriquement dans tout le domaine des fréquences.

Ces deux cas sont donc fondamentalement différents, mais une autre distinction, d'ordre purement mathématique, apparaît dans le calcul de (13), selon le signe de $\gamma - \gamma'$.

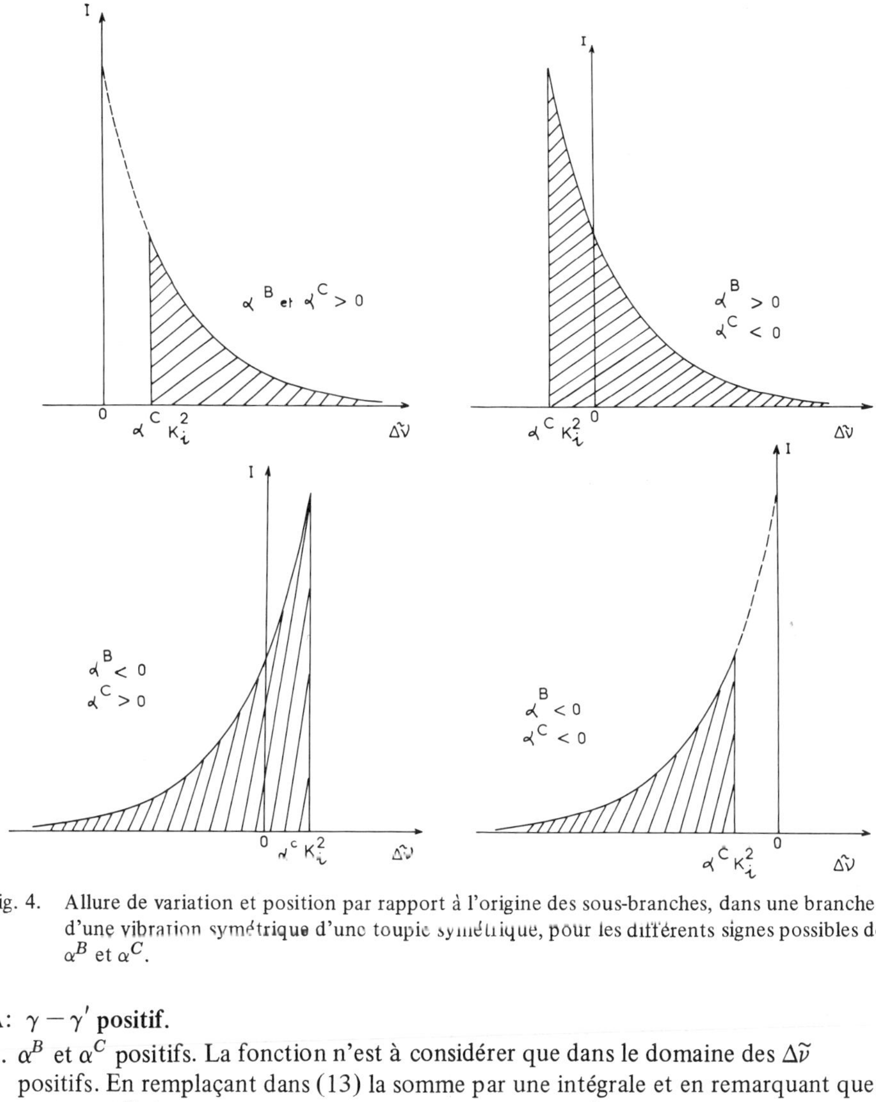

Fig. 4. Allure de variation et position par rapport à l'origine des sous-branches, dans une branche Q d'une vibration symétrique d'une toupie symétrique, pour les différents signes possibles de α^B et α^C.

A: $\gamma - \gamma'$ positif.

1. α^B et α^C positifs. La fonction n'est à considérer que dans le domaine des $\Delta\tilde{\nu}$ positifs. En remplaçant dans (13) la somme par une intégrale et en remarquant que, pour un $\Delta\tilde{\nu}$ donné, il existe les sous-branches correspondant à $K = 0, 1, 2, \ldots (\Delta\tilde{\nu}/\alpha^C)^{1/2}$, il vient

$$I(\Delta\tilde{\nu}) \propto \exp - \left[\frac{\sigma}{\alpha^B}\Delta\tilde{\nu}\right] \operatorname{erf}\left[\frac{\sigma(\gamma-\gamma')}{\alpha^C}\Delta\tilde{\nu}\right]^{1/2} \tag{14}$$

2. α^B positif et α^C négatif. La fonction est à considérer pour toutes les fréquences mais il faut distinguer deux domaines, selon le signe de $\Delta\tilde{\nu}$. Pour les $\Delta\tilde{\nu}$ négatifs, la fonction sera

$$I(\Delta\tilde{\nu}) \propto \exp - \left[\frac{\sigma}{\alpha^B}\Delta\tilde{\nu}\right] \operatorname{erf} c\left[\frac{\sigma(\gamma-\gamma')}{\alpha^C}\Delta\tilde{\nu}\right]^{1/2} \tag{15}$$

Elle devient, dans le domaine des $\Delta\tilde{\nu}$ positifs où l'on somme sur les K de 0 à l'infini en tout point

$$I(\Delta\tilde{\nu}) \propto \exp - \left[\frac{\sigma}{\alpha^B} \Delta\tilde{\nu} \right] \qquad (16)$$

le coefficient de proportionnalité étant le même que dans (15).

B: $\gamma - \gamma'$ **négatif.**

On peut remarquer que

$$\gamma - \gamma' = \frac{C}{B} - \frac{\alpha^C}{\alpha^B}$$

et que si α^B et α^C sont de signes opposés, $\gamma - \gamma'$ ne peut être négatif. On n'a donc à considérer que le cas où α^B et α^C sont positifs. Comme précédemment la fonction n'est à considérer que dans le domaine positif des $\Delta\tilde{\nu}$ et sa présentation mathématique, différente de (14), sera

$$I(\Delta\tilde{\nu}) \propto \exp - \left[\frac{\sigma}{\alpha^B} \Delta\tilde{\nu} \right] \int_{x=0}^{A\Delta\tilde{\nu}^{1/2}} \exp x^2 \, dx \qquad (17)$$

où

$$A = [\sigma(\gamma - \gamma')/\alpha^C]^{1/2}$$

C: $(\gamma - \gamma') = 0$.

Cette éventualité recouvre deux cas particuliers importants:

1. γ et γ' sont différents de zéro et c'est le cas de la toupie symétrique plane (benzène, trifluorure de bore etc.) où $\gamma = \gamma' = -0{,}5$ pour les vibrations totalement symétriques.
2. γ et γ' sont tous deux égaux à zéro; c'est le cas de la toupie sphérique.

Dans ces deux cas, on voit que, selon (12), toutes les sous-branches ont le même facteur d'intensité, égal à 1. La sommation de (13) se ramène à une multiplication par K soit $(\Delta\tilde{\nu}/\alpha^C)^{1/2}$. Pour la toupie symétrique plane il s'agit d'une sommation sur des sous-branches physiquement existantes; pour la toupie sphérique cette sommation correspond à la dégénérescence. On peut voir que le résultat est bien identique à (8).

Aucune analyse de profils expérimentaux ne semble avoir été faite, pour les toupies symétriques et sphériques; certaines sont en cours dans notre laboratoire. Ces études pourraient être utilisées pour obtenir des informations sur l'anharmonicité des vibrations actives en diffusion seulement. Mais il ne faut pas dissimuler les difficultés qu'apportent les bandes chaudes non résolues, que l'on met en évidence par un comportement anormal du profil en fonction de la température, mais qui peuvent interdire toute mesure de α.

CONTOURS DES BANDES DEGENEREES

Toupies sphériques

Deux types de bandes sont à considérer selon la symétrie des niveaux mis en jeu:

1. les bandes correspondant aux transitions $E_g \leftarrow A_{1g}$;

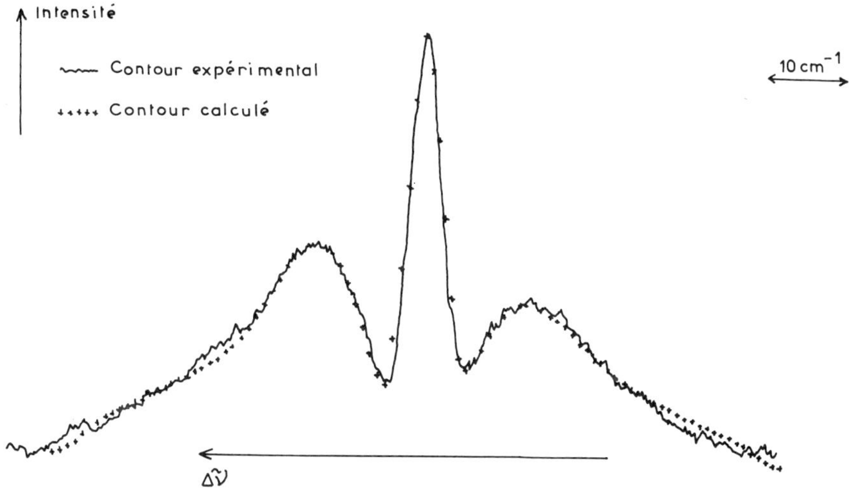

Fig. 5. Bande dégénérée de type E de CF_4. Comparaison des contours calculé et observé. $T = 290$ K, $B = 0{,}1950$ cm^{-1}, $\alpha = 0{,}0004$ cm^{-1}.

2. celles correspondant aux transitions $F_{2g} \leftarrow A_{1g}$.

La structure des premières, appelées dans la suite bandes E, est la plus simple, car il n'y a pas d'interaction de Coriolis du premier ordre dans un niveau E d'une toupie sphérique. En fait, la distribution d'intensité dans les ailes de rotation OP et RS est la même dans ce cas que dans les ailes d'une bande symétrique d'une toupie symétrique, dont les moments d'inertie seraient tous égaux, et ce cas a été traité par Placzek et Teller.[1]

Dans les états de vibration F_2, chaque niveau de rotation est partagé en trois sous-niveaux par l'interaction de Coriolis, de telle sorte que, dans une bande F_2, les règles de sélection n'excluant aucune transition, les cinq branches O, P, Q, R et S sont décomposées en quinze sous-branches. Ces sous-branches sont affectées de facteurs d'intensité différents qui ont été donnés par Teller[5] puis légèrement corrigés par Herranz et Stoicheff.[15] On peut noter que les bandes E peuvent aussi être considérées comme un cas particulier des bandes F_2 où $\zeta = 0$.

Masri et Fletcher les premiers[16] ont calculé des contours par sommation des raies individuelles, donnant les séparations $\Delta\tilde{\nu}_{OP-RS}$ en fonction de ζ et en comparant les prévisions du calcul avec des résultats expérimentaux de la littérature.

Les expressions analytiques dans le cas limite classique sont très simples[17, 18] lorsqu'on ne tient compte que de ζ. Si on veut introduire la constante d'interaction de vibration–rotation α, il faut adopter la méthode de sommation de raies, et on peut remarquer alors[18] que, tant que la valeur de α demeure dans des limites qui ne paraissent pas être dépassées dans l'expérience, la séparation $\Delta\tilde{\nu}_{OP-RS}$ n'est pas affectée de façon significative par α, mais seulement le rapport des intensités des maximums des ailes. Ainsi, si la séparation $\Delta\tilde{\nu}_{OP-RS}$ est sensible à la valeur de ζ seulement, la répartition d'intensité dans la bande est sensible à α et il a été possible de proposer dans certains cas les valeurs de ζ et de α qui permettent de retrouver exactement le tracé expérimental (Figs. 5 et 6).

Clark et Rippon[19] ont donné indépendamment de nous les expressions analytiques des contours, dans le cas limite classique et dans l'approximation d'ordre un du cas

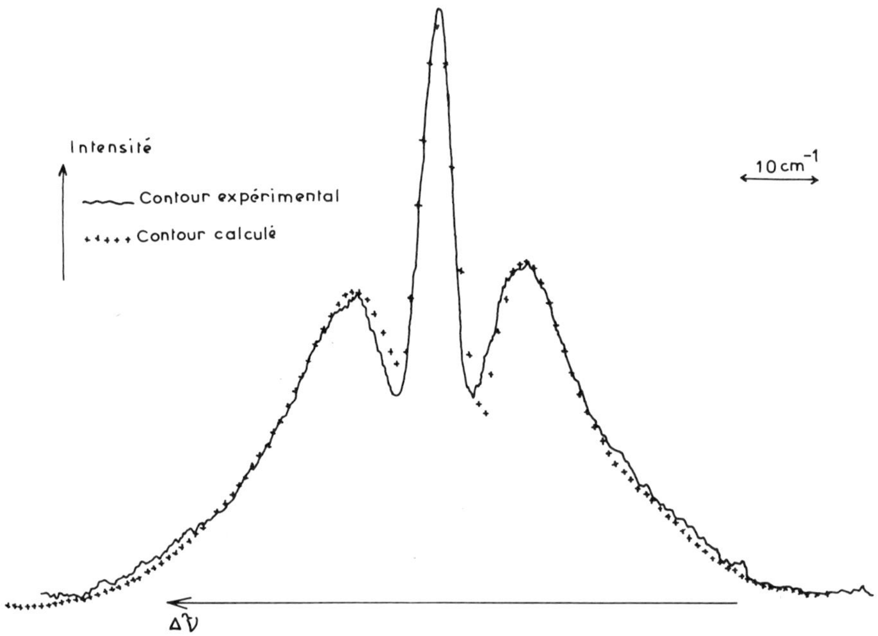

Fig. 6. Band dégénérée de type F_2 de SiF_4 (ν_4). Comparaison des contours calculés et observés. $T = 293$ K, $B = 0{,}1385$ cm^{-1}, $\alpha = -0{,}0002$ cm^{-1}, $\zeta = -0{,}10$.

quantique, selon la méthode déjà employée par Placzek et Teller.[1] Ils ont déterminé la valeur de ζ pour des bandes F_2 de plusieurs tétrahalogénures. Pour ces molécules, il y a quatre fréquences fondamentales et cinq constantes indépendantes dans l'expression du potentiel harmonique, de telle sorte que la connaissance des ζ est nécessaire pour le calcul du potentiel.

Toupies symétriques

Les règles de sélection des transitions mettant en jeu un niveau dégénéré de vibration dans la toupie symétrique ne sont pas simples. A la règle générale sur les J ($\Delta J = 0, \pm 1, \pm 2$) vient s'ajouter la règle sur les K, qui dépend en fait de la nature des éléments de matrice de polarisabilité non nuls dans la transition considérée. Pour les transitions où ces éléments sont la paire dégénérée (α_{yz}, α_{zx}), on a $\Delta K = \pm 1$; si c'est la paire ($\alpha_{xx} - \alpha_{yy}, \alpha_{xy}$) on a $\Delta K = \pm 2$.

Pour les molécules de symétrie telle que ces deux paires dégénérées appartiennent à des types de symétrie différents, il y aura des bandes dégénérées avec $\Delta K = \pm 1$, et d'autres avec $\Delta K = \pm 2$. Pour illustrer cela, prenons l'exemple du groupe D_{3h}, dans lequel la paire (α_{yz}, α_{zx}) est de symétrie E'', et la paire ($\alpha_{xx} - \alpha_{yy}, \alpha_{xy}$) de symétrie E'. Pour les transitions $E'' \leftarrow A_1'$, on a $\Delta K = \pm 1$, et pour $E' \leftarrow A_1'$, $\Delta K = \pm 2$. Lorsqu'on rencontre cette situation, il est possible de calculer le contour de la bande Raman, lequel dépend non seulement du rapport des moments principaux d'inertie, mais aussi de la valeur de ζ. Aucune expression analytique n'a été établie à notre connaissance, mais des calculs par sommation de raies individuelles ont été faits, lesquels ne couvrent que superficiellement le domaine des valeurs possibles du rapport des moments d'inertie. Ainsi Masri[20] a traité avec quelques détails le cas des toupies aplaties; l'un de nous[13] a donné quelques valeurs complémentaires des séparations $\Delta\tilde{\nu}_{OP-RS}$ en fonction de ζ et du rapport des moments.

En fait les molécules de symétrie telle que les paires $(\alpha_{xx} - \alpha_{yy}, \alpha_{xy})$ et $(\alpha_{yz}, \alpha_{zx})$ sont de types différents sont rares. Leurs groupes de symétrie sont C_p, D_p, C_{pv} avec $p \geqslant 4$, C_{ph}, D_{ph} avec $p \geqslant 3$, D_{pd} avec $p = 2, 4, 5$ etc., S_{2pu} avec $p = 2, 4, 6$ etc. Mais la plupart des toupies symétriques sont de symétrie C_{3v}. Dans ce cas, comme dans tout autre où les deux paires dégénérées d'éléments de matrice de polarisabilité sont de même symétrie, les bandes dégénérées sont en fait la superposition de deux systèmes ($\Delta K = \pm 1$ et $\Delta K = \pm 2$), dont les répartitions d'intensité sont différentes et qui sont affectés de facteurs d'intensité vibrationnels différents, ce qui semble interdire tout calcul de contour. Toutefois Masri[21] a pu montrer que dans le cas de la ν_8 de CF_3–CN, le système dû à $\Delta K = \pm 2$ correspondait à un contour très aplati, de telle sorte que les maximums des ailes étaient attribuables, dans ce cas précis, au système dû à $\Delta K = \pm 1$, ce qui lui a permis de déterminer ζ.

On doit rappeler enfin dans ce cadre le travail déjà ancien de Mathai, Shepherd et Welsh[7] qui ont analysé les contours de certaines bandes du cyclopropane, où les bandes avec $\Delta K = \pm 1$ et $\Delta K = \pm 2$ sont distinctes (symétrie D_{3h}).

CONCLUSION

Alors que la spectroscopie Raman des gaz est devenue expérimentalement très accessible, l'étude des contours semble devoir apporter des renseignements précieux dans l'analyse des spectres Raman en phase gazeuse, en général, et plus particulièrement dans les problèmes d'attribution, de mise en évidence de bandes chaudes. Les profils des branches Q peuvent conduire à des valeurs assez précises de la constante α pour les vibrations symétriques, alors que la séparation des ailes des bandes dégénérées permet d'atteindre la valeur de ζ, dans les cas favorables.

Ces données peuvent compléter heureusement celles que, avec des considérations analogues, on obtient par absorption infrarouge.

REFERENCES

1 G. Placzek et E. Teller, *Z. Phys.* **81**, 209 (1933).
2 B. P. Stoicheff, *Advances in Spectroscopy,* Vol. 1, (Ed. H. W. Thompson), Interscience, New York, 1959, p. 91.
3 B. P. Stoicheff, *Can. J. Phys.* **36**, 218 (1958).
4 I. M. Mills, *J. Mol. Spectry.* **5**, 334 (1960).
5 E. Teller, *Hand- und Jahrbuch Chem. Phys. Vol. 9* **11**, 43 (1934).
6 M Lapp, L. M. Goldman et C. M. Penney, *Science* **175**, 1112 (1972).
7 P. M. Mathai, G. G. Shepherd et H. L. Welsh, *Can. J. Phys.* **34**, 1448 (1956).
8 S. Brodersen, *J. Mol. Spectry.* **6**, 511 (1961).
9 R. Gaufrès et S. Sportouch, *C. R. Acad. Sci., Paris* **272**, 995 (1971).
10 Y. Morino et T. Nakagawa, *J. Mol. Spectry.* **26**, 496 (1968).
11 A. E. Douglas, C. K. Møller et B. P. Stoicheff, *Can. J. Phys.* **41**, 1174 (1963).
12 M. A. A. Clyne et J. A. Coxon, *J. Mol. Spectry.* **33**, 381 (1970).
13 S. Sportouch, *Thèse de doctorat ès-Sciences,* Montpellier, 1972.
14 R. Gaufrès et S. Sportouch, *J. Mol. Spectry.* **39**, 527 (1971).
15 J. Herranz et B. P. Stoicheff, *J. Mol. Spectry.* **10**, 448 (1963).
16 F. N. Masri et W. H. Fletcher, *J. Chem. Phys.* **52**, 5759 (1970).

17 S. Sportouch, C. Lacoste et R. Gaufrès, *J. Mol. Structure* **9**, 119 (1971).
18 S. Sportouch et R. Gaufrès, *J. Chim. Phys.* **69**, 470 (1972).
19 R. J. H. Clark et D. M. Rippon, *Chem. Commun.* 1295 (1971).
20 F. N. Masri, (communication privée – à paraître).
21 F. N. Masri, (communication privée – à paraître).

63. VAPOUR PHASE RAMAN SPECTRA OF SPHERICAL TOP MOLECULES

R. J. H. Clark and D. M. Rippon
Christopher Ingold Laboratories, University College, London, England

INTRODUCTION

It is well known that the Raman and infrared band contours of the triply degenerate fundamentals of a spherical top molecule depend markedly on the Coriolis forces in the molecule. If the rotational fine structure of these bands can be resolved in either the infrared or the Raman, the Coriolis constants may be evaluated directly from the line spacing in the rotational branches. However, the B values of most molecules are such that the line spacings are too small currently to be resolved and hence only band contours can in general be observed.

In the case of the infrared spectra of spherical top molecules, the Coriolis constants may be evaluated from a knowledge of the separations between the P and R branch maxima, $\Delta\nu_{PR}$, by use of the equation[1, 2]

$$\Delta\nu_{PR} = 4(BkT/hc)^{1/2}(1 - \zeta)$$

where B, the rotational constant is $h/8\pi^2 cI_B$ (in cm^{-1}), T is the temperature in °K, and ζ is the Coriolis constant for the triply degenerate fundamental under study. This equation has already been applied successfully to the determination of Coriolis constants for many spherical top molecules, including the tetrafluorides CF_4, SiF_4 and GeF_4 and the tetrachlorides $SiCl_4$, $GeCl_4$ and $SnCl_4$.

Experimental difficulties, principally associated with the lack of suitable window materials in the low frequency infrared region, have so far precluded the recording of the vapour phase infrared spectra of metal tetrabromides and metal tetraiodides, and hence no Coriolis constants are available for these molecules. However, with the development of the powerful Ar^+ and Kr^+ laser sources and of photon counting techniques, it is now possible to study the Raman band contours of heavy spherical top molecules. Apart from the intensity of colour of the metal tetraiodides, the tetrabromides and tetraiodides pose no particular problem in the study of their vapour phase Raman spectra, because silica containers may be used at most exciting frequencies.

Because of the less restrictive selection rules in the Raman effect, which allow changes in the rotational quantum number ΔJ of 0, ± 1, and ± 2, the band contours are

Table 1. Raman shifts of the Q branches and the $\Delta\nu_{OP, RS}$ branch separations for the fundamentals of each tetrahalide (cm^{-1}).

Halide	Sample Temp. °C	$\nu_1(a_1)$ Max	$\nu_2(e)$ Max	$\nu_2(e)$ $\Delta\nu_{OP, RS}$	$\nu_3(t_2)$ Max	$\nu_3(t_2)$ $\Delta\nu_{OP, RS}$	$\nu_4(t_2)$ Max	$\nu_4(t_2)$ $\Delta\nu_{OP, RS}$	$\nu_2(e)\Delta\nu_{OP, RS}$ Expt. 300 K	$\nu_2(e)\Delta\nu_{OP, RS}$ Calc. 300 K
CF_4	20	908.4	434.5	27.9	1283.0	26.7	631.2	12.8	28.2	27.9
SiF_4	22	800.8	264.2	23.8	1029.6	14.5	388.7	23.0	24.0	23.7
GeF_4	20	735.0	202.9	22.0	800.1	23.7	273.1	8.6, 21.6	22.3	22.0
CCl_4	140	460.0	214.2	17.7	792.0	–	313.5	9.8	15.1	15.2
$SiCl_4$	65	423.1	145.2	14.4	616.3	–	220.3	12.6	13.6	13.3
$GeCl_4$	100	396.9	125.0	14.7	459.1	–	171.0	14.8	13.2	12.7
$SnCl_4$	97	369.1	95.2	13.5	408.2	13.2	126.1	14.0	12.2	11.8
$TiCl_4$	65	389	114	12.8	498	–	136	13.6	12.1	12.4
$ZrCl_4$	380	377	98	16 ± 1	418	–	113	–	10.8	11.6
$HfCl_4$	380	382	101.5	15 ± 1	390	–	112	–	10.2	11.5
$SiBr_4$	210	246.7	84.8	10.7	494	–	133.6	3.1, 8.5	8.4	8.3
$GeBr_4$	135	235.7	74.7	9.3	332	–	111.1	8.7	8.0	7.8
$SnBr_4$	100	222.1	59.4	8.3	284	–	85.9	8.9	7.4	7.4
$TiBr_4$	125	231.5	68.5	9.3	393	–	88	8.9	8.0	7.8
$ZrBr_4$	380	225.5	60.0	11.0	315	–	72	–	7.5	7.4
$HfBr_4$	380	235.5	63.0	10.3	273	–	71	–	7.0	7.4
SiI_4	210	165.9	57.4	7.7	404	–	88.8	–	6.1	5.8
GeI_4	270	156.0	51.6	8.1	273	–	77.3	6.6	6.0	5.7
SnI_4	260	147.7	42.3	7.2	~210	–	62.8	7.2	5.4	5.4
ZrI_4	420	158	43	8.0	254	–	55	–	5.3	5.4
HfI_4	420	158	55	–	224	–	63	–	–	5.4

more complicated than are infrared band contours. The original theoretical treatment of Raman band contours, due to Placzek and Teller,[3] was confined to the totally symmetric stretching modes of symmetric top molecules of which spherical top molecules are a limiting case. The treatment was developed for the case in which the Coriolis constant is zero, and is therefore applicable herein only to the doubly degenerate fundamental of spherical top molecules. The present results, which relate to the OP, RS branch maxima separations, $\Delta\nu_{OP, RS}$, of the $\nu_2(e)$ fundamental of tetrahedral molecules, provide a general test for this theory.

A method for the determination of Coriolis constants from the Raman band contours of t_2 fundamentals has been developed[4] and compared with that of Masri and Fletcher.[5] A related analysis has been developed by Sportouch, Lacoste and Gaufrès.[6] Where ζ values are also available from infrared data, these agree well with the presently determined Raman values.

RESULTS

Tetrahedral molecules

In accordance with the selection rules for tetrahedral molecules of the type MX_4, the $\nu_1(a_1)$ fundamental consists solely of a Q branch in each case. Each of the other fundamentals, however, has OPQRS structure.

For fundamentals of e symmetry, the Placzek–Teller treatment leads (for the

Table 2. Branch-maxima separations $\Delta\nu_{OP, RS}$ transformed to 300 K for each halide, and the derived Coriolis constants.

$\nu_4(t_2)$ fundamental				
Halide	$B(cm^{-1})$	$\Delta\nu_{OP, RS}$	ζ_4(expt.)	ζ_4(calc.)[a]
CF_4	0.1919	12.9_5	−0.36	−0.43
SiF_4	0.1385	23.2	−0.08	−0.09
GeF_4	0.1193	8.7	0.35	0.24
		21.8_5[c]	0.2 ± 0.2[b]	0.24
CCl_4	0.0572	8.3_5	−0.47	−0.45
$SiCl_4$	0.0437	11.9	−0.20	−0.22
$GeCl_4$	0.0399	13.2_5	0.2 ± 0.2[b]	0.09
$SnCl_4$	0.0334	12.6	0.2 ± 0.2[b]	0.21
$TiCl_4$	0.0379	13.6	0.2 ± 0.2[b]	0.03
$SiBr_4$	0.0171	2.4	−0.27	−0.37
		6.6_5[c]	−0.29	−0.37
$GeBr_4$	0.0151	7.4_5	−0.11	−0.15
$SnBr_4$	0.0133	8.0	0.2 ± 0.2[b]	−0.01
GeI_4	0.0080	4.9	−0.24	−0.27
SnI_4	0.0071	5.4	0.00	−0.14

$\nu_3(t_2)$ fundamental				
Halide	$\Delta\nu_{OP, RS}$	ζ_3(expt.)	ζ_3(calc.)[a]	$\zeta_3 + \zeta_4$ (Raman, expt.)
CF_4	27.0	0.83	0.93	0.47
SiF_4	14.6	0.63	0.59	0.55
GeF_4	23.9_5	0.2 ± 0.2	0.26	*ca* 0.55[d]
$SnCl_4$	11.9	0.2 ± 0.2	0.29	*ca* 0.4[d]

[a] S. J. Cyvin, *Molecular Vibrations and Mean Square Amplitudes*, Elsevier, Amsterdam, 1968.
[b] The value of $\Delta\nu_{OP, RS}$ leads to a value of Δx which lies in a part of the Δx versus ζ plot which is insensitive to ζ.
[c] Two pairs of maxima occur on either side of the Q branch, as expected for the ζ values concerned.
[d] Very uncertainly established *cf.* footnote *b*.

limiting situation in which $B = 0$) to the expression

$$\Delta\nu_{OP, RS} = 4.412\,(BkT/hc)^{1/2}$$

For finite B values, the constant term is reduced slightly, e.g.

$$\Delta\nu_{OP, RS} = 4.401\,(BkT/hc)^{1/2} \text{ for } B = 2.085 \text{ cm}^{-1}$$

and the $\Delta\nu_{OP, Q} < \Delta\nu_{Q, RS}$ i.e. some frequency (and also intensity) asymmetry develops. The experimental and calculated branch separations of the *e* fundamentals of various metal tetrahalides are given in Table 1. In all but two cases, the two quantities agree to within ± 0.5 cm^{-1}; this is a convincing demonstration of the validity of the analysis.

The separations of the principal branch maxima of the $\nu_3(t_2)$ and $\nu_4(t_2)$ fundamentals of each molecule, are, in general, different from each other, and also from that of the $\nu_2(e)$ fundamental of the same molecule on account of Coriolis coupling. This alters the spacing between the rotational lines so that the rotational energies of the three sub-levels of a triply degenerate fundamental are then different and given, in first order, by the expressions

$$F^+ = BJ(J+1) + 2B\zeta(J+1)$$
$$F^0 = BJ(J+1)$$
$$F^- = BJ(J+1) - 2B\zeta J$$

where the dependence of B on the vibrational quantum number has been ignored. As each of the permitted Raman transitions $\Delta J = 0, \pm 1, \pm 2$, goes to all three of the upper states, there are fifteen permitted sub-branches each with a different relative intensity. For $1 > \zeta > 0$, the high frequency wing is therefore a composite of the Q^+, R^+, R^0, R^-, S^-, S^0, and S^+ sub-branches while the low frequency wing is a composite of the Q^-, P^+, P^0, P^-, O^+, O^0 and O^- sub-branches. Expressions for the intensity of each sub-branch have been developed, those for each wing summed and the overall expression differentiated in order to locate the maximum. The branch separations (Δx) are then expressed as a function of the Coriolis constant, where Δx is a reduced frequency defined by

$$\Delta x = (hc/BkT)^{1/2} \Delta\nu_{OP,\, RS}$$

Secondary maxima also occur for certain values of ζ. The zeta values so derived are given in Table 2. The Coriolis constants so obtained have been used to help to define better the force fields of the molecules. Similar analyses have been carried out for OsO_4, P_4, As_4, and the molecules $M(CH_3)_4$ where M = C, Si, Ge, Sn or Pb.

Octahedral molecules

Corresponding analyses have been carried out for the octahedral molecules SF_6, SeF_6, TeF_6, MoF_6, WF_6 and UF_6. The spectra of these molecules will also be presented and discussed.

REFERENCES

1 W. F. Edgell and R. E. Moynihan, *J. Chem. Phys.* **27**, 155 (1957).

2 R. S. McDowell, *J. Chem. Phys.* **41**, 2557 (1964); **43**, 319 (1965).

3 G. Placzek and E. Teller, *Z. Phys.* **81**, 209 (1933).

4 R. J. H. Clark and D. M. Rippon, *Chem. Commun.* 295 (1971); *J. Mol. Spectry.* (in press); Y. M. Bosworth, R. J. H. Clark and D. M. Rippon, *J. Mol. Spectry.* (submitted for publication).

5 F. N. Masri and W. H. Fletcher, *J. Chem. Phys.* **52**, 5759 (1970).

6 S. Sportouch, C. Lacoste and R. Gaufrès, *J. Mol. Structure* **9**, 119 (1971); *J. Chim. Phys.* (in press).

64. ETUDE EXPERIMENTALE DE L'ELARGISSEMENT PAR LA PRESSION DES RAIES DE ROTATION DU SPECTRE RAMAN DE L'ACIDE CHLORHYDRIQUE

D. Fabre, G. Widenlocher, M. M. Thiery,
H. Vu et B. Vodar
Laboratoire des Hautes Pressions, C.N.R.S.,
Bellevue, France

ABSTRACT

An experimental study of the pressure perturbations of pure rotational Raman lines is presented for pure HCl and for HCl perturbed by argon. For pure HCl, the pressure range lies between 1 and 35 bar, and for HCl perturbed by argon the pressure range is 0–275 bar. The true linewidth is deduced from the apparent experimental linewidth after corrections taking into account both slit function and eventual line overlapping. The true linewidth is proportional to the perturber pressure and the dependence of the broadening coefficient on the *J* quantum number is sensitive to the nature of the perturber. These experimental results are in good agreement with the predictions of Gray's theory based on the effects of anisotropic intermolecular forces.

L'étude de l'élargissement des raies spectrales sous l'effet de la pression dans les spectres Raman des gaz présente de l'intérêt par les informations qu'elle peut apporter sur la nature des forces intermoléculaires. Différentes théories ont été proposées pour relier le phénomène d'élargissement et de déplacement des raies d'absorption, quand la densité d'un gaz augmente, aux interactions dont il est le siège; une extension de ces théories aux phénomènes de diffusion Raman a été faite par Fiutak et van Kranendonk,[1,2] puis Gordon,[3] May,[4] Gray,[5,6]. Plusieurs études expérimentales (travaux de l'école russe, résumés par Bazhulin et Pinter;[7,8] travaux de l'école canadienne résumés par Jammu *et al.*[9] et Gray et Welsh[10]) donnent l'élargissement par la pression des raies rotationnelles de molécules simples, généralement non polaires ou faiblement polaires; on constate le plus souvent une variation du coefficient d'élargissement suivant le nombre quantique de rotation *J*. Nous nous sommes proposé d'étudier expérimentalement le cas d'une molécule très polaire, l'acide chlorhydrique, à la température ambiante; la même étude a été faite parallèlement, dans des conditions légèrement différentes, par Rich et Welsh[11] à Toronto. Il était intéressant de comparer à l'expérience les prévisions théoriques de Gray.[6] Ce dernier, en effet, à la suite d'une étude générale de l'élargissement par la pression des raies spectrales (infrarouge et Raman) de gaz peu denses, sous l'action des forces intermoléculaires anisotropes, détermine, pour la diffusion Raman, dans le cas particulier de HCl, les sections efficaces optiques et les coefficients d'élargissement à partir du calcul des quatre interactions multipolaires d'ordre inférieur; il en déduit pour le coefficient d'autoélargissement et sa dépendance de *J*, un comportement particulier qu'on ne trouve pas dans le cas des molécules non polaires. Il calcule également le coefficient d'élargissement dans le cas d'une perturbation de HCl par une pression d'argon, en prenant en considération l'interaction de dispersion.

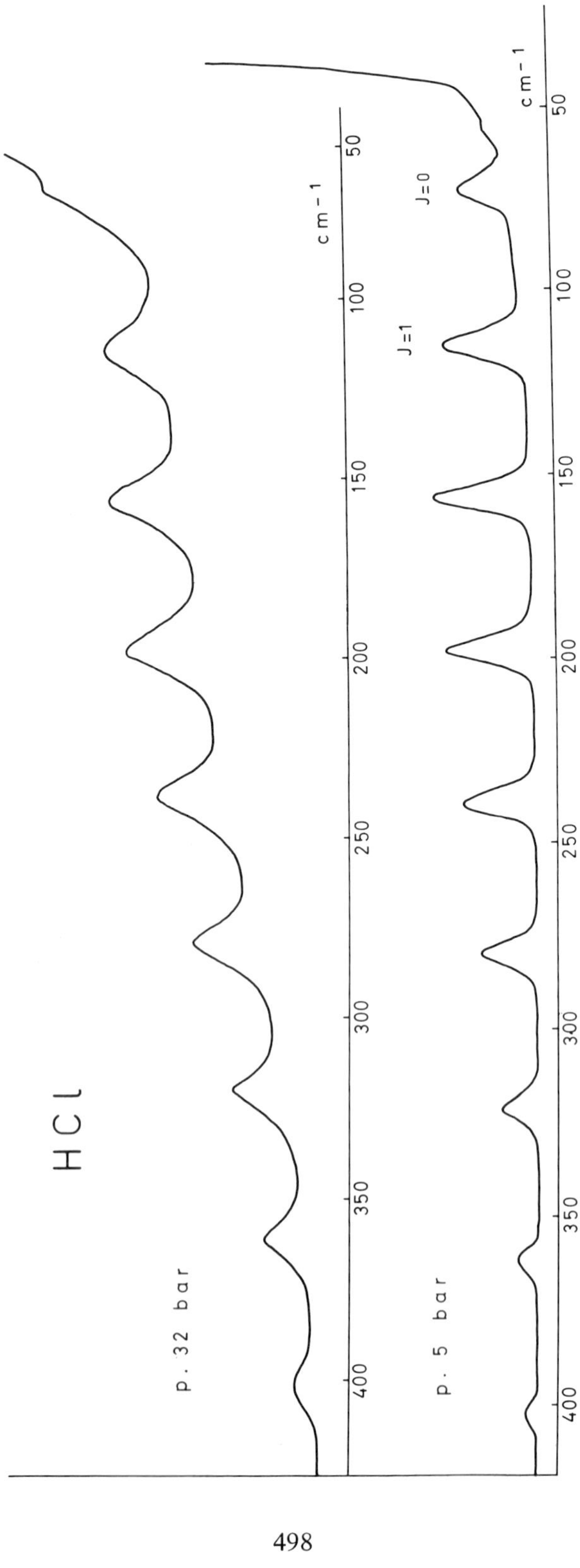

Fig. 1. Spectres de rotation pure de HCl pour des pressions de gaz (HCl) de 5 bar et 32 bar.

Le dispositif expérimental utilisé pour cette étude a été décrit par ailleurs[12]; la cuve est en acier inoxydable à fenêtres de saphir; on a fait varier la pression du gaz entre environ 1 et 35 bar dans le cas de HCl pur; au-delà, le recouvrement des raies de rotation élargies par la pression, rend les mesures de largeur trop imprécises; dans le cas de HCl perturbé par l'argon, la pression de HCl était de 5 bar et on a fait varier la pression d'argon entre 0 et 275 bar. Le rayonnement excitateur (6.328 Å) est fourni par un laser à He–Ne de 150 mW, et la lumière diffusée par le gaz, analysée dans un monochromateur double, à réseaux de 1200 traits par millimètre, équipé d'un photomultiplicateur refroidi. Les composantes du rayonnement diffusé n'ont pas été séparées: cependant, la proportion de la composante dépolarisée est prédominante dans le signal obtenu, car dans le domaine spectral balayé, le monochromateur transmet avec un meilleur rendement le rayonnement polarisé horizontalement, c'est-à-dire perpendiculairement à la direction de polarisation de la radiation excitatrice. La Fig. 1 montre deux spectres de rotation pure de HCl pour des pressions de respectivement 5 et 32 bar de HCl.

Le rayonnement Raman diffusé étant faible, il a fallu utiliser des fentes relativement larges en regard des largeurs de raies mesurées; il a, par conséquent, été nécessaire d'effectuer une correction de fonction d'appareil. D'autre part, le recouvrement des raies lorsque la pression augmente et les élargit, contribue également à donner à chaque raie, sur les spectres enregistrés, un profil apparent différent de celui qu'elle aurait si elle était isolée. Dans nos conditions de travail, la fonction de fentes, évaluée par enregistrement de la raie monchromatique donnée par le laser, peut être considérée comme une fonction triangle de largeur 6,3 cm^{-1}. Pour des pressions inférieures à une dizaine de bars, les raies rotationnelles sont suffisamment isolées pour qu'on puisse considérer leur profil expérimental comme donné simplement par le produit de convolution de leur profil vrai (de dispersion) par la fonction de fentes; au-delà, on a tenu compte du recouvrement des raies en faisant une correction dont le calcul a été simplifié par les approximations suivantes:

1. Etant donné l'espacement moyen relativement grand des raies ($d \sim 41{,}6$ cm^{-1}), on admet qu'à chaque raie s'ajoute seulement la contribution de sa plus proche voisine de chaque côté.
2. On attribue à ces deux raies même intensité et même largeur qu'à la raie étudiée.
3. On suppose qu'elles conservent chacune, lorsque la pression augmente et que le recouvrement est amorcé, le profil de Lorentz caractéristique de la raie isolée. Sauf dans le cas d'une raie $J = 0$, qui n'a qu'une raie adjacente, le profil expérimental d'une raie est alors donné par

$$J(\nu') = I_0\gamma^2 \int_{\nu'-\delta}^{\nu'+\delta} \left[\frac{1}{\nu^2+\gamma^2} + \frac{1}{(\nu+d)^2+\gamma^2} + \frac{1}{(\nu-d)^2+\gamma^2}\right] \left[1 - \frac{|\nu'-\nu|}{\delta}\right] d\nu$$

où $J(\nu')$ est l'intensité expérimentale à la fréquence ν', l'origine des fréquences étant prise au centre de la raie; I_0 et γ sont respectivement l'intensité maximale et la mi-largeur à mi-hauteur de la raie isolée idéale; d est l'écartement moyen entre deux raies de rotation consécutives; et δ est la largeur spectrale des fentes égales du monochromateur. $J(d/2)$ représente la valeur de l'intensité minimale de la raie (à mi-distance du maximum suivant)

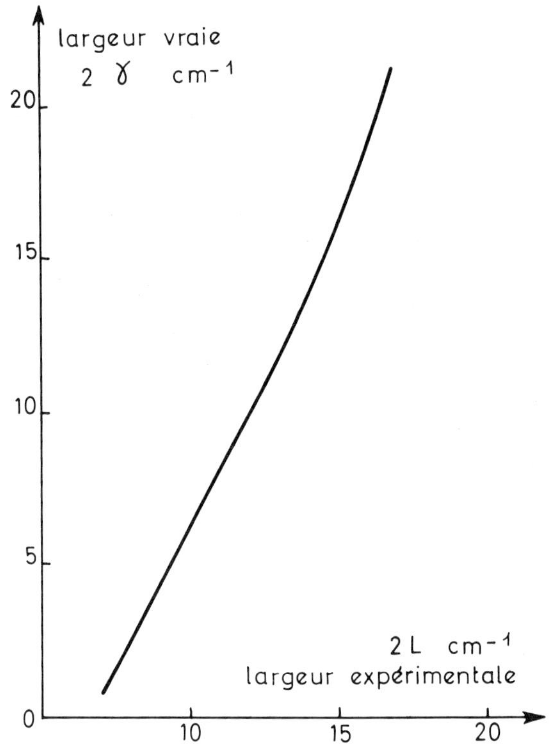

Fig. 2. Courbe de correction, donnant les largeurs vraies 2γ des raies de rotation en fonction de leur largeur apparente $2L$.

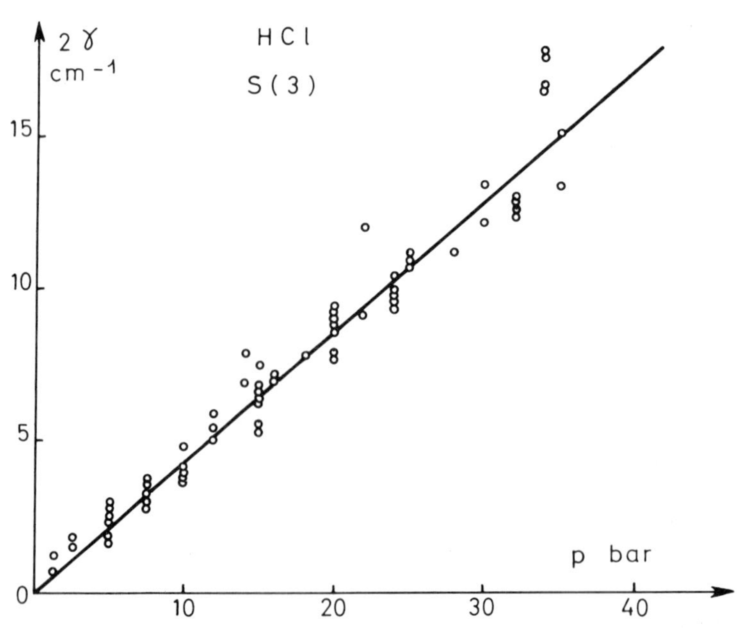

Fig. 3. Variations de la largeur (vraie) 2γ de la raie de rotation $S(3)$, en fonction de la pression de HCl.

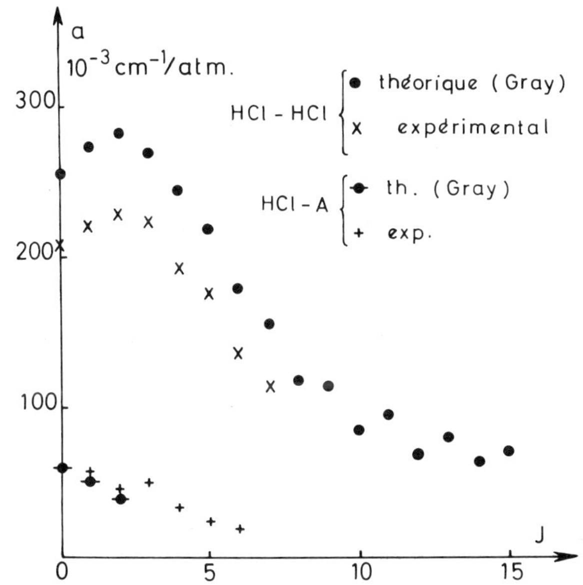

Fig. 4. Coefficient d'élargissement par la pression des raies Raman rotationnelles de HCl, suivant les valeurs du nombre quantique J. $a = \gamma/p$; γ mi-largeur à mi-hauteur de la raie; Cas perturbateur: HCl. • Valeurs théoriques (Gray) × Valeurs expérimentales; Cas perturbateur: Argon. -•- Valeurs théoriques (Gray) + Valeurs expérimentales.

et $J(0)$ son maximum; la demie largeur apparente L est alors donnée par

$$J(L) = \tfrac{1}{2}[J(0) + J(d/2)]$$

cette égalité fournit une relation implicite entre les largeurs apparente et réelle $2L$ et 2γ; elle a été résolue numériquement sur une calculatrice Pallas avec $d = 41{,}6$ cm^{-1} et $\delta = 6{,}3$ cm^{-1}; la Fig. 2 montre la correspondance entre $2L$ et 2γ. On voit que pour les raies de largeur inférieure à environ 14 cm^{-1}, l'influence de la largeur de la fente est prédominante, d'où il résulte que la largeur apparente de la raie est supérieure à sa largeur réelle; tandis que lorsque les raies sont plus larges, l'effet du recouvrement sur le profil apparent de la raie devient prépondérant, ce qui se traduit par un rétrécissement apparent de la raie.

Le calcul n'a pas été fait dans le cas de $J = 0$, car dans le domaine de pressions rendant indispensable une correction (au moins approchée) de recouvrement de raies, l'élimination du fond continu sous-jacent, dû à l'élargissement de la composante anisotrope de la raie Rayleigh, devient incertaine et rend de toutes façons la mesure de largeur de la raie trop imprécise; le coefficient d'élargissement pour la raie $J = 0$, dans le cas HCl–HCl, a été déduit de mesures faites sous des pressions inférieures à 15 bar, avec uniquement une correction de fentes; nous n'avons pu le mesurer dans le cas de HCl–A.

Les variations avec la pression de la largeur des différentes raies de rotation $S(J)$ suivent, dans la limite de précision expérimentale, une loi linéaire; on peut en déduire un coefficient d'élargissement $a = \gamma/p$; dans le cas HCl pur, la droite représentative, extrapolée par une pression nulle, passe par l'origine. La Fig. 3 montre, à titre d'exemple, les résultats obtenus, dans ce cas, pour la raie $S(3)$.

Les valeurs obtenues pour a, Fig. 4, diffèrent suivant la valeur du nombre quantique de rotation, conformément à la théorie de Gray[6] qui prévoit que:

1. **Cas HCl–HCl.** Par suite du fort moment dipolaire de HCl ($\mu \sim 1D$) et de sa grande constante rotationnelle ($B_e \sim 10$ cm^{-1}) les interactions dipolaires doivent apporter une grande contribution à l'élargissement, et les collisions résonantes jouent le rôle principal dans le phénomène; une collision ayant d'autant plus de probabilité d'être résonante que la molécule rayonnante est dans un niveau J très peuplé, ce sont les raies les plus intenses qui doivent être aussi les plus élargies (en ce qui concerne l'interaction dipôle–dipôle); on doit donc s'attendre à un coefficient d'auto-élargissement maximal vers $J = 3$, alors que pour les molécules non polaires, dont l'élargissement résulte principalement des interactions quadrupolaires et de dispersion, le coefficient décroit de façon monotone quand J augmente. Les valeurs expérimentales, Fig. 4, qui passent par un maximum pour $J = 2$, confirment ces prévisions; on remarque cependant qu'elles sont inférieures aux valeurs calculées par Gray; en revanche, elles sont supérieures à celles de Rich et Welsh[11] qui mesurent l'auto-élargissement de HCl dans la gamme de pression 3–14 atmos. Pour les grandes valeurs de J, l'élargissement est principalement dû aux interactions dipôle–quadrupôle, dont Gray montre qu'elles conduisent à une alternance des intensités du coefficient a suivant la parité de J, comme le montrent les valeurs théoriques de la Fig. 4. Malheureusement la faible intensité des raies de nombre quantique supérieur à 8 ne nous permet pas de mesures de largeur assez précises pour vérifier cette prédiction théorique.

2. **Cas HCl–A.** L'absence de collisions résonantes conduit à des sections efficaces optiques beaucoup plus petites que dans le cas HCl–HCl, donc à un coefficient d'élargissement plus faible, et l'on doit s'attendre à une décroissance monotone de a quand J augmente. C'est ce que l'on trouve expérimentalement, avec cependant un léger écart à la loi de décroissance de a pour la raie $S(3)$; dans l'ensemble, nos valeurs expérimentales montrent le même ordre de grandeur que celles calculées par Gray pour les trois premières raies de rotation (Fig. 4).

En conclusion, les résultats expérimentaux semblent, en première approximation, confirmer la théorie; il serait donc intéressant de reprendre les mesures dans de meilleures conditions expérimentales pour pouvoir suivre l'évolution du coefficient d'élargissement pour les grandes valeurs de J et comparer avec plus de précision les valeurs expérimentales et théoriques.

REMERCIEMENTS

Nous remercions le Dr. C. Gray d'avoir bien voulu nous communiquer avant parution, son article sur les spectres Raman des gaz.[10]

REFERENCES

1 J. Fiutak et J. van Kranendonk, *Can. J. Phys.* **40**, 1085 (1962).
2 J. Fiutak et J. van Kranendonk, *Can. J. Phys.* **41**, 21 (1963).
3 R. G. Gordon, *J. Chem. Phys.* **44**, 3083 (1966).
4 A. D. May et J. D. Poll, *Can. J. Phys.* **43**, 1836 (1965).
5 C. G. Gray et J. van Kranendonk, *Can. J. Phys.* **44**, 2411 (1966).
6 C. G. Gray, *Thesis*, University of Toronto, 1967; C. G. Gray, *Chem. Phys. Letters* **8**, 528 (1971).

7 P. A. Bazhulin, *Soviet Phys. Usp.* **5**, 661 (1962).
8 F. Pinter, *Opt. Spectry.* **17**, 428 (1964).
9 K. S. Jammu, G. E. St. John et H. L. Welsh, *Can. J. Phys.* **44**, 797 (1966).
10 C. G. Gray, H. L. Welsh, in *Essays in Structural Chemistry*, (Eds. D. A. Long, A. J. Downs et L. A. K. Stavely), Macmillan, London, 1971.
11 N. H. Rich et H. L. Welsh, *Chem. Phys. Letters* **11**, 292 (1971).
12 R. Aumont, E. Dayan, B. Vodar, H. Vu et G. Widenlocher, *C.R. Acad. Sci. Paris* **268**, 1565 (1969).

65. PURE ROTATION AND VIBRATION–ROTATION RAMAN SPECTRA OF FLUORINE AND TRITIUM

H. G. M. Edwards, D. A. Long and R. Love
School of Chemistry, University of Bradford, Bradford, England

INTRODUCTION

This chapter gives an account of experimental studies of the pure rotational Raman spectra of tritium and fluorine and part of the vibration–rotation Raman spectra of fluorine and tritium. There are no reports of previous studies of the spectra of tritium. The spectra of fluorine have been studied previously but the present work yields the first experimental value of the centrifugal stretching constant D_0 and a more accurate value of B_0 and hence r_0.

It had been hoped to employ direct photoelectric recording of the spectra in these studies. As far as signal-to-noise was concerned, spectra of good quality could be directly recorded using an argon ion laser (about 1.0–1.5 W output at 4880 Å) for excitation, simple sampling arrangements with the sample at the focus of the laser beam, a Spex 1400 double monochromator for dispersion and photon counting for detection. However, it was found that the lead screw of the monochromator did not enable the necessary precision in frequency measurement to be achieved. Although many preliminary studies were carried out using direct recording and some spectra obtained in this way will be reproduced here, all frequency measurements were obtained from photographically recorded Raman spectra on which were superimposed iron arc calibration spectra.

EXPERIMENTAL PROCEDURE

A diagram of the simple arrangements for illumination of the sample and collection of the scattered radiation is shown in Fig. 1. A 5 cm focal length lens focuses the laser radiation into the sample. The scattered radiation is collected at a numerical aperture of 1.0 by a 6 cm focal length lens and matched to the numerical aperture of the spectrograph.

The spectrograph used for these studies was of the Czerny-Turner type and had been constructed in our laboratories. The collimator and camera mirrors have focal lengths of 165 cm and the effective numerical aperture of each is about 16. The spectrograph uses a Bausch and Lomb diffraction grating, 830 lines mm^{-1}, size 102 mm × 102 mm, and blazed at 8500 Å. For the work reported here it was used in the second order ($n = -2$) giving a reciprocal linear dispersion of 9.9 cm^{-1} mm^{-1} at 4880 Å.

The Spectra Physics model 140 and model 165/03 argon ion lasers, with outputs in the range 1.0–1.5 W at 4880 Å were used throughout this work.

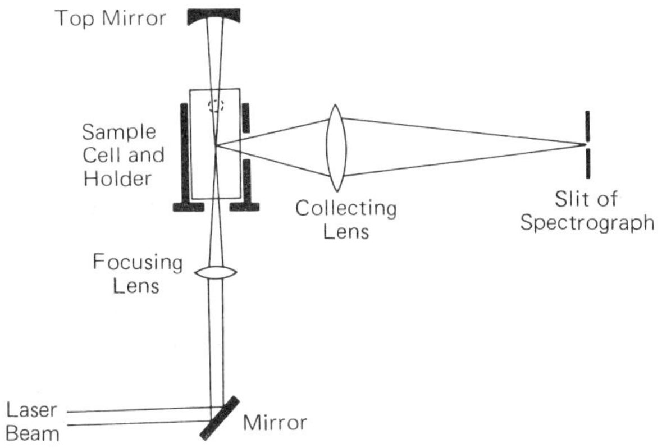

Fig. 1. Sample illumination and collecting system.

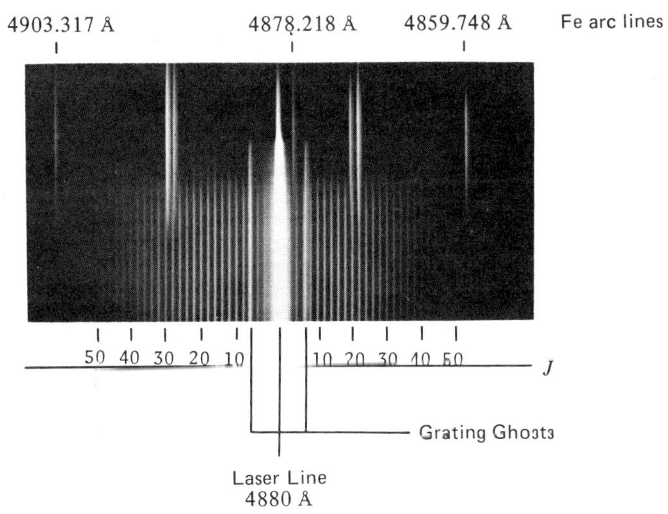

Fig. 2. Pure rotational Raman spectrum of carbon dioxide.

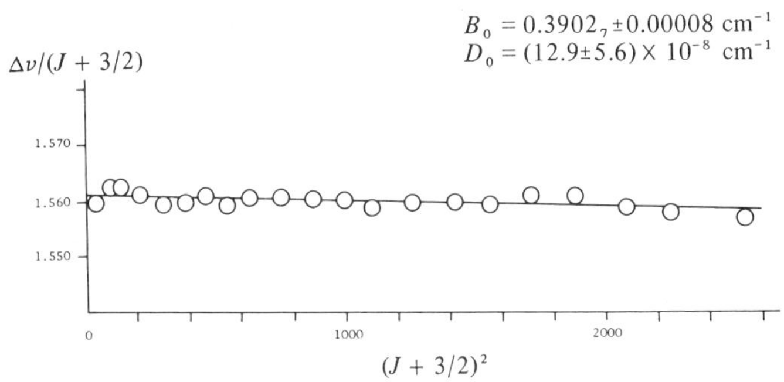

Fig. 3. Plot of $\Delta\nu/(J + 3/2)$ against $(J + 3/2)^2$ for carbon dioxide.

Since Rayleigh scattering is highly polarized, whereas pure rotational Raman scattering is completely depolarized, the plane of the electric vector of the incident plane polarized laser radiation can be chosen so that the scattered radiation seen by the spectrograph includes a strong component of rotational Raman scattering but the weak component of Rayleigh scattering. This arrangement gives useful discrimination against Rayleigh scattering and was used throughout this work for studies of rotational spectra. Vibration–rotation spectra are however polarized and the observation of the strong component requires the plane of the incident electric vector to be that which gives the strong component of the Rayleigh scattering. The resulting lack of discrimination against Rayleigh scattering is unimportant for vibration–rotation spectra since they are well removed from the Rayleigh line.

Rotational spectra were recorded on Ilford Zenith Astronomical plates which have maximum sensitivity around 4880 Å. Stokes vibration–rotation spectra were recorded on Kodak 103aF plates which have maximum sensitivity at longer wavelengths.

Frequency calibration was achieved using an iron arc spectrum which overlapped the Raman spectrum. Measurements of distances on the plate were generally made with a Hilger and Watts L76/3 measuring microscope. This can be read to ± 0.001 mm although the absolute accuracy is only ± 0.002 mm. The frequency of a given Raman line was determined by linear interpolation between the two nearest iron arc lines, one below and one above the unknown frequency.

In general, frequencies were derived from measurements by several observers of more than one plate. The details of such measurments and their statistical treatment are considered separately for each system studied.

Pure rotational Raman spectrum of carbon dioxide

The pure rotational Raman spectrum of carbon dioxide has been studied by Stoicheff,[1] by Tyulin *et al.*[2] and by Barrett and Weber.[3] Barrett and Weber used laser excitation and recorded the spectra photographically using a grating spectrograph with a reciprocal linear dispersion of 2 cm^{-1} mm^{-1}. Because of this high dispersion they were able to obtain very accurate values of B_0 and D_0. Since the rotational constants of carbon dioxide are so well established, as part of the present work the pure rotational Raman spectrum of carbon dioxide was photographed and measured to check on the accuracy of our procedure.

Figure 2 shows a photographically recorded pure rotational Raman spectrum of CO_2 obtained in the present work from 1.5 ml of gas at 1 atmos pressure, using about 1.2 W of 4880 Å laser radiation. The spectral slit width was 0.5 cm^{-1} and the exposure time 15 h. Rotational lines out to $J = 48$ were observed in both the Stokes and anti-Stokes spectra.

The pure rotation Raman shifts ($\Delta\nu$ cm^{-1}) for a linear molecule are related in general to the rotational constants B_0 and D_0 by the formula

$$\Delta\nu = (4B_0 - 6D_0)(J + 3/2) - 8D_0(J + 3/2)^3 \quad (J = 0, 1, \ldots) \qquad (1)$$

where $B_0 = h/(8\pi^2 cI)$, I is the moment of inertia and D_0 is the centrifugal stretching constant. Hence

$$\frac{\Delta\nu}{(J + 3/2)} = (4B_0 - 6D_0) - 8D_0(J + 3/2)^2 \qquad (2)$$

Table 1. B_0, D_0 and r_0 values for CO_2.

Source	B_0 cm^{-1}	10^8 D_0 cm^{-1}	r_0 Å
Stoicheff[1] (1955)	0.39041	–	
Tyulin *et al.*[2] (1963)	0.39013 ± 0.00015	10 ± 15	–
Barrett and Weber[3] (1970)	0.39027 ± 0.000007	12.9 ± 0.3	–
Present work	0.39027 ± 0.00008	12.9 ± 5.6	1.1618

and so by plotting $\Delta\nu/(J + 3/2)$ against $(J + 3/2)^2$, B_0 and D_0 can be evaluated. Such a plot for CO_2 is shown in Fig. 3, and a least squares analysis gives $B_0 = 0.39027 \pm 0.00008$ cm^{-1} and $D_0 = (12.9 \pm 5.6) \times 10^{-8}$ cm^{-1}. It is encouraging to note from Table 1, in which the results of various workers are compared, that the present results overlap with those of Barrett and Weber. The uncertainty in the present results is larger than that in the results of Barrett and Weber but this is to be expected since Barrett and Weber used a spectrograph with reciprocal linear dispersion five times smaller than that used here.

Rotational and vibration–rotation Raman spectra of fluorine

The first observation of the pure rotational Raman spectrum of fluorine was reported by Andrychuk[4] in 1951 and represented a considerable experimental achievement at that time. Andrychuk used a 1 litre sample of fluorine in a quartz cell illuminated with Toronto mercury arcs. The spectra were recorded photographically with a Zeiss three-prism spectrograph having a reciprocal linear dispersion of 58 cm^{-1} mm^{-1} at 4047 Å. The spectral slit widths used ranged from 0.8 to 1.2 cm^{-1}. The rotational spectra were excited with 4047 Å Hg radiation and showed the expected intensity alternation consistent with a nuclear spin of ½ for ^{19}F. However, only the strong lines of odd J number could be satisfactorily measured and Andrychuk's analysis of the rotational spectrum was based on measurements of nine Stokes and anti-Stokes pairs (odd J values from 7–23 inclusive) and four Stokes lines (J = 5, 25, 27 and 29). The values of the frequency displacements from the exciting line used in the analysis were based on the means of eight readings on each of three plates. An iron arc spectrum was used for frequency calibration. However, the plot of $\Delta\nu/(J + 3/2)$ against $(J + 3/2)^2$ revealed that the accuracy of the measurements was not sufficient to determine the slope (which is equal to $-8D_0$) with any accuracy. Andrychuk therefore calculated a value of $-8D_0 = -0.000027$ cm^{-1} from the approximate relationship

$$D_0 = \frac{4B_0^3}{\omega_0^2} \tag{3}$$

using preliminary values of B_0 and ω_0, the vibration frequency. He then determined the straight line of slope -0.000027 cm^{-1} which fitted best the experimental points. This line had an intercept of 3.5310, whence $B_0 = 0.8828 \pm 0.001_0$ cm^{-1} and $r_0 = 1.417_7 \pm 0.001_5$ Å.

In 1969 Claassen *et al.*[5] reported that they had recorded directly the pure rotational spectrum of fluorine gas at 25°C and 4 atmos pressure using a 50 × 10^{-3} W Spectra

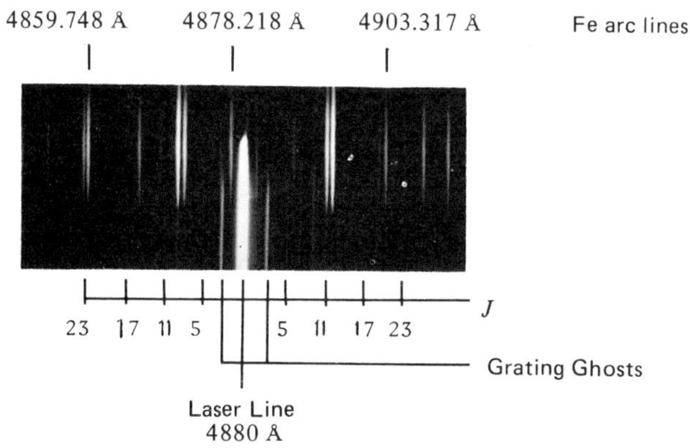

Fig. 4. Pure rotational Raman spectrum of fluorine.

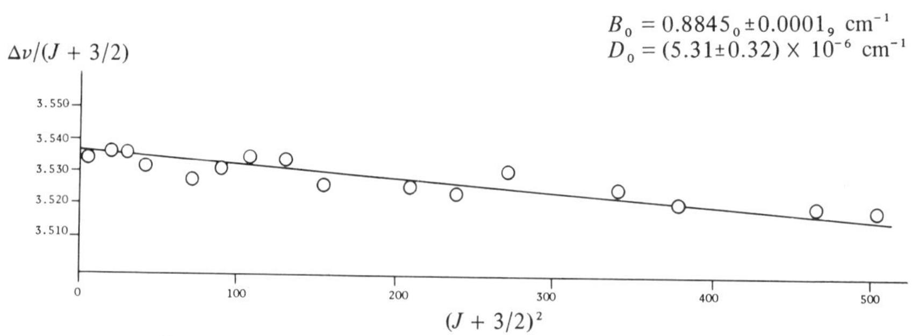

Fig. 5. Plot of $\Delta\nu/(J + 3/2)$ against $(J + 3/2)^2$ for fluorine.

Physics 125 helium–neon laser and a Spex 1400-II double monochromator. The spectrum they obtained was of good quality with the weak (even J) and strong (odd J) rotational lines clearly visible from $J = 1$ to $J = 27$ on both the Stokes and the anti-Stokes sides. They ran the pure rotational Raman spectrum of oxygen immediately after running the fluorine spectrum and used the known frequencies of the pure rotational lines of oxygen which have been determined by Weber and McGinnis[6] to $\pm$ 0.05 cm^{-1} for calibration. Spectra were run on four different days and measured separately by three observers. However, plots of $\Delta\nu/(J + 3/2)$ against $(J + 3/2)^2$ gave values of D_0 which differed by a factor of more than two for two experiments. Hence Claassen *et al.* assumed a value of $D_0 = 3.47 \times 10^{-6}$ cm^{-1} (or $8D_0 = 2.8 \times 10^{-5}$ cm^{-1}) calculated from Eqn. (3). Using this assumed value of D_0, the measured values of the frequency shifts for the seven lines from $J = 19$ to $J = 25$ for both Stokes and anti-Stokes regions, for spectra run on four different days, were used to calculate independent values of B_0. This procedure gave an average value of $B_0 = 0.8841 \pm 0.0006$ cm^{-1} and a corresponding value of $r_0 = 1.4168 \pm 0.0005$ Å.

In view of the present authors' experience it should be noted that with their Spex double monochromator Claassen *et al.* found that they could achieve the same accuracy by using the Spex wavelength marker and averaged values of Stokes and anti-Stokes shifts without using oxygen calibration lines.

In the present work the fluorine was contained in a small silica cell of volume 1.5 ml. The cell had taps fitted with Teflon plugs and was carefully dried and purged with dry argon before being filled with highly purified fluorine to a pressure of 1 atmos. Excitation was with the 4880 Å line of an argon ion laser and the power at the sample was about 3 W.

The pure rotational Stokes and anti-Stokes spectra, lying within ± 100 cm^{-1} of the exciting line and the Stokes vibration–rotation spectrum centred at 5100 Å were investigated in separate sets of experiments.

The pure rotational spectra were recorded photographically using a spectral slit width of 0.5 cm^{-1} and exposure times of from 17–36 h. The spectra were of good quality and the weak even J value lines and the stronger odd J value lines were clearly visible out to $J = 23$ in both Stokes and anti-Stokes spectra. A photographic print of the Raman spectrum of fluorine is shown in Fig. 4.

The analysis of the spectra was based on measurements of 17 Stokes and anti-Stokes pairs; the remaining lines were rejected for measurement because of their weakness or the interference of grating ghosts. Two pure rotational spectra, each recorded on a separate plate were used for the analysis. One plate was measured twice and the other plate three times by separate observers. A typical plot of $\Delta\nu/(J + 3/2)$ against $(J + 3/2)^2$ for fluorine is shown in Fig. 5. Each set of measurements was subjected to a least squares analysis to determine the best value of B_0 and D_0 and the error in each. It was found that the values of D_0 did not show the gross variation found by Claassen *et al.* The five values of B_0 and D_0 so obtained were then combined together according to the principle of least error to give a final best estimate of $B_0 = 0.8845_0 \pm 0.0001_9$ cm^{-1} and $D_0 = (5.31 \pm 0.32) \times 10^{-6}$ cm^{-1}. This value of B_0 yields a value of $r_0 = 1.4165_7 \pm 0.0001_5$ Å. The results for fluorine are summarized in Table 2 which also includes earlier values for comparison.

Attempts were made to observe the Stokes vibration–rotation spectrum of fluorine, but even after 72 h exposure and using a spectral slit width of 1.0 cm^{-1} only the unresolved Q branch for the $0 \rightarrow 1$ vibrational transition was observed. There was no trace of features assignable to the O and S branches. The frequency of the unresolved Q branch was determined as $\Delta\nu = 892.02 \pm 0.03$ cm^{-1}. This overlaps with Andrychuk's value of 891.85 ± 0.4 cm^{-1} but the uncertainty in the present determination is less because of the lower reciprocal linear dispersion of the spectrograph used.

Rotational and vibration–rotation Raman spectra of tritium

As far as we are aware no work has been published previously on the rotational or vibration–rotation Raman spectrum of tritium, T_2. A possible reason for this could be the prohibitive cost of the large volume of tritium required for the study of rotational Raman spectra using mercury arc excitation.

In the present work, the pure rotational Stokes and anti-Stokes spectra, covering a region of ± 550 cm^{-1} about the exciting line, and the Stokes Q branch associated with the $0 \rightarrow 1$ vibrational transition centred at 5550 Å were investigated in separate sets of experiments. The sample volume was 1.5 ml and the laser power at the sample about 3 W at 4880 Å.

Figure 6 shows the directly recorded rotational Raman spectrum of tritium but all

Table 2. B_0 and D_0 values for fluorine.

Source	B_0 cm^{-1}	$10^6 D_0$ cm^{-1}	r_0 Å	ν_0 cm^{-1}
Andrychuk[4]	$0.882_8 \pm 0.001$	[3.33]	$1.417_7 \pm 0.001$	891.85 ± 0.4
Claassen[5]	0.8841 ± 0.0006	[3.47]	1.4158 ± 0.0005	–
Present work	$0.8845_0 \pm 0.0001_9$ [a]	$5.3_1 \pm 0.3_2$ [a]	$1.4164_7 \pm 0.0001_5$	892.02 ± 0.03

[a] Obtained by combining together the following five independent measurements according to the principle of least error:

	B_0 (cm^{-1})	$10^6 D_0$ (cm^{-1})
(i)	$0.8846_0 \pm 0.0006$	6.10 ± 0.91
(ii)	$0.8843_4 \pm 0.0006$	5.36 ± 0.75
(iii)	$0.8844_5 \pm 0.0003$	3.78 ± 0.31
(iv)	$0.8840_5 \pm 0.0004$	5.68 ± 0.63
(v)	$0.8849_5 \pm 0.0004$	4.83 ± 0.49

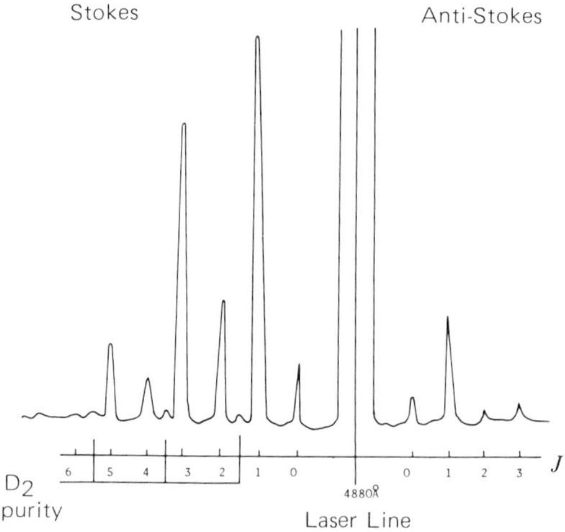

Fig. 6. Pure rotational Raman spectrum of tritium – direct recording.

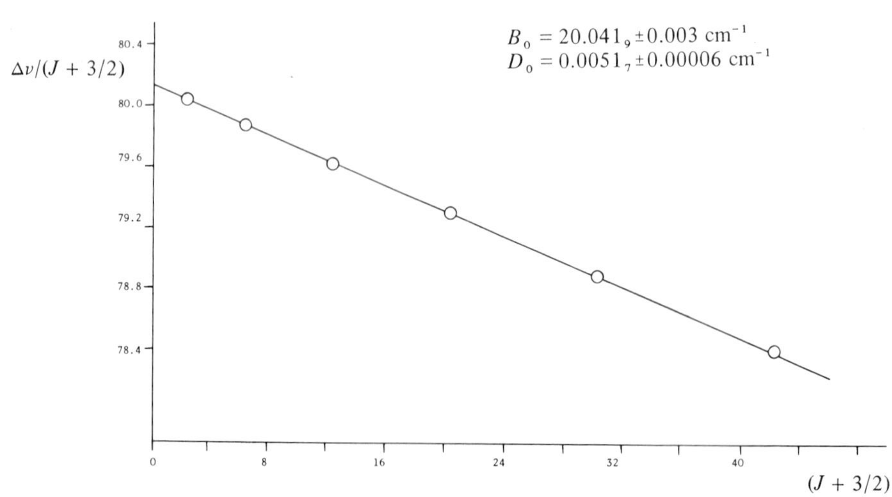

Fig. 7. Plot of $\Delta\nu/(J + 3/2)$ against $(J + 3/2)^2$ for tritium.

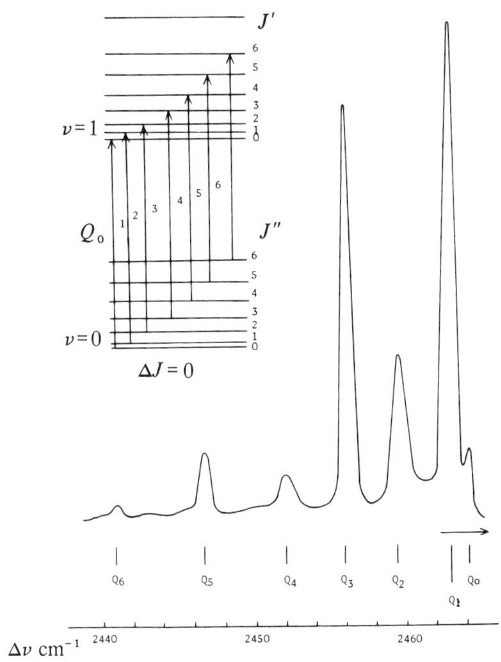

Fig. 8. Vibration–rotation spectrum of tritium; Q branch of the $\nu = 0$ to $\nu = 1$ transition.

Table 3. B_0, D_0, ν_0, B_1 and D_1 values for tritium.

B_0 (cm^{-1})	=	$20.041_9 \pm 0.003^a$
D_0 (cm^{-1})	=	$0.0051_7 \pm 0.00006^a$
ν_0 (cm^{-1})	=	2464.08_3
B_1 (cm^{-1})	=	$19.463_2 \pm 0.001_7$
D_1 (cm^{-1})	=	$0.0050_0 \pm 0.00004$

[a] Obtained by combining together the following three independent measurements according to the principle of least error:

	B_0 (cm^{-1})	D_0 (cm^{-1})
(i)	$20.0378 \pm 0.006_1$	$0.0049_0 \pm 0.0001_3$
(ii)	$20.041_2 \pm 0.008_8$	$0.0051_0 \pm 0.0001_4$
(iii)	$20.045_1 \pm 0.006_0$	$0.0052_8 \pm 0.0001_0$

frequency measurements were derived from spectra photographically recorded on Zenith Astronomical plates using a spectral slit width of 0.8 cm^{-1} and exposure times of 30–90 h. As Fig. 6 clearly shows, the even and odd J value lines have intensities in the ratio 1 to 3. This intensity alternation is consistent with a nuclear spin of ½ for the ^{3}H nucleus and shows that tritium is a Fermi gas. Figure 6 also shows that deuterium is present as an impurity to the extent of a few percent.

The analysis of the rotational spectra was based on measurements of 6 Stokes and anti-Stokes pairs ($J = 0$ to $J = 5$).

Three separate spectra were recorded and each was measured twice by two separate observers. A typical plot of $\Delta\nu/(J + 3/2)$ against $(J + 3/2)^2$ for tritium is shown in

Table 4. B_v, D_v, r_v, α and β values for tritium and deuterium.

Source	Molecule	State	B_v cm^{-1}	D_v cm^{-1}	r_v Å	α_e cm^{-1}	β_e
Present Work	T_2	$\nu = 0$	20.0419 ± 0.003	0.00517 ± 0.00006	0.74650 ± 0.00006		
		$\nu = 1$	19.4632 ± 0.002	0.00500 ± 0.00004	0.75755 ± 0.00003		
		Equilibrium configur-ation	20.3313	0.00526	0.74120	0.5787	−0.00017
Jones (8)[a]	T_2	Equilibrium configur-ation	20.3243	0.00519	–	0.5922	
Stoicheff (7)	D_2	$\nu = 0$	29.9105	0.01134	0.74820		
		$\nu = 1$	28.8482	0.01075			
		Equilibrium configur-ation	30.442 ± 0.01	0.01164	0.74165 ± 0.00006	1.0623	−0.00059
	H_2	$\nu = 0$	59.3392	0.04599	0.75105		
		$\nu = 1$	56.3640	0.04323			
		Equilibrium configur-ation	60.841 ± 0.01	0.04737	0.74173 ± 0.00006	2.9752	−0.00276

[a] Calculated from H_2 and D_2

Fig. 7. Each set of measurements was subjected to a least squares analysis to determine the best value of B_0 and D_0 and the errors. The values obtained were then combined together according to the principle of least error to give a final best estimate of B_0 and D_0 and the errors therein. The individual values of B_0 and D_0, and the final best estimates with errors, are presented in Table 3.

Figure 8 shows the directly recorded Stokes Q branch associated with the $0 \rightarrow 1$ vibrational transition. Seven lines in the Q branch corresponding to $J = 0$ to $J = 6$ were observed in the directly recorded spectrum. These lines show the same intensity alternation as observed in the pure rotation spectrum. All frequency measurements were derived from spectra photographically recorded on Kodak 103aF plates with spectral slit widths 0.8–1.0 cm^{-1} and exposure times up to 108 h. Even in the longest exposures no lines attributable to the O and S branches associated with the $0 \rightarrow 1$ vibrational transition were observed.

Although five lines ($J = 1, 2, 3, 4$ and 5) were observed photographically, the analysis of the Q branch was based on measurements of only three Stokes Q branch lines ($J = 1, 3$ and 4). The $J = 2$ and 5 lines were too weak for satisfactory measurement.

The frequency shifts for the Q branch lines are given by

$$\nu_Q = \nu_0 + (B_1 - B_0)J(J+1) - (D_1 - D_0)J^2(J+1)^2 \tag{4}$$

Assuming the values of B_0 and D_0 already obtained from analysis of the pure rotational spectrum, this equation involves three unknown molecular parameters, ν_0, B_1 and D_1. Since the frequencies of three Q branch lines were determined, there exist three linear equations in the three unknowns. The values of ν_0, B_1 and D_1 calculated in this way are also given in Table 3.

Since

$$B_v = B_e - \alpha_e(v + ½) \tag{5}$$

and

$$D_v = D_e + \beta_e(v + ½) \tag{6}$$

the results obtained here enable α_e, β_e, B_e (and hence r_e) and D_e to be calculated. The values of all the parameters for T_2 are summarized in Table 4 which also includes for comparison corresponding values for H_2 and D_2 taken from the work of Stoicheff[7] and predicted values for T_2 calculated by Jones.[8]

ACKNOWLEDGEMENTS

We are indebted to the Science Research Council for substantial grants for equipment for Raman spectroscopy, and to Dr. A. J. Edwards of the University of Birmingham for preparing the fluorine used in this work. One of use (R. Love) gratefully acknowledges the award of a postgraduate studentship from the University of Bradford.

REFERENCES

1 B. P. Stoicheff, *Advances in Spectroscopy,* (Ed. H. W. Thompson), Vol. 1, Wiley Interscience, New York, 1959, p. 91.

2 V. I. Tyulin and V. M. Tatevskii, *Opt. Spectry.* **15**, 18 (1963).

3 J. J. Barrett and A. Weber, *J. Opt. Soc. Amer.* **60**, 70 (1970).

4 D. Andrychuk, *Can. J. Phys.* **29**, 15 (1951).

5 H. H. Claassen, H. Selig and J. Shamir, *Appl. Spectry.* **23**, 8 (1969).
6 A. Weber and E. A. McGinnis, *J. Mol. Spectry.* **4**, 195 (1960).
7 B. P. Stoicheff, *Can. J. Phys.* **35**, 730 (1957).
8 W. J. Jones, *J. Chem. Phys.* **16**, 1077 (1948).

66. STRUCTURE ET LARGEUR DES RAIES RAMAN νOH ET νNH DES CRISTAUX D'ACIDE ACETIQUE, DE TRIAZOLE ET D'IMIDAZOLE ETUDIES ENTRE 300 ET 20 K

R. Foglizzo et A. Novak

Laboratoire de Chimie Physique, C.N.R.S.,
Thiais, France

ABSTRACT

The Raman spectra of crystalline acetic acid, triazole, imidazole and some deuterated derivatives were studied between 1800 and 3600 cm^{-1}, at several temperatures down to 20 K. The corresponding νOH, νOD, νNH and νND bands were observed for the first time in scattering. At room temperature, they are broad and exhibit numerous secondary maxima, as do the corresponding infrared absorption bands. The frequency and breadth of the νAH bands decrease with temperature and the secondary maxima are better resolved. In the cases of triazole and imidazole, the intensity of the main maximum increases strongly with respect to that of the secondary ones. This effect is striking at 20 K for the νNH bands and is already observed at 90 K for the νND band.

From the results obtained, it can be concluded that the breadth and structure of the νAH band of a hydrogen bond AH . . . B of moderate strength, are due to the anharmonicity of the A–H oscillator. The internal vibrations νAH would be due to the structure and the external ones, which concern the A . . . B length are due to the breadth.

INTRODUCTION

Si les études en absorption des bandes ν AH d'un groupment AH engagé dans une liaison hydrogène AH . . . B sont nombreuses,[1,2] il n'en est pas de même en diffusion. En effect, aucun travail correspondant en spectroscopie Raman n'est signalé dans le livre récent sur la liaison hydrogène.[2] Nous avons entrepris une étude par spectrométrie Raman de quelques composés contenant des liaisons hydrogène de force et de type différents[3-5] afin d'examiner en particulier l'intensité et le profil des bandes dues aux vibrations de valence ν AH. Dans ce mémoire, nous présentons les spectres Raman de 3600 à 1800 cm^{-1} des cristaux d'acide acétique, de triazole, d'imidazole et de quelques dérivés deutériés à différentes températures jusqu'à 20 K.

PARTIE EXPERIMENTALE

Produits

Le triazole (Synthèse et Recherche) et l'imidazole (Eastman Kodak) étaient purifiés par sublimation sous pression réduite. L'acide acétique CH_3COOD a été préparé par décomposition de l'anhydride acétique par l'eau lourde en présence de traces d'anhydride phosphorique. Le triazole (D_1)-1 a été préparé à partir du triazole par deux échanges à température ambiante en présence d'eau lourde. Après évaporation de l'eau, le produit

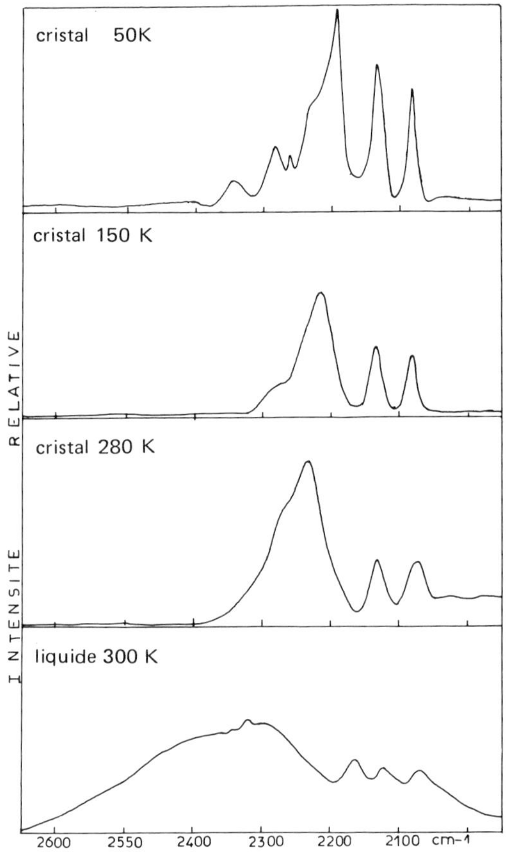

Fig. 1. Spectres Raman de l'acide acétique CH_3COOD à différentes températures.

obtenu était séché et sublimé sous vide. Le taux isotopique des dérivés deutériés était de l'ordre de 95%.

Spectres

Les spectres ont été enregistrés sur un spectromètre Coderg PH0 équipé d'un laser à argon ionisé CRL modèle 52, dont la raie à 4880 Å, utilisée comme excitatrice, avait une puissance de 1 W. Aucune correction tenant compte de la réponse du photomultiplicateur en fonction de la longueur d'onde n'a été apportée à la mesure des intensités relatives des raies. Les fréquences étaient mesurées à ± 1 cm^{-1} près. La largeur spectrale de fente était comprise entre 8 et 1 cm^{-1} suivant l'intensité des raies considérées.

Les composés étaient examinés en tubes scellés sous vide. Le cryostat utilisé pour atteindre 20 K était constitué essentiellement d'une tête réfrigérante modèle 'Cryodyne' construite par Cryogenic Technology Inc. utilisant l'hélium gazeux en circuit fermé comme réfrigérant; la température était contrôlée à ± 5 K au moyen d'un thermocouple.

RESULTATS

Acide acétique

Dans le cristal d'acide acétique, les molécules sont associées en chaines infinies par l'intermédiaire de liaisons hydrogène OH . . . O dont la distance O . . . O est de 2,625 Å.[6] En absorption, la bande ν OH, située vers 2850 cm^{-1}, est très intense, large et structurée;[7] en diffusion par contre, elle est de faible intensité et en partie masquée par les bandes du groupement méthyle. Nous avons donc choisi l'acide CH_3COOD pour étudier la bande ν OD située dans un domaine qui ne présente pas d'autre vibration fondamentale.

Le spectre Raman, de 2600 à 2000 cm^{-1}, du cristal d'acide acétique CH_3COOD à différentes températures est présenté sur la Fig. 1 avec celui de l'acide à l'état liquide pur, enregistré pour comparaison. La bande νOD du cristal à 280K est observée à 2228 cm^{-1}; sa largeur à mi-hauteur $\nu_{1/2}$ est voisine de 80 cm^{-1}. Les deux bandes de plus faible intensité, à 2133 et 2078 cm^{-1}, sont attribuées, comme en absorption,[7] aux combinaisons 2 δ OD et ρ CH_3 + δ OD respectivement.

Lorsqu'on refroidit le cristal, la fréquence et la demi-largeur de la bande ν OD diminuent tandis que l'intensité de sous bandes vers 2133 et 2078 cm^{-1} augmente par rapport à celle de la bande ν OD. A 50 K par exemple, la composante principale de la bande ν OD située à 2194 cm^{-1} n'a que 20 cm^{-1} comme demi-largeur et l'intensité des combinaisons 2 δ OD et δ OD + ρ CH_3 devient comparable à celle de la raie fondamentale. Ce comportement confirme l'interprétation des sous-bandes du massif ν OD de l'acide acétique observées en absorption:[7] les niveaux de la vibration ν OD et des combinaisons se rapprochent avec l'abaissement de la température et l'intensité des combinaisons est renforcée par résonance de Fermi. Les combinaisons les plus intenses possèdent probablement un certain caractère ν OD, de même que l'harmonique 2 δ CH du chloroforme, de type A_1, présente un caractère ν CH.[8]

En liquide pur à 300 K, la bande ν OD a une fréquence beaucoup plus élevée, 2320 cm^{-1}; sa demi-largeur, $\nu_{1/2}$, voisine de 290 cm^{-1}, est au moins trois fois plus grande que pour le cristal à une température peu différente, 280 K (Fig. 1). Cette différence de largeur s'explique surtout par un désordre de structure important à l'état liquide où coexistent des dimères cycliques et des oligomères.[7] En outre, une nouvelle sous-bande est observée à 2164 cm^{-1} qui est due à l'harmonique 2 δ OD de type A_g de la vibration δ OD de type A_g du dimère cyclique; cette dernière, située à 1092 cm^{-1}, est de fréquence plus élevée que la déformation δ OD de type B_u, identifiée à 1055 cm^{-1}. Rappelons que l'harmonique 2 δ OD du dimère cyclique est active uniquement en diffusion.

Triazole

Le cristal de triazole comporte également des chaines infinies de molécules associées par liaisons hydrogène NH . . . N assez fortes dont les distances N . . . N sont voisines de 2,82 Å.[9] Les spectres Raman des cristaux de triazole et de triazole (D_1) à différentes températures sont réunis sur la Fig. 2.

Massif NH. A 300 K, le massif νNH s'étend de 3150 à 2400 cm^{-1}, très large et centré vers 2780 cm^{-1}; il s'oppose à la raie très intense et fine des deux vibrations νCH vers 3120 cm^{-1}. Il présente de nombreux sous-maximums et son intensité intégrée parait importante; sa largeur à mi-hauteur $\nu_{1/2}$ est estimée à 400 cm^{-1}. Les intensités des sous-maximums sont comparables.

Lorsqu'on abaisse la température, la fréquence du centre de gravité du massif diminue, le nombre de sous-maximums augmente mais leurs intensités baissent par rapport à celle du pic le plus intense, de telle façon que la largeur à mi-hauteur $\nu_{1/2}$ de la bande ν NH diminue: à 20 K la fréquence du pic le plus intense est 2660 cm^{-1} et la rapport des intensités des sous-maximums à 3090 cm^{-1} et le maximum à 2660 cm^{-1} est

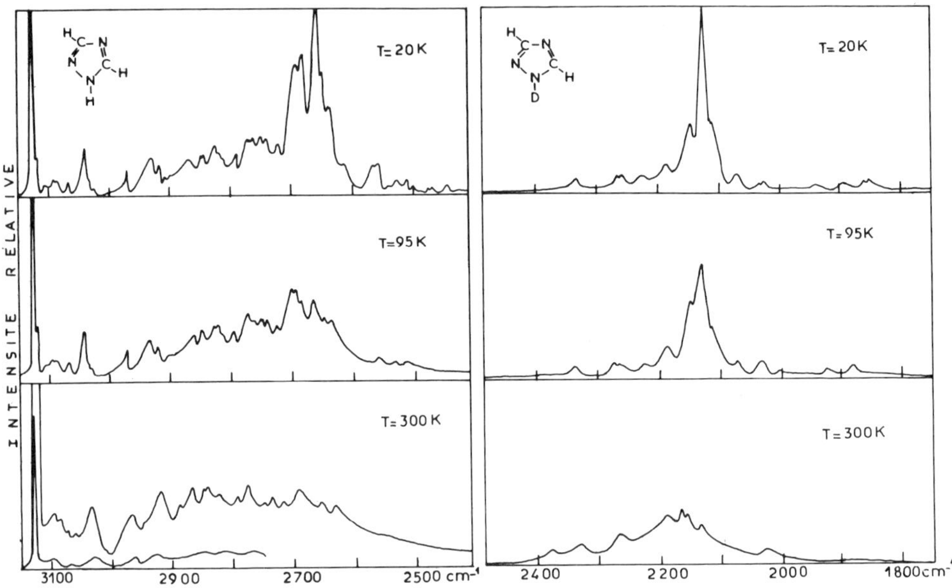

Fig. 2. Spectres Raman du triazole et du triazole (D_1)-1 à différentes températures; sur le spectre du triazole à 300 K l'intensité relative est multipliée par 5.

10 fois plus petit qu'à 300 K. Enfin, la bande centrale est constituée de cinq composantes entre 2700 et 2625 cm^{-1}; la demi-largeur $\nu_{1/2}$ de l'ensemble ne dépasse pas 60 cm^{-1}.

Massif ND. Le massif ν ND du triazole (D_1) est beaucoup moins large que le massif ν NH du triazole; à 300 K il s'étend de 2400 à 2000 cm^{-1} environ, sa fréquence est voisine de 2175 cm^{-1} et sa demi-largeur $\nu_{1/2}$ estimée à 160 cm^{-1}. Il ne comporte que six sous-maximums dont quatre d'intensité comparable à 2192, 2166, 2157 et 2134 cm^{-1}.

Comme pour le massif ν NH, l'intensité de la bande principale augmente aux dépens de celles des sous-maximums lorsqu'on diminue la température. A 20 K, la bande à 2125 cm^{-1}, simple et fine, est environ 10 fois plus intense que le sous-maximums à 2192 cm^{-1}: sa demi-largeur $\nu_{1/2}$ est voisine de 15 cm^{-1}. Les autres sous-bandes sont également plus fines et mieux résolues.

Imidazole

La même analyse a été faite du massif ν ND de l'imidazole cristallisé qui, comme le triazole, possède une liaison hydrogène NH . . . N moyennement forte (2,86 Å) associant les molécules en chaines infinies.[10] Le massif ν ND à la température ambiante (Fig. 3) présente déjà une raie sensiblement plus intense que l'ensemble des autres sous-maximum mais sa largeur à mi-hauteur est encore importante: $\nu_{1/2} = 116\ cm^{-1}$. A 50 K, la fréquence νND s'abaisse de 2140 à 2110 cm^{-1}; elle est constituée de deux composantes seulement, mal résolues, et la demi-largeur de l'ensemble ne dépasse pas 36 cm^{-1}. Les trois autres sous-maximums* par contre ne sont pratiquement plus observés.

*La bande à 2350 cm^{-1} est due à la vibration νCD des impuretés isotopiques de l'imidazole.

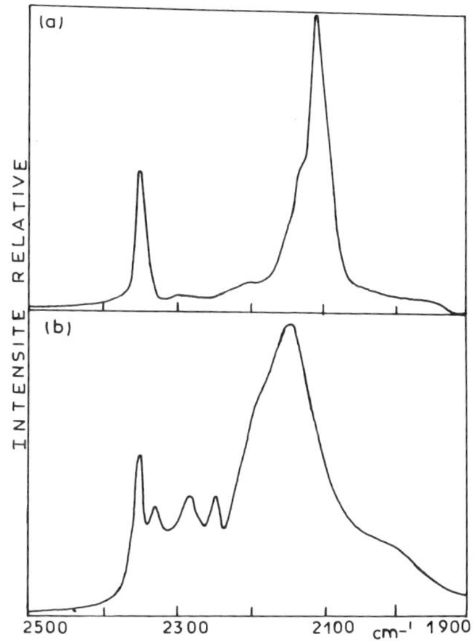

Fig. 3. Spectres Raman de l'imidazole (D_1)-1 à 50 K (a) et 300 K (b).

DISCUSSION

Structure de la bande ν AH

Les nombreux sous-maximums de la bande ν AH ou ν AD peuvent correspondre aux combinaisons ayant un certain caractère ν AH par analogie avec le chloroforme.[8] Il existe cependant une différence importante: la vibration δ CH du chloroforme est localisée et ne donne qu'une harmonique 2 δ CH intense en absorption tandis que les vibrations δ OH et δ NH sont habituellement réparties parmi plusieurs vibrations. C'est ainsi que les déformations δ OD et ρ CH_3 paraissent couplées et les combinaisons 2 δ OD et ρ CH_3 + δ OD possèdent un certain caractère δ OD. Leurs fréquences sont inférieures à celle de la vibration νOD et par abaissement de la température, les bandes correspondantes deviennent plus intenses (Fig. 1).

La vibration δ NH du triazole de l'imidazole, d'autre part, se répartit notamment parmi les vibrations de squelette désignées par R_1, R_2, R_3, R_4 dont les fréquences se situent entre 1540 et 1200 cm^{-1}.[11] Les combinaisons résultantes auraient donc un caractère ν NH et se comporteraient toutes comme la vibration fondamentale vis à vis de la liaison hydrogène. Pour la plupart, leurs fréquences sont supérieures à la fréquence ν NH et donnent lieu aux sont-maximums de la bande νNH. Lorsque la bande ν NH décroit en fréquence et devient moins large les sous-maximums s'affaiblissent ou disparaissent parce que les niveaux de Fermi s'éloignent.

Des considérations analogues peuvent être émises pour la bande νND qui présente moins de sous-maximums; ceci s'explique par le fait que le niveau νND est moins large et que le nombre de combinaisons à caractère ν ND, et dont les fréquences se situent entre 2400 et 2000 cm^{-1}, est plus faible.

Largeur de la bande ν AH

Une diminution importante de la demi-largeur de la bande ν AH a déjà été observée en absorption pour des liaisons hydrogène relativement faibles telles que celles du pyrrole[3] ou bien des alcools[12] dont les bandes ν OH n'ont pas de structure. C'est par contre la première fois qu'un comportement est mis en évidence dans le cas de liaisons hydrogène relativement fortes, caractérisées par des déplacements relatifs de la fréquence ν AH de l'ordre de 20 à 25%. L'interprétation de cette diminution doit toutefois être fondamentalement la même pour les deux cas.

Etant donné le désordre plus faible dans les cristaux que dans les liquides, la diminution de la largeur avec l'abaissment de la température doit s'expliquer d'une part par une réduction de l'anharmonicité, d'autre part par une modification de la distribution de Boltzmann des vibrateurs mettant en jeu la distance A . . . B.[13] En effet, les vibrations ν OH et ν NH des acides carboxyliques et des bases azotées comportant des liaisons hydrogène sont très anharmoniques; leurs fréquences sont très sensibles aux distances intermoléculaires O . . . O ou N . . . N. L'abaissement de la température augmente l'ordre dans les cristaux: il dépeuple les états excités des vibrations externes, situées entre 200 et 50 cm^{-1} et qui mettent en jeu les distances intermoléculaires, et il diminue l'anharmonicité. Sur le spectre Raman, il en résulte, à 20 K, une bande ν AH dont la composante principale, beaucoup plus intense qu'à 300 K, a une largeur à mi-hauteur de 20 cm^{-1} seulement au lieu de quelques centaines de cm^{-1} à température ambiante. Plus la vibration ν AH est anharmonique et plus la modulation par les vibrations externes est importante: les bandes νAD sont beaucoup moins larges que les bandes νAH correspondantes. D'autre part, l'anharmonicité importante du vibrateur νNH est mise en évidence dans le cas du triazole par les cristallographes[9] selon lesquels l'agitation thermique la plus notable, et de loin, est celle de l'atome d'hydrogène tautomère (du groupement NH).

CONCLUSION

En conclusion, les résultats obtenus permettent de penser que la largeur et la structure de la bande ν AH sont dues principalement à l'anharmonicité du vibrateur ν AH: les combinaisons à caractère ν AH des vibrations internes seraient responsables de la structure, tandis que les vibrations externes qui mettent en jeu la distance A . . . B le seraient de la largeur.

REFERENCES

1 G. C. Pimentel et A. L. McClellan, *The Hydrogen Bond,* Freeman, San Francisco, 1960.

2 S. N. Vinogradov et R. H. Linell, *Hydrogen bonding*, Van Nostrand Reinhold, New York, 1971.

3 A. Lautie et A. Novak, *J. Chem. Phys.* **56**, 2479 (1972).

4 J. de Villepin et A. Novak, *Spectrochim. Acta* **27 A**, 1259 (1971).

5 J. de Villepin et A. Novak, *Spect. Letters,* **4**, 1 (1971).

6 I. Nahringbauer, *Acta Chem. Scand.* **24**, 453 (1970).

7 M. Haurie et A. Novak, *Spectrochim. Acta* **21**, 1217 (1965); *J. Chim. Phys.* **62**, 120 (1965).

8 W. E. Thompson et G. C. Pimentel, *Z. Elektrochem.,* **64**, 748 (1960).

9 P. Goldstein, J. Ladell et G. Abowitz, *Acta Cryst.* **25**, 135 (1969).

10 S. Martinez-Carrera, *Acta Cryst.* **20**, 783 (1966).

11 R. Foglizzo, B. Saint Roch et A. Novak, (à paraitre); B. Saint Roch, *Thèse,* Bordeaux, 1972.

12 C. Sandorfy, *Can. J. Spectry.*, **17**, 24 (1972).
13 J. N. Finch et E. R. Lippincott, *J. Chem. Phys.* **24**, 908 (1956); *J. Phys. Chem.* **61**, 894 (1957).

SECTION VIII
MOLECULAR STRUCTURES

67. RAMAN SPECTROSCOPY AND MOLECULAR STRUCTURES

F. A. Miller

Department of Chemistry, University of Pittsburgh, Pittsburgh, Pennsylvania, USA

INTRODUCTION

It is well known that Raman spectroscopy is a useful technique for determining molecular structures, especially when combined with infrared measurements. This review will summarize some recent results of such studies. They are either as yet unpublished, or they seem to the author to be unusually interesting. Such a large amount of work is being done that a review of this length cannot possibly mention all of the significant recent contributions. The treatment will therefore be selective, and the emphasis will be on very new information. The details of *how* Raman spectroscopy is used will be omitted because they are widely known. This review will concentrate on results and conclusions.

There are three levels of refinement in determining molecular structures by Raman (and by infrared) spectroscopy. They provide answers to the following three questions.

1. *What groups of atoms, or what kinds of bonds, are present?* This application depends heavily on the use of characteristic group frequencies. It is the most primitive level of structure determination, but it is an important first step for large molecules. For most molecules, in fact, this is as far as Raman and infrared spectroscopy can take one.
2. *What is the molecular symmetry?* This generally depends on the use of Raman and infrared selection rules, and a comparison of what is expected for a given assumed symmetry with what is observed. It is the intermediate level of refinement.
3. *What are the bond distances and angles?* This is the highest level of refinement, and of course is the ultimate goal. It depends on determining the moments of inertia and then deducing the structure from them. Raman (and infrared) spectroscopy cannot compete with microwave spectroscopy where the latter is applicable. However if a molecule has no permanent dipole moment, it usually has no microwave spectrum. (Oxygen is the most notable exception.) Therefore there are a number of small, symmetrical molecules for which vibrational spectroscopy can make important contributions.

This classification will provide the basic framework for this review. Examples for each of the three categories will be discussed.

APPLICATIONS TO DETERMINING GROUPS OF ATOMS OR TYPES OF BONDS

This application depends on the existence of characteristic group frequencies. A very simple example is a problem which was presented to the author by Professor Henry Gilman several years ago. A crystalline sample was known to have the composition C_4X_6, where $X = Si(CH_3)_3$. Was it an acetylene derivative (I) or a butadiene one (II)? The

$$X_3C{-}C{\equiv}C{-}CX_3$$

I

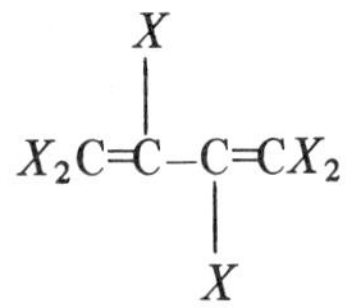

II

infrared spectrum showed no absorption between 1400 and 2900 cm^{-1}, which was evidence for I but by no means proof of it. (The antisymmetric C=C stretch of II is formally infrared-allowed, but its intensity is unpredictable.) However the Raman spectrum had a strong band near 2135 cm^{-1} and nothing else down to 1420 cm^{-1}. This was convincing proof for I and against II.

The field of characteristic Raman group frequencies has not proved to be as lively as many had expected. This is probably because infrared, and later NMR, spectroscopy were brought to a high stage of development first and have skimmed off the cream of the problems, leaving only rather marginal contributions amenable to Raman work.

A new book is in preparation which will bring together the large amount of widely scattered information on Raman group frequencies, and add a considerable amount of new material.[1] It should be a very useful contribution to this subject.

DETERMINATIONS OF MOLECULAR SYMMETRY

This is certainly the most active of our three levels of structure determination. Several specific examples will now be described.

Trisilylamine, $N(SiH_3)_3$ (ref. 2)

Most XY_3 compounds, where $X = N$, P, As or Sb, are pyramidal. $N(SiH_3)_3$ seems to be an exception. Earlier studies of its electron diffraction[3, 4] and infrared and Raman spectra[5-7] led to the conclusion that it has a trigonal planar NSi_3 skeleton. This is a remarkable finding, as is shown in Table 1. It is so remarkable that an additional line of evidence would be welcome, particularly since the arguments based on infrared and Raman selection rules are not rigorous proofs. It occurred to us that Raman spectroscopy could supply such evidence. Consider the totally symmetric NSi_3 stretching mode and its $^{14}N-^{15}N$ isotopic frequency shift. The amplitude of the nitrogen atom in this vibration will depend upon the extent that the skeleton deviates from planarity. If the skeleton is planar, symmetry requires that the nitrogen does not move, and the frequency will be independent of the mass of the nitrogen atom. If the skeleton is pyramidal the nitrogen will have a considerable amplitude, and there will be an isotopic frequency shift. A crude calculation suggests that as the NSi_3 bonds change from planar to tetrahedral, the isotopic frequency shift goes from 0 to about 11 cm^{-1}.

This mode, which we shall call ν_6, can be assigned with confidence to a highly polarized Raman band at 493 cm^{-1}. Fortunately for our purposes, it is relatively strong and sharp in the Raman spectrum of condensed states, so that it is fairly well suited to careful frequency measurement. Nearly isotopically pure ^{14}N and ^{15}N samples of trisilylamine were prepared by Dr. G. A. Gibbon and Miss Beverly Swisshelm of Chatham

Table 1. Structures of some $X(YH_3)_3$ molecules.

	$(CH_3)_3$	$(SiH_3)_3$	$(GeH_3)_3$
N	Pyramidal[a]	*Planar*	
P	Pyramidal[a]	Pyramidal[b]	Pyramidal[c, d]
As	Pyramidal[a]	Pyramidal[b]	Pyramidal?[e]
Sb		Pyramidal[c]	Pyramidal?[e]

a *Tables of Interatomic Distances and Configurations in Molecules and Ions* and supplement, The Chemical Society, London. Special Publications Nos. 11 and 18. pp. M156-7, M100s-101s.
b B. Beaghley, A. G. Robiette and G. M. Sheldrick, *J. Chem. Soc.* A, 3002, 3006 (1968).
c D. W. H. Rankin, A. G. Robiette, G. M. Sheldrick, B. Beaghley and T. G. Hewitt, *J. Inorg. Nucl. Chem.* **31**, 2351 (1969).
d S. Cradock, E. A. V. Ebsworth, G. Davidson and L. A. Woodward, *J. Chem. Soc.* A, 1229 (1967).
e E. A. V. Ebsworth, D. W. H. Rankin and G. M. Sheldrick, *J. Chem Soc.* A, 2828 (1968).

Table 2. Isotope shift for ν_6 of $N(SiH_3)_3$ (in cm^{-1}).

State	$\nu(^{14}N)$	$\nu(^{15}N)$	$\nu(^{14}N) - \nu(^{15}N)$
Gas	493.1 ± 0.1 (2)	493.15 ± 0.55 (2)	− 0.05
Liquid	495.86 ± 0.24 (5)	495.39 ± 0.11 (5)	+ 0.47
Solid	501.25 ± 0.05 (2)	499.75 ± 0.05 (2)	+ 1.50

Figures in parenthesis = number of measurements.
± gives the total spread of all measurements – i.e. *all* the results fell within this range.

College, and the Raman work was done in our laboratory. The equipment has been described elsewhere.[8] Raman spectra were excited by Ar^+ 4880 Å for the gas at room temperature (about 290 torr), for the liquid at 218 K, and for the solid at 123 K. The argon laser has a non-lasing discharge line about 35 cm^{-1} away from the Raman line which was used as a frequency reference. The procedure was to remove the spike filter so that the laser discharge line was detectable, and to include it and ν_6 in each scan. The spectral slit width was 1 cm^{-1} for liquid and solid and 4 cm^{-1} for the vapour. The ^{14}N and ^{15}N samples were alternated between successive scans.

The measured values for ν_6 are summarized in Table 2. Measurements were made to the nearest 0.1 cm^{-1}. There are not enough measurements to warrant a full statistical treatment, so we shall use the total spread of observed values as a measure of precision. This is conservative because the precision is certainly better than the total scatter.

For the gas the band is broader and weaker than for the condensed states, and the reproducibility for the ^{15}N compound is not what it should be. There are also only two measurements for each isotope. The difference between the average values, $\nu(^{14}N) - \nu(^{15}N)$, is -0.05 cm^{-1}. Since the shift is unlikely to be negative, we conclude that it is really zero and that this represents experimental error. (Negative isotope shifts do occur, although rarely, as a result of Fermi resonance. There is no evidence that this is likely here, however.) A zero isotope shift means a planar NSi_3 skeleton, so we believe a third line of evidence has been added for this which is quite independent of the previous two. We do, however, wish that the spread for the ^{15}N compound had been smaller. Unfortunately the sample

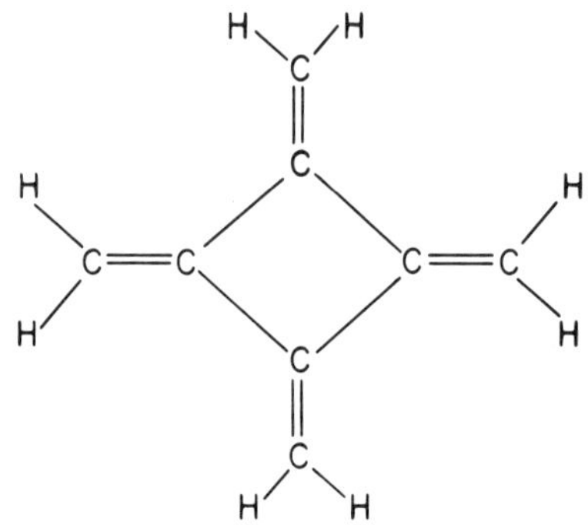

Fig. 1. [4]-radialene, or tetramethylenecyclobutane. For perchloro-[4]-radialene, each hydrogen atom is replaced by a chlorine.

was not available for additional measurements.

For both the liquid and the solid there is a small but definite isotope shift which is outside the total spread of the two measurements. This means that the molecule is distorted in some way from planar trigonal symmetry so that the nitrogen atom has some amplitude of vibration in the ν_6 mode. The distortion may be out of the plane, or it may be in the plane in such a way as to make one of the N–Si bonds slightly different from the other two. We believe that this is an effect of nearby neighbours in the condensed states. While it could be a permanent deformation in the crystal due to a site symmetry lower than D_{3h}, in the liquid it is probably a transitory effect. The important conclusion is that the isolated molecule in the gas does have a planar trigonal NSi_3 skeleton.

In contrast to ν_6, the degenerate NSi_3 stretch at 996 cm^{-1} has an isotope shift of 19 cm^{-1} (Raman spectrum of the solid).

[4]-radialene (tetramethylene cyclobutane)

This molecule has the chemical structure shown in Fig. 1. We have recently isolated it, and also its perdeuterated derivative, for the first time.[9] They are white crystalline solids at $-22°C$ which rapidly polymerize upon warming to room temperature. The vapour at room temperature (a few torr) has decomposed enough in 30 min so that the products can be seen visually. Nonetheless it was possible to obtain infrared and Raman data, mainly at low temperatures. They are completely compatible only with a planar D_{4h} structure. The reader is referred to ref. 9 for details.

Perchloro-[4]-radialene

This molecule is [4]-radialene with all the hydrogens replaced by chlorine atoms. It was first prepared by Heinrich and Roedig.[10] An X-ray crystal structure determination[11] showed that the molecule is not planar D_{4h} like the parent hydrocarbon. Instead it is deformed to a D_{2d} structure by (a) puckering the C_4 ring, (b) tilting each $=CCl_2$ group out of the plane defined by the three adjacent C–C–C ring atoms, and (c) tilting each CCl_2 so that the $=C\langle^{Cl}_{Cl}$ bonds are no longer co-planar.

Since it is known that symmetries are sometimes different in crystals from in gases,

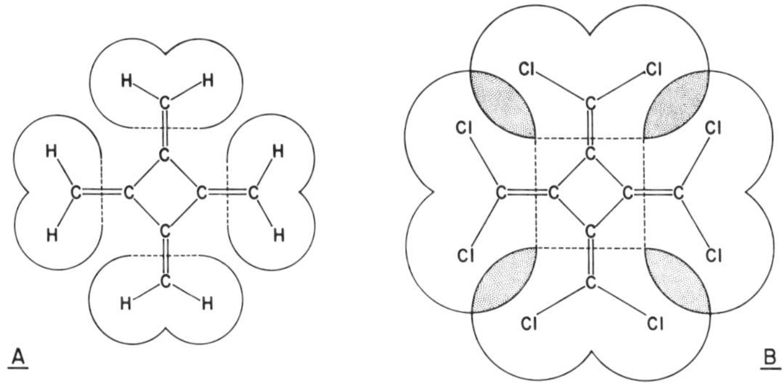

Fig. 2. Scale drawing to illustrate steric interference in planar structures of: (A). [4]-radialene (no interference, permitting a planar D_{4h} structure). (B). Perchloro-[4]-radialene (serious interference, producing a non-planar D_{2d} structure).

liquids or solutions because of packing requirements and crystal energies, it seemed worthwhile to study the vibrational spectrum of perchloro-[4]-radialene in various states. We also wanted to confirm that the distortions in the solid really did result in relaxing the selection rules. It has long been a puzzle as to how far a molecule must deviate from an ideal symmetry before this is apparent in the spectrum. We have found that the spectra are essentially the same for crystal, liquid and solution, showing only the small frequency shifts that are expected for change of phase.[12] Thus the symmetry does not change with phase. Furthermore the data are definitely not compatible with D_{4h}, but do fit D_{2d} nicely.

[4]-Radialene and perchloro-[4]-radialene have different symmetries because steric hindrance in the planar form is negligible in the first but is severe in the second. This is illustrated in Fig. 2. These are scale drawings, assuming a planar structure and using accepted values of bond distances and van der Waal's radii. The serious overlap in the perchloro compound is relieved by distorting the molecule so that successive $C{=}CCl_2$ groups are alternatively above and below the plane of the figure.

Tetrafluorodiphosphine, P_2F_4

For some time we have been interested in the inorganic molecules X_2Y_4, where X is N or P and Y is H or halogen, because they show an interesting change of conformation. This is summarized in Table 3. Hydrazine is entirely *gauche*, whereas N_2F_4 consists of both *gauche* and *trans* forms with an energy difference of only 100–200 cal/mole, and P_2Cl_4 is entirely *trans*. The reason for this change is not completely clear, although the sizes of the atoms must be one important factor. The conformations of two other molecules, P_2H_4 and P_2F_4, have not been firmly established. Since they are expected to be near the cross-over from *gauche* to *trans*, it would be useful to know the actual conformations. We have just completed a study of P_2F_4.[13] This material was first prepared and studied by Rudolph, Taylor and Parry,[14] who concluded that the molecule may have a centre of symmetry. However they remark that a more detailed spectroscopic study will be necessary to establish the structure unequivocally. Our results are more extensive, and support their conclusion completely. Furthermore there is no evidence for any isomer but the *trans* one. Details will not be given here, but Raman spectroscopy was a vital part of the work, as it always is when one is trying to find whether or not a centre of symmetry is present.

Table 3. Some inorganic X_2Y_4 molecules.

Compound	State[a]	Conformer	References
N_2H_4	g	*gauche*	*b, c*
N_2F_4	g, l, s	*gauche* and *trans*	*d, e, f, g*
P_2H_4	g	*gauche*?	*h, i*
P_2H_4	s	*trans*	*j*
P_2F_4	g, l, s	*trans*	*k, l*
P_2Cl_4	g, l, s	*trans*	*m*
P_2I_4	s, soln.	*trans*	*n, o*

[a] g = gas, l = liquid, s = solid.
[b] A. Yamaguchi, I. Ichishima, T. Shimanouchi and S. I. Mizushima, *Spectrochim. Acta* **16**, 1471 (1960).
[c] T. Kasuya and T. Kojima, *J. Phys. Soc. Japan* **18**, 364 (1963); *Sci. Papers Inst. Phys. Chem. Res. (Tokyo)* **56**, 1 (1956).
[d] C. B. Colburn, F. A. Johnson and E. Haney, *J. Chem. Phys.* **43**, 4526 (1965).
[e] F. A. Johnson, B. F. Aycock, C. Haney and C. B. Colburn, *J. Mol. Spectry.* **31**, 66 (1969).
[f] J. R. Durig and J. W. Clark, *J. Chem. Phys.* **48**, 3216 (1968).
[g] D. F. Koster and F. A. Miller, *Spectrochim. Acta* **24A**, 1487 (1968).
[h] E. R. Nixon, *J. Phys. Chem.* **60**, 1054 (1956).
[i] B. Beaghley, A. R. Conrad, J. M. Freeman, J. J. Monaghan, B. G. Norton and G. C. Holywell, *J. Mol. Structure* **11**, 371 (1972).
[j] S. G. Frankiss, *Inorg. Chem.* 7, 1931 (1968).
[k] R. W. Rudolph, R. C. Taylor and R. W. Parry, *J. Amer. Chem. Soc.* **88**, 3729 (1966).
[l] K. H. Rhee, A. M. Snider and F. A. Miller (unpublished research).
[m] S. G. Frankiss and F. A. Miller, *Spectrochim. Acta* **21**, 1235 (1965).
[n] Y. C. Leung and J. Waser, *J. Phys. Chem.* **60**, 539 (1956).
[o] S. G. Frankiss, F. A. Miller, H. Stammreich and T. T. Sans, *Spectrochim. Acta* **23A**, 543 (1967).

We have learned that L. S. Bartell and H. L. S. Hodges have recently made an electron diffraction study of P_2F_4, and have also concluded that it is *trans*.[15]

Hexafluoroazomethane, $F_3C{-}N{=}N{-}CF_3$

Electron diffraction results should not always be believed, however! This molecule

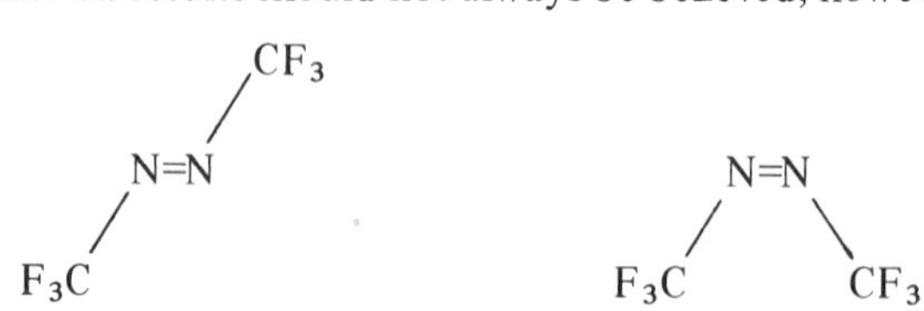

trans *cis*

provides an example. It is expected to exist in either the *trans* or the *cis* form, or as a mixture of the two. As a result of an electron diffraction study, it was concluded that only the *cis* form is present.[16] Recently a study of the vibrational spectrum has been completed by Fateley, Hayden and Tuazon[17] which led to exactly the opposite result: that the molecule is entirely *trans*. Raman and infrared spectra were obtained on the gas phase as well as on various condensed states. The evidence for a centre of symmetry is overwhelming; out of 32 bands below 1650 cm^{-1}, there is not a single Raman–infrared coincidence! The sample cannot be all *cis*, and if it is a mixture it does not contain enough *cis* to be detected in either the vibrational or NMR spectra.

Table 4. Some observed bands for LiO_2.

	Neutral O_2	O_2^- ^{6}Li	O_2^- ^{7}Li
Raman			
20% $^{16}O_2$	1548	1093	1093
50% $^{16}O-^{18}O$	1506	1063	1063
30% $^{18}O_2$	1461	1032	1032
Infrared			
		1099.0 w	1098.5 w
		720 s	626 s
		490 s	474 s

Table 5. LiO_2: summary of fundamentals.

	$^6Li^{16}O_2$ Raman	$^6Li^{16}O_2$ Infrared	Description
ν_1	1093	1099 w	$[O-O]^-$ stretch
ν_2	–	720 s	Symmetric Li–O stretch
ν_3	–	490 s	Antisymmetric Li–O stretch

The *trans* and *cis* forms may be separable and stable, analogous to *trans* and *cis* dichloroethylene. Could Chang *et al.* have worked with a *cis* sample, and Fateley *et al.* with a *trans* one? This is possible, but I think improbable. In the first place both bought their samples from the same source: Merck, Sharp and Dohme of Canada. (The one used by Fateley *et al.* had about 4% CF_3I impurity, which was effectively removed with molecular sieves.) Secondly, the sample of Fateley *et al.* had the three bands at 1584, 1561, and 1532 cm^{-1} which were present in that of Chang *et al.* (Incidentally, these are all weak and are readily explained as binary combinations, so they are not evidence for *cis* isomer as is implied by Chang *et al.*) Although one cannot be absolutely positive, it seems very likely that both experiments were done on the same isomer, and that the electron diffraction results are wrong. The latter work should be re-done. This is also indicated by the fact that work from the same laboratory has been incorrect in several other instances.[18]

Lithium superoxide, LiO_2

Papers are now starting to appear on the use of matrix isolation in Raman spectroscopy. One interesting study of this type which led to a determination of a molecular structure will be cited as an example. It is the work of Hatzenbuhler and Andrews on LiO_2.[19] They co-deposited lithium atoms (from a Knudsen cell) with oxygen molecules at 15 K. Some of the observed bands are given in Table 4. In the Raman spectrum, the band of $^{16}O_2$ at 1553 cm^{-1} was observed, and in addition a new band at 1093 cm^{-1}. The latter frequency was independent of whether ^{6}Li or ^{7}Li was used. When the matrix was a mixture of 20% $^{16}O_2$, 50% $^{16}O-^{18}O$, and 30% $^{18}O_2$, the three lines for the neutral oxygen molecules were found at 1548, 1506 and 1461 cm^{-1} respectively. Now there were three new bands, at 1093, 1063 and 1032 cm^{-1}, with the same relative intensities as for the neutral molecules. This indicates that the substance producing the new bands contains two oxygen atoms. The central band at 1063 cm^{-1} is sharp, and has the same half width (8 cm^{-1}) as the other two members. This is evidence that the two oxygen atoms are equivalent. In earlier infrared work with mixed lithium isotopes, it was shown that the absorbing entity contains only one lithium atom. Hence the substance has the formula LiO_2, and the two oxygen atoms are equivalent.

In the infrared spectrum new bands are observed at 1099.0 (w), 720 (s), and 490 (s) cm^{-1} for ^{6}Li (Table 4). For ^{7}Li they are at 1098.5 (w), 626 (s), and 474 (s) cm^{-1}

It is concluded from these data that LiO_2 is best represented as

$$\begin{matrix}[O{-}O]^{-} \\ Li^{+}\end{matrix}$$

The fundamentals are summarized in Table 5. The 1093 cm^{-1} band is in the right region for an $[O{-}O]^{-}$ stretch, and it is strong in the Raman spectrum but weak in the infrared as expected. Furthermore it has almost no lithium isotope shift. The other two bands are strong in the infrared and missing in the Raman spectrum as is expected of ionic bonds, and they show a considerable Li isotope shift – very large for ν_2, the symmetric Li–O stretch. This constitutes a nice structure determination for an experimentally difficult system. The Raman experiment was independently duplicated, and the results confirmed by Hüber and Ozin.[20] Very recently the same structure has been reported for the Na, K, Rb, and Cs superoxides.[21, 22]

$C(HgX)_4$, with X = Cl, Br, or I

These molecules have been studied recently by Fateley and Pensenstadler.[23] The results confirm expectations – the molecules are very strong Raman scatterers, and are tetrahedral. I mention them becuase of an amusing point. As far as we are aware, the substance $C(HgI)_4$ has the lowest weight percent of carbon of any known organic compound – 0.91%!

$B_2S_3X_2$ compounds

We have been carrying out a collaborative study of the compounds

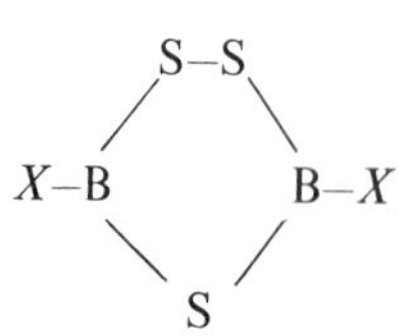

, X = Cl, Br, I, CH_3

with Dr. Robert E. Sacher of the Army Materials and Mechanics Research Center and with Dr. Walter Siebert of the University of Würzburg.[24] Various kinds of evidence indicate that these are ring compounds, but it is still necessary to decide whether the ring is planar (C_{2v}), folded (C_s), or twisted (C_2). Usually it is difficult to distinguish between these three point groups, but for the above molecules the data definitely indicate planar C_{2v} symmetry. The structure is thus analogous to that of BH–O–O–BH–O.[25, 26]

Cyclobutadiene iron tricarbonyl, $C_4H_4Fe(CO)_3$

Among the many novel metallo-organic compounds that have been prepared in the

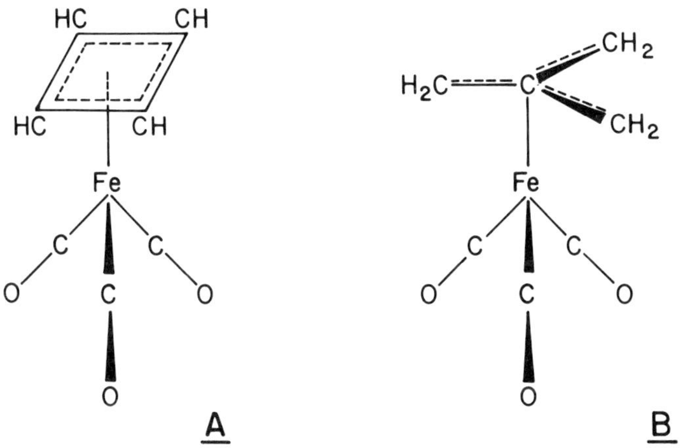

Fig. 3. (A). Structure of cyclobutadiene iron tricarbonyl. (B) Structure of trimethylenemethane iron tricarbonyl.

last few years, this is one of the most interesting. Cyclobutadiene, which has never been isolated, is stabilized in this substance by its linkage to iron. Furthermore it has been shown that the bonds of the cyclobutadiene portion are all equivalent, so that the structure is as shown in Fig. 3(A). The spectrum is worth studying because the C_4H_4 system is so unusual. We have been doing so in our laboratory,[27] and recently a paper by Andrews and Davidson has appeared.[28] The problem is the reverse of the usual case; here the structure is known in advance, and one hopes to obtain information about an unusual kind of bond.

The relative orientation of the two ends of the molecule is not known. Since one end is three-fold symmetric and the other four-fold symmetric about the same axis, the potential function for the torsion is twelve-fold symmetric. The barrier will therefore almost certainly be very small, so that it makes little difference what the relative orientation of the two ends may be. Under this circumstance it is reasonable to use the concept of local symmetry to interpret the spectra. Thus one regards the C_4H_4Fe portion as a square pyramid, and the $Fe(CO)_3$ portion as a trigonal pyramid, and then treats the central portion as though it were a linear M–Fe–N. Andrews and Davidson did this, and so do we. While we do not agree with them in all details of data and interpretation, they are correct in the main. The totally symmetric ring stretch is around 1230 cm^{-1}. Thus this mode changes from ~ 1000 cm^{-1} for C_6H_6 to ~ 1100 cm^{-1} for C_5H_5 to ~ 1230 cm^{-1} for C_4H_4 when these groups are in π-bonded complexes with metals. The highest C–C stretch in $C_4H_4Fe(CO)_3$ is near 1330 cm^{-1}, so the bonds are intermediate between single and double links.

Trimethylenemethane iron tricarbonyl, $(CH_2)_3CFe(CO)_3$

This is another example of an unstable molecule $[C(CH_2)_3]$ which is stabilized by forming its iron tricarbonyl complex (see Fig. 3(B)). The preparation of this material has recently been described,[29] and Mr. D. F. Finseth in our laboratory has made some and is studying its vibrational spectrum. The symmetry situation is a little simpler here than for butadiene iron tricarbonyl because the over-all symmetry is probably D_{3d} (staggered), or

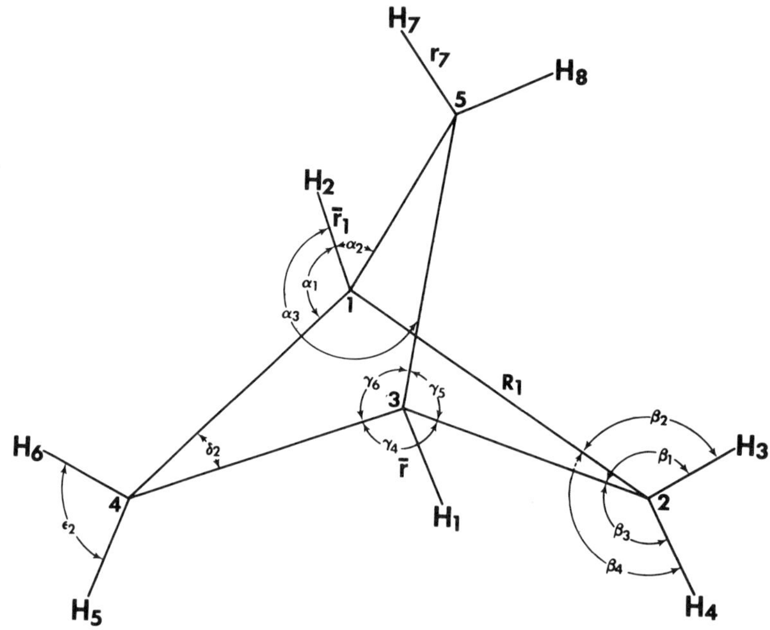

Fig. 4. Structure of bicyclo-[1.1.1]-pentane. (Figure by courtesy of Dr. Ira W. Levin.)

possibly D_{3h} (eclipsed). Again the C–C stretching frequencies are intermediate between those for double and single bonds.

bis-Cyclopropylidene

This molecule, which has the interesting bond structure

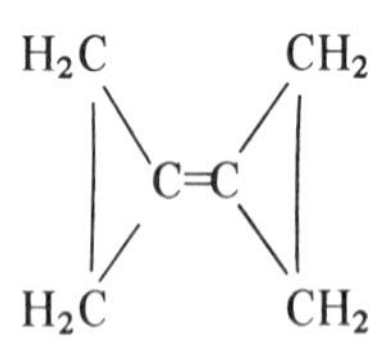

is currently under study by Lord and Wurrey.[30] They have shown that it has D_{2h} symmetry. It provides a striking exception to the rule that C=C stretching modes are always intense in the Raman spectrum (providing they are symmetry-allowed). In this molecule the C=C stretch is very weak – about the same intensity as the overtone of a ring mode which comes about 50 cm^{-1} higher. The reason for this is not known. The C=C stretch is also very high – but this is expected for a double bond exocyclic to a three-membered ring. Finally, it is interesting to note that the 'butterfly' mode occurs near 180 cm^{-1} in the infrared.

Bicyclo-[1.1.1]-pentane

This novel molecule has the structure shown in Fig. 4. It contains three four-membered rings. The Raman and infrared spectra of it and two of its deuterated deriva-

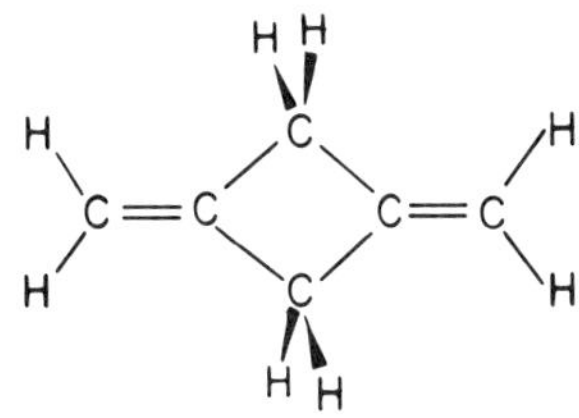

Fig. 5. 1,3-dimethylene cyclobutane.

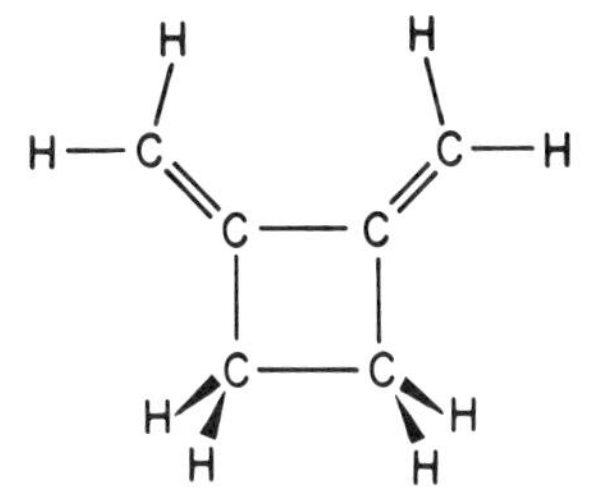

Fig. 6. 1,2-dimethylene cyclobutane.

tives have been studied by Levin *et al.*[31] The data provide good evidence for D_{3h} symmetry.

Molecular structures from ring puckering modes

The puckering modes of four-membered rings are very anharmonic. They have received a great deal of study in the infrared spectrum, and now papers are beginning to appear reporting gas phase Raman data. Although the primary interest is not molecular structure in the usual sense of the term, structural information is sometimes an incidental by-product. For example the first gas-phase Raman results of this type were for cyclobutane.[32, 33] The double-minimum potential was determined quantitatively, and from the positions of the two minima it was possible to conclude that the equilibrium dihedral angle of the molecule is 145°.

A somewhat different type of information was obtained in our studies of the dimethylene cyclobutanes.[34] The vibrational spectrum of 1,3-dimethylene cyclobutane (Fig. 5) shows that it definitely has a centre of symmetry and therefore a planar C_4 ring. The symmetry of 1,2-dimethylene cyclobutane (Fig. 6) is much more difficult to establish. The highest it can be is C_{2v}, which demands a planar C_4 ring. If the ring is puckered the symmetry is C_2. C_s and C_1 are also possible but far less likely. For a molecule of this size it is usually very difficult to distinguish unequivocally between these point groups on the basis of selection rule arguments. If a series of the puckering mode transitions can be observed, there is another line of evidence which can be used. If the ring is planar at equilibrium, the potential function is a pure quartic one to a good approximation. (In Eqn. (1), $A = 0$.)

$$V = AQ^2 + BQ^4 \tag{1}$$

The spacing between successive energy levels then increases steadily as they get higher in

Table 6. Raman transitions for the ring puckering modes of trimethylene oxide, cyclobutane, and cyclobutane-d_8. (In general, the intensities diminish as one goes downward in the table).

C_3H_6O cm^{-1}	Δ	C_4H_8 cm^{-1}	Δ	C_4D_8 cm^{-1}	Δ
143.1		199.4		157.1	
	51.2		22.3		15.8
194.3		177.1		141.3	
	28.4		17.8		8.4
222.7		159.3		132.9	
	24.8		40.1		4.7
247.5		119.2		128.2	
	20.2				23.2
267.7				105	
	18.4				13
286.1				92	
	16.0				
302.1					
	14.9				
317.0					
planar ring		puckered ring		puckered ring	

the potential well. Thus the $1 \leftarrow 0$ transition, usually the most intense, is at smallest wavenumber, and $2 \leftarrow 1$, $3 \leftarrow 2$, etc. come at successively larger wavenumbers in a fairly smooth pattern. This is illustrated by the impressive Raman results of Kiefer *et al.* for trimethylene oxide,[35] summarized in Table 6. The barrier is known to be only 15 cm^{-1}, which is below the lowest level, so the ring is effectively planar and the potential function almost pure quartic.

Suppose, however, that the ring is puckered by a potential barrier for the planar form so that the potential function has a symmetrical double minimum. (Now A in Eqn. (1) is non-zero and negative. We shall not consider cases which may have unsymmetrical potential functions.) If this barrier can be crossed readily, the pattern of bands is very different. The splitting of the levels due to tunnelling, and the very anharmonic potential near the top of the barrier, lead to an erratic pattern of levels so that the observed bands are spaced very irregularly. The intensities are also erratic. The data for cyclobutane and cyclobutane-d_8 in Table 6 illustrate the frequency behaviour.

The conclusion is that *if* a series of the puckering mode transitions can be observed, one can frequently tell merely by looking at them whether a ring is planar or folded. If the bands trail off to higher wavenumbers with a fairly smooth pattern of positions and intensities, the ring is planar. If they are erratic in position and intensity, the ring is folded.

Unfortunately we could not detect the puckering fundamental of 1,2-dimethylene cyclobutane in the infrared spectrum, and could not resolve it in the Raman spectrum, but it was observed as a series of combination tones in the mid infrared. Because of the erratic frequency pattern, we have come to the tentative conclusion that the ring is puckered. If true, this eliminates C_{2v} symmetry and leaves C_2 as most probable. This is a surprising finding, and we shall do more work to try to strengthen the evidence.

Table 7. Results of some high-resolution Raman work.

	Authors	B_0 (cm^{-1})	Method
A.	N_2		
	Stoicheff (1954)	1.9897_3	Raman
	Wilkinson (1957)	1.9898	Vacuum ultraviolet
	Vanderslice (1965)	1.9896	Vacuum ultraviolet
	Butcher *et al.* (1971)	1.98950_6	Raman
B.	O_2		
	Babcock and Herzberg (1948)	1.43777	Electronic[b]
	West and Mizushima (1966)	1.43767[a]	Microwave[b]
	McNight and Gordy (1968)	1.4376807[a]	Microwave[b]
	Butcher *et al.* (1971)	1.437682	Raman
C.	CS_2, 00°0 state		
	Stoicheff (1958)	0.10910	Raman
	Rank *et al.* (1958)	0.10910	Infrared
	Plyler *et al.* (1962)	0.10912 0.10907 0.10904	Infrared
	Walker and Weber (1971)	0.10912	Raman
D.	CS_2, 01^10 state		
	Rank *et al.* (1958)	0.109321	Infrared
	Plyler *et al.* (1962)	0.10940	Infrared
	Walker and Weber (1971)	0.10935	Raman

[a] These values were originally given in MHz. The value of c used for conversion to cm^{-1} was $2.997925 \times 10^{+10}$ cm/s.
[b] Magnetic dipole allowed.

DETERMINATION OF MOMENTS OF INERTIA

The Raman data for this application come from very high resolution studies of either pure rotation spectra or rotation–vibration spectra. Only a few people are engaged in such work. One thinks of the groups led by Professor H. L. Welsh at Toronto, Professor A. Weber at Fordham, Dr. W. J. Jones at Cambridge, and Dr. J. J. Barrett when he was with Perkin-Elmer. A few published results will be mentioned here to give an impression of the accuracy that has been achieved. Weber and Schlupf have described the goals and the difficulties succinctly: 'The aim of high resolution Raman spectroscopy of gases is to obtain from the resolved rotation–vibration spectrum of a molecule the rotation, centrifugal distortion, coriolis, and anharmonic constants for the various vibrational states, and to obtain these constants with accuracies that qualify them for inclusion with the constants derived from high-resolution infrared spectroscopy in a unified study of the molecular properties. In view of this criterion the demands that are placed on Raman spectroscopy in terms of resolving power, wavenumber accuracy, and sensitivity are quite enormous.'[36]

Butcher, Willetts, and Jones[37] studied the pure rotation spectrum of O_2 and N_2 by crossing a short focal length spectrograph with a Fabry-Perot etalon to obtain both high resolution and high light throughput. The sample was inside the cavity of an argon ion laser. Their results for the rotational constant B_0 are given in Table 7, and are compared there with other measurements. It will be seen that the Raman results are among the best available, with values quoted to about one part in 10^7.

Walker and Weber have measured the rotational constant of CS_2 in the ground state and in the 01^10 excited state.[38] These results are also included in Table 7. While they are not obviously superior to those obtained by other means, they are still impressive. The later paper by Weber and Schlupf[36] describes the use of an argon ion laser in single-mode operation to narrow the exciting line. A resolution of $\sim 0.07\ cm^{-1}$ was observed for the pure rotation spectrum of hexafluorobenzene, which is the best reported so far for 90° scattering.

A recent paper by Fast and Welsh describes a high resolution Raman study of acetylene and its two deuterated derivatives.[39] The pure rotation spectra were measured, and also vibration–rotation structure for a total of five bands. These data were combined with infrared results to give the best values extant for the fundamental frequencies, the B's, D's, x_{ij}'s, and α's. From these B_e and r_e values were calculated. It is a thorough, careful piece of work which illustrates the unified use of both Raman and infrared data referred to earlier.

Finally, I wish to mention that two separate surveys of high resolution Raman spectroscopy by Professor Weber will appear soon.[40, 41]

CONCLUSION

This concludes our survey of the recent uses of Raman spectroscopy in determining molecular structures. To all those workers who have published interesting and significant work that has not been mentioned here, I express apologies. I have assumed that my audience would rather hear about work which has not yet appeared than to have a review of papers which might already be known to them. To those who have so generously given me their results in advance of publication, I am most appreciative.

An excellent review chapter on Raman spectroscopy has recently been written by Durig and Harris.[42] It includes many examples of molecular structure determinations. Since they are all different from the ones described here, the reader is referred to their review for more cases.

In closing, I should like to draw your attention to a new book by Fateley *et al.* entitled *Infrared and Raman Selection Rules for Molecular and Lattice Vibrations* which will be out shortly.[43] It is an elaboration of their paper on the Correlation Method[44] which many of you may already have found useful.

ACKNOWLEDGEMENT

Our recent work in Raman spectroscopy has been supported by grants GP-9260 and GP-24727 from the National Science Foundation. I am most appreciative to the Organizing Committee of the Conference for the opportunity to present this material.

REFERENCES

1 F. R. Dollish, W. G. Fateley and F. F. Bentley, *Characteristic Raman Frequencies of Organic Compounds*, Wiley-Interscience, New York, 1973.

2 F. A. Miller, J. Perkins, G. A. Gibbon and B. A. Swisshelm (unpublished research). From the Ph.D. thesis of J. Perkins, University of Pittsburgh, 1971.

3 K. Hedberg, *J. Amer. Chem. Soc.* **77**, 6491 (1955).

4 B. Beaghley and A. R. Conrad, *Trans. Faraday Soc.* **66**, 2740 (1970).

5 E. A. V. Ebsworth, J. R. Hall, M. J. Mackillop, D. C. McKean, N. Sheppard and

L. A. Woodward, *Spectrochim. Acta* **13**, 202 (1958).
6 D. W. Robinson, *J. Amer. Chem. Soc.* **80**, 5924 (1958).
7 V. H. Kriegsmann and W. Forster, *Z. Anorg. Allgem. Chem.* **298**, 212 (1959).
8 K. H. Rhee and F. A. Miller, *Spectrochim. Acta* **27A**, 1 (1971).
9 F. A. Miller, F. R. Brown and K. H. Rhee, *Spectrochim. Acta* **28A**, 1467 (1972).
10 B. Heinrich and A. Roedig, *Angew. Chemie* **80**, 367 (1968); *Angew. Chem. Intern. Ed. Engl.* **7**, 375 (1968).
11 F. P. van Remoortere and F. P. Boer, *J. Amer. Chem. Soc.* **92**, 3355 (1970).
12 F. A. Miller and D. F. Finseth (unpublished research).
13 K. H. Rhee, A. M. Snider and F. A. Miller (unpublished research).
14 R. W. Rudolph, R. C. Taylor and R. W. Parry, *J. Amer. Chem. Soc.* **88**, 3729 (1966).
15 L. S. Bartell and H. L. S. Hodges, University of Michigan (private communication).
16 C. H. Chang, R. F. Porter and S. H. Bauer, *J. Amer. Chem. Soc.* **92**, 5313 (1970).
17 W. G. Fateley, R. A. Hayden, and E. C. Tauzon, Carnegie-Mellon University (unpublished research).
18 (a) N_2F_4. See D. F. Koster and F. A. Miller, *Spectrochim. Acta* **24A**, 1487 (1968) and references therein. (b) 1,4-Cyclohexadiene. See J. Laane and R. C. Lord, *J. Mol. Spectry.* **39**, 340 (1971). (c) Perfluoropropene. A. B. Harvey and L. Y. Nelson, *J. Chem. Phys.* **55**, 4145 (1971).
19 D. A. Hatzenbuhler and L. Andrews, *J. Chem. Phys.* **56**, 3398 (1972).
20 H. Hüber and G. A. Ozin, *J. Mol. Spectry.* **41**, 595 (1972).
21 L. Andrews, paper I-3 at Symposium on Molecular Structure and Spectroscopy, Ohio State University, June, 1972.
22 R. R. Smardzewski and L. Andrews, *J. Chem. Phys.* **57**, 1327 (1972).
23 W. G. Fateley and D. F. Pensenstadler, Carnegie-Mellon University (unpublished research).
24 R. E. Sacher, W. Siebert, J. Nakovich and F. A. Miller (unpublished research).
25 W. V. F. Brooks, C. C. Costain and R. F. Porter, *J. Chem. Phys.* **47**, 4186 (1967).
26 F. A. Grimm and R. F. Porter, *Inorg. Chem.* **8**, 731 (1969).
27 D. F. Finseth, K. H. Rhee and F. A. Miller (unpublished research).
28 D. C. Andrews and G. Davidson, *J. Organometal. Chem.* **36**, 349 (1972).
29 K. Ehrlich and G. F. Emerson, *J. Amer. Chem. Soc.* **94**, 2464 (1972).
30 R. C. Lord and C. J. Wurrey, Mass. Inst. Technology (private communication).
31 I. W. Levin, T. P. Lewis and K. B. Wiberg (private communication from Dr. Ira W. Levin, National Institutes of Health). An abbreviated discussion will appear in a chapter by I. W. Levin in *Advances in Chemoreception*, Vol. II Meredith, New York.
32 J. M. R. Stone and I. M. Mills, *Mol. Phys.* **18**, 631 (1970).
33 F. A. Miller and R. J. Capwell, *Spectrochim. Acta* **27A**, 947 (1971).
34 F. R. Brown and F. A. Miller (unpublished research).
35 W. Kiefer, H. J. Bernstein, M. Danyluk and H. Wieser, *Chem. Phys. Letters* **12**, 605 (1972).
36 A. Weber and J. Schlupf, *J. Opt. Soc. Amer.* **62**, 428 (1972).
37 R. J. Butcher, D. V. Willetts and W. J. Jones, *Proc. Roy. Soc.* **A324**, 231 (1971).
38 W. J. Walker and A. Weber, *J. Mol. Spectry.* **39**, 57 (1971).
39 H. Fast and H. L. Welsh, *J. Mol. Spectry.* **41**, 203 (1972).
40 A. Weber, Recent Developments in High Resolution Raman Spectroscopy of Gases, *Developments in Applied Spectroscopy*, Vol. 10, Plenum Press, New York, (in press).

41 A. Weber, 'High Resolution Raman Studies of Gases', Chap. 10 of *The Raman Effect*, Vol. 2, (Ed. A. Anderson), Dekker, New York, 1972.

42 J. R. Durig and W. C. Harris, in *Physical Methods of Chemistry*, Vol.1, (Eds. A. Weissburger and B. Rossiter), Wiley, New York, 1972, part 3B.

43 W. G. Fateley, F. R. Dollish, N. T. McDevitt and F. F. Bentley, *Infrared and Raman Selection Rules for Molecular and Lattice Vibrations*, Wiley-Interscience, New York, 1972.

44 W. G. Fateley, N. T. McDevitt and F. F. Bentley, *Appl. Spectry.* **25**, 155 (1971).

68. RAMAN SPECTRA OF HALOGEN AND INTERHALOGEN CATIONS

R. J. Gillespie
Laboratoire des Acides Mineraux,
Université de Montpellier, France
and
M. J. Morton and J. M. Sowa
Department of Chemistry, McMaster University,
Hamilton, Ontario, Canada

INTRODUCTION

The halogens and the interhalogens may be oxidized with several oxidizing agents such as $S_2O_6F_2$, SbF_5 and AsF_5 to a variety of cationic species such as I_2^+, Br_2^+, I_3^+, Br_3^+, Cl_3^+, Cl_2F^+, ICl_2^+ and I_2Cl^+. These cations have been obtained as stable species in suitable very weakly basic solvents such as HSO_3F and SO_2 and in some cases in the form of crystalline salts with weakly basic anions such as $SbCl_6^-$, $AlCl_4^-$, SbF_6^-, AsF_6^- and $Sb_2F_{11}^-$. Four of these cations, Br_2^+, ICl_2^+, ClF_2^+ and BrF_2^+,[1-5] have had their structures determined by X-ray crystallography. In principle, useful structural information may also be obtained by vibrational spectroscopy but the dark colours of the cations containing iodine and bromine and the great sensitivity to moisture of the cations, particularly those containing fluorine, have inhibited the study of their infrared and Raman spectra until the last few years. The advent of laser Raman spectroscopy has enabled the vibrational spectra of these cations to be studied much more easily although the fact that some of them are very strongly absorbing in the visible region of the spectrum still presents some problems which we will discuss in detail in particular cases.

The diatomic cations I_2^+ and Br_2^+

The I_2^+ cation may be prepared[6] in solution in HSO_3F by oxidizing iodine with $S_2O_6F_2$ according to the equation

$$2I_2 + S_2O_6F_2 \rightarrow 2I_2^+ + 2SO_3F^-$$

It may also be prepared simply by dissolving iodine in 65% oleum[7] in which iodine is oxidized to I_2^+ by SO_3 according to the equation

$$2I_2 + 5SO_3 + H_2S_4O_{13} \rightarrow 2I_2^+ + 2HS_4O_{13}^- + SO_2$$

These solutions of I_2^+ have an intense blue colour and have strong absorption bands with maxima at 640, 490 and 410 nm. Because of this strong absorption the first attempts to observe the Raman spectrum of I_2^+[8] were unsuccessful due to the complete

Table 1. Vibrational spectra of X_2^+ cations.

	I_2^+	I_2[a]	Br_2^+	Br_2[a]	Cl_2^+	Cl_2[a]
Stretching frequency (cm^{-1})	238	213	360	318	645	554
Bond length Å	–	2.66	2.13	2.27	–	1.98
Force constant (mdyn $Å^{-1}$)	2.12	1.70	3.05	2.38	4.29	3.16

[a] Ref. 13.

Table 2. Vibrational frequencies and force constant for the triatomic halogen cations X_3^+.

	ν_1	ν_2 (cm^{-1})	ν_3	f (mdyn $Å^{-1}$)	d
I_3^+	207	114	233	1.7	0.31
Br_3^+	290	–	–	–	–
Cl_3^+	490	225	508	2.5	0.36

absorption of the exciting radiation. However, using 6328 Å He–Ne laser light and by carefully controlling the concentration of the solution contained in a very narrow capillary tube it was possible to obtain a resonance Raman spectrum of I_2^+. This spectrum which is very intense and consists of a fundamental band accompanied by a large number of overtones is discussed in more detail elsewhere.[9,10] From this spectrum the stretching frequency was found to be 238 cm^{-1}. The increase in the vibrational frequency from 213 cm^{-1} for I_2 is consistent with the loss of an anti-bonding electron, giving a formal bond order of 1.5.

The $Sb_3F_{16}^-$ salt of Br_2^+ was prepared by Edwards and his coworkers[1] by the reaction of bromine with bromine pentafluoride and antimony pentafluoride

$$9Br_2 + 2BrF_5 + 30SbF_5 \rightarrow 10(Br_2^+Sb_3F_{16}^-)$$

It is scarlet red paramagnetic solid that is immediately decomposed by moisture. An X-ray crystallographic study[1] showed the solid to consist of the Br_2^+ cation and the *trans* $Sb_3F_{16}^-$ ion. The bond length in Br_2^+ was found to be 2.13 Å compared with 2.27 Å in the neutral Br_2 molecule and a stretching frequency of 368 cm^{-1} was observed in the Raman spectrum of the solid compared with the frequency of 318 cm^{-1} for the Br_2 molecule. The increase in the stretching frequency and the decrease in the bond length are again consistent with an increase in bond order resulting from the loss of an anti-bonding electron.

Solutions of Br_2^+ may be prepared by oxidizing Br_2 with $S_2O_6F_2$ in solution in the super acid HSO_3F–SbF_5–SO_3. The solutions have a characteristic deep cherry-red colour and a strong absorption with a maximum at 510 nm. A resonance Raman effect was observed for the Br_2^+ ion using the 5145 Å radiation of an argon ion laser. The stretching frequency in solution was found to be 360 cm^{-1}.

There is no reliable evidence for the existence of Cl_2^+ in the form of stable salts or even in solution[12] although its electronic spectrum has been recorded in the gas phase.

From this spectrum the stretching frequency has been found to be 645 cm^{-1}. The stretching frequencies, bond lengths and force constants of the diatomic cations Cl_2^+, Br_2^+ and I_2^+ and the neutral halogen molecules are summarized in Table 1.

TRIATOMIC HALOGEN CATIONS

The tri-iodine cation I_3^+

Solutions of this cation may be prepared by the reaction of iodine and iodic acid in 100% sulphuric acid[14] according to the equation

$$HIO_3 + 7I_2 + 8H_2SO_4 \rightarrow 5I_3^+ + 3H_3O^+ + 8HSO_4^-$$

Solutions in fluorosulphuric acid may also be prepared by the reaction

$$3I_3 + S_2O_6F_2 \rightarrow 2I_3^+ + 2SO_3F^-$$

These solutions have a deep red-brown colour and have absorption maxima at 305 and 470 nm. A very dark, almost black, solid may be prepared by the reaction of iodine with antimony pentafluoride in liquid sulphur dioxide according to the equation

$$3I_2 + 3SbF_6 \rightarrow 2I_3^+SbF_6^- + SbF_3$$

When this compound is dissolved in 100% H_2SO_4 the same red-brown solution is obtained.

The solutions of I_3^+ in H_2SO_4 gave Raman spectra which in addition to the solvent peaks had bands at 114, 207 and 233 cm^{-1} which may be assigned as the ν_2, ν_1 and ν_3 vibrations of an angular molecule (Table 2). As expected the 114 and 207 cm^{-1} bands were found to be polarized. The study of these intensely coloured solutions was facilitated by the use of a rotating cell.

In initial attempts to obtain the spectrum with 6328 Å radiation an intense resonance Raman spectrum of I_2^+ was obtained which obscured the 233 cm^{-1} peak of I_3^+. This I_2^+ was presumably formed by slight solvent oxidation of I_3^+. This difficulty was overcome however, by the use of the 5145 Å radiation of the Ar^+ laser. It may be noted that the average stretching frequency of 220 cm^{-1} in the I_3^+ molecule is appreciably lower than the stretching frequency of I_2^+ at 238 cm^{-1} and it is indeed closer to the frequency of 213 cm^{-1} for the neutral molecule. This is consistent with the fact that in I_3^+ both bonds have a formal bond order of 1.0 compared with 1.5 for the I_2^+ molecule. Using a simple valency force field and a bond angle of 114°, a stretching force constant $f = 1.7$ mdyn $Å^{-1}$ and a bending force constant $d = 0.31$ mdyn $Å^{-1}$ were obtained.

The tribromine cation Br_3^+

This ion may be formed quantitatively by oxidation of bromine with $S_2O_6F_2$ in the superacid medium $HSO_3F–SbF_5–SO_3$ [11]

$$3Br_2 + S_2O_6F_2 \rightarrow 2Br_3^+ + 2SO_3F^-$$

These solutions are brown in colour and have an absorption band with a maximum at 300 nm and a shoulder at 375 nm. The only Raman band in addition to the solvent and

Table 3. Raman spectra of $Cl_3^+AsF_6^-$ and $Cl_2F^+AsF_6^-$.

$\Delta\nu(cm^{-1})$	Assignments Cl_3^+	Assignments AsF_6^-	$\Delta\nu(cm^{-1})$	Assignments Cl_2F^+	Assignments AsF_6^-
170 (20)			293, 299 (20)	ν_3	
225 (75)	ν_2		375 (12)		ν_5
370 (25)			528, 535 (100)	ν_2	
394 (6)		ν_5	563 (16)		ν_2
485 (75)	ν_1		581 (12)		(cont.)
493 (100)	(cont.)		685 (20)		ν_1
508 (60)	ν_3		744 (78)	ν_1	
571 (18)		ν_2			
674 (50)		ν_1			

anion peaks that has so far been observed for these solutions is at 290 cm^{-1}. Presumably this is the ν_1 symmetrical stretching mode. Further investigation of the Raman spectra of these solutions is in progress.

The trichlorine cation Cl_3^+

The yellow solid $Cl_3^+AsF_6^-$ has been prepared from chlorine, chlorine monofluoride and arsenic pentafluoride at $-76°C$. The Raman spectrum of the solid (Table 3) at $-80°$ shows the active bands of the AsF_6^- ion together with three relatively intense bands at 490, 225 and 508 cm^{-1} which may be assigned as ν_1, ν_2 and ν_3 respectively of the bent Cl_3^+ cation. The assigned frequencies are very close to the vibrational frequencies 514, 208 and 535 cm^{-1} of the isoelectronic SCl_2 molecule which has a bond angle of 103°. A simple valency force field treatment of the Cl_3^+ molecule gave good agreement with the observed frequencies for a bond angle of 100°, a stretching force constant of 2.5 mdyn $Å^{-1}$ and a bending force constant of 0.36 mdyn $Å^{-1}$.

Triatomic interhalogen cations

If it is assumed that the most electronegative atom always occupies the central position, there are 16 possible interhalogen triatomic cations and of these only ClF_2^+, BrF_2^+, IF_2^+, Cl_2F^+, I_2Cl^+ and ICl_2^+ have so far been established with certainty and crystal structure determinations have been published for ICl_2^+, BrF_2^+ and ClF_2^+ which all have the expected angular structure.[2–5]

Adducts of ClF_3 and BrF_3 with a number of fluoride ion acceptors such as AsF_5, SbF_5 and BF_3 have been known for some time and several recent crystal structure determinations have provided information on the geometry of the halogen cation.[3–5] In $ClF_2^+SbF_6^-$ the ClF_2^+ ion has a bond angle of 96° and a bond length of 1.58 Å whereas for $ClF_2^+AsF_6^-$ the rather different values of 103° and 1.54 Å were obtained. In $BrF_2^+SbF_6^-$ the BrF_2^+ cation was found to have a bond angle of 93.5° and a bond length of 2.29 Å. In each case a very approximately square coordination is achieved around the central halogen by the formation of two rather long fluorine bridges to neighbouring anions. The frequencies observed in the Raman spectra are shown in Table 4.[20,21]

The structures of the adducts $AsF_5.2ClF$ and $BF_3.2ClF$ have been established from their infrared[16] and Raman spectra[12] which show that they are ionic compounds con-

Table 4. Raman spectra of $ClF_2^+AsF_6^-$ and $BrF_2^+AsF_6^-$.

$\Delta\nu(cm^{-1})$	Assignments ClF_2^+	AsF_6^-	$\Delta\nu(cm^{-1})_2$	Assignments BrF_2^+	AsF_6^-
373 (14)		ν_5	308 (5)	ν_2	
384 (14)	ν_2		360 (18)		ν_5
544 (26) }		ν_2	382		ν_4
602 (60) }			519 (15)		?
693 (77)		ν_1	572 (10) }		ν_2
806 (100) }	ν_1		600 (8) }		
809 (90) }			691 (44)		ν_1
821 (51)	ν_3		703 (30)	ν_3	
			706 (100)	ν_1	

Table 5. Raman spectra of $ICl_2^+SbCl_6^-$ and $I_2Cl^+SbCl_6^-$.

$\Delta\nu(cm^{-1})$	Assignments ICl_2^+	$SbCl_6^-$	$\Delta\nu(cm^{-1})$	Assignments I_2Cl^+	$SbCl_6^-$
148	ν_2		124	ν_3(bend)	
175		ν_5	184	ν_2(I–I str.)	
250 }		ν_2	264 }		ν_2
269 }			298 }		
323		ν_1	335		ν_1
354	$\nu_1(I^{37}Cl_2^+)$		351	$\nu_1(I^{37}Cl)$	
362	$\nu_1(I^{35}Cl^{37}Cl^+)$		358	$\nu_1(I^{35}Cl)$	
370	$\nu_1(I^{35}Cl_2^+)$				

taining the Cl_2F^+ cation which has the unsymmetrical angular structure

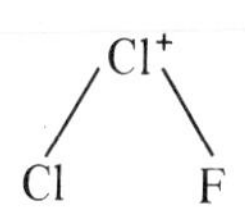

The vibrational frequencies obtained from the Raman spectrum are given in Table 3. As in the case of $ClF_2^+AsF_6^-$ and $BrF_2^+AsF_6^-$ the splitting of ν_2 of AsF_6^- can reasonably be attributed to a lowering of the symmetry of the AsF_6^- anion by fluorine bridging.

From their studies of the system I_2–$SbCl_5$ both Ruff[17] in 1915 and more recently Fialkov and his coworkers[18] have reported evidence for the compound $SbCl_5.2ICl$ which in view of the evidence for other interhalogen cations may reasonably be formulated as $I_2Cl^+SbCl_6^-$. Evidence for the formation of I_2Cl^+ in solution has also been given by Garrett *et al.*[14] from their conductimetric and cryoscopic studies of HIO_3–I_2–ICl solutions in 100% H_2SO_4.

We have prepared the compound $SbCl_5.2Cl$ by crystallization from a hot solution of ICl in excess $SbCl_5$. Sublimation of the crude product yielded well-formed very dark brown crystals. The frequencies observed in the Raman spectrum of the crystalline solid are given in Table 5. The observation of frequencies that can clearly be assigned as I–I and I–Cl stretching modes confirms the expected unsymmetrical structure

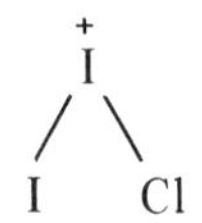

X-ray crystallographic studies[2] of the adducts of ICl_3 with $AlCl_3$ and $SbCl_5$ have shown that they may be regarded as ionic compounds containing the angular ICl_2^+ cation with the following dimensions:

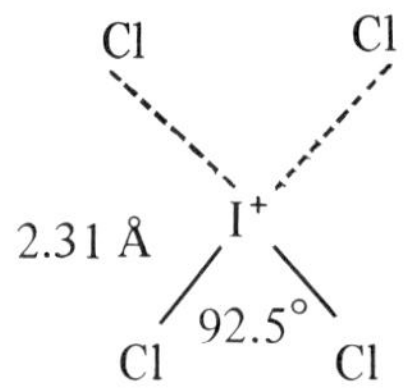

$ICl_2^+SbCl_6^-$

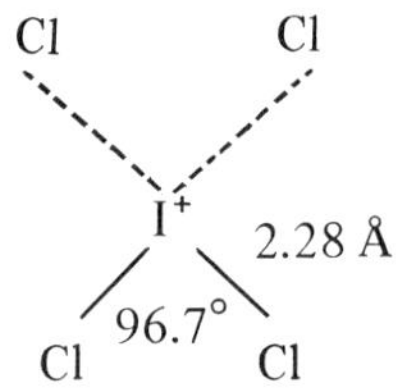

$ICl_2^+AlCl_4^-$

In both cases an approximately square planar coordination around the iodine is completed by chlorine bridges to the $SbCl_6^-$ ion. It has also been demonstrated that ICl_3 ionizes to the ICl_2^+ cation in solution in disulphuric acid and that it may be formed by the reaction of I_2, Cl_2 and HIO_3 or I_2, ICl_3 and HIO_3 in solution in H_2SO_4.[17] The Raman spectrum of the orange solid $ICl_2^+.SbCl_6^-$ gave, in addition to bands which may be attributed to the anion, a band at 148 cm^{-1} which may be assigned as the ν_2 bending mode and bands at 354, 362 and 370 cm^{-1} which may be assigned as the symmetric ν_1 stretching mode for $I^{37}Cl_2^+$, $I^{35}C^{37}Cl_2^+$ and $I^{35}Cl_2^+$ respectively (Table 5). The asymmetric stretching mode which is expected to have a smaller isotopic splitting must lie beneath the bands due to ν_1. Thus ν_1 and ν_3 must have very similar frequencies and this is consistent with a bond angle of close to 90° as has been found by X-ray crystallography. Using a simple valency force field treatment, a stretching force constant of 2.2 mdyn Å^{-1} and a bending force constant of 0.19 mdyn Å^{-1} were obtained.

In sulphuric acid solutions of ICl_2^+ prepared according to the equations

$$2I_2 + HIO_3 + 5Cl_2 + 8H_2SO_4 \rightarrow 5ICl_2^+ + 3H_3O^+ + 8HSO_4^-$$

$$I_2 + 3HIO_3 + 10ICl_3 + 24H_2SO_4 \rightarrow 15ICl_2^+ + 9H_3O^+ + 24HSO_4^-$$

two bands were observed at 148 and 387 cm^{-1}. These must be the bending mode ν_2 and the unresolved stretching modes ν_1, ν_3 of the ICl_2^+ cation although the stretching frequencies are somewhat higher than in the solid state.

REFERENCES

1 A. J. Edwards, G. R. Jones and R. J. C. Sills, *Chem. Commun.* 1527 (1968).
2 C. G. Vonk and E. H. Wiebenga, *Acta Cryst.* **12**, 859 (1959).
3 A. J. Edwards and R. J. C. Sills, *J. Chem. Soc.* 2697 (1970).

4 H. Lynton and J. Passmore, *Can. J. Chem.* **49**, 2539 (1971).
5 A. J. Edwards and G. R. Jones, *J. Chem. Soc.* 1467 (1969).
6 R. J. Gillespie and J. B. Milne, *Inorg. Chem.* **5**, 1577 (1966).
7 R. J. Gillespie and K. C. Malhotra, *Inorg. Chem.* **8**, 1751 (1969).
8 R. D. W. Kemmitt, M. Murray, V. M. McRae, R. D. Peacock and M. C. R. Symons, *J. Chem. Soc.* 862 (1968).
9 R. J. Gillespie and M. J. Morton, *J. Mol. Spectry.* **30**, 178 (1969).
10 M. R. Booth, M. J. Morton and R. J. Gillespie, in *Advances in Raman Spectroscopy*, Vol. 1 (Ed. J. P. Mathieu), Heyden & Son, London, 1973, p. 364.
11 R. J. Gillespie and M. J. Morton, *Inorg. Chem.* **11**, 586 (1972).
12 R. J. Gillespie and M. J. Morton, *Inorg. Chem.* **11**, 591 (1972).
13 W. Holzer, W. F. Murphy and H. J. Bernstein, *J. Chem. Phys.* **52**, 399 (1970).
14 R. A. Garrett, R. J. Gillespie and J. B. Senior, *Inorg. Chem.* **4**, 563 (1965).
15 R. J. Gillespie and J. B. Milne, *Inorg. Chem.* **5**, 1236 (1966).
16 K. O. Christie and W. Sawodny, *Inorg. Chem.* **8**, 212 (1969).
17 O. Ruff, *Chem. Ber.* **48**, 2068 (1915).
18 Y. A. Fialkov and I. L. Abarbarchuk, *Ukr. Khim. Zh.* **15**, 372 (1949).
19 J. B. Senior and J. L. Grover, *Can. J. Chem.* **49**, 2688 (1971).
20 K. O. Christie and C. J. Schack, *Inorg. Chem.* **9**, 2296 (1970).
21 T. Sarles, H. H. Hyman, L. A. Quarterman and A. I. Popov, *Inorg. Chem.* **9**, 2726 (1970).

69. RAMAN INTENSITIES OF THE $\nu_1(A_1)$ LINE OF VO_4^{3-}, CrO_4^{2-}, MoO_4^{2-}, WO_4^{2-}, ReO_4^-, OsO_4^- AND WS_4^{2-}

H. Schulze* and A. Müller
Institut für Chemie der Universität Dortmund, Germany

INTRODUCTION

The determination of the bond polarizability derivative $\bar{\alpha}'_{XY}$ from Raman intensities is sometimes useful in explaining bond properties. Yoshino and Bernstein[1] have shown that in the case of hydrocarbon gases $\bar{\alpha}'_{XY}$ is directly proportional to the bond order.

Some authors studied the intensity of the $\nu_1(A_1)$ line of tetrahedral molecules and ions. The results were also partly discussed with respect to bond properties. Chantry and Plane[2,3] measured the intensity of the $\nu_1(A_1)$ line and found the following trend: $PO_4^{3-} < SO_4^{2-} < ClO_4^- < IO_4^-$, Witt and Hammaker[4] have found the trend: $ClO_4^- < BrO_4^- < IO_4^-$, and Clark and Willis[5] $MCl_4 < MBr_4 < MI_4$ (M = C, Si, Ge, Sn, Ti) and $CX_4 < SiX_4 < GeX_4 < SnX_4 < TiX_4$ (X = Cl, Br, I). All these measurements were carried out for molecules and ions with a main group element as the central atom. In these cases the Raman intensity increases when the central atom becomes heavier.

We have now measured the intensities of some transition metal compounds like VO_4^{3-}, CrO_4^{2-}, MoO_4^{2-}, WO_4^{2-}, ReO_4^-, OsO_4, WS_4^{2-} and discuss the results.

EXPERIMENTAL SECTION

Chemicals. All chemicals were used as received (Merck). Osmium tetroxide was a gift from Fa. Degussa (Hanau).

Spectrometer. The Raman spectra were recorded with a Coderg Raman Spectrometer PH1 with a He/Ne laser (~200 mW, 6328 Å) and an argon laser (~2 W, 5145 and 4880 Å). The spectral slit width was in the range of 2–12 cm^{-1} and the scanning speed was 3 cm^{-1}/min.

Intensity measurements. Following the procedure of Chantry and Plane[2] the Raman intensities were measured using, in several cases, the $\nu_1(A_1)$ line of the perchlorate ion as internal standard. Where possible, we have taken the ions under investigation also as internal standards to get more possibilities to compare the results. We have measured the intensities for the following pairs: ClO_4^-/CrO_4^{2-}, ClO_4^-/MoO_4^{2-}, CrO_4^{2-}/MoO_4^{2-}, CrO_4^{2-}/WO_4^{2-}, MoO_4^{2-}/WO_4^{2-}, VO_4^{3-}/MoO_4^{2-} and VO_4^{3-}/WO_4^{2-}.

The intensity of the $\nu_1(A_1)$line of the ReO_4^- ion was determined using only MoO_4^{2-}

---

* Present address: Anorganisch-Chemisches Institut der Universität Göttingen, W. Germany.

as internal standard and the WS_4^{2-} ion was measured against SO_4^{2-}. OsO_4 was directly measured in CCl_4 solution. The intensity values for ClO_4^- and SO_4^{2-} were taken from ref. 2. Band areas were determined by planimeter. In the case of WS_4^{2-} slight decomposition occurred at the concentrations used even in alkaline solutions. The concentrations used were ~ 1 M.

Calculations. The observed intensity for a totally symmetric $\nu_1(A_1)$ Raman line for a tetrahedral molecule is given by

$$I = \frac{KM(\nu_0 - \nu_1)^4}{\nu_1(1 - \exp(-h\nu_1/kT))} 45 \left(\frac{\delta\bar{\alpha}}{\delta Q_1}\right)^2 \tag{1}$$

In this equation, K is a constant, M the molar concentration of the scattering species, ν_0 the exciting frequency, ν_1 the frequency of normal coordinate Q_1 and $\bar{\alpha}$ the mean molecular polarizability. T, k and h have their usual meanings.

To relate the measured intensity to bond properties, we have to transform the polarizability derivative with respect to Q_1 to a derivative with respect to the internal coordinate $r(\mathrm{A-B})$. For a tetrahedral molecule of the type AB_4, we obtain

$$\bar{\alpha}'_{r(\mathrm{A-B})} = \frac{\sqrt{M_\mathrm{B}}}{2} \bar{\alpha}'_Q \tag{2}$$

M_B is the mass of the atom B.

The δ function potential model, proposed by Long and Plane,[6] enables us to estimate bond orders from experimentally determined polarizability derivatives

$$\bar{\alpha}'_{r(\mathrm{A-B})} = 2/3\,(g\sigma/Za_0)(n/2)\,r(\mathrm{A-B})^3 \tag{3}$$

where $n/2$ is the bond order, $r(\mathrm{A-B})$ the internuclear distance, a_0 the Bohr radius, g the δ function strength, σ is the covalent bond character as defined by Pauling and Z the effective nuclear charge. The δ function strength, g, is here taken as $(X_\mathrm{A}^{1/2} X_\mathrm{B}^{1/2})^{1/2}$, where X_α is electronegativity of the atom α. The covalent bond character is given by $\exp[(-1/4)(X_\mathrm{A} - X_\mathrm{B})^2]$ and the effective nuclear charge, Z, is taken to be the geometric mean of Z_A and Z_B, $(Z_\mathrm{A} Z_\mathrm{B})^{1/2}$, where Z is the atomic number minus the number of inner shell electrons.

RESULTS AND DISCUSSIONS

In Table 1 the results of the intensity measurements are tabulated along with the absorption frequencies, the frequencies of the first allowed electronic transition and the internuclear distances. Molecular polarizability derivatives were calculated using Eqns. (1) and (2). These calculations include all correction factors as given by Long, Milner and Thomas.[7] The molecular polarizability derivatives were placed on an absolute scale[2] using $\bar{\alpha}'_{\mathrm{C-Cl}} = 2.04$ Å^2 for CCl_4. For the calculation of bond orders from Eqn. (3) we have used electronegativity values from Pauling,[8] Allred[9] and from Jørgensen[10] (in this case values for O^{2-}, S^{2-}, V^{5+}, Cr^{6+}, Mo^{6+}, W^{6+}, Re^{7+} and Os^{8+}), but only the values of Jørgensen gave plausible results.

An enhancement of the Raman intensity by the resonance Raman effect could not be excluded for all species. Following the procedure of Clark and Willis,[5] we have measured the Raman intensity with three different exciting lines (6328 Å, 5145 Å and 4880 Å) and

Table 1. Wavenumbers of the $\nu_1(A_1)$ line, the longest wavelength transitions in the electronic spectrum ν_e, bond distance r, $\bar{\alpha}'$ values, calculated bond orders $n/2$ and stretching force constants f_r.

	ν_1 (cm^{-1})	ν_e $(kK)^a$	R^b	$\bar{\alpha}'$ (Å^2)	$n/2^d$	$n/2^e$	r^f (Å)	f_r^g (mdyn/Å)
CCl_4	459		1.00	2.04				
VO_4^{3-}	824	36.9	1.74 ± 0.05	2.67 ± 0.07	2.1	1.3	1.71	4.59
CrO_4^{2-}	847	26.8	$(1.89)^c$	2.80 ± 0.30	2.2	1.5	1.66	5.58
MoO_4^{2-}	897	43.2	1.38 ± 0.05	2.51 ± 0.09	2.4	1.8	1.77	5.87
WO_4^{2-}	931	50.3	1.11 ± 0.05	2.31 ± 0.10	2.7	1.9	1.79	6.42
ReO_4^-	972	43.7	1.24 ± 0.05	2.51 ± 0.10	2.6	2.2	1.76	7.55
OsO_4	965	33.8	1.30 ± 0.03	2.55 ± 0.06	2.2	2.2	1.75	7.98
WS_4^{2-}	459	25.5	$(2.48)^c$	3.15 ± 0.40	1.3	1.8	2.17	3.57

[a] A. Müller and E. Diemann[13]
[b] R is the molar intensity relative to $R[\nu_1(CCl_4)] = 1.00$; $R_i = \dfrac{I_i \cdot M_{std}}{I_{std} \cdot M_i} \cdot \dfrac{I_{std} \cdot M_{CCl_4}}{I_{CCl_4} \cdot M_{std}}$
[c] calculated from extrapolated $\bar{\alpha}'$ values (see Eqn. (4))
[d] electronegativity values from Jørgensen
[e] using the bond order calculated with the formula of Siebert[14,15]
[f] values for XO_4^{n-} from Kalman[16] and for WS_4^{2-} from Sasvari[17]
[g] A. Müller and B. Krebs[18]

plotted the $\bar{\alpha}'$ values against A given by the equation [11]

$$A = \frac{1 + (\nu_0/\nu_e)^2}{[1 - (\nu_0/\nu_e)^2]^2} \tag{4}$$

Here ν_0 is the exciting frequency and ν_e is the frequency of the first allowed electronic transition. Extrapolation of A versus 1 ($\nu_0 \to 0$; $\nu_e \to \infty$) gives the corrected $\bar{\alpha}'$ values. Only in the case of the coloured ions CrO_4^{2-} and WS_4^{2-}, could we fine remarkable contribution of the resonance Raman effect.

It can be seen from Table 1 that the derivative of the mean molecular polarizability increases in the isoelectronic rows when the central atom becomes heavier (see for instance WO_4^{2-}, ReO_4^- and OsO_4). Both the bond order in the definition of Siebert[14,15] and the valence force constant are found to increase in the same direction.[12] Here we got the same result as in the case of isoelectronic ions of the main group elements (e.g. PO_4^{3-}, SO_4^{2-} and ClO_4^-).[2,3]

In the case of ions where the central atoms belong to vertical rows of the periodic system (such as CrO_4^{2-}, MoO_4^{2-}, WO_4^{2-}) it is rather difficult to predict general rules, because the change in the type of chemical bonding is very large. According to our opinion it is physically more reasonable to relate $\bar{\alpha}'$ values to bond orders only for molecules containing identical bonds (as for instance the C–H bond in hydrocarbons). The measurements show that the Raman intensity increases when the central atom is a main group element and becomes heavier.[4,5] But if the central atom is a transition metal, the Raman intensity decreases when the transition metal becomes heavier, though the π bond order and the valence force constants increase.[12]

It should be mentioned that the calculated bond orders have only qualitative meaning (the most plausible are those calculated from electronegativity data of Jørgensen). But the results show unambiguously that π bonds exist in all investigated ions. The same conclusion was earlier drawn by us from the valence force constants,[19,20] electronic spectra,[13,20,21] MO calculations[22,23] and the bond distances[24,25] for the investigated and similar ions.

ACKNOWLEDGEMENT

The authors wish to thank the "Deutsche Forschungsgemeinschaft" and the "Fonds der Chemischen Industrie" for financial support and "Degussa", Hanau, for a gift of OsO_4.

REFERENCES

1 T. Yoshino and H. J. Bernstein, *Spectrochim. Acta* **14**, 127 (1959).
2 G. W. Chantry and R. A. Plane, *J. Chem. Phys.* **32**, 319 (1960).
3 G. W. Chantry and R. A. Plane, *J. Chem. Phys.* **34**, 1268 (1961).
4 J. D. Witt and R. M. Hammaker, *Inorg. Chem.* **10**, 1093 (1971).
5 R. J. H. Clark and C. J. Willis, *Inorg. Chem.* **10**, 1118 (1971).
6 T. V. Long II and R. A. Plane, *J. Chem. Phys.* **43**, 457 (1965).
7 D. A. Long, D. C. Milner and A. G. Thomas, *Proc. Roy. Soc. (London)* **A237**, 197 (1956).
8 L. Pauling, *The Nature of the Chemical Bond*, Cornell University Press, 1960.
9 A. L. Allred, *J. Inorg. Nucl. Chem.* **17**, 215 (1961).
10 Chr. K. Jørgensen, *Progr. Inorg. Chem.* **12**, 101 (1970).
11 J. Behringer in *Raman Spectroscopy*, (Ed. H. A. Szymanski), Plenum Press, New York, 1967, p.168.
12 A. Müller and B. Krebs, *Spectrochim. Acta* **23A**, 1591 (1967).
13 A. Müller and E. Diemann, *Chem. Phys. Letters* **9**, 369 (1971).
14 H. Siebert, *Anwendungen der Schwingungsspektroskopie in der anorganischen Chemie,* Springer-Verlag, Berlin, Heidelberg, New York, 1966.
15 H. Siebert, *Z. Anorg. Allgem. Chem.* **275**, 225 (1954).
16 A. Kalman, *J. Chem. Soc. A* 1857 (1971).
17 K. Sasvari, *Acta Cryst.* **16**, 719 (1963).
18 B. Krebs, A. Müller and A. Fadini, *J. Mol. Spectry.* **24**, 198 (1967).
19 A. Müller, K. H. Schmidt, K. H. Tytko, J. Bouwma and F. Jellinek, *Spectrochim. Acta* **28A**, 381 (1972).
20 A. Müller, *Chimia* **24**, 346 (1970); A. Müller, E. Diemann and C. K. Jørgensen, *Structure and Bonding,* Springer-Verlag, Berlin, 1973.
21 A. Müller, W. Rittner and G. Nagarajan, *Z. Phys. Chem.* **54**, 229 (1967).
22 R. Kebabcioglu and A. Müller, *Chem. Phys. Letters* **8**, 59 (1971).
23 R. Kebabcioglu, A. Müller and W. Rittner, *J. Mol. Structure* **9**, 207 (1971).
24 A. Müller, B. Krebs and H. Beyer, *Z. Naturforsch.* **23b**, 1537 (1968).
25 B. Krebs, A. Müller and E. Kindler, *Z. Naturforsch.* **25b**, 222 (1970).

70. RAMAN SPECTRA, SPECIES FORMATION AND MELT STRUCTURE IN $MeCl_3$– Alk Cl (Me = Al, Ga, In; Alk = Li, K, Cs)

E. Rytter and H. A. Øye

Institute of Inorganic Chemistry, The Technical University of Norway, Trondheim, Norway

and

P. Klaboe

Department of Chemistry, University of Oslo, Norway

INTRODUCTION

Raman spectroscopy is one of the most powerful tools for identifying complexes in molten salts. The spectra are easily recorded at higher temperatures and in the low frequency region whereas infrared spectroscopy applied to molten salts imposes many experimental difficulties. When interpreting the Raman melt spectra two factors should be recognized:

1. the possibility of phonons arising from the lattice-like structure of the melt;
2. the spectral observation time ($< 10^{-12}$s) is small compared with the lifetime of a melt configuration ($\sim 10^{-11}$s) and the spectra merely reveal that a configuration is statistically dominant without necessarily being kinetically stable.

The present $MeCl_3$–AlkCl molten systems ($AlCl_3$–LiCl, $AlCl_3$–KCl, $AlCl_3$–CsCl, $GaCl_3$–CsCl, $InCl_3$–KCl) are well suited for studying models of melt structures. Between 50 and 100 mole % $MeCl_3$ the properties of the melt change from a mixture of ionic species $MeCl_4^-$ and Alk^+ to a covalent Me_2Cl_6 structure. At intermediate compositions there is indirect evidence for the formation of the species $Al_2Cl_7^-$ from ultraviolet spectra,[1-3] vapour phase data[4,5] and EMF studies.[6,7] Raman investigations[8-14] have confirmed the existence of halogen bridged $Me_2Cl_7^-$ species and reveal a tendency towards formation of higher polymers $Me_nCl_{3n+1}^-$ ($n \geqslant 3$).

When the polarizing effect of the trivalent metal ion decreases from Al^{3+} to In^{3+} a reduced tendency to complex formation is expected. On the other hand, changing the alkali counter ion from Cs^+ to Li^+ causes an increased interaction between the complex and the alkali ion.

RESULTS

In Fig. 1 the Raman spectra of 50.0, 58.3, 66.7, 75.0 and 100.0 mole % $AlCl_3$ with LiCl are given as an example. The whole concentration range cannot be obtained for the $AlCl_3$–AlkCl system because of an immiscibility gap between 80 and 99 mole % $AlCl_3$. Various spurious peaks due to noise are removed from the spectra. Particularly for the $AlCl_3$ systems the spectra were disturbed by non-periodic noise due to microparticles which made slow scanning speeds and high time constants (20–30 s) necessary. Each

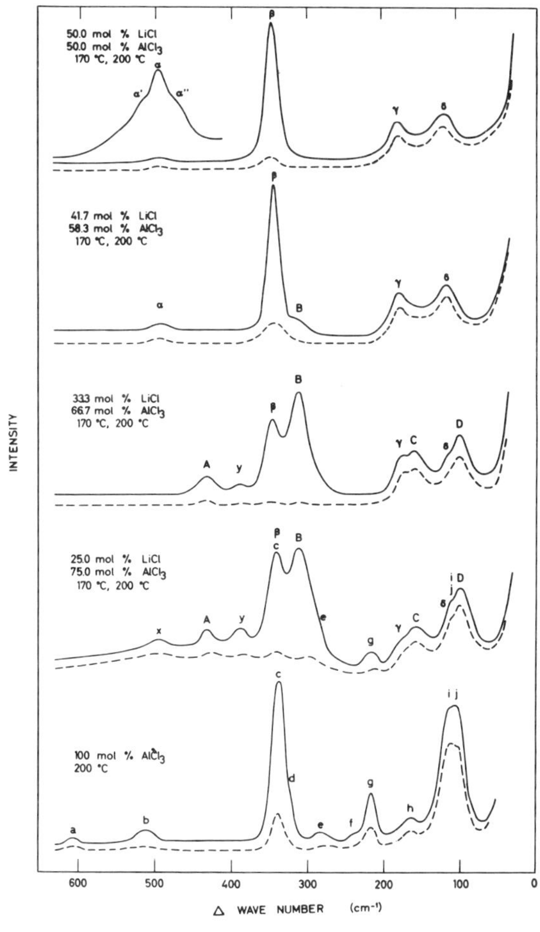

Fig. 1. Raman spectra of molten $AlCl_3$–LiCl. —— = the intensity $I(\parallel)$ (obs. ($\perp$). – – – = the intensity $I(\perp)$ (obs. ($\parallel$)). The absolute intensity scale for each composition is arbitrary. Frequency assignation: $AlCl_4^-$: α–δ; $Al_2Cl_7^-$: A–D; Al_nCl_{3n+1}, $n \geqslant 3$: w–z; Al_2Cl_6 : a–j.

band was determined from 6–13 spectra, except for pure $AlCl_3$ for which only two spectra were recorded. The complete set of spectra and lists of frequencies for the various systems can be found in various recent papers.[9,11–14]

INTERPRETATION

As is apparent from Fig. 1, the spectral changes observed when $MeCl_3$ was added to $Alk^+MeCl_4^-$ cannot be described by a *gradual* shift of frequencies. The variations can be well described by intensity variations in certain bands with composition change. However, the relative intensities of one set of frequencies do not change significantly, indicating existing species equilibria in the melt rather than a lattice-like structure.

$MeCl_4^-$. The spectra of all the melts consisting of 50 mole % $MeCl_3$ can definitely be interpreted in terms of $MeCl_4^-$ species of tetrahedral symmetry which can now be considered well established.[9,10]

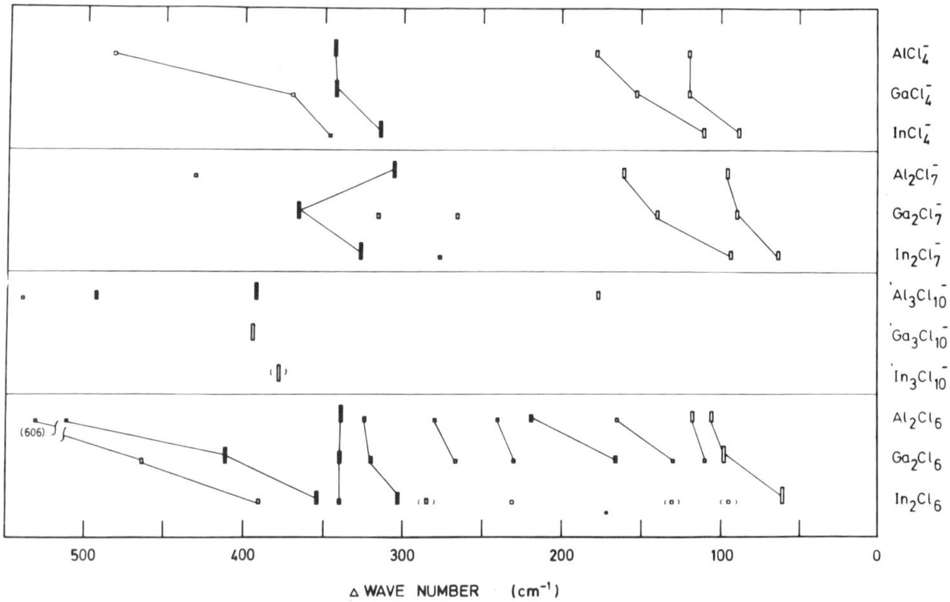

Fig. 2. Correlation diagram for $MeCl_4^-$, $Me_2Cl_7^-$, '$Me_3Cl_{10}^-$' and Me_2Cl_6 (Me = Al, Ga, In) frequencies; intensities are indicated by the length of the column. Filled columns, polarized; open columns, depolarized or uncertain polarization ratios. Data from $AlCl_3$–CsCl,[11] $GaCl_3$–CsCl[13], and $InCl_3$–KCl[14] systems.

Me_2Cl_6. The bridged structure (D_{2h}) of pure $AlCl_3$ are discussed by various authors (see ref. 9 and references therein). Pure molten $GaCl_3$ gives a Raman spectrum in excellent correspondence with Al_2Cl_6 (Fig. 2) and has been attributed to D_{2h} structure.[12,13] In $InCl_3$ melts the spectra (Fig. 2) are not clear-cut; a dimer structure seems probable although the relatively high electrical conductance[16] indicates a certain dissociation into ionic species.

$Me_2Cl_7^-$. Raman bands appear upon addition of $MeCl_3$ to AlkCl, not corresponding with the Al_2Cl_6 spectra, and they have been attributed to the ion $Al_2Cl_7^-$. This dimer is inferred from earlier data[1-7] and from the existence of a crystalline compound $CsGa_2Cl_7$.[17] The double tetrahedron $Me_2Cl_7^-$ sharing a bridging Cl^- ion provides a link between the isolated tetrahedra $MeCl_4$ and the double-bridged Me_2Cl_6. The bands assigned to $Me_2Cl_7^-$ are very similar for $Al_2Cl_7^-$, $Ga_2Cl_7^-$ and $In_2Cl_7^-$ and will be discussed later.

Higher polymers. Two peaks in the $AlCl_3$–LiCl (four in $AlCl_3$–CsCl[11]) system (Fig. 2) are enhanced relative to the $Me_2Cl_7^-$ peaks when the $MeCl_3$ content is increased beyond 66.7 mole %. They do not belong to Al_2Cl_6 and are attributed to higher polymers $Me_nCl_{3n+1}^-$($n \geqslant 3$). The tendency to form higher polymers is reduced with the polarizing effect of the cation from Al^{3+} to In^{3+}.

Frequency correlations. When the trivalent metal cation is changed from Al^{3+} to In^{3+} a general decrease in the Raman frequencies were observed for the $MeCl_4^-$ and Me_2Cl_6 species. These shifts are expected from the increased mass of the metal ions, whereas the calculated force constants do not change significantly.[11] Less straightforward shifts

Table 1. Variation in frequencies (cm^{-1}) with cation for the complexes $AlCl_4^-$, $Al_2Cl_7^-$ and 'higher polymers'. Values for the system $AlCl_3$–NaCl are taken from Torsi *et al.*[16]

	$AlCl_4^-$					
	α'	α	α''	β	γ	δ
Li^+	<u>512</u>	↑498	<u>473</u>	348	182	119
Na^+		490		351	184	121
K^+		487		↑350	182	122
Cs^+		483		344	178	120
	$Al_2Cl_7^-$					
	A	*B*	*C*	*D*		
Li^+	432	312	159	100		
Na^+	433	312	160	97		
K^+	432	↑311	161	97		
Cs^+	431	306	161	96		
	Higher Polymers					
	w	*x*	*y*	*z*		
Li^+		493	391			
Na^+		495	388			
K^+		495	393			
Cs^+	<u>~540</u>	493	392	<u>177</u>		

Significant effects of cations are indicated by arrows to the left of the column. Bands appearing only for some cations are underlined.

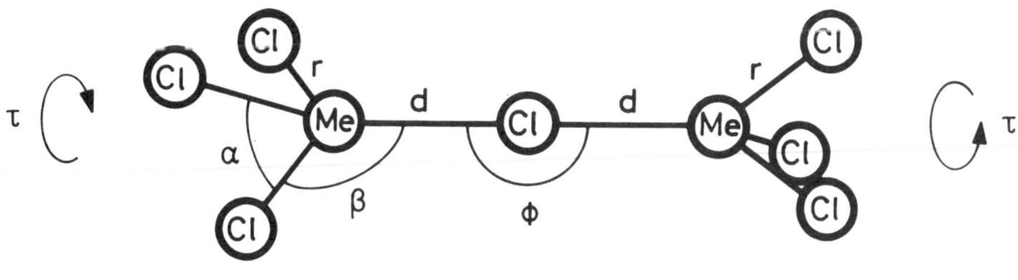

Fig. 3. Model for $Me_2Cl_7^-$ with symmetry D_{3d}, the letters indicating valence coordinates.

were found in the $Me_2Cl_7^-$ ions for a polarized Raman line in $Ga_2Cl_7^-$. Øye and Bues[13] could, however, account for this apparently erratic shift by a vibrational analysis.

Vibrational interaction of the alkali cations. The variations in the vibrational frequencies of $AlCl_4^-$, $Al_2Cl_7^-$ and higher polymers with larger alkali cations are listed in Table 1. Significant shifts to higher frequencies with smaller alkali ions were observed for $AlCl_4^-$ (α and β) and $Al_2Cl_7^-$ (*B*). Moreover, in the $LiAlCl_4$ system a splitting of the T_2 mode (α) at 512, 498 and 473 cm^{-1} appeared (Fig. 1). From force constant calculations[11] this splitting can be explained as a breakdown of the $AlCl_4^-$ T_d symmetry by the highly polarizing Li^+ ions forming Cl–Li interactions. The calculations resulted in a splitting of the T_2 α-mode with practically no splitting of the other degenerate modes in accordance with the observations. Independent indications of Al–Cl ...... Li interaction are provided by decreased electrical conductivity from molten $NaAlCl_4$ to $LiAlCl_4$[18] and the preponderance of $AlCl_4^-$ and Al_2Cl_6 at the cost of the polymers $Al_nCl_{3n+1}^-$ ($n > 2$) in the $AlCl_3$–LiCl system.

CHEMICAL EQUILIBRIA

Addition of AlkCl to $MeCl_3$ can be described in terms of the successive equilibria

$$n/2\ Me_2Cl_6 + Cl^- = Me_nCl_{3n+1}^-\ (n \gtreqless 3)$$

$$2\ Me_3Cl_{10}^- + Cl^- = 3\ Me_2Cl_7^-$$

$$Me_2Cl_7^- + Cl^- = 2\ MeCl_4^-$$

The change in relative intensities with composition for one Raman band belonging to each species was employed to calculate the dissociation constant K. An approximate calculation of the dissociation constant for: $2\,Al_2Cl_7^- = 2\,AlCl_4^- + Al_2Cl_6$ neglecting higher polymers gives a value $K = 0.8 \cdot 10^{-2}$ M for the $AlCl_3$–KCl system[9] compared with $K \approx 10^{-2}$ M found from vapour pressure data[4,5] for $AlCl_3$–NaCl.

The Raman spectra of $AlCl_3$–AlkCl[11] reveal that $AlCl_4^-$ persists at higher $AlCl_3$ contents and, moreover, Al_2Cl_6 is more predominant at intermediate compositions for the LiCl system. Thus, the species $AlCl_4^-$ and Al_2Cl_6 are here relatively more stable than the polymers. Generally, the formation of polymers increases in the order LiCl < KCl < CsCl. Furthermore, in the $AlCl_3$–CsCl system two additional bands are assigned to the higher polymers (Table 1).

THE STRUCTURE OF $Me_2Cl_7^-$

Normal coordinate analysis for the $Me_2Cl_7^-$ ions were performed in terms of the point groups D_{3d}, D_{3h}, D_3, C_{2v}, C_s as well as free rotation about the bridging chlorine,[8,9,11,13] transferring force constants from $MeCl_4^-$. Because of the small number of Raman bands observed, a diheder group and particularly D_{3d} (Fig. 3) with staggered structure appeared most likely. This model predicts only six Raman active bands ($3\,A_{1g}$ + $3\,E_g$). The calculations give a good correspondence with the observations except for one predicted totally symmetric band around 100 cm^{-1}. However, this band could be hidden by the depolarized band in this region.

A bent Me–Cl–Me bridge cannot be ruled out in view of the observed C_s symmetry for $Al_2Cl_7^-$ in $Te_4(Al_2Cl_7)_2$ ($\angle$Al–Cl–Al = 110°),[19] but the 21 Raman-active fundamentals expected for this model are in sharp contrast with the number of observed lines.[11]

REFERENCES

1 R. H. Moore, J. R. Morrey and E. E. Voiland, *J. Phys. Chem.* **67**, 744 (1963).
2 J. R. Morrey and R. H. Moore, *J. Phys. Chem.* **67**, 748 (1963).
3 H. A. Øye and D. M. Gruen, *Inorg. Chem.* **4**, 1173 (1965).
4 E. W. Dewing, *J. Amer. Chem. Soc.* **77**, 2639 (1955).
5 H. A. Øye and D. M. Gruen, *Inorg. Chem.* **3**, 836 (1964).

6 B. Tremillon and G. J. Letisse, *Electroanal. Chem.* **17**, 371 (1968).
7 A. Torsi and G. Mamantov, *Inorg. Chem.* (in press).
8 S. J. Cyvin, P. Klaboe, E. Rytter and H. A. Øye, *J. Chem. Phys.* **52**, 2776 (1970).
9 H. A. Øye, E. Rytter, P. Klaboe and S. J. Cyvin, *Acta Chem. Scand.* **25**, 559 (1971)
10 A. Torsi, G. Mamantov and G. M. Begun, *Inorg. Nucl. Chem. Letters* **6**, 663 (1970).
11 E. Rytter, H. A. Øye, S. J. Cyvin, B. N. Cyvin and P. Klaboe, *J. Inorg. Nucl. Chem.* (in press).
12 H. A. Øye and W. Bues, *Inorg. Nucl. Chem. Letters* **8**, 31 (1972).
13 H. A. Øye and W. Bues, *Inorg. Chem.* (in press).
14 H. A. Øye, E. Rytter and P. Klaboe, (to be published).
15 J. Kendall, E. D. Crittenden, and H. K. Miller, *J. Amer. Chem. Soc.* **45**, 963 (1923).
16 W. Klemin, *Z. Anorg. Allgem. Chem.* **152**, 255 (1926).
17 P. J. Fedorov and V. V. Tsimbalist, *Russ. J. Inorg. Chem.* **9**, 908 (1964)
18 R. H. Moss, *Thesis*, University of Connecticut (1955).
19 T. W. Couch, D. A. Lokken and J. D. Corbett, *Inorg. Chem.* **11**, 357 (1972).

71. ETUDE PAR SPECTROSCOPIE RAMAN DU COMPORTEMENT DE PCl_5 FONDU ET EN SOLUTION DANS HCl LIQUIDE

P. V. Huong et B. Desbat

Université de Bordeaux, Talence, France

ABSTRACT

The Raman spectra of phosphorus pentachloride without impurities have been recorded in its solid state, in its molten-salt state and in solution in various non-aqueous solvents. The covalent and ionic characters of this compound are described and the vibrational frequencies of PCl_5, PCl_4^+ and PCl_6^- species are definitely assigned.

The possibility of obtaining solutions containing selectively or simultaneously the PCl_4^+ and PCl_6^- ions is also shown.

INTRODUCTION

Il est bien connu, qu'à l'état gazeux, le pentachlorure de phosphore existe sous forme covalente PCl_5 et possède une symétrie D_{3h}.[1] Par contre, à l'état solide, il se présente sous forme ionique $PCl_4^+PCl_6^-$, le cation possédant une structure tétraédrique Td et l'anion une structure octaédrique O_h.[2] Les méthodes spectroscopiques ont permis ensuite d'attribuer les différentes fréquences de vibrations de ces espèces selon leur symétrie.[3,4]

L'analyse spectroscopique a aussi permis de constater le caractère covalent de PCl_5 dans le sulfure de carbone et dans le benzène.[5] Dans les autres milieux non aqueux, si les espèces ioniques ont été parfois signalées, leurs fréquences ont été toujours mélangées à celles d'autres produits réactionnels ou des impuretés telles que PCl_3 et $POCl_3$.[4,6,7] Quant au pentachlorure de phosphore fondu, son spectre Raman, bien qu'incomplet, indique que le produit se trouve sous forme covalente.[8]

Dans cette communication, nous présenterons l'étude par spectroscopie Raman du comportement de PCl_5 à différents états physiques et en solution dans des milieux divers. Nos conditions opératoires nous ont permis d'obtenir le spectre complet en l'absence totale d'impureté, ce qui conduit à une détermination sans ambiguité des fréquences vibrationnelles. Nous montrerons ensuite la possibilité d'obtenir des solutions contenant sélectivement l'ion PCl_4^+ ou PCl_6^-.

PARTIE EXPERIMENTALE

Les produits chimiques utilisés dans cette étude sont PCl_5, BCl_3, $(CH_3)_4N^+Cl^-$, HCl, CS_2 et CH_3NO_2. Tous ces composés ont été distillés et manipulés soit en enceinte parfaite-

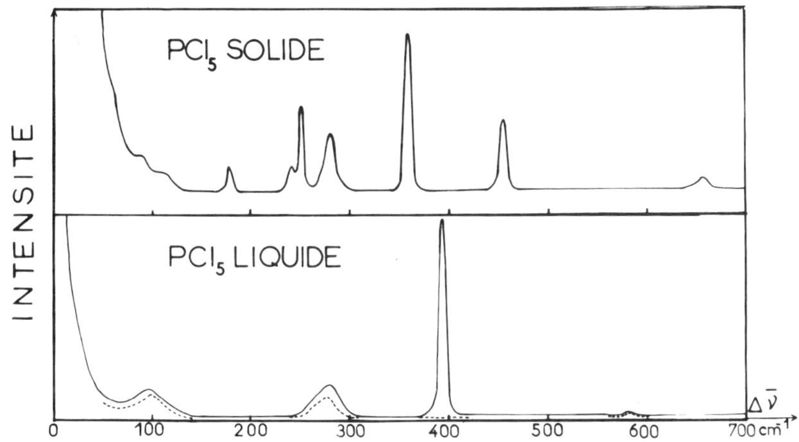

Fig. 1. Spectres Raman du pentachlorure de phosphore solide à 20°C et liquide fondu à 170°C. — : $I_{//}$; . . . . : $I_{\perp}$.

Tableau 1. Fréquences en cm^{-1} des vibrations fondamentales du pentachlorure de phosphore covalent.

Vibration et activité			Gaz	Gaz piégé	Solution	Liquide fondu	
							ce travail
			a	*b*	*c*	*d*	*e*
ν_1	a_1'	R, pol.			395	392	394
ν_2	a_1'	R, pol.			370 ?		385
ν_3	a_2''	IR	465	446	441	449 ?	444[f]
ν_4	a_2''	IR		295	301		299[f]
ν_5	e'	R, depol. IR	592	591	581		580
ν_6	e'	R, depol. IR		276	281	271 ?	278
ν_7	e'	R, depol. IR		110	100	100	98
ν_8	e''	R, depol.			261		261

[a] fréquences infrarouges, réf. 3
[b] fréquences infrarouges, réf. 4
[c] fréquences IR et R dans CS_2 et C_6H_6, réf. 5
[d] fréquences Raman, réf. 8
[e] fréquences Raman
[f] fréquences infrarouges dans CS_2

ment desséchée, soit sous rampe à vide.

Les mélanges étaient faits en tube cylindrique de verre scellé, de 3 mm de diamètre intérieur, supportant, à la température ambiante, une pression de 60 atmosphères; ces tubes sont placés avec leur axe perpendiculaire au faisceau laser.

Les tubes de sels fondus ont été chauffés à 170°C dans une enceinte chauffante électrique. Les spectres Raman ont été enregistrés sur un spectromètre Coderg PH 1 équipé de sources laser He-Ne et argon ionisé, Spectra Physics.

Tableau 2. Fréquences vibrationnelles en cm^{-1} du cation PCl_4^+.

Vibration et activité			Solide		Solution		
			$-PCl_6^-$	$-AlCl_4^-$ [a]	HCl	HCl/BCl_4^-	CH_3NO_2
ν_1	a_1	R	456	459	458	456	458
ν_2	e	R	177	172	178	175	178
ν_3	f_2	R, IR	657	556	662	660	masquée
ν_4	f_2	R, IR	252	252	255	250	255

[a] réf. 4

Tableau 3. Fréquences vibrationnelles en cm^{-1} de l'anion PCl_6^-.

Vibration et activité			Solide		Solution	
			$-PCl_4^+$	$-Et_4N^+$ [a]	CH_3NO_2	CH_3NO_2/Me_4N^+
ν_1	a_{1g}	R	359	360	354	353
ν_2	e_g	R	280	283	266	266
ν_3	f_{1u}	IR	443	444		
ν_4	f_{1u}	IR	283	285		
ν_5	f_{2g}	R	240	238	241	240
ν_6	f_{2u}	inactive				

[a] fréquences Raman et infrarouge, réf. 7

RESULTATS ET DISCUSSION

Spectre vibrationnel du pentachlorure covalent PCl_5 et ionique $PCl_4^+PCl_6^-$

1. Le spectre Raman du pentachlorure de phosphore *solide* présente toutes les raies caractéristiques des ions PCl_4^+ et PCl_6^-.[4] L'échantillon recristallisé à partir du sel fondu fait apparaître (Fig. 1) en outre, des raies à 60, 90 et 110 cm^{-1} dues aux vibrations du réseau cristallin.

2. A l'état liquide *fondu* à 170°C, ou en solution dans un solvant de basse constante diélectrique comme le sulfure de carbone, le pentachlorure de phosphore présente par contre un spectre vibrationnel caractéristique de l'espèce covalente. Signalons que, dans la littérature, les résultats reportés comportent soit un nombre incomplet de raies,[8] soit un nombre élevé de raies dues à la présence d'impuretés.[4, 5] Les spectres de nos échantillons, absents d'impuretés (Fig. 1), ont permis une détermination sans ambiguité des fréquences vibrationnelles. En particulier, nous ne retrouvons pas à 370 cm^{-1} la raie signalée antérieurement;[5] par contre, nous trouvons, vers le côté de basse fréquence de la bande intense ν_1 à 394 cm^{-1}, un épaulement à 385 cm^{-1} attribuable à la vibration ν_2. Il nous semble raisonnable que les fréquences ν_1 et ν_2 soient si proches l'une de l'autre car elles représententent toutes les deux l'élongation symétrique PCl'_2Cl_3, Cl′ étant les atomes de chlore axiaux dans la structure D_{3h} du pentachlorure de phosphore.[9]

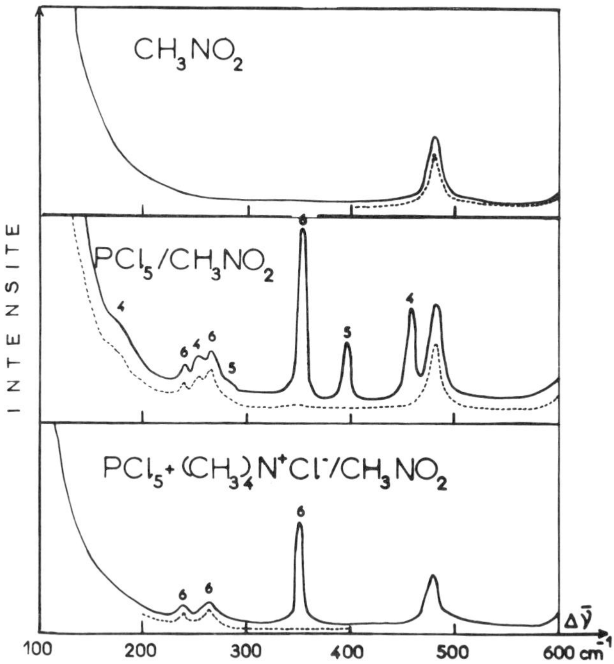

Fig. 2. Spectres Raman des solutions de pentachlorure de phosphore dans le nitrométhane et dans le nitrométhane en présence du chlorure de tétraméthylammonium. — : $I_{//}$; . . . : $I_{\perp}$. Les raies marquées 4, 5, 6 appartiennent respectivement à PCl_4^+, PCl_5 et PCl_6^-.

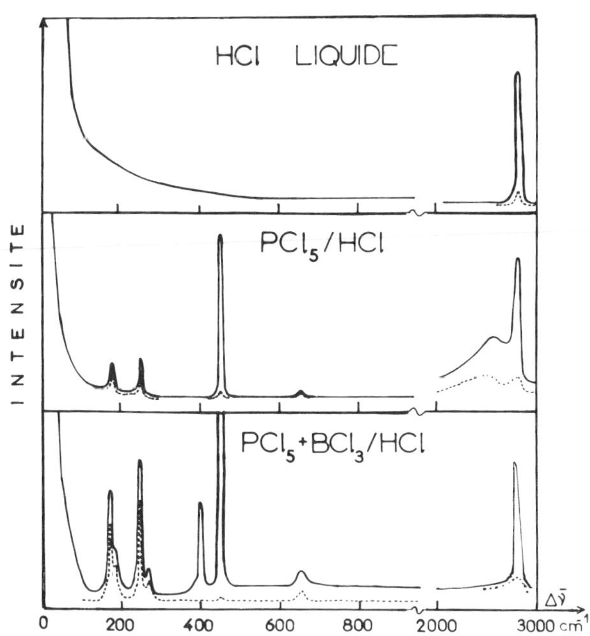

Fig. 3. Spectres Raman des solutions de pentachlorure de phosphore dans HCl liquide a 20° C et dans HCl liquide en présence de BCl_3.

Solutions contenant à la fois PCl_5, PCl_4^+ et PCl_6^-

Le spectre Raman du pentachlorure de phosphore dans un solvant ionisant de forte constante diélectrique comme le nitrométhane ($\epsilon_{30°} = 36$) présente les caractéristiques des trois espèces PCl_5, PCl_4^+ et PCl_6^- (Tableaux 2 et 3). Il y a donc l'équilibre d'ionisation

$$2\,PCl_5 \rightleftharpoons PCl_4^+ + PCl_6^-$$

Solutions contenant selectivement l'ion PCl_4^+ ou PCl_6^-

Ayant constaté que le pentachlorure de phosphore peut s'autoioniser en milieu liquide et connaissant les caractéristiques spectrales des espèces covalente PCl_5 et ioniques PCl_4^+ et PCl_6^-, il nous a paru intéressant de provoquer le déplacement de l'équilibre d'ionisation ou d'éliminer une des deux espèces ioniques formées en changeant le solvant ou en additionnant un ion adéquat. Ces séparations peuvent présenter un intérêt dans la conduite des réactions chimiques sélectives.

1. En effet, si, à la solution de pentachlorure de pentachlorure de phosphore dans le nitrométhane, on additionne un chlorure de tétraalkylammonium, le spectre Raman du mélange montre la présence exclusive des ions PCl_6^- et NR_4^+ (Fig. 2). Tout se passe comme si l'on était en présence de la réaction

$$PCl_5 + Cl^-(CH_3)_4N^+ \rightarrow PCl_6^- + (CH_3)_4N^+$$

Cette réaction est totale si l'on en croît l'absence de raies attribuables au PCl_5.

2. En *solution dans HCl liquide* anhydre à 20°C, le pentachlorure de phosphore s'ionise complètement. Le spectre de ce mélange présente en effet les quatre raies caractéristiques de l'ion PCl_4^+ (Fig. 3). On voit d'autre part l'apparition, vers le côté basse fréquence de la bande ν(ClH) liquide, d'un large massif attribuable à l'ion Cl^- solvaté. Cette bande est identique à celle observée avec une solution de chlorure de tétraméthylammonium dans HCl liquide. On assiste donc à l'ionisation suivante

$$PCl_5 + nHCl \rightarrow PCl_4^+ + (ClH)_nCl^-$$

Ces résultats sont à rapprocher de la forte conductivité électrique trouvée par Waddington[10] avec les mêmes solutions à −100°C.

Nous notons d'une part un déplacement du maximum du massif vers 2600 cm^{-1} selon la concentration du mélange et d'autre part l'absence des raies caractéristiques de l'ion triatomique $(ClHCl)^-$.[11] Cet ensemble de faits montre que l'ion Cl^- formé est solvaté par un nombre de molécules de HCl varié et supérieur à 2.

Nous avons pu déplacer cet ion Cl^- en ajoutant dans la solution précédente un capteur de Cl^- qui est BCl_3. Sur le spectre résultant, si les bandes correspondant à l'ion PCl_4^+ persistent, la bande de l'ion Cl^- solvaté vers 2600 cm^{-1} disparaît complètement; il en est de même pour les bandes caractéristiques de la molécule BCl_3. On voit par contre apparaître quatre raies à 190, 273 et 402, 705 cm^{-1} que nous attribuons respectivement aux vibrations ν_2, ν_4, ν_1 et ν_3 de l'ion BCl_4^-; certaines fréquences de cet ion étaient déjà connues[12–14] en particulier celle à 705 cm^{-1} active à la fois en infrarouge et en Raman.[12] Signalons qu'avec une concentration élevée en sel, on obtient un précipité dont l'analyse spectrale donne les caractéristiques du sel solide $PCl_4^+\,BCl_4^-$, ce qui confirme notre interprétation.

CONCLUSION

Nous avons montré que la spectroscopie Raman constitue une méthode commode pour l'étude du comportement du pentachlorure de phosphore à l'état fondu ou en solution dans divers milieux non aqueux.

Nous avons indiqué, à cette occasion, la possibilité d'obtenir des solutions contenant sélectivement ou simultanément les espèces ioniques PCl_4^+ et PCl_6^- provenant de l'ionisation du pentachlorure de phosphore.

REFERENCES

1 L. M. Roualt, *Ann. Physik.* Leipzig, **14**, 78 (1940).
2 D. Clark, H. M. Powell et A. F. Wells, *J. Chem. Soc.* 642 (1942).
3 J. K. Wilmshurst et H. J. Bernstein, *J. Chem. Phys.* **27**, 661 (1957).
4 G. L. Carlson, *Spectrochim Acta.* **19**, 1291 (1963).
5 M. J. Taylor et L. A. Woodward, *J. Chem. Soc.* 4670 (1963).
6 I. R. Beatie et M. Webster, *J. Chem. Soc.* 38 (1963).
7 I. R. Beatie, T. Gilson, K. Livingston, V. Fawcett et G. A. Ozin, *J. Chem. Soc. A* 712 (1967).
8 H. Moureu, M. Magat et G. Wetroff, *Compt. Rend.* **205**, 276 (1937).
9 K. Nakamoto, *Infrared Spectra of Inorganic and Coordination Compounds,* Wiley, New York, 1970.
10 T. C. Waddington, *Non-aqueous Solvent Systems*, Academic Press, London, 1965.
11 J. C. Evans et G. Y-S. Lo, *J. Phys. Chem.* **70**, 11 (1966).
12 W. Kynaston, B. E. Larcombe et H. S. Turner, *J. Chem. Soc.* 1772 (1960).
13 H. Gerding et J. C. Duinker, *Rev. Chim. Minerale* **3**, 815 (1966).
14 J. A. Creighton, *J. Chem. Soc.* 6589 (1965).

72. METAL ISOTOPE EFFECT ON RAMAN SPECTRA OF COORDINATION COMPOUNDS

K. Nakamoto
Department of Chemistry, Marquette University, Milwaukee, Wisconsin, USA

SUMMARY

We have shown previously that the metal isotope technique is highly useful in assigning metal–ligand vibrations in infrared spectra. If a molecule contains one metal atom at the centre of symmetry, no metal isotope shifts are expected for Raman-active fundamentals. Thus, the lack of metal isotope shifts in Raman spectra proves the presence of a centre of symmetry. The *trans* structure of $SnCl_4$(pyridine)$_2$ has been confirmed by this method. The $Zn(NH_3)_4^{2+}$ ion exhibits two Zn–N stretching bands in the Raman spectrum: the 412 cm^{-1} band is shifted whereas the 432 cm^{-1} band is not shifted by the ^{64}Zn–^{68}Zn substitution. Thus, these two bands can be assigned to the antisymmetric (F_2) and totally symmetric (A_1) Zn–N stretching modes, respectively. In molecules such as ZnX_2(pyridine)$_2$, both Zn–N and Zn–X stretching modes are metal isotope sensitive. In this case, the distinction of these two modes can be made by the deuteration of the ligand or the substitution of the halogen atom (X).

73. INTENSITES RAMAN DE METAUX CARBONYLE: DETERMINATION DES $\overline{\alpha}'_{C-O}$ ET INTERPRETATION

M. F. Koenig et M. Bigorgne
Laboratoire de Chimie de Coordination,
E.N.S.C.P., Paris, France

ABSTRACT

The intensities of the Raman scattering bands of the complexes $Ni(CO)_{4-n}L_n$ ($n = 0; n = 1,2$ for $L = PMe_3$; $n = 1, 2, 3$ for $L = P(OMe)_3$) and of acetone have been measured using a laser excitation spectrometer. The derivatives $\overline{\alpha}'_p = \partial\overline{\alpha}/\partial Q_p$ have been deduced. It is shown that to obtain the values of $\overline{\alpha}'_{C-O} = \partial\overline{\alpha}/\partial r_{C-O}$ one needs only consider the A_1 M–C and C–O bands.

Using some assumptions, the signs of corresponding $\overline{\alpha}'_p$ have been ascertained and the value $\overline{\alpha}'_{C-O}$ of each compound has thus been obtained.

The results indicate that the electrons of the multiple C–O bond play a preponderant role which is analogous to that played by the electrons of multiple C–C bonds. They also point out the important negative contribution of the free electrons on oxygen and chiefly on the carbon atom, which is responsible for the observed increase of $\overline{\alpha}'_{C-O}$ in the complex as compared with free CO. Finally it seems that the electrons of the σ and π M–C bonds play no important role in $\overline{\alpha}'_{C-O}$.

INTRODUCTION

Un intérêt croissant s'est récemment manifesté pour l'étude et l'interprétation des intensités des bandes de diffusion Raman. Dans le domaine des composés de coordination, il faut notamment citer les travaux de Kettle, Paul et Stamper[1,2] et ceux de Terzis et Spiro[3] portant tous deux sur les métaux carbonyle non substitués.

Dans ces études apparaît une difficulté, car les intensités sont liées aux carrés des dérivées $\overline{\alpha}'_p = \partial\overline{\alpha}/\partial Q_p$ de la polarisabilité moléculaire moyenne par rapport à la coordonnée normale Q_p du mode p considéré, mais le signe de ces grandeurs est inconnu. Le but de notre étude est d'apporter une solution au cas particulier de certains métaux carbonyle et de leurs dérivés de substitution, en utilisant leur propriété de former des séries homogènes, et en les comparant à des composés voisins quant à la nature de leur liaison C–O (acétone, ion CO_3^{2-}, CO libre) pour lesquels cette ambiguité de signe a été levée.

PARTIE EXPERIMENTALE

Les composés étudiés sont présentés dans le Tableau 1. Les fréquences des vibrations C–O et les constantes de force calculées indiquent que l'ordre de liaison C–O diminue quand le degré de substitution augmente. Tous ces composés sont des liquides incolores ou jaunes.

L'appareil utilisé pour la mesure des intensités des bandes de diffusion est un spectro-

Tableau 1. Composés étudiés.
(a) Dérivés $Ni(CO)_{4-n}L_n$.

n	L	$\nu_{C-O}(A_1)$ cm^{-1}	k_{C-O} mdyn/Å
0		2127	17,82
1	$P(OMe)_3$	2080	15,86
	PMe_3	2069	15,65
2	$P(OMe)_3$	2025	15,25
	PMe_3	1993	14,71
3	$P(OMe)_3$	1957	14,63

(b) Acétone: $\nu_{C-O} = 1710\ cm^{-1}$.

mètre Raman Coderg muni d'une source laser hélium–néon (6328 Å). Il est équipé pour mesurer avec précision les taux de dépolarisation et nous avons déterminé la variation de sa sensibilité de détection en fonction de la fréquence.

Les taux de dépolarisation ρ_p et les intensités I_p ont été mesurés pour les bandes des vibrations totalement symétriques des composés étudiés. Les valeurs obtenues, corrigées des facteurs d'appareil, ont été placées sur une échelle absolue au moyen d'une méthode d'étalon externe utilisant pour référence la bande 313 cm^{-1} du tétrachlorure de carbone. Dans l'échelle de Shorygin on donne à cette bande l'intensité 100. Les valeurs de $\bar{\alpha}'_p$ sont exprimées en $Å^2.g^{-1/2}$ en utilisant la valeur

$$\gamma'_{CCl_4\,313} = 0{,}548\ Å^2.g^{-1/2} \qquad \text{(ref. 4)}$$

où γ' désigne l'invariant anisotropie du tenseur $(\partial\alpha/\partial Q)$.

RESULTATS

Les mesures effectuées sur les bandes A_1 conduisent aux valeurs des $(\partial\bar{\alpha}/\partial Q_p)^2$. Si on connaît avec leur signe les valeurs $\bar{\alpha}'_p$ pour toutes les bandes A_1 et la matrice L reliant coordonnées normales A_1 et coordonnées symétriques A_1

$$S = LQ$$

on peut calculer les dérivées $\bar{\alpha}'_i = \partial\bar{\alpha}/\partial r_i$ (en $Å^2$) de la polarisabilité moléculaire moyenne par rapport à l'une quelconque des n_i coordonnées internes du type i. On a en effet

$$\bar{\alpha}'_i = \frac{1}{\sqrt{n_i}} \sum_p L^{-1}_{p_i} \bar{\alpha}'_p$$

avec

n_i = nombre de coordonnées internes équivalentes du type i à partir desquelles est construite la coordonnée symétrique S_i;

$L^{-1}_{p_i}$ = élément de la matrice ${}^tL^{-1}$ (t désignant la matrice transposée);

la somme est étendue à tous les modes normaux p de symétrie A_1.

Tableau 2. Valeurs de I_p, ρ_p et $|\bar{\alpha}'_p|$ pour les bandes C–O et Ni–C.

Composé	I_p Echelle de Shorygin C–O	I_p Echelle de Shorygin Ni–C	ρ_p C–O	ρ_p Ni–C	$|\bar{\alpha}'_p| = |\partial\bar{\alpha}/\partial Q_p|$ ($Å^2 \cdot g^{-1/2}$) C–O	$|\bar{\alpha}'_p|$ Ni–C
$Ni(CO)_4$	53,8	408	0	0	1,04	0,88
$Ni(CO)_3PMe_3$	47,4	225	0,21	0,04	0,73	0,67
$Ni(CO)_2[PMe_3]_2$	59,8	267[a]	0,60	0,04	0,36	0,77
$Ni(CO)_3P(OMe)_3$	76,1	511[a]	0,18	0,05	0,98	0,99
$Ni(CO)_2[P(OMe)_3]_2$	55,8	208	0,45	0,04	0,54	0,68
$Ni(CO)[P(OMe)_3]_3$	38,5	72,8[a]	0,73	0,05	0,11	0,42

[a] Valeur entachée d'erreur par suite de la présence d'une bande non A_1 qui n'est pas séparée; l'erreur ne se reporte pas sur $\bar{\alpha}'_p$.

Tableau 3. Valeurs de $\bar{\alpha}'_{C-O}$ selon le choix de la combinaisons de signes.

Composé	$\bar{\alpha}'_{C-O} = \partial\bar{\alpha}/\partial r_{C-O}$ ($Å^2$) Combinaison de signes $\left(\frac{\partial\bar{\alpha}}{\partial Q_{C-O}}, \frac{\partial\bar{\alpha}}{\partial Q_{Ni-C}}\right)$ (+,+)	(–,+)
$Ni(CO)_4$	2,70	−0,03
$Ni(CO)_3PMe_3$	2,19	−0,02
$Ni(CO)_2[PMe_3]_2$	2,10	0,76
$Ni(CO)_3P(OMe)_3$	3,07	0,12
$Ni(CO)_2[P(OMe)_3]_2$	2,25	0,26
$Ni(CO)[P(OMe)_3]_3$	1,35	0,75

Notre but étant d'étudier les liaisons C–O dans le nickel carbonyle et ses dérivés, nous cherchons plus particulièrement à obtenir les valeurs de $\bar{\alpha}'_{C-O} = \partial\bar{\alpha}/\partial r_{C-O}$ qui sont plus faciles à interpréter. Des calculs complets de $\bar{\alpha}'_{C-O}$ ont été effectués pour le nickel carbonyle et ses substitués par la triméthylphosphine en considérant tous les modes p A_1 et toutes les combinaisons possibles de signes des $\bar{\alpha}'_p$. Les résultats montrent que, *pour obtenir les valeurs de $\bar{\alpha}'_{C-O}$, on peut se limiter à l'étude des deux bandes A_1* C–O *et* Ni–C.

Le Tableau 2 rassemble les valeurs obtenues pour I_p, ρ_p et $\bar{\alpha}'_p$ pour les bandes C–O et Ni–C des composés étudiés. La précision de ces paramètres est estimée à 10%.

Comme $\partial\bar{\alpha}/\partial Q_{C-O}$ et $\partial\bar{\alpha}/\partial Q_{Ni-C}$ peuvent avoir, chacun indépendamment de l'autre, le signe + ou le signe –, il y a quatre combinaisons de signes possibles pour les $\bar{\alpha}'_p$ et donc quatre solutions deux à deux opposées pour $\bar{\alpha}'_{C-O}$; deux d'entre elles sont reportées dans le Tableau 3 avec indication de la combinaison de signes correspondante.

DISCUSSION

Le problème qui nous intéresse est de trouver parmi les combinaisons de signes possibles celle qui correspond à la réalité physique; ce problème revient à déterminer si pour le nickel carbonyle la valeur de $\bar{\alpha}'_{C-O}$ est ± 2,7 Å^2 ou 0 (Tableau 3). Pour cela, nous cherchons à comparer les ordres de grandeur de $\bar{\alpha}'_{C-O}$ dans un dérivé de métal carbonyle et dans une cétone. On sait que dans un dérivé trisubstitué l'ordre de la liaison C–O se rapproche de 2, c'est-à-dire que la liaison C–O tend à devenir cétonique.

Les mesures que nous avons effectuées sur l'acétone conduisent à deux types de valeurs possibles pour $\bar{\alpha}'_{C-O}$ dans ce composé: ± 1,28 Å^2 ou ± 0,08 Å^2. Or, pour l'ion CO_3^{2-} la valeur de $\bar{\alpha}'_{C-O}$ est 1,04 Å^2,[5] et une formule semi-empirique de Long *et al.*[6] indique que $\partial\bar{\alpha}/\partial r$ augmente avec l'ordre de la liaison correspondante. Nous conservons donc la valeur

$$\bar{\alpha}'_{C-O} = +\ 1{,}28\ \text{Å}^2 \text{ dans } CH_3{-}CO{-}CH_3$$

Les dérivées $\bar{\alpha}'_{C-O} = \partial\bar{\alpha}/\partial r_{C-O}$ représentent la variation infinitésimale de la polarisabilité moléculaire moyenne au cours d'une extension unité (infiniment petite) de la liaison C–O. Cette variation de polarisabilité peut provenir de différents facteurs et nous admettrons que, en première approximation, $\bar{\alpha}'_{C-O}$ peut être décomposé en une somme de termes indépendants dont seul le premier est relatif à la liaison C–O elle-même; les autres dépendent de l'environnement de cette liaison (liaison Ni–C, doublets libres sur O et éventuellement C)

$$\frac{\partial\bar{\alpha}}{\partial r_{C-O}} = \frac{\partial\bar{\alpha}_{\text{liaison C-O}}}{\partial r_{C-O}} + \frac{\partial\bar{\alpha}_C}{\partial r_{C-O}} + \frac{\partial\bar{\alpha}_O}{\partial r_{C-O}} + \frac{\partial\bar{\alpha}_{Ni-C\,\sigma}}{\partial r_{C-O}} + \frac{\partial\bar{\alpha}_{Ni-C\,\pi}}{\partial r_{C-O}}$$

$$= A + A_C + A_O + A_\sigma + A_\pi$$

A représente la contribution des électrons de la liaison C–O et devrait croître avec l'ordre de cette liaison;

A_C et A_O représentent les contributions des électrons du carbone et de l'oxygène non engagés dans les liaisons (doublets libres);

A_σ et A_π représentent les contributions des électrons de la liaison Ni–C voisine du C–O.

Nous formulons alors *l'hypothèse fondamentale* suivante: *les électrons des liaisons* C–O *se comportent de la même manière que ceux des liaisons* C–C *du point de vue de la variation de* $\partial\bar{\alpha}/\partial r$ *avec l'ordre de la liaison*: les résultats de Yoshino et Bernstein[7] indiquent que $\partial\bar{\alpha}/\partial r_{C-C}$ positif varie sensiblement proportionnellement à l'ordre de la liaison C–C. En comparant les valeurs obtenues pour $\bar{\alpha}'_{C-O}$ dans le nickel carbonyle, le dérivé trisubstitué et l'acétone, nous arrivons alors à la conclusion que: *il y a une seule combinaison de signes acceptable*: la combinaison(+, +); c'est-à-dire que $\partial\bar{\alpha}/\partial Q_{C-O}$ *et* $\partial\bar{\alpha}/\partial Q_{Ni-C}$ sont tous les deux positifs.

Les valeurs conservées sont alors

$$\bar{\alpha}'_{C-O} = +\ 2{,}70\ \text{Å}^2 \text{ dans } Ni(CO)_4$$

$$\bar{\alpha}'_{C-O} = +\ 1{,}35\ \text{Å}^2 \text{ dans } Ni(CO)\,[P(OMe)_3]_3$$

cette dernière valeur est très proche de celle obtenue pour l'acétone pour un ordre de liaison voisin.

Tableau 4. Valeurs des différentes contributions à $\bar{\alpha}'_{C-O}$.

Nature de la contribution à $\bar{\alpha}'_{C-O}$		Composés dans lesquels elle intervient	Valeur de la contribution ($Å^2$)
A	ordre de liaison ≅ 3	$Ni(CO)_4$, CO	+ 3,1
	ordre de liaison ≅ 2	$Ni(CO)[P(OMe)_3]_3$, $CH_3–CO–CH_3$	+ 2,1
A_C		CO	− 1
A_O		CO,$Ni(CO)_4$, $CH_3–CO–CH_3$,$Ni(CO)[P(OMe)_3]_3$	− 0,6
A_σ		$Ni(CO)_4$, $Ni(CO)[P(OMe)_3]_3$	+ 0,2
A_π		$Ni(CO)[P(OMe)_3]_3$	+ 0,3

Pour l'oxyde de carbone CO libre, les mesures de Yoshino et Bernstein[8] fournissent la valeur $\bar{\alpha}'_{C-O} = +\ 1{,}5\ Å^2$. Comme les ordres de liaison C–O de l'oxyde de carbone et du nickel carbonyle sont très voisins (de l'ordre de trois), la différence observée entre leurs valeurs de $\bar{\alpha}'_{C-O}$ ne peut provenir que de la différence entre les contributions des électrons qui constituent le doublet libre du carbone dans l'oxyde de carbone et qui sont engagés pour former la liaison Ni–C dans le nickel carbonyle. On peut alors calculer les différentes contributions à $\bar{\alpha}'_{C-O}$. Les valeurs obtenues sont reportées dans le Tableau 4.

CONCLUSIONS

Les conclusions à tirer de ces résultats sont les suivantes:

1. $\bar{\alpha}'_{C-O}$ dans un dérivé de métal carbonyle dépend en premier lieu de l'ordre de la liaison C–O et varie dans le même sens. Ceci explique que $\bar{\alpha}'_{C-O}$ diminue lorsque le degré de substitution augmente et que l'ordre de la liaison C–O passe progressivement de trois à deux. *La contribution à $\bar{\alpha}'_{C-O}$ des électrons de la liaison multiple* C–O (contribution A) *est du même ordre que la contribution à $\bar{\alpha}'_{C-C}$ des électrons des liaisons multiples* C–C (+3 $Å^2$ pour les liaisons triples C–O et C–C).
2. *Les doublets électronique libres sur le carbone et sur l'oxygène apportent une contribution négative à $\bar{\alpha}'_{C-O}$* particulièrement importante dans le cas du carbone. L'augmentation de $\bar{\alpha}'_{C-O}$ de 1,5 à 2,7 $Å^2$ en passant de CO libre à $Ni(CO)_4$ (pour lequel la liaison C–O est voisine) est essentiellement due au fait que l'on supprime la contribution négative du doublet du carbone en formant par coordination la liaison Ni–C.
3. *Les liaisons* Ni–C *fournissent une contribution* ($A_\sigma + A_\pi$) qui est *faible* par rapport à la contribution de la liaison C–O et de même signe qu'elle.

Le résultat auquel nous sommes parvenus, à savoir que $\partial\bar{\alpha}/\partial Q_{C-O}$ et $\partial\bar{\alpha}/\partial Q_{Ni-C}$ sont positifs, est le même que celui obtenu par Terzis et Spiro,[3] avec une méthode différente, dans l'étude des métaux hexacarbonyle.

La conclusion concernant les modifications de $\partial\bar{\alpha}/\partial r$ produites par la transformation du doublet du carbone par coordination est en parfait accord avec les observations de Chantry et Plane concernant les complexes cyanure:[9] la valeur de $\partial\bar{\alpha}/\partial r_{C-N}$ passe de 1,7 $Å^2$ dans l'ion cyanure libre à 2,2 $Å^2$ dans $Hg(CN)_4^{2-}$ (liaison σ partielle) et 2,8 $Å^2$ dans

$Cu(CN)_4^{3-}$ (liaison $\sigma + \pi$). Cependant, il ne semble pas que la coordination de la triméthylphosphine PMe_3 produise une variation de $\partial\bar{\alpha}/\partial r_{P-C}$ par suite de la transformation du doublet du phosphore.

REFERENCES

1 S. F. A. Kettle, I. Paul et P. J. Stamper, *Chem. Commun.* 1724 (1970).
2 S. F. A. Kettle, I. Paul et P. J. Stamper, *Chem. Commun.* 235 (1971).
3 A. Terzis et T. G. Spiro, *Inorg. Chem.* **10**, 643 (1971).
4 D. A. Bahnick et W. B. Person, *J. Chem. Phys.* **48**, 1251 (1968).
5 D. A. Long, D. C. Milner et A. G. Thomas, *Proc. Roy. Soc.* **A237**, 197 (1956).
6 D. A. Long, R. B. Gravenor et D. C. Milner, *Trans. Faraday. Soc.* **59**, 46 (1963).
7 T. Yoshino et H. J. Bernstein, *Spectrochim. Acta* **14**, 127 (1959); H. W. Schrötter et H. J. Bernstein, *J. Mol. Spectry.* **12**, 1 (1964).
8 T. Yoshino et H. J. Bernstein, *J. Mol. Spectry.* **2**, 213 (1958); W. F. Murphy, W. Holzer et H. J. Bernstein, *Appl. Spectry.* **23**, 211 (1969).
9 G. W. Chantry et R. A. Plane, *J. Chem. Phys.* **35**, 1027 (1961).

74. RAMAN INTENSITY AND CONJUGATION

E. D. Schmid and B. Brosa
Institut für physikalische Chemie der Universität Freiburg, Freiburg, Germany

SUMMARY

We have measured the Raman intensities of the phenyl ring vibrations near 1600 cm^{-1} and the C=C stretching vibrations of the polyene chain of four diphenyl-polyenes Ph-$(CH{=}CH)_n$-Ph with n varying from 1–4. All measurements have been carried out in sufficiently diluted CCl_4 solutions. The results we obtained can be interpreted quantitatively in terms of a simple relation by which the Raman intensity is given by $S = \text{const}\, f_B\, K$. By definition, the f_B factor describes quantitatively the resonance Raman effect on the observed intensities whereas the value of the K factor is a measure of the conjugation effect. This relation can be justified empirically for the present investigation. The influence of the conjugation effect on the relative Raman intensities of the phenyl ring vibrations is particularly large. In contrast the conjugation effect on the relative intensities of the C=C stretching vibrations of the polyene chain is extremely small. Both vibrations exhibit clearly a near-resonance type behaviour which turns out to be uniform.

75. RAMAN SPECTROSCOPIC STUDIES OF UREA/n-PARAFFIN CLATHRATES

V. Fawcett and D. A. Long
School of Chemistry, University of Bradford, Bradford, England

INTRODUCTION

In this chapter we report experimental studies of the Raman spectra of orientated single crystals of urea/n-paraffin clathrates and discuss vibrational assignments of the urea host and the guest n-paraffin in the light of these results. Urea clathrates containing the following n-paraffins have been studied: n-decane, n-tetradecane, n-hexadecane and n-undecane. In the space available we can give a detailed analysis of only one clathrate, urea/n-decane, but reference will also be made to features of special interest in the spectra of the other clathrates studied.

Before presenting the experimental results we shall consider briefly the advantages expected to result from orientated single crystal studies of these clathrates. In essence, these compounds are unique devices for *orientation* studies of *single* n-paraffin molecules in *one* configuration, the extended planar zig-zag configuration. In the urea/n-paraffin clathrate, the urea lattice has an hexagonal structure[1] in which there are long channels whose cross sectional dimensions are sufficient to accommodate the n-paraffin chain in an extended planar zig-zag configuration with small lattice–paraffin and paraffin–paraffin interactions. Single crystals of n-paraffins of adequate size are very difficult to grow and there have been no single crystal studies using infrared or Raman spectroscopy. Even if such studies were possible the spectra are likely to be more complicated than those obtained from the clathrates for although n-paraffins do assume the planar form in the crystal there is usually more than one molecule per unit cell and the intermolecular interactions can be large.

IDENTIFICATION OF HOST AND GUEST FREQUENCIES IN THE CLATHRATE

To distinguish between those vibrational frequencies arising from the urea host and those from the paraffin guest in the vibrational spectrum of a given urea/n-paraffin clathrate, the spectra of the urea/n-paraffin clathrate and the corresponding urea-d_4 clathrate were studied and compared. The frequency shifts on deuteration lead to an almost unambiguous identification of the urea host vibrational frequencies.

RELEVANT THEORY FOR INTERPRETATION OF ORIENTATION STUDIES

Orientation studies of the clathrate crystals enable the symmetry classes of the vibrations of the host and the guest molecules to be determined as the following analysis will indicate.

Host lattice

In the clathrate, X-ray studies[1] have shown that the urea host lattice has an hexagonal form, space group $P6_1 22$ with six molecules per unit cell. The urea molecules are arranged in a hydrogen bonded spiral forming hexagonal shaped channels along the crystal *c* axis. An isolated urea molecule, which is planar and belongs to the point group C_{2v}, has 18 internal vibrations: $7 A_1 + 2 A_2 + 3 B_1 + 6 B_2$. A full factor group treatment for the host lattice with 6 molecules of urea per unit cell and factor group D_6 gives 141 optical vibrations of which 58 would be Raman active. However since the observed Raman spectrum of the clathrate contains far fewer bands associated with the urea host, a full factor group treatment is inappropriate. In the present work the analysis has been based on a factor group D_{6h}. This would be the true factor group if the site symmetry of the urea molecule were C_{2v}. Although the true site symmetry of the urea molecule is C_2 it is very close to C_{2v}.

For the factor group D_{6h} there should be 58 Raman active vibrations, 45 internal and 13 external with the following symmetry classification: internal, $9 A_{1g} + 18 E_{1g} + 18 E_{2g}$ and external, $2 A_{1g} + 5 E_{1g} + 6 E_{2g}$.

The Raman scattering tensors for this factor group are

$$\underset{A_{1g}}{\begin{bmatrix} a & & \\ & a & \\ & & b \end{bmatrix}} \quad \underset{E_{2g}}{\begin{bmatrix} c & & \\ & -c & \\ & & \end{bmatrix} \begin{bmatrix} & c & \\ c & & \\ & & \end{bmatrix}} \quad \underset{E_{1g}}{\begin{bmatrix} & & d \\ & & \\ d & & \end{bmatrix} \begin{bmatrix} & & \\ & & d \\ & d & \end{bmatrix}}$$

The crystals are uniaxial with the optic axis lying along the *c* axis of the crystal. Orientation measurements were made by illumination along the *b* axis and observation at right angles to this axis. Using the Porto nomenclature the following four orientation measurements were made: $b(cc)a$, $b(cb)a$, $b(ab)a$ and $b(ac)a$ It can be seen that only A_{1g} modes will be observed in $b(cc)a$ measurements; only E_{2g} modes in $b(ab)a$ measurements; and only E_{1g} modes will be observed in either $b(cb)a$ or $b(ac)a$ measurements. Thus the symmetry of all Raman active modes of the host molecule can be determined unambiguously.

Guest n-paraffin

In the clathrate the n-paraffin assumes the planar extended zig-zag configuration. For *n* even, the molecular point group is C_{2h} and the Raman active vibrations belong to the symmetry classes A_g and B_g. The Raman scattering tensors are

$$\underset{A_g}{\begin{bmatrix} a & d & \\ d & b & \\ & & c \end{bmatrix}} \quad \underset{B_g}{\begin{bmatrix} & & e \\ & & f \\ e & f & \end{bmatrix}}$$

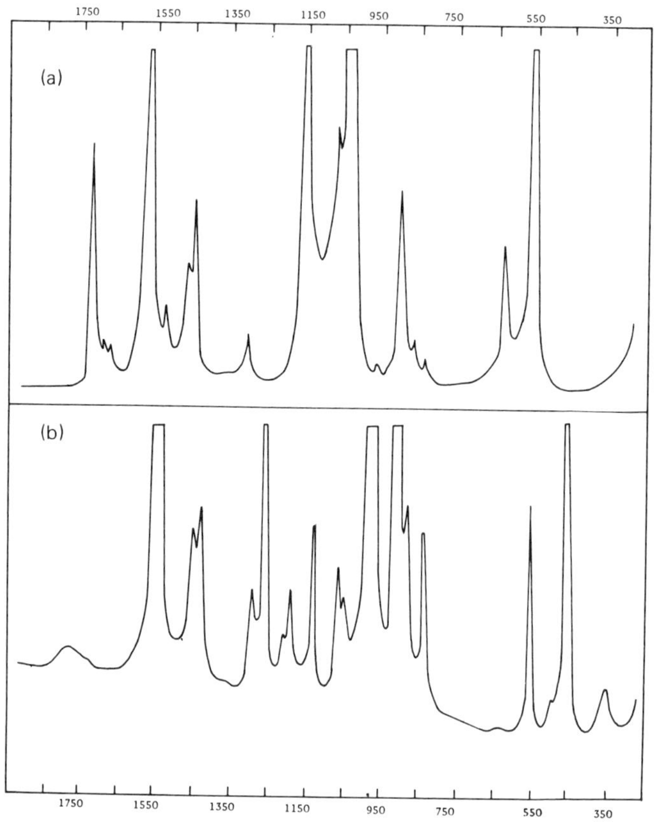

Fig. 1. (a) Raman spectrum of n-decane/urea clathrate in orientation $b(cc)a$; (b) Raman spectrum of n-decane/urea-d_4 clathrate in orientation $b(cc)a$.

The following four orientation measurements were made: $b(cc)a$, $b(cb)a$, $b(ab)a$ and $b(ac)a$. Because of essentially free rotation about the molecular x axis (which is coincident with the crystal c axis), the orientation measurements $b(cb)a$ and $b(ac)a$ should be equivalent and indeed were found to be so. Nevertheless the A_g and B_g modes can be distinguished. The A_g modes will be observed in $b(cc)a$ and $b(ab)a$ orientation measurements and the B_g modes will be observed in $b(cb)a$ and $b(ac)a$ orientation measurements. For n-decane, the n-paraffin considered in most detail here, there will be 26 A_g modes and 19 B_g modes. No details will be given here of the analysis for n odd.

RESULTS

To illustrate the changes in the Raman spectrum of the clathrate on complete deuteration of the urea host Fig. 1 shows the $b(cc)a$ Raman spectra of urea/n-decane and urea-d_4/n-decane in the frequency range 400–1800 cm^{-1}. To show the changes in the Raman spectra with orientation, Fig. 2 compares the $b(cc)a$, $b(cb)a$, $b(ab)a$ and $b(ac)a$ Raman spectra of urea/n-decane in the frequency range 0–1000 cm^{-1}.

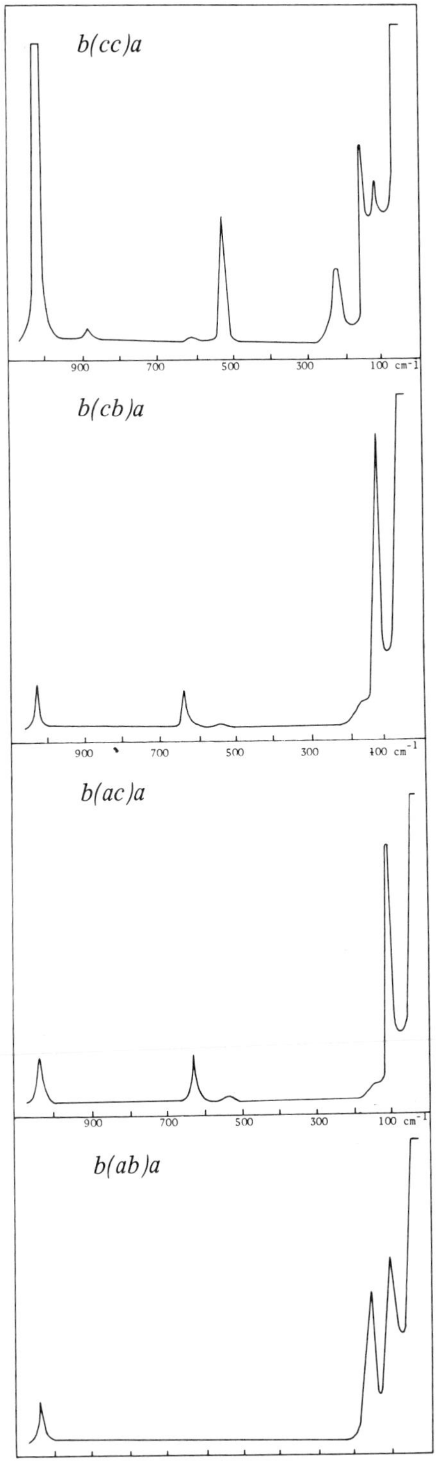

Fig. 2. Raman spectra of n-decane/urea clathrate in various orientations.

Table 1. Observed Raman active frequencies and assignments for hexagonal urea.

Hexagonal urea			Tetragonal urea	
Observed ν(cm^{-1})	Symmetry[a]	Assignment	Observed[b] ν(cm^{-1})	Calculated[c] ν(cm^{-1})
63	A_{1g}	lattice mode	63	–
101	E_{1g}	lattice mode	101	–
133	A_{1g}	lattice mode	132	–
167	E_{2g}	lattice mode	167	–
535	A_{1g}	skeletal deformation	556	555
611	E_{1g}	skeletal deformation	570	558
1024	A_{1g}	CN stretch	1000	1012
1065	E_{1g}	NH_2 rock	1150	1152
1142	A_{1g}	NH_2 rock	1150	1143
1515	E_{1g}	CN stretch	1464	1461
1562	A_{1g}	CO stretch	1603	1615
1604	E_{1g}	NH_2 bend	–	–
1664	E_{1g}	NH_2 bend	1629	1639
1679	E_{1g}	NH_2 bend	–	–
1706	A_{1g}	NH_2 bend	1686	1648
3232	A_{1g}	NH stretch	–	–
3276	A_{1g}	NH stretch	–	–
3317	E_{1g}	NH stretch	–	–
3389	A_{1g}	NH stretch	–	–
3402	A_{1g}	NH stretch	–	–

[a] As determined from orientation studies in the present work.
[b] Lattice frequencies taken from Arnau and Giguère,[3] other frequencies taken from Yamaguchi *et al.*[2]
[c] Calculated by Yamaguchi *et al.*[2]

DISCUSSION

The observed Raman spectrum of urea/n-decane contains 30 distinct bands in the frequency range 0–3500 cm^{-1}. On complete deuteration of the urea host, 10 of these remain unshifted in frequency and 2 more show a reduction in intensity. It follows that 12 bands belong to the n-decane and 20 to the urea host with 2 bands for the n-decane being accidentally coincident in frequency with 2 bands of the urea. The assignments of the bands of the urea host and the n-paraffin will now be considered in turn.

The urea host

Table 1 shows the frequencies of the 20 bands attributed to urea and also the symmetries of the bands as deduced from orientation studies. 16 of these bands have frequencies above 200 cm^{-1} and are internal vibrations. The other 4 bands have frequencies below 200 cm^{-1} and are lattice vibrations. The assignments in the free urea molecules, urea and urea-d_4 are on the whole quite well established as a result of a series of spectroscopic studies and normal coordinate calculations for the in-plane modes.[2] The correlation between the factor group D_{6h} and the molecular point group C_{2v} gives the following inter-relationships: A_{1g}–A_1, E_{1g}–A_2 and B_2, E_{2g}–A_1 and B_1. Using these correlations, the 16 internal frequencies of the urea host may be fairly readily assigned by comparison with the established assignments for urea and urea-d_4.

There are 5 bands in the frequency region 3200–3400 cm^{-1} associated with N–H stretching modes. Four fundamental N–H stretching frequencies would be expected, 2 of A_{1g} symmetry correlating with the 2 A_1 modes (symmetric N–H stretches) in free urea, and 2 of E_{1g} symmetry correlating with the 2 B_2 modes (antisymmetric N–H stretches) in free urea. Only one band of E_{1g} symmetry is observed with frequency 3389 cm^{-1}; 4 bands are observed with A_{1g} symmetry with frequencies 3402 cm^{-1}, 3232 cm^{-1} 3317 cm^{-1} and 3276 cm^{-1}. The A_{1g} components will arise from Fermi resonance between N–H stretching vibrations and overtones of NH_2 bending vibrations. The absence of the other E_{1g} band can be explained if it is assumed to be weak and to lie under a strong A_{1g} band.

In tetragonal urea the A_1 symmetric NH_2 bending mode is found at 1686 cm^{-1} and the B_2 antisymmetric NH_2 bending mode is found at 1629 cm^{-1}. There are four bands in this region in the urea host, one of A_{1g} symmetry at 1706 cm^{-1} and three of E_{1g} symmetry at 1679 cm^{-1}, 1664 cm^{-1} and 1604 cm^{-1} respectively. All four have been assigned to NH_2 bending modes.

In tetragonal urea the A_1 CO stretch is found at 1603 cm^{-1} and the band at 1562 cm^{-1} of A_{1g} symmetry is assigned to this mode in hexagonal urea. The lower frequency in the hexagonal form may be attributed to hydrogen bonding.

In tetragonal urea the C–N stretching modes are found at 1464 cm^{-1} (antisymmetric B_2) and 1000 cm^{-1} (symmetric A_1). In hexagonal urea the bands at 1515 cm^{-1} (E_{1g}) and 1024 cm^{-1} (A_{1g}) are assigned to antisymmetric and symmetric CN stretches respectively. It should be noted that the band at 1024 cm^{-1} which is a very strong band in orientation measurement *b*(*cc*)*a* forms a good test for the purity of the clathrate since any tetragonal urea is indicated by the presence of the band at 1000 cm^{-1}. The slightly higher frequency for the symmetric CN stretch in hexagonal urea may be interpreted in terms of a slight shortening of the C–N bond.

The NH_2 rocking modes in tetragonal urea (symmetric A_1 and antisymmetric B_2) appear to be accidentally coincident in frequency and have both been assigned to a band at 1150 cm^{-1}. The bands are quite distinct in hexagonal urea and are found at 1142 cm^{-1} (A_{1g}) and 1065 cm^{-1} (E_{1g}). These two bands overlap two much weaker paraffin bands but since the urea bands are so much stronger, the orientation measurements are still significant for the urea bands.

The skeletal vibrations in tetragonal urea are found at 556 cm^{-1} (symmetric A_1) and 570 cm^{-1} (antisymmetric B_2). The equivalent bands in hexagonal urea occur at 535 cm^{-1} (A_{1g}) and 611 cm^{-1} (E_{1g}).

In tetragonal urea four strong bands occur below 200 cm^{-1} due to external or lattice vibrations. On the basis of the small shifts in frequency with temperature, Giguère and Arnau[3] suggested the following assignments: 63 cm^{-1} (E), 101 cm^{-1} (A_1), 132 cm^{-1} (B_1) and 167 cm^{-1} (E). In hexagonal urea four bands at virtually the same frequencies are also observed. The orientation measurements lead to the following unambiguous assignments for these bands in hexagonal urea: 63 cm^{-1} (A_{1g}), 101 cm^{-1} (E_{1g}), 133 cm^{-1} (A_{1g}), 167 cm^{-1} (E_{2g}).

The frequencies observed in hexagonal urea-d_4 are presented in Table 2. No hexagonal urea-d_4 orientation measurements were made. However reasonable assignments of the urea-d_4 bands may be based on the hexagonal urea assignments and the expected frequency shifts on deuteration as indicated by the calculation of Yamaguchi *et al.*[2]

In the N–D stretching region three bands were observed at 2400 cm^{-1}, 2500 cm^{-1} and 2556 cm^{-1} as compared with five in the N–H stretching region of hexagonal urea. Fermi resonance contributions are less important in the N–D stretching region.

Table 2. Observed Raman active frequencies and assignments for hexagonal urea-d_4.

Hexagonal urea-d_4			Tetragonal urea-d_4	
Observed ν (cm^{-1})	Symmetry[a]	Assignment	Observed[b] ν (cm^{-1})	Calculated[b] ν (cm^{-1})
52	A_{1g}	lattice mode	–	–
93	E_{1g}	lattice mode	–	–
127	A_{1g}	lattice mode	–	–
159	E_{2g}	lattice mode	–	–
454	A_{1g}	skeletal deformation	475	476
556	E_{1g}	skeletal deformation	512	523
836	E_{1g}	ND_2 rock	887	893
912	A_{1g}	CN stretch	887	874
983	A_{1g}	ND_2 rock	1001	986
1047	E_{1g}	ND_2 bend	–	–
1200	E_{1g}	ND_2 bend	1154	1174
1273	A_{1g}	ND_2 bend	1245	1267
1384	E_{1g}	ND_2 bend	–	–
1565	A_{1g}	CO stretch	1610	1597
2400	A_{1g}	ND stretch	–	–
2500	A_{1g}	ND stretch	–	–
2556	E_{1g}	ND stretch	–	–

[a] Inferred from correlation with hexagonal urea.
[b] Taken from Yamaguchi *et al.*[2]

Table 3. Observed Raman active frequencies and assignments for n-decane in urea clathrate.

Observed ν (cm^{-1})	Symmetry[a]	Assignment	Shimanouchi and Mizushima[b]
234	A_g	skeletal bend	231
890	A_g	CH_3 rock	886
1064	B_g	CH_2 rock	1060
1142	A_g	skeletal stretch	1136
1301	B_g	CH_2 twist	1295
1430	A_g	CH_2 bend	1447
1459	A_g	CH_2 bend	1475
2850	A_g	CH_2 stretch	2843
2871	A_g	CH_3 stretch	2877
2901	B_g	CH_2 stretch	–
2931	A_g	CH_2 stretch	2935
2960	A_g	CH_3 stretch	2964

[a] As determined from orientation studies in the present work.
[b] See reference 8.

In tetragonal urea-d_4 the C–O stretching mode is found at 1610 cm^{-1}. The only band in the spectrum of hexagonal urea-d_4 which can be assigned to a C–O stretching mode is that at 1565 cm^{-1}. Again following the assignment in tetragonal urea-d_4 the antisymmetric CN stretching mode should have a frequency around 1500 cm^{-1}. There is no evidence of a band in this region in the spectrum of hexagonal urea-d_4 so presumably the antisymmetric CN stretch is too weak to be observed. This leaves four bands above 1000 cm^{-1}, at 1047 cm^{-1}, 1200 cm^{-1}, 1273 and 1384 cm^{-1}. It seems likely that these can all be assigned to ND_2 bending vibrations. In tetragonal urea-d_4 these have frequencies of 1154 cm^{-1} and 1250 cm^{-1}.

The five frequencies in the range 400–1000 cm^{-1} are fairly readily assigned. The band at 912 cm^{-1} can be attributed to the symmetric CN stretch (observed at 887 cm^{-1} in tetragonal urea-d_4). The bands at 836 cm^{-1} and 983 cm^{-1} can be assigned to the ND_2 rocking vibrations (observed at 1001 cm^{-1} and 887 cm^{-1} in tetragonal urea-d_4). The bands at 454 cm^{-1} and 556 cm^{-1} can be assigned to the skeletal deformation modes (observed at 475 cm^{-1} (A_1) and 512 cm^{-1} (B_2) in tetragonal urea-d_4).

As may be expected the lattice vibrations shift only very slightly on deuteration. Four strong bands are observed below 200 cm^{-1} in hexagonal urea-d_4 at 159 cm^{-1}, 127 cm^{-1}, 93 cm^{-1} and 52 cm^{-1}. There can be little doubt that there is a one-to-one correspondence between these bands and the bands of corresponding frequency in hexagonal urea.

The n-decane guest

Table 3 shows the frequencies and the symmetries, as deduced from orientation studies, of all the bands attributed to n-decane in a urea clathrate. There have been many vibrational spectroscopic studies of n-paraffins[4–7] in the liquid state and in the solid state but the symmetries of bands have not previously been determined for a normal paraffin in a specific configuration. The symmetries and assignments of the bands determined here will be discussed and for bands below 1500 cm^{-1} compared with the assignments for n-decane previously made by Mizushima and Shimanouchi,[8] these are included in Table 3.

A single molecule of n-decane will have 90 internal vibrations of which 45 would be Raman active. These vibrations may be classified according to their symmetry and the type of motion involved, namely methyl modes, methylene modes or skeletal modes.[4] The number of modes and the range of frequencies for each symmetry and each type of motion may be calculated.[9,10] The results of such calculations for n-decane are summarized in Table 4. Usually the several bands associated with one type of motion of a particular symmetry form an array in which the intensity falls off very rapidly. Consequently only the strongest member and exceptionally the next strongest member of an array will be observed. If only the strongest member of each array is counted the Raman spectrum of n-decane would contain only 18 bands; 12 bands are observed for n-decane in the urea clathrate.

Five bands are observed in the C–H stretching region of the Raman spectrum between 2800 and 3000 cm^{-1}. There should be 3 methyl stretching modes ($2\,A_g + 1\,B_g$) and 8 methylene stretching modes ($4\,A_g + 4\,B_g$). From studies of $(CH_3)(CH_2)_{10}(CH_3)$ and $(CD_3)(CH_2)_{10}(CD_3)$, Jones and Ripley[11] identified the methyl stretching modes in n-dodecane at 2961 cm^{-1} and 2876 cm^{-1}. The two bands of A_g symmetry at 2960 cm^{-1} and 2871 cm^{-1} in n-decane have therefore been assigned to the two A_g methyl stretching modes. The broad band centred at 2901 cm^{-1} is of B_g symmetry and is assigned to one or more of the expected four antisymmetric Raman active methylene stretching vibrations. The other two bands at 2850 cm^{-1} and 2931 cm^{-1} have A_g symmetry and are difficult to explain other than as a Fermi resonance interaction of a methylene symmetric stretching mode and an overtone of a methylene bending mode.

Theory predicts four Raman active CH_2 bending vibrations of A_g symmetry

Table 4. Classification of vibrational modes in n-decane.[a]

Description of mode	Number of modes Symmetry A_g	Symmetry B_g	Expected Frequency Range
Methyl modes:			
Asymmetric stretch	1	1	
Asymmetric bend	1	1	
Asymmetric rock	1	1	
Symmetric stretch	1	0	
Symmetric bend	1	0	
Torsion	0	1	
Methylene modes:			
Asymmetric stretch	0	4	
Symmetric stretch	4	0	
Bend	4	0	1450–1480 cm^{-1}
Wag	4	0	1175–1415 cm^{-1}
Twist	0	4	1170–1300 cm^{-1}
Rock	0	4	720–1050 cm^{-1}
Skeletal modes:			
Stretch	5	0	970–1140 cm^{-1}
Bend	4	0	
Torsion	0	3	
TOTAL	26	19	

[a] Based on Snyder.[4]

Table 5. Comparison of observed and calculated accordion mode frequencies.

Paraffin	Order *m*	ν (cm^{-1}) calculated[a]	ν (cm^{-1}) observed
C_8H_{18}	1	277.8	284
	3	637.6	
$C_{10}H_{22}$	1	225.4	234
	3	522.5	(obscured)
$C_{12}H_{26}$	1	189.3	198
	3	473.2	
$C_{14}H_{30}$	1	163.4	174
	3	431.1	
$C_{16}H_{34}$	1	143.6	136
	3	392.3	392
	5	535.7	(obscured)
$C_{11}H_{24}$	1	205	212
	2	383	(very weak)
	3	495	

[a] Calculated using the Schaufele formula:[12]

$$\nu\ (cm^{-1}) = A(m/n) + B(m/n)^2 + C(m/n)^3 + \ldots + F(m/n)^6$$

where $A = 2495$, $B = -5.867 \times 10^3$, $C = 6.253 \times 10^4$, $D = -3.485 \times 10^5$, $E = 7.329 \times 10^5$, $F = -4.724 \times 10^5$, *m* is the order of the mode and *n* the number of carbon atoms in the chain.

calculated to lie in the 1400–1500 cm^{-1} region and the two bands observed at 1430 cm^{-1} and 1459 cm^{-1} (both of A_g symmetry) are assigned to methylene bending vibrations.

There should be four Raman active methylene twisting modes of B_g symmetry calculated to lie in the frequency range 1170–1300 cm^{-1}. The one band with the correct symmetry observed in this region at 1301 cm^{-1} is assigned as a methylene twisting mode.

Of the remaining methyl group modes only the A_g methyl rocking mode is observed in the Raman spectrum. The strong band at 890 cm^{-1} is assigned to this mode, which is found in this region in most paraffins.

Although there should be four Raman active A_g methylene wagging modes predicted to lie in the range 1175–1415 cm^{-1}, no bands attributable to these modes were observed.

There remain to be considered: five A_g skeletal stretching modes, four A_g skeletal bending modes, four B_g methylene rocking modes and three B_g skeletal torsions. The band of A_g symmetry at 1142 cm^{-1} is assigned to one of the symmetric skeletal stretching vibrations which are calculated to lie in the range 970–1140 cm^{-1}. The band at 1064 cm^{-1} of B_g symmetry is assigned to one of the methylene rocking modes. Finally the band at 234 cm^{-1} is associated with one of the array of skeletal bending modes often described as longitudinal accordion modes. Their frequencies vary inversely as the chain length of the n-paraffin. Table 5 shows the accordion mode frequencies observed in all the n-paraffin clathrates studied and compares these with values calculated using the formula due to Schaufele.[12] There is good general agreement. In the case of hexadecane two modes of the accordion array were observed.

ACKNOWLEDGEMENT

We are indebted to the Science Research Council for substantial grants for equipment for Raman spectroscopy.

REFERENCES

1 A. E. Smith, *Acta Cryst.* **5**, 224 (1952).
2 A. Yamaguchi, T. Miyazawa, T. Shimanouchi and S. Mizushima, *Spectrochim. Acta* **10**, 170 (1957).
3 J. L. Arnau and P. A. Giguère, *C. R. Acad. Sci.* **268**, 245 (1969).
4 R. G. Snyder, *J. Mol. Spectry.* **4**, 411 (1960).
5 J. R. Nielsen and A. H. Woollet, *J. Chem. Phys.* **26**, 1391 (1957).
6 T. P. Lin and K. L. Koenig, *J. Mol. Spectry.* **9**, 228 (1962).
7 M. Tasumi, T. Shimanouchi and T. Miyazawa, *J. Mol. Spectry.* **9**, 261 (1962).
8 T. Shimanouchi and S. Mizushima, *J. Chem. Phys.* **17**, 1102 (1949).
9 R. G. Snyder and J. H. Schachtschneider, *Spectrochim. Acta.* **19**, 85 (1963).
10 R. G. Snyder and J. H. Schachtschneider, *Spectrochim. Acta.* **19**, 117 (1963).
11 R. N. Jones and R. A. Ripley, *Can. J. Chem.* **42**, 305 (1964).
12 R. F. Schaufele and T. Shimanouchi, *J. Chem. Phys.* **47**, 3605 (1967).

76. VAPOUR PHASE RAMAN SPECTROSCOPY OF NON-RIGID MOLECULES

J. Laane
Department of Chemistry, Texas A & M University, College Station, Texas, USA

INTRODUCTION

Considerable work on non-rigid ring molecules using far-infrared spectroscopy has been reported.[1,2] These studies on molecules such as cyclobutane, cyclopentane, cyclopentene, 1,4-cyclohexadiene and their derivatives have made it possible to analyse the ring-puckering vibration and to determine the configuration of each molecule. With the exception of cyclobutane[3,4] and trimethylene oxide[5] the vapour phase Raman spectra of very few non-rigid molecules have been reported. In the work to be discussed, we have used a high powered argon ion laser to study the Raman spectra of several non-rigid molecules in the vapour phase. The spectra of cyclopentene will be analysed in detail and compared to the far-infrared spectra discussed previously.[6]

EXPERIMENTAL

Vapour-phase Raman spectra of cyclopentene and other samples were obtained using a Cary 82 spectrometer, a Coherent Radiation Model 53 argon ion laser, and the standard Cary multipass system for gases. A vapour pressure of 300 torr of cyclopentene was used inside the standard Cary gas cell, which has Brewster angle windows. The laser line at 5145 Å with a power of about 6 W was used as the exciting source. A spectral bandwidth ranging from 1 to 3 cm^{-1} was used for the scans, and a sodium lamp was used for calibration.

RESULTS AND DISCUSSION – CYCLOPENTENE

The Raman spectrum of cyclopentene vapour between 90 and 260 cm^{-1} is shown in Fig. 1. The frequencies, which correspond to ring-puckering transitions with $\Delta v = 2$, are tabulated in Table 1. The observed values are in excellent agreement with those calculated from far-infrared data.[6] As is the case for trimethylene oxide,[5] only the totally symmetric transitions ($\Delta v = 2$) are observed even though the $\Delta v = 1, 3$, etc. are also Raman active. It is somewhat surprising that the observed transitions all have relatively high depolarization ratios of about 0.7 as compared to 0.2 for trimethylene oxide. Nevertheless, these apparently correspond to $A_1 \rightarrow A_1$ or $B_2 \rightarrow B_2$ transitions in C_{2v} symmetry ($A' \longleftrightarrow A'$ or $A'' \longleftrightarrow A''$ in C_s symmetry).

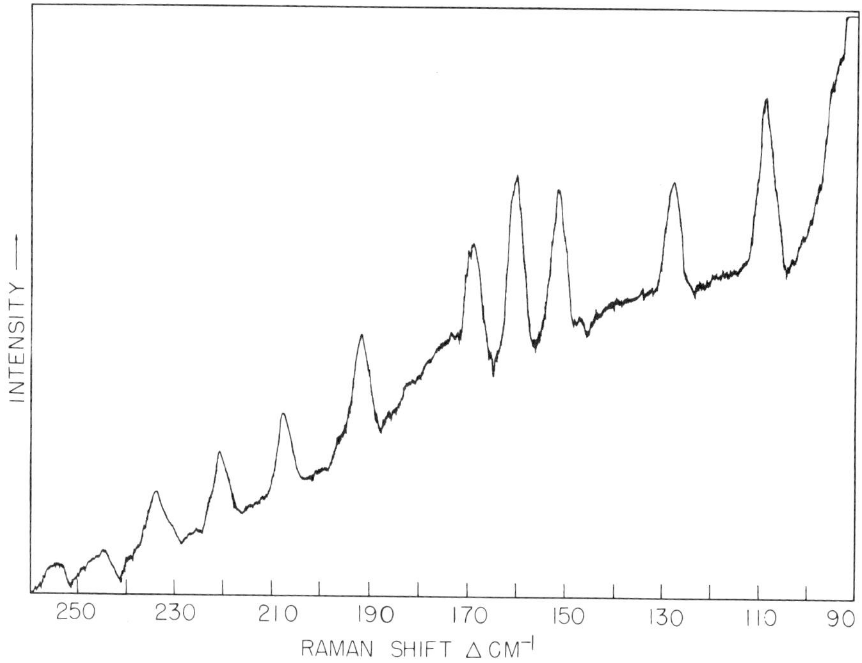

Fig. 1. Raman spectrum of cyclopentene vapour, 90–260 cm^{-1}. Conditions as for Fig. 2, curve a.

Table 1. Ring-puckering transitions of cyclopentene vapour (Raman frequency cm^{-1}).

Transition	Observed	Calculated[a]	Relative intensity
0 → 2	128.5	128.0	0.6
1 → 3	152.7	152.0	1.0
2 → 4	109.5	108.3	1.0
3 → 5	160.9	159.7	1.0
4 → 6	168.5	168.6	0.5
5 → 7	191.6	191.8	0.5
6 → 8	208.0	207.3	0.4
7 → 9	220.8	220.7	0.3
8 → 10	234.8	232.7	0.3
9 → 11	245.8	245.5	0.2
10 → 12	~ 256	258.7	0.1

[a] From far-infrared transitions ($\Delta v = 1$) in ref. 6.

The other low frequency ring vibration in cyclopentene, which is essentially a C=C twist of A_2 symmetry, gives rise to a Raman band centred at 385 cm^{-1}. It shows no Q branches but resembles a type B band in an infrared spectrum.

Ueda and Shimanouchi have previously reported a band progression in the infrared spectrum of cyclopentene vapour[7] with the 2861 cm^{-1} frequency as the reference band. In the Raman spectrum of the vapour we have been able to identify one difference band

Table 2. Raman spectrum of cyclopentene vapour, 2800–3250 cm^{-1}.

Frequency (cm^{-1})	Relative intensity	Assignments[a]
2806.0	10	$\nu_p(13 \to 11)$
2813.5	9	$\nu_p(12 \to 10)$
2820.9	10	$\nu_p(11 \to 9)$
2835.0	16	$\nu_p(10 \to 8)$
2860.4	1000	$\nu_3(A_1)$
2865.0	94	ν_H
2867.4	9	ν_H
2876.7	16	$\nu_p(7 \to 5)$
2877.9	4	ν_H
2879.7	66	ν_H
2884.6	61	$2\nu_6(\mathrm{HCH}, A_1)$
2903.2	285	$\nu_{28}(B_2)$
2907.6	19	$\nu_p(5 \to 3)$
2914.6	5	ν_H
2916.9	47	$\nu_p(3 \to 1)$
2923.1	121	$\nu_2(A_1)$
2926.8	13	ν_H
2932.2	187	$\nu_{19}(B_1)$
2938.9	37	ν_H
2942.1	13	$\nu_p(2 \to 0)$
2944.3	33	ν_H
2956.3	667	$\nu_{27}(B_2)$
2959 sh	w	ν_H
2964.1	66	$\nu_p(4 \to 2)$
2971.2	150	$\nu_{12}(A_2)$
2982.3	8	ν_H
2986.2	16	ν_H
2990.7	17	ν_H
2994.9	14	ν_H
~3005	9	ν_H
3054.6	5	$\nu_4 + \nu_6$
3061.6	65	ν_H
3065.5	47	$\nu_{18}(B_1)$
3070.4	613	$\nu_1(A_1)$
3082.4	2	$\nu_4 + \nu_5$
3090.4	16	ν_H
~3223	42	$2\nu_4(\mathrm{C{=}C}, A_1)$
~3232 broad	6	ν_H

[a] ν_p is the puckering difference band with 3070.4 cm^{-1} as reference; each is highly polarized ($\rho < 0.1$). ν_H: hot band resulting from excited puckering state; or, possibly, an overtone or combination band.

progression with the A_1 olefinic C–H stretching band at 3070.4 cm^{-1} as the reference line. Surprisingly, no sum band progression was observed. For trimethylene oxide,[5] both sum and difference bands were present.

Figure 2 shows the Raman spectrum and Table 2 lists the frequencies observed for cyclopentene vapour in the region 2800–3210 cm^{-1} along with the observed relative intensities and the proposed assignment for each band. Of the 32 observed frequencies, 8 correspond to fundamentals and 9 to the difference bands. Each of these A_1 differ-

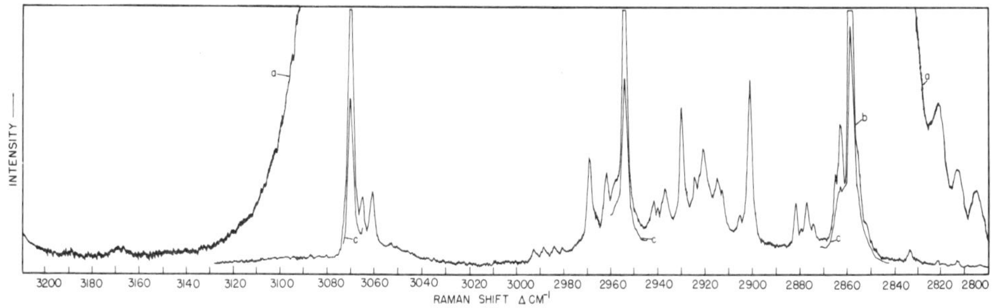

Fig. 2. Raman spectrum of cyclopentene vapour, 2800–3210 cm^{-1}; pressure 300 torr, argon 5145 Å exciting line, 6 W; (a) slit 3 cm^{-1}, sensitivity 100 counts/s; (b) 1.5 cm^{-1}, 1000 counts/s; (c) 2.0 cm^{-1}, 4000 counts/s.

Table 3. Raman difference bands with origin at 3070.4 cm^{-1} (frequency in cm^{-1}).

Puckering transition	Observed	Difference	Pure[a]	Separation in excited state (cm^{-1})[b]
$2 \rightarrow 0$	2942.1	128.3	128.5	128.3
$3 \rightarrow 1$	2916.9	153.5	152.7	153.5
$4 \rightarrow 2$	2964.1	106.3	109.4	106.0
$5 \rightarrow 3$	1907.6	162.8	160.9	162.7
$6 \rightarrow 4$	*c*	–	168.5	–
$7 \rightarrow 5$	2876.7	193.7	191.6	190.6
$8 \rightarrow 6$	*c*	–	208.0	–
$9 \rightarrow 7$	*c*	–	220.8	–
$10 \rightarrow 8$	2835.0	235.4	234.8	–
$11 \rightarrow 9$	2820.9	249.5	245.8	–
$12 \rightarrow 10$	2813.5	256.9	256	254.2
$13 \rightarrow 11$	2806.0	264.4	–	260.2

[a] Raman frequency.
[b] Calculated for excited state of 3070.4 cm^{-1} vibration using far-infrared frequencies[6] and the difference bands.
[c] Obscured by fundamental.

ence bands is highly polarized. The other 15 lines apparently result from hot bands similar to those found for trimethylene oxide,[5] or else are due to overtones or other combination bands. Table 3 compares the difference frequencies with those observed directly in the 100–260 cm^{-1} region. The agreement is good although it is clear that the excited state of the $\nu_1(A_1)$ vibration has a somewhat different puckering potential than the ground state.

CONCLUSION

It has been demonstrated that the use of a high powered laser can lead to Raman spectra which arise from the low-frequency skeletal motions of molecules. Since these vibrational motions are very sensitive to the molecular configurations, the Raman spectra can be used to determine the configuration of each molecule.

ACKNOWLEDGEMENTS

Figures 1 and 2 have been reproduced with the permission of *Chem. Phys. Letters*.

The author acknowledges financial support from the Robert A. Welch Foundation.

REFERENCES

1 C. S. Blackwell and R. C. Lord, 'Far-Infrared Spectra of Four-Membered Ring Compounds' in *Vibrational Spectra and Structure*, Vol. 1, (Ed. J. R. Durig), Marcel Dekker, 1972.

2 J. Laane, 'Pseudorotation of Five-Membered Rings' in *Vibrational Spectra and Structure*, Vol. 1, (Ed. J. R. Durig), Marcel Dekker, 1972.

3 J. M. R. Stone and I. M. Mills, *Mol. Phys.* **18**, 631 (1970).

4 F. A. Miller and R. J. Capwell, *Spectrochim. Acta* **27A**, 947 (1971).

5 W. Kiefer, H. J. Bernstein, M. Danyluk and H. Wieser, *Chem. Phys. Letters* **12**, 605 (1972).

6 J. Laane and R. C. Lord, *J. Chem. Phys.* **47**, 4941 (1967).

7 T. Ueda and T. Shimanouchi, *J. Chem. Phys.* **47**, 5018 (1967).

77. RECHERCHES SUR LA STRUCTURE DE SOLUTIONS AQUEUSES CONCENTREES DE SELS METALLIQUES

M. Alves-Marques,* B. Oksengorn et B. Vodar

Laboratoire des Hautes Pressions, C.N.R.S., Bellevue, France

ABSTRACT

The pressure and temperature shifts of some Raman bands attributed to the vibrations of hydrates of metallic cations and to the vibrations of chlorocomplexes of zinc cations in concentrated aqueous solutions of their inorganic salts, were measured. The more representative intermolecular forces acting in the neighbouring of cations were used in numerical calculation of these band shifts: the agreement between calculated and experimental values is rather good. However the validity of the assumptions advanced in these calculations was discussed.

INTRODUCTION

L'hydratation des ions permet d'expliquer de nombreuses propriétés des solutions électrolytiques. Des recherches spectrographiques, par l'effet Raman spontané, ont mis en évidence la structure définie des hydrates de quelques cations métalliques dans les solutions aqueuses liquides de leurs sels inorganiques.[1]

On décrit dans cet article les modifications subies par les bandes Raman des hydrates de quelques cations métalliques en fonction de la pression et de la température. On a aussi étudié un autre type de complexe de coordination électrostatique dans les solutions de chlorure de zinc.

DESCRIPTION DES EXPERIENCES

Les solutions aqueuses ont été enfermées dans une petite cuve cylindrique en Pyrex de 8 mm de diamètre intérieur, et 4 cm^3 de capacité. Les parois étaient rodées sur 3 cm pour laisser glisser librement le bouchon transparent lorsque la pression augmentait; elles étaient transparentes à mi-hauteur pour permettre la sortie de la lumière diffusée à 90°.

La cuve de verre était placée à l'intérieur d'une enceinte à haute pression comportant deux fenêtres colinéaires pour l'entrée et la sortie du faisceau laser, et une troisième fenêtre à 90° pour l'observation de la lumière diffusée.

On a utiliseé des cylindres en Pyrex de 12,5 mm de diamètre et 9 mm d'épaisseur, collées à l'araldite sur les faces polies des bouchons en acier, percés au centre d'un trou de 4 mm. La compression était réalisée avec de l'argon, jusqu'à 2.500 atmos, au moyen d'un générateur huile-mercure-gaz. On a vérifié que les quantités d'argon qui se dissol-

* Boursier de l'Instituto de Alta Cultura, Portugal.

Tableau 1. Résultats obtenus en faisant varier la pression.

Solutions aqueuses	Mode de vibration		ν cm^{-1}	Δp atmos	$\partial\nu/\nu\partial p$ 10^{-2}
SO_4Be conc. 4,3 N	$Be(OH_2)_6^{2+}$	ν'_3	351	2.500	+ 1,9
		ν_1	533	2.500	~ 0
	$Be(OH_2)_4^{2+}$	ν_1	765	2.500	~ 0?
$(NO_3)_2Be$ conc. 5,2 N	$Be(OH_2)_6^{2+}$	ν'_3	348	2.000	+ 1,9
		ν_1	528	2.000	~ 0
Cl_3Al conc. 9,1 N	$Al(OH_2)_6^{3+}$	ν_3	344	2.500	~ 0
		ν_1	532	2.500	~ 0
$(NO_3)_4Th$ conc. 7,0 N	$Th(OH_2)_8^{4+}$	ν_1	423	2.500	+ 0,6
Cl_2Zn conc. 12,2 N	$ZnCl_4^{2-}$	ν_1	289	2.500	− 0,8

Tableau 2. Résultats obtenus en faisant varier la température.

Solutions aqueuses	Mode de vibration		ν cm^{-1}	ΔT °C	$\partial\nu/\nu\partial T$ 10^{-2}
$(NO_3)_2Be$ conc. 5,2 N	$Be(OH_2)_6^{2+}$	ν'_3	348	63	+ 4
		ν_1	528	63	+ 1
Cl_3Al conc. 9,1 N	$Al(OH_2)_6^{3+}$	ν_3	344	90	+ 1,5
		ν_1	532	90	~ 0

ν représente la fréquence de la bande Raman mesurée à la température de 30° C et à la pression atmosphérique, Δp est la variation de pression, ΔT est la variation de température, $\partial\nu/\nu\partial p$ indique la variation relative de la fréquence de la bande rapportée à une variation de pression de 1.000 atmosphères, $\partial\nu/\nu\partial T$ indique la variation relative de la fréquence rapportée à une variation de température de 100° C, ν indique toujours un mode de vibration totalement symétrique. La forme des vibrations ν_3 et ν'_3 est indiquée sur la Fig. 1. L'erreur expérimentale sur le coefficient de pression est de l'ordre de grandeur de $0,4 \times 10^{-2}$ et de 10^{-2} sur le coefficient de température.

vaient dans la solution étaient pratiquement insignifiantes.

On a utilisé la raie 4.880 Å d'un laser à argon ionisé C.G.E., dont la puissance sur cette raie était de l'ordre de 1 W au moment des expériences.

Les solutions aqueuses ont été ultrafiltrées sur des membranes de collodium fabriquées d'après les techniques déjà indiquées.[1] La lumière diffusée a été observée au moyen d'un spectrographe Huet très lumineux, ouvert à *f*/6; avec des plaques Kodak 103 a G les poses duraient entre 1 heure et 12 heures.

Les Tableaux 1 et 2 donnent les déplacements avec l'augmentation de pression, ainsi que ceux obtenus par l'élévation de la température pour quelques bandes Raman attribuées soit à des modes de vibration des hydrates des cations, soit aux complexes de coordination des anions Cl^- autour du cation Zn^{2+}.[2]

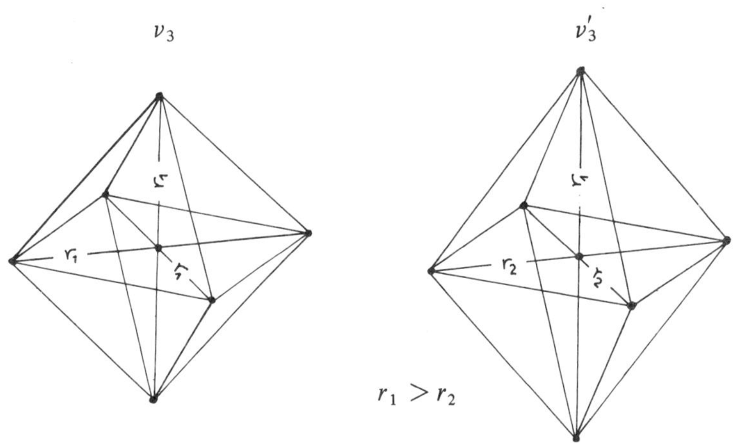

Fig. 1. Vibrations dégénérées des complexes de six molécules d'eau autour du cation.

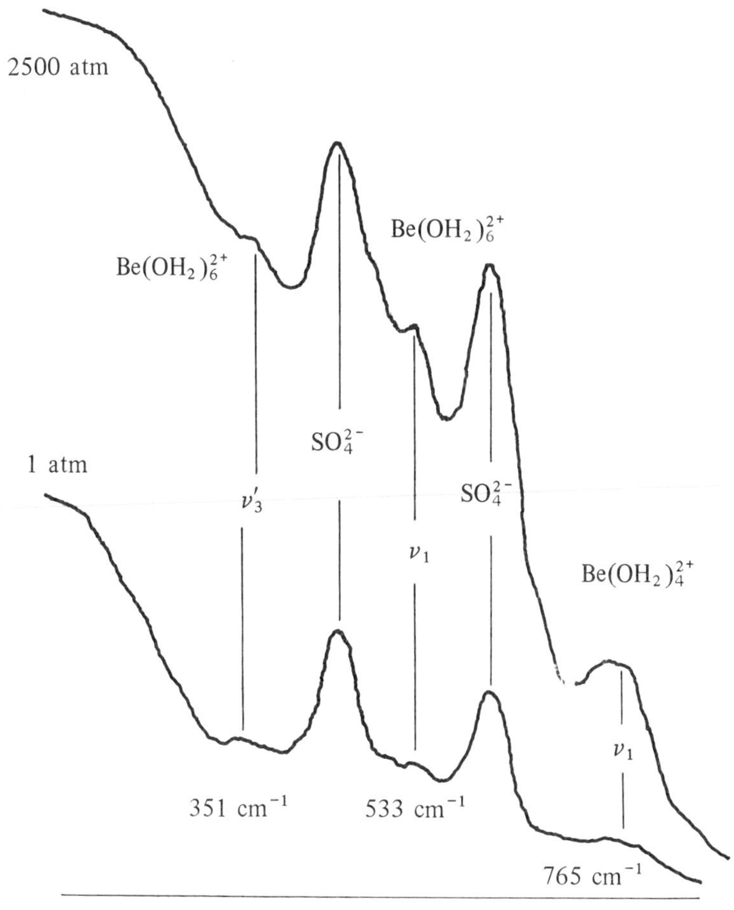

Fig. 2. Spectre Raman jusqu'à 900 cm^{-1} d'une solution aqueuse 4,3 N de sulfate de béryllium sous pression, comparé au spectre Raman à la pression normale.

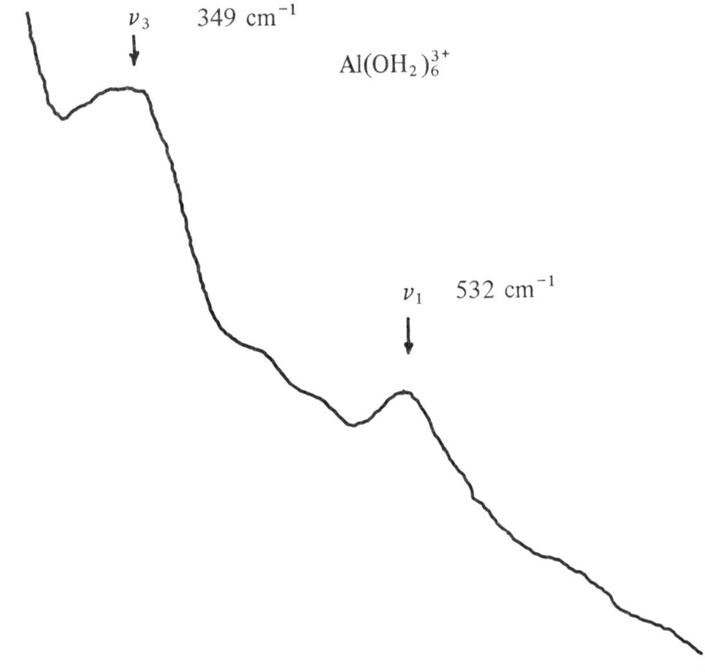

Fig. 3 Spectre Raman jusqu'à 700 cm^{-1} d'une solution aqueuse saturée de chlorure d'aluminium à la température de 120°C.

Des tables qui donnent les concentrations des solutions aqueuses à partir de leurs densités[3] ont été utilisées pour déterminer les concentrations des solutions observées, et on a mesuré les densités au picnomètre.

On a fait des mesures de déplacement sur les bandes les plus intenses des hydrates des cations, car en général, le spectre Raman de ces complexes est faible avec des largeurs de bandes atteignant dans certains cas de valeurs de l'ordre de 70 cm^{-1}.

Sur la Fig. 2, on donne le spectre Raman d'une solution aqueuse 4, 3 N de sulfate de béryllium à 2.500 atmos et à 30° C; on remarque que les bandes attribuées aux hydrates de béryllium se détachent bien du fond continu.

La Fig. 3 représente le spectre Raman d'une solution aqueuse saturée de chlorure d'aluminium à la pression normale et à 120°C.

CALCUL DES DEPLACEMENTS DES BANDES RAMAN

On a essayé de determiner par des calculs numériques sur ordinateur le changement de fréquence pour les vibrations des hydrates des cations à partir des modifications de configuration des modèles d'hydratation à plusieurs couches de molécules d'eau utilisés par ailleurs[4] (Fig. 4). Ces configurations, variables avec la pression ou avec la température, correspondent à une énergie potentielle minimale pour les forces intermoléculaires qu'on a considérées comme les plus représentatives des interactions au voisinage du cation. Dans le calcul de l'énergie d'interaction, on a fait intervenir des forces électrostatiques entre le cation considéré comme une charge électrique ponctuelle et les molécules d'eau,

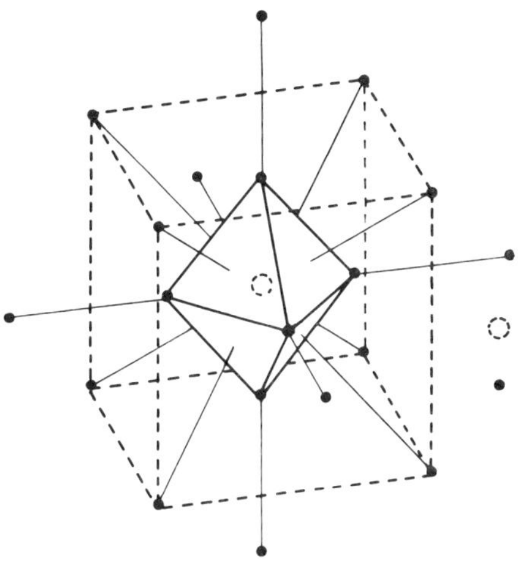

Fig. 4. Configuration des trois couches de molécules d'eau dans un des modèles d'hydratation du cation qu'on a utilisés. Les molécules d'eau de la couche la plus proche du cation sont aux sommets d'une bipyramide tétragonale; les molécules d'eau de la deuxième couche sont aux sommets d'un cube et les autres six molécules sont situées devant les centres des faces du cube, aux sommets d'une deuxième bipyramide tétragonale (non représentée) coaxiale à la première.

polarisées, représentées par des dipôles centrés sur leurs atomes d'oxygène, ainsi que des forces centrales de courte portée. On a admis l'additivité de toutes ces forces.

L'expression de l'énergie potentielle de l'aggrégat peut être donnée par

$$V = -\sum \frac{ze\mu_i}{r_i^2} + \sum \frac{D_{ik}\mu_i\mu_k}{r_{ik}^3} + \sum \frac{(\mu_i - \mu_0)^2}{2\alpha} + \sum \frac{a_c}{r_i^{n_c}} + \sum \frac{a}{r_{ik}^{\ n}} \qquad (1)$$

ze est la valeur de la charge de l'ion central, e celle de l'électron, μ_i les moments dipolaires des molécules d'eau coordonnées, r_i les distances entre les diverses particules.

Dans le deuxième terme de l'expression (1) qui représente l'énergie de répulsion électrostatique des molécules d'eau, D_{ik} sont des coefficients qui dépendent de l'orientation relative des dipôles μ_i et μ_k. La valeur de la polarisabilité α de la molécule d'eau, qu'on a fait intervenir dans les calculs, est un peu inférieure à la valeur de la polarisabilité moyenne mesurée dans la vapeur d'eau, ce qu'on justifie ultérieurement; on a admis la valeur $\alpha = 1{,}1.10^{-24}$ cgs.[4] Dans le troisième terme, qui représente l'énergie élastique de polarisation, μ_0 indique le moment dipolaire permanent de la molécule d'eau, égal à $1{,}84.10^{-18}$ cgs. Les exposants de Born pour les forces répulsives entre le cation et les molécules d'eau (n_c), et entre les molécules d'eau elles-mêmes (n) sont voisins des valeurs indiquées par Pauling[5] pour des ions avec une structure électronique semblable à celle des atomes des gaz rares (Tableau 3). Les coefficients a_c et a ont été fixés de telle manière qu'on obtienne une bonne concordance entre les valeurs calculées et les valeurs expérimentales pour les distances entre le cation et les molécules d'eau ainsi que pour les fréquences des bandes Raman qu'on a attribuées aux hydrates des cations.[4]

L'influence de la pression extérieure s'ajoute à la pression d'origine électrostatique

Tableau 3. Forces répulsives.

Interactions	n_c, n	a_c, a (cgs)
$Be^{2+} \ldots OH_2$	6,5	$9{,}2 \times 10^{-63}$
$Al^{3+} \ldots OH_2$	7,0	$3{,}0 \times 10^{-66}$
$Th^{4+} \ldots OH_2$	9,5	$1{,}5 \times 10^{-84}$
$OH_2 \ldots OH_2$	8,0	$6{,}7 \times 10^{-74}$

Tableau 4. Valeurs calculées du coefficient de pression.

Mode de vibration		$\partial\nu/\nu\partial p$ 10^{-2}
$Be(OH_2)_6^{2+}$	ν'_3	+ 2
	ν_1	< 0,4
$Be(OH_2)_4^{2+}$	ν_1	+ 0,7
$Al(OH_2)_6^{3+}$	ν_3	< 0,4
	ν_1	< 0,4
$Th(OH_2)_8^{4+}$	ν_1	+ 0,8

qui s'exerce toujours sur les couches de molécules d'eau qui entourent le cation; il en résulte naturellement une réduction des distances d'équilibre des molécules d'eau au cation central. Les calculs ont été faits avec un modèle qui ne tient pas compte de l'agitation thermique; en effet, l'ordre de grandeur qu'on cherche ne change pas si on utilise un modèle d'aggrégat un peu aplati.

Le Tableau 4 donne les résultats obtenus pour la variation relative $\partial\nu/\nu\partial p$ de la fréquence des modes de vibration attribués aux bandes les plus intenses du spectre Raman des hydrates des cations, lorsque la pression extérieure atteint la valeur de 1,000 atmos.

On a trouvé qu'au-dessous de 10^4 atmos, les variations de fréquences pour les vibrations des hydrates sont proportionnelles à la valeur de la compression réalisée.

Un calcul analogue à celui qu'on a fait pour les hydrates de cations, ne semble pas présenter d'intérêt pour les complexes des anions Cl^- coordonnés autour du cation Zn^{2+}. En effet, trop d'interactions moléculaires jouent dans l'entourage du cation, et il n'est pas possible d'en négliger certaines en face des autres. L'expression de l'énergie potentielle de l'aggrégat constitué autour du cation comporterait un si grand nombre de paramètres inconnus, que les résultats expérimentaux pourraient difficilement fixer leurs valeurs. Toutefois, quelques considérations qualitatives pourront être utiles pour l'interprétation des résultats obtenus. En effet, l'intervention d'une compression extérieure perturbe l'équilibre statistique de l'entourage des complexes et modifie les valeurs des distances des anions aux cations. La force s'exerçant directement sur les anions rend le complexe plus compact. Au contraire, les molécules d'eau localisées en face du centre des faces du tétraèdre et se trouvant à une distance du cation de l'ordre de grandeur du rayon du complexe, ont tendance à agrandir ce dernier. La prépondérance d'une de ces deux influences fournit des renseignements sur le degré d'association des molécules d'eau aux anions, et sur celui des molécules d'eau entre elles-mêmes. Si l'on admet que les ions Cl^- sont faiblement associés aux molécules d'eau de la première couche autour du complexe, et que les molécules d'eau plus proches du cation Zn^{2+} sont fortement liées aux autres molécules d'eau de la solution, on peut espérer que l'effet résultant sera une augmentation du rayon du complexe, ce qui aura pour conséquence une diminution de la valeur de la fréquence de la vibration totalement symétrique, conformément à l'expérience.

DISCUSSION

D'après nos calculs, on voit que les résultats expérimentaux sont en assez bon accord avec ceux calculés à partir des modèles d'hydrates d'ions qu'on vient de proposer. Toutefois, les modèles ne sont pas satisfaisants à tous les points de vue. En effet, l'absence des liaisons hydrogène est peut-être une des plus graves de leurs imperfections, et la précision sur les valeurs calculées pour l'énergie de formation d'un hydrate n'est pas supérieure à 3%.

Par ailleurs, on a fait l'approximation de la molécule d'eau parfaitement symétrique du point de vue des interactions à courte portée, ce qui conduit à des distances trop petites entre les molécules d'eau de couches d'hydratation successives, incompatibles avec les résultats obtenus par diffraction des rayons X.[6] On peut lever cette contradiction en multipliant par un facteur de l'ordre de *trois* les énergies d'interaction de courte portée entre les molécules d'eau de la première couche d'hydratation et celles de la deuxième couche d'hydratation, ce qui ne change pratiquement pas les configurations d'équilibre de l'hydrate.

Les simplifications qu'on a pratiquées sur le modèle électrostatique de la molécule d'eau: dipôle centré sur le noyau d'oxygène, isotropie de polarisabilité dipolaire, absence des polarisabilités d'ordre plus élevé, ne changent pas l'ordre de grandeur des variations de la fréquence pour les modes de vibration étudiés. En outre, l'adoption, pour la polarisabilité de la molécule d'eau, d'une valeur un peu inférieure à celle de la polarisabilité moyenne mesurée dans la vapeur d'eau, trouve une justification dans le fait que les molécules d'eau étant situées dans les hydrates très près des cations qui les attirent fortement, les régions extérieures des nuages électroniques ne jouent plus un rôle important pour la polarisabilité.

Par ailleurs, on n'a pas tenu compte d'une façon explicite des forces de Van der Waals. Ceci n'est pas très grave, car les forces électrostatiques sont très intenses dans ce type d'aggrégats moléculaires, et les coefficients *a* des forces répulsives dues au recouvrement des couches électroniques ont été déterminés en faisant intervenir les rayons des hydrates obtenus par la diffraction des rayons X. Ainsi, les valeurs des termes représentatifs de ces forces répulsives qui varient comme r^{-8} ou r^{-10}, sont modifiés implicitement par les forces de Van der Waals.

L'adoption d'un calcul de forces additives a été suggérée par la vérification éxpérimentale du fait que les valeurs des fréquences des vibrations internes de la molécule d'eau ne se trouvent apparemment pas changées par rapport à leurs valeurs dans l'eau liquide. Un calcul de perturbation serait donc suffisant pour rendre compte des légères modifications de la structure interne des molécules.

Enfin, des recherches récentes sur la structure des solutions aqueuses par diffraction des rayons X[6] révèlent que les anions se disposent de manière symétrique autour des cations, en dehors de la première couche d'hydratation, ce qui a permis d'estimer la compression supplémentaire correspondant à leur présence dans le voisinage de l'hydrate et d'en tenir compte au cours des calculs.

CONCLUSION

Ces études sur l'effet de pression ont fourni une preuve expérimentale supplémentaire de l'existence d'une coordination du cation Be^{2+} différente de la coordination tétraèdrique, qui est à l'origine des bandes Raman très sensibles à la pression. En effet, les bandes de vibration des hydrates de quatre molécules d'eau ne se déplacent pratiquement pas, d'après les calculs effectués sur les modèles utilisés avec une compression de 2.500 atmos.

Pour le cation Th^{4+}, on a vérifié par les calculs qu'une coordination octaédrique de six molécules d'eau pourrait rendre compte des vibrations dont les fréquences ne varieraient pratiquement pas avec une compression extérieure du modèle adopté, de l'ordre de 3.000 atmos. Au contraire, avec une coordination cubique de huit molécules d'eau autour du cation, on obtient une vibration totalement symétrique dont la fréquence varie sensiblement avec les pressions utilisées. Nos expériences ont donc confirmé l'existence des hydrates $Th^{4+}(OH_2)_8$, hypothèse déjà avancée par l'un de nous.[4]

En conclusion, les recherches dont on vient de rendre compte semblent adaptées à la résolution de problèmes de structure dans les solutions liquides, contenant des complexes de coordination dont les modes de vibration sont actifs en diffusion Raman.

REFERENCES

1 A da Silveira et E. Bauer, *C.R. Acad. Sci. Paris* **195**, 416 (1932); A. da Silveira, *C.R. Acad. Sci. Paris* **197**, 1033 (1933); *J. Chem. Phys.* **7**, 380 (1939); J. P. Mathieu, *C.R. Acad. Sci. Paris* **231**, 896 (1950); R. Lafont, *C.R. Acad. Sci. Paris* **244**, 1481 (1957); A. da Silveira, M. A. Marques et N. M. Marques, *C.R. Acad. Sci. Paris* **252**, 3983 (1961); *Mol. Phys.* **9**, 271 (1965); D. E. Irish, B. McCarroll et T. F. Young, *J. Chem. Phys.* **39**, 3436 (1963); R. Hester et R. A. Plane, *J. Inorg. Chem.* **3**, 768 (1964).

2 M. L. Dewaulle, *C. R. Acad. Sci. Paris* **240**, 2132 (1955).

3 *International Critical Tables,* Vol. III, 1928, p. 54.

4 M. A. Marques, à paraître 1972.

5 L. Pauling, *The Nature of the Chemical Bond,* Oxford University Press, 1948.

6 M. A. Marques et M.I.B. Marques, à paraître 1972.

78. THE CONFIGURATION OF WEAKLY BOUND COMPLEXES IN SOLUTION

J. A. Creighton
University Chemical Laboratory, University of Kent, Canterbury, Kent, England

Vibrational spectroscopy has been widely used for identifying the symmetry of molecular species, but more detailed structural information on the *sense* of a distortion of a highly symmetric molecule as a result of the formation of a complex may sometimes be deduced by evaluating the resulting perturbation of the force field.

This method has been used here to investigate the structure of the complex $CBr_4.Cl^-$, which exists in solutions of CBr_4 and tetralkylammonium chlorides in aprotic polar solvents. The ν_1 vibration of this complex (260 cm^{-1}) is well resolved from ν_1 of free CBr_4 (267 cm^{-1}), and Job plots and equilibrium studies using Raman intensities confirm the 1:1 stoichiometry, with formation constants at room temperature of 2.03 l $mole^{-1}$ in CH_2Cl_2 and 4.78 l $mole^{-1}$ in CH_3CN. In addition to the shift in ν_1 there are also changes in the other stretching band $\nu_3(t_2)$ of CBr_4 on complexing with chloride ions. Measurements on solutions in CH_3CN containing virtually no free CBr_4 showed that this band of the complex has an asymmetric contour (see Fig. 1), with the unresolved weaker component lying on the low frequency side and having a depolarization ratio $\rho < 6/7$. Within experimental erro there were no shifts in the bending frequencies $\nu_2(e)$ and $\nu_4(t_2)$.

The 1:1 stoichiometry requires that the symmetry of CBr_4 is lower than T_d, and possible configurations are those in which a chloride ion is bound to a face, edge or apex of a CBr_4 tetrahedron, whose symmetry is then C_{3v}, C_{2v}, or C_{3v},respectively. The effect of a C_{2v} distortion is to split ν_3 into three components $[t_2(T_d) \longrightarrow a_1 + b_1 + b_2\ (C_{2v})]$, and it may be shown that for a small perturbation the a_1 component lies approximately centrally between the b_1 and b_2 components (this is exactly true to a first approximation if the change in the stretch–stretch interaction force constant connecting inequivalent bonds in C_{2v} symmetry is equal to the mean of those connecting equivalent bonds). For a small C_{2v} perturbation the contour of ν_3 therefore remains symmetric. A C_{3v} distortion, however, splits ν_3 into two components $[t_2(T_d) \rightarrow a_1 + e\ (C_{3v})]$, and since for a small effect the intensity of the e is approximately twice that of the a_1 component, the overall contour of ν_3 is then asymmetric. As shown in Fig. 1, the observed contour is thus consistent with a C_{3v} perturbation but not in accord with C_{2v}.

Assuming C_{3v} symmetry the ν_3 band was decomposed into two bands of the same half-width and with heights in the ratio 1:2 as shown in Fig. 1, giving ν_1 = 260 cm^{-1},

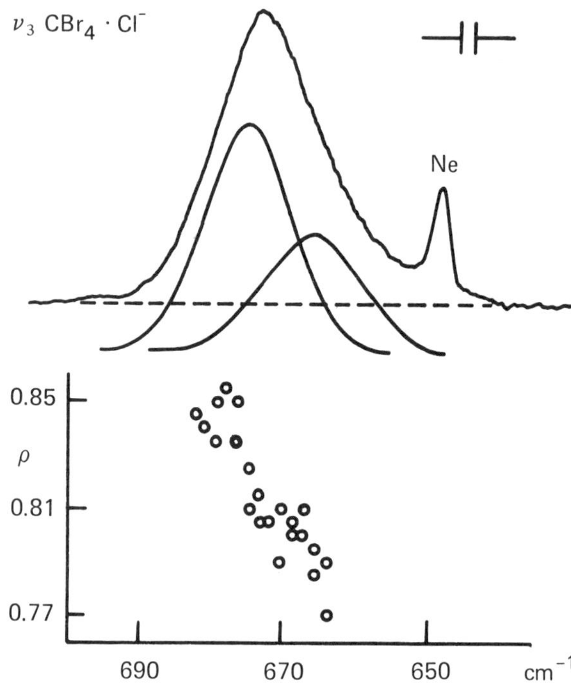

Fig. 1. The ν_3 Raman band of CBr_4Cl^-, and its depolarization ratio, from a solution of Bu_4NCl (1.6 M) and CBr_4 (1.3 M) in CH_3CN.

ν_3 (a_1) = 666 cm^{-1}, ν_3 (e) = 675 cm^{-1}. These frequencies may be compared with the stretching frequencies of CBr_4 in the same solvent CH_3CN of ν_1 = 267 cm^{-1}, and ν_3 = 675 cm^{-1}.

In order to investigate whether the chloride ion is bound to a face or an apex of CBr_4, the shifts in the stretching frequencies were used to calculate the changes in the valence force constants of CBr_4. If the same symmetry coordinates are used for both free and complexed CBr_4, we may define $\mathscr{F}_1 = \mathscr{F}(CBr_4.Cl^-) - \mathscr{F}(CBr_4)$, and the changes in the valence force constants are then the elements of $F_1 = \tilde{U}\mathscr{F}_1 U$, where all the components of the degenerate symmetry coordinates must be included in U. To simplify the calculations, the mixing of stretching with bending motion was neglected. This seemed justified as a first approximation since ν_3 and ν_4 are well separated in CBr_4, and the small interaction between them is likely to be little affected by the small frequency shifts which occur on complexing. There are only three C–Br stretching frequencies from which to calculate four force constants of $CBr_4.Cl^-$ and it was necessary to use one force constant as a variable parameter. The off-diagonal element $\mathscr{F}_{12}$ of the a_1 block of the $\mathscr{F}$ matrix of $CBr_4.Cl^-$ was chosen for this purpose, since it must be small for a small perturbation, and its value may therefore be used to define the limits of acceptable solutions. Figure 2 shows the fractional changes α_{VF} and **a** in the bond stretching and stretch–stretch interaction constants involving the bond along the C_3 axis, and the corresponding changes γ_{VF} and **c** in the valence force constants involving only the other three bonds, as a function of $\mathscr{F}_{12}$ within the range of values of $\mathscr{F}_{12}$ of roughly ± 5% of the diagonal elements of

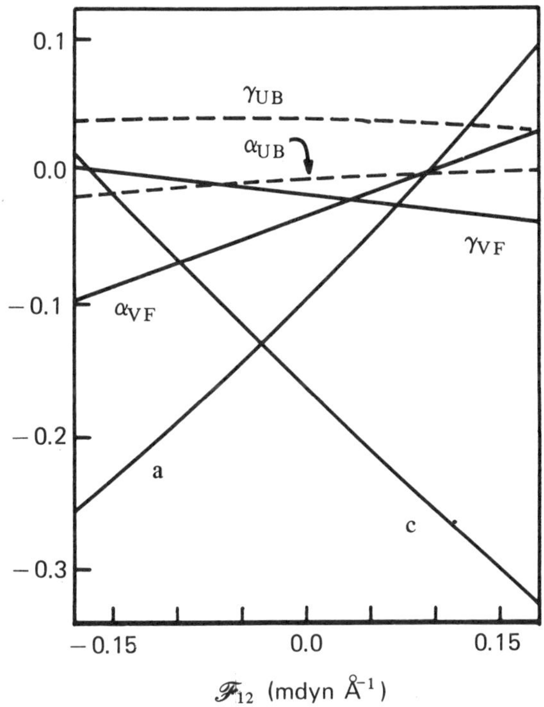

Fig. 2. The fractional changes in the force constants of CBr_4 on complexing with chloride ions in solution in CH_3CN.

It may be seen that close to $\mathcal{F}_{12}$ – 0.09 mdyn Å^{-1} there are solutions with $\alpha_{VF} \triangleq \mathbf{a} \triangleq 0$ and γ_{VF} and **c** negative, consistent with face-centred coordination of CBr_4, but also near $\mathcal{F}_{12} = -0.16$ mdyn Å^{-1} are solutions consistent with apical coordination with $\gamma_{VF} \triangleq \mathbf{c} \triangleq 0$ and α_{VF} and **a** negative. Two arguments favour the solutions in the range consistent with the facial configuration. Firstly, off-diagonal elements of $\mathcal{F}$ matrices such as $\mathcal{F}_{12}$, are normally positive. Secondly it has been suggested that a Urey–Bradley force field provides a better description of the forces in CBr_4 than a valence field, and α_{VF} and γ_{VF} were therefore analysed in terms of the fractional changes α_{UB} and γ_{UB} in the Urey–Bradley bond stretching constants K and the fractional changes **a** and **c** in the non-bonded repulsion constant F. Using the relationships for a tetrahedral molecule XY_4

$$f_r = K + 1.9\,F; \quad 3\,f_{rr} = 2.1\,F$$

we find

$$\mathrm{a_{UB}} = \mathrm{a_{VF}} = \mathrm{a}; \quad \mathrm{c_{UB}} = \mathrm{c_{VF}} = \mathrm{c}$$

$$K\alpha_{UB} = f_r\alpha_{VF} - 1.9\,F\mathrm{a}$$

$$K\gamma_{UB} = f_r\gamma_{VF} - 1.9\,F\,(\mathrm{a} + 2\mathrm{c})/3$$

As shown in Fig. 2, solutions near $\mathcal{F}_{12} = 0.09$ mdyn Å^{-1} have $\alpha_{UB} \triangleq \mathbf{a} \triangleq 0$ and are consistent with face-centred coordination also in a Urey–Bradley field, whereas there is no solution with small values of both γ_{UB} and **c**. Indeed $|\gamma_{UB}| > |\alpha_{UB}|$ throughout the whole range of Fig. 2, and the small values of γ_{VF} near $\mathscr{F}_{12} = -0.16$ mdyn Å^{-1} are interpreted in the Urey--Bradley field as being due to fairly large but opposing values of γ_{UB} and **a**.

Clearly a detailed interpretation of the force constant changes depends on the type of force field assumed, but the valence-force and Urey–Bradley fields represent two extremes in which respectively none and all the stretch–stretch interaction is attributed to non-bonded repulsions. It thus seems clear that with intermediate force fields, as with the two extremes, the force constant changes indicate a face-centred structure as the more probable configuration of $CBr_4.Cl^-$ in solution. It is interesting that a similar structure has recently been suggested[1] from NMR and infrared evidence for the 1:1 halide ion complex of CHI_3 in solution, in which the halide ion appears to be bound to the three polarizable iodine atoms rather than participating in a hydrogen bond.

A considerable simplification of the method used here may be made if the changes in the interaction force constants are negligibly small. It may be shown that the frequency shifts are then given to a first order approximation by

$$\Delta\lambda_1 = \mu_y f_r(\alpha_{VF} + 3\gamma_{VF})/4$$
$$\Delta\lambda_3(a_1) = (\mu_y + 4/3\,\mu_x) f_r\,(3\alpha_{VF} + \gamma_{VF})/4$$
$$\Delta\lambda_3(e) = (\mu_y + 4/3\,\mu_x) f_r\,\gamma_{VF}$$

where $\lambda_i = 4\pi^2\nu_i^2$ and μ_x and μ_y are the inverse masses, and f_r is the valence stretching force constant of the XY_4 molecule. If $\nu_3(a_1) < \nu_3(e)$ then $\alpha < \gamma$, i.e. if the effect of complexing is to reduce the stretching force constants, then $\nu_3(a_1) < \nu_3(e)$ is consistent with apical coordination. Conversely $\nu_3(a) > \nu_3(e)$ indicates facial coordination. Eliminating α_{VF} and γ_{VF} from the above equations gives

$$3\Delta\lambda_1\,(\mu_y + 4/3\,\mu_x) = [\Delta\lambda_3(a_1) + 2\Delta\lambda_3(e)]\,\mu_y$$

and the validity of this expression may be used to test whether the interaction force constant changes are indeed small.

A modification of this approach has recently been used to discuss the shifts in the CO stretching frequencies on forming the ion pair $Na^+[Co(CO)_4^-]$ in tetrahydrofuran solution, and since ν(CO)asym.$(a_1) < \nu$(CO)asym.(e) an apical ion pair was established.[2] It is noteworthy that in $CBr_4.Cl^-$, $\nu_3(a_1)$ is also less than $\nu_3(e)$, but consideration also of the interaction constant changes leads to the conclusion that in this complex the configuration is probably facial.

REFERENCES

1 R. D. Green and J. S. Martin, *J. Amer. Chem. Soc.* **90**, 3659 (1968).
2 W. F. Edgell and J. Lyford, *J. Amer. Chem. Soc.* **93**, 6407 (1971).

79. LOW TEMPERATURE RAMAN SPECTROSCOPY OF MOLECULAR SOLIDS

W. C. Harris
Department of Chemistry, Furman University, Greenville, South Carolina, USA

and

I. W. Levin
National Institutes of Health, Bethesda, Maryland, USA

INTRODUCTION

In attempting to establish vibrational assignments for large polyatomic molecules, one generally utilizes a wide variety of data including vapour phase band contours, depolarization ratios and isotopic shift factors. It also appears that spectra recorded at cryogenic temperatures may be of significant value since the spectral line widths are intrinsically more narrow. Thus, in principle, it is possible both to resolve overlapping features that often occur in spectra recorded at room temperature and to identify weak transitions. This additional information provides further data with which to confirm a vibrational assignment, for example, through a force field calculation. We note that most laboratories have had low temperature infrared capabilities for some time; however, it has only been since the introduction of the laser excitation source that similar experiments could routinely be attempted by the Raman spectroscopist. In addition, the feasibility of Raman matrix isolation studies has been demonstrated by several groups.[1–4] For instance, the measurement of depolarization ratios[1] as well as establishing limits for the stretch–bend interaction force constant in carbon tetrachloride have been reported[2] using the experimental data from Raman matrix procedures.

We have initiated a study of large, fused ring systems and have found the low temperature Raman data to be essential for suggesting vibrational descriptions upon which force field calculations can be based. Similarly, cryogenic studies have been extremely useful in the designation of the weak methyl rocking and torsional modes of several smaller molecules. It is noted that, with the exception of acetone oxime, all of the 'smaller' molecules to be summarized in this study have C_{2v} symmetry. Thus, direct observation of the a_2 and b_2 methyl torsional fundamentals permits evaluation of both the internal rotational barrier and the importance of top–top coupling.

The observation of bands resulting from methyl torsions is difficult in the vapour and liquid states since the population of excited levels leads to broad, diffuse bands. At reduced temperatures, however, identification of the $1 \leftarrow 0$ transition appears more definite. Thus, our emphasis will be on the torsional spectra obtained at liquid nitrogen temperatures with the resulting barriers corresponding to those representative of the solid state. The following discussion of the respective molecules will be considered in three separate sections:

1. norbornane and norbornadiene,
2. acetone and acetone oxime,
3. isobutene and 3-methyl-1,2-butadiene.

EXPERIMENTAL

The Raman spectra of the commercial samples were obtained on Cary Model 81 and Jarrell-Ash Model 500[5] spectrophotometers equipped with argon ion laser sources. Sublimed films were observed at liquid nitrogen temperatures in a cell viewed at 90° to the incident beam.[6] The spectra of the gaseous and liquid samples were recorded in standard cells with multi-passing capability. The spectral resolution varied from 3 to 5 cm^{-1} and the reported frequencies are accurate to ± 2 cm^{-1}.

Norbornane and norbornadiene

The study of bicyclic molecules has occupied the attention of chemists for some time.[7–10] Particular emphasis has centred on the chemical reaction of bicyclo [2.2.1.] heptane and its derivatives since their reactions often take an unexpected course or display unusual rates. The observed ambiguities have been explained by differences in internal strain or by the presence of low energy intermediates with special electronic structures (i.e. non-classical ions). It should be noted that the interpretation of these reaction phenomena is still controversial; however, it is generally accepted that the distinguishing characteristics of the bicyclic systems depend critically on their detailed molecular geometry. For this reason, a number of recent electron diffraction studies of bicyclic systems, have been undertaken.[11–13] Vibrational studies on such strained ring systems are seldom attempted because of the obvious spectral complexity resulting from the large number of available fundamental and combination bands.

It is apparent[14–16] that vibrational assignments and subsequent force field calculations are becoming increasingly necessary for strained ring systems since attempts are currently being made to determine their structures through theoretical calculations. At present, the lack of experimental data necessitates the transference of empirical force constants of dubious validity. Ideally, of course, a potential function compatible with the geometry and the vibrational frequencies of a molecule is desired. Therefore, infrared and Raman investigations, with the objectives of assigning fundamental transitions and of determining intramolecular force fields, would complement and extend the current efforts in calculating structures and energies.

By selecting bicyclic systems which have similar vibrational motions and which possess a reasonable degree of symmetry, it should be possible to suggest spectral assignments upon which force field calculations can be initiated. The present study of norbornane and norbornadiene was undertaken for these reasons. Since isotopic derivatives were unavailable, the spectra are assigned on the basis of depolarization ratios, band positions and intensities. It is noted that the large number of fundamentals predicted for the fingerprint spectral regions for norbornane and norbornadiene encourages the utilization of low temperature techniques.

The results of electron diffraction studies of norbornane[12] and norbornadiene[11,17] have recently been published and are consistent with structures having C_{2v} symmetry. The $C_2(Z)$ axis in each molecule is perpendicular to the non-bonded distance between the

Table 1. Summary of the fundamental vibrations of norbornane.

Vibration number	Symmetry species and approx. description of normal modes	ν^a (cm^{-1})	Vibration number	Symmetry species and approx. description of normal modes	ν^a (cm^{-1})
	a_1			b_2	
1	CH_2 Stretching	2964	27	CH_2 Stretching	2964
2	CH Bridgehead		28	CH Bridgehead	
	Stretching	2952		Stretching	2952
3	CH_2 Stretching	2916	29	CH_2 Stretching	2916
4	CH_2 Stretching	2916	30	CH_2 Scissors	1450
5	CH_2 Scissors	1463	31	CH_2 Bending	1400
6	CH_2 Scissors	1450	32	CH_2 Bending	1238
7	CH_2 Bending	1259	33	CH Bridgehead Bending	1145
8	CH Bridgehead Bending	1207	34	CH_2 Bending	1071
9	CH_3 Bending	1120	35	CH_2 Bending	1031
10	Skeletal Stretching	993	36	Skeletal Stretching	949
11	Skeletal Stretching	923	37	Skeletal Stretching	789
12	Skeletal Stretching	874	38	Skeletal Bending	455
13	CH_2 Bending	754	39	Skeletal Torsion	342
14	Skeletal Bending	540		b_2	
15	Skeletal Bending	412	40	CH_2 Stretching	2948
	a_2		41	CH_2 Stretching	2916
16	CH_2 Stretching	2928	42	CH_2 Stretching	2916
17	CH_2 Stretching	2891	43	CH_2 Scissors	1450
18	CH_2 Scissors	1442	44	CH Bridgehead Bending	1301
19	CH Bridgehead Bending	1317	45	CH_2 Bending	1207
20	CH_2 Bending	1217	46	CH_2 Bending	1031
21	CH_2 Bending	1071	47	Skeletal Stretching	949
22	CH_2 Bending	1031	48	Skeletal Stretching	889
23	CH_2 Bending	958	49	CH_2 Bending	754
24	Skeletal Stretching	787	50	CH_2 Bending	701[b]
25	Skeletal Bending	485	51	Skeletal Bending	342
26	Skeletal Torsion	342			

[a] Reported frequencies are those for the low temperature spectra.
[b] Calculated frequency

two bridgehead carbons, with the two symmetry planes occurring coincident (σ_{xz}) and normal (σ_{yz}), respectively, to this line. The irreducible representations for the 51 normal modes of norbornane are: $15a_1 + 11a_2 + 13b_1 + 12b_2$; correspondingly, the 39 modes of norbornadiene may be listed as $12a_1 + 8a_2 + 10b_1 + 9b_2$. All of the normal vibrations should be Raman active, but only the vibrations characteristic of symmetry species a_1, b_1 and b_2 are infrared active.

The vibrational spectra of norbornane and norbornadiene have been interpreted in detail[18] and the assignments suggested for the respective internal modes are summarized in Tables 1 and 2. A number of overtone and combination bands in Fermi resonance with specific fundamentals have also been observed and assigned. Actually, complicating spectral features, such as combination bands, are expected for these bicyclic molecules because of the associated large number of vibrational degrees of freedom. Raman spectra, recorded of both the liquid and solid states, were more useful than the corresponding infrared studies[18] (see Figs. 1 and 2). Thus, the low temperature Raman data were essential for the location of numerous a_2 vibrations and, also, for the analysis of certain broad, diffuse regions observed in the spectra of the liquids. The interesting low frequency

Table 2. Summary of the fundamental vibrations of norbornadiene.

Vibration number	Symmetry species and approximate description of normal modes	ν^a (cm^{-1})
	a_1	
1	=CH Stretching	3068
2	CH Bridgehead Stretching	2986
3	CH_2 Stretching	2935
4	Skeletal Stretching	1572
5	CH_2 Scissors	1445
6	CH Bridgehead Bending	1248
7	=CH Bending	1227
8	Skeletal Stretching	1101
9	Skeletal Stretching	933
10	=CH Bending	773
11	Skeletal Bending	502
12	Skeletal Bending	428
	a_2	
13	=CH Stretching	3058
14	=CH Bending	1408
15	CH Bridgehead Bending	1271
16	CH_2 Bending	1113
17	Skeletal Stretching	913
18	=CH Bending	711
19	Skeletal Bending	549
20	Skeletal Bending	446
	b_1	
21	=CH Stretching	3082
22	CH Bridgehead Stretching	3000
23	=CH Bending	1445
24	=CH Bending, CH_2 Bending	1345
25	CH Bridgehead Bending	1208
26	Skeletal Stretching	1062
27	Skeletal Stretching, CH_2 Bending	955
28	=CH Bending	797
29	Skeletal Bending, Skeletal Stretching	549
30	Skeletal Bending	446
	b_2	
31	=CH Stretching	3118
32	CH_2 Stretching	2973
33	Skeletal Stretching	1543
34	CH Bridgehead Deformation, Skeletal Stretching	1309
35	=CH Bending	1149
36	CH_2 Bending	895
37	Skeletal Stretching	733
38	=CH Bending	667
39	Bridgehead Deformation	428

[a] Reported frequencies are those from the low temperature spectra. We note that these tentative assignments which may be changed upon completion of our normal coordinate studies.

skeletal bending fundamentals, for example, are also more conveniently identified by the Raman technique. It is interesting to note that the twisting fundamentals of norbornane have only slightly higher frequencies (i.e. ~ 50 cm^{-1}) than their counterparts in simple

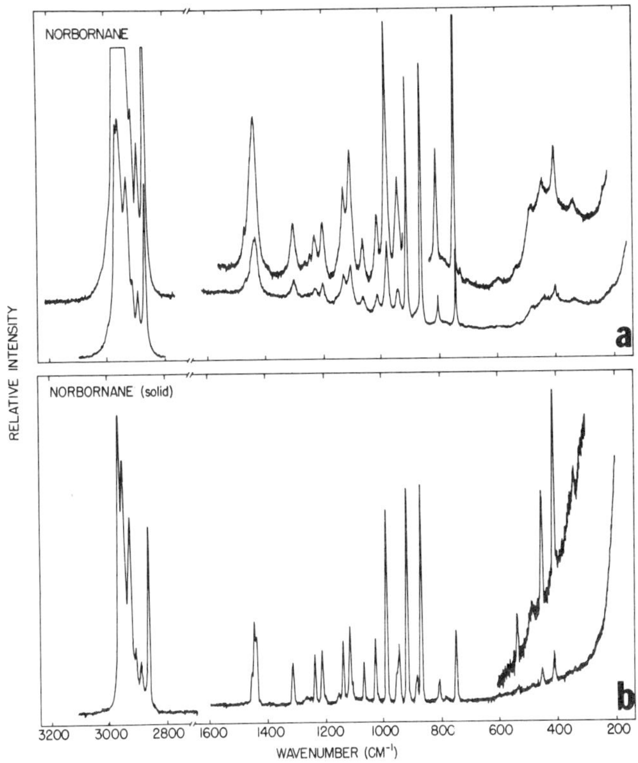

Fig. 1. (a) Raman spectrum of liquid (melt) norbornane. (b) Raman spectrum of solid norbornane.

saturated and unsaturated five-membered rings. Although a majority of the fundamentals can be assigned with confidence, we recognize that certain ambiguities are inherent in initial studies of such complex molecules.

Acetone and acetone oxime

Although there have been a number of earlier studies on the vibrational spectra of acetone,[19-26] certain ambiguities persist in the assignment of several of the methyl and skeletal fundamentals. For instance, assignment[19] of the low frequency a_1 skeletal bending fundamental to a band of 385 cm^{-1} does not appear consistent with recent depolarization measurements,[20] which suggests that this band originates from the b_2 C=O deformation. In addition, the a_2 modes, which are Raman active only, are often difficult to detect since they usually give rise to weak lines or are concealed by other fundamentals at approximately the same frequencies. Because of these difficulties, the frequencies of the a_2 modes in acetone and acetone-d_6 were necessarily estimated on the basis of normal coordinate calculations[19] and have not been experimentally verified. We examine in the present study the applicability of the low temperature Raman technique to acetone in an effort to determine a detailed vibrational assignment and to obtain a value for the methyl rotational barrier of the solid system.[27]

Definitive assignments of methyl torsional vibrations are difficult since such motions typically give rise to weak features in both the infrared and Raman spectra. In order to

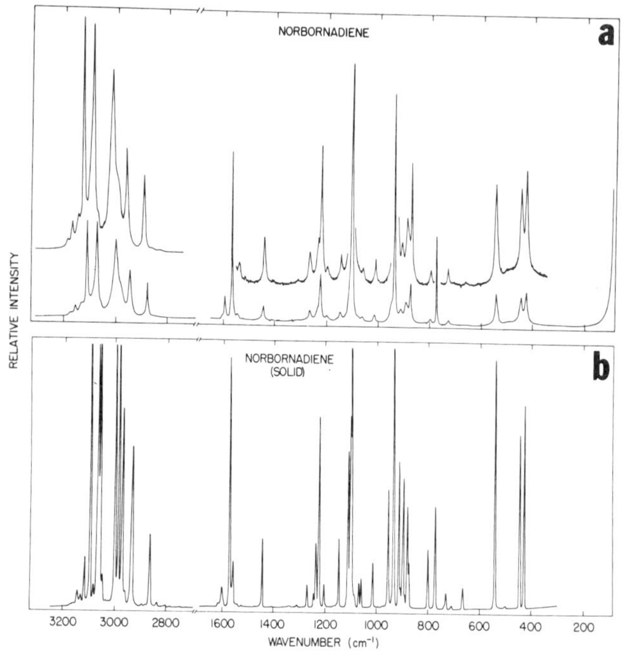

Fig. 2. (a) Raman spectrum of liquid norbornadiene. (b) Raman spectrum of solid norbornadiene.

Table 3. Observed torsional frequency splittings for several two-top molecules.

Molecule	b_2 (cm^{-1})	a_2 (cm^{-1})	Frequency splitting (cm^{-1})	Reference
$CH_3-C(=O)-CH_3$	130	112	18	Present study
CH_3-O-CH_3	268	244	24	14
$CH_3-C(=CH_2)-CH_3$	239	209	30	10
$CH_3-CH_2-CH_3$	271	216	55	17

ensure that one is observing transitions of a torsional fundamental rather than difference bands, for example, a simultaneous investigation of deuterium analogues is required. Furthermore, recent studies[28-32] have demonstrated the advantages associated with acquiring torsional spectral data at liquid nitrogen temperatures. At these reduced temperatures, excited state levels are effectively depopulated and only sharp, well-defined $1 \leftarrow 0$ transitions are observed. Acetone, of course, has two internal rotational degrees of freedom and it is of interest not only to calculate the rotational barrier, but to determine the presence or absence of rotor–rotor interactions. Since we have experienced

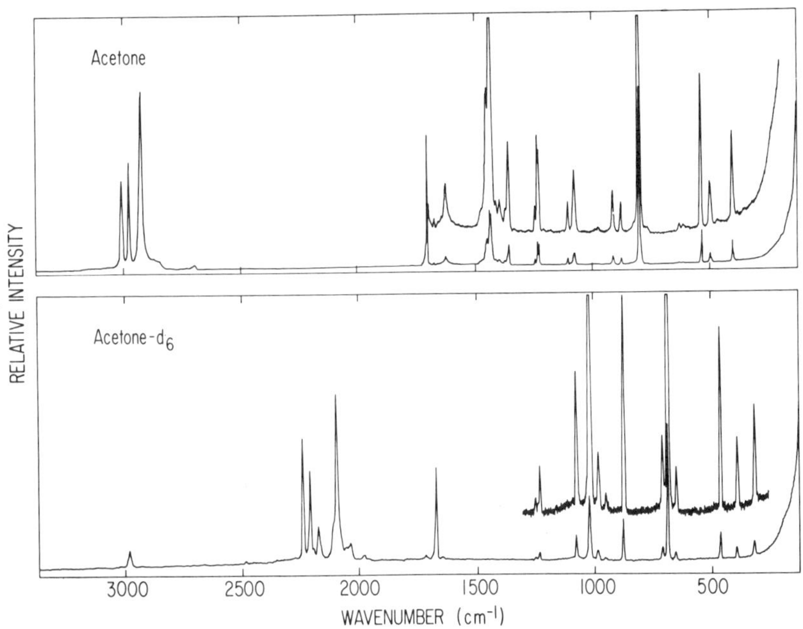

Fig. 3. Raman spectra of solid acetone (top) and acetone-d_6 (bottom).

Table 4. Summary of the possible low frequency torsional-skeletal bending fundamental combination bands.

Skeletal fundamental (Δcm^{-1})	Torsional fundamental (Δcm^{-1})	Combination band (Δcm^{-1})
533	94	627
533	76	609
497	130	627
497	112	609
402	225	627
402	207	609

difficulty in measuring weak spectra for polycrystalline samples below 150 cm^{-1}, the location of Raman spectral features originating directly from the a_2 and b_2 torsional modes of acetone was considered unlikely with our present instrumentation. Thus, inspection of the more accessible mid-frequency region for combination bands, from which torsional frequencies could be determined, offered an alternative approach. The possibility of utilizing actual 'mid-infrared' data to locate the unobserved far-infrared frequencies has been demonstrated in numerous studies[33–36] of ring-puckering vibrations in small ring compounds.

Although no direct evidence for the a_2 fundamental in acetone has been found, a splitting of 15–30 cm^{-1} might be anticipated on the basis of recent studies of 'two-top'

Table 5. Summary[a] of the fundamental vibrations for acetone and acetone-d_6.

Species	Acetone (Δcm^{-1})		Acetone-d_6 (Δcm^{-1})	Approximate description
			SKELETAL MODES	
ν_3 (a_1)	1697		1689	C=O stretch
ν_7 (a_1)	796		703	CC_2 symmetrical stretch
ν_8 (a_1)	497		410	CC_2 scissors
ν_{17} (b_1)	1229	$\nu_{15}(b_1)$	1251	CC_2 antisymmetrical stretch
ν_{19} (b_1)	533		480	C=O in–plane deformation
ν_{23} (b_2)	402		335	C=O out–of–plane deformation
			METHYL STRETCHING MODES	
ν_1 (a_1)	3004		2224	CH_3 (D_3) asymmetrical stretch
ν_2 (a_1)	2920		2109	CH_3 (D_3) symmetrical stretch
ν_9 (a_2)	2972		2255	CH_3 (D_3) asymmetrical stretch
ν_{13} (b_1)	3004		2224	CH_3 (D_3) asymmetrical stretch
ν_{14} (b_1)	2920		2109	CH_3 (D_3) symmetrical stretch
ν_{20} (b_2)	2972		2255	CH_3 (D_3) asymmetrical stretch
			METHYL DEFORMATIONS	
ν_4 (a_1)	1431	ν_5 (a_1)	1033	CH_3 (D_3) asymmetrical deformation
ν_5 (a_1)	1351	ν_4 (a_1)	1092	CH_3 (D_3) symmetrical deformation
ν_{10} (a_2)	1426		1033	CH_3 (D_3) asymmetrical deformation
ν_{15} (b_1)	1408	$\nu_{17}(b_1)$	999	CH_3 (D_3) asymmetrical deformation
ν_{16} (b_1)	1366	$\nu_{16}(b_1)$	1045	CH_3 (D_3) symmetrical deformation
ν_{21} (b_2)	1444		1045	CH_3 (D_3) asymmetrical deformation
			METHYL ROCKING MODES	
ν_6 (a_1)	1072		890	CH_3 (D_3) rock
ν_{11} (a_2)	872		668	CH_3 (D_3) rock
ν_{18} (b_1)	905		725	CH_3 (D_3) rock
ν_{22} (b_2)	1098		890	CH_3 (D_3) rock

[a] The a_2 (ν_{12}) and b_2 (ν_{24}) torsional modes are not listed in this table since they were not directly observed.

molecules (see Table 3 for typical values). Inspection of Fig. 3, however, reveals weak lines at 627 and 609 cm^{-1} which presumably originate from bands involving fundamentals below 550 cm^{-1}. If the 627 – 609 cm^{-1} doublet is assumed to originate from either the 402 or 533 cm^{-1} lines and the two torsions, the predicted $1 \leftarrow 0$ torsional transitions (see Table 4) are not in accord with recent studies of molecular solids in which only relatively small (< 25%), positive solid-vapour frequency shifts have been measured.[37] Frequencies of 130 and 112 cm^{-1} can be deduced for the acetone torsional modes, however, if the 627 – 609 cm^{-1} doublet is *assumed* to result from a combination of the two torsions and the a_1 skeletal bending fundamental (CC_2 scissor motion) at 497 cm^{-1}.

A summary of the estimated torsional frequencies in solid acetone is given in Table 3. It is noted that the b_2 torsional fundamental and b_2 C=O deformation mode may be

Table 6. Summary of the fundamental vibrations for acetone oxime.

Fundamental	Approximate description	$(CH_3)_2CNOH$ (cm^{-1})
	a' species	
1	O–H stretch	3653^a
2	CH_3 asymmetrical stretch	2996
3	CH_3 asymmetrical stretch	2996
4	CH_3 symmetrical stretch	2920
5	CH_3 symmetrical stretch	2920
6	C=N stretch	1675
7	CH_3 asymmetrical deformation	1426
8	CH_3 asymmetrical deformation	1426
9	CH_3 symmetrical deformation	1378
10	CH_3 symmetrical deformation	1364
11	O–H deformation	1344^a
12	CC_2 antisymmetric stretch	1272
13	CH_3 rock	1071
14	CH_3 rock	976
15	N–O stretch	925^a
16	CC_2 symmetric stretch	826
17	Skeletal deformation	600
18	Skeletal deformation	491
19	Skeletal deformation	353
	a'' species	
20	CH_3 asymmetrical stretch	2960
21	CH_3 asymmetrical stretch	2960
22	CH_3 asymmetrical deformation	1426
23	CH_3 asymmetrical deformation	1426
24	CH_3 rock	1083
25	CH_3 rock	966
26	Skeletal deformation	388
27	O H deformation	369^a
28	Skeletal deformation	305
29	CH_3 torsion	217
30	CH_3 torsion	217

^{a} Indicates gas phase frequency; others taken from low temperature spectra.

mixed: therefore, the 'apparent' $a_2 - b_2$ frequency splitting is due in part to both top–top coupling and the mixing of the two b_2 fundamentals. Using the average of the solid state torsional frequencies, an effective internal rotational barrier for acetone at liquid nitrogen temperature is estimated to be 1014 cal/mole. Finally, a complete vibrational assignment for the 24 normal modes of acetone, based upon group frequency arguments, depolarization ratios and isotopic shift factors, is summarized in Table 5.

The infrared and Raman spectra of acetone oxime and acetone oxime–O–*d* have been interpreted in detail and a vibrational assignment, consistent with C_s symmetry, has been proposed (see Table 6).[38] It is important to point out the utility of the low temperature infrared and Raman spectra for resolving complicated spectral patterns and for detecting weak transitions in molecules such as $(CH_3)_2CNOH$. In this respect, the assignment of the low frequency skeletal deformations and methyl rocking modes was

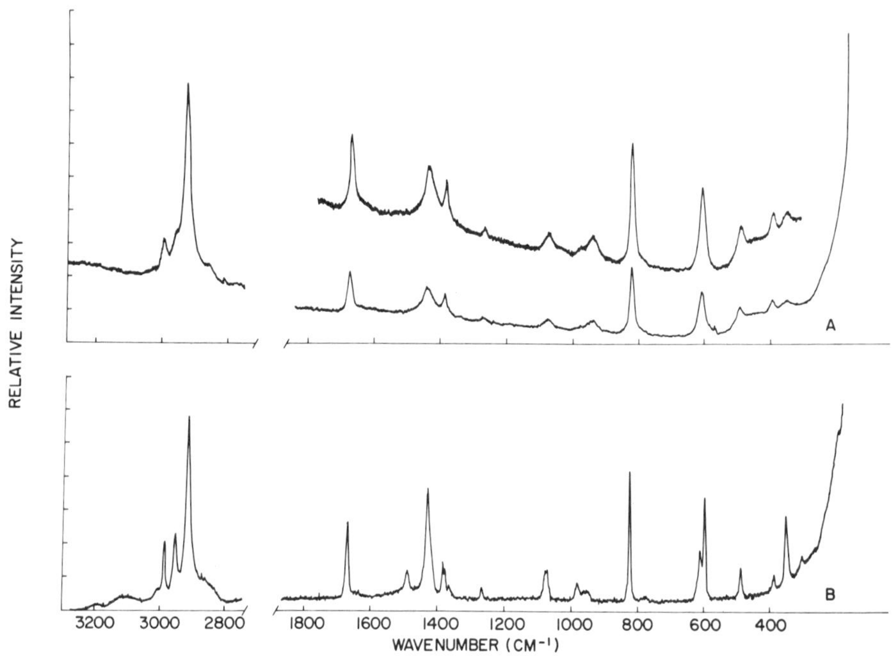

Fig. 4. (a) Raman spectrum of liquid $(CH_3)_2CNOH$. (b) Raman spectrum of polycrystalline $(CH_3)_2CNOH$.

greatly facilitated (see Fig. 4(a) and 4(b)). Furthermore, it was possible to observe the CH_3 torsional fundamentals in the low temperature far infrared spectrum.[38] The close correspondence of the vibrational assignments for the analogous modes in acetone oxime and acetone can be seen by inspection of Tables 5 and 6.

The two a'' methyl torsional modes of solid acetone oxime were assigned to an infrared band at 217 cm^{-1}. Since no top–top splitting was measured, an internal rotational barrier of 3.0 kcal/mole was estimated[38] on the basis of the Mathieu equation.

Isobutene and 3-methyl-1, 2-butadiene

For molecular systems containing two or more methyl groups bonded to a common atom, ambiguities usually arise in the vibrational assignments for the carbon–hydrogen motions as a consequence of the near degeneracies that occur for many of these fundamental modes. The situation is substantially complicated for isobutene $[(CH_3)_2C{=}CH_2]$ and 3-methyl-1, 2-butadiene $[(CH_3)_2C{=}C{=}CH_2]$ since the terminal methylene group contributes six additional vibrations that must be differentiated from the methyl motions. Thus, using the group frequency concept as a guide for locating spectral positions, eight carbon–hydrogen stretching motions are expected near 3000 cm^{-1}, seven carbon–hydrogen bending modes in the 1400 cm^{-1} region and six rocking and wagging motions about 950 cm^{-1}.

For isobutene-d_0, the vibrational assignment has been moderately well established,[39–41] although the assignments for the deuterium isotopes remain less clear.[40,41] In particular, Pathak and Fletcher[40] observed only nineteen of the thirty fundamentals expected for isobutene-d_8. The situation for isobutene-d_6 is better as Shimanouchi and Abe[41] have apparently been able to assign most of the infrared active modes using both

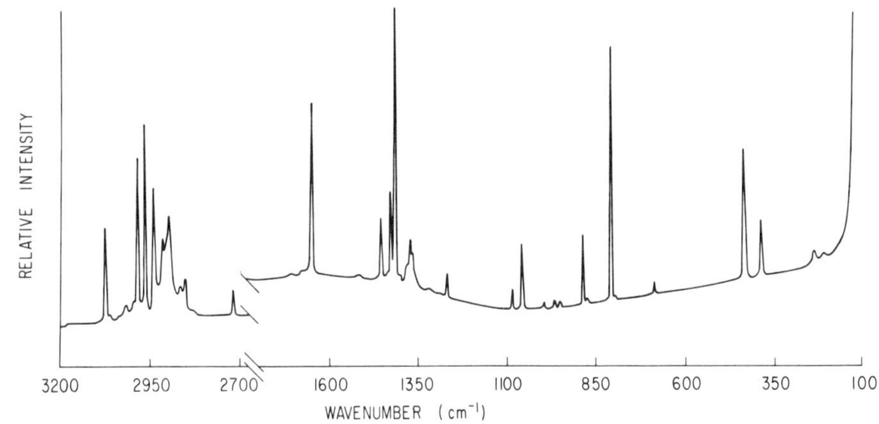

Fig. 5. Raman spectrum of solid isobutene-d_0.

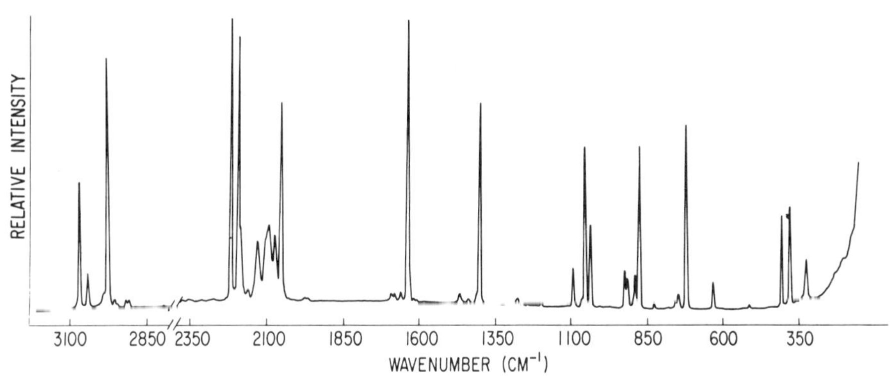

Fig. 6. Raman spectrum of solid isobutene-d_6.

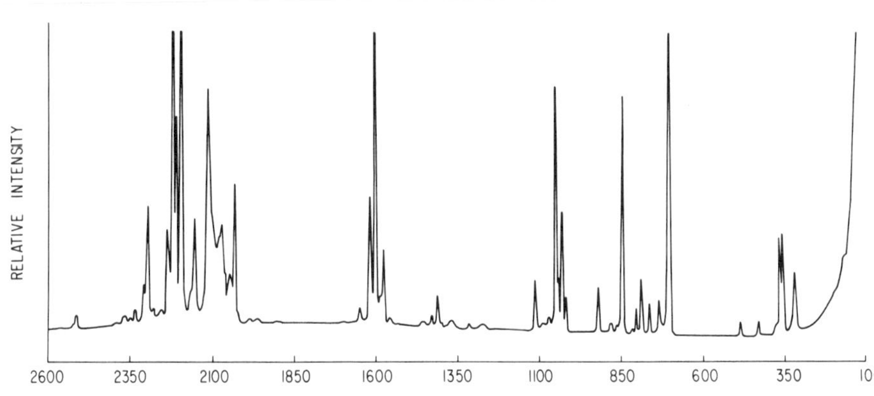

Fig. 7. Raman spectrum of solid isobutene-d_8.

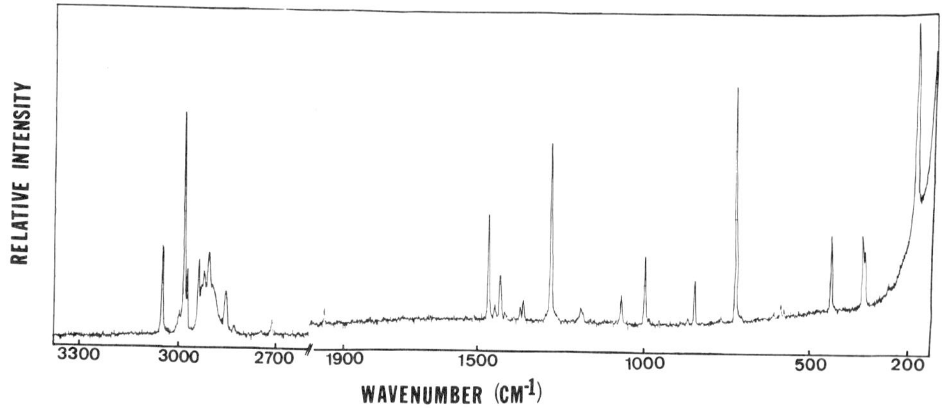

Fig. 8. Raman spectrum of solid 3-methyl-1, 2-butadiene.

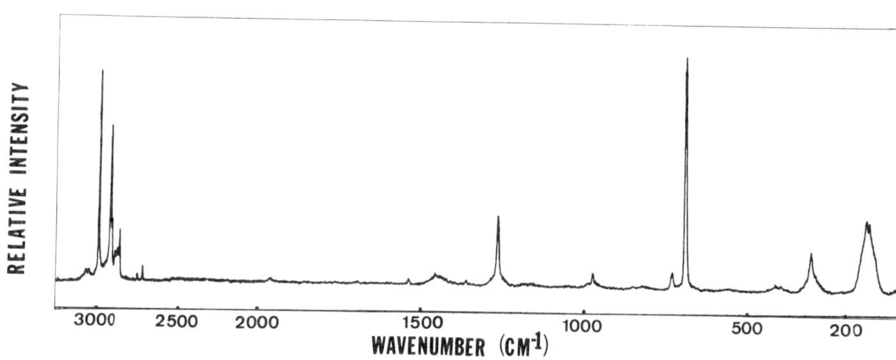

Fig. 9. Raman spectrum of gaseous 3-methyl-1, 2-butadiene.

infrared band contours and infrared solid frequencies. Since no Raman data were recorded by these authors for the -d_6 species, the infrared inactive a_2 modes remained unobserved. Also, many of the a_2 modes for isobutene-d_0 and -d_8 were not detected in the Raman spectra of the liquid,[39-41] although estimates were given for the unobserved modes in force field analyses.[40,41] No previous infrared and Raman study could be found for 3-methyl-1, 2-butadiene.

In an effort to resolve more definitively the individual components contributing to the congested areas of the spectrum, we examined the laser Raman spectra of isobutene-d_0 and two isotopes, isobutene-d_6 and isobutene-d_8, as polycrystalline films at liquid nitrogen temperatures[42] (see Figs. 5 to 7). In addition, we complement this study with a detailed analysis of the infrared and Raman spectra of 3-methyl-1, 2-butadiene[43] (see Fig. 8). It is noted that the capability of observing the entire spectrum, including the usually inaccessible 'far infrared' region, and of detecting the infrared inactive a_2 vibrations encourages the use of the Raman technique. Finally, as a consequence of the high signal–to–noise ratio obtained in the Raman spectra of the solid, many of the weaker methyl rocking and torsional modes become clearly distinguishable. Gas phase Raman spectra of 3-methyl-1, 2-butadiene have also been measured and appear most

Table 7. Summary of the torsional fundamentals for isobutene and 3-methyl-1,2-butadiene.

	$(CH_3)_2C{=}CH_2$ (cm^{-1})	$(CD_3)_2C{=}CH_2$ (cm^{-1})	$(CD_3)_2C{=}CD_2$ (cm^{-1})	$(CH_3)_2C{=}C{=}CH_2$ (cm^{-1})
$\nu(b_2)$	239	172	172	235
$\nu(a_2)$	209	148	148	185

useful in the assignment of the a_1 modes (see Fig. 9).

In summary, group frequency information and isotopic shift data have been used to propose a complete vibrational assignment for isobutene. These data were then applied as a guide toward formulating a vibrational description for the more complex allene derivative $(CH_3)_2C{=}C{=}CH_2$ for which deuterium isotopes were unavailable. We recognize that the use of group frequencies implies approximate descriptions for the normal modes which are probably only qualitatively correct in cases where strong couplings are dominant. In any event, the Raman spectra of polycrystalline films lead to a clarification of the spectral data, which, in conjunction with Raman liquid and infrared gas phase information, enable us to assign, in particular, features to the carbon–hydrogen motions with greater certainty.

The frequencies assigned to the methyl torsional modes for each molecule are summarized in Table 7. Although there is little question about the assignment of the a_2 and b_2 torsions for isobutene, the analogous description for 3-methyl-1, 2-butadiene is more speculative since isotopic derivatives are not available. However, since methyl torsional vibrations characteristically give rise to very weak Raman lines, one can differentiate between the low frequency skeletal bending modes and the a_2 and b_2 torsional fundamentals on the basis of both the low temperature far infrared spectrum and the relative Raman intensities of the low frequency region.[43]

ACKNOWLEDGEMENT

Acknowledgement is made to the donors of the Petroleum Research Fund, administered by the American Chemical Society, for support of this research (W.C.H.).

REFERENCES

1 J. W. Nibbler and D. A. Coe, *J. Chem. Phys.* **55**, 5133 (1971).

2 I. W. Levin and W. C. Harris, *Raman Spectra of Matrix Isolated Molecules: Application to the Force Field for CCl_4*, *J. Chem. Phys.* (in press).

3 H. Huber, G. A. Ozin and A. Vander Voet, *J. Mol. Spectry.* **40**, 421 (1971).

4 H. H. Claassen and G. L. Huston, *J. Chem. Phys.* **55**, 1505 (1971).

5 Purchased with funds from a National Science Foundation Grant, GP–32802.

6 I. W. Levin, *Spectrochim. Acta* **25A**, 1157 (1969).

7 T. P. Neven, E. DeSalas and C. L. Wilson, *J. Chem. Soc.* 1188 (1939).

8 J. March, *Advanced Organic Chemistry: Reactions, Mechanisms and Structure*, McGraw-Hill, New York, 1968.

9 P. D. Bartlett, *Nonclassical Ions*, W. A. Benjamin, New York, 1965.

10 G. D. Sargent, *Quart. Rev (London)* **20**, 301 (1966).

11 Y. Morino, K. Kuchitsu and A. Yokozeki, *Bull. Chem. Soc. Japan* **40**, 1552 (1967).

12 J. F. Chiang, C. F. Wilcox, Jr. and S. H. Bauer, *J. Amer. Chem. Soc.* **90**, 3149 (1968).
13 J. F. Chiang and S. H. Bauer, *J. Amer. Chem. Soc.* **92**, 1614 (1970).
14 N. L. Allinger, J. A. Hirsch, M. Miller, I. J. Tyminski and F. A. Van-Catledge, *J. Amer. Chem. Soc.* **90**, 1199 (1968).
15 N. L. Allinger, M. T. Tribble, M. Miller and D. H. Wertz, *J. Amer. Chem. Soc.* **93**, 1637 (1971).
16 J. E. Williams, P. J. Stang and P. von R. Schleyer, *Ann. Rev. Phys. Chem.* **19**, 531 (1968).
17 G. Dallinger and L. H. Toneman, *Rec. Trav. Chim.* **87**, 805 (1968).
18 I. W. Levin and W. C. Harris, *Infrared and Raman Spectra and Strained Ring Systems: Nobornane and Norbornadiene* (to be published).
19 G. Dellepiane and J. Overend, *Spectrochim. Acta* **22A**, 593 (1966).
20 J. R. Allkins and E. R. Lippincott, *Spectrochim. Acta* **25A**, 761 (1969).
21 M. T. Forel and M. Fouassier, *Spectrochim. Acta* **23A**, 1977 (1967).
22 P. Mirone, *Spectrochim. Acta* **20A**, 1646 (1964).
23 J. Overend and J. R. Scherer, *Spectrochim, Acta* **16A**, 773 (1960).
24 P. Mirone and P. Chivroboli, *Ann. Chim.* **50**, 1095 (1960).
25 P. Cossee and J. H. Schachtschneider, *J. Chem. Phys.* **44**, 97 (1966).
26 T. Miyazawa, *Nippon Kagaku Zasshi* **74**, 915 (1953).
27 W. C. Harris and I. W. Levin, *J. Mol. Spectry.* **43**, 117 (1972).
28 W. C. Harris and I. W. Levin, *J. Chem. Phys.* **54**, 3227 (1971).
29 J. R. Durig, S. M. Craven and J. Bragin, *J. Chem. Phys.* **51**, 5663 (1969).
30 J. R. Durig, S. M. Craven and J. Bragin, *J. Chem. Phys.* **52**, 2046 (1970).
31 J. R. Durig, C. M. Player, Jr. and J. Bragin, *J. Chem. Phys.* **54**, 460 (1971).
32 C. M. Player, Jr., M. S. Dissertation, University of South Carolina, August, 1970.
33 A. B. Harvey, J. R. Durig and A. C. Morrissey, *J. Chem. Phys.* **47**, 4864 (1967).
34 W. H. Green, *J. Chem. Phys.* **52**, 2156 (1970).
35 T. Ueda and T. Shimanouchi, *J. Chem. Phys.* **47**, 5018 (1967).
36 J. F. Blanke, T. H. Chao and J. Laane, *J. Chem. Phys.* **38**, 483 (1971).
37 J. R. Durig, S. M. Craven, K. K. Lau and J. Bragin, *J. Chem. Phys.* **54**, 479 (1971).
38 W. C. Harris and S. F. Bush, *J. Chem. Phys.* **56**, 6147 (1972).
39 W. Luttke and S. Braun, *Ber. Bunsenges, Physik. Chem.* **71**, 34 (1967).
40 C. M. Pathak and W. H. Fletcher, *J. Mol. Spectry.* **31**, 32 (1969).
41 T. Shimanouchi and Y. Abe, *Vibrational Spectra and Molecular Force Field of Isobutylene and Isobutylene-d_6* (private communication).
42 W. C. Harris and I. W. Levin, *J. Mol. Spectry.* **39**, 441 (1971).
43 W. C. Harris and C. T. Longshore, *Vibrational Spectra and Structure of 3-methyl-1, 2-butadiene* (to be published).

80. CALCUL DES FREQUENCES ET INTENSITES APPROCHEES DANS LES SPECTRES INFRAROUGE ET RAMAN DES HEXAMETHYLBENZENE, HEXAMETHYLPRISMANE, ET HEXAMETHYLBENZENE DEWAR*

D. Bougeard, P. Bleckman et B. Schrader
Universität Dortmund, Germany

ABSTRACT

A method derived from the GF-Matrix method of Wilson was used for the calculation of the vibrations of hexamethylbenzene and two of his non-planar isomers: hexamethylbenzene Dewar and hexamethylprismane.

Two models, which use the calculated forms of the vibrations, were used to calculate the approximate intensities of the infrared and Raman spectra. The second one, which uses the CNDO method of Pople, gave the best results for infrared spectra.

The precision for frequency calculations was about 1%. Interpretation of force constants gave information about intramolecular phenomena: repulsion of methyl groups, overlap of π-orbitals, resonance and hybridization. The calculation for hexamethylprismane gave indices of a possible torsion of the carbon skeleton.

INTRODUCTION

Jusqu'à ces dernières années l'étude quantitative complète du spectre de vibration des molécules était limitée par la taille des ordinateurs. On était alors contraint à traiter les molécules d'une certaine taille soit de façon qualitative (utilisation des fréquences et intensités caractéristiques, utilisation du degré de polarisation), soit avec des approximations (traitement des groupes méthyle ou phényle comme des points de masse 15 ou 77, séparation totale des vibrations dans le plan de la molécule et de celles hors de ce plan). Le traitement du spectre de vibration était alors limité à des molécules de l'ordre de dix atomes. Depuis une dizaine d'années, les ordinateurs se sont dévelopées et on a maintenant la possibilité d'étudier des molécules plus importantes. Des groupes de travail ont développé des études sur les champs de forces pour des séries de molécules semblables (Schachtschneider et Snyder[1] pour les parafines, Scherer[2] ou Neto et Scrocco[3] pour les hydrocarbures aromatiques). Le but de ce travail était de tenter de délimiter quelles informations on peut obtenir à partir d'une étude quantitative complète des spectres de vibration des molécules organiques d'une certaine taille.

Pour le calcul des fréquences, nous avons utilisé une variante de la méthode de la matrice GF de Wilson.[4] Le programme utilisé est une version modifiée du programme mis au point par Shimanouchi à Tokio. Les données sont entrées en coordonnées cartésiennes ce

---

* L'exposé détaillé des résultats obtenus est prévu dans trois publications en préparation pour *Z. Naturforsch.*

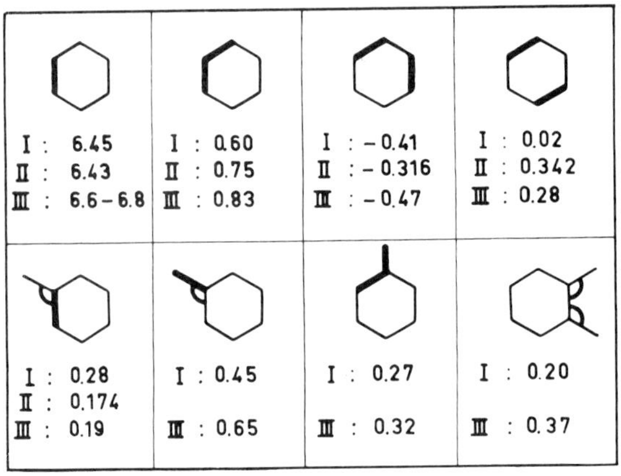

Fig. 1. Constantes de force du benzene. I – nos resultats; II – Califano; III – Scherer.

qui permet d'obtenir en plus des fréquences des vibrations aussi les formes de celles-ci. Nous avons utilisé les formes des vibrations pour le calcul des intensités des bandes infrarouge et Raman. L'information que constitue l'intensité d'une bande est très importante pour l'attribution dans le cas de grosses molécules où plusieurs attribution sont possibles et où les considérations de symétrie ne suffisent plus.

Deux méthodes de calcul des intensités ont été utilisées. Dans la première, on attribue à chaque liaison de la molécule un moment dipolaire et un tenseur de polarisabilité ainsi que leurs dérivés par rapport à la longueur de la liaison $\partial\mu/\partial r$ et $\partial\alpha/\partial r$ (voir Averbukh et Mayants[5,6]) et on suppose que ces propriétés physiques sont additives pour une molécule. Si Q est une des coordonnées normales, l'intensité en infrarouge est proportionelle à $(\partial\mu/\partial Q)_0^2$ et l'intensité en Raman à $5b^2 + 13g^2$, où b et g sont la trace et l'anisotropie du tenseur $(\partial\alpha/\partial Q)_0$. On peut développer comme suit

$$\left(\frac{\partial\mu}{\partial Q}\right)_0 = \sum_{\text{liaisons}} \left(\frac{\partial\mu}{\partial r}\right)_0 \frac{\partial r}{\partial Q}$$

et

$$\left(\frac{\partial\alpha}{\partial Q}\right)_0 = \sum_{\text{liaisons}} \left(\frac{\partial\alpha}{\partial r}\right)_0 \frac{\partial r}{\partial Q}$$

Les $(\partial\mu/\partial r)_0$ et $(\partial\alpha/\partial r)_0$ sont les paramètres introduits plus haut; les $\partial r/\partial Q$ sont les éléments de la matrice qui décrit les formes des vibrations. Dans la deuxième méthode que nous n'avons pour le moment appliquée qu'au spectre infrarouge, on calcule à l'aide des méthodes de la chimie quantique (dans notre cas la méthode CNDO/2 de Pople[7]) la courbe $\mu = f(Q)$. La dérivée de cette courbe pour chaque vibration aux environs de l'état d'équilibre conduit alors directement à l'intensité de cette vibration.

Nous avons cherché pour notre étude des molécules qui soit assez différentes les unes des autres et qui présentent des géométries suffisamment particulières pour donner de gros effets: réhybridisation, répulsion, etc. D'autre part, nous voulions ne pas avoir trop

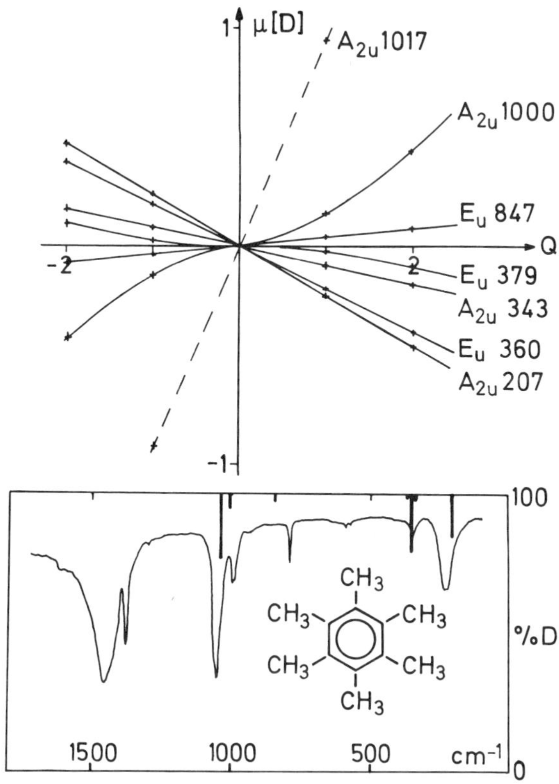

Fig. 2. Intensité infrarouge par CNDO

de paramètres libres; c'est pourquoi nous avons choisi des composés contenant six groupes méthyle dont les constantes de force sont bien connues. Les composés choisis sont: l'hexaméthylbenzène (HMB), l'hexaméthylbenzène Dewar (HMDB), et l'hexaméthyl-prismane (HMP). Ces substances one été proposées le siècle dernier comme alternatives aux structures de Kékulé du benzène mais n'étaient envisagées à cette époque que comme des formules planes. Depuis cinq ans la chimie des isomères non plans du benzène s'est considérablement développée et a permis de synthétiser photochimiquement ces deux dérivés. Ceux-ci sont relativement stables à température ambiante malgré des énergies de tension assez considérables: 44,5 kcal/mole pour HMDB et 116 kcal/mole pour HMP. Nous avons essayé d'étudier par spectroscopie infrarouge et Raman comment cette tension des squelettes carbonés se traduit dans les constantes de force et de comparer ces résultats avec ceux d'autres méthodes d'étude.

HEXAMETHYLBENZENE

Tout d'abord, nous allons voir les résultats de l'HMB qui est la substance qui posait le moins de problèmes. L'attribution a été faite à l'aide des spectres de monocristaux pour les vibrations observées en Raman. Ces spectres correspondent au groupe ponctuel D_{6h} bien que Hamilton[8] ait trouvé un groupe ponctuel D_{3d} par la dispersion des neutrons. Pour le calcul nous avons utiliser pour les groupes méthyle les constantes de force calculées par Schachtschneider et Snyder.[1] Les plus importantes constantes de force sont rapportées dans la Fig. 1. Quelques remarques s'imposent. La constante de force pour la liaison C–CH_3 est 6,45 mdyn/Å, soit environ 10% de plus qu'une liaison C–C «normale». Nous pensons que cela doit être attribué à l'état d'hybridisation sp_2 des carbones du noyau benzènique. La présence des constantes d'interaction *ortho, méta, para,* correspond aux résultats d'autres travaux. La valeur assez faible de l'interaction de liaisons en position *para* est sans doute un signe que la résonance est moins forte que dans le benzène lui-même. Cela peut provenir du fait mentionné par Hamilton, que les groupes méthyle sont un peu déplacés hors du plan du noyau benzènique. Une autre observation vient appuyer cette hypothèse: il nous a été nécessaire pour obtenir un ajustement convenable de la raie 456 du spectre Raman d'introduire une interaction entre deux groupes méthyle en position *ortho*. La valeur de cette constante de force est 0,62 mdyn/Å.

Les intensités des bandes du spectre infrarouge ont été calculées par les deux méthodes exposées plus haut. Le résultat du calcul pour le domaine 0–1000 cm^{-1} par la méthode CNDO est exposé dans la Fig. 2. L'accord avec l'expérience est très satisfaisant. Le calcul par la première méthode conduit, pour toutes les vibrations où les groupes méthyle ne participent pas, à une intensité négligeable. L'examen du calcul CNDO montre que la plus grosse part de l'intensité est causée par des phénomènes de réhybridisation. Or ces phénomènes ne sont pas reproduits par la première méthode qui ne prend en considération que le changement du moment dipolaire avec la longueur ou l'orientation de la liaison correspondante. Les principaux résultats apportés par ce calcul sont donc:

1. Les groupes méthyle sont peu troublés par le composé où ils sont fixés mais se gênent quand ils sont en position *ortho*.
2. Les phénomènes de résonance et de réhybridisation sont très importants et responsables d'une très grande part de l'intensité en infrarouge.
3. Le premier modèle pour le calcul de l'intensité infrarouge n'est pas suffisant pour un système aromatique.

HEXAMETHYLBENZENE DEWAR

Nous en venons maintenant à l'HMDB où nous allons retrouver quelques uns des résultats de l'HMB et voir quelles informations on peut obtenir à propos d'un squelette carboné. La structure de cette substance a été étudiée par diffraction des électrons. Les études sont en accord sur tous les paramètres sauf sur la liaison *para* qui a été évaluée à 1,63 et 1,58 Å. Notre calcul s'est avéré le meilleur pour une liaison de 1,60 Å. L'angle dihédral mesure environ 120°. Les autres liaisons et angles sont «normaux». Le calcul a été fait dans le groupe ponctuel C_2 bien que les spectres révèlent que la symétrie est vraisemblablement C_{2v}. La Fig. 3 montre les résultats du calcul pour les liaisons C–C du squelette carboné. Les liaisons C–C simples ne montrent pas de comportement particulier. La longue liaison *para* a un ordre qui correspond à ce l'on attend d'après les études de Coulson sur la corrélation entre ordre et longueur d'une liaison. Les résultats pour les doubles liaisons sont un peu plus étonnants, étant donné que l'ordre de ces liaisons descend à 1,64. A la suite de cette constation, nous avons effectué un calcul des orbitales moléculaires de HMDB par la méthode CNDO. Ce calcul a montré que le recouvrement des orbitales p_y et p_z de deux atomes de carbone appartenant à des doubles liaisons différentes est du même ordre de grandeur que le recouvrement des orbitales 1*s* de l'hydrogène et 2*s* du carbone pour une liaison CH. Il semble donc que l'on soit en présence

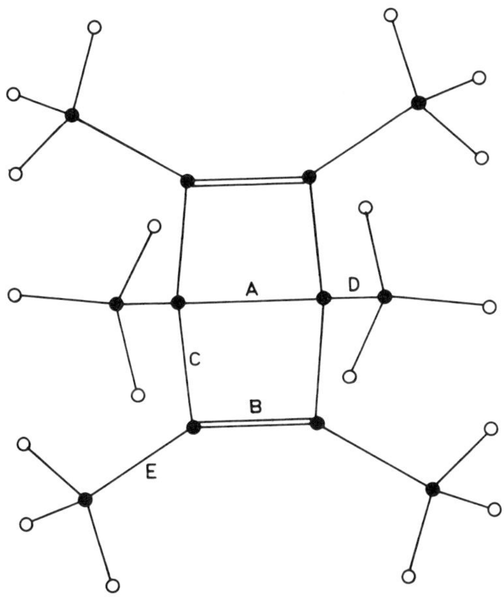

	Longueur Å	Constante de Force mdyn/Å	Ordre
A	1,60	3,63	0,87
B	1,35	8,20	1,64
C	1,52	4,40	0,99
D	1,54	4,40	1.00
E	1,50	4,60	1,02

Fig. 3. Propriétés des liaisons CC de HMDB

d'une sorte d'homoconjugaison, c'est-à-dire d'une «liaison à travers l'espace» où est localisée une partie des électrons que l'on attend dans la double liaison.

Le calcul des intensités infrarouge et Raman a été fait par la méthode de Averbukh et Mayants. Les résultats se trouvent sur la Fig. 4 en comparaison avec les spectres observés. On peut se rendre compte ici des faiblesses de cette méthode pour des grosses molécules. Dans le spectre infrarouge il y a quelques bandes (735 et 785 cm^{-1}) qui apparaîssent dans le calcul par suite du facteur $(\partial\mu/\partial r)$ pour les liaisons $C-CH_3$. Or ces bandes n'apparaîssent pas dans le spectre. Cela veut dire que d'autres processus annulent au cours de la vibration la participation à l'intensité apportée par $(\partial\mu/\partial r)\,C-CH_3$. Dans le spectre Raman on peut se rendre compte que les vibrations où seules les liaisons sont en mouvement (577 et 1684 cm^{-1}) apparaissent relativement bien ainsi que celles où les déformations ne sont pas trop importantes (465, 492, 516 cm^{-1}). La reproduction du spectre devient mauvaise pour les vibrations de déformation pures et en particulier pour les torsions. Celles-ci se produisent sans changement de la longueur des liaisons si bien que l'intensité devient nulle. On pourrait supprimer ces difficultés par l'introduction de paramètres supplémentaires mais ce serait au dépend de la compréhension de la signification physique de ces paramètres.

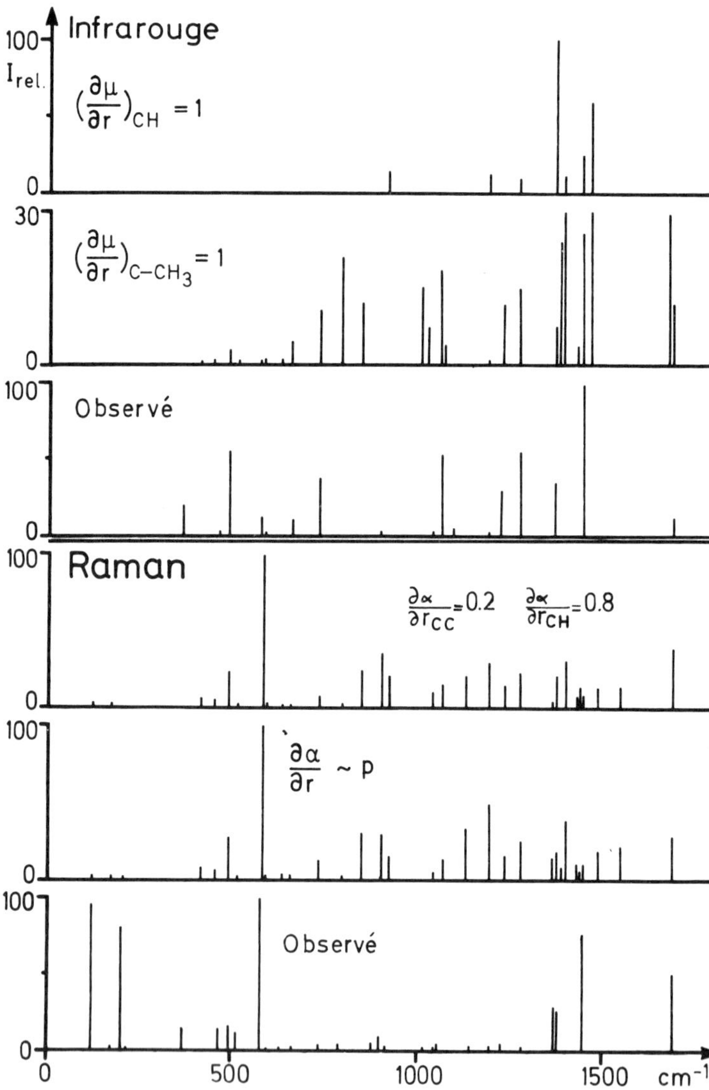

Fig. 4. Intensités d'après Averbukh et Mayants.

Le calcul CNDO nous a montré que l'analyse des spectres permet de déceler des phénomènes telles que l'homoconjugaison et de recueillir des renseignements sur les liaisons des molécules. Il s'avère de nouveau ici que la première méthode de calcul des intensités n'est pas suffisante pour une étude poussée.

HEXAMETHYLPRISMANE

Nous en venons maintenant à l'HMP. Cette molécule est celle qui pose le plus de problèmes car la structure n'est pas encore connue. Nous avons déterminée cette structure par le calcul du minimum de l'énergie de la molécule à l'aide du programme

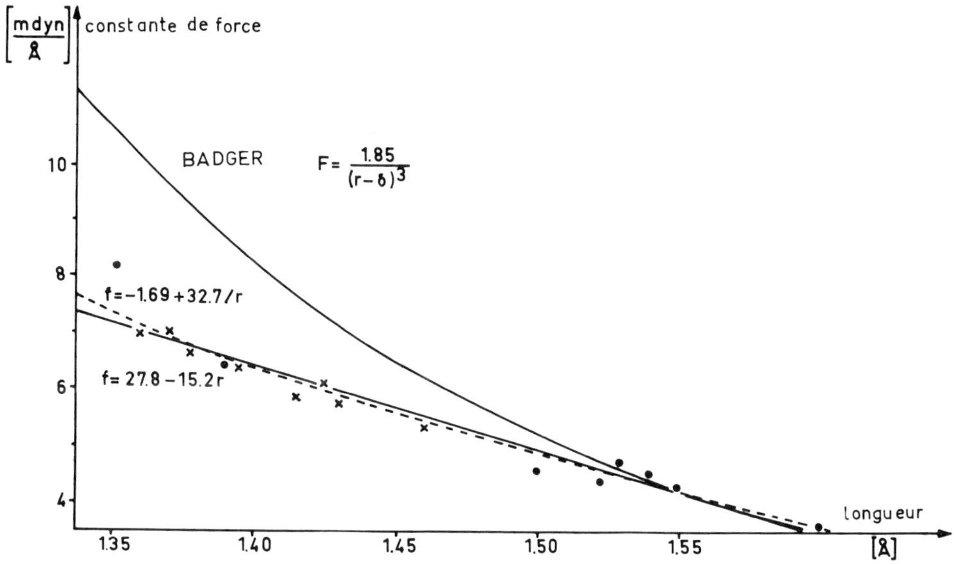

Fig. 5. Corrélation entre constante de force et longueur des liaisons C–C.

CNDO. La structure finalement utilisée pour le calcul est:

r: liaison dans le cyclopropane: 1,53 Å
R: liaison dans le cyclobutane: 1,55 Å
θ: angle entre C–C du cyclobutane et C–CH_3: 130°

Il semble que cette structure soit encore éloignée de la structure réelle (l'énergie de la molécule est plus faible que celle de HMB). Nous avons calculé avec cette structure car un meilleur ajustement n'était pas possible (trois paramètres: r, R, θ et une valeur de référence: l'énergie). La molécule appartient au groupe ponctuel D_{3h}. Le calcul des fréquences donne une précision de l'ordre de 1,3%. Mais pour atteindre cette précision il faut attribuer certaines bandes à des vibrations qui, d'après les règles de sélection, ne devraient pas être observées. Nous ne pensons pas que cela signifie que la structure supposée est complètement fausse, mais plutôt un peu déformée. Les spectres observés sont des spectres de poudre. Il est possible que le champ cristallin trouble les règles de sélection, mais aussi que l'interaction des molécules entre elles entraine un déplacement des CH_3 de la structure idéale ou une torsion du squelette carboné. Il est à remarquer que l'ordre des liaisons est proche de 1, c'est à dire que la torsion des liaisons n'apparaît pas dans les constantes de force. Il est possible que la torsion apparaîsse dans les constantes d'interaction, mais celles-ci sont difficiles à évaluer du fait que les vibrations qui en dépendent fortement ne sont pas observées.

Nous avons fait le tour des renseignements propres à chaque molécule. Pour terminer, nous allons étudier quelques résultats plus généraux. Tout d'abord en ce qui concerne les constantes de force. La Fig. 5 rassemble nos constantes de force pour les liaisons C–C (ronds), ainsi que quelques valeurs rassemblées par Neto (croix).[9] Les points montrent une corrélation assez nette qui peut être exprimée avec presque la même précision par une droite ou par une hyperbole. L'accord avec la règle de Badger

$$F\,(\text{mdyn/Å}) = 1{,}85/(r-\delta)^3$$

où δ dépend des atomes de la liaison, est assez bon dans le domaine des liaisons simples. On peut remarquer que l'approximation avec une hyperbole correspond aux résultats de Lippincott[10] qui calcule différentes propriétés physiques des molécules simples à l'aide de fonctions δ. Cette représentation des constantes de force en fonction de la longueur de la liaison est de toute façon trop simple puisque nous avons vu que d'autres phénomènes entrent en jeu (homoconjugaison, liaisons courbées). On devrait essayer de tenir compte aussi de facteurs tels que les orbitales de recouvrement.

Par suite de la participation de plusieurs angles à une même vibration de déformation, il est difficile d'estimer les constantes de force pour les angles. Il ne nous a pas été possible de retrouver une corrélation entre la valeur des constantes de force pour les angles et d'autres paramètres physiques de la molécule.

L'observation des formes des vibrations montre que les vibrations caractéristiques des groupes méthyle se retrouvent dans des régions de fréquence connues. Mais on peut se rendre compte que le passage par exemple de δ_s à ρ se fait par des vibrations pour lesquelles une part de l'énergie potentielle vient des vibrations des liaisons C–C. On peut d'autre part se rendre compte que les groupes méthyle se conduisent au-dessous de 600 cm^{-1} comme des solides indéformables. C'est donc seulement au-dessous de 600 cm^{-1} que l'on peut utiliser l'approximation des groupes méthyle comme points de masse 15. Une dernière remarque concernant le spectre Raman. On observe dans les trois spectres que la bande la plus intense (exceptées les vibrations de valence des CH_3) se trouve vers 575 cm^{-1} (HMB: 559, HMDB: 577, HMP: 580) et qu'elle est très polarisée. Les formes de ces vibrations montrent que cette bande correspond pour chacune des molécules à la vibration dans laquelle on a une expansion des cycles.

CONCLUSION

Cet exposé a tenté de montrer à l'aide de trois exemples quelles informations on peut espérer obtenir à l'aide des spectres de vibration. Une des difficultés reste l'obtention de constantes de force transposables d'une molécule à une autre. Nous avons aussi essayer de montrer comment on peut utiliser l'information contenue dans l'intensité des bandes. Le problème est loin d'être complètement résolu mais on peut espérer dans un avenir proche pouvoir utiliser plus complètement les possibilités des spectres infrarouge et Raman.

REFERENCES

1 J. H. Schachtschneider, et R. G. Snyder, *Spectrochim. Acta* **19**, 117 (1963).
2 J. R. Scherer, *Spectrochim. Acta* **17**, 719 (1961); *Spectrochim. Acta* **20**, 345 (1964); *Spectrochim. Acta* **23A**, 1489 (1967).
3 N. Neto et M. Scrocco, *Spectrochim. Acta* **22**, 1981 (1966).
4 E. B. Wilson, J. C. Decius et P. C. Cross, *Molecular Vibrations,* McGraw-Hill, New York, 1951.
5 B. S. Averbukh, *Opt. Spectry.* **23**, 28 (1967).
6 L. S. Mayants, *J. Mol. Spectry.* **22**, 197 (1967).
7 J. A. Pople, *Beveridge Approximate Molecular Orbital Theory,* McGraw-Hill, New York, 1970.
8 W. C. Hamilton, J. W. Edmonds et A. Tippe, Oxford Congress, 1968.
9 N. Neto et C. diLauro, *Spectrochim. Acta* **26A**, 1175 (1970).
10 E. R. Lippincott, *J. Chem. Phys.* **26**, 1678 (1957).

81. SPECTRE DE DIFFUSION RAMAN ET LUMINESCENCE DE LA BENZOPHENONE CRISTALLINE

P. Ranson et P. Peretti

Département des Recherches Physiques, Laboratoire associé au C.N.R.S., Université de Paris VI, France

et

J. L. Laporte et Y. Rousset

Laboratoire d'Optique Moléculaire, Université de Reims, Reims, France

ABSTRACT

A comparison between the phosphorescence and Raman spectra of benzophenone crystal at very low temperatures is given.

Some vibronic bands of the phosphorescence spectrum provides a coupling between the lattice and the intramolecular motion.

INTRODUCTION

Les spectres de diffusion Raman de la benzophénone cristalline ont été effectués dans deux régions distinctes, l'une concernant les vibrations internes de la molécule, l'autre les vibrations du réseau cristallin. Le but de cette étude est de montrer, l'apport de la diffusion Raman dans l'analyse vibrationnelle des émissions de phosphorescence, dans lesquelles en particulier un certain nombre de bandes ne s'interprètent que par des couplages entre le réseau et les mouvements intramoléculaires. Ces interactions sont d'autant plus fortes que les électrons responsables de la transition optique participent aux liaisons intermoléculaires.

STRUCTURE

La molécule de benzophénone $(C_6H_5)_2CO$ appartient au groupe de symétrie C_2,[1] l'axe de symétrie C_2 étant porté par la liaison carbonyle C=O. La répulsion entre les deux atomes d'hydrogène H (2) et H (2′) (Fig. 1) entrainant une rotation de même sens[1, 2] des cycles benzéniques d'un angle $\alpha = 38°$,[2] détruit en effet la symétrie C_{2v} à laquelle on aurait pu s'attendre.

Les données radiocristallographiques[3, 4] de la forme stable de la benzophénone solide sont les suivantes

groupe spatial (D_2^4), $P\,2_12_12_1$
symétrie de site C_1 [5, 6]

Le cristal appartient au système orthorhombique, les dimensions de la cellule élémentaire étant[4]

$$a = 10{,}26\ \text{Å}$$
$$b = 12{,}09\ \text{Å}$$
$$c = 7{,}88\ \text{Å}$$

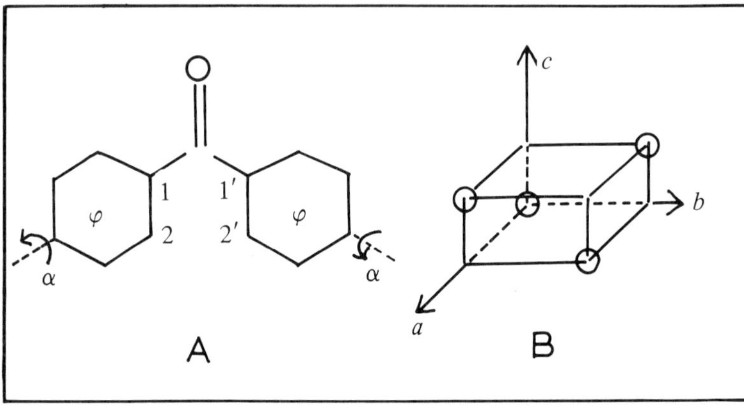

Fig. 1. A – Schéma de la molécule de benzophénone; B – Position des molécules dans la maille cristalline.

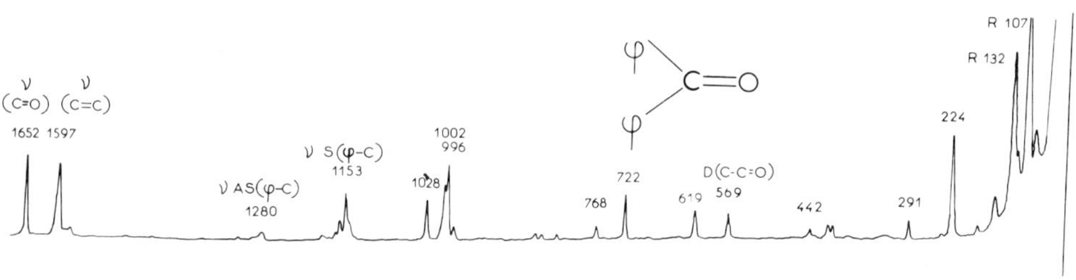

Fig. 2. Spectre Raman de la benzophénone en poudre, à 77 K. D = déformation; R = libration; νS = vibration symétrique; νAS = vibration antisymétrique.

Le diagramme de corrélation déduit des tables de caractère[7] prévoit que chaque vibration de la molécule isolée donnera dans le cristal quatre composantes de type de symétrie A, B_1, B_2 et B_3, les axes **x**, **y** et **z** étant les axes **a**, **b**, **c** de la maille orthorhombique.

Pour les vibrations externes, nous pouvons nous attendre à 21 vibrations correspondant aux branches optiques[8, 9] soit: 12 vibrations de pivotement (librations) toutes actives en Raman dont

3 du type A
3 du type B_1 soit S A A
3 du type B_2 soit A A S
3 du type B_3 soit A S A

et 9 vibrations de translation, toutes actives en Raman dont

3 du type A
2 du type B_1
2 du type B_2
2 du type B_3

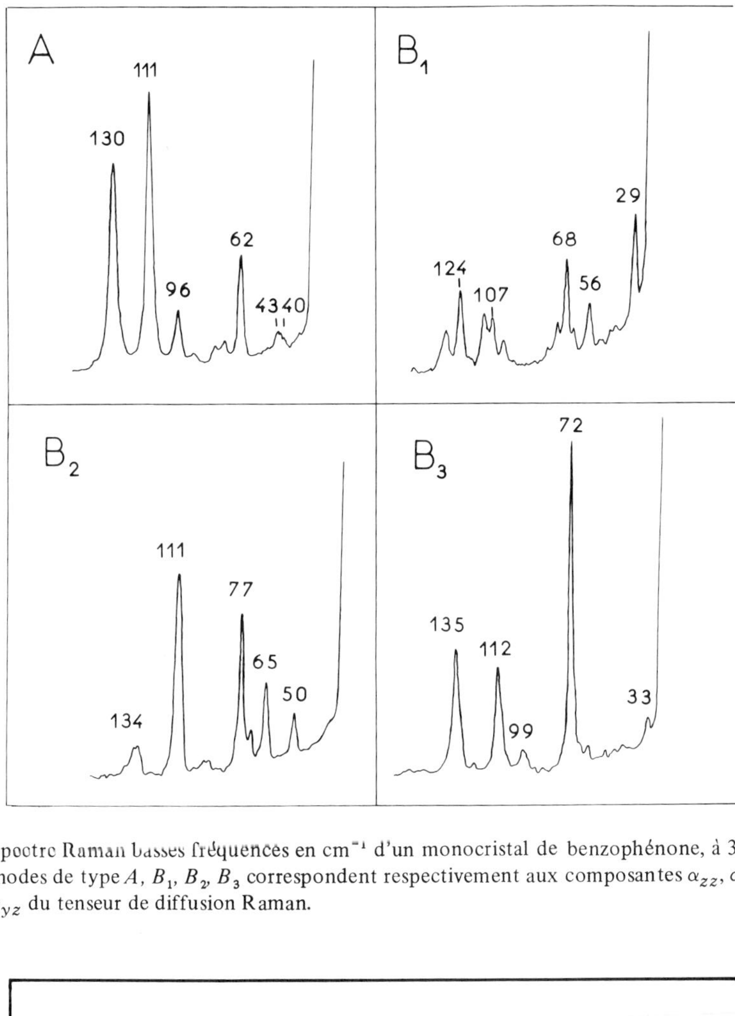

Fig. 3. Spectre Raman basses fréquences en cm^{-1} d'un monocristal de benzophénone, à 30 K. Les modes de type A, B_1, B_2, B_3 correspondent respectivement aux composantes α_{zz}, α_{xy}, α_{xz} et α_{yz} du tenseur de diffusion Raman.

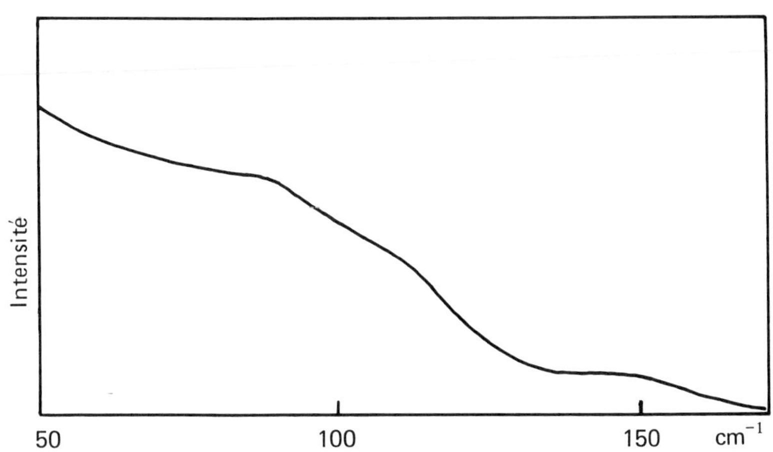

Fig. 4. Spectre Raman basses fréquences de la benzophénone en solution dans CCl_4, à 300 K. La courbe est corrigée du fond de diffusion du solvant.

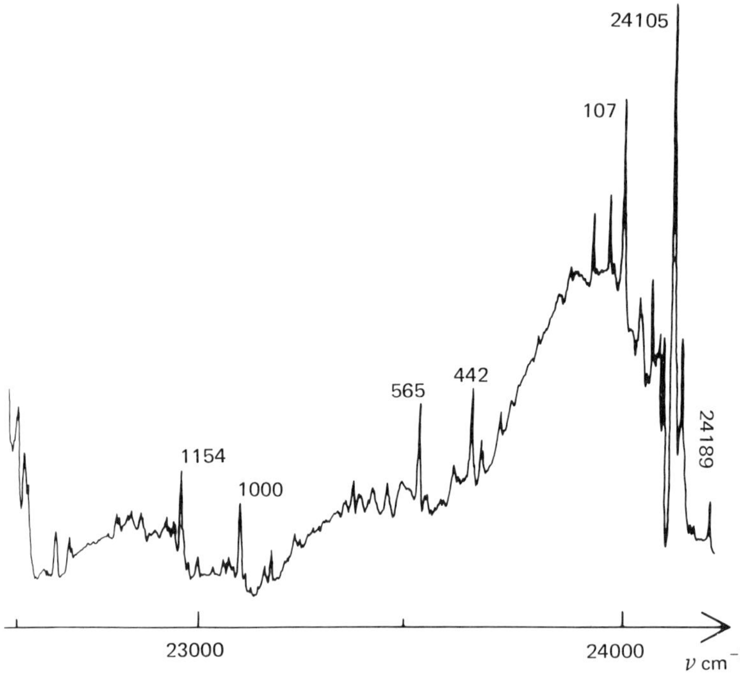

Fig. 5. Première séquence du spectre de phosphorescence de la benzophénone en poudre, à 7 K.

Tableau 1.

Emission de longue durée de vie $\tau = 2 \times 10^{-2}$ s $\bar{\nu}$ (cm^{-1})	$\Delta\bar{\nu}$ (cm^{-1})	Spectre Raman (cm^{-1})
24,156	−51	
24,147	−42	
24,128	−19	
24,105	0	
24,087	18	
24,079	26	29
24,074	31	33
24,063	42	41
24,055	50	50
24,034	71	72
24,010	95	96–99
23,998	107	107
23,979	126	124
23,967	138	134–135

DIFFUSION RAMAN

La Fig. 2 montre la totalité du spectre de diffusion Raman de la benzophénone en poudre à 77 K. La Fig. 3 correspond à la partie basses fréquences du spectre de diffusion Raman d'un monocristal, à 30 K.[10]

Tableau 2.

Emission de longue durée de vie $\tau = 2 \times 10^{-2}$ s $\bar{\nu}$ (cm^{-1})	$\Delta\bar{\nu}$ (cm^{-1})	Spectre Raman (cm^{-1})	Spectre infrarouge (cm^{-1})	Combinaison dans le spectre de phosphorescence
24,105	0			
23,950	155	159		
23,930	175			71 + 107
23,885	220	224		
23,810	295	291		
23,726	379	383		
23,864	418	415		
23,663	442	442		
23,625	480		488	
23,555	550			442 + 107
23,540	565	569		
23,485	620	619		
23,460	645	641		
23,433	672			565 + 107
23,400	705	707	699	
23,378	727	722		
23,258	847	852	846	
23,239	866	864		
23,185	920		921	
23,168	937		935	
23,120	985	986		
23,105	1,000	996–1,002	1,000–1,005	
23,075	1,030	1,028	1,028–1,032	
23,065	1,040		1,040	
22,999	1,106			1,000 + 107
22,951	1,154	1,153	1,160	
22,937	1,168	1,167		
22,914	1,191	1,190		
22,841	1,164			1,154 + 107
22,818	1,287	1,282–1,285		
22,790	1,315	1,320		
22,655	1,450	1,452	1,445	
22,614	1,491	1,494	1,490	
22,536	1,572	1,580	1,575	
22,512	1,593	1,597	1,592	
22,494	1,611		1,608	
22,455	1,650	1,652		

Il est à noter que le spectre Raman de la benzophénone à l'état liquide ou en solution dans CCl_4, présente deux bandes très larges (de l'ordre de 30 cm^{-1}), l'une située vers 110 cm^{-1}, l'autre vers 70 cm^{-1} (Fig. 4).

La comparaison avec les résultats déduits du spectre Raman du monocristal[10] nous montre que ces bandes, devant correspondre à des mouvements de torsion des noyaux benzéniques autour des liaisons $C–C_1$ et $C–C_{1'}$ apparaissent sous forme de librations à l'état solide.

Tableau 3.

Emission de l'exciton durée de vie $\tau = 2 \times 10^{-4}$ s	
$\bar{\nu}$ (cm^{-1})	$\Delta\bar{\nu}$ (cm^{-1})
24,189	0
24,181	108
23,850	439
23,625	564
23,181	1,004
23,033	1,156
22,536	1,653

PHOSPHORESCENCE DE LA BENZOPHENONE

L'analyse spectrale et la résolution temporelle de la phosphorescence à très basse température de la benzophénone cristallisée, montrent qu'il existe plusieurs émissions distinctes de phosphorescence. Deux d'entre elles sont constituées de bandes très fines. La première de courte durée de vie ($\tau = 2 \times 10^{-4}$ s) commence à 24,189 cm^{-1}. La seconde débutant à 24,105 cm^{-1} a une durée de vie beaucoup plus longue ($\tau = 2 \times 10^{-2}$ s). Une troisième émission correspondant au fond important du spectre est constituée de bandes larges. A partir de l'évolution spectrale en fonction de la température, nous avons pu montrer l'existence d'ilôts vitreux au sein du cristal.[11] La troisième émission provient des molécules formant ces ilôts vitreux. L'émission d'origine 24,105 cm^{-1} est due aux molécules perturbées situées à la frontière limitant les ilôts vitreux du cristal régulier. La coïncidence de la bande (0, 0) de la première émission avec celle du spectre d'absorption[12] montre que l'émission commençant à 24,189 cm^{-1} est celle de l'exciton triplet.[13–15]

Examinons la première séquence du spectre, obtenu à 7 K, de l'émission de longue durée de vie (Fig. 5). Les bandes de basses fréquences ($\Delta\bar{\nu} < 150$ cm^{-1}) caractéristiques des vibrations du réseau cristallin sont reportées dans le Tableau 1. Nous constatons qu'il y a coïncidence entre le spectre de diffusion Raman[10] et le spectre de luminescence bien que celui-ci soit relatif aux molécules perturbées de benzophénone. Il semble que cette perturbation soit insuffisante pour empêcher ces molécules d'entrer en résonnance avec les fréquences de vibration du réseau cristallin.[16]

Les bandes de faible intensité placées en avant de la bande (0, 0) (24,105 cm^{-1}) peuvent être attribuées à des transitions à partir d'un niveau vibrationnel excité de l'état triplet et correspondent encore aux fréquences de vibration du réseau. Pour les bandes de fréquences supérieures à 150 cm^{-1} (Tableau 2), nous observons le bon accord avec les fréquences de vibrations internes des molécules déduites du spectre Raman et infrarouge. (Refs. 5, 6, 8, 9, 17, and 18).

Remarquons qu'il apparaît dans la phosphorescence de la benzophénone, correspondant à la transition $^3\Gamma_{n\pi^*} \longrightarrow {}^1\Gamma_1$, un assez grand nombre de raies de combinaison de fréquences $\Delta\bar{\nu} + 107$ cm^{-1} (Tableau 2). Ceci peut résulter d'un fort couplage vibronique[12] entre les états $^1\sigma\pi^*$ du carbonyle et $^1\pi\pi^*$ du noyau aromatique, la fréquence 107 cm^{-1} étant associée au mode torsionnel (mouvement relatif des deux noyaux benzéniques) de la molécule de benzophénone. La couplage spin-orbitre entre les états $^3n\pi^*$ et $^1\sigma\pi^*$ du carbonyle permet ensuite la transition électronique $^3\Gamma_{n\pi^*} \longrightarrow {}^1\Gamma_1$.

Les principales bandes du spectre d'émission de l'exciton sont reportées dans le Tableau 3. Nous retrouvons dans ce spectre les principales fréquences de vibrations internes et externes de la benzophénone, ce qui montre que cet exciton est de type moléculaire.[19]

REFERENCES

1 C. Menard et A. Mellier, *C. R. Acad. Sci., Paris* **271 B**, 1181 (1970).
2 R. Hoffmann et J. R. Swenson, *J. Phys. Chem.* **74**, 415 (1970).
3 E. B. Vil et G. M. Lobanova, *Soviet Phys. Cryst.* **12**, 233 (1967).
4 G. M. Lobanova, *Soviet Phys. Cryst.* **13**, 856 (1969).
5 K. Mallikarjuna Rao et C. K. Narayanaswamy, *Indian J. Pure Appl. Phys.* **7**, 243 (1969).
6 K. Mallikarjuna Rao et C. K. Narayanaswamy, *Indian J. Pure Appl. Phys.* **8**, 100 (1970).
7 E. B. Wilson, J. C. Decius et P. C. Cross, *Molecular Vibrations*, McGraw-Hill, New York, 1955.
8 B. Wyncke, A. Hadni, D. Chanal et E. Decamps, *Ann. Phys.* **2**, 123 (1967).
9 B. Wyncke et A. Hadni, *Spectrochim. Acta* **27 A**, 1929 (1971).
10 P. Ranson, P. Peretti, Y. Rousset et J. A. Koningstein, *Chem. Phys. Letters* **16**, 396 (1972).
11 J. L. Laporte, G. Nouchi et Y. Rousset, *J. Chem. Phys.* **57**, 1767 (1972).
12 S. Dym, R. M. Hochstrasser et M. Schafer, *J. Chem. Phys.* **48**, 646 (1968).
13 V. J. Melnik et M. T. Shpak, *Opt. Spectry.* **29**, 326 (1970).
14 A. Nakahara, M. Koyanagi et Y. Kanda, *J. Chem. Phys.* **50**, 552 (1969).
15 J. L. Laporte, R. Lochet, G. Nouchi et Y. Rousset, *C. R. Acad. Sci., Paris* **273**, 124 (1971).
16 R. M. Hochstrasser et P. N. Prasad, *J. Chem. Phys.* **56**, 2814 (1972).
17 T. M. K. Nedungadi, *Proc. Indian Acad. Sci.* **13A**, 161 (1941).
18 M. S. Mathur, C. A. Frenzel et E. B. Bradley, *Spectrochim. Acta* **26A**, 451 (1970).
19 J. L. Laporte, Y. Rousset, G. Nouchi, P. Ranson et P. Peretti, *C. R. Acad. Sci., Paris* **275**, 17 (1972).

R 7/9/73

82. LIGHT SCATTERING FROM OXIDE SURFACES*

G. Careri, V. Mazzacurati,† M. Sampoli and G. Signorelli
Istituto di Fisica, Università di Roma, Italy

INTRODUCTION

The occurrence of a very intense and broad scattering background (S.B.) in Raman spectra from biomaterials, plastics and oxides is still of controversial origin. This S.B. is found to decay when the sample is irradiated with the laser beam. The S.B. intensity is so high that an explanation in terms of Raman scattering or impurity fluorescence is incompatible with the available data.

In the past S.B. has been observed in the Raman study of absorbed species on oxide[1] and silica surfaces,[2] and on Vycor[3] where it was attributed to hydrocarbons contamination. We have noticed S.B. in solutions of liquid proteins,[4] in solid pellets[5] and we have attributed it to water bound to the protein, in contrast with other workers in the field of bio-polymers.[6] The purpose of this paper is to show that S.B. observed on oxide surfaces correlates with the presence of water molecules tightly hydrogen-bonded close to some Lewis sites, and to consider some further physical aspects of this somewhat mysterious phenomenon.

EXPERIMENTAL

Our Raman laser apparatus has already been described.[1] This apparatus has been calibrated by help of a standard emitter and the raw data have been corrected to take into account resolution and transmission in order to obtain the scattered power density. The light scattering spectra recorded from oxide pellets show an intense and broad Stokes band, centred around 3000 cm^{-1}, with a shape somewhat dependent on the exciting wavelength. An anti-Stokes band is also observed, and it is less intense than the Stokes one, roughly consistent with a Boltzman factor.

The most peculiar and intriguing aspect of both the Stokes and anti-Stokes bands is the laser intensity dependence of the scattered light. This feature has been already reported by our group[2] in solid protein pellets, and it is observed in oxide pellets as well, both in vacuum and in air. The time dependence of the emitted radiation is a complex phenomenon. It shows a rapid fall in the first few minutes and a subsequent slow decrease in the following hours.[7] The recovery time in the dark is much larger, and final equilibrium can be obtained after days or weeks in air. No recovery effects in a few hours are noticed

* Work partially supported by C.N.R. and by E.N.I.
† Present address: Istituto di Fisica, Università di Camerino.

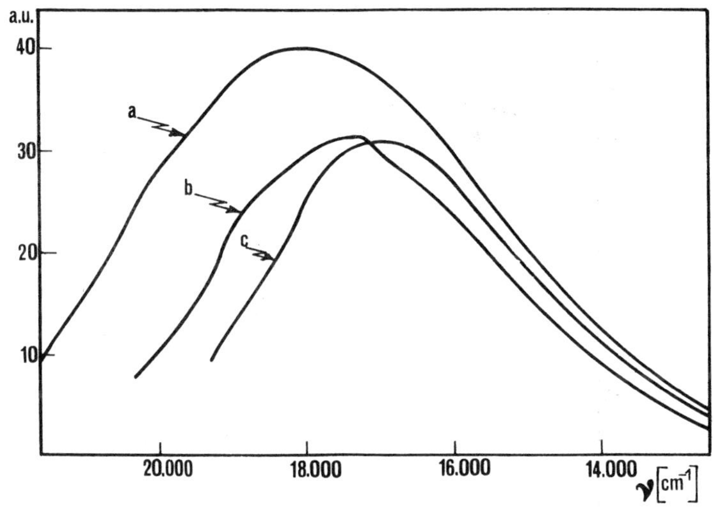

Fig. 1. Scattered intensity per unit incident power density (arbitrary units) plotted versus absolute wavelength, from a pressed disc of γ-Al_2O_3 excited at: (a) 4579, (b) 4880 (c) 5145 Å.

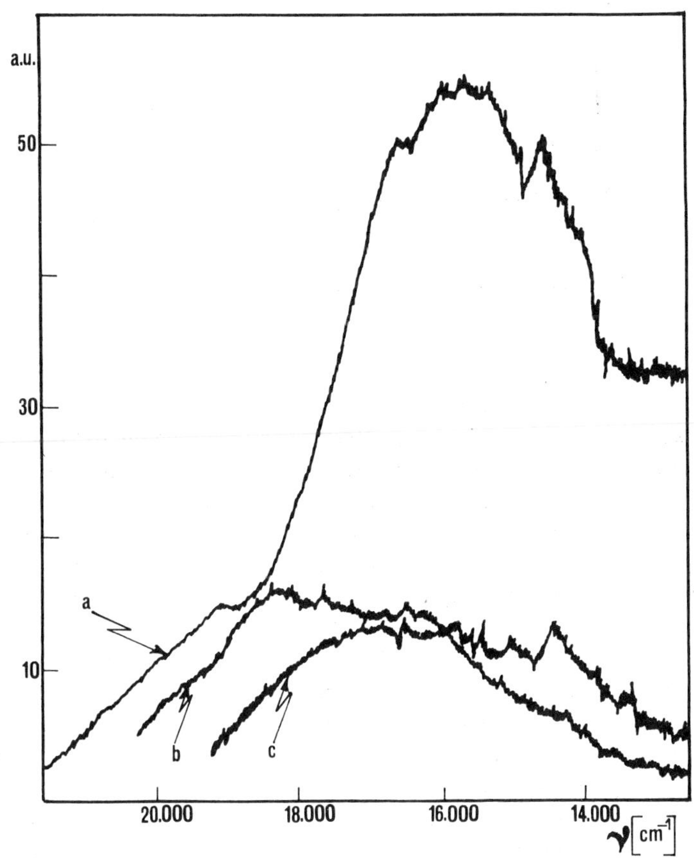

Fig. 2. The same plot as Fig. 1, after a delay time of 150 μs.

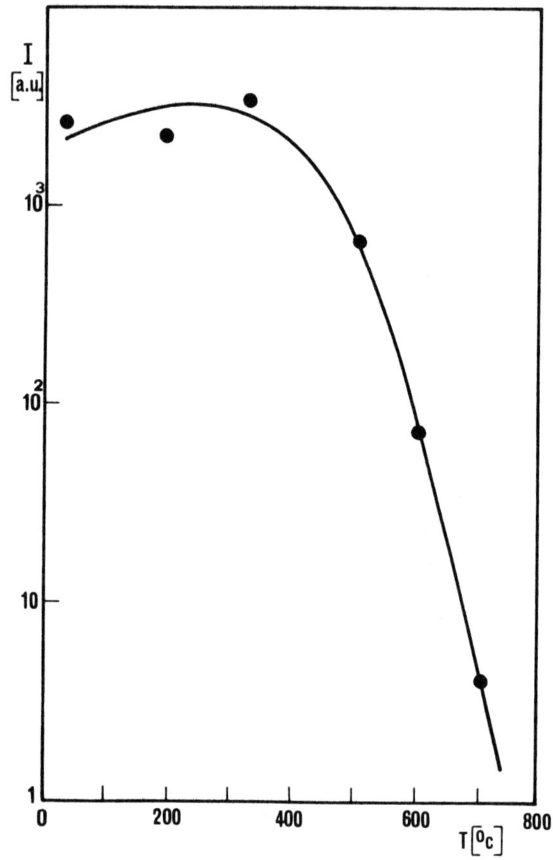

Fig. 3. Trend of the light scattering intensity in γ-Al_2O_3 heated in high vacuum, plotted versus temperature.

when the laser beam intensity is suddenly reduced.

The Stokes band intensity yields a scattering coefficient about 10^4 times larger than the one of a typical Raman band (i.e. OH-stretching mode of H_2O). Such a large coefficient makes the interpretation of this band as a simple Raman scattering process quite doubtful. Furthermore the luminescence spectrum recorded after a delay time of 150 μs reveals a somewhat similar band whose characteristics are strongly dependent on the exciting wavelength, with an intensity up to 20% of the 'synchronous' ordinary spectrum. In this case the presence of a long time constant fluorescence band ($\tau = 10^{-3}$s) is clearly demonstrated (see Figs. 1 and 2). The presence of a short time fluorescence ($\tau = 10^{-8}$ s) remains to be detected; this experiment is in progress at the present time.

All the above reported features were observed in a large number of oxide pellets under similar irradiation conditions and are reported here (a more detailed account on the chemical aspects, is given elsewhere[8]).

The first experiment concerns the behaviour of a sample of γ-Al_2O_3 under different conditions of hydration in vacuum. Upon heating up to 800°C in vacuum, the S.B. integrated intensity was reduced, as shown in Fig. 3, while the overall shape was preserved. The

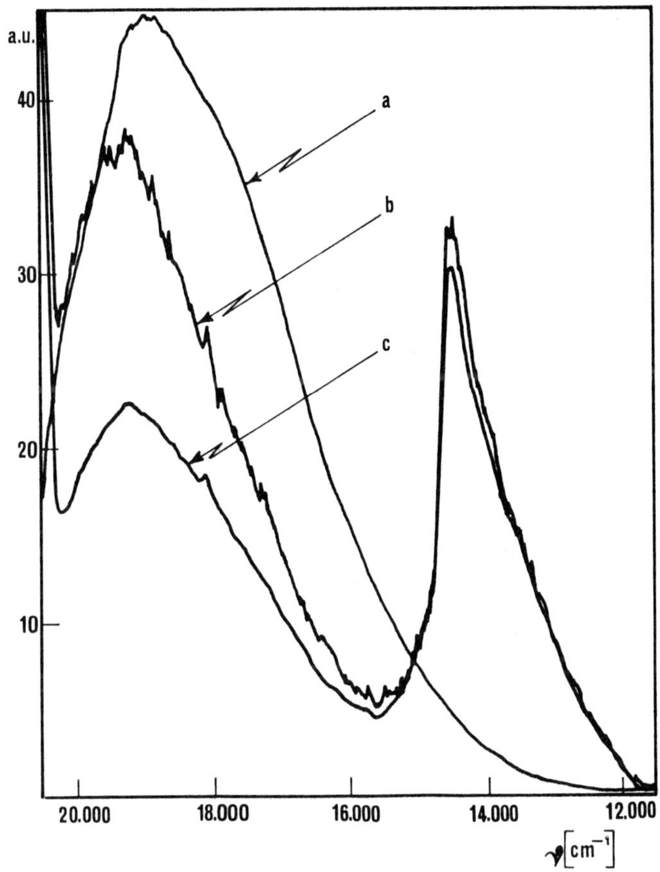

Fig. 4. (a) Raw data of the spectrum reported in Fig. 1(b). (b) and (c) refer to the same sample poisoned by Na, and (c) has been irradiated at laser power 10 times larger than (b) or (a). (a) has been reduced by a factor of 125.

original intensity was then recovered, after admission of moist air at room temperature for several weeks. When the same experiment was carried out in air at temperatures up to 600°C, the S.B. essentially disappeared and the spectrum revealed a narrow peak in the 14.000 cm^{-1}. This band is in fact easily attributed to a fluorescence peak of so formed α-Al_2O_3.

A second group of experiments was directed to investigate some significant oxide surfaces:

1. By adding Na to γ-Al_2O_3, the S.B. intensity was lowered by a factor of 100. Moreover the spectra revealed the presence of both a moderate S.B. and a newly formed fluorescent band around 14.000 cm^{-1}, the two bands displaying quite a different dependence upon the irradiation power (see Fig. 4).
2. The S.B. integrated intensity has been measured in a variety of oxides, with the following results (in arbitrary units): γ-Al_2O_3 (44.5), silica alumina (9.0), 13X-zeolite (0.42), MgO (0.16), silica (0.04), α-Al_2O_3 (0.31), ZnO (0.00). When silica alumina was activated under vacuum at 500°C, the intensity increased by a factor of 25. When

the decationated 13X-zeolite was heated in vacuum at 500°C, and then cooled to room temperature and exposed to open air, the intensity increased by a factor of 50.

DISCUSSION

The nature of S.B. is quite intriguing. There are two basic questions to be answered: the first one concerns the chemical species which are responsible for this rather widespread phenomenon, and the second one the physical process which takes place. Obviously, the two questions are interconnected.

Concerning the first question, we believe that the very similar features observed in a reproducible way in a large variety of substances, should rule out its origin in terms of fluorescent hydrocarbons impurities. Moreover there is no way to explain, in every sample, the recovery to its original S.B. level by readsorption of 'impurities'. This point of view is further supported by the following facts:

1. The occurrence of S.B. in polywater,[9] where the amount of impurities has been measured.
2. The striking pH dependence of S.B. in polymethacrilic acid solutions,[10] where nothing but NaOH was added.
3. The nearly complete absence of S.B. on some solid hydrophobic surfaces, for instance polyethylene oxide (work in progress in this laboratory).

This being the situation, one is forced to look for another possible source. We have already suggested that water bound to solid proteins could have been the origin of S.B. We believe that the three points quoted above and the experimental results presented here are consistent with this hypothesis.

As a matter of fact, our results can be explained by attributing this S.B. to the water molecules on the surface, which are tightly hydrogen-bonded close to some Lewis sites. Evidence for this statement can be offered as follows (for more details see ref. 8):

1. Our data reported in Fig. 3 on one hand the infrared spectra of partially hydrated γ-Al_2O_3 reported by Peri and Hannan [11] on the other hand suggest a parallel trend between the intensity of the S.B. and that of the low frequency tail of the OH-band (already assigned to the OH-stretching mode of tightly hydrogen bonded water).
2. The S.B. decrease in the γ-Al_2O_3 samples poisoned by sodium led one to conclude that some acidic sites on the surface must be involved.
3. Chemisorption of water on γ-Al_2O_3 occurs on sites where Al^+O^{2-} are available. It is easy to think this water to be bonded[11] to O^{2-} and coordinated to the Al^+ Lewis acidic site as well. Moreover it is known that quite a similar mechanism of water chemisorption holds for silica alumina and for the decationated zeolites, while in silica the water is chemisorbed in quite a different way.

The second question, namely the nature of the optical process which takes place, is much more difficult to answer. Actually, we believe that oxides are not the best system to investigate in order to answer this question, and simpler systems are under consideration in our laboratory at the present time. Nevertheless there are some features that seem sufficiently clear to be stated, as follows:

1. The observation of a long time fluorescence band similar to S.B. indicates that a real optical absorption process takes place. This absorption is also suggested by the typical fluorescent behaviour of the S.B. in the region below 16.000 cm^{-1} (see Fig. 1), because of the independence on the exciting wavelength.

2. On the other hand the S.B. exhibits a Raman-like behaviour for displacements lower than 2000 cm^{-1} from the exciting line. A wide fluorescent band similar to this S.B. has been described in retinyl-polyenes.[13] Both Stokes and anti-Stokes bands are explained in these substances by the existence of a vibrational broad band in the ground state.[14] But in our case this part of the spectrum is similar in both the 'synchronous' and the delayed scattering, and these facts are in conflict with the usual idea of fluorescence.

To conclude, we believe that some chemisorbed water, by virtue of its altered electronic structure, gives rise to a complex scattering process and that its physical understanding requires further study. This chemisorbed water seems to be in a state which is easily damaged by irradiation and is difficult to restore afterwards.

REFERENCES

1 P. J. Hendra, J. R. Horder and E. J. Loader, *Chem. Commun.* 563 (1970).
2 E. J. Loader, *J. Catalysis* **22**, 41 (1971).
3 T. A. Egerton, A. H. Hardin, Y. Kozirovski and N. Sheppard, *Chem. Commun.* 887 (1971).
4 G. Careri, V. Mazzacurati and G. Signorelli, *Phys. Letters* **31A**, 425 (1970).
5 G. Careri, V. Mazzacurati, M. Sàmpoli and G. Signorelli, *Phys. Letters* **32A**, 495 (1970).
6 J. L. Koenig and P. L. Sutton, *Biopolymers* **8**, 167 (1969).
7 The failure to observe SB in protein solution by Lord *et al.* (*Biochim. Biophys. Acta* **257**, 280 (1972)), is due to the high laser power employed and to the long time of irradiation used by these authors before recording their spectra.
8 G. Careri, V. Mazzacurati, M. Sampoli and G. Signorelli, *J. Catalysis*, (in press).
9 D. L. Rousseau and S. P. S. Porto, *Science* **167**, 1715 (1971).
10 M. C. Tobin, *J. Chem. Phys.* **50**, 4551 (1968).
11 J. B. Peri and R. B. Hannan, *J. Phys. Chem.* **64**, 1526 (1960).
12 H. De Boer, J. M. H. Fortuin, B. C. Lippens and W. H. Meljs, *J. Catalysis* **2**, 1 (1963).
13 A. J. Thomson, *J. Chem. Phys.* **51**, 4106 (1969).
14 B. I. Stepanov and V. P. Gribkovskii, *Theory of Luminescence,* Iliffe, London, 1968.

CONTRIBUTORS INDEX

SUBJECT INDEX